The Revolution Continues . . .

The first edition of Stephen Marshak's *Essentials of Geology* was warmly embraced by the geology community and practically overnight became a best-selling introductory text. Geology instructors have enthusiastically responded to the integration of the theories of plate tectonics and Earth systems science; the spectacular art and photographs; the innovative pedagogy, highlighted by "What a Geologist Sees" figures; and the remarkably inviting writing style. Here is just a sampling of some of the comments that we have received:

From Students

"I originally chose to take a geology class because I had a relatively small interest in Earth Science and it was an interesting way to satisfy my lab credit requirement. Upon starting the class, and reading into your textbook, I have decided to make geology (specifically tectonics) my life's work.

"Without your textbook, I am not so sure I would have found my calling. I look forward to refining my knowledge in geology as my college career moves on."
—email from a student at *Nassau Community College* to Stephen Marshak

"The geology course I'm enrolled in uses your text, and I just wanted to tell you how fantastic it is. The book truly astonishes me. It's very well done, and the illustrations are marvelous. I just thought I would let you know."
—email from a *University of Cincinnati* student to Stephen Marshak

"Professor Marshak, I'm a theatre major at UW Madison taking geology to fulfill science requirements; however, after studying for my midterm, something struck me about your textbook—*Essentials of Geology*. I think it's the best textbook I've ever had in my 3 years of school here. I knew nothing coming into the course and I feel like I've actually learned the material I've read, as opposed to having memorized it. It's not as 'preachy' as some textbooks are. Everything is explained thoroughly. Overall, it has a very down-to-earth and conversational feel. I find myself recommending this course to others using the phrase, 'You should take it, the text book is wonderful!' So, from a college student who has seen her fair share of wretched textbooks, I felt that I had to write you a thank-you. Thank you for understanding what students need."
—email from a *University of Wisconsin, Madison* student to Stephen Marshak

From Adopters

"The book is extremely well written and very easy to understand. Marshak has a gift for presenting material in a clear, easy-to-understand fashion that I envy. I find I can assign a reading on a topic and count on the students having a pretty good understanding before they come to class. This is definitely not true of most books."
—Tom Juster, *University of South Florida*

"Lucid, thorough writing style and great graphics. . . . Marshak is a fabulous writer. I can't say enough about how good this text is."
—David Osleger, *University of California, Davis*

"I very much like the structure of the book, with plate tectonics as the 'hinge' to understand geology and geological processes."
—Charly Bank, *Colorado College*

"Marshak's book is written from the perspective of the Earth as a dynamic system with multiple interrelated components. Each chapter makes sense both from a component perspective and as an integral piece of the overall system."
—Michael Bradley, *Eastern Michigan University*

"This is a clear and extremely well written textbook. . . . It has excellent illustrations, and my students have praised it for its readability."
—Peter Buseck, *Arizona State University*

"About Marshak—I love it. I found it very complete and understandable."
—James Woodhead, *Occidental College*

"There is simply no competition."
—Torbjorn Tornqvist, *University of Illinois at Chicago*

The next few pages show how the text works and what's been added to the Second Edition.

MODERN ORGANIZATION AND APPROACH

Essentials of Geology, Second Edition, introduces plate tectonics early—Chapter 2—after an important chapter on the formation and structure of the Earth (Chapter 1). From this point on, the theory of plate tectonics, the unifying theory of the Earth sciences, is used to connect all the appropriate subjects, especially volcanoes, earthquakes, mountain formation, and the history of the Earth.

Likewise, Earth System science, the framework that stresses the interconnectivity of the Earth's landmasses, bodies of water, and atmosphere, informs the entire narrative. It is particularly evident in the chapters on oceans and coasts, deserts, glaciers, and global warming.

In short, the text consistently emphasizes the fact that several powerful cycles are at play in nearly every process that drives our Earth.

For the Second Edition, Stephen Marshak has added a great deal of new material on comparative planetary geology, content that is informed by the successful satellite explorations of Mars, Saturn, Venus, and Jupiter.

There is also extensive coverage of Hurricane Katrina and the December 2004 Indian Ocean Tsunami.

Of course, all of the chapters are self-contained and can be taught in any order the instructor wants.

Comprehensive chapter on the formation
and structure of the Earth; it sets the stage
for the chapter on Plate Tectonics

Plate Tectonics presented
early

Coverage of the
December 2004
Indian Ocean
Tsunami

Full chapter on the history
of life as it relates to the
history of the Earth

Coverage of
Hurricane Katrina

Important chapter
on global warming

BRIEF CONTENTS

Preface		xvii
Prelude	*And Just What Is Geology?*	1
Chapter 1	*The Earth in Context*	8
Chapter 2	*The Way the Earth Works: Plate Tectonics*	35
Chapter 3	*Patterns in Nature: Minerals*	78
Interlude A	*Rock Groups*	95
Chapter 4	*Up from the Inferno: Magma and Igneous Rocks*	102
Chapter 5	*A Surface Veneer: Sediments, Soils, and Sedimentary Rocks*	121
Chapter 6	*Metamorphism: A Process of Change*	153
Interlude B	*The Rock Cycle*	174
Chapter 7	*The Wrath of Vulcan: Volcanic Eruptions*	180
Chapter 8	*A Violent Pulse: Earthquakes*	206
Interlude C	*Seeing Inside the Earth*	240
Chapter 9	*Crags, Cracks, and Crumples: Crustal Deformation and Mountain Building*	248
Interlude D	*Memories of Past Life: Fossils and Evolution*	273
Chapter 10	*Deep Time: How Old Is Old?*	283
Chapter 11	*A Biography of Earth*	306
Chapter 12	*Riches in Rock: Energy and Mineral Resources*	329
Interlude E	*An Introduction to Landscapes, and the Hydrologic Cycle*	364
Chapter 13	*Unsafe Ground: Landslides and Other Mass Movements*	373
Chapter 14	*Streams and Associated Flooding: The Geology of Running Water*	391
Chapter 15	*Restless Realm: Oceans and Coasts*	418
Chapter 16	*A Hidden Reserve: Groundwater*	450
Chapter 17	*Dry Regions: The Geology of Deserts*	473
Chapter 18	*Amazing Ice: Glaciers and Ice Ages*	493
Chapter 19	*Global Change in the Earth System*	525
Appendix	*Scientific Background: Matter and Energy*	A-1
Metric Conversion Chart		A-12
Glossary		G-1
Credits		C-1
Index		I-1

TOTALLY UP-TO-DATE

Complete Coverage of Hurricane Katrina

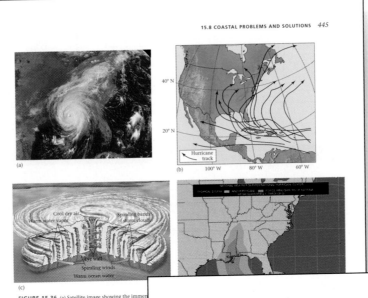

(a)

(b)

(c)

FIGURE 15.36 (a) Satellite image showing the imme... Hurricane Katrina as it crossed over the Gulf coast of the... States. (b) The tracks of typical hurricanes during the pa... (c) A three-dimensional sketch illustrating the architectu... hurricane. (d) A wind-swath map of Hurricane Katrina. T... area was affected by hurricane-force winds, and the gold... affected by tropical-storm-force winds. Note how the hu... grew after entering the Gulf of Mexico.

- *Wind:* Winds of weaker hurricanes tear off branche... smash windows. Stronger hurricanes uproot trees,... roofs, and collapse walls.
- *Waves:* Winds shearing across the sea surface durin...

In the three-dimensional sketch labels: Cool dry air, Warm water vapor, Spiraling bands of storm clouds, Eye, Eye wall, Spiraling winds, Warm ocean water.

40° N
20° N
100° W   80° W   60° W
Hurricane track

- *Rain, Stream Flooding, and Landslides:* Rain drenches the Earth's surface beneath a hurricane. In places, half a meter or more of rain falls in a single day. Rain causes streams to flood, even far inland, and by saturating the ground, can trigger landslides, especially in coastal areas.
- *Disruption of Social Structure:* When the storm passes, the hazard is not over. By disrupting transportation and com... networks, breaking water mains, and ...age-treatment plants, hurricane... ...vere obstacles to search and rescue, and ...read of disease, fire, and looting.

...icanes that reach the coast cause death ...ut some are truly catastrophic. ...cyclone making landfall on the low-... Bangladesh led to an estimated 500,000 ...rge from a hurricane that struck ...n 1900 caused up to 12,000 deaths, and ...n 1992 leveled extensive areas of south-...ed over $30 billion in damage, and left ...omeless. Hurricane Katrina, in 2005, ...expensive hurricane to strike the United ...me. Because of where Katrina passed, ...prediction scenarios came to be reality, ...istoric city of New Orleans. Let's look at ...

...Katrina came into existence over the ...d west. Just before landfall in southeast-...trengthened and the storm became Hur-...hurricane sliced across the southern tip of

Florida, causing several deaths and millions of dollars in damage. It then entered the Gulf of Mexico and passed directly over the Loop Current, an eddy (circular flow) of summer-heated water from the Caribbean that had entered the Gulf of Mexico (see Fig. 15.36d). Water in the Loop Current reaches temperatures of 32°C (90°F), and thus stoked the storm, injecting it with a burst of energy sufficient for the storm to morph into a Category 5 monster whose swath of hurricane-force winds reached a width of 325 km (200 miles). When it entered the central Gulf of Mexico, Katrina turned north and began to bear down on the Louisiana-Mississippi coast. The eye of the storm passed just east of New Orleans, and then across the coast of Mississippi. Storm surges broke records, in places rising 7.5 m (25 ft) above sea level, and they washed coastal communities off the map along a broad swath of the Gulf Coast (▶Fig. 15.37a, b). In addition to the devastating wind and surge damage, Katrina led to the drowning of New Orleans.

To understand what happened to New Orleans, we must consider the city's environmental history. New Orleans grew on the Mississippi Delta, between the banks of the Mississippi River on the south and Lake Pontchartrain (actually a bay of the Gulf of Mexico) on the north. The oldest part of the town, a neighborhood called the French Quarter, was built on the relatively high land of the Mississippi's natural levee. Younger parts of the city, however, spread out over the lower delta plain. As the decades passed, people modified the surrounding delta landscape by draining wetlands, by constructing artificial levees that confined the Mississippi River, and by extracting groundwater. Sediment beneath the delta compacted,

...An air photo, looking straight down at the coast of Mississippi, following ...e distribution of debris and the near-total destruction of buildings shows the ...m surge washed inland. (b) Close-up of Hurricane Katrina storm damage.

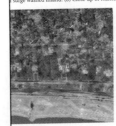

(b)

FIGURE 15.38 (a) Even before Hurricane Katrina, maps clearly displayed the vulnerability of New Orleans to flooding. The red areas are below sea level, and stayed dry only because of levees, as emphasized by the cross-section shown in (b). (b) Three breaches occurred in the levee system. This map shows the levee system, and the region near downtown New Orleans that rising waters submerged.

USA

LAKE PONTCHARTRAIN

Swamp   Gulf of Mexico

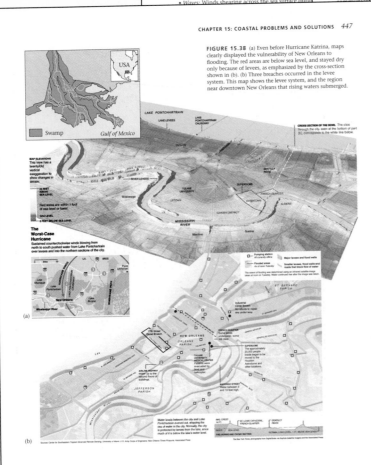

(a)

(b)

Complete Coverage of the December 2004 Indian Ocean Tsunami

(a)

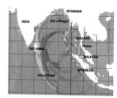

FIGURE 8.33 A computer model showing the 2004 Indian Ocean tsunami about 2 hours after the earthquake. The yellow areas have the greatest wave height.

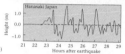

(b)

FIGURE 8.32 (a) In the aftermath of a tsunami that hit Kodiak, Alaska, after the 1964 earthquake, boats rest amid the rubble of a coastal town. (b) Tidal gauge at Hanasaki, Japan, twelve hours later.

determine whether a particular earthquake has generated a tsunami. If observers detect a tsunami, they flash warnings to authorities around the Pacific. In recent years, to help in the effort, researchers have used computer models that predict how tsunamis propagate (▶Fig. 8.33). Unfortunate... warning system existed in the Indian Ocean, and though Hawaiian observers detected the earthquake a... ized that a tsunami was likely, they did not have the me... contact many local authorities. Even if they had, a... towns had no evacuation plans in place. A multina... team has begun work to remedy this situation.

Disease

Once the ground shaking and fires have stopped, ... may still threaten lives in an earthquake-damaged ... Earthquakes cut water and sewer lines, destroying ... water supplies and exposing the public to bacteria, an...

when and where an earthquake will happen, so people could evacuate dangerous buildings, turn off gas and electricity, and, with more warning, build stronger structures.

Can seismologists predict earthquakes? The answer depends on the time frame of the prediction. With our present understanding of the distribution of seismic zones and the frequency at which earthquakes occur, we can make long-term predictions (on the time scale of decades to centuries). For example, with some certainty, we *can* say that a major earthquake will rattle California during the next one hundred years, and that a major earthquake won't strike central ...

stretches of Indian Ocean coast. Far-field (or distant) tsunamis crossed the ocean and struck Sri Lanka 2.5 hours after the earthquake, the coast of India half an hour later, and the coast of Africa, on the west side of the Indian Ocean, 5.5 hours after the earthquake (▶Fig. 8.30d, e). Coastal towns vanished, fishing fleets sank, and beach resorts collapsed into rubble. In the end, more than 220,000 people died that day.

Japanese word that translates literally as ... apt name because tsunamis can be par... harbor towns. This name replaces the ... wave." Tsunamis have no relationship to ... them tidal waves can be misleading. ... aused by undersea earthquakes but, as we ... 13, they can also be set off by large ...

cause, tsunamis are very different from fa... storm waves. Large wind-driven waves ... of 10 to 30 meters in the open ocean. But ... rs have wavelengths of only tens of me... y affect the surface realm of the ocean. ... wave moves down to a depth of about half ... a submarine cruising at 100 m depth will ... m water, even if a storm rages above.) In ... a tsunami in deep water may cause a rise ... ost only a few tens of centimeters—a ship ... ldn't even notice—tsunamis have wave... hundreds of kilometers and thus affect ... of the ocean. In simpler terms, we can ... of a tsunami, in map view, as being more ... e width of a wind-driven wave. Because of ... storm wave and a tsunami have very dif... they strike the shore.

wave approaches the shore, friction be... the wave and the sea floor slows the bot... so the back of the wave catches up to the ... ed volume of water builds the wave higher ... top of the wave may fall over the front of ... e a breaker. In the case of a wind-driven ... may be tall when it washes onto the ... ve is so narrow that it doesn't contain ... , the wave makes it only part way up the ... ns out of water, friction slows it to a stop, ... s the water to spill back seaward. In the ... the wave is so wide that, as friction slows ... into a plateau that can be tens of meters ... kilometers wide. Thus, when a tsunami ... ontains so much water that it crosses the ... ps on going (▶Fig. 8.31b, c).

ian Ocean event remains etched in peo... se of the immense death toll and the ... erage. But it is not unique. Tsunamis gen... t Chilean earthquake of 1960 (M_w = 9.5) ... towns of South America and crossed the

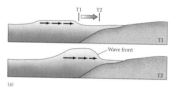

(a)

Wind-driven waves contain a small volume of water, and do not submerge higher areas.

(b)

Tsunamis are so wide (measured perpendicular to shore) that, like a plateau of water, they submerge the land.

(c)

FIGURE 8.31 (a) As a tsunami approaches the shore, friction between water and the sea floor slows the wave, allowing the back part of the wave to catch up with the front part. As a consequence, the wave becomes much higher than it was in the open ocean as the front moves from its position at Time 1 to its position at Time 2. (b) Wind-driven waves may be quite high, but they are so narrow that they contain relatively little water, and they only go part-way up the beach until the water stops moving landward. The water in such a wave has a circular motion. (c) When a tsunami washes ashore, the water moves landward in a straight line. A tsunami is like a plateau of water—it contains so much water that the water keeps moving inland and submerges a broad area.

Pacific, causing a 10.7-m-high wall of water to strike Hawaii 15 hours later. When, 21 hours after the earthquake, the tsunami reached Japan, it flattened coastal villages and left 50,000 people homeless. The tsunami following the 1964 Good Friday earthquake in Alaska destroyed ports at Valdez and Kodiak (▶Fig. 8.32a). Typically, an earthquake generates several tsunamis that arrive on distant shores as much as an hour apart (▶Fig. 8.32b).

Because of the danger of tsunamis, predicting their arrival can save thousands of lives. A tsunami warning center in Hawaii keeps track of earthquakes around the Pacific and uses data relayed from tide gauges and sea-floor pressure gauges to

(a)

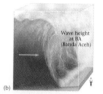

Wave height at BA (Banda Aceh)

(b)

(c)

(d)

(e)

(f)

(g)

FIGURE 8.30 The Indian Ocean tsunami disaster of 2004. (a) A tourist took this photo of the tsunami approaching the shore of Sumatra. (b) At its highest, the wave front was many times the height of a person. (c, d) Satellite photos of the Indonesian province of Aceh before and after the tsunami struck. Note that the city was washed away, and that the beach vanished. (e) A satellite photo of the tsunami submerging land up to 500 m inland of the coastline of Sri Lanka; (f) A map of the Indian Ocean showing the position of the wave at various times after the earthquake. The enlargement of the boxed area (g) shows the configuration of plate boundaries near the epicenter.

STATE-OF-THE-ART ILLUSTRATIONS

One of Stephen Marshak's goals in *Essentials of Geology* was to develop art that conveys the dynamic way that geologic processes work. Marshak worked directly with a team of top artists to create figures that are clear and simple enough for students to understand but realistic enough to provide a reference framework. These spectacular 3-D illustrations are accompanied by photographs to help students visualize real geology. All of these illustrations are available to instructors as PowerPoint slides on the Norton Media Library CD-ROM that accompanies the text.

> Many new figures have been added for the Second Edition, and many from the First Edition have been enhanced.

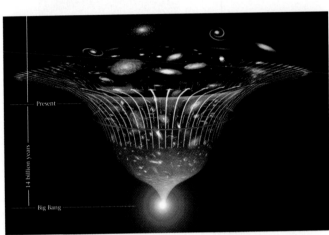

FIGURE 1.6 An artist's rendition of the big bang, followed by expansion of the Universe. The horizontal direction represents size and the vertical direction represents time. Recent work suggests that the rate of expansion has changed as time passes.

reached an age of 3 minutes, its temperature had fallen below 1 billion degrees, and its diameter had grown to about 100 billion km (60 billion miles). Under these conditions, nuclei of new atoms began to form through the collision and fusion (sticking together) of hydrogen atoms. Formation of new nuclei by fusion reactions at this time is called "big bang nucleosynthesis" because it happened *before* any stars existed. Big bang nucleosynthesis could produce only small atoms, meaning ones containing a small number of protons, and it happened very rapidly. In fact, virtually all of the new atomic nuclei that would form by big bang nucleosynthesis had formed by the end of 5 minutes. Eventually, the Universe became cool enough for chemical bonds to bind atoms of certain elements together in molecules—most notably, two hydrogen atoms could join to form molecules of H_2. As the Universe continued to expand and cool further, atoms and molecules slowed down and accumulated into **nebulae**, patchy clouds of gas separated from one another by the vacuum of space. Gas mak-

ing up the earliest nebulae of the Universe consisted almost entirely of the two smallest atoms, namely, hydrogen (74%, by volume) and helium (24%).

Forming the First Stars

When the Universe reached its 200 millionth birthday, it contained immense, slowly swirling, dark nebulae separated by vast voids of empty space (▶Fig. 1.7). The Universe could not remain this way forever, however, because of the invisible but persistent pull of gravity. Eventually, gravity began to remold the Universe pervasively and permanently.

All matter exerts gravitational pull—a type of force—on its surroundings, and as Isaac Newton first pointed out, the amount of pull depends on the amount of mass. Somewhere in the young Universe, the gravitational pull of an initially denser region of a nebula began to suck in surrounding gases and, in a grand example of "the rich getting richer," grew in mass and, therefore, density (mass per

The Effects of Urbanization and Agriculture on Flooding

Human society has had a major impact on the world's rivers. People have introduced pollution into streams, have restricted flow by building dams, and have diverted flow into canals and ditches that carry it to farms and cities. Notably, overuse of water by people has resulted in the long-term drying up of many rivers around the world. But although the consumption of water for agricultural and industrial purposes decreases the overall supply of river water, urbanization may actually increase the short-term supply. Cities cover the ground with impermeable concrete or blacktop, so rainfall does not soak into the ground but rather runs into storm sewers and then into streams, causing local flooding. Stream discharge during a rainfall thus increases much more rapidly than it would without urbanization. Similarly, although damming rivers decreases the amount of silt a river carries downstream, agriculture may increase the sediment supply: agriculture decreases the vegetative cover on the land, so that when it rains, soil washes into streams.

FIGURE 14.35 The Central Arizona Project canal, shunting water from the Colorado River to Phoenix.

CHAPTER SUMMARY

- Streams are bodies of water that flow down channels and drain the land surface. Channels develop when sheetwash cuts into the substrate and concentrates the water flow; they grow by headward erosion.

- Drainage networks consist of many tributaries that flow into a trunk stream.

- Permanent streams exist where the water table lies above the bed of the channel. Where the water table lies below the channel bed, streams are ephemeral.

- The discharge of a stream is the volume of water passing a point in a second.

- Streams erode the landscape by scouring, lifting, abrading, and dissolving. Sediment in a stream consists of dissolved, suspended, and bed loads. The total quantity of sediment carried by a stream is its capacity. Competence is the maximum particle size a stream can carry. When stream water slows, it deposits alluvium.

- Longitudinal profiles of streams are concave up, for a stream has steeper gradients at its headwaters than near its mouth. Streams cannot cut below the base level.

- Streams cut valleys or canyons. Where a stream has a bed littered with large rocks, rapids develop. Where a stream plunges off a vertical face, a waterfall forms.

- Meandering streams wander back and forth across a floodplain. Such a stream erodes its outer bank and builds out sediment into a point bar on the inner bank. Eventually,

THE VIEW FROM SPACE The Ganges River drains the base of the Himalaya Mountains and carries sediment into the Indian Ocean. The river divides into many meandering channels and has built out a huge delta. These low-lying lands are home to millions of people, but many of these areas flood during the monsoon season, or during typhoons.

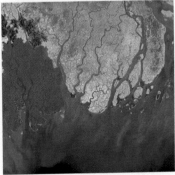

> Over a dozen dramatic satellite photographs have been added to the Second Edition.

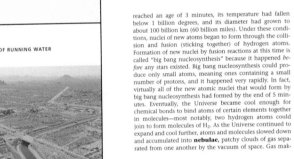

WHAT A GEOLOGIST SEES

Stephen Marshak has developed a particularly clever device to show students how thoughtful observation can reveal a large amount of information. Using simple drawings, he conveys what a trained geologist sees when viewing a photograph of a landscape. The untrained eye sees only a pretty picture—a geologist sees a page of Earth history.

FIGURE 9.17 (a) An anticline, exposed in a road cut near Kingston, New York. (b) A syncline, exposed in a road cut near Sideling Hill, in Maryland. (c) A train of folds exposed in sea cliffs in eastern Ireland. Note that the axial planes of these folds are not vertical. (d) What a geologist sees in the previous photo.

Eight new "What a Geologist Sees" figures have been added to the Second Edition.

Folds develop in two principal ways. During formation of flexual folds, a stack of layers bends, and slip occurs between the layers. The same phenomenon happens when you bend a deck of cards—to accommodate the change in shape, the cards slide with respect to each other. Flow folds form when the rock, overall, is so soft that it behaves like weak plastic and slowly flows; these folds develop simply because different parts of the rock body flow at different rates.

Folds develop for a variety of reasons. Some layers wrinkle up, or buckle, in response to end-on compression (▶Fig. 9.19a). Others form where shear stress gradually moves one part of a layer up and over another part (▶Fig. 9.19b). Still others develop where rock layers move up and over bends in a fault and must curve to conform with the fault's shape (▶Fig. 9.19c). Finally, some folds form when a block of basement moves and bends the overlying sedimentary layers (▶Fig. 9.19d).

9.7 TECTONIC FOLIATION IN ROCKS

In an undeformed sandstone, the grains of quartz are roughly spherical, and in an undeformed shale, clay flakes press together into the plane of bedding so that shales tend to split parallel to the bedding. During ductile deformation, however, internal changes take place in a rock that gradually modify the original shape and arrangement of grains. For example, quartz grains may transform into cigar shapes, elongate ribbons, or tiny pancakes, and clay flakes may recrystallize or reorient so that they lie at an angle to the bedding. Overall, deformation can produce inequant grains and can cause them to align parallel to each other, thereby generating metamorphic foliation in the rock. We refer to layering created by the alignment of deformed and/or reoriented grains as tectonic foliation (▶Fig. 9.20a).

TWO-PAGE SYNOPTIC PAINTINGS

These comprehensive paintings by the most respected geology artist in the world—Gary Hincks—are designed to encapsulate a number of topics. Each of the paintings is a complete synopsis of a major part of a chapter. By studying them carefully, students can see interconnections among subtopics covered in the chapter.

Three new paintings were commissioned for the Second Edition. They appear in the chapters on energy, mineral resources, and oceans and coasts.

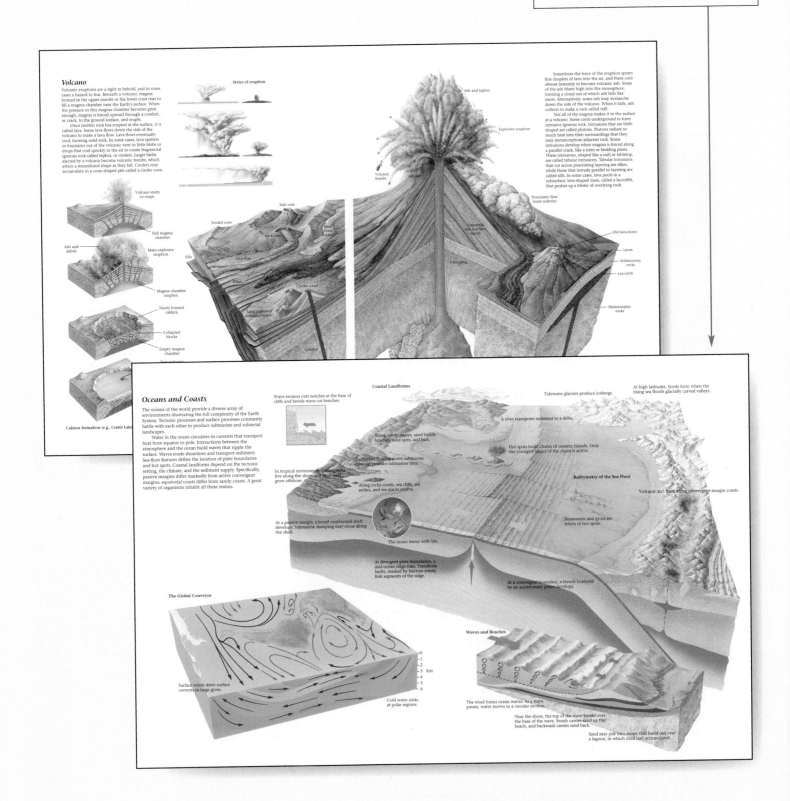

CAPTURES STUDENT INTEREST

CHAPTER 5

A Surface Veneer: Sediments, Soils, and Sedimentary Rocks

5.1 INTRODUCTION

In the 1950s, the government of Egypt decided to build the Aswan High Dam to trap water of the Nile River in a huge reservoir before the water could reach the Mediterranean Sea. To identify a good site for the dam's foundation, geologists drilled holes into the ground to find the depth to bedrock. They discovered that the present-day Nile River flows on the surface of a 1.5-km-thick layer of gravel, sand, and mud that fills what was once a canyon as large as the Grand Canyon (▶Fig. 5.1). How could the river once have carved a canyon 1.5 km deep, and why did this canyon later fill with sediment?

The origin of the pre-Nile canyon remained a mystery until the summer of 1970, when geologists had the opportunity to drill holes into the floor of the Mediterranean Sea to find out what lay beneath. They expected the sea floor to be covered with shells of plankton (tiny floating organisms) that had settled out of the water, or with clay that rivers had carried to the

These cliffs, near Bryce Canyon (Utah), expose beds of sedimentary rock deposited in lakes, and by streams, over 40 million years ago. Present-day erosion has produced aprons of debris at the base of the cliffs.

when a slurry composed of sand and/or gravel mixed with cement and water is allowed to harden. The hardening takes place when a complex assemblage of minerals grows by chemical reactions in the slurry; these minerals bind together the grains of sand or gravel in mortar or concrete. (Note that the word *mortar* refers to the substance that holds bricks or stone blocks together, whereas *concrete* refers to the substance that workers shape into roads or walls by spreading it out into a layer or by pouring it into a form.) The **cement** in mortar or concrete starts out as a powder composed of lime (CaO), quartz (SiO_2), aluminum oxide (Al_2O_3), and iron oxide (Fe_2O_3). Typically, lime accounts for 66% of cement, silica for 25%, and the remaining chemicals for about 9%.

It appears that the ancient Romans were the first to use cement—they made it from a mixture of volcanic glass and limestone. During recent centuries, most cement has been produced by heating specific types of limestone (which happened to contain calcite, clay, and quartz in the correct proportions) in a kiln up to a temperature of about 1,450°C; the heating releases CO_2 gas and produces "clinker," chunks consisting of lime and other oxide compounds. Manufacturers crush the clinker into cement powder and pack it in bags for transport. But natural limestone with the exact composition of cement is fairly rare, so most cement used today is Portland cement, made by intentionally mixing limestone, sandstone, and shale in just the right proportions to provide the proper chemical makeup. Isaac Johnson, an English engineer, came up with the recipe for Portland cement in 1844, and he named it after the town of Portland, England, because he thought it resembled rock exposed there.

Nonmetallic Minerals in Your Home

We use an astounding variety of nonmetallic geologic resources (▶Table 12.1) without ever realizing where they come from. Consider the materials in a house or apartment. The concrete foundation consists of cement, made from limestone mixed with sand or gravel. The bricks in the exterior walls originated as clay, formed from the chemical weathering of silicate rocks and perhaps dug from the floodplain of a stream. To make bricks, workers mold wet clay into blocks, which they then bake. Baking drives out water and causes metamorphic reactions that recrystallize the clay.

The glass used to glaze windows consists largely of silica, formed by first melting and then freezing pure quartz sand from a beach deposit or a sandstone formation. Gypsum board (drywall), used to construct interior walls, is a sandwich of gypsum powder between sheets of paper. Gypsum ($CaSO_4 \cdot 2H_2O$) occurs in evaporite strata precipitated from seawater or saline lake water. Evaporites provide other useful minerals as well, such as halite and borax. Truly, without the geologic resources of the Earth, modern society would grind to a halt.

12.15 GLOBAL MINERAL NEEDS

How Long Will Resources Last?

The average citizen of an industrialized country uses 25 kilograms (kg) of aluminum, 10 kg of copper, and 550 kg of iron and steel in a year's time (▶Fig. 12.37). If you combine

Water on Mars?

In 1877, an Italian astronomer named Giovanni Schiaparelli studied the surface of Mars with a telescope and announced that long, straight *canali* criss-crossed the planet's surface. *Canali* should have been translated into the English word "channel," but perhaps because of the recent construction of the Suez Canal, newspapers of the day translated the word into the English "canal," with the implication that the features had been constructed by intelligent beings. An eminent American astronomer began to study the "canals" and suggested that they had been built to carry water from polar ice caps to Martian deserts.

Late-twentieth-century satellite mapping of Mars showed that the "canals" do not exist—they were simply optical illusions. There are no lakes, oceans, rainstorms, or flowing rivers on the surface of Mars today. The atmosphere of Mars has such low density, and thus exerts so little pressure on the planet's surface, that any liquid water released at the surface in recent time would quickly evaporate. Thus, Mars has no hydrologic cycle the way the Earth does. But three crucial questions remain: Does liquid water ever form, even for short periods of time, on the Martian surface today? Did Mars ever have a significant amount of running water or standing water in the past? If the planet once had significant water, where is the water now? The question of the presence of water lies at the heart of the even more basic question: Given that the simplest life as we know it requires water, is there, or was there, life on Mars?

Many planetary geologists believe that the case for liquid water on Mars is quite strong. Much of the evidence comes from comparing landforms on the planet's surface with landforms of known origin on Earth. High-resolution images of Mars reveal a number of landforms that look as if they formed in response to flowing water. Examples include networks of channels resembling river networks on Earth (▶Fig. E.5), scour features, deep gullies, and streamlined deposits of sediment.

Studies by the *Odyssey* satellite in 2003, and by Mars rovers (*Spirit* and *Opportunity*) that landed on the planet in 2004, have added in-

triguing new data to the debate. *Odyssey* detected hints that hydrogen, an element in water, exists beneath the surface of the planet over broad regions, and the Mars rovers have documented the existence of hematite and gypsum minerals that form in the presence of water. The rovers have also found sedimentary deposits that appear to have been deposited in water. Researchers speculate that Mars was much wetter in its past, perhaps billions of years ago. But since the atmosphere became less dense, the water evaporated and now lies hidden underground or trapped in polar ice caps.

FIGURE E.5 A riverlike network of channels on Mars, as photographed by a satellite. Small tributary channels join a large trunk channel.

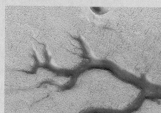

Mars has no mountain belts or volcanic arcs. In fact, most landscape features on Mars, with the exception of wind-related ones, are over three billion years old. Thermal activity also led to the eruption of gargantuan hot-spot volcanoes, such as the 22-km-high Olympus Mons (▶Fig. E.6). Mars also boasts the largest known canyon, the Valles Marineris, a gash over 3,000 km long and 8 km deep—no comparable feature exists on Earth. Because Mars has no vegetation and no longer has rain, its surface does not weather and erode like that of Earth, so it still bears the scars of impact by swarms of meteors earlier in the history of the solar system, scars that have long since disappeared on Earth.

Venus is closer in size to the Earth and may still have operating mantle plumes. Virtually the entire surface of Venus was covered by volcanic deposits about 300 to 1,600 Ma,

making the planet's surface much younger than those of the Moon and Mars. Further, Venus has a dense atmosphere that protects it from impacts by smaller objects. Because there has been relatively little cratering since the resurfacing event, volcanic and tectonic features dominate the landscape of Venus (▶Fig. E.7). Satellites have used radar to reveal a variety of volcanic constructions (such as shield volcanoes, lava flows, calderas). Rifting on Venus produced faults. Liquid water cannot survive the scalding temperatures of Venus's surface, so no hydrologic cycle operates there and no life exists. Because of the density of the atmosphere, winds are too slow to cause much erosion or deposition.

After this brief side trip to other planets, let's now return to Earth. Our planet has the greatest diversity of landscapes in the solar system.

The Sidewalks of New York

Untold tons of concrete have gone into the construction of New York City. In fact, with the exception of a few city parks, most of the walking space in the city consists of concrete. And concrete skyscrapers tower above this concrete plain. Where does all this concrete come from?

Much of the sand used in New York concrete was deposited during the last ice age. As vast glaciers moved southward over 14,000 years ago, they ground away the igneous and metamorphic rocks that constituted central and eastern Canada. These ancient rocks contained abundant quartz, and since quartz lasts a long time (it does not undergo chemical weathering easily), the sediment transported by the glaciers retained a large amount of quartz. Glaciers deposited this sediment in huge piles. As the glaciers melted, fast-moving rivers of meltwater washed the sediment, sorting sand from mud and pebbles. The sand was deposited by the meltwater rivers.

What about the cement? Cement contains a mixture of lime, derived from limestone, and other elements derived from shale and sandstone. The bedrock of New York, though, consists largely of schist

and gneiss, not sedimentary rocks. Fortunately, a source of rocks appropriate for making cement lies up the Hudson River. A rock unit called the Rosendale Formation, which naturally contains exactly the right mixture of lime and silica needed to make a durable cement, crops out in low ridges just to the west of the river. Beginning in the late 1820s, workers began quarrying the Rosendale Formation for cement, creating a network of underground caverns. They then dumped the excavated rock into nearby kilns and roasted it to produce cement. The resulting powder was packed into barrels, loaded onto barges, and shipped downriver to New York. As demand for cement increased, operators eventually dug open-pit quarries from which they excavated other limestone and shale units, mixing them together in the correct proportion to make Portland cement.

The rocks making up limestone consist of cemented-together shell fragments and small, reeflike colonies of organisms. In other words, the lime in the concrete of New York sidewalks was originally extracted from seawater by living organisms over 350 million years ago.

ANIMATIONS

Developed by Stephen Marshak to illustrate dynamic Earth processes, over sixty original animations emphasize plate tectonics, geologic hazards, and Earth systems science concepts. Approximately half of these animations are new to the Second Edition, and many are based on "What a Geologist Sees" figures from the text.

Designed for use by instructors and students, animations can be enlarged to full-screen view for lecture display. VCR-like controls make it easy for instructors to control the pace of the animations during lectures.

Animations are available on the Norton Media Library Instructor's CD-ROM, and on the Student CD-ROM and web site at wwnorton.com/earth. See pp. *xix–xx* for a complete list and descriptions of supplements for students and instructors.

"The Instructor CD contains some of the more imaginative and detailed figures and animations I have found."—Jeff Knott, *California State University, Fullerton*

"The Student CD and Web site were a welcome surprise when I first adopted the book. The animations are very good. . . . I use many of them during class."—Tom Juster, *University of Southern Florida*

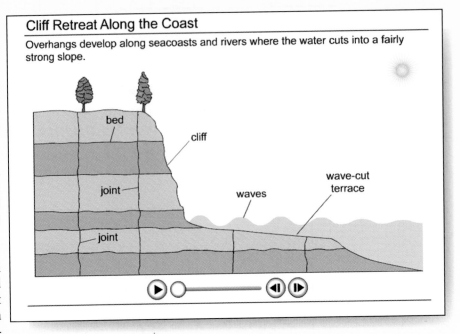

Cliff Retreat Along the Coast

Overhangs develop along seacoasts and rivers where the water cuts into a fairly strong slope.

What A Geologist Sees—Thrust Fault

ESSENTIALS OF GEOLOGY

Second Edition

Essentials of
Geology

Second Edition

STEPHEN MARSHAK

University of Illinois

W. W. NORTON & COMPANY

NEW YORK   LONDON

W. W. Norton & Company has been independent since its founding in 1923, when William Warder Norton and Mary D. Herter Norton first published lectures delivered at the People's Institute, the adult education division of New York City's Cooper Union. The Nortons soon expanded their program beyond the Institute, publishing books by celebrated academics from America and abroad. By mid-century, the two major pillars of Norton's publishing program—trade books and college texts—were firmly established. In the 1950s, the Norton family transferred control of the company to its employees, and today—with a staff of four hundred and a comparable number of trade, college, and professional titles published each year—W. W. Norton & Company stands as the largest and oldest publishing house owned wholly by its employees.

Composition by TSI Graphics
Manufacturing by Courier
Illustrations for the Second Edition by Precision Graphics

Editor: Jack Repcheck
Project editor: Thomas Foley
Copy editor: Barbara Gerr
Electronic media editor: April Lange
Director of manufacturing: Christopher Granville
Photography editor: Stephanie Romeo
Editorial assistant: Mik Awake
Book designer: Joan Greenfield

Library of Congress Cataloging-in-Publication Data

Marshak, Stephen, 1955–
    Essentials of geology / Stephen Marshak.—2nd ed.
        p. cm.
    Includes bibliographical references and index.

    ISBN 13: 978-0-393-92815-0 (pbk.)
    ISBN 10: 0-393-92815-2 (pbk.)

    1. Geology—Textbooks. I. Title.

QE28.M3415 2006
550—dc22   2006045754

W. W. Norton & Company, Inc., 500 Fifth Avenue, New York, N.Y. 10110
www.wwnorton.com

W. W. Norton & Company Ltd., Castle House, 75/76 Wells Street, London W1T 3QT

2 3 4 5 6 7 8 9 0

BRIEF CONTENTS

Preface		xvii
Prelude	*And Just What Is Geology?*	1
Chapter 1	*The Earth in Context*	8
Chapter 2	*The Way the Earth Works: Plate Tectonics*	35
Chapter 3	*Patterns in Nature: Minerals*	78
Interlude A	*Rock Groups*	95
Chapter 4	*Up from the Inferno: Magma and Igneous Rocks*	102
Chapter 5	*A Surface Veneer: Sediments, Soils, and Sedimentary Rocks*	121
Chapter 6	*Metamorphism: A Process of Change*	153
Interlude B	*The Rock Cycle*	174
Chapter 7	*The Wrath of Vulcan: Volcanic Eruptions*	180
Chapter 8	*A Violent Pulse: Earthquakes*	206
Interlude C	*Seeing Inside the Earth*	240
Chapter 9	*Crags, Cracks, and Crumples: Crustal Deformation and Mountain Building*	248
Interlude D	*Memories of Past Life: Fossils and Evolution*	273
Chapter 10	*Deep Time: How Old Is Old?*	283
Chapter 11	*A Biography of Earth*	306
Chapter 12	*Riches in Rock: Energy and Mineral Resources*	329
Interlude E	*An Introduction to Landscapes, and the Hydrologic Cycle*	364
Chapter 13	*Unsafe Ground: Landslides and Other Mass Movements*	373
Chapter 14	*Streams and Associated Flooding: The Geology of Running Water*	391
Chapter 15	*Restless Realm: Oceans and Coasts*	418
Chapter 16	*A Hidden Reserve: Groundwater*	450
Chapter 17	*Dry Regions: The Geology of Deserts*	473
Chapter 18	*Amazing Ice: Glaciers and Ice Ages*	493
Chapter 19	*Global Change in the Earth System*	525
Appendix	*Scientific Background: Matter and Energy*	A-1
Metric Conversion Chart		A-12
Glossary		G-1
Credits		C-1
Index		I-1

"WHAT A GEOLOGIST SEES"

Shiprock, Fig. 4.11c — 111
Basalt Sill in Antarctica, Fig. 4.11e — 111
Sierra Nevada, Fig. 4.13c — 112
Grand Canyon, Fig. 5.2 — 123
Channel Shape, Fig. 5.29d — 145
Geologist's Interpretation of Outcrop, Fig. 6.9b — 161
Mt. Vesuvius, Fig. 7.1b — 181
Displacement on Fault, Fig. 8.5b — 209

San Andreas Fault, Fig. 9.13b — 256
A Normal Fault, Fig. 9.13d — 256
Thrust Fault, Fig. 9.15b — 258
Train of Folds, Fig. 9.17d — 260
Shear Movement, Fig. 9.20d — 262
Siccar Point Unconformity, Fig. 10.5b — 289
Stromatolite Deposit, Fig. 11.4c — 309

TWO-PAGE SYNOPTIC PAINTINGS

The Birth of the Earth-Moon System, Chapter 1 — 16–17
The Earth, from Surface to Center, Chapter 1 — 30–31
The Theory of Plate Tectonics, Chapter 2 — 72–73
The Formation of Igneous Rocks, Chapter 4 — 114–15
The Formation of Sedimentary Rocks, Chapter 5 — 148–49
Environments of Metamorphism, Chapter 6 — 168–69
Rock-Forming Environments and the Rock Cycle, Interlude B — 178–79
Volcano, Chapter 7 — 188–89
Faulting in the Crust, Chapter 8 — 216–17
The Collision of India with Asia, Chapter 9 — 264–65
The Record in Rocks: Reconstructing Geologic History, Chapter 10 — 298–99

The Evolution of Earth, Chapter 11 — 326–27
Power from the Earth, Chapter 12 — 348–49
Forming and Processing Earth's Mineral Resources, Chapter 12 — 360–61
The Hydrologic Cycle, Interlude E — 368–69
Mass Movement, Chapter 13 — 386–87
River Systems, Chapter 14 — 410–11
Oceans and Coasts, Chapter 15 — 440–41
Caves and Karst Landscape, Chapter 16 — 466–67
The Desert Realm, Chapter 17 — 486–87
Glaciers and Glacial Landforms, Chapter 18 — 510–11
The Earth System, Chapter 19 — 538–39

CONTENTS

PREFACE   xvii

Prelude
And Just What Is Geology?   1

P.1   In Search of Ideas   1
P.2   The Nature of Geology   2
P.3   Themes of This Book   4

BOX P.1   SCIENCE TOOLBOX: *The Scientific Method*   6

Chapter 1
The Earth in Context   8

1.1   Introduction   8
1.2   The Modern Image of the Universe   9
1.3   Forming the Universe   9
1.4   Making Order from Chaos   12
1.5   We Are All Made of Stardust   14

Featured painting: The Birth of the Earth-Moon System   16–17

1.6   Welcome to the Earth System and Its Neighborhood   20
1.7   The Atmosphere   21
1.8   Land and Oceans   22
1.9   What Is the Earth Made of?   23
1.10   Discovering That the Earth Has Layers   25
1.11   What Are the Layers Made of?   27

BOX 1.1   THE REST OF THE STORY: *Meteors and Meteorites*   28

1.12   The Lithosphere and the Asthenosphere   29

Featured painting: The Earth, from Surface to Center   30–31

Chapter 2
The Way the Earth Works: Plate Tectonics   35

2.1   Introduction   35
2.2   Wegener's Evidence for Continental Drift   36
2.3   Paleomagnetism and Apparent Polar-Wander Paths   39

BOX 2.1   SCIENCE TOOLBOX: *Some Fundamentals of Magnetism*   40

2.4   Setting the Stage for the Discovery of Sea-Floor Spreading   45

2.5   Harry Hess and His "Essay in Geopoetry"   48
2.6   Marine Magnetic Anomalies: Evidence for Sea-Floor Spreading   48
2.7   Deep-Sea Drilling: Further Evidence   53
2.8   What Do We Mean by Plate Tectonics   53
2.9   Divergent Plate Boundaries and Sea-Floor Spreading   56
2.10   Convergent Plate Boundaries and Subduction   59
2.11   Transform Plate Boundaries   63
2.12   Special Locations in the Plate Mosaic   64
2.13   The Birth and Death of Plate Boundaries   68
2.14   What Drives Plate Motion?   69
2.15   The Velocity of Plate Motions   71

Featured Painting: The Theory of Plate Tectonics   72–73

Chapter 3
Patterns in Nature: Minerals   78

3.1   Introduction   78
3.2   What Is a Mineral?   79

BOX 3.1   SCIENCE TOOLBOX: *Some Basic Definitions from Chemistry*   80

3.3   Beauty in Patterns: Crystals and Their Structure   81
3.4   How Can You Tell One Mineral From Another?   85
3.5   Organizing Our Knowledge: Mineral Classification   89
3.6   Something Precious—Gems!   91

BOX 3.2   THE REST OF THE STORY: *Where Do Diamonds Come From?*   92

Interlude A
Rock Groups   95

A.1   Introduction   95
A.2   What Is Rock?   96
A.3   Rock Occurrences   97
A.4   The Basis of Rock Classification   97
A.5   Studying Rock   100

Chapter 4

*Up from the Inferno: Magma and
Igneous Rocks* 102

4.1	Introduction	102
4.2	The Formation of Magma	104
4.3	What Is Magma Made Of?	105
4.4	The Movement of Magma and Lava	107

BOX 4.1 THE REST OF THE STORY:
Bowen's Reaction Series 108

4.5	Extrusive Versus Intrusive Environments	109
4.6	Transforming Magma into Rock	111

*Featured Painting: The Formation of Igneous
Rocks* 114–15

4.7	Igneous Rock Textures	116
4.8	Classifying Igneous Rocks	117

Chapter 5

*A Surface Veneer: Sediments, Soils, and
Sedimentary Rocks* 121

5.1	Introduction	121
5.2	Weathering: The Formation of Sediment	122
5.3	Soil	128
5.4	Classes of Sedimentary Rocks	131
5.5	Sedimentary Structures	138
5.6	Sedimentary Environments	143
5.7	Sedimentary Basins	147

*Featured Painting: The Formation of
Sedimentary Rocks* 148–49

5.8	Diagenesis	150

Chapter 6

Metamorphism: A Process of Change 153

6.1	Introduction	153
6.2	What Is a Metamorphic Rock?	154
6.3	Causes of Metamorphism	154
6.4	Types of Metamorphic Rocks	158
6.5	The Intensity of Metamorphism	161

BOX 6.1 THE REST OF THE STORY:
Metamorphic Facies 163

6.6	Metamorphism in the Context of Plate Tectonics	164

BOX 6.2 THE HUMAN ANGLE: *Pottery Making—
An Analog for Thermal Metamorphism* 167

*Featured Painting: Environments of
Metamorphism* 168–69

6.7	Bringing Metamorphic Rocks Back to the Earth's Surface	170
6.8	Where Do You Find Metamorphic Rocks?	171

Interlude B

The Rock Cycle 174

B.1	Introduction	174
B.2	A Case Study of the Rock Cycle	175
B.3	Rates of Movement Through the Rock Cycle	175
B.4	What Drives the Rock Cycle in the Earth System?	177

*Featured Painting: Rock-Forming Environments
and the Rock Cycle* 178–79

Chapter 7

The Wrath of Vulcan: Volcanic Eruptions 180

7.1	Introduction	180
7.2	The Products of Volcanic Eruptions	182
7.3	The Architecture and Shape of Volcanoes	186

Featured Painting: Volcano 188–89

7.4	Eruptive Styles: Will It Flow, or Will It Blow?	190

BOX 7.1 THE HUMAN ANGLE:
Volcanic Explosions to Remember 192

7.5	Plate Boundary and Rift Eruptions	194
7.6	Hot-Spot Eruptions	194
7.7	Volcanoes in the Landscape	197
7.8	Beware: Volcanoes Are Hazards!	198
7.9	Protection from Vulcan's Wrath	201
7.10	Volcanoes and Climate	202
7.11	Volcanoes on Other Planets	203

Chapter 8

A Violent Pulse: Earthquakes 206

8.1	Introduction	206
8.2	Causes of Earthquakes	207
8.3	Seismic Waves	212
8.4	Measuring and Locating Earthquakes	212

Featured Painting: Faulting in the Crust 216–17

8.5	Where and Why do Earthquakes Occur?	220
8.6	Damage from Earthquakes	226
8.7	Predicting the "Big One"	234
8.8	Earthquake Engineering and Zoning	237

Interlude C
Seeing Inside the Earth   240

C.1   Introduction   240

C.2   The Movement of Seismic Waves Through the Earth   240

C.3   The Reflection and Refraction of Wave Energy   241

C.4   Discovering the Crust-Mantle Boundary   242

C.5   Defining the Structure of the Mantle   243

C.6   Discovering the Core-Mantle Boundary   244

C.7   Discovering the Nature of the Core   244

C.8   Fine-Tuning Our Image of the Earth's Layers   245

C.9   An Integrated View of the Earth: New Discoveries   246

Chapter 9
Crags, Cracks, and Crumples: Crustal Deformation and Mountain Building   248

9.1   Introduction   248

9.2   Mountain Belts and the Concept of Orogeny   249

9.3   Rock Deformation in the Earth's Crust   249

9.4   Joints: Natural Cracks in Rocks   254

BOX 9.1   THE REST OF THE STORY: *Describing the Orientation of Structures*   254

9.5   Faults: Fractures on Which Sliding Has Occurred   256

9.6   Folds: Curving Rock Layers   258

9.7   Tectonic Foliation in Rocks   260

9.8   Uplift and the Formation of Mountain Topography   261

Featured Painting: The Collision of India with Asia   264–65

9.9   Causes of Mountain Building   267

9.10   Cratons and the Deformation within Them   268

9.11   Measuring Mountain Building in Progress   270

Interlude D
Memories of Past Life: Fossils and Evolution   273

D.1   The Discovery of Fossils   273

D.2   Fossilization   274

D.3   Classifying Life   278

D.4   Classifying Fossils   278

D.5   The Fossil Record   279

D.6   Is the Fossil Record Complete?   280

D.7   Evolution and Extinction   280

Chapter 10
Deep Time: How Old Is Old?   283

10.1   Introduction   283

10.2   The Concept of Geologic Time   284

10.3   Physical Principles for Defining Relative Age   284

10.4   Adding Fossils to the Story: Fossil Succession   285

10.5   Unconformities: Gaps in the Record   287

10.6   Stratigraphic Formations and Their Correlation   289

10.7   The Geologic Column   293

10.8   Numerical Age and the Radiometric Clock   295

Featured Painting: The Record in Rocks: Reconstructing Geologic History   298–99

BOX 10.1   THE REST OF THE STORY: *Carbon-14 Dating*   300

10.9   Adding Numerical Ages to the Geologic Column   301

10.10   The Age of the Earth   302

10.11   Picturing Geologic Time   304

Chapter 11
A Biography of Earth   306

11.1   Introduction   306

11.2   The Hadean Eon: Hell on Earth?   307

11.3   The Archean Eon: The Birth of the Continents and the Appearance of Life   308

11.4   The Proterozoic Eon: Transition to the Modern World   310

11.5   The Phanerozoic Eon: Life Diversifies, and Today's Continents Form   313

11.6   The Paleozoic Era: From Rodinia to Pangaea   313

BOX 11.1   SCIENCE TOOLBOX: *Stratigraphic Sequences and Sea-Level Change*   315

11.7   The Mesozoic Era: When Dinosaurs Ruled   319

11.8   The Cenozoic Era: The Final Stretch to the Present   322

Featured Painting: The Evolution of Earth   326–27

Chapter 12
Riches in Rock: Energy and Mineral Resources 329

12.1	Introduction	329
12.2	Sources of Energy in the Earth System	330
12.3	Oil and Gas	330
12.4	Making an Oil Reserve	331
12.5	Oil Exploration and Production	334

BOX 12.1 THE REST OF THE STORY: *Types of Oil and Gas Traps* 335

12.6	Alternative Reserves of Hydrocarbons	336
12.7	Coal: Energy from the Swamps of the Past	338
12.8	Nuclear Power	341
12.9	Other Energy Sources	343
12.10	Energy Choices, Energy Problems	345

Featured Painting: Power from the Earth 348–49

12.11	Introducing Mineral Resources	350
12.12	Metals, Ores, and Ore Deposits	350
12.13	Ore-Mineral Exploration and Production	355
12.14	Nonmetallic Mineral Resources	356
12.15	Global Mineral Needs	357

BOX 12.2 THE HUMAN ANGLE: *The Sidewalks of New York* 357

Featured Painting: Forming and Processing Earth's Mineral Resources 360–61

Interlude E
An Introduction to Landscapes, and the Hydrologic Cycle 364

E.1	Introduction	364
E.2	Shaping the Earth's Surface	364
E.3	Factors Controlling Landscape Development	366
E.4	The Hydrologic Cycle	367

Featured Painting: The Hydrologic Cycle 368–69

E.5	Landscapes of Other Planets	370

BOX E.1 GEOLOGIC CASE STUDY: *Water on Mars?* 371

Chapter 13
Unsafe Ground: Landslides and Other Mass Movements 373

13.1	Introduction	373
13.2	Types of Mass Movement	374

BOX 13.1 GEOLOGIC CASE STUDY: *The Storegga Slide and the North Sea Tsunamis* 377

13.3	Why Do Mass Movements Occur?	380
13.4	How Can We Protect Against Mass-Movement Disasters?	384

Featured Painting: Mass Movement 386–87

Chapter 14
Streams and Associated Flooding: The Geology of Running Water 391

14.1	Introduction	391
14.2	Draining the Land	392
14.3	Discharge and Turbulence	395
14.4	The Work of Running Water	396
14.5	How Streams Change Along Their Length	398
14.6	Streams and Their Deposits in the Landscape	400
14.7	The Evolution of Drainage	404
14.8	Raging Waters	408

Featured Painting: River Systems 410–11

14.9	Rivers: A Vanishing Resource?	414

Chapter 15
Restless Realm: Oceans and Coasts 418

15.1	Introduction	418
15.2	Landscapes Beneath the Sea	419
15.3	Ocean Water and Currents	423

BOX 15.1 SCIENCE TOOLBOX: *The Coriolis Effect* 425

15.4	The Tides Go Out . . . The Tides Come In . . .	426
15.5	Wave Action	427
15.6	Where Land Meets Sea: Coastal Landforms	431
15.7	Causes of Coastal Variability	438
15.8	Coastal Problems and Solutions	439

Featured Painting: Oceans and Coasts 440–41

Chapter 16
A Hidden Reserve: Groundwater 450

16.1	Introduction	450
16.2	Where Does Groundwater Reside?	451
16.3	Permeability: The Ease of Flow	452
16.4	The Water Table	452
16.5	Groundwater Flow	454
16.6	Tapping the Groundwater Supply	456

16.7   Hot Springs and Geysers   457

BOX 16.1   THE HUMAN ANGLE: *Oases*   459

16.8   Groundwater Usage Problems   461

16.9   Caves and Karst: A Spelunker's Paradise   464

        Featured Painting: Caves and Karst Landscapes   466–67

Chapter 17

Dry Regions: The Geology of Deserts   473

17.1   Introduction   473

17.2   What Is a Desert?   474

17.3   Types of Deserts   475

BOX 17.1   THE REST OF THE STORY: *Convection in the Atmosphere and Generation of Prevailing Wind*   476

17.4   Weathering and Erosional Processes in Deserts   478

17.5   Depositional Environments in Deserts   482

17.6   Desert Landscapes   485

        Featured Painting: The Desert Realm   486–87

17.7   Desert Problems   489

Chapter 18

Amazing Ice: Glaciers and Ice Ages   493

18.1   Introduction   493

18.2   The Nature of Glaciers   494

18.3   Carving and Carrying by Ice   502

18.4   Deposition Associated with Glaciation   505

        Featured Painting: Glaciers and Glacial Landforms   510–11

18.5   Other Consequences of Continental Glaciation   512

18.6   Periglacial Environments   515

18.7   The Pleistocene Ice Ages   516

BOX 18.1   THE HUMAN ANGLE: *So You Want to See Glaciation?*   518

18.8   The Causes of Ice Ages   520

18.9   Will There be Another Glacial Advance?   522

Chapter 19

Global Change in the Earth System   525

19.1   Introduction   525

19.2   Unidirectional Changes   526

19.3   Physical Cycles   528

19.4   Biogeochemical Cycles   529

19.5   Global Climate Change   530

BOX 19.1   THE HUMAN ANGLE: *Global Climate Change and the Birth of Legends*   533

19.6   Anthropogenic Changes in the Earth System   535

        Featured Painting: The Earth System   538–39

19.7   The Future of the Earth: A Scenario   543

Appendix   Scientific Background: Matter and Energy   A-1

Metric Conversion Chart   A-12

Glossary   G-1

Credits   C-1

Index   I-1

The modern science of geology (or geoscience), the study of the Earth, began in the late eighteenth century. So in comparison with other sciences, geology is a young subject. Nevertheless, over the past two centuries, thousands of geologists have provided answers to a wide range of questions: Why do earthquakes and volcanoes happen? What causes mountains to rise? How do Earth's varied landscapes develop and change through time? How has the climate changed through time? When did our planet form, and by what process? Where do we dig to find valuable ore, and where do we drill to find oil? Indeed, a look at almost any natural feature leads to a new question, and new questions fuel the need for new research. Thus, geoscience remains an active and exciting field today.

Before the mid-twentieth century, geoscientists studied each of the questions listed above on its own, without considering its relation to other issues. But since 1960, there have been two "paradigm-shifting" ideas that have unified thinking about the Earth and its features. The first idea, called the theory of plate tectonics, shows that the Earth's outer shell, rather than being static, consists of discrete plates that constantly move very slowly, so that the map of our planet constantly changes. We now understand that plate interactions cause earthquakes and volcanoes, build mountains, provide gases for the atmosphere, and affect the distribution of life on Earth. The second idea, the concept of Earth systems science, emphasizes that the planet's water, land, atmosphere, and living inhabitants are dynamically interconnected. Earth materials constantly cycle among various living and nonliving reservoirs on, above, and within the planet, and the history of life is intimately linked to the history of the physical Earth.

Essentials of Geology is an introduction to geology that weaves the theory of plate tectonics and the concept of Earth systems science into its narrative from the beginning, and thus strives to create a modern, coherent image of our planet.

NARRATIVE THEMES

To develop a complete understanding of the Earth, students must go beyond vocabulary and be aware of fundamental concepts, or narrative themes, that explain how the Earth works. These themes provide a peg-board on which to hang observations and ideas, and allow students to make connections between them. Several narrative themes (discussed more fully in the Prelude) are emphasized throughout the text:

1. The Earth is a complex system in which the solid Earth, the oceans, the atmosphere, and life are interconnected to yield a planet unique in the solar system.
2. Most geological processes can be understood in the context of plate tectonics theory.
3. The Earth is a planet, formed like other planets from dust and gas. But, in contrast to other planets, Earth continues to undergo significant change.
4. The Earth is very old—about 4.57 billion years old. During this time, the map of the planet and its surface features have changed, and life has evolved.
5. Internal processes (driven by Earth's internal heat) and external processes (driven by heat from the Sun) interact at the Earth's surface to create our landscapes.
6. Natural hazards—earthquakes, volcanoes, landslides, foods—can be understood. In some cases, people can reduce their effects.
7. Energy and mineral resources come from the Earth and are formed by geologic phenomena.
8. Physical features of the Earth are linked to life processes.
9. Science comes from observation, and people make scientific discoveries.
10. The study of geology provides an excellent means to improve science literacy.

ORGANIZATION

The topics covered in this book have been arranged so that students can build their knowledge of geology on a foundation of basic concepts. Thus, the book starts with cosmology and the formation of the Earth, and then introduces the architecture of our planet, from surface to center. With this background, we can delve into plate tectonics theory. Plate tectonics appears early so that students can relate content in subsequent chapters to this concept. Knowing about plate tectonics, for example, helps students understand the suite of chapters on minerals, rocks, and the rock cycle. A knowledge of plate tectonics and rocks together then provides a basis for learning about volcanoes, earthquakes, and mountains. And with this background, we can see how the map of the Earth has changed through the vast expanse of geologic time, and how energy and mineral resources have developed.

The final chapters of the book address processes and problems occurring at or near the Earth's surface, from the unstable slopes of hills, down the course of rivers, to the shores of the sea and beyond. This section concludes with a topic of growing concern in society—global change, particularly climate change.

SPECIAL FEATURES

Broad Application

Essentials of Geology provides concise coverage of topics used in an introductory geology course. It has been shortened, relative to its progenitor (*Earth: Portrait of a Planet*), by removal of topics less frequently covered in an introductory geology course, and by simplification of the treatment of the remaining topics.

Flexible Organization

Though the sequence of chapters was chosen for a reason, this book is designed to be flexible enough for instructors to choose their own strategies for teaching geology. Thus, each chapter is largely self-contained, reiterating relevant material or at least referring to other chapters where certain topics can be reviewed. This apparent redundancy in some parts of the text is intentional, for geology is a nonlinear subject: the individual topics are so interrelated that there is not always a single best way to order them.

Societal Issues

Geology's practical applications are addressed in several chapters. Students will learn about such topics as energy resources, mineral resources, global change, and mass wasting. Further, chapters on earthquakes, volcanoes, and landscapes highlight geological hazards. And students are encouraged to apply their geological understanding to environmental issues, where relevant.

Boxed Inserts

Throughout the text, boxes expand on specific topics by giving further scientific background, additional detail, or related information that's just plain interesting.

Detailed Illustrations

It's hard to understand features of the Earth system without being able to see them. To help students visualize topics, this book is lavishly illustrated, with figures that attempt to give a realistic context for a geologic feature without overwhelming students with extraneous detail. The talented artists who worked on the book have pushed the envelope of modern computer graphics, and the result is the most realistic pedagogical art ever provided by a geoscience text.

Photographs from around the world (many by the author) have been assembled for this book. Where appropriate, they are accompanied by annotated sketches labeled "What a geologist sees," to help students discover what the photos show.

Featured Paintings

In addition to individual figures, British painter Gary Hincks has provided spectacular two-page spreads for most chapters. These paintings illustrate key concepts introduced in the chapters and visually emphasize the relationships between components of the Earth system.

CHANGES IN THE SECOND EDITION

The Second Edition of *Essentials of Geology* is not simply a cosmetic modification of the First Edition. Though the basic organization remains the same, the text has been intelligently modified and thoroughly updated. Key changes include:

1. Incorporation of New Discoveries. Recent research has yielded fundamental new understandings that belong even in introductory books. While basic geoscience remains the backbone of *Essentials of Geology,* we have incorporated discussions of these new discoveries, and many others, into the text.

2. New Examples. To keep the discussion of natural hazards current, we have updated the treatment to include events that have happened since publication of the First Edition.

3. The View from Space. Satellites orbiting the Earth provide spectacular imagery of our planet's surface. We have added numerous satellite images to the Second Edition.

4. New Synoptic Art. Famed geologic artist Gary Hincks has provided new, spectacular two-page paintings that provide an overview of key concepts. The drama of these paintings appeals to students' intuition.

5. New Figures and Photos. Numerous figures have been added to the book, and over 200 figures from the First Edition have been updated and improved. Further, many photographs have been added.

6. New WAGS. The "What a Geologist Sees" feature, which provides an annotated sketch of a photograph so that students can better see what the picture shows, have proven to be particularly helpful. Numerous WAGS have been added to this edition.

7. <u>Clarifications</u>. To help ensure that *Essentials of Geology* is as accurate and up-to-date as possible, each chapter of the First Edition was sent to an expert reviewer for comment. In addition, we revised the text on a line-by-line basis to ensure that it is as clear as possible and have made many of the more complex discussions even more readable.

8. <u>Key Terms and Case Studies</u>. We have reduced the number of bold-faced terms, highlighting only those that are crucial to understanding a topic. We have added "case studies" to illustrate the practical side of the geosciences, especially as regards environment-related issues.

SUPPLEMENTS

Highlights:

Earth: Videos of a Planet Instructor's DVD
This outstanding lecture resource features 24 short film clips, carefully selected from USGS archives.

Animations
Developed by Professor Marshak to illustrate dynamic Earth processes, this rich collection of animations emphasizes plate tectonics, geological hazards, and earth systems science concepts. Approximately half of these animations are new to the Second edition, and many are based on groundbreaking "What a Geologist Sees" figures from the text.

Zoomable Art
Explore Gary Hinck's spectacular synoptic paintings in vivid detail using the zoomable art feature. Ideal for lecture use, these figures are included among the resources on the Instructor's CD-Rom.

NortonEbook
Essentials of Geology, Second Edition is now available from W. W. Norton in a value-priced electronic edition. For details visit nortonebooks.com

For Instructors:

1. Norton Media Library Instructor's CD-ROM

This instructor CD-ROM offers a wealth of easy-to-use multimedia resources, all structured around the text and designed for use in lecture presentations, including:

- editable PowerPoint lecture outlines by Ron Parker of Earlham College

- all of the art and all photographs from the text

- additional photographs from Stephen Marshak's own archives

- over sixty animations unique to *Earth: Portrait of a Planet,* Second Edition and *Essentials of Geology,* Second Edition

- "Zoomable" versions of the synaptic paintings by Gary Hincks

2. Instructor's Manual and Test Bank

Prepared by John Werner of Seminole Community College, Terry Engelder of Pennsylvania State University, and Stephen Marshak, this manual offers useful material to help instructors as they prepare their lectures. It includes over 1,200 multiple-choice and true-false test questions. The test bank is available both in printed form and electronically in Exam-view and WebCT- or Blackboard-ready formats.

3. Transparencies

Over 200 figures from the text are available as color acetates.

4. Supplemental Slide Set

This collection of 35mm slides supplements the photographs in the text with additional images from Stephen Marshak's own photo archives. These images are also available as PowerPoint slides on the Norton Media Library CD-ROM.

For Students

1. Student CD-ROM and Web Site

wwnorton.com/earth
Developed specifically for *Essentials of Geology,* Second Edition, this free online study guide offers materials that reinforce core concepts from the text and animations to help students visualize dynamic processes. A CD-ROM version of this resource is available as an optional package item. It includes:

- **Review Materials**—Chapter overviews, key-term crosswords, and multiple-choice quizzes help students get the most from their reading assignments. A new grade book utility makes it easy for instructors to collect scores from web site quiz assignments.

- **Animations**—Over sixty animations developed by Stephen Marshak, half of them new to this edition, emphasize plate tectonics, geologic hazards, and Earth systems science. Many of the new animations are based on the "What a Geologist Sees" art from the text.

- **Earth Science News**—Weekly updates feature news about geologic events, recent advances, and contemporary environmental issues.

ACKNOWLEDGMENTS

I am very grateful for the assistance of many people in bringing this book from the concept stage to the shelf in the first place, and for helping to provide the momentum needed to make the Second Edition take shape. First and foremost, I wish to thank my family. My wife, Kathy, carried out the immense task of cutting content from *Earth: Portrait of a Planet,* Second Edition to produce the template of *Essentials of Geology,* Second Edition. She also edited new text, rationalized the art list, and cross-checked all sets of proofs. Without her efforts, creation of *Essentials of Geology,* Second Edition would not have been possible. My daughter, Emma, located and scanned photographs and searched for resources on the Web. My son, David, helped me keep the project in perspective and highlighted places where the writing could be improved. During the initial development of the First Edition, I greatly benefited from discussions with Philip Sandberg, and during later stages in the development of the First Edition, Donald Prothero contributed text, editorial comments, and end-of-chapter material.

The publisher, W. W. Norton & Company, has been incredibly supportive of my work and has been very generous in their investment in this project. Steve Mosberg signed the First Edition, and Rick Mixter put the book on track. Jack Repcheck bulldozed aside numerous obstacles and brought the First Edition to completion. He has continued to be a fountain of sage advice and an understanding friend throughout the development of the Second Edition. Jack has provided numerous innovative ideas that have strengthened the book and brought it to the attention of the geologic community. Under Jack's guidance, the First Edition became one of the most widely used geology texts worldwide.

April Lange has expertly coordinated development of the ancillary materials. She has not only managed their development, but she also introduced innovative approaches and wrote part of the material. Her contributions have set a standard of excellence. Thom Foley and Christopher Granville have expertly and efficiently handled the task of managing production for the Second Edition. They have calmly managed all the back and forth involved in developing a book and in keeping it on schedule. Susan Gaustad did an outstanding job of copy editing the First Edition. This tradition continued through the efforts of Barbara Gerr on the Second Edition. Stephanie Romeo did an excellent job of photo research and of obtaining permissions, Mik Awake, editorial assistant, was a great help in tying up any and all loose ends.

Production of the illustrations has involved many people. I am particularly grateful to Joanne Brummett and Stan Maddock, who helped create the overall style of the figures, produced a great many of them for both editions, and have worked closely with me on improvements. I would also like to thank the many artists of Precision Graphics in Champaign, Illinois, who, joined Joanne and Stan to draw all of the new art for this edition. Joanne ably coordinated and supervised the art team at Precision Graphics; Terri Hamer has done an excellent job as production manager; and Jon Prince has creatively programmed the animations. It has been a delight to interact with the artists, production staff, and management of Precision Graphics over the past several years. It has also been great fun to interact with Gary Hincks, who painted the incredible two-page spreads, in part using his own designs and geologic insights. Some of Gary's paintings originally appeared in *Earth Story* (BBC Worldwide, 1998) and were based on illustrations jointly conceived by Simon Lamb and Felicity Maxwell, working with Gary. Others were developed specifically for *Earth: Portrait of a Planet.* Some of the chapter quotes were found in *Language of the Earth,* compiled by F. T. Rhodes and R.O. Stone (Pergamon, 1981).

In developing the First Edition, I was helped by insightful review and discussions of the manuscript by many geologists. The development of the Second Edition benefited greatly from input by expert reviewers for specific chapters, by new general reviewers of the entire book, and by comments from faculty and students who have used the First Edition and were kind enough to contact me by e-mail. The list of people whose comments were incorporated includes:

Jack C. Allen, Bucknell University
David W. Anderson, San Jose State University
Philip Astwood, University of South Carolina
Eric Baer, Highline University
Victor Baker, University of Arizona
Keith Bell, Carleton University
Mary Lou Bevier, University of British Columbia
Daniel Blake, University of Illinois
Michael Bradley, Eastern Michigan University
Sam Browning, Massachusetts Institute of Technology
Rachel Burks, Towson University
Peter Burns, University of Notre Dame
Katherine Cashman, University of Oregon
George S. Clark, University of Manitoba
Kevin Cole, Grand Valley State University
Patrick M. Colgan, Northeastern University
John W. Creasy, Bates College
Norbert Cygan, Chevron Oil, retired
Peter DeCelles, University of Arizona
Carlos Dengo, ExxonMobil Exploration Company
John Dewey, University of California, Davis
Charles Dimmick, Central Connecticut State University
Robert T. Dodd, State University of New York at Stony Brook
Missy Eppes, University of North Carolina, Charlotte
Eric Essene, University of Michigan
James E. Evans, Bowling Green State University

Leon Follmer, Illinois Geological Survey
Nels Forman, University of North Dakota
Bruce Fouke, University of Illinois
David Furbish, Vanderbilt University
Grant Garvin, John Hopkins University
Christopher Geiss, Trinity College, Connecticut
Gayle Gleason, SUNY Cortland
William D. Gosnold, University of North Dakota
Lisa Greer, William & Mary College
Henry Halls, University of Toronto at Mississuaga
Bryce M. Hand, Syracuse University
Tom Henyey, University of South Carolina
Paul Hoffman, Harvard University
Neal Iverson, Iowa State University
Donna M. Jurdy, Northwestern University
Thomas Juster, University of Southern Florida
Dennis Kent, Lamont Doherty/Rutgers
Jeffrey Knott, California State University, Fullerton
Ulrich Kruse, University of Illinois
Lee Kump, Pennsylvania State University
David R. Lageson, Montana State University
Robert Lawrence, Oregon State University
Craig Lundstrom, University of Illinois
John A. Madsen, University of Delaware
Jerry Magloughlin, Colorado State University
Paul Meijer, Utrecht University (NL)
Alan Mix, Oregon State University
Robert Nowack, Purdue University
Charlie Onasch, Bowling Green State University
David Osleger, University of California, Davis
Lisa M. Pratt, Indiana University
Mark Ragan, University of Iowa
Bob Reynolds, Central Oregon Community College
Joshua J. Roering, University of Oregon
Eric Sandvol, University of Missouri
William E. Sanford, Colorado State University
Doug Shakel, Pima Community College
Angela Speck, University of Missouri
Tim Stark, University of Illinois (CEE)
Kevin G. Stewart, University of North Carolina at Chapel Hill
Don Stierman, University of Toledo
Barbara Tewksbury, Hamilton College
Thomas M. Tharp, Purdue University
Kathryn Thornbjarnarson, San Diego State University
Basil Tickoff, University of Wisconsin
Spencer Titley, University of Arizona
Robert T. Todd, State University of New York at Stony Brook
Torbjörn Törnqvist, University of Illinois, Chicago
Jon Tso, Radford University
Alan Whittington, University of Illinois
Lorraine Wolf, Auburn University
Christopher J. Woltemade, Shippensburg University

I apologize if I inadvertently left anyone off this list.

ABOUT THE AUTHOR

Stephen Marshak is currently head of the Department of Geology at the University of Illinois, Urbana-Champaign. He holds an A.B. from Cornell University, an M.S. from the University of Arizona, and a Ph.D. from Columbia University. Steve's research interests lie in the fields of structural geology and tectonics. Over the years, he has explored geology in the field on several continents. Since 1983, Steve has been on the faculty of the University of Illinois, where he teaches courses in introductory geology, structural geology, tectonics, and field geology, and has won the university's highest teaching award. In addition to research papers and *Essentials of Geology,* Steve has authored or co-authored *Earth: Portrait of a Planet; Earth Structure: An Introduction to Structural Geology and Tectonics;* and *Basic Methods of Structural Geology.*

THANKS!

I am very grateful to the faculty who selected the First Edition for use in their classes and to the students who engaged so energetically with it. The response was very gratifying. I particularly appreciate the comments from readers that helped to make improvements in the Second Edition. I continue to welcome your comments and can be reached at smarshak@uiuc.edu.

Stephen Marshak

*To see the world in a grain of sand
and heaven in a wild flower.
To hold infinity in the palm of the hand
and eternity in an hour.*

—William Blake (British poet, 1757–1827)

ESSENTIALS OF GEOLOGY

Second Edition

And Just What Is Geology?

Civilization exists by geological consent, subject to change without notice.
—WILL DURANT (1885–1981)

P.1 IN SEARCH OF IDEAS

In the glow of the midnight sun, our C-130 Hercules transport plane rose from a smooth ice runway on the frozen sea surface at McMurdo Station, Antarctica, and we were off to spend a month studying unusual rocks exposed on a cliff about 250 km (kilometers) away. As we climbed past the smoking summit of Mt. Erebus, Earth's southernmost volcano, we had one nagging thought: no aircraft had ever landed at our destination, so the ground conditions there were unknown; if deep snow covered the landing site, the massive plane might get stuck and would not be able to return to McMurdo. Because of this concern, the flight crew had added a crate of rocket canisters to the pile of snowmobiles, sleds, tents, and food in the plane's cargo hold. "If the turboprops can't lift us, we can clip a few canisters to the tail, light them, and rocket out of the snow," they claimed.

For the next hour, we flew along the Transantarctic Mountains, a ridge of rock that divides the continent into two parts, East Antarctica and West Antarctica (▶Fig. P.1). A vast ice sheet, in places over 3 km thick, covers East Antarctica—the surface of this ice sheet forms a high plain known as the Polar Plateau. Rivers of ice from the Polar Plateau slowly flow down valleys cut through the Transantarctic

We can see the Earth System at a glance near the Maroon Bells, a row of mountains in Colorado. Here, sunlight, air, water, rock, and life all interact.

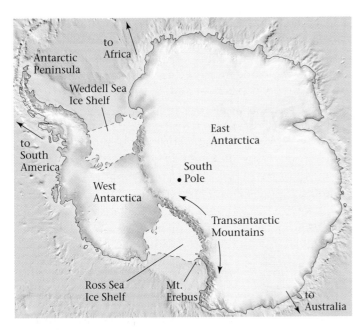

FIGURE P.1 Map of Antarctica.

Mountains. (Ice sheets and ice rivers are called glaciers.) From the plane's windows, we admired long stripes of rock debris that had been shed from mountains onto the glaciers in these valleys. The stripes move with the ice and highlight the direction of flow. Suddenly, we heard the engines slow.

As the plane descended, it lowered its ski-equipped landing gear. The loadmaster shouted an abbreviated reminder of the emergency alarm code: "If you hear three short blasts of the siren, hold on for dear life!" Roaring toward the ground, the plane touched the surface of our first choice for a landing spot, the ice at the base of the rock cliff we wanted to study. *Wham, wham, wham, wham!!!!* Frozen snowdrifts rippled the ice surface, and as the plane's skis slammed into them at about 180 km an hour, it seemed as though a fairy-tale giant was shaking the plane. Seconds later, the landing aborted, we were airborne again, looking for a softer runway above the cliff. Finally, we landed in a field of deep snow, unloaded, and bade farewell to the plane. The Hercules trundled for kilometers through the snow before gaining enough speed to take off, but fortunately did not need to use the rocket canisters. When the plane passed beyond the horizon, the silence of Antarctica hit us—no trees rustled, no dogs barked, and no traffic rumbled in this stark land of black rock and white ice. It would take us a day and a half to haul our sleds of food and equipment down to our study site (▶Fig. P.2). All this to look at a few dumb rocks?

Geologists, scientists who study the Earth, explore remote regions like Antarctica almost routinely. Such efforts often strike people in other professions as a strange way to make a living. Scottish poet Walter Scott (1771–1832), when describing geologists at work, said: "Some rin uphill and down dale, knappin' the chucky stones to pieces like sa' many roadmakers run daft. They say it is to see how the warld was made!" Indeed—to see how the world was made, to see how it continues to evolve, to find its valuable resources, to prevent contamination of its waters and soils, and to predict its dangerous movements. That is why geologists spend months at sea drilling holes in the ocean floor, why they scale mountains (▶Fig. P.3), camp in rain-drenched jungles, and trudge through scorching desert winds. That is why geologists use electron microscopes to examine the atomic structure of minerals, use mass spectrometers to define the composition of rock and water, and use supercomputers to model the paths of earthquake waves. For over two centuries, geologists have pored over the Earth—in search of ideas to explain the processes that form and change our planet.

P.2 THE NATURE OF GEOLOGY

Geology, the study of the Earth, focuses on describing our planet's composition, behavior, and history. In fact, because geology covers such a diversity of subjects, researchers commonly use the term "geoscience" as a name for the discipline.

Not only do geologists address academic questions, such as the formation and composition of the Earth, the causes of earthquakes and ice ages, and the evolution of life, they also address practical problems, such as how to prevent groundwater contamination, how to find oil and minerals, and how to stabilize slopes. And in recent years, geologists have contributed to the study of global climate change, for the long-term record of Earth's past climate lies in layers of sediment and rock. When news reports begin with "Scientists say . . ." and then continue with "an earthquake occurred today off

FIGURE P.2 Geologists sledding to a field area in Antarctica. The sleds contain a month's worth of food, sample bags, rock hammers, and notebooks as well as tents and clothes (and a case of frozen beer).

FIGURE P.3 A geologist studying exposed rocks on a mountain slope on the desert island of Zabargad, in the Red Sea, off the coast of Egypt.

Oil, coal, and uranium are energy resources whose distribution is controlled by geologic processes.

• Do you ever wonder about where the copper in your home's wires come from? Metals come from geologic materials—ore deposits—found by geologists.

• Have you seen fields of green crops surrounded by desert and wondered where the water to irrigate the crops comes from? Most likely, the water comes from underground, where it fills cracks and pores in geologic materials.

• Would you like to buy a dream house on a coastal sandbar (a ridge of sand just offshore)? The surroundings look beautiful, but geologists suggest that, on a time scale of centuries, sandbars are temporary landforms, and your investment might disappear in the next storm.

Clearly, all citizens of the twenty-first century, not just professional geologists, will need to make decisions concerning Earth-related issues. And they will be able to make more reasoned decisions if they have a basic understanding of geologic phenomena. History is full of appalling stories of people who ignored geologic insight and paid a horrible price for their ignorance. Your knowledge of geology may help you to avoid building your home on a hazardous floodplain or fault zone, on an unstable slope, or along a rapidly eroding coast. With a basic understanding of groundwater, you may be able to save money when drilling an irrigation well. With knowledge of the geologic controls on resource distribution, you may be able to invest more wisely in the resource industry.

Second, the study of geology gives you a perspective on the planet that no other field can. As you will see, the Earth is a complicated system; its living organisms, climate, and solid

Japan," or "landslides will threaten the city," or "contaminants from the proposed toxic waste dump are destroying the town's water supply," or "there's only a limited supply of oil left," the scientists referred to are geologists.

The fascination of geology attracts many people to careers in this science. Thousands of geologists work for oil, mining, water, engineering, and environmental companies, while a smaller number work in universities, government geological surveys, and research laboratories. Nevertheless, since the majority of students reading this book will not become professional geologists, it's fair to ask the question "Why should people, in general, study geology?"

First, geology may be one of the most practical subjects you can learn, for geologic phenomena and issues affect our daily lives, sometimes in unexpected ways:

• Do you live in a region threatened by landslides, volcanoes, earthquakes, or floods (▶Fig. P.4)? These are geologic natural hazards that destroy property and take lives.

• Are you worried about the price of energy or about whether there will be a war in an oil-supplying country?

FIGURE P.4 Human-made cities cannot withstand the vibrations of a large earthquake. These apartment buildings collapsed during an earthquake in Turkey.

rock interact with one another in a great variety of ways. Geologic study reveals Earth's antiquity (it's about 4.57 billion years old) and demonstrates how the planet has changed profoundly during its existence. What was the center of the Universe to our ancestors becomes, with the development of geologic perspective, our "island in space" today, and what was an unchanging orb originating at the same time as humanity becomes a dynamic planet that existed long before people did.

Third, the study of geology puts human achievements and natural disasters in context. On the one hand, our cities seem to be no match for the power of an earthquake, and a rise in sea level may swamp all major population centers. But on the other hand, we are now changing the face of the land worldwide at rates that far exceed those resulting from natural geologic processes. By studying geology, we develop a frame of reference for judging the extent and impact of changes.

Finally, when you finish reading this book, your view of the world will be forever colored by geologic curiosity. When you walk in the mountains, you will think of the many forces that shape and reshape the Earth's surface. When you hear about a natural disaster, you will have insight into the processes that brought it about. And when you next go on a road trip, the rock exposures next to the highway will no longer be gray, faceless cliffs, but will present complex puzzles of texture and color telling a story of Earth's history.

P.3 THEMES OF THIS BOOK

A number of narrative themes appear (and reappear) throughout this text. These themes, listed below, can be viewed as the book's "take-home message."

1. *The Earth is a unique, evolving system.* Geologists increasingly recognize that the Earth is a complicated system; its interior, solid surface, oceans, atmosphere, and life forms interact in many ways to yield the landscapes and environment in which we live. Within this **Earth System,** chemical elements pass in cycles between different types of rock, between rock and sea, between sea and air, and between all of these entities and life. Aside from the residue of occasional collisions with fragments of rock or ice from space (asteroids and comets), all the material involved in these cycles originates in the Earth itself. Our planet is truly an island in space.

2. *Plate tectonics is a unifying idea that explains Earth processes.* Like other planets, Earth is not a homogeneous ball, but rather consists of concentric layers: from center to surface, Earth has a core, mantle, and crust. We live on the surface of the crust, where it meets the atmosphere and the oceans. In the 1960s, geologists recognized that the crust together with the uppermost part of the underlying mantle form a 100- to 150-km-thick semirigid shell. Large cracks separate this shell into discrete pieces, called **plates,** which move very slowly relative to one another (▶Fig. P.5). The

FIGURE P.5 Simplified map of the Earth's principal plates. The arrow on each plate indicates the direction the plate moves, and the length of the arrow indicates the plate's velocity (the longer the arrow, the faster the motion). We discuss the types of plate boundaries in Chapter 2.

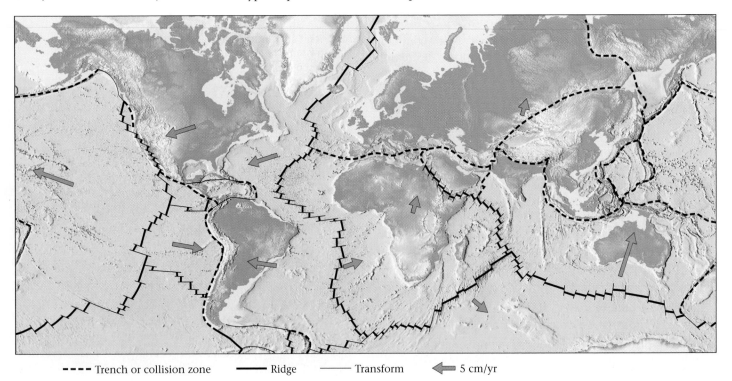

- - - - Trench or collision zone ———— Ridge ———— Transform ⬅ 5 cm/yr

theory that describes this movement and its consequences is now known as the **theory of plate tectonics,** and it is the foundation for understanding most geologic phenomena. Although plates move very slowly, generally less than 10 cm (centimeters) a year, their movements yield earthquakes, volcanoes, and mountain ranges, and cause the distribution of continents to change over time.

3. *The Earth is a planet.* Despite the uniqueness of Earth's system and inhabitants, Earth fundamentally can be viewed as a planet, formed like the other planets of the solar system from dust and gas that encircled the newborn Sun. Though Earth resembles the other inner planets (Mercury, Venus, and Mars), it differs from them in having plate tectonics, an oxygen-rich atmosphere and liquid-water ocean, and abundant life. Further, because of the dynamic interactions among various aspects of the Earth System, our planet is constantly changing; the other inner planets are static.

4. *The Earth is very old.* Geologic data indicate that the Earth formed 4.57 billion years ago—plenty of time for geologic processes to generate and destroy features of the Earth's surface, for life forms to evolve, and for the map of the planet to change. Plate-movement rates of only a few centimeters per year, if those movements continue for hundreds of millions of years, can move a continent thousands of kilometers. In geology, we have time enough to build mountains and time enough to grind them down, many times over. To define intervals of this time, geologists developed the **geologic time scale** (▶Fig. P.6). Geologists call the last 542 million years the **Phanerozoic** Eon, and all time before that the **Precambrian.** They further divide the Precambrian into three main intervals named, from oldest to youngest, the **Hadean,** the **Archean,** and the **Proterozoic** Eons, and the Phanerozoic Eon into three main intervals named, from oldest to youngest, the **Paleozoic,** the **Mesozoic,** and the **Cenozoic** Eras. (Chapter 10 provides further details about geologic time.)

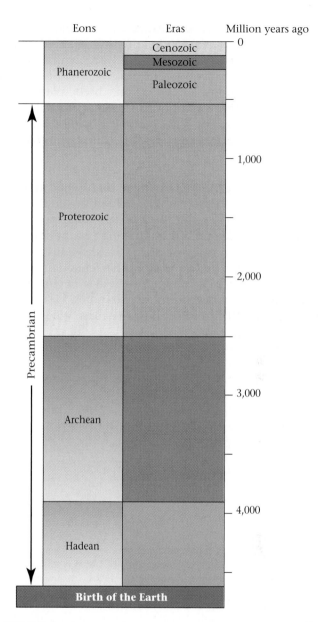

FIGURE P.6 The major divisions of the geologic time scale.

5. *Internal and external processes interact at the Earth's surface.* Internal processes are those phenomena that ultimately are driven by heat from inside the Earth. Plate movement is an example, and since plate movements cause mountain building, earthquakes, and volcanoes, we call all of these phenomena internal processes as well. External processes are those phenomena that ultimately are driven by heat supplied by radiation coming to the Earth from the Sun. This heat drives the movement of air and water, which grinds and sculpts the Earth's surface and transports the debris to new locations, where it accumulates. The interaction between internal and external processes form the landscapes of our planet.

6. *Geologic phenomena affect our environment.* Volcanoes, earthquakes, landslides, floods, and even more subtle processes such as groundwater flow and contamination or depletion of oil and gas reserves are of vital interest to every inhabitant of this planet. They are often a matter of life and death. Linkages between geology and the environment are therefore stressed throughout the book.

7. *Physical aspects of the Earth System are linked to life processes.* All life on this planet depends on such physical features as the minerals in soil; the temperature, humidity, and composition of the atmosphere; and the flow of surface and subsurface water. And life in turn affects and

alters these same physical features. For example, the atmosphere's oxygen comes primarily from plant photosynthesis, a life activity, and it is this oxygen that permits complex animals to survive. The oxygen also affects chemical reactions among air, water, and rock. Without the physical Earth, life could not exist, but without life, this planet's surface might have become a frozen wasteland like that of Mars, or enshrouded in acidic clouds like that of Venus.

8. **Science comes from observation, and people make scientific discoveries.** Science is not a subjective guess or an arbitrary dogma, but rather a consistent set of objective statements resulting from the application of the **scientific method** (▶Box P.1). Every scientific idea must be constantly subjected to testing and possible refutation, and can be accepted only when supported by documented observations. Further, scientific ideas do not appear out of nowhere, but are the result of human efforts. Wherever possible, this book shows where geologic ideas came from, and tries to answer the question "How do we know that?"

9. **The study of geology can increase general science literacy.** Studying geology provides an ideal opportunity to learn basic concepts of chemistry and physics, because these concepts can be applied directly to understanding tangible phenomena. Thus, in this book, where appropriate, basic concepts of physical science are introduced in boxed features called "Science Toolboxes." Also, the Appendix provides a systematic introduction to matter and energy, for those readers who have not learned this information previously or who need a review.

As you read this book, please keep these themes in mind. Don't view geology as a list of words to memorize, but rather as an interconnected set of concepts to digest. Most of all, enjoy yourself as you learn about what may be the most fascinating planet in the Universe.

BOX P.1
SCIENCE TOOLBOX

The Scientific Method

Sometime during the past 200 million years, a large block of rock or metal, which had previously been orbiting the Sun, crossed the path of Earth's orbit. In seconds, it pierced the atmosphere and slammed into our planet (thereby becoming a meteorite) at a site in what is now the central United States, a landscape of flat cornfields. The impact released more energy than a nuclear bomb—a cloud of shattered rock and dust blasted skyward, and once-horizontal layers of rock from deep below the ground sprang upward and tilted on end in the gaping hole left by the impact. When the dust had settled, a huge crater surrounded by debris marked the surface of the Earth at the impact site. Later in Earth history, running water and blowing wind wore down this jagged scar, and some 15,000 years ago, sediments (sand, gravel, and mud) carried by a vast glacier buried what remained, hiding it entirely from view (▶Fig. P.7a, b). Wow! So much history beneath a cornfield. Have you ever wondered how geologists come up with such a story? It takes scientific investigation.

The movies often portray science as a dangerous tool, capable of creating Frankenstein's monster, and scientists as warped or nerdy characters with thick glasses and poor taste in clothes. In reality, **science** is simply the use of observation, experiment, and calculation to explain how nature operates, and **scientists** are people who study and try to understand natural phenomena. Scientists carry out their work using the *scientific method,* a sequence of steps for systematically analyzing scientific problems in a way that leads to verifiable results. Let's see how geologists employed the steps of the scientific method to come up with the meteorite-impact story.

1. *Recognizing the problem:* Any scientific project, like any detective story, begins by identifying a mystery. The cornfield mystery came to light when water drillers discovered limestone, a rock typically made of shell fragments, just below the 15,000-year-old glacial sediment. In surrounding regions, the rock at this depth consists of sandstone, made of cemented-together sand grains, which differs greatly in composition from limestone. Since limestone can be used to build roads, make cement, and produce the agricultural lime used in treating soil, workers stripped off the glacial sediment and began excavating the limestone. They were amazed to find that rock layers exposed in the quarry tilted steeply and had been shattered by large cracks. In the surrounding regions, all rock layers are horizontal like the layers in a birthday cake, the limestone layer lies underneath the sandstone, and the rocks contain relatively few cracks. Curious geologists came to investigate, and soon realized that the geologic features of the land just beneath the cornfield presented a problem to be explained: What phenomena had brought limestone up close to the Earth's surface, tilted the layering in the rocks, and shattered the rocks?

2. *Collecting data:* The scientific method proceeds with the collection of observations or clues that point to an answer. Geologists studied the quarry and determined the age of its rocks, measured the orientation of rock layers, and *documented* (made a written or photographic record of) the fractures that broke up the rocks.

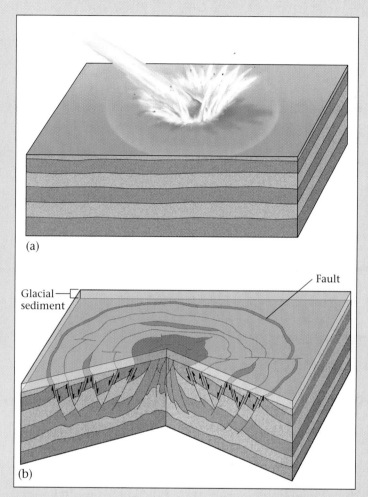

(a)

Glacial sediment

Fault

(b)

FIGURE P.7 (a) The site of an ancient meteorite impact in the American Midwest, before impact. Note the horizontal layers of rock below the ground surface. The thin lines represent boundaries between the successive layers. During impact, a large crater, surrounded by debris, forms. (b) The site of the impact today. The crater and the surface debris were eroded away. Relatively recently, the area was buried by glacial sediments. Underground, the impact disrupted layers of rock by tilting them and by generating faults (fractures on which sliding occurs).

3. *Proposing hypotheses:* A scientific **hypothesis** is merely a possible explanation, involving only naturally occurring processes, that can explain a set of observations. Scientists propose hypotheses during or after their initial data collection. The geologists working in the quarry came up with two alternative hypotheses. First, the features in this region could result from a volcanic explosion; and second, they could result from a meteorite impact.

4. *Testing hypotheses:* Since a hypothesis is no more than an idea that can be either right or wrong, scientists must put hypotheses through a series of tests to see if they work. The geologists at the quarry compared their field observations with published observations made at other sites of volcanic explosions and

meteorite impacts, and studied the results of experiments designed to simulate such events. They learned that if the geologic features visible in the quarry were the result of volcanism, the quarry should contain rocks made of frozen lava (the melt that flows from a volcano). But no such rocks were found. If, however, the features were the consequence of an impact, the rocks should contain **shatter cones,** small, cone-shaped cracks formed only by meteorite impact (▶Fig. P.8). Shatter cones can easily be overlooked, so the geologists returned to the quarry specifically to search for them, and found them in abundance. The impact hypothesis passed the test!

Theories are scientific ideas supported by an abundance of evidence; they have passed many tests and have failed none. Scientists have much more confidence in a theory than they do in a hypothesis. Continued study in the quarry eventually yielded so much evidence for impact that the impact hypothesis came to be viewed as a theory. Scientists continue to test theories over a long time. Successful theories withstand these tests and are supported by so many observations that they come to be widely accepted. (As you will discover in Chapter 2, geologists consider the idea that continents drift around the surface of the Earth to be a theory, because so much evidence supports it.) However, some theories may eventually be disproven, to be replaced by better ones.

Some scientific ideas must be considered absolutely correct, for if they were violated, the natural Universe as we know it would not exist. Such ideas are called **scientific laws,** and examples include the law of gravity.

FIGURE P.8 Shatter cones in limestone. These cone-shaped fractures, formed only by severe impact, open up in the direction away from the impact. At this locality, the cones open up downward, indicating that the impact came from above.

CHAPTER 1

The Earth in Context

When the Hubble Space Telescope looks into what, to the naked eye, appears to be the black void of the night sky, it reveals a spectacle of disks and spirals of hazy light. Each of these is a distant galaxy, a cluster of as many as 300 billion stars. This is the fabric of space.

1.1 INTRODUCTION

Sometime in the distant past, perhaps more than 100,000 generations ago, humans developed the capacity for complex, conscious thought. This amazing ability, which distinguishes our species from all others, brought with it the gift of curiosity, an innate desire to understand and explain the workings of ourselves and all that surrounds us—our **Universe.** Questions that we ask about the Universe differ little from questions a child asks of a playmate: Where do you come from? How old are you? Such musings first spawned legends in which heroes, gods, and goddesses used supernatural powers to mold the planets and sculpt the landscape. Increasingly, science, the systematic analysis of natural phenomena, has provided insight into these questions. However, the development of **cosmology,** the study of the overall structure of the Universe, has proven to be a rough one, booby-trapped with tempting but flawed approaches and cluttered with misleading prejudices.

In this chapter, we begin by presenting a brief sketch of modern cosmological science—we characterize the basic architecture of the Universe, introduce the big bang theory for the formation of the Universe, and discuss scientific ideas concerning the birth of the Earth. Then we outline the basic characteristics of our home planet by building an image of its surroundings, surface, and interior. Our high-speed tour of the Earth provides a reference frame for the remainder of this book.

1.2 THE MODERN IMAGE OF THE UNIVERSE

Beginning around 600 B.C.E., Greek philosophers began to argue about the structure of the Universe. Some advocated the geocentric Universe concept (▶Fig. 1.1a), in which the Earth sat motionless in the center of the heavens while other bodies orbited around it. A few centuries later, around 250 B.C.E., the heliocentric Universe concept, in which all heavenly objects including the Earth orbited the Sun (▶Fig. 1.1b), was proposed, but this idea found little favor. The argument about the position of the Earth in the Universe continued until Ptolemy (100–170 C.E.), an influential Egyptian mathematician, completed calculations that seemed to predict the wanderings of the planets successfully, in the context of the geocentric concept. The church hierarchy of Europe adopted the geocentric hypothesis as dogma, because it effectively certified that the Earth occupied the most important place in the Universe and implied that humans were the Universe's most important creatures. With the fall of Rome in 476 C.E., Europe and the Middle East entered the Middle Ages. For the next millennium, most scientific study in the Western world ceased, and those who disagreed with the Ptolemaic view of the Universe risked charges of heresy.

Then came the Renaissance. The very word means "rebirth" and "revitalization," and in fifteenth-century Europe bold thinkers spawned a new age of exploration and discovery. The burst of scientific discovery during the Renaissance forced people to change their view of Earth's central place in the Universe. Eventually, they realized that the Earth is but one of nine planets that orbit the Sun and that the Sun is but one of a vast number of stars. Stars are not randomly scattered through the Universe; gravity holds them together in immense groups, called **galaxies.** The Sun and over 300 billion stars together form the Milky Way galaxy; more than 100 billion galaxies constitute the visible Universe (see chapter opening photo). Galaxies are so far away that, to the naked eye, they look like stars in the night sky. The nearest galaxy to ours, Andromeda, lies over 2.2 million light years away. A light year is the distance light travels in one year—about 10 trillion kilometers (6 trillion miles).

If we could view the Milky Way from a great distance, it would look like a flattened spiral, 100,000 light years across, with great curving arms slowly swirling around a glowing, disk-like center (▶Fig. 1.2a, b). Presently, our solar system lies near the outer edge of one of these arms and rotates around the center of the galaxy about once every 250 mil-

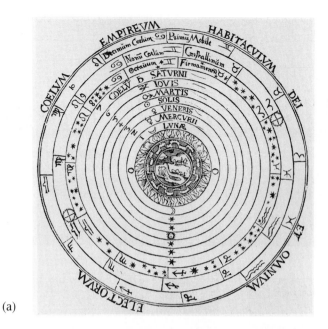

(a)

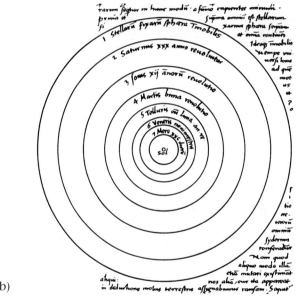

(b)

FIGURE 1.1 (a) The geocentric image of the Universe. Earth, at the center, is surrounded by air and fire and the Moon, Mercury, Venus, the Sun, Mars, Jupiter, and Saturn. Everything lies within the globe of the stars. (b) The heliocentric view of the Universe, as illustrated in this woodcut from Copernicus's *De revolutionibus*.

lion years. We hurtle through space, relative to an observer standing outside the galaxy, at about 200 km per second.

1.3 FORMING THE UNIVERSE

Do galaxies move with respect to other galaxies? Does the Universe become larger or smaller with time? Has the

(a)

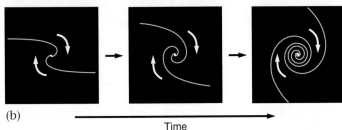

(b)

Time

FIGURE 1.2 (a) An image of what the Milky Way galaxy might look like if viewed from outside. Note that the galaxy consists of spiral arms around a central cluster. Our Sun lies at the edge of one of these arms. (b) The spiral shape is due to rotation of the galaxy. Here we see three stages in a computer simulation of spiral formation.

Universe always existed? Answers to these fundamental questions came from an understanding of a phenomenon called the Doppler effect. Though the term may be unfamiliar, the phenomenon it describes is an everyday experience.

Waves and the Doppler Effect

When a train whistle screams, the sound you hear has moved through the air from the whistle to your ear in the form of sound waves. Waves are disturbances that transmit energy from one point to another by causing periodic motions. As each wave passes, air alternately compresses, then expands. We refer to the distance between successive waves as the wavelength, and the number of waves that pass a point in a given time interval as the frequency of the wave. If the wavelength decreases, more waves pass a point in a given time interval, so the frequency increases. The pitch of the sound, meaning its note in the musical scale, depends on the frequency of the sound waves. Now imagine that as you are standing on the station platform, the train moves toward you. The sound of the whistle gets

louder as the train approaches, but its pitch remains the same. Then, the instant the train passes, the pitch abruptly changes; it sounds like a lower note in the musical scale. An Austrian physicist, C. J. Doppler (1803–1853), first interpreted this phenomenon, and thus it is now known as the **Doppler effect.** When the train moves toward you, the sound has a higher frequency (the waves are closer together so the wavelength is smaller), because the sound source, the whistle, has moved slightly closer to you between the instant that it emits one wave and the instant that it emits the next (▶Fig. 1.3a, b). When the train moves away from you, the sound has a lower frequency (the waves are farther apart), because the whistle has moved slightly farther from you between the instant it emits one wave and the instant it emits the next.

Light energy also moves in the form of waves. We can represent light waves symbolically by a periodic succession of crests and troughs (▶Fig. 1.4a, b). Visible light comes in many colors—the colors of the rainbow. The color of light you see depends on the frequency of the light waves, just as the pitch of a sound you hear depends on the frequency of sound waves. Red light has a longer wavelength (lower frequency) than blue light (Fig. 1.4a, b). The concept of the Doppler effect also applies to light but can be noticed only if the light source moves very fast (at least a few percent of the speed of light). If a light source moves away from you, the light you see becomes redder (as the light shifts to longer wavelength or lower frequency), and if the source moves toward you, the light you see becomes bluer (as the light shifts to higher frequency). We call these changes the red shift and the blue shift, respectively (▶Fig. 1.4c).[1]

Red Shifts and the Expanding Universe Theory

In the 1920s, astronomers such as Edwin Hubble (after whom the Hubble Space Telescope was named) braved many a frosty night beneath the open dome of a mountaintop observatory in order to aim telescopes into deep space. These researchers had begun a search for distant galaxies. At first, they documented only the location and shape of newly discovered galaxies. But then, one astronomer began an additional project to study the wavelength of light produced by the distant galaxies. The results yielded a surprise that would forever change humanity's perception of the Universe.

Astronomers found, to their amazement, that the light of distant galaxies displayed red shifts relative to the light of nearby stars. Hubble pondered this mystery and, around 1929, realized that the red shifts must be a consequence of

---

[1]In detail, the cause of red and blue shifts for light is quite different than the cause of frequency shifts for sound. To explain this difference would require discussion of Einstein's theory of relativity.

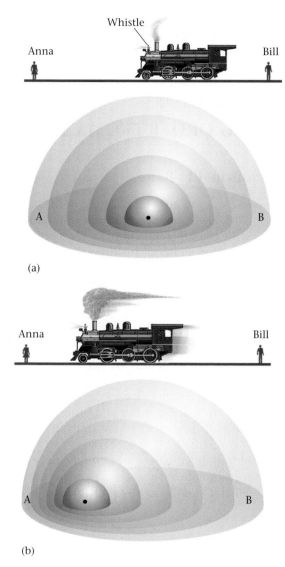

FIGURE 1.3 Wavelength is the spacing of waves. Wavelength, and, therefore, frequency (the number of waves passing a point in an interval of time) are affected by the speed of the source. Frequency determines pitch. (a) Sound emanating from a stationary source has the same wavelength in all directions (the circular shells represent the waves), and observers at points A and B (Anna and Bill) hear the same pitch. (b) If the source is moving toward Anna, she hears a shorter-wavelength sound than does Bill. Therefore, Anna hears a higher-pitched (higher-frequency) sound than does Bill.

the Doppler effect—and thus that the distant galaxies must be moving away from Earth at an immense velocity. At the time, astronomers thought the Universe had a fixed size, so Hubble initially assumed that if some galaxies were moving away from Earth, others must be moving toward Earth. But this was not the case. On further examination, Hubble concluded that the light from *all* distant galaxies, regardless of their direction from Earth, exhibits a red shift. In other words, *all* distant galaxies are moving rapidly away from us.

How can all galaxies be moving away from us, regardless of which direction we look? Hubble puzzled over this question and finally recognized the solution, one that actually was implied by calculations that Albert Einstein had made. The whole Universe must be expanding! To picture the expanding Universe, imagine a ball of bread dough with raisins scattered throughout. As the dough bakes and expands into a loaf, each raisin moves away from its neighbors, in every direction (▶Fig. 1.5a, b). This idea came to be known as the **expanding Universe theory.**

Hubble's expanding Universe theory marked a revolution in thinking. No longer could we view the Universe as being fixed in dimension, with galaxies locked in position. Now we see the Universe as an expanding bubble, in which galaxies race away from each other at incredible speeds. This image immediately triggers the key question of cosmology: Did the expansion begin at some specific time in the past? If it did, then this instant would mark the beginning of the Universe, the beginning of space and time.

FIGURE 1.4 Light waves resemble ocean waves in shape, but physically they are quite different. (a) Blue light has a relatively short wavelength (higher frequency). (b) Red light has a relatively long wavelength (lower frequency). (c) The shift in light frequency that an observer sees depends on whether the source is moving toward or away from the observer.

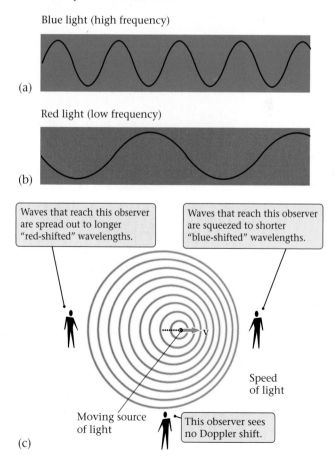

The Big Bang

Most astronomers have concluded that expansion did indeed begin at a specific time, with a cataclysmic explosion called the **big bang.** According to the big bang theory, all matter and energy—everything that now constitutes the Universe—was initially packed into an infinitesimally small point. For reasons that no one fully understands, the point exploded, according to current estimates, 13.7 (± 1%) billion years ago.

(To learn what may have happened next, you may first need to review some terms from basic physics and chemistry; please study the Appendix.) Since the big bang, the Universe has been continually expanding (▶Fig. 1.5c).

1.4 MAKING ORDER FROM CHAOS

Aftermath of the Big Bang

Of course, no one was present at the instant of the big bang, so no one actually saw it happen. But by combining clever calculations with careful observations, researchers have developed a consistent model of how the Universe evolved beginning an instant after the explosion (▶Fig. 1.6). The calculations come from applying the laws of physics. The observations come from examining the edge of the Universe with large telescopes, for when we look at very distant objects in space we are seeing the distant past. The most distant objects yet observed lie about 13 billion light years away, so looking at them provides a snapshot of the Universe when it was very young indeed.

According to the contemporary model of the big bang, profound change happened at a fast and furious rate at the outset. During the first instant (10^{-43} seconds) of existence, the Universe was so small, so dense, and so hot (10^{28} degrees centigrade) that it consisted entirely of energy—atoms, or even the smallest subatomic particles that make up atoms, could not even exist. By the time the Universe

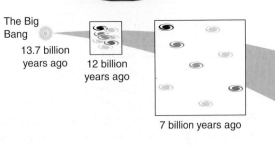

FIGURE 1.5 (a) At the dough stage, raisins in raisin bread are relatively close to one another. (b) During baking the bread expands, and the raisins (like galaxies in the expanding Universe) have all moved away from each other. Notice that *all* raisins move away from their neighbors, regardless of direction. (c) The concept of the expanding Universe; the spirals represent galaxies.

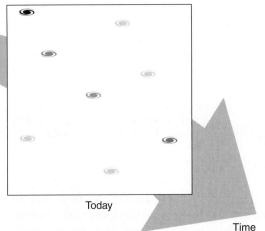

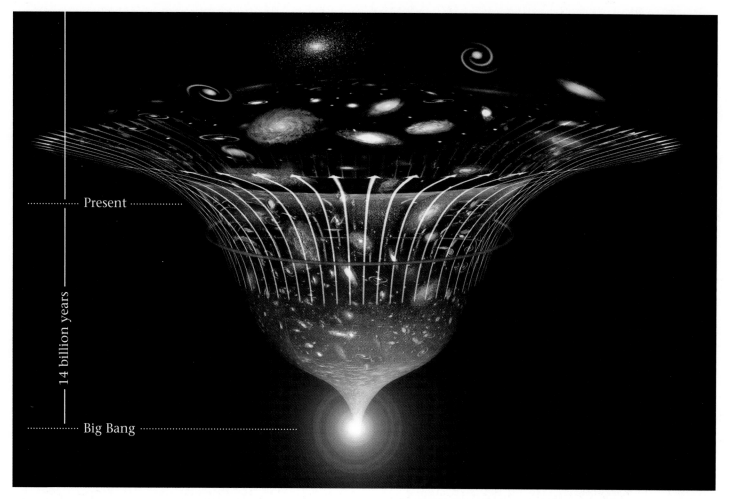

FIGURE 1.6 An artist's rendition of the big bang, followed by expansion of the Universe. The horizontal direction represents size and the vertical direction represents time. Recent work suggests that the rate of expansion has changed as time passes.

reached an age of 3 minutes, its temperature had fallen below 1 billion degrees, and its diameter had grown to about 100 billion km (60 billion miles). Under these conditions, nuclei of new atoms began to form through the collision and fusion (sticking together) of hydrogen atoms. Formation of new nuclei by fusion reactions at this time is called "big bang nucleosynthesis" because it happened *before* any stars existed. Big bang nucleosynthesis could produce only small atoms, meaning ones containing a small number of protons, and it happened very rapidly. In fact, virtually all of the new atomic nuclei that would form by big bang nucleosynthesis had formed by the end of 5 minutes. Eventually, the Universe became cool enough for chemical bonds to bind atoms of certain elements together in molecules—most notably, two hydrogen atoms could join to form molecules of H_2. As the Universe continued to expand and cool further, atoms and molecules slowed down and accumulated into **nebulae,** patchy clouds of gas separated from one another by the vacuum of space. Gas mak-

ing up the earliest nebulae of the Universe consisted almost entirely of the two smallest atoms, namely, hydrogen (74%, by volume) and helium (24%).

Forming the First Stars

When the Universe reached its 200 millionth birthday, it contained immense, slowly swirling, dark nebulae separated by vast voids of empty space (▶Fig. 1.7). The Universe could not remain this way forever, however, because of the invisible but persistent pull of gravity. Eventually, gravity began to remold the Universe pervasively and permanently.

All matter exerts gravitational pull—a type of force—on its surroundings, and as Isaac Newton first pointed out, the amount of pull depends on the amount of mass. Somewhere in the young Universe, the gravitational pull of an initially denser region of a nebula began to suck in surrounding gases and, in a grand example of "the rich getting richer," grew in mass and, therefore, density (mass per

FIGURE 1.7 Gases clump to form distinct nebulae, which look like clouds in the sky. In this Hubble Space Telescope picture, new stars are forming at the top of the nebula on the left. Stars that have already formed light up the nebulae from behind.

unit volume). As this denser region attracted progressively more gas, more matter compacted into a smaller region, and the initial swirling movement of gas transformed into a rotation around an axis that became progressively faster and faster. (The same phenomenon occurs when a spinning ice skater pulls her arms inward and speeds up.) Because of rotation, the condensing portion of the nebula evolved into a spinning disk-shaped mass of gas called an accretion disk (see art on pp. 16–17). Eventually, the gravitational pull of the accretion disk became great enough to trigger wholesale inward collapse of the surrounding nebula. With all the additional mass available, gravity aggressively pulled the inner portion of the accretion disk into a dense ball. The energy of motion (kinetic energy) of gas falling into this ball transformed into heat (thermal energy) when it landed on the ball. Moreover, the squeezing together of jostling gas atoms and molecules in the ball further increased the gas's temperature. Eventually, the central ball of the accretion disk became hot enough to glow, and at this point it became a **protostar.**

A protostar continues to grow, by pulling in successively more mass, until its core becomes very dense and its temperature reaches about 10 million degrees. Under such conditions, hydrogen nuclei fuse (stick together), in a series of steps, to form helium nuclei. Astronomers refer to this type of nuclear fusion as "hydrogen burning," because it produces huge amounts of energy and makes a star into a

fearsome furnace. When the first nuclear fusion reactions began in the first protostar, the body "ignited" and the first true star formed, and when this happened, the first starlight pierced the newborn Universe, perhaps only 800 million years after the big bang. This process would soon happen again and again, and many first-generation stars began to burn.

First-generation stars tended to be very massive (as much as 100 times the mass of the Sun). Astronomers have shown that the larger the star, the hotter it burns and the faster it runs out of fuel and dies. A huge star may survive only a few million years to a few tens of millions of years before it explodes to form a supernova. (This word comes from the Latin word *nova*, which means "new"; when the light of a supernova explosion reaches Earth, it looks like a very bright new star in the sky.) Thus, not long after the first generation of stars formed, the Universe began to be peppered with the first generation of supernova explosions.

1.5 WE ARE ALL MADE OF STARDUST

Element Factories

Nebulae from which the first-generation stars formed consisted entirely of small atoms (elements with atomic numbers smaller than 5; "atomic number" refers to the number of protons in the nucleus), because only these small atoms were generated by big bang nucleosynthesis. In contrast, the Universe of today contains ninety-two naturally occurring elements (see Appendix). Where do the other eighty-seven elements come from? In other words, how did elements such as carbon, sulfur, silicon, iron, gold, and uranium form? These elements, which are common on Earth, have large atomic numbers (carbon has an atomic number of 6, and uranium has an atomic number of 92). Physicists now realize that these elements didn't form during or immediately after the big bang. Rather, they formed later, during the life cycle of stars, by the process of "stellar nucleosynthesis." Because of stellar nucleosynthesis we can consider stars to be element factories, constantly fashioning larger atoms out of smaller atoms.

The specific reactions that take place during stellar nucleosynthesis depend on the mass of the star. Low-mass stars, like our Sun, burn slowly and may survive for 10 billion years. Nuclear reactions in these stars produce elements up to an atomic number of 6 (carbon). High-mass stars (10 to 100 times the mass of the Sun) burn hotter and more quickly, and may survive for only 20 million years. They mostly produce elements up to an atomic number of 26 (iron).

Most very large atoms (atoms with atomic numbers greater than that of iron) require even more violent circumstances to form than can occur within a high-mass star.

These atoms form most efficiently during a supernova explosion. Supernova explosions produce large quantities of neutrons; many collide and attach to nearby atoms, quickly building very large atoms.

What happens to the atoms formed in stars? Some escape from a star into space during the star's lifetime, simply by moving fast enough to overcome the star's gravitational pull. The stream of atoms emitted from a star during its lifetime is a stellar wind (▶Fig. 1.8). Some escape upon the death of a star. Specifically, a low-mass star (like our Sun) releases a large shell of gas as it dies, whereas a high-mass star blasts matter into space during a supernova explosion (▶Fig.1.9). Once in space, atoms form new nebulae or mix back into existing nebulae.

In effect, when the first generation of stars died, they left a legacy of new elements that mixed with residual gas from the big bang. A second generation of stars and associated planets formed out of the new, compositionally more diverse nebulae. Second-generation stars lived and died, and contributed elements to third-generation stars. Succeeding generations of stars and planets contain a greater proportion of heavier elements. Because different stars live for varied periods of time, at any given moment the Universe contains many different generations of stars, including small stars that have been living for a long time and large stars that have only recently arrived on the scene. The mix of elements we find on Earth includes relics of primordial gas from the big bang as well as the disgorged guts of dead stars. Think of it—the elements that make up your body once resided inside a star!

The Nature of Our Solar System

We've just discussed how stars form from nebulae. Can we apply this model to our own solar system—that is, the Sun plus all the objects that orbit it? The answer is yes, but before we do so, let's define the components that make up our solar system.

Our Sun does not sail around the Milky Way in isolation. In its journey it holds, by means of gravitational "glue," many other objects. Of these, the largest are the nine planets: Mercury, Venus, Earth, Mars, Jupiter, Saturn, Uranus, Neptune, and Pluto (named in order from the closest to the Sun to the farthest). A **planet** is a sizable solid object orbiting a star, and it may itself travel with a moon or even many moons. By definition, a **moon** is an object locked in orbit around a planet; all but two of the planets have them. For example, Earth has one moon (the Moon) whereas Jupiter has at least sixteen, of which four are as big as or bigger than our Moon. In the past several years astronomers have documented planets in association with dozens of other stars. In addition to planets and moons, our solar system also includes a belt of asteroids (relatively small chunks of rock and/or metal) between the orbit of Mars and the orbit of Jupiter, and

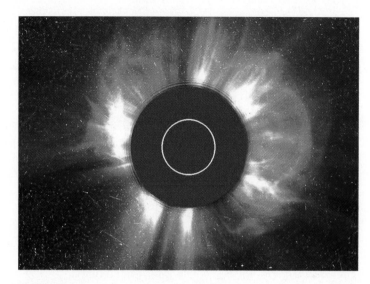

FIGURE 1.8 In this image, a black disk hides the Sun so we can see the stellar wind that the Sun produces.

perhaps a trillion comets (relatively small blocks of "ice" orbiting the Sun; ice, in this context, means the solid version of materials that could be gaseous under Earth's surface conditions) in clouds that extend very far beyond the orbit of Pluto. Even though there are many objects in the solar system, 99.8% of the solar system's mass resides in the Sun. The next largest object in the Solar system—the planet Jupiter—accounts for 99.5% of all non-solar mass in the solar system.

FIGURE 1.9 Very heavy elements form during supernova explosions. Here, we see the rapidly expanding shell of gas ejected into space from an explosion in 1054 C.E. This shell is called the Crab Nebula.

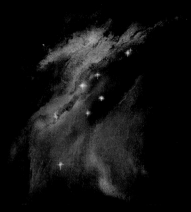

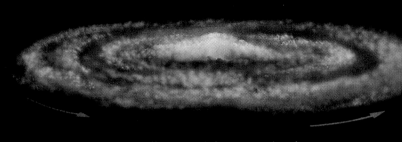

Gravity pulls gas and dust inward to form an accretion disk. Eventually a glowing ball--the proto-Sun--forms at the center of the disk.

Forming the solar system, according to the nebula hypothesis: A nebula forms from hydrogen and helium left over from the big bang, as well as from heavier elements that were produced by fusion reactions in stars or during explosions of stars.

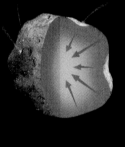

Gravity reshapes the proto-Earth into a sphere. The interior of the Earth differentiates into a core and mantle.

Forming the planets from planetesimals: Planetesimals grow by continuous collisions. Gradually, an irregularly shaped proto-Earth develops. The interior heats up and becomes soft.

Soon after Earth forms, a small planet collides with it, blasting debris that forms a ring around the Earth.

The Moon forms from the ring of debris.

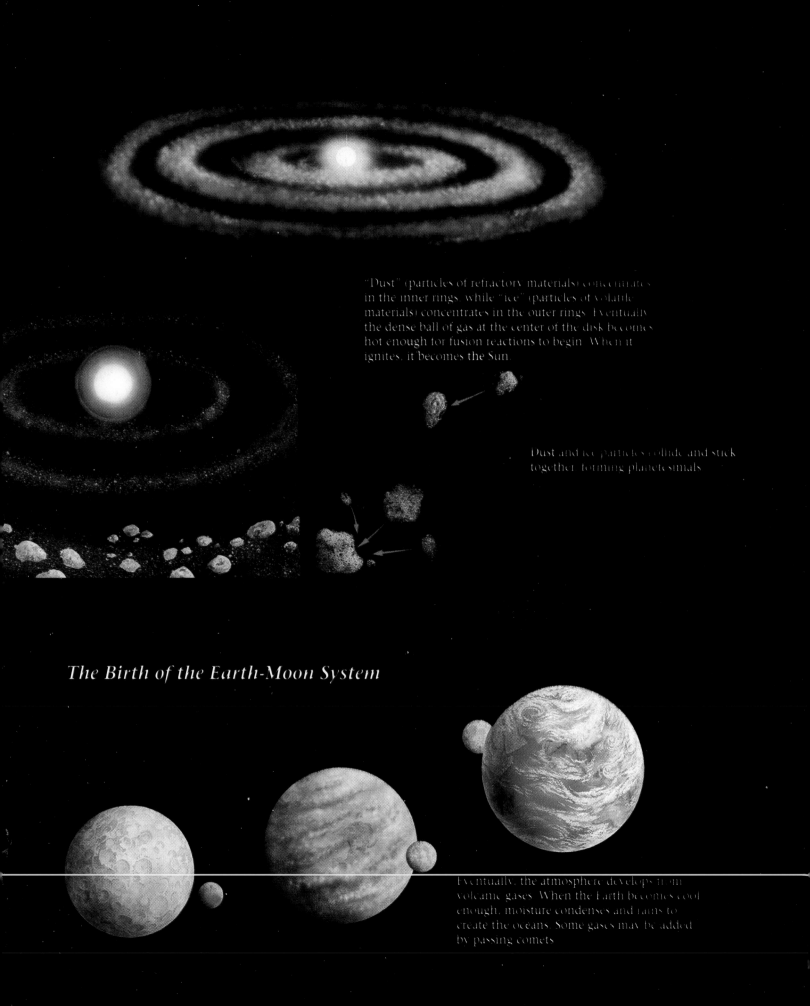

"Dust" (particles of refractory materials) concentrates in the inner rings, while "ice" (particles of volatile materials) concentrates in the outer rings. Eventually the dense ball of gas at the center of the disk becomes hot enough for fusion reactions to begin. When it ignites, it becomes the Sun.

Dust and ice particles collide and stick together, forming planetesimals.

The Birth of the Earth-Moon System

Eventually, the atmosphere develops from volcanic gases. When the Earth becomes cool enough, moisture condenses and rains to create the oceans. Some gases may be added by passing comets.

Of the nine planets in our solar system, the inner four (Mercury, Venus, Earth, and Mars) are called the inner planets or the **terrestrial planets** (Earth-like planets) because they consist of a shell of rock surrounding a core of iron alloy, as does the Earth. The next four (Jupiter, Saturn, Uranus, and Neptune) have been traditionally called the outer planets, the Jovian planets (Jupiter-like planets), or the **gas-giant planets.** The adjective "giant" certainly seems appropriate for these planets (▶Fig. 1.10). Jupiter, for example, has a mass that is about 318 times that of Earth and a diameter 11.2 times larger. The ninth planet, Pluto, is so small (about two-thirds the size of Earth's moon) that astronomers didn't find it until 1930.

Forming the Solar System

Earlier in this chapter we presented ideas about how the first stars formed, during the early history of the Universe, from very light gases (mostly hydrogen) produced during the big bang. To keep things simple, we didn't mention the formation of planets, moons, asteroids, or comets in that discussion.

Now, let's develop a model to explain where these other objects came from (see art on pp. 16–17). Since this model involves the formation of our solar system out of a nebula, it is called the **nebula theory** of planet formation.

Our solar system formed at about 4.57 Ga (Ga means giga annum or a billion years ago), over 9 billion years after the big bang. Thus our Sun is probably a third- or fourth- or fifth-generation star (no one can say for sure) created from a nebula that, while still predominantly composed of hydrogen, contained all ninety-two elements. Recall that the other elements formed in stars or supernovas.

The materials in nebulae such as the one from which our solar system formed can be divided into two classes. Volatile materials—such as hydrogen, helium, methane, ammonia, water, and carbon monoxide—are materials that could exist as gas at the Earth's surface. In the pressure and temperature conditions of space, some volatile materials remain in gaseous form, but others condense or freeze to form different kinds of ice. Refractory materials are those that melt only at high temperatures, and they condense to form solid soot-sized particles of "dust" in

FIGURE 1.10 (a) The relative sizes of the planets of our solar system. (b) The planets' orbits. Note that the orbits of all planets except Pluto lie in the same plane. Pluto may be an errant moon or a large comet; some astronomers do not call it a planet.

(a)

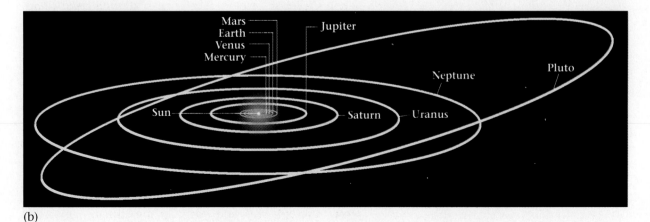

(b)

the coldness of space. Thus, when astronomers refer to dust, they mean specks of solid rocky or metallic material, or clumps of the molecules that make up rock or metal.

We saw earlier that the first step in the formation of a star is the development of an accretion disk. When our solar system formed, this accretion disk contained not only hydrogen and helium gas, but also other gases as well as ice and dust. Such an accretion disk can also be called a proto-planetary disk, because it contains the raw materials from which planets form. With time, the central ball of our proto-planetary disk developed into the proto-Sun, and the remainder evolved into a series of concentric rings. A protoplanetary disk is hotter toward its center than toward its rim, so volatile elements in the central region of the disk evaporated, and migrated outward. Thus, the warmer, inner rings of the disk ended up with higher concentrations of dust, whereas the cooler, outer rings ended up with higher concentrations of ice.

Even before the proto-Sun ignited, the material of the surrounding rings began to coalesce, or "accrete." Recall that accretion refers to the process by which smaller pieces clump and bind together to form larger pieces. First, soot-sized particles merged to form sand-sized grains. Then, these grains clumped together to form grainy basketball-sized blocks, which in turn collided. The fate of these blocks depended on the speed of the collision—if the collision was slow, blocks stuck together or simply bounced apart; if the collision was fast, one or both of the blocks shattered, producing smaller fragments that had to recombine all over again.

Eventually, enough blocks accreted to form **planetesimals,** bodies whose diameter exceeded about 1 km (▶Fig. 1.11). Because of their mass, planetesimals exert enough gravity to attract and pull in other objects that are nearby. Figuratively, planetesimals acted like vacuum cleaners, sucking in debris—small pieces of dust and ice as well as smaller planetesimals—into their orbit, and in the process they grew progressively larger. Eventually, victors in the competition to attract mass grew into **protoplanets,** bodies almost the size of today's planets. Once the protoplanets had succeeded in incorporating virtually all the debris near their orbits, so that their growth nearly ceased and they were the only inhabitants of their orbits, they became the planets that exist today.

Early stages in the accretion process probably occurred very quickly—some computer models suggest that it may have taken only a few hundred thousand years to go from the dust and gas stage to the large planetesimal stage. Estimates for the growth of planets from planetesimals range from 10 million years (m.y.) to 200 m.y. Recent studies favor faster growth.

The nature of the planet resulting from accretion depended on distance from the proto-Sun. In the inner orbits, where the protoplanetary disk consisted mostly of dust, small terrestrial planets composed of rock and metal formed. In the outer part of the protoplanetary disk, where in addition to gas and dust significant amounts of ice existed, larger protoplanets—as big as 15 times the size of Earth—grew. These, in turn, pulled in so much gas and ice that they grew into the gas-giant planets. Rings of dust and ice orbiting these planets became their moons.

When the Sun ignited, toward the end of the time when planets were forming, it generated a strong stellar wind (in this case, the solar wind) of particles traveling rapidly into space. The solar wind blew any remaining gases out of the inner portion of the newborn solar system. But the wind was too weak to blow away the atmospheres of the gas-giant planets.

Differentiation of the Earth and Formation of the Moon

When they first developed, larger planetesimals and protoplanets had a fairly uniform distribution of material throughout, because the smaller pieces from which they formed all had much the same composition and collected together in no particular order. But large planetesimals did not stay homogeneous for long. As they formed, they began to heat up. The heat came primarily from two sources: the transformation of kinetic energy into thermal energy during collisions, and the decay of radioactive elements. In bodies whose temperature rose sufficiently to cause melting, denser iron alloy separated out and sank to the center of the body, whereas lighter rocky materials remained in a shell surrounding the center. By this process, called **differentiation,** protoplanets and large planetesimals developed

FIGURE 1.11 Cross section of a particular type of meteorite (a fragment of solid material that fell from space and landed on Earth) thought to display the texture of a small planetesimal.

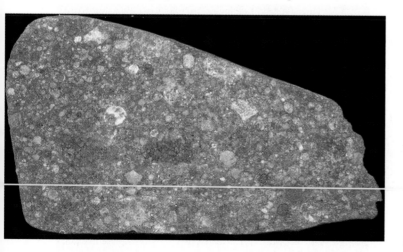

internal layering early in their history. As we will see later, the central ball of iron alloy constitutes the body's core and the outer shell constitutes its mantle. Eventually, even partially molten planetesimals cooled and largely solidified.

In the early days of the solar system, planets continued to be bombarded by **meteorites** (solid objects falling from space that land on Earth) even after the Sun had ignited and differentiation had occurred. Heavy bombardment in the early days of the solar system pulverized the surfaces of planets and eventually left huge numbers of craters. It also contributed to heating the planets.

Constrained by the age of Moon rocks, most geologists have concluded that sometime between Earth's accretion (about 4.57 Ga) and the oldest known rock on the Moon (4.47 Ga), a Mars-sized protoplanet slammed into Earth. In the process, the colliding body disintegrated, along with a large part of the Earth's mantle. As much as 65% of the objects melted and some of their mass may even have vaporized. A ring of debris formed around the Earth, which at the time would have been covered with a sea of molten rock, and quickly (perhaps in as little as 100,000 years) accreted to form the Moon. Of note, not all moons in the solar system necessarily formed in this manner. Some may be protoplanets or comets captured by a larger planet's gravity.

Why Is the Earth Round?

Planetesimals were jagged or irregular in shape, and asteroids today have irregular shapes. Planets, on the other hand, are essentially spheres. Why are planets spherical? Simply put, when a protoplanet gets big enough, gravity can change its shape. To picture how, take a block of cheese outside on a hot summer afternoon and place it on a table. As the Sun warms the cheese, it gets softer and softer, and eventually gravity causes the cheese to spread out in a pancake-like blob on the table. This model shows that gravitational force alone can cause material to change shape if the material is soft enough. Now let's apply this model to planetary growth.

The rock composing a small planetesimal is cool and strong enough so that the force of gravity is not sufficient to cause the rock to flow. But once a planetesimal grows beyond a certain critical size, the insides of the planet become warm and the rock becomes soft enough to flow in response to gravity; also, the gravitational force becomes stronger. As a consequence, protrusions are pulled inward toward the center, and the planetesimal re-forms into a shape that permits the force of gravity to be the same at all points on its surface. This special shape is a sphere because in a sphere the distribution of mass around the center has evened out.

1.6 WELCOME TO THE EARTH SYSTEM AND ITS NEIGHBORHOOD

So far in this chapter, we've described scientific ideas about how the Universe, and then the solar system, formed. Now, let's focus on our home planet and develop an image of Earth's overall architecture. To do this, imagine that we are explorers from outside our solar system on a visit to the Earth for the first time. We will see that our planet consists of several complex components—the atmosphere (Earth's gaseous envelope), the hydrosphere (Earth's surface and near-surface water), the biosphere (Earth's great variety of life forms), the lithosphere (the outer, rigid shell of the Earth), and the interior (the softer solid and molten material inside the Earth). These systems interact in complex ways, comprising the **Earth System.** This planet remains a dynamic place, because heat inside the Earth, and the heat of the Sun, provide energy.

Our journey begins in interstellar space, somewhere between the nearest star and Pluto. Compared with the air we breathe at sea level, interstellar space is a **vacuum,** meaning that it contains very little matter in a given volume. In interstellar space, there is only about 1 atom in every 1 to 10 liters. As our rocket approaches the Earth, and we see its beautiful bluish glow (▶Fig. 1.12), our instruments detect the planet's magnetic field, like a signpost shouting, "Approaching Earth!" A **magnetic field,** in a general sense, is the region affected by the force emanating from a magnet. This force, which grows progressively stronger as you approach the magnet, can attract or repel another magnet and can cause charged particles to move. Earth's magnetic field, like the familiar magnetic field around a bar magnet, is largely a dipole, meaning it has a north pole and a south pole. We can portray the magnetic field by drawing magnetic field lines, the trajectories or paths

FIGURE 1.12 A view of the Earth as seen from space. Oceans, clouds, and land stand out clearly.

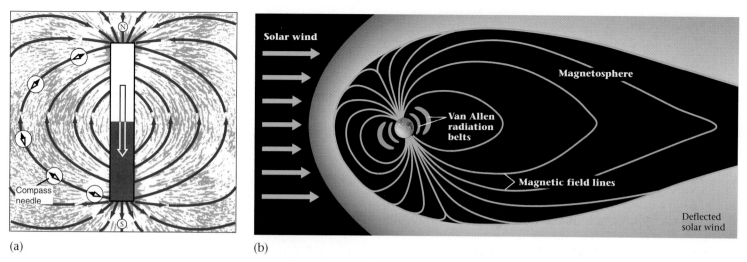

(a)   (b)

FIGURE 1.13 (a) The magnetic field around a bar magnet can be displayed by sprinkling iron filings on a sheet of paper lying over the magnet. The filings define curving trajectories—these are the field lines. Charged particles placed in the field would flow along these lines, following the arrows, and compass needles would align with them. Note that the magnet has a north pole and a south pole. (b) The magnetic field of the Earth interacts with the solar wind—the wind distorts the field so that it tapers away from the Sun, and the field isolates the Earth from most of the wind. Note the Van Allen belts near the Earth.

along which magnetic particles would align, or charged particles would flow, if placed in the field (▶Fig. 1.13a).

The solar wind interacts with Earth's magnetic field, distorting it into a huge teardrop pointing away from the Sun.

FIGURE 1.14 (a) Satellite view showing the glowing ring of the aurora borealis as it appears superimposed on a map of North America. (b) The aurora as seen from the ground in Alaska.

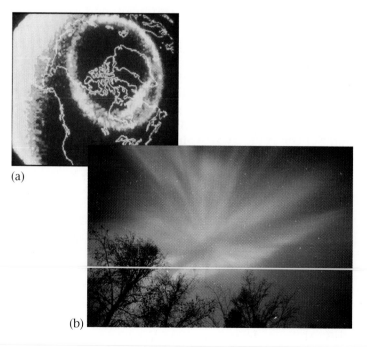

(a)

(b)

Fortunately, the magnetic field deflects most of the wind, so that most of the particles in the wind do not reach Earth's surface. In this way, the magnetic field acts like a shield against the solar wind; the region inside this magnetic shield is called the "magnetosphere" (▶Fig. 1.13b).

Though it protects the Earth from most of the solar wind, the magnetic field does not stop our rocket ship, and we continue to speed toward the planet. At distances of about 3,000 km and 10,500 km out from the Earth, we encounter two distinct belts called the Van Allen radiation belts, named for the physicist who first recognized them in 1959. These consist of solar wind particles as well as cosmic rays (nuclei of atoms, which may have been emitted from supernova explosions, and that bombard the Earth from deep space) that were moving so fast they were able to penetrate the weaker outer part of the magnetic field and were then trapped by the stronger magnetic field closer to the Earth. By trapping cosmic rays, the Van Allen belts protect life on Earth from dangerous radiation. Some charged particles make it past the Van Allen belts and are channeled along magnetic field lines to the polar regions of Earth. When they interact with gas atoms in the upper atmosphere they cause the gases to glow, like the gases in neon signs, creating spectacular aurorae (▶Fig. 1.14a, b).

1.7 THE ATMOSPHERE

Approaching still closer to the Earth, we enter its **atmosphere,** an envelope of gas consisting overall of 78% nitrogen (N_2) and 21% oxygen (O_2), with minor amounts

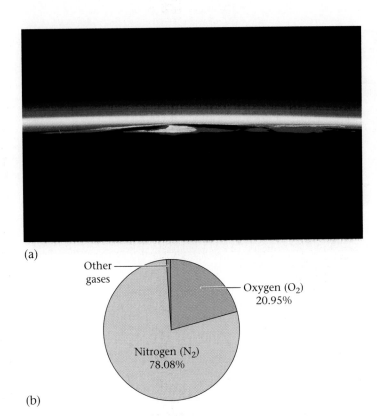

(a)

(b)

FIGURE 1.15 (a) Sunset view from the space shuttle. The gases and dust of the atmosphere reflect light and absorb certain wavelengths of light, creating a glowing palette of color. The vacuum of space is always black. (b) Nitrogen and oxygen constitute most of the gas in the atmosphere.

(1% total) of argon, carbon dioxide (CO_2), neon, methane, ozone, carbon monoxide, and sulfur dioxide, that surrounds the Earth (▶Fig. 1.15a, b). Other terrestrial planets have atmospheres, but none of them are like Earth's.

The weight of overlying air squeezes on the air below, and thus pushes gas molecules in the air below closer together (▶Fig. 1.16a, b). Thus, both the density (mass per unit volume) of air and the air pressure (the amount of push that the air exerts on material beneath it) increase closer to the Earth's surface (▶Fig. 1.16c). Technically, we specify pressure in units of force, or push, per unit area. (Such units include atmospheres, abbreviated "atm," and "bars." 1 atm = 1.04 kilograms per square centimeter, or 14.7 pounds per square inch. Atmospheres and bars are almost the same, for 1 atm = 1.01 bars.) At sea level, average air pressure is 1 atm, whereas on the peak of Mt. Everest, 8.85 km above sea level, air pressure is only 0.3 atm. In the Earth's atmosphere, 99% of the gas lies below 50 km, and most of the remaining 1% lies between 50 and 500 km. Where the space shuttle orbits the Earth, an altitude of about 400 km (about 250 miles), air pressure is only 0.0000001 atm. Keep in mind that humans cannot live for long at elevations greater than about 4.5–5.5 km. If your climb to the top of Mt. Everest and down again lasts more than a few hours, you'll need to breathe bottled oxygen to survive.

The nature of the atmosphere changes with distance from the Earth. Because of these changes, atmospheric scientists divide the atmosphere into layers. Most winds and clouds develop only in the lowest layer, the troposphere, where air undergoes convection (hence the prefix *tropos*, Greek for "turning"). "Convection" refers to the circulation that occurs when different parts of a fluid have different densities—less dense portions of a fluid rise, while denser portions sink (see Appendix). Winds develop when sinking cool air moves in to replace rising warm air. Clouds, mists of tiny droplets, form when rising air contains moisture, for as this air rises, it cools, and the moisture condenses. The layers of the atmosphere that lie above the troposphere are named, in sequence from base to top: the stratosphere, the mesosphere, and the thermosphere (▶Fig. 1.16d). Boundaries between layers are called "pauses." Specifically, the boundary between the troposphere and the overlying stratosphere is the tropopause; the boundary between the stratosphere and the mesosphere is the stratopause; and the boundary between the mesosphere and the thermosphere is the mesopause. The pauses are defined as elevations at which temperature stops decreasing and starts increasing, or vice versa.

1.8 LAND AND OCEANS

Now we've gone into orbit around the Earth, and we've set about mapping the planet—what obvious features should we put on the map? To start with, Earth's surface looks totally different from that of any other terrestrial planet or moon in the solar system. Earth has three distinctly different types of surface: dry land (continents and islands), the oceans, and ice-covered areas. Land (including ice-covered land) covers about 30% of the Earth's surface, while the oceans cover the remaining 70%. Lakes and rivers cover a small portion of the land. Geologists sometimes refer to the surface water of Earth (lakes, rivers, and oceans), along with groundwater (water that fills pores, tiny cracks or holes, in rock and sediment underground) and water vapor in the atmosphere, as the **hydrosphere.**

While orbiting the Earth, our instruments record variations in elevation—the **topography**—of the solid Earth, and show diverse landscapes such as mountains, valleys, and plains on the continents (▶Fig. 1.17). The instruments can also detect variations in the elevation of the sea floor, so we can recognize submarine plains, oceanic ridges, and deep trenches, or troughs. A graph plotting surface elevation on the vertical axis and the percentage of the Earth's surface on the horizontal axis shows that a relatively small proportion of the Earth's surface occurs at very high elevations (mountains) or at great depths (deep trenches). In fact,

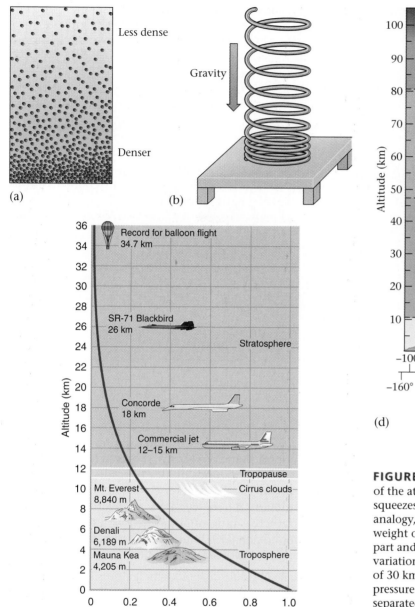

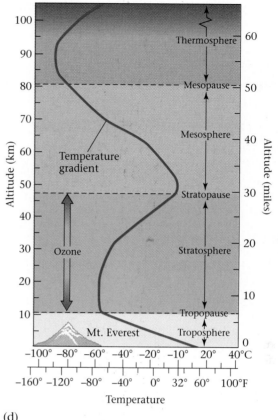

FIGURE 1.16 (a) Atmospheric density increases toward the base of the atmosphere because the weight of the upper atmosphere squeezes together gas molecules in the lower atmosphere. (b) By analogy, if you place a spring on a table in a gravity field, the weight of the upper part of the spring pushes down on the lower part and causes it to squeeze together. (c) A graph displaying the variation of air pressure with elevation shows that by an elevation of 30 km, atmospheric pressure is less than 1% of the atmospheric pressure at sea level. (d) The principal layers of the atmosphere are separated from each other by pauses. At a pause, the temperature gradient in the atmosphere changes direction.

most of the land surface lies just within a kilometer of sea level, while most of the sea floor is between 4 and 5 km deep (▶Fig. 1.18). A slight change in sea level would dramatically change the amount of dry land by flooding it.

1.9 WHAT IS THE EARTH MADE OF?

Elemental Composition

At this point, we leave our fantasy space voyage and turn our attention to the materials that make up the solid Earth, which we need to know about before proceeding to the Earth's interior. Earlier in this chapter, we learned that the atoms that make up the Earth consist of a mixture of elements left over from the big bang as well as elements produced by fusion reactions in stars and during supernova explosions. During the birth of the solar system, solar wind blew volatile materials ("ices") away from the region in which Earth was forming, like wind separating chaff from wheat. Earth and other terrestrial planets formed from the heavier materials left behind. Iron (35%), oxygen (30%), silicon (15%), and magnesium (10%) make up most of Earth's mass. The remaining 10% consists of the other eighty-eight naturally occurring elements.

FIGURE 1.17 This map of the Earth shows variations in elevation of both the land surface and the sea floor. Darker blues are deeper water in the ocean. Greens are lower elevation on land.

Categories of Earth Materials

The elements making up the Earth combine to form a great variety of materials. We can organize these into several categories.

FIGURE 1.18 This graph shows the proportions of the Earth's solid surface at different elevations. Two principal zones—the continents and adjacent continental shelf areas (the submerged margins of continents), and the ocean floor—account for most of Earth's area. Mountains and deep trenches cover relatively little area.

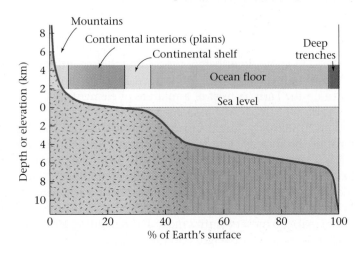

- *Organic chemicals:* Carbon-containing compounds that either occur in living organisms, or have characteristics that resemble compounds in living organisms, are called organic chemicals. Examples include oil, protein, plastic, fat, and rubber. Of note, certain simple carbon-containing materials, such as pure carbon (C), carbon dioxide (CO_2), carbon monoxide (CO), and calcium carbonate ($CaCO_3$), are *not* considered organic.

- *Minerals:* A solid substance in which atoms are arranged in an orderly pattern is called a **mineral.** Almost all minerals are inorganic (not organic) chemicals. Minerals grow either by the cooling and freezing of a liquid or by precipitation out of a water solution. Precipitation occurs when atoms that had been dissolved in water come together and form a solid. For example, solid salt forms by precipitation out of seawater when the water evaporates. A single coherent sample of a mineral that grew to its present shape and has smooth, flat faces is a crystal, while an irregularly shaped sample, or a fragment derived from a once-larger crystal or group of crystals, is a grain.

- *Glasses:* A solid in which atoms are not arranged in an orderly pattern is called glass. Glass forms when a liquid freezes so fast that atoms do not have time to organize into an orderly pattern.

- *Rocks:* Aggregates of mineral crystals or grains, and masses of natural glass, are called **rocks.** Geologists recognize three main groups of rocks. (1) Igneous rocks develop when hot molten (melted) rock cools and freezes solid. (2) Sedimentary rocks form from grains that break off preexisting rock and become cemented together, or from minerals that precipitate out of a water solution. An accumulation of loose mineral grains (grains that have not stuck together) is called **sediment.** (3) Metamorphic rocks form when preexisting rocks undergo changes, such as the growth of new minerals in response to heat and pressure.

- *Metals:* Solids composed of metal atoms (such as iron, aluminum, copper, and tin) are called metals. In a metal, outer electrons are able to flow freely. An alloy is a mixture containing more than one type of metal atom (for example, bronze is a mixture of copper and tin).

- *Melts:* **Melts** form when solid materials become hot and transform into a liquid. Molten rock is a type of melt— geologists distinguish between magma, which is molten rock beneath the Earth's surface, and lava, molten rock that has flowed out onto the Earth's surface.

- *Volatiles:* Materials that easily transform into gases at the relatively low temperatures found at the Earth's surface are called volatiles.

Notably, the most common minerals in the Earth contain silicon and oxygen mixed in varying proportions with other elements (typically iron, magnesium, aluminum, calcium, potassium, and sodium). These minerals are called silicate minerals, and, no surprise, rocks composed of silicate minerals are silicate rocks. Geologists distinguish four classes of igneous silicate rocks based, in essence, on the proportion of silicon to iron and magnesium. In order, from greatest to least proportion of silicon to iron and magnesium, these classes are: *felsic* (or *silicic*), *intermediate, mafic,* and *ultramafic.* As the proportion of silicon in a rock increases, the density (mass per unit volume) decreases. Thus, felsic rocks are less dense than mafic rocks.

Within each class, there are many different rock types, each with a name, that differ from one another in terms of composition (chemical makeup) and crystal size. These will be discussed in detail in Chapters 4–6, but for now we use only four rock names:

- *granite:* a felsic rock with large grains

- *gabbro:* a mafic rock with large grains

- *basalt:* a mafic rock with small grains

- *peridotite:* a mafic rock with large grains

1.10 DISCOVERING THAT THE EARTH HAS LAYERS

The world's deepest mine shaft penetrates gold-bearing rock that lies about 3.5 km (2 miles) beneath South Africa. Though miners seeking this gold must begin their workday by plummeting straight down a vertical shaft for almost ten minutes aboard the world's fastest elevator, the shaft represents little more than a pinprick on Earth's surface, when compared with the planet's radius; the distance from the center to the surface is 6,371 km. Even the deepest well ever drilled, a 12-km-deep hole in northern Russia, penetrates only the upper 0.03% of the Earth. We literally live on the thin skin of our planet, its interior forever inaccessible to our wanderings.

People have speculated about the Earth's interior since ancient times. What is the source of incandescent lavas spewed from volcanoes, of precious gems and metals, of sparkling spring water, and of the mysterious powers that shake the ground and topple buildings? Without the ability to observe the Earth's interior firsthand, pre-twentieth-century authors dreamed up fanciful images of it. For example, the English poet John Milton (1608–1674) described the underworld as a "dungeon horrible, on all sides round, as one great furnace flamed" (▶Fig. 1.19). Perhaps his image was inspired by volcanoes in the Mediterranean. In the eighteenth and nineteenth centuries, some European writers thought that the Earth's interior resembled a sponge, containing open caverns variously filled with molten rock, water, or air. In fact,

FIGURE 1.19 A literary image of the Earth's insides: *The Fallen Angels Entering Pandemonium, from* [Milton's] *"Paradise Lost," Book 1,* by the English painter John Martin (1789–1854).

in the popular 1864 novel *Journey to the Center of the Earth*, by the French author Jules Verne, three explorers find a route through interconnected caverns to the Earth's center.

Our present image of the Earth's interior, one made up of distinct layers, is the end product of many discoveries made during the past two hundred years. Specifically, researchers measured the mass of the whole Earth, and from this information calculated the *average* density (mass per unit volume) of the whole Earth. They discovered that our planet's average density far exceeds the density of common rocks found at its surface. Thus, the interior of the Earth must contain denser material than its outermost layer. Then, by studying the shape of the Earth, researchers determined that most of the Earth's mass had to be concentrated closer to the center of the planet. Thus, they recognized that the Earth resembled a hard-boiled egg, in that it had three principal layers: a not-so-dense **crust** (like an eggshell; composed of granite, basalt, gabbro, and other rocks), a denser solid **mantle** in between (the white; composed of a then-unknown material), and a very dense **core** (the yolk; composed of a different unknown material). Clearly, many questions remained. How thick are the layers? Are the boundaries between layers sharp or gradational? And what exactly are the layers composed of?

Clues from the Study of Earthquakes: Refining the Image

One day in 1889, a physicist in Germany noticed that the pendulum in his lab began to move without having been touched. He reasoned that the pendulum was actually standing still, because of its inertia (the tendency of an object at rest to remain at rest, and of an object in motion to remain in motion), and that the Earth was moving under it. A few days later, he read in a newspaper that a large **earthquake** (ground shaking due to the sudden breaking of rocks in the Earth) had taken place in Japan minutes before the movement of his pendulum began. The physicist deduced that vibrations due to the earthquake had traveled through the Earth from Japan and had jiggled his laboratory in Germany. The energy in such vibrations moves in the form of waves, called either "seismic waves" or "earthquake waves," that resemble the shock waves you feel with your hands when you snap a stick (▶Fig. 1.20). The breaking of rock during an earthquake either produces a new fracture on which sliding occurs or causes sliding on a preexisting fracture. A fracture on which sliding occurs is called a **fault.**

Geologists immediately realized that the study of seismic waves traveling through the Earth might provide a tool for exploring the Earth's insides, much as ultrasound helps doctors study a patient's insides. Specifically, laboratory measurements demonstrated that earthquake waves travel at different velocities (speeds) through different materials. Thus, by detecting depths at which velocities suddenly change, geoscien-

tists pinpointed the boundaries between layers and even recognized subtler boundaries within layers (see Interlude C).

Pressure and Temperature Inside the Earth

In order to keep underground tunnels from collapsing under the pressure created by the weight of overlying rock, mining engineers must design sturdy support structures. It is no surprise that deeper tunnels require stronger supports: the downward push from the weight of overlying rock increases with depth, simply because the mass of the overlying rock layer increases with depth. At the Earth's center, pressure probably reaches about 3,600,000 atm.

Temperature also increases with depth in the Earth. Even on a cool winter's day, miners who chisel away at gold veins exposed in tunnels 3.5 km below the surface swelter in temperatures of about 53°C (127°F). We refer to the rate of change in temperature with depth as the **geothermal gradient.** In the upper part of the crust, the geothermal gradient averages between 15° and 50°C per km. At greater depths, the rate decreases (to 10°C per km or less). Thus, 35 km below the surface of a continent, the temperature reaches 400° to 700°C. No one has ever directly measured the temperature at the Earth's center, but recent calculations suggest it may exceed 4,700°C.

FIGURE 1.20 When the rock inside the Earth suddenly breaks and slips, forming a fracture called a fault, it generates shock waves that pass through the Earth and shake the surface (creating an earthquake), much as the sound waves from a stick snapping travel to you and make your eardrum vibrate.

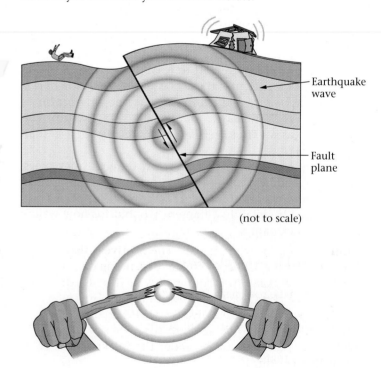

Earthquake wave

Fault plane

(not to scale)

1.11 WHAT ARE THE LAYERS MADE OF?

We noted earlier that the material composing the Earth's insides are denser than familiar surface rocks like granite and basalt. To discover what this material consists of, geologists

- conducted laboratory experiments to determine what kinds of materials inside the Earth could serve as a source for magma;
- studied unusual chunks of rock that have been carried up from the mantle in magma;
- conducted laboratory experiments to measure densities in samples of known rock types, so that they could compare these with observed densities in the Earth; and
- estimated which elements would be present in the Earth if the Earth had formed out of planetesimals similar in composition to meteorites (chunks of rock and/or metal alloy that fell from space and landed on Earth; ▶Box 1.1).

As a result of this work, we have a pretty clear sense of what the layers inside the Earth are made of. Let's now look at the properties of individual layers, starting with the Earth's surface.

The Crust

When you stand on the surface of the Earth, you are standing on the top of its outermost layer, the crust. The crust is our home and the source of all our resources. How thick is this all-important layer? Or, in other words, what is the depth to the crust-mantle boundary? An answer came from the studies of Andrija Mohorovičić, a researcher working in Zagreb, Croatia. In 1909, Mohorovičić discovered that the velocity of earthquake waves suddenly increased tens of kilometers beneath the Earth's surface, and he suggested that this increase was caused by an abrupt change in the properties of rock (see Interlude C for further detail). This change can be found most everywhere around our planet, though it occurs at different depths in different locations—it's deeper beneath continents than beneath oceans. Geologists now consider the change to be the crust-mantle boundary, and they refer to it as the **Moho** in Mohorovičić's honor. The relatively shallow depth of the Moho (7–70 km, depending on location) as compared to the radius of the Earth (6,371 km) emphasizes that the crust is very thin indeed. In fact, the crust is only about 0.1% to 1.0% of the Earth's radius, so if the Earth were the size of a balloon, the crust would be about the thickness of the balloon's skin.

Geologists distinguish between two fundamentally different types of crust—oceanic crust, which underlies the sea floor, and continental crust, which underlies continents (▶Fig. 1.21a). The crust is not simply cooled mantle, like the skin on chocolate pudding, but rather consists of a variety of rocks that differ in composition (chemical makeup) from mantle rock.

Oceanic crust is only 7 to 10 km thick. At highway speeds (100 km per hour), you could drive a distance equal to the thickness of the oceanic crust in about five minutes. We have a good idea of what oceanic crust looks like in cross section, because geologists have succeeded in drilling down through its top few kilometers and have found places where slices of oceanic crust, known as ophiolites, have been incorporated in mountain belts and therefore have been exposed on dry land. Studies of such examples show that oceanic crust consists of fairly uniform layers. At the top, we find a blanket of sediment, generally less than 1 km thick, composed of clay and tiny shells that settled like snow out of the

FIGURE 1.21 (a) This simplified cross section illustrates the differences between continental crust and oceanic crust. Note that the thickness of continental crust can vary greatly. (b) Oceanic crust is denser than continental crust.

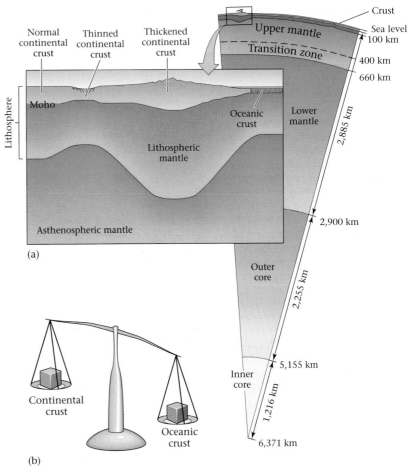

sea. Beneath this blanket, the oceanic crust consists of a layer of basalt and, below that, a layer of gabbro.

Most continental crust is about 35 to 40 km thick—about four to five times the thickness of oceanic crust—but its thickness varies significantly. Continents stretch like taffy in narrow bands called rifts, where crust can thin to only 25 km thick. Or it can squash together in mountain belts, where crust can increase to 70 km thick. In contrast to oceanic crust, continental crust contains a great variety of rock types, ranging from mafic to felsic in composition. On average, continental crust is less mafic than oceanic crust—it has a felsic (granite-like) to intermediate composition—so a block of average continental crust weighs less than a same-size block of oceanic crust (▶Fig. 1.21b). By analyzing the composition of rocks, geologists can estimate the proportions of chemicals making up the Earth's crust overall (▶Table 1.1). A glance at Table 1.1 shows that oxygen is the most abundant element in the crust.

TABLE 1.1 A table illustrating the abundance of elements in the Earth's crust.

Element	Symbol	Percentage by weight	Percentage by volume	Percentage by atoms
Oxygen	O	46.6	93.8	60.5
Silicon	Si	27.7	0.9	20.5
Aluminum	Al	8.1	0.8	6.2
Iron	Fe	5	0.5	1.9
Calcium	Ca	3.6	1	1.9
Sodium	Na	2.8	1.2	2.5
Potassium	K	2.6	1.5	1.8
Magnesium	Mg	2.1	0.3	1.4
All others	—	1.5	0.01	3.3

BOX 1.1
THE REST OF THE STORY

Meteors and Meteorites

During the early days of the solar system, the Earth collided with and incorporated almost all of the planetesimals and smaller fragments of solid material lying in its path. Intense bombardment ceased about 4.0 Ga, but even today collisions with space objects continues, and over 1,000 tons of material (rock, metal, dust, and ice) fall to Earth, on average, every year. The vast majority of this material consists of fragments derived from comets and asteroids sent careening into the path of the Earth after billiard ball–like collisions with each other out in space, or because of the gravitational pull of a passing planet. Some of the material, however, consists of chips of the Moon or Mars, ejected into space when large objects collided with these bodies.

Astronomers refer to any object from space that enters the Earth's atmosphere as a meteoroid. Meteoroids move at speeds of up to 75 km/s, so fast that when they reach an altitude of about 150 km, friction with the atmosphere causes them to begin to evaporate, leaving a streak of bright, glowing gas. The glowing streak, an atmospheric phenomenon, is a **meteor** (also known colloquially, though incorrectly, as a "falling star"). Most visible meteoroids completely evaporate by an altitude of about 30 km. But dust-sized ones may slow down sufficiently to float to Earth, and larger ones (fist-sized or bigger) can survive the heat of entry to reach the surface of the planet. Objects that strike the Earth are called **meteorites**. Most are asteroidal or planetary fragments, for the icy material of small cometary bodies is too fragile to survive the fall. In some cases, meteoroids explode in brilliant fireballs; such particularly luminous objects are called bolides. In fact, the explosion of a cometary bolide over Siberia in 1908 flattened a region of forest with a force equivalent to that of 2,000 atomic bombs.

Scientists did not realize that meteors were the result of solid objects falling from space until 1803, when a spectacular meteor shower (the occurrence of a large number of meteors during a short time) lit the sky over Normandy, France, and over 3,000 meteorites were subsequently found on the ground. In the succeeding two centuries, many meteorites have been collected and studied in detail. From this work, researchers recognize three basic classes of meteorites: iron (made of iron-nickel alloy), stony (made of silicate rock), and stony iron (rock embedded in a matrix of metal). Of all known meteorites, about 93% are stony and 6% are iron. From their composition, researchers have concluded that some meteors (a special subcategory of stony meteorites called carbonaceous chondrites, because they contain carbon and small spherical nodules called chondrules) are asteroids derived from planetesimals that never underwent differentiation into a core and mantle. Others (other stony meteorites and all iron meteorites) are asteroids derived from planetesimals that differentiated into a metallic core and a rocky mantle early in solar system history but later shattered into fragments during collisions with other planetesimals. Most meteorites have yielded radiometric dates of 4.54 Ga, but carbonaceous chondrites are as old as 4.57 Ga, the oldest known material ever measured. Meteorites are important to geologists because they represent matter from the earliest days of the solar system. Rocks that formed on Earth, in contrast, melted or were otherwise modified directly or indirectly as a result of plate tectonics.

Although almost all meteors are small and have not caused notable damage on Earth in historic time, a very few have smashed through houses, dented cars, and bruised people. During the longer term of Earth history, however, there have been some catastrophic collisions that left huge craters. As we will see later in the book, the largest collisions probably caused mass extinctions of life forms on our planet. It is likely that such collisions may happen again in the future—a large asteroid, for example, passed within 3 million miles of Earth in 2004, fortunately without incident.

Finally, it is important to note that most rock in the crust contains pores (tiny open spaces), and in much of the upper several kilometers of the crust the pores are filled with liquid water. This subsurface water, or groundwater, provides the water that farmers pump out of wells for irrigation and that cities pump out for their water supplies.

The Mantle

The mantle of the Earth forms a 2,885-km-thick layer surrounding the core. In terms of volume, it is the largest part of the Earth. In contrast to the crust, the mantle consists entirely of an ultramafic rock called peridotite. This means that peridotite, though rare at the Earth's surface, is actually the most abundant rock in our planet! On the basis of the occurrence of changes in the velocity of earthquake waves, geoscientists divide the mantle into three sublayers: the upper mantle, down to a depth of 400 km, the transition zone, from there down to a depth of 660 km, and the lower mantle, from there down to the core-mantle boundary.

Almost all of the mantle is solid rock. But even though it's solid, mantle rock below a depth of 100 to 150 km is so hot that it's soft enough to flow. This flow, however, takes place extremely slowly—at a rate of less than 15 centimeters a year. "Soft" here does *not* mean liquid; it simply means that over long periods of time mantle rock can change shape, like soft wax, without breaking. Note that we said *almost* all of the mantle is solid—in fact, up to a few percent of the mantle has melted. This melt occurs in films or bubbles between grains in the mantle at depths of between 100 and 200 km beneath the ocean floor. The presence of melt causes seismic waves to slow down, so geologists refer to the interval containing melt as the "low-velocity zone."

Though overall the temperature of the mantle increases with depth, temperature also varies significantly with location even at the same depth. The warmer regions are less dense, while the cooler regions are denser. The blotchy pattern of warmer and cooler mantle indicates that the mantle convects like water in a simmering pot. Warm mantle gradually flows upward, while cooler, denser mantle sinks.

The Core

Early calculations suggested that the core had the same density as gold, so for many years people held the fanciful hope that vast riches lay at the heart of our planet. Alas, geologists eventually concluded that the core consists of a far less glamorous material, iron alloy (iron mixed with lesser amounts of other elements). They arrived at this conclusion, in part, by comparing the properties of the core with the properties of metallic (iron) meteorites (▶ Box 1.1).

Studies of how earthquake waves bend as they pass through the Earth, along with the discovery that certain types of seismic waves cannot pass through the outer part of the core (see Interlude C), led geoscientists to divide the core into two parts, the outer core (between 2,900 and 5,155 km deep) and the inner core (from a depth of 5,155 km down to the Earth's center at 6,371 km). The outer core consists of *liquid* iron alloy. It can exist as a liquid because the temperature in the outer core is so high that even the great pressures squeezing the region cannot lock atoms into a solid framework. Because it is a liquid, the iron alloy of the outer core can flow (see art pp. 30–31); this flow generates Earth's magnetic field.

The inner core, with a radius of about 1,220 km, is a *solid* iron-nickel alloy that may reach a temperature of over 4,700°C. Even though it is hotter than the outer core, the inner core is a solid because it is deeper and is subjected to even greater pressure. The pressure keeps atoms from wandering freely, so they pack together tightly in very dense materials. The inner core probably grows over time, at the expense of the outer core, as the Earth slowly cools and the lower part of the outer core solidifies. Recent data suggests the inner core rotates slightly faster than the rest of the Earth because of the force applied to it by the Earth's magnetic field.

1.12 THE LITHOSPHERE AND THE ASTHENOSPHERE

So far, we have identified three major layers (crust, mantle, and core) inside the Earth that differ compositionally from each other. Earthquake waves travel at different velocities through these layers. An alternate way of thinking about Earth layers comes from studying the degree to which the material making up a layer can flow. In this context we distinguish between "rigid" materials, which can bend but cannot flow, and "plastic" materials, which are relatively soft and can flow.

Let's apply this concept to the outer portion of the Earth's shell. Geologists have determined that the outer 100 to 150 km of the Earth is relatively rigid; in other words, the Earth has an outer shell composed of rock that cannot flow easily. This outer layer is called the **lithosphere,** and it consists of the crust plus the uppermost part of the mantle. We refer to the portion of the mantle within the lithosphere as the lithospheric mantle. Note that the terms "lithosphere" and "crust" are not synonymous—the crust is just part of the lithosphere. The lithosphere lies on top of the **asthenosphere,** which is the portion of the mantle in which rock can flow. Notice that the asthenosphere is entirely in the mantle and lies below a depth of 100–150 km. We can't assign a specific depth to the base of the asthenosphere because

The Earth, from Surface to Center

If we could remove all the air that hides much of the solid surface from view, we would see that both the land areas and the sea floor have plains and mountains.

If we could break open the Earth, we would see that its interior consists of a series of concentric layers, called (in order from the surface to the center) the crust, the mantle, and the core. The crust is a relatively thin skin (7–10 km beneath oceans, 25–70 km beneath the land surface). Oceanic crust consists of basalt (mafic rock), while the average continental crust is intermediate to silicic. The mantle, which overall has the composition of ultramafic rock, can be divided into three layers: upper mantle, transition zone, and lower mantle. The core can be divided into an outer core of liquid iron alloy and an inner core of solid iron alloy. Temperature increases progressively with depth, so at the Earth's center the temperature may approach that of the Sun's surface.

Within the mantle and outer core, there is swirling, convective flow. Flow within the outer core generates the Earth's magnetic field.

When discussing plate tectonics, it is convenient to call the outer part of the Earth a relatively rigid shell composed of the crust and uppermost mantle, the lithosphere, and the underlying warmer, more plastic portion of the mantle the asthenosphere. These are not shown in this painting.

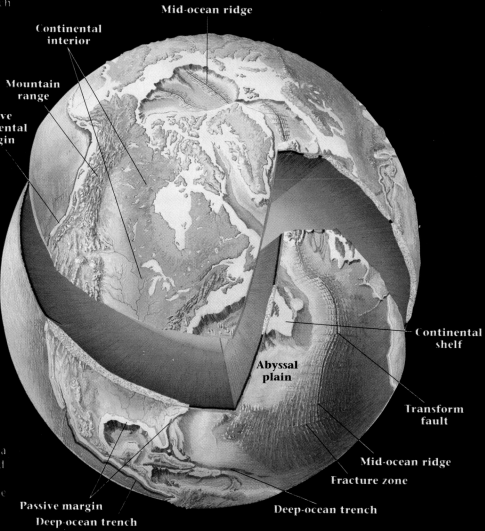

Continental interior

Mid-ocean ridge

Mountain range

Active continental margin

Continental shelf

Abyssal plain

Transform fault

Mid-ocean ridge

Fracture zone

Passive margin

Deep-ocean trench

Deep-ocean trench

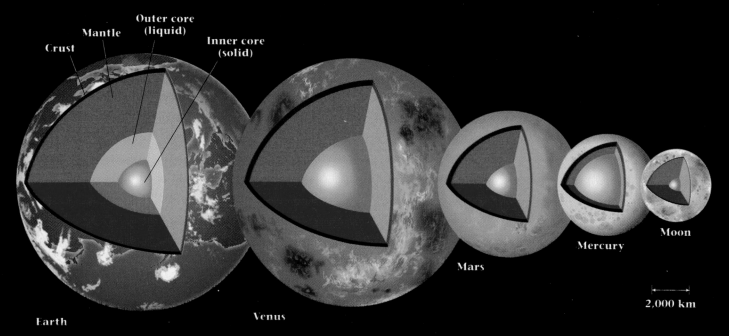

Crust

Mantle

Outer core (liquid)

Inner core (solid)

Earth

Venus

Mars

Mercury

Moon

2,000 km

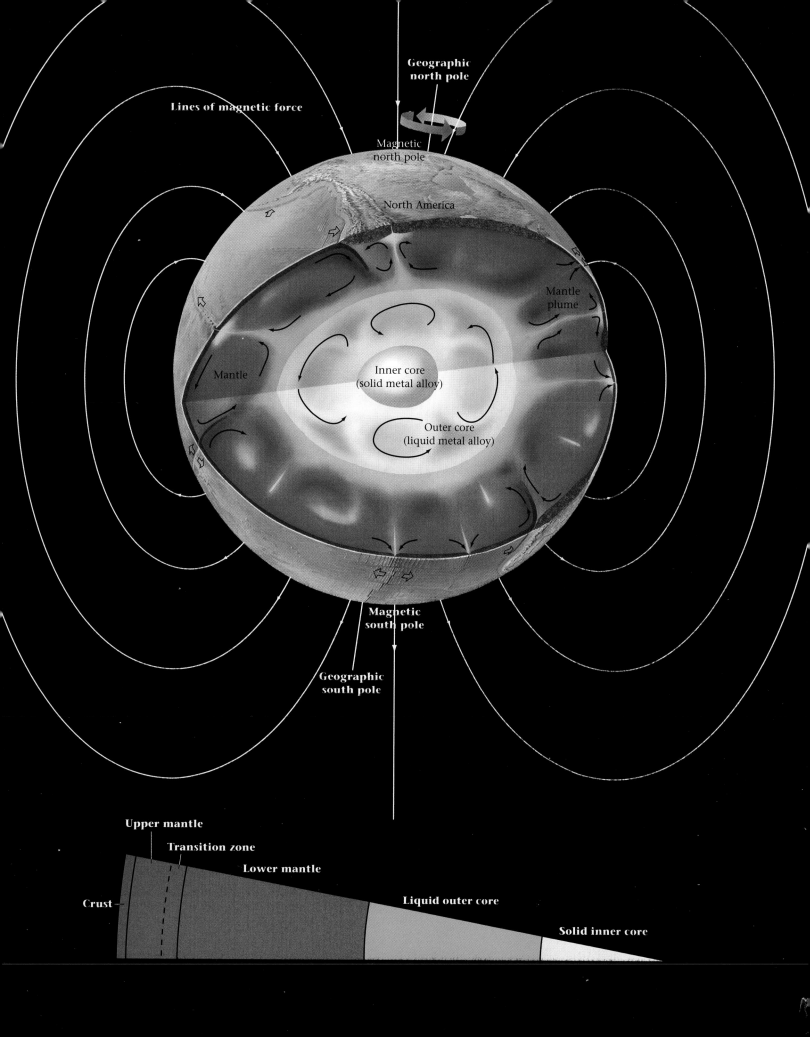

all of the mantle below 150 km can flow, but for convenience, some geologists place the base of the asthenosphere at the upper mantle/transition zone boundary (a depth of 400 km). One final point: even though the asthenosphere can flow, do not think of it as a liquid. It is not. Rather, the asthenosphere is largely solid. (A small amount of melt occurs only in the low-velocity zone.) At its fastest, the asthenosphere flows at rates of 10–15 cm/year.

Oceanic lithosphere and continental lithosphere are somewhat different (▶Fig. 1.22). Oceanic lithosphere, topped by oceanic crust, generally has a thickness of about 100 km. In contrast, continental lithosphere, topped by continental crust, generally has a thickness of about 150 km.

The boundary between the lithosphere and asthenosphere occurs where the temperature reaches about 1,280°C, for at this temperature mantle rock becomes soft enough to flow. To see how temperature affects the ability of a material to flow, take a cube of candle wax and place it in the freezer. The wax becomes very rigid and can maintain its shape for long periods of time; in fact, if you were to drop the cold wax, it would shatter. But if you take another block of wax and place it in a warm (not hot) oven, it becomes soft, so that you can easily mold it into another shape. Rock behaves somewhat similarly to wax. When rock is cool, it is quite rigid, but at high temperatures, rocks become soft and can flow, though much more slowly than wax. This ability to flow slowly can occur at a temperature much lower than is necessary to cause rocks to melt. Rock of the lithosphere is cool enough to behave rigidly, whereas rock of the asthenosphere is warm enough to flow easily.

Now, with an understanding of Earth's overall architecture at hand, we can discuss geology's grand unifying theory—plate tectonics. The next chapter introduces this key topic.

FIGURE 1.22 A cross section of the lithosphere, emphasizing the difference between continental and oceanic lithosphere.

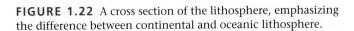

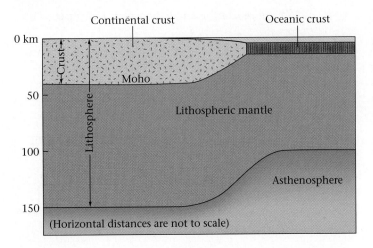

CHAPTER SUMMARY

- Most Greek philosophers favored a geocentric Universe concept, which placed the Earth at the center of the Universe. The heliocentric (Sun-centered) Universe concept was not widely accepted. The Renaissance brought a revolution in scientific thought, and Copernicus reintroduced the heliocentric concept.

- The Earth is one of nine planets orbiting the Sun. Our solar system lies on the outer edge of a slowly revolving galaxy, the Milky Way, which is composed of about 300 billion stars. The Universe contains at least hundreds of billions of galaxies.

- The red shift of light from distant galaxies, a manifestation of the Doppler effect, indicates that all distant galaxies are moving away from the Earth. This observation leads to the expanding Universe theory. Most astronomers agree that this expansion began after the big bang, a cataclysmic explosion about 13.7 billion years ago.

- The first atoms (hydrogen and helium) of the Universe developed before about 1 million years after the big bang. These atoms accumulated in vast gas clouds, called nebulae.

- Heavier elements form during fusion reactions in stars; the heaviest are probably made during supernova explosions. The Earth and the life forms on it contain elements that could only have been produced during the life cycle of stars. Thus, we are all made of stardust.

- Gravity caused clumps of gas in the nebulae to coalesce into revolving balls. As these balls of gas collapsed inward, they evolved into flattened disks with bulbous centers. The protostars at the center of these disks eventually became dense and hot enough that fusion reactions began in them. When this happened, they became true stars, emitting heat and light.

- Planets developed from the rings of gas and "dust," the planetary nebulae, that surrounded protostars. The gas condensed into planetesimals that then clumped together to form protoplanets, and finally true planets. The rocky and metallic balls in the inner part of the solar system did not acquire huge gas coatings; they became the terrestrial planets. Outer rings grew into gas-giant planets.

- The Moon formed from debris blasted free from Earth when our planet collided with a Mars-sized planet early during the history of the solar system.

- A planet the size of the Moon or larger will assume a near-spherical shape because the warm rock inside is so soft that gravity can smooth out irregularities.

- The Earth has a magnetic field, which shields it from solar wind. Closer to Earth, the field creates the Van Allen belts, which trap cosmic rays.

- A layer of gas surrounds the Earth. This atmosphere (78% nitrogen, 21% oxygen, 1% other) can be subdivided

THE VIEW FROM SPACE A group of newborn stars occurs in a cluster 12 billion light years from Earth, as viewed by the Hubble Space Telescope. The light from the stars makes the gas and dust surrounding them glow. Our solar system may have formed from such a cloud.

into distinct layers. All weather occurs in the troposphere, the layer we live in. Air pressure decreases with elevation, so 50% of the gas in the atmosphere resides below 5.5 km.

• The surface of the Earth can be divided into land (30%) and ocean (70%). Most of the land surface lies within 1 km of sea level.

• The Earth consists of organic chemicals, minerals, glasses, rocks, metals, melts, and volatiles. Most rocks on Earth contain silica (SiO_2) and thus are called silicate rocks. We distinguish between felsic, intermediate, mafic, and ultramafic rocks based on the proportion of silica.

• The Earth's interior can be divided into three compositionally distinct layers, named in sequence from the surface down: the crust, the mantle, and the core. The first recognition of this division came from studying the density and shape of the Earth.

• Pressure and temperature both increase with depth in the Earth. At the center, pressure is 3.6 million times greater than at the surface, and the temperature reaches over 4,700°C. The rate at which temperature increases as depth increases is the geothermal gradient.

• Studies of seismic waves have revealed the existence of sublayers in the core (outer core and inner core) and mantle (upper mantle, transition zone, and lower mantle).

• The crust is a thin skin that varies in thickness from 7–10 km (beneath the oceans) to 25–70 km (beneath the continents). Oceanic crust is mafic in composition, while average continental crust is felsic to intermediate. The mantle is composed of ultramafic rock. The core consists of iron alloy and consists of two parts—the outer core is liquid, and the inner core is solid. Flow in the outer core generates the magnetic field.

• The crust plus the upper part of the mantle constitute the lithosphere, a relatively rigid shell. The lithosphere lies over the asthenosphere, mantle that is capable of flowing.

KEY TERMS

asthenosphere (p. 29)
atmosphere (p. 21)
big bang (p. 12)
core (p. 26)
cosmology (p. 8)
crust (p. 26)
differentiation (p. 19)
Doppler effect (p. 10)
earthquake (p. 26)
Earth System (p. 20)
expanding Universe theory (p. 11)
fault (p. 26)
galaxy (p. 9)
gas-giant planet (p. 18)
geothermal gradient (p. 26)
hydrosphere (p. 22)
lithosphere (p. 29)
magnetic field (p. 20)
mantle (p. 26)

melts (p. 25)
meteor (p. 28)
meteorite (pp. 20, 28)
mineral (p. 24)
Moho (p. 27)
moon (p. 15)
nebula (p. 13)
nebular theory (p. 18)
planet (p. 15)
planetesimal (p. 19)
protoplanet (p. 19)
protostar (p. 14)
rock (p. 24)
sediment (p. 25)
terrestrial planet (p. 18)
topography (p. 22)
Universe (p. 8)
vacuum (p. 20)

REVIEW QUESTIONS

1. Contrast the geocentric and heliocentric Universe concepts.

2. Imagine you hear the main character in a cheap science-fiction movie say he will "return ten light years from now." What's wrong with his usage of the term "light year"? What are light years actually a measure of?

3. Describe how the Doppler effect works.

4. What does the red shift of distant galaxies tell us about their motion with respect to the Earth?

5. Briefly describe the steps in the formation of the Universe according to the big bang theory.

6. How does the composition of the solar system, in terms of the elements making up the Sun and the planets, differ

from that of the nebulae that existed a million years after the birth of the Universe? Explain the difference.

7. Why do the inner planets consist mostly of rock and metal, but the giant outer planets mostly of gas?

8. Why isn't the Earth homogeneous?

9. Describe how the Moon was formed.

10. Why is the Earth round?

11. What is the Earth's magnetic field? Draw a representation of the field on a piece of paper; your sketch should illustrate the direction in which charged particles would flow if placed in the field.

12. How does the magnetic field interact with solar wind? Be sure to consider the magnetosphere, the Van Allen radiation belts, and the aurorae.

13. What is the Earth's atmosphere composed of? Why would you die of suffocation if you were to eject from a fighter plane at an elevation of 12 km without taking an oxygen tank with you?

14. Describe the major categories of materials constituting the Earth.

15. What are the principal layers of the Earth? What happens to earthquake waves when they reach the boundary between the layers?

16. How do temperature and pressure change with increasing depth in the Earth? Be sure to explain the geothermal gradient.

17. What is the Moho, and how was it first recognized? Describe the differences between continental crust and oceanic crust.

18. What is the mantle composed of? What are the three sublayers within the mantle? Is there any melt within the mantle?

19. What is the core composed of? How do the inner core and the outer core differ from each other? We can't sample the core directly, but geologists have studied samples of material that are probably very similar in composition to the core. Where do these samples come from?

20. What is the difference between the lithosphere and the asthenosphere? Be sure to consider material differences and temperature differences. Which layer is softer and flows more easily? At what depth does the lithosphere/asthenosphere boundary occur? Is this above or below the Moho?

SUGGESTED READING

Allegre, C. 1992. *From Stone to Star.* Cambridge, Mass.: Harvard University Press.

Bolt, B. A. 1982. *Inside the Earth.* San Francisco: W.H. Freeman.

Brown, G. C., and A. E. Mussett. 1993. *The Incredible Earth.* London: Chapman and Hall.

Freedman, R. A., and W. J. Kaufmann III. 2001. *Universe,* 6th ed. New York: Freeman.

Hawking, S. 2005. *A Brief History of Time.* New York: Bantam.

Helffrich, G. R., and B. J. Wood. 2001. The Earth's Mantle. *Nature* 412: 501–507.

Hester, J., et al. 2002. *21st Century Astronomy.* New York: W. W. Norton.

Hoyle, F., G. Burbridge, and J. W. Narlikar. 2000. *A Different Approach to Cosmology: From a Static Universe through the Big Bang towards Reality.* Cambridge, UK: Cambridge University Press.

Kirshner, R. P. 2002. *The Extravagant Universe: Exploding Stars, Dark Energy, and the Accelerating Cosmos.* Princeton: Princeton University Press.

Liddle, A. 2003. *An Introduction to Modern Cosmology,* 2nd ed. New York: John Wiley & Sons.

Stein, S., and M. Wysession. 2003. *An Introduction to Seismology, Earthquakes, and Earth Structure.* London: Blackwell.

The Way the Earth Works: Plate Tectonics

2.1 INTRODUCTION

In September 1930, fifteen explorers led by a German meteorologist, Alfred Wegener, set out across the endless snowfields of Greenland to resupply two weather observers stranded at a remote camp. The observers were planning to spend the long polar night recording wind speeds and temperatures on Greenland's polar plateau. At the time, Wegener was well known, not only to researchers studying climate but also to geologists. Some fifteen years earlier, he had published a small book, *The Origin of the Continents and Oceans,* in which he had dared to challenge geologists' long-held assumption that the continents had remained fixed in position through geologic time (the time since the formation of the Earth). Wegener proposed, instead, that the continents had once fitted together like pieces of a giant jigsaw puzzle, to make one vast supercontinent. He suggested that this supercontinent, which he named **Pangaea** (pronounced Pan-jee-ah; Greek for "all land"), later fragmented into separate continents that then drifted apart, moving slowly to their present positions (▶Fig. 2.1). This idea came to be known as the **continental drift** hypothesis.

Wegener presented many observations in favor of the hypothesis, but he met with strong resistance. Drifting

Astronauts in the space shuttle Endeavor *could see the consequences of plate tectonics right outside their window. This photo from space shows the Sinai Peninsula, separated from Egypt to the west and the Arabian Peninsula to the east by rifts, narrow belts where the crust has stretched and broken apart. Note the green stripe in the middle—this is the Nile River Valley. The river flows north into the Mediterranean Sea. The green triangle is the Nile Delta.*

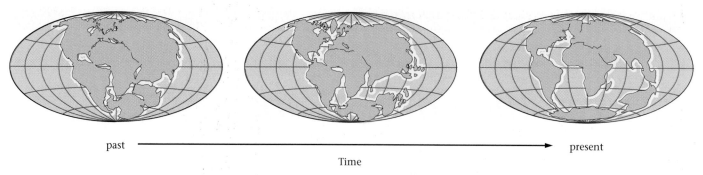

past ——————————————————————————————————→ present

Time

FIGURE 2.1 Wegener's image of Pangaea and its subsequent breakup and dispersal.

continents? Absurd! Or so proclaimed the leading geologists of the day. At a widely publicized 1926 geology conference in New York City, a crowd of celebrated American professors scoffed: "What force could possibly be great enough to move the immense mass of a continent?" Wegener's writings didn't provide a good answer, so despite all the supporting observations he had provided, most of the meeting's participants rejected continental drift.

Now, four years later, Wegener faced his greatest challenge. As he headed into the interior of Greenland, the weather worsened and most of his party turned back. But Wegener felt he could not abandon the isolated observers, and with two companions he trudged forward. On October 30, 1930, Wegener reached the observers and dropped off enough supplies to last the winter. Wegener and one companion set out on the return trip the next day, but they never made it home.

Had Wegener survived to old age, he would have seen his hypothesis become the foundation of a scientific revolution. Today, most geologists accept Wegener's ideas and take it for granted that the map of the Earth constantly changes; continents indeed waltz around its surface, variously combining and breaking apart through geologic time. The revolution began in 1960, when Harry Hess, a Princeton University professor, proposed that as continents drift apart, new ocean floor forms between them by a process that his contemporary Robert Dietz also described and labeled **sea-floor spreading.** Hess also suggested that continents move toward each other when the old ocean floor between them sinks back down into the Earth's interior, a process now called **subduction.** By 1968, geologists had developed a fairly complete model describing continental drift, sea-floor spreading, and subduction. In this model, Earth's lithosphere, its outer, relatively rigid shell, consists of about twenty distinct pieces, or plates, that slowly move relative to each other. Because we can confirm this model by many observations, it has gained the status of a theory (see Prelude), which we now call the theory of **plate tectonics,** from the Greek word *tekton,* which means "builder"; plate movements "build" regional geologic features. Geologists view plate tectonics as the grand

unifying theory of geology, because it can successfully explain a great many geologic phenomena, as we will see.

In this chapter, we introduce the observations that led Wegener to propose his continental drift hypothesis. Then we look at paleomagnetism, the record of Earth's magnetic field in the past, which provides a key proof of continental drift. Next, we learn how observations about the sea floor made by geologists during the mid-twentieth century led Harry Hess to propose the concept of sea-floor spreading. We conclude by describing modern plate tectonics theory, the foundation for discussing all geology.

2.2 WEGENER'S EVIDENCE FOR CONTINENTAL DRIFT

Before Wegener, geologists viewed the continents and oceans as immobile—fixed in position throughout geologic time. Wegener, however, suggested that a vast supercontinent, Pangaea, existed until the Mesozoic Era (the interval of geologic time, commonly known as the "age of dinosaurs," that lasted from 251 to 65 million years ago). During the Mesozoic, Pangaea broke apart to form the continents we see today; these continents then drifted away from each other. (Geologists now realize that supercontinents formed and dispersed at least a few times during Earth's history—the name Pangaea applies only to the most recent supercontinent.) Let's look at some of Wegener's arguments and see why he came to this conclusion.

The Fit of the Continents

Almost as soon as maps of the Atlantic coastlines became available in the 1500s, scholars noticed the fit of the continents. The northwestern coast of Africa could tuck in against the eastern coast of North America, and the bulge of eastern South America could nestle cozily into the indentation of southwestern Africa. Australia, Antarctica, and India could all connect to the southeast of Africa, while Greenland,

Europe, and Asia could pack against the northeastern margin of North America. In fact, all the continents could be joined, with remarkably few overlaps or gaps, to create Pangaea. (Modern plate tectonics theory can now even explain the misfits.) Wegener concluded that the fit was too good to be coincidence.

Locations of Past Glaciations

Wegener was an Arctic meteorologist by training, so it's no surprise that he had a strong interest in glaciers, the rivers or sheets of ice that slowly flow across the land surface. He realized that glaciers form mostly at high latitudes, and that by studying the past locations of glaciers, he might be able to determine the past locations of continents. We'll look at glaciers in detail in Chapter 18, but we need to know something about them here to understand Wegener's arguments.

When a glacier moves, its ice carries sediment (pebbles, boulders, sand, silt, and mud). Some sediment freezes into the base of the glacier, so the glacier then becomes like a rasp and grinds exposed rock beneath it. In fact, rocks protruding from the base of the ice carve striations (scratches) into the underlying rock, and these striations indicate the direction in which the ice flowed. When the glacier eventually melts, the sediment collects on the ground and creates a distinctive layer of sediment called glacial till, a mixture of mud, sand, pebbles, and larger rocks. Later on, the till may be buried and preserved. Today, glaciers are found only in polar regions and in high mountains. But by studying the distribution and age of ancient till, geoscientists have determined that at several times during Earth's history, glaciers covered large areas of continents. We refer to these times as ice ages. One of the major ice ages occurred about 260–280 million years ago, near the end of the Paleozoic Era (the interval of geologic time between 542 and 251 million years ago).

Why was the study of ancient glacial till important to Wegener? When he plotted the locations of late Paleozoic till, he found that glaciers of this time interval occurred in southern South America, southern Africa, southern India, Antarctica, and southern Australia. These places are now widely separated from one another and, with the exception of Antarctica, do not currently lie in cold polar regions (▶Fig. 2.2a). To Wegener's amazement, all late Paleozoic glaciated areas lie adjacent to each other on his map of Pangaea (▶Fig. 2.2b). Furthermore, when he plotted the orientation of glacial striations, they all pointed roughly outward from a location in southeastern Africa. In other words, Wegener determined that the distribution of glaciations at the end of the Paleozoic Era could easily be explained if the continents had been united in Pangaea, with the southern part of Pangaea lying beneath the center of a huge ice cap. This distribution of glaciation could not be explained if the continents had always been in their present positions.

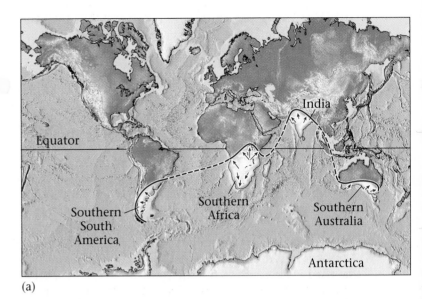

(a)

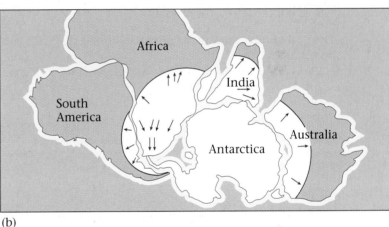

(b)

FIGURE 2.2 (a) The distribution of late Paleozoic glacial deposits on a map of the present-day Earth. The arrows indicate the orientation of striations. (b) The distribution of these glacial deposits on a map of the southern portion of Pangaea. Note that the glaciated areas fit together to define a polar ice cap.

The Distribution of Equatorial Climatic Belts

If the southern part of Pangaea had straddled the South Pole at the end of the Paleozoic Era, then during this same time interval, southern North America, southern Europe, and northwestern Africa would have straddled the equator and would have had tropical or subtropical climates. Wegener searched for evidence that this was so by studying sedimentary rocks that were formed at this time, for the material making up these rocks can reveal clues to the past climate. Specifically, in the swamps and jungles of tropical regions, thick deposits of plant material accumulate, and when deeply buried, this material transforms into coal. Further, in the clear, shallow seas of tropical regions, large reefs develop offshore. Finally, subtropical regions, on either side of the tropical belt, contain deserts, an environment for sand-dune formation and for the accumulation of salt from evaporating seawater or salt lakes. Wegener speculated that the distribution of late Paleozoic coal, sand-dune deposits, and salt deposits could define climate belts on Pangaea.

Sure enough, in the belt of Pangaea that Wegener expected to be equatorial, late Paleozoic sedimentary rock layers include abundant coal and the relicts of reefs; and in the portions of Pangaea that Wegener predicted would be subtropical, late Paleozoic sedimentary rock layers include relicts of desert dunes and of salt (▶Fig. 2.3). On a present-day map of our planet, these deposits are scattered around the globe at a variety of latitudes—including high latitudes, where they cannot possibly have formed. However, in Wegener's Pangaea, the deposits align in continuous bands that occupy appropriate latitudes.

The Distribution of Fossils

Today, different continents provide homes for different species. Kangaroos, for example, live only in Australia. Similarly, many kinds of plants grow only on one continent

FIGURE 2.3 Map of Pangaea, showing the distribution of coal deposits and reefs (tropical environments), and sand-dune deposits and salt deposits (subtropical environments). Deposits now on different continents align in distinct belts.

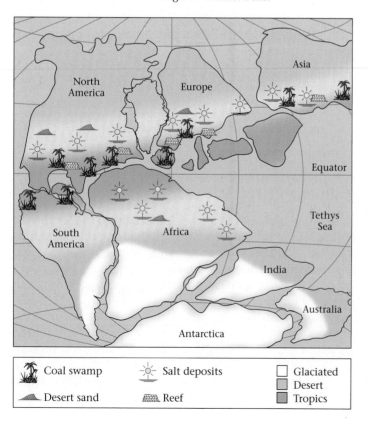

Coal swamp · Salt deposits · ☐ Glaciated · ☐ Desert · ☐ Tropics
Desert sand · Reef

and not on others. Why? Because land-dwelling species of animals and plants cannot swim across vast oceans, and thus they evolved independently on different continents. During a period of Earth history when all continents were in contact, however, land animals and plants conceivably could have lived on many continents.

With this concept in mind, Wegener plotted locations of fossils of land-dwelling species that existed during the late Paleozoic and early Mesozoic Eras (between about 300 and 210 million years ago) and found that they had indeed populated several continents (▶Fig. 2.4). For example, an early Mesozoic land-dwelling reptile called *Cynognathus* lived in both southern South America and southern Africa. *Glossopteris*, a species of seed fern, flourished in regions that now constitute South America, Africa, India, Antarctica, and Australia. *Mesosaurus*, a freshwater reptile, inhabited portions of what is now South America and Africa. *Lystrosaurus*, another land-dwelling reptile, wandered through present-day Africa, India, and Antarctica. None of these species could have traversed a large ocean. Thus, Wegener argued, the distribution of these species required the continents to have been adjacent to one another in the late Paleozoic and early Mesozoic Eras.

Matching Geologic Units

In the same way that an art historian can identify a Picasso painting and an architect a Victorian design, a geoscientist can identify a distinctive group of rocks. Wegener found that the same distinctive Precambrian (the interval of geologic time before 542 million years ago) rock assemblages occurred on the eastern coast of South America and the western coast of Africa, regions now separated by an ocean (▶Fig. 2.5a). If the continents had been joined to create Pangaea in the past, then these matching rock groups would have been adjacent to each other, and thus could have composed continuous blocks. Wegener also noted that belts of rocks in the Appalachian Mountains of the United States and Canada resembled belts of rocks in mountains of southern Greenland, Great Britain, Scandinavia, and northwestern Africa (▶Fig. 2.5b), regions that would have lain adjacent to each other in Pangaea. Wegener thus demonstrated that not only did the coastlines of continents match, the rocks adjacent to the coastlines did too.

Criticism of Wegener's Ideas

Wegener's model of a supercontinent that later broke apart explained the distribution of glaciers, coal, sand dunes, distinctive rock assemblages, and fossils we find today. Clearly, he had compiled a strong case for continental drift. But Wegener, as noted earlier, could not adequately explain how or why continents drifted. He left on his final expedition to Greenland having failed to convince his peers,

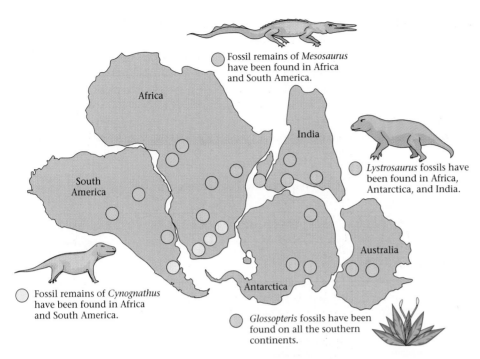

FIGURE 2.4 This map shows the distribution of terrestrial (land-based) fossil species. Note that creatures like *Lystrosaurus* could not have swum across the Atlantic to reach Africa. Sample locations are approximate.

FIGURE 2.5 (a) Distinctive areas of rock assemblages on South America link with those on Africa, as if they were once connected and later broke apart. (b) If the continents are returned to their positions in Pangaea by closing the Atlantic, mountain belts (shown in brown) of the Appalachians lie adjacent to similar-age mountain belts in Greenland, Great Britain, Scandinavia, and Africa. "Archean" is the older part of the Precambrian, and "Proterozoic" is the younger part.

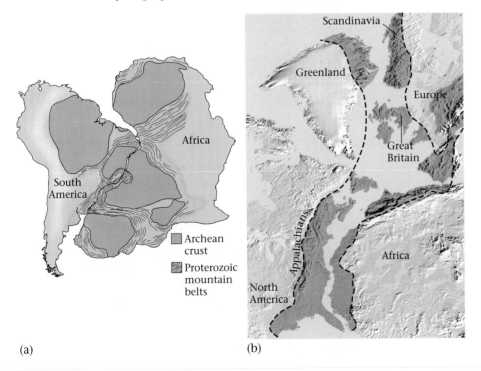

and he died in the icy wasteland never knowing that his ideas would lie dormant for decades before being reborn as the basis of the broader theory of plate tectonics.

In effect, Wegener was ahead of his time. It would take three more decades of research before geologists had obtained sufficient data to test Wegener's hypotheses properly. Collecting this data required instruments and techniques that had not existed in Wegener's day. Geologists learned to determine the age of rock assemblages, greatly improved our image of the Earth's interior, and produced detailed maps of the ocean floor and of its physical characteristics. But of the many geologic discoveries made in the middle decades of the twentieth century, the key that ultimately proved continental drift and sea-floor spreading, and which opened the door to plate tectonics, came from the discovery of a phenomenon called paleomagnetism.

2.3 PALEOMAGNETISM AND APPARENT POLAR-WANDER PATHS

In 1853, an Italian physicist noticed that volcanic rock behaved like a very weak magnet, and proposed that this rock became magnetic when it solidified from melt. It took a century, however, before instruments became available that could routinely measure such weak magnetization. Researchers in England used these instruments to study magnetization in *ancient* rocks. Their work showed that rocks preserve a record of Earth's past magnetic field. The record of ancient magnetism preserved in rock is called **paleomagnetism.** In order to understand paleomagnetism and how it enabled geologists to test the continental drift hypothesis, and later the sea-floor-spreading hypothesis, we must first look at the nature of Earth's magnetic field in a bit more depth than we did in Chapter 1. If you need to review background information on magnetism in general, please see ▶ Box 2.1.

Earth's Magnetic Field: Some Details

In Chapter 1, we learned that Earth has a magnetic field that deflects the solar wind and traps cosmic rays. Why does this field exist? Geologists do not yet have a complete answer, but they have hypothesized that the field results from the circulation of liquid iron alloy in the Earth's outer core—in other words, the outer core behaves like a "dynamo" that produces a magnetic field due to the motion of an electrical conductor (▶Fig. 2.7a). (A physics text provides additional detail on this phenomenon.) For convenience, however, *we can picture* the planet as a giant bar magnet, with a north magnetic pole and a south magnetic pole (▶Fig. 2.7b). The

---

BOX 2.1
SCIENCE TOOLBOX

Some Fundamentals of Magnetism

If you hold a magnet over a pile of steel paper clips, it will lift the paper clips against the force of gravity. The magnet exerts an attractive force that pulls on the clips. A magnet can also create a repulsive force that pushes an object away. For example, when oriented appropriately, one magnet can levitate another. The push or pull exerted by a magnet is a magnetic force; this force creates an invisible magnetic field around the magnet. Magnetic forces can be generated by a permanent magnet, a special material that behaves magnetically for a long time all by itself. Magnetic forces can also be produced by an electric current passing through a wire. An electrical device that produces a magnetic field is an electromagnet. The stronger the magnet, the greater its magnetization. When other magnets, special materials (such as iron), or electric charges enter a magnetic field, they feel a magnetic force. The strength of the pull that an object feels when placed in a magnet's field depends on the magnet's magnetization and on the object's distance from the magnet.

Compass needles are simply magnetic needles that can pivot freely and that align with Earth's magnetic field. Recall from Chapter 1 that you can symbolically represent a magnetic field by a pattern of curving lines, known as magnetic field lines. You can see the form of these lines by sprinkling iron filings on a sheet of paper placed over a bar magnet; each filing acts like a tiny magnetic compass needle and aligns itself with the magnetic field lines (Fig. 1.13).

All magnets have two **magnetic poles,** a north pole at one end and a south pole at the other. Opposite poles attract, but like poles repel. The imaginary line through the magnet that connects one pole to another represents the magnet's dipole. Physicists specify the dipole by an arrow that points from the north to the south pole. The **polarity** of a magnet refers to the direction the arrow points; the dipoles of magnets with opposite polarity are represented by arrows with arrowheads at opposite ends. Because of the dipolar nature of magnetic fields, we can draw arrowheads on magnetic field lines oriented to form a continuous loop through the magnet.

An electron, which is a spinning, negatively charged particle that orbits the nucleus of an atom, behaves like a tiny electromagnet because its movement produces an electric current. Most of the magnetism is due to the electron's spin, but a little may come from its orbital motion (▶Fig. 2.6a). Each atom, therefore, can be pictured as a little dipole (▶Fig. 2.6b). But even though all materials consist of atoms, not all materials behave like strong, permanent magnets. In fact, most materials (wood, plastic, glass, gold, tin, etc.) are essentially nonmagnetic. That's because the atomic dipoles in the materials are randomly oriented, so overall the dipoles of the atoms cancel each other out (▶Fig. 2.6c). In a permanent magnet, however, atomic dipoles lock into alignment with one another. When this happens, the magnetization of each atom adds to that of its neighbor, so the material as a whole becomes magnetic (▶Fig.2.6d).

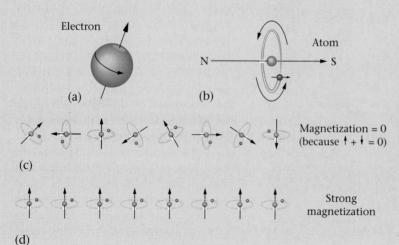

(a)   (b)

(c)   Magnetization = 0 (because ↑ + ↓ = 0)

(d)   Strong magnetization

FIGURE 2.6 (a) A spinning electron creates an electric current. (b) The magnetic dipole of an atom can be represented by an arrow that points from north to south. (c) In a nonmagnetic material, atoms tilt all different ways, so the dipoles cancel each other out, yielding a net magnetization of 0. (d) In a permanent magnet, the dipoles lock into alignment, shown here symbolically, so that they add to each other and produce a strong magnetization.

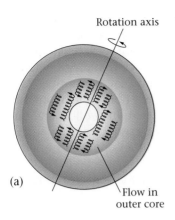

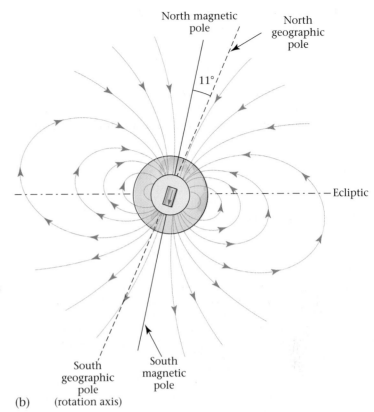

FIGURE 2.7 (a) The convective flow of liquid iron alloy in the Earth's outer core creates an electric current that in turn generates a magnetic field. Recent studies suggest the flow spirals up spring-like coils. (b) Earth's magnetism creates magnetic lines of force in space. We can picture Earth's magnetism by imagining that it contains a giant bar magnet. The dipole of this magnet points presently from the north magnetic pole to the south magnetic pole, and it pierces the Earth at the magnetic poles. Today, the magnetic poles do not coincide exactly with the Earth's geographic poles; they are 11° apart.

FIGURE 2.8 The projection of magnetic field lines in North America at present. Recall that lines of longitude run north-south, so in most places a compass needle will not parallel longitude. For example, a compass needle at New York would make an angle of about 14° to the west of true north. Note that along the line that passes through both magnetic north and geographic north, the magnetic declination = 0°.

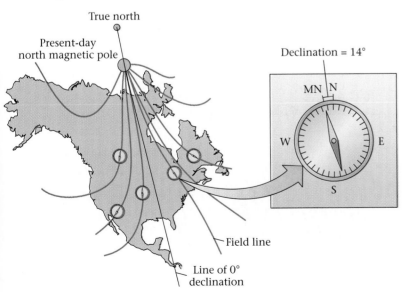

north-seeking end of a compass points toward the north magnetic pole, while the south-seeking end points toward the south magnetic pole. We define the **dipole** of the Earth as an imaginary arrow that points from the north magnetic pole to the south magnetic pole, and passes through the planet's center.

Presently, Earth's dipole tilts at about 11° to the planet's rotational axis (the imaginary line through the center of the Earth around which Earth spins). Therefore, the geographic poles of the planet, the places where the rotational axis intersects the Earth's surface, do not coincide exactly with the magnetic poles. For example, the north magnetic pole currently lies in the Arctic Ocean off the north coast of Canada. As a consequence, the north-seeking end of a compass needle in New York points about 14° west of north. The angle between the direction that a compass needle points at a given location and the direction to "true" (geographic) north is called the magnetic declination (►Fig. 2.8). Measurements show that magnetic poles migrate very slowly over time, probably never straying more than about 15° of latitude from the geographic pole. In fact, the magnetic declination of a compass changes by 0.2° to 0.5° per year. Notably, when averaged over about 10,000 years, the magnetic poles are thought to coincide roughly with the geographic poles.

Figure 2.7b illustrates the magnetic field lines in space around the Earth, as seen in cross section (without the warping caused by solar wind). Note that close to the Earth, the lines parallel Earth's surface at the equator; the lines tilt at an angle to the surface at mid-latitudes; and the lines are perpendicular to the surface at the magnetic poles. Thus, if we traveled to the equator and set up a magnetic needle such that it could pivot up and down freely, the needle would be horizontal. If we took the needle to mid-latitudes, it would tilt at an angle to Earth's surface. And at a magnetic pole, the needle would point straight down. The needle's angle of tilt (which, as ▶Fig. 2.9 shows, depends on latitude) is called the magnetic inclination. Note that a regular compass needle does not indicate inclination, because it cannot tilt—a compass needle aligns parallel to the shadow of a magnetic field line on Earth's surface.

How Do Rocks Develop Paleomagnetism?

More than 1,500 years ago, Chinese sailors discovered that an elongated piece of lodestone suspended from a thread "magically" pivots until it points north, and thus that this rock could help guide their voyages. We now know that lodestone exhibits this behavior because it consists of magnetite, an iron-rich mineral that acts like a permanent magnet (see Box 2.1). Small crystals of magnetite or other magnetic minerals occur in many rock types. Each crystal produces a tiny magnetic force. The sum of the magnetic forces produced by all these crystals makes the rock, as a whole, weakly magnetic.

FIGURE 2.9 An illustration of magnetic inclination. A magnetic needle that is free to rotate around a horizontal axis aligns with magnetic field lines (here depicted in cross section). Because magnetic field lines curve in space, this needle is horizontal at the equator, tilts at an angle at mid-latitudes, and is vertical at the magnetic pole. Therefore, the angle of tilt depends on the latitude.

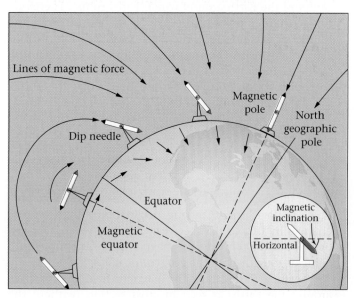

To see how magnetic rocks preserve a record of Earth's past magnetic field, let's examine the development of magnetization in one type of rock, basalt. Basalt is a magnetite-containing igneous rock that forms when lava, flowing out of a volcano, cools and solidifies. When lava first comes out of a volcano, it is very hot (up to about 1,200°C), and thermal energy makes its atoms wobble and tumble chaotically. The magnetic force exerted by one atom cancels out the force of another with an oppositely oriented dipole, so the lava as a whole is not magnetic (▶Fig. 2.10a). However, as the temperature of the lava decreases to below the melting

FIGURE 2.10 The formation of paleomagnetism. (a) At high temperatures (greater than 350°–550°C), thermal vibration makes atoms have random orientations; the dipoles thus cancel each other out and the sample has no overall magnetic dipole. (b) As the sample cools to below 350°–550°C, the atoms slow down and their dipoles lock into alignment with the Earth's field.

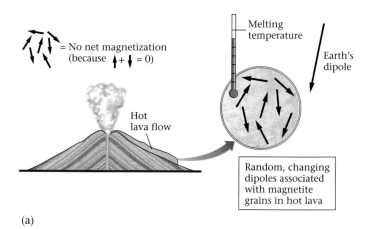

(a)

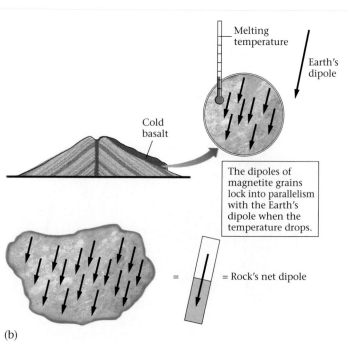

(b)

temperature, basalt starts to solidify. As the magnetite crystals in the basalt form and cool, their iron atoms slow down. The dipoles of all the atoms gradually become parallel with each other and with the Earth's magnetic field lines at the location where the basalt cools. Finally, at temperatures below 350°–550°C, the dipoles lock into position, pointing in the direction of the magnetic pole, and the basalt becomes a permanent magnet (▶Fig. 2.10b). Since this alignment is permanent, the basalt provides a record of the orientation of the Earth's magnetic field lines, relative to the rock, at the time the rock cooled. This record is paleomagnetism. Paleomagnetism can also develop in other rock types, for a variety of reasons.

Interpreting *Apparent* Polar-Wander Paths: Evidence for Continental Drift

When geologists measured paleomagnetism in samples of basalt that had formed millions of years ago, they were surprised to find that the dipole representing this paleomagnetism did not point to the present-day magnetic poles of the Earth (▶Fig. 2.11). At first, they interpreted this observation to mean that the magnetic poles of Earth moved through time and thus had different locations in the past, a phenomenon that they called polar wander, and they introduced the term **paleopole** to refer to the supposed position of the

Earth's magnetic pole at a time in the past. Geologists next measured paleopoles in a variety of different rocks of different ages from near the same location, and plotted the results on a map. The line connecting the positions of paleopoles as a function of time, for a given location, came to be known as a polar-wander path (▶Fig. 2.12a, b). Geologists at first thought a polar-wander path showed how the position of Earth's magnetic pole changes as time passes. Polar-wander paths end near the present North Pole because, geologists thought at first, younger rocks became magnetized when the Earth's magnetic pole was close to its present position. As we now see, the initial interpretation of polar-wander paths we've just described is wrong!

In the late 1950s, geologists determined the polar-wander path for Europe. When this path was first plotted, they did not accept the notion of continental drift, and assumed that the position of the continents was fixed. Thus, they *interpreted* the path to represent how the position of Earth's north magnetic pole migrated through time. Were they in for a surprise! When geologists then determined the polar-wander path for North America, they found that North America's path differed from Europe's (▶Fig. 2.13). In fact, when paths were plotted for all continents, they all turned out to be different from one another. The hypothesis that the continents are fixed and the magnetic poles move simply *cannot* explain this observation. If the magnetic poles really moved while the continents stayed fixed, then all continents should have the same polar-wander paths.

Geologists suddenly realized that they were looking at polar-wander paths in the wrong way. *It's not the pole that moves with respect to fixed continents, but rather the continents that move with respect to a fixed pole* (▶Fig. 2.14a, b)! Further, if the pole is fixed, then in order for each continent to have its own unique polar-wander path, the continents must move, or "drift," in relation to each other. Poles do *not* wander, but the continents do drift while the pole stays fixed. To emphasize this point, we now refer to the curve passing through a succession of paleopoles as an **apparent polar-wander path.**

Having finally understood the meaning of apparent polar-wander paths, geologists looked once again at the paths for Europe and North America, and realized that they had the same shape between about 280 and 180 million years ago, the period during which the continents linked together to form part of Pangaea (▶Fig. 2.15). This makes sense—during times when continents move with respect to one another, they develop different apparent polar-wander paths; but when continents are stuck together, they develop the same apparent polar-wander path.

This discovery led to a renewal of support for Wegener's model of continental drift. But how could continental drift take place? The stage was set for the proposal of the seafloor-spreading hypothesis, an idea that ultimately explains how continental drift occurs.

FIGURE 2.11 A rock sample can maintain paleomagnetization for millions of years. In this example, the dipole representing the paleomagnetism in this rock outcrop, from a village on the equator, does not parallel the Earth's present field. Note that I (inclination) is not 0°, as it would be today for rock forming near the equator. D is the declination.

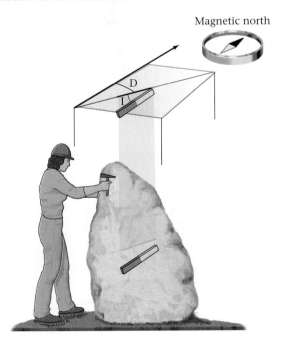

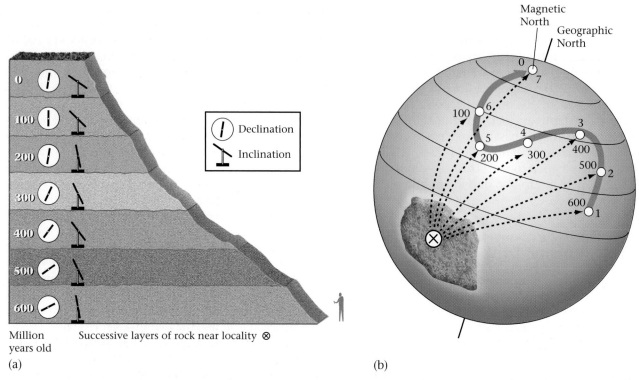

(a)

(b)

FIGURE 2.12 (a) A cliff at location X exposes a succession of dated lava flows. A geologist measures the orientation of the dipole in the rock. (The arrowheads aren't shown, as they don't matter here.) The paleopoles are the places where the dipole intersects the surface of the Earth. (b) The succession of paleopole positions through time for location X defines the apparent polar-wander path for the location. The path ends at the position of the present-day magnetic pole, near the North Pole.

FIGURE 2.13 Apparent polar-wander paths of North America, Europe, and Africa for the past several hundred million years, plotted on a present-day map of the Earth.

FIGURE 2.14 The two alternative explanations for an apparent polar-wander path. (a) In a "true polar-wander" model, the continent is fixed, so to explain polar-wander paths, the magnetic pole must move substantially. (In reality, the magnetic pole does move a little, but it never strays very far from the geographic pole.) (b) In a continental drift model, the magnetic pole is fixed near the geographic pole, and the continent drifts relative to the pole.

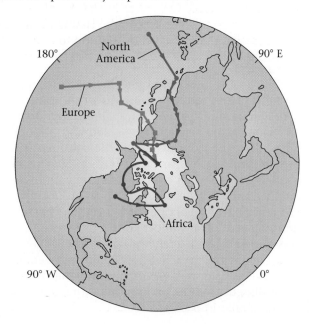

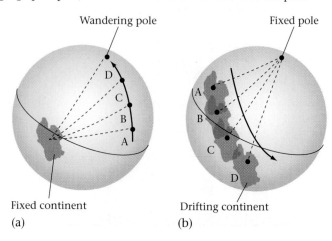

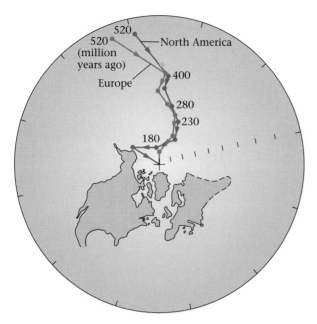

FIGURE 2.15 The apparent polar-wander paths for North America and Europe would have coincided with one another from about 280 to 180 million years ago, because Europe and North America moved together as a unit when both were part of Pangaea. When Pangaea broke up, the two began to develop separate apparent polar-wander paths.

2.4 SETTING THE STAGE FOR THE DISCOVERY OF SEA-FLOOR SPREADING

New Images of Sea-Floor Bathymetry

Military needs during World War II gave a boost to sea-floor exploration, for as submarine fleets grew, navies required detailed maps of bathymetry that shows variations in the depth of the sea floor. The invention of echo sounding (sonar) per-

mitted such maps to be made. Echo sounding uses the same phenomenon that a bat uses to navigate. A sound pulse emitted from a ship travels down through the water, bounces off the sea floor, and returns up as an echo through the water to a receiver on the ship. Since sound waves travel at a known velocity, the time between the sound emission and the detection of the echo indicates the distance between the ship and the sea floor (velocity = distance/time, so distance = velocity × time). As the ship moves, echo sounding permits observers to obtain a continuous record of the depth of the sea floor; the resulting cross section showing depth plotted against location is called a bathymetric profile (▶Fig. 2.16a, b). By cruising back and forth across the ocean many times, investigators obtained a series of bathymetric profiles and from these constructed maps of the sea floor. Bathymetric maps revealed several important features of the ocean floor.

- *Mid-ocean ridges:* The floor beneath all major oceans includes **abyssal plains,** which are broad, relatively flat regions of the ocean that lie at a depth of about 4–5 km below sea level; and **mid-ocean ridges,** elongate submarine mountain ranges whose peaks lie only about 2–2.5 km below sea level (▶Figs. 2.17, 2.18a). Geologists call the crest of the mid-ocean ridge the ridge axis. All mid-ocean ridges are roughly symmetrical—bathymetry on one side of the axis is nearly a mirror image of bathymetry on the other side.

- *Deep-ocean trenches:* Along much of the perimeter of the Pacific Ocean, and in a few other localities as well, the ocean floor reaches depths of 8–12 km—deep enough to swallow Mt. Everest. These deep areas define elongate troughs that are now referred to as **trenches.** Trenches border **volcanic arcs,** curving chains of active volcanoes.

- *Seamount chains:* Numerous volcanic islands poke up from the ocean floor: for example, the Hawaiian Islands lie in the middle of the Pacific. In addition to islands that rise above sea level, echo sounding has detected many

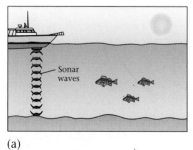

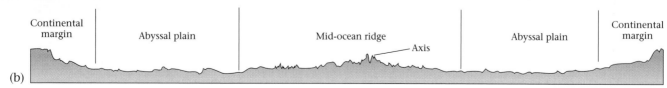

FIGURE 2.16 (a) To make a bathymetric profile, researchers use sonar. (b) An east-west bathymetric profile of the Atlantic Ocean.

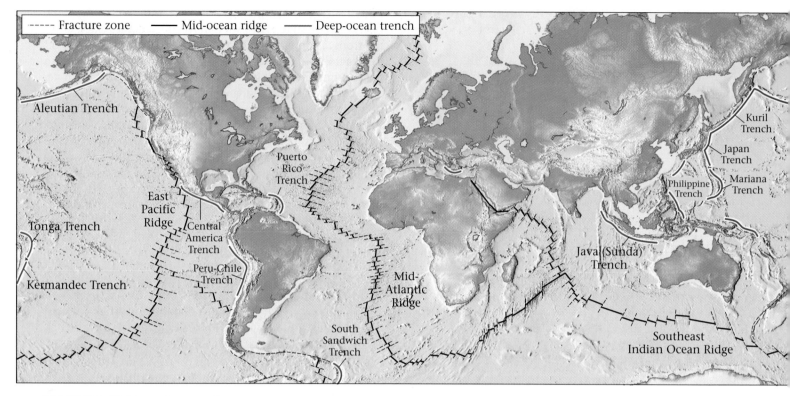

FIGURE 2.17 The mid-ocean ridges, fracture zones, and principal deep-ocean trenches of today's oceans.

seamounts (isolated submarine mountains), which were once volcanoes but no longer erupt. Volcanic islands and seamounts typically occur in chains, but in contrast to the volcanic arcs that border deep-ocean trenches, only one island at the end of a seamount chain remains capable of erupting today.

• *Fracture zones:* Surveys reveal that the ocean floor is diced up by narrow bands of vertical fractures. These fracture zones lie roughly at right angles to mid-ocean ridges, effectively segmenting the ridges into small pieces (►Fig. 2.18b).

New Observations on the Nature of Oceanic Crust

By the mid-twentieth century, geologists had discovered many important characteristics of the sea-floor crust. These discoveries led them to realize that oceanic crust is quite different from

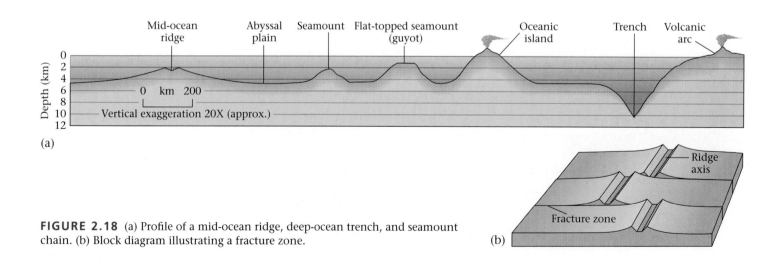

FIGURE 2.18 (a) Profile of a mid-ocean ridge, deep-ocean trench, and seamount chain. (b) Block diagram illustrating a fracture zone.

continental crust, and further that bathymetric features of the ocean floor provide clues to the origin of the crust. Specifically:

• A layer of sediment composed of clay and the tiny shells of dead plankton covers much of the ocean floor. This layer becomes progressively thicker away from the mid-ocean ridge axis. But even at its thickest, the sediment layer is too thin to have been accumulating for the entirety of Earth history.

FIGURE 2.19 In a mid-ocean ridge, heat from the mantle flows up through the crust; heat flow decreases away from the ridge axis.

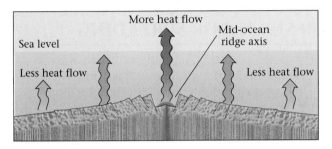

• By dredging up samples, geologists learned that oceanic crust is fundamentally different in composition from continental crust. Beneath its sediment cover, oceanic crust bedrock consists of basalt—it does not display the great variety of rock types found on continents.

• Heat flow, the rate at which heat rises from the Earth's interior up through the crust, is not the same everywhere in the oceans. Rather, more heat seems to rise beneath mid-ocean ridges than elsewhere (▶Fig. 2.19). This observation led geologists to speculate that magma might be rising into the crust just below the mid-ocean ridge axis, for hot molten rock could bring heat into the crust.

• When maps showing the distribution of earthquakes in oceanic regions became available in the years after World War II, geologists realized that earthquakes do not occur randomly, but rather define distinct belts (▶Fig. 2.20). Some belts follow trenches, some follow mid-ocean ridge axes, and others lie along portions of fracture zones. Since earthquakes define locations where rocks break and move, geologists realized that these bathymetric features are places where movements of the crust take place.

FIGURE 2.20 A 1953 map showing the distribution of earthquake locations in the ocean basins. Note that earthquakes occur in belts.

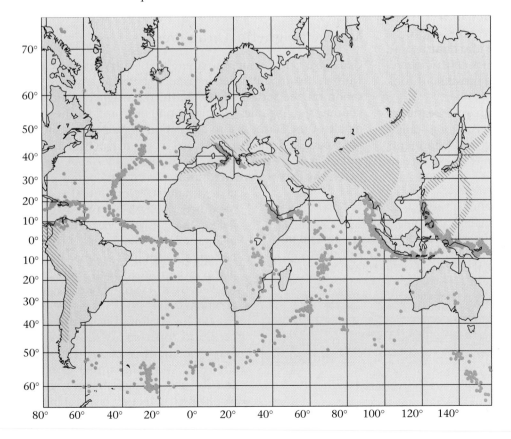

Now let's see how Harry Hess used these observations to come up with the hypothesis of sea-floor spreading.

2.5 HARRY HESS AND HIS "ESSAY IN GEOPOETRY"

In the late 1950s, Harry Hess, after studying the observations described above, realized that the overall thinness of the sediment layer on the ocean floor meant that the ocean floor might be much younger than the continents, and that the progressive increase in thickness of the sediment away from mid-ocean ridges could mean that the ridges themselves were younger than the deeper parts of the ocean floor. If this was so, then somehow new ocean floor must be forming at the ridges, and thus an ocean could be getting wider with time. But how? The association of earthquakes with mid-ocean ridges suggested to him that the sea floor was cracking and splitting apart at the ridge. The discovery of high heat flow along mid-ocean ridge axes provided the final piece of the puzzle, for it suggested the presence of molten rock beneath the ridges. In 1960, Hess suggested that molten rock rose upward beneath mid-ocean ridges and that this material solidified to form oceanic crust. The new sea floor then moved away from the ridge, a process we now call **sea-floor spreading.** Hess realized that old ocean floor must be consumed somewhere, or the Earth would have to expand. He suggested that deep-ocean trenches might be places where the sea floor sank back into the mantle, and that earthquakes at trenches were evidence of this movement, but he didn't understand how the movement took place (▶Fig. 2.21).

Hess and his contemporaries realized that the sea-floor-spreading hypothesis instantly provided the long-sought explanation of how continental drift occurs. Continents passively move apart as the sea floor between them spreads at mid-ocean ridges, and they passively move together as the sea floor between them sinks back into the mantle at trenches. (Geologists now realize that it is the lithosphere that moves, not just the crust.) Thus, sea-floor spreading proved to be an important step on the route to plate tectonics—the idea seemed so good that Hess referred to his description of it as "an essay in geopoetry." But other key discoveries would have to take place before the whole theory of plate tectonics came together.

2.6 MARINE MAGNETIC ANOMALIES: EVIDENCE FOR SEA-FLOOR SPREADING

For a hypothesis to become a theory, there must be proof. The proof of sea-floor spreading emerged from two discoveries. First, geologists found that the measured strength of Earth's magnetic field is not the same everywhere in the ocean basins; the variations are now called marine magnetic anomalies. Second, they found that Earth's dipole reverses direction every now and then; such sudden reversals of the Earth's polarity are now called magnetic reversals. As we see below, the discovery of reversals provides a way to explain the formation of anomalies, and the pattern of anomalies requires that sea-floor spreading takes place.

Marine Magnetic Anomalies

Geologists can measure the strength of Earth's magnetic field with an instrument called a magnetometer. At any given location on the surface of the Earth, the magnetic field that you measure includes two parts: one produced by the main dipole of the Earth, due to circulation of molten iron in the outer core, and another produced by the magnetism of near-surface rock. A **magnetic anomaly** is the difference between the expected strength of the Earth's main dipole field at a certain location and the *actual* measured strength of the magnetic field at that location. Places where the field strength is stronger than expected are positive anomalies, and places where the field strength is weaker than expected are negative anomalies.

Geologists towed magnetometers back and forth across the ocean to map variations in magnetic field strength. As a ship cruised along its course, the magnetometer's gauge would first detect a strong signal (a positive anomaly) and then a weak

FIGURE 2.21 Harry Hess's basic concept of sea-floor spreading. New sea floor forms at the mid-ocean ridge axis. As a result, the ocean grows wider. Old sea floor sinks into the mantle at a trench. Earthquakes occur at ridges and trenches. Hess implied, incorrectly, that only the crust moved. We will see that this sketch is an oversimplification.

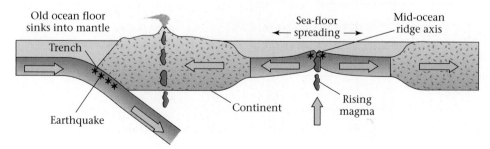

signal (a negative anomaly). A graph of signal strength versus distance along the traverse, therefore, has a sawtooth shape (►Fig. 2.22a). When geologists compiled data from many cruises on a map, these **marine magnetic anomalies** defined distinctive, alternating bands. If we color positive anomalies dark and negative anomalies light, the pattern made by the anomalies resembles the stripes on a candy cane (►Fig. 2.22b). The mystery of the marine magnetic anomaly pattern, however, remained unsolved until geologists recognized the existence of magnetic reversals.

Magnetic Reversals

When geologists measured the paleomagnetism of a succession of rock layers that had accumulated over a long period of time, they found that the polarity (which end of a magnet points north and which end points south; see Box 2.1) of the paleomagnetic field preserved in some layers was the same as that of Earth's present magnetic field, while in other layers it was the opposite. Recall that Earth's magnetic field can be represented by an arrow, representing the dipole, that points from north to south; in some of the rock layers, the paleomagnetic dipole pointed south (these layers have normal polarity), but in others the dipole pointed north (these layers have reversed polarity) (►Fig. 2.23).

At first, observations of reversed polarity were largely ignored, thought to be the result of lightning strikes or of chemical reactions between rock and water. But when repeated measurements from around the world revealed a systematic pattern of alternating normal and reversed polarity in rock layers, geologists realized that reversals were a global, not a local, phenomenon. At various times during Earth history, the polarity of Earth's magnetic field has suddenly reversed! In other words, sometimes the Earth has normal polarity, as it does today, and sometimes it has reversed polarity (►Fig. 2.24a, b). Times when the Earth's field flips from normal to reversed polarity, or vice versa, are called **magnetic reversals.** When the Earth has reversed polarity, the south magnetic pole lies near the north geographic pole, and the north magnetic pole lies near the south geographic pole. If you were to use a compass during periods when the Earth's magnetic field was reversed, the north-seeking end of the needle would point to the south geographic pole.

In the 1950s, about the same time geologists discovered polarity reversals, they developed a technique that permitted them to define the age of a rock in years, by measuring the decay of radioactive elements in the rock. The technique

FIGURE 2.22 (a) A ship towing a magnetometer through the ocean detects a positive anomaly and then a negative one, then a positive one, then a negative one. (b) Magnetic anomalies on the sea floor off the northwestern coast of the United States. The dark bands are positive anomalies, the light bands negative anomalies. Note the distinctive stripes of alternating anomalies. A positive anomaly overlies the crest of the Juan de Fuca Ridge (a small mid-ocean ridge).

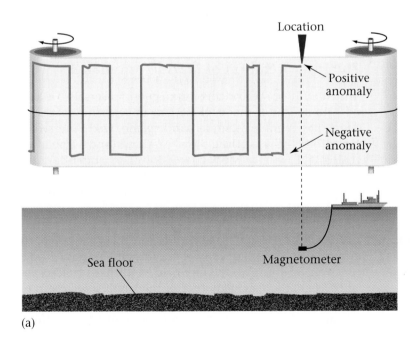

(a)

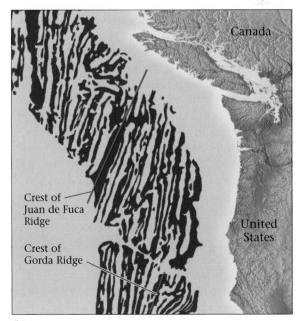

(b)

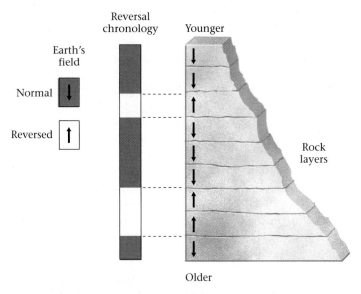

FIGURE 2.23 In a succession of rock layers on land, different flows exhibit different polarity (indicated here by whether the arrow points up or down). When these reversals are plotted on a time column, we have a magnetic-reversal chronology.

is called radiometric dating (it will be discussed in detail in Chapter 10). Geologists applied the technique to determine the ages of rock layers in which they obtained their paleo-magnetic measurements, and thus determined *when* the magnetic field of the Earth reversed. With this information, they constructed the "history" of magnetic reversals, the magnetic-reversal chronology. A diagram representing the Earth's magnetic-reversal chronology (▶Fig. 2.25) shows that reversals do not occur regularly, so the lengths of different polarity chrons, the time intervals between reversals, are different. For example, we have had a normal-polarity chron for about the last 700,000 years. Before that, there was a reversed-polarity chron. Geologists named the youngest four polarity chrons (Brunhes, Matuyama, Gauss, and Gilbert) after scientists who had made important contributions to the study of rock magnetism. As more measurements became available, investigators realized that there were some short-duration reversals (less than 200,000 years long) within the chrons, and called these shorter reversals polarity subchrons. Radiometric dating methods are not accurate enough to date reversals that are older than about 4.5 million years, because for rocks older than that, the uncertainty in a date exceeds the duration of a polarity subchron. (As we see later in this chapter, there are other ways to determine the ages of older anomalies.) Of note, very detailed studies demonstrate that reversals may take place in less than 1,000 years. Recent computer models suggest that reversals happen because of changes in the flow pattern of liquid iron alloy in the outer core.

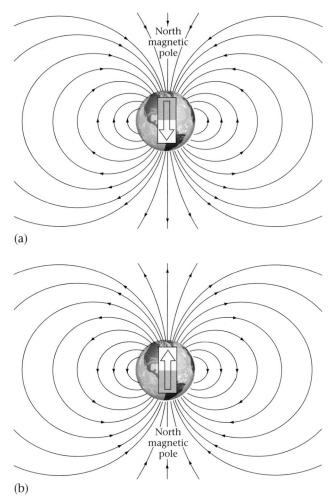

FIGURE 2.24 The magnetic field of the Earth has had reversed polarity at various times during Earth history. (a) If the dipole points from north to south, Earth has normal polarity. (b) If the dipole points from south to north, Earth has reversed polarity.

The Interpretation of Marine Anomalies

Armed with our knowledge of magnetic reversals, we can now understand the explanation for marine magnetic anomalies. A graduate student in England, Fred Vine, working with his adviser, Drummond Mathews, and a Canadian geologist, Lawrence Morley (working independently), discovered a solution to this riddle. Simply put, a positive anomaly occurs over areas of sea floor where basalt has normal polarity. In these areas, the magnetic force produced by the basalt *adds* to the force produced by Earth's dipole and creates a stronger magnetic signal than expected, as measured by the magnetometer. A negative anomaly occurs over regions of sea floor where basalt has reversed polarity. Here, the magnetic force of the basalt *subtracts* from the force produced by the dipole and results in a weaker magnetic signal (▶Fig. 2.26a). Sea floor yielding positive

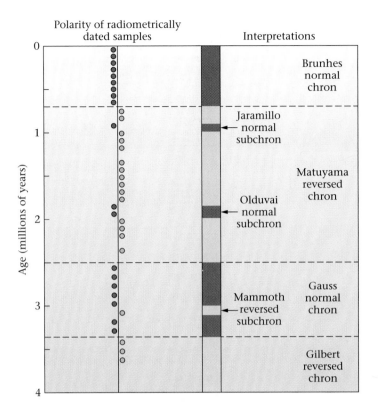

Polarity of radiometrically dated samples   Interpretations

Age (millions of years)

Brunhes normal chron

Jaramillo normal subchron

Matuyama reversed chron

Olduvai normal subchron

Gauss normal chron

Mammoth reversed subchron

Gilbert reversed chron

FIGURE 2.25 Radiometric dating of lava flows allows us to determine the age of magnetic reversals during the past 4 million years. Major intervals of a given polarity are referred to as polarity chrons, and are named after scientists who contributed to the understanding of Earth's magnetic field. Shorter-duration reversals are called subchrons.

By relating the stripes on the sea floor to magnetic reversals found in basalt of known age (►Fig. 2.27b), geologists dated the sea floor back to an age of 4.5 million years, and they found that the relative widths of anomaly stripes on the sea floor exactly corresponded to the relative durations of polarity chrons in the magnetic-reversal chronology. The relationship between anomaly-stripe width and polarity-chron duration provides the key for determining the rate (velocity) of sea-floor spreading, for it indicates that the rate of spreading has been constant for the last 4.5 million years. Remember that velocity = distance/time. In the North Atlantic Ocean, 4.5-million-year-old sea floor lies 45 km away from the ridge axis. Therefore, the velocity (v) at which the sea floor moves away from the ridge axis is 1 cm

FIGURE 2.26 (a) The explanation of marine anomalies. The sea floor beneath positive anomalies has the same polarity as Earth's field and therefore adds to it. The sea floor beneath negative anomalies has reversed polarity and thus subtracts from Earth's field. (b) The symmetry of the magnetic anomalies measured across the Mid-Atlantic Ridge south of Iceland. Note that individual anomalies are somewhat irregular, because the process of forming the sea floor, in detail, happens in discontinuous pulses along the length of the ridge.

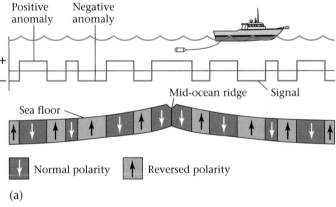

Positive anomaly   Negative anomaly

Mid-ocean ridge   Signal

Sea floor

↓ Normal polarity   ↑ Reversed polarity

(a)

anomalies formed at times when the Earth had normal polarity, while sea floor yielding negative anomalies formed when the Earth had reversed polarity.

The sea-floor-spreading model easily explains why the magnetic anomaly pattern on one side of a mid-ocean ridge is a mirror image of the anomaly pattern on the other (►Fig. 2.26b). To see how, please refer to ►Figure 2.27a. At time 1 (sometime in the past), a time of normal polarity, the dark stripe of sea floor forms. The tiny dipoles of magnetite grains in basalt making up this stripe align with the Earth's field. As it forms, the rock in this stripe migrates away from the ridge axis, half to the right and half to the left. Later, at time 2, the field has reversed, and the light-gray stripe forms with reversed polarity. As it forms, it too moves away from the axis, and still younger crust begins to develop along the axis. As the process continues over millions of years, many stripes form. A positive anomaly exists along the ridge axis today because this represents sea floor that has developed during the most recent interval of time, a chron of normal polarity. The magnetism of the rock along the ridge adds to the magnetism of the Earth's field.

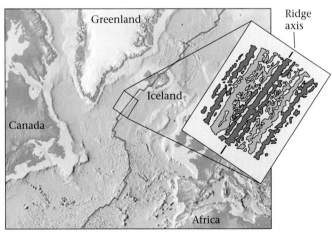

Greenland

Iceland

Canada

Africa

Ridge axis

(b)

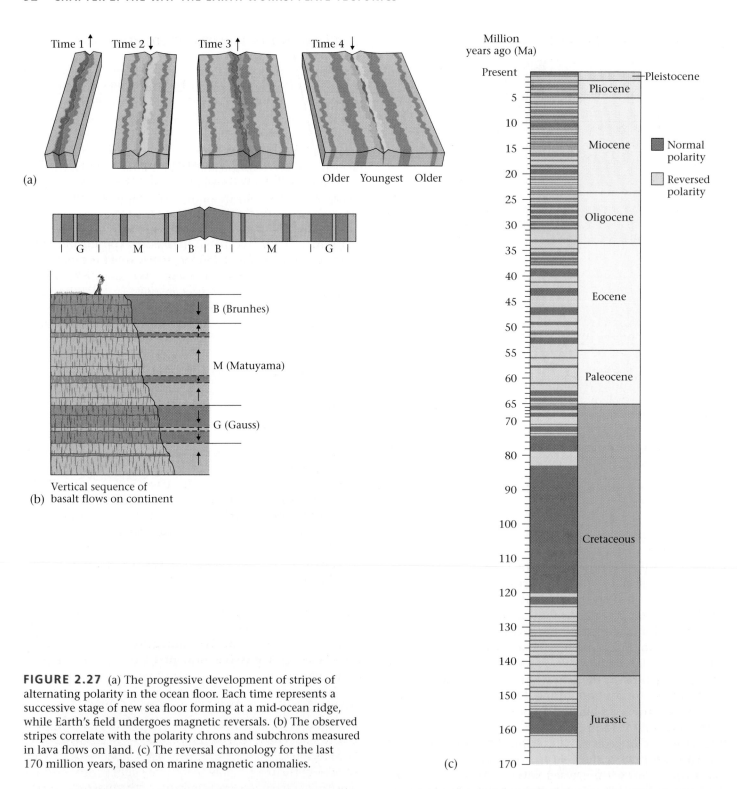

FIGURE 2.27 (a) The progressive development of stripes of alternating polarity in the ocean floor. Each time represents a successive stage of new sea floor forming at a mid-ocean ridge, while Earth's field undergoes magnetic reversals. (b) The observed stripes correlate with the polarity chrons and subchrons measured in lava flows on land. (c) The reversal chronology for the last 170 million years, based on marine magnetic anomalies.

per year. Thus, a point on one side of the ridge moves away from a point on the other side by 2 cm per year. We call this number the spreading rate. At a spreading rate of 2 cm/y, a 2,000-km-wide ocean can form in 100 million years.

Geologists also realized that if they could assume that the rate of sea-floor spreading has remained fairly constant for a long time, then they could date the ages of magnetic-field reversals further back in Earth history, simply by measuring the distance of successive magnetic anomalies from the ridge axis, because: time = distance/velocity. Such analysis eventually defined the magnetic-reversal chronology back to about 170 million years ago (▶Fig. 2.27c).

2.7 DEEP-SEA DRILLING: FURTHER EVIDENCE

In the late 1960s, a drilling ship called the *Glomar Challenger* set out to sail around the ocean drilling holes into the sea floor. This amazing ship could lower enough drill pipe to drill in 5-km-deep water and could continue to drill until the hole reached a depth of about 1.7 km (1.1 miles) below the sea floor. Drillers brought up cores of rock or sediment that geoscientists then studied on board.

On one of its early cruises, the *Glomar Challenger* drilled a series of holes through sea-floor sediment to the basalt layer. These holes were spaced at progressively greater distances from the axis of the Mid-Atlantic Ridge. If the model of sea-floor spreading was correct, then the sediment layer should be progressively thicker away from the axis, and the age of the oldest sediment just above the basalt should be progressively older away from the axis. When the drilling and the analyses were complete, the predictions were confirmed.

The drilling results, together with the interpretation of marine magnetic anomalies, effectively proved sea-floor spreading. So, by the early 1960s, it had become clear that Wegener had been right all along—continents do drift. But, though the case for drift had been greatly strengthened by the discovery of apparent polar-wander paths, it really took the proposal and proof of sea-floor spreading to make believers of most geologists. Very quickly, these ideas became the basis of the theory of plate tectonics.

2.8 WHAT DO WE MEAN BY PLATE TECTONICS?

The discovery of sea-floor spreading and the paleomagnetic proof of continental drift set off a "scientific revolution" in geology. By 1968, geologists realized that many of their existing interpretations of global geology, based on the premise that the positions of continents and oceans remain fixed in position through time, were simply wrong! Researchers dropped what they were doing and turned their attention to studying the broader implications of sea-floor spreading and continental drift. From these efforts the theory of plate tectonics evolved. This theory states that the outer layer of the Earth, the lithosphere, consists of separate pieces, or "plates," that move with respect to each other. New studies clarified the meaning of a plate, defined the types of plate boundaries, constrained plate motions, related plate tectonics to earthquakes and volcanoes, showed how plate motions generate mountain belts and seamount chains, and outlined the history of past plate motions. To begin our explanation of the key elements of plate tectonics theory, we describe lithosphere plates and their boundaries in more detail.

The Concept of a Lithosphere Plate

We learned earlier that geoscientists divide the outer part of the Earth into two layers. The **lithosphere** consists of the crust plus the top (cooler) part of the upper mantle. It behaves relatively rigidly, meaning that when a force pushes or pulls on it, it does not flow but rather bends or breaks (▶Fig. 2.28). The lithosphere floats on a relatively soft, or "plastic," layer called the **asthenosphere,** composed of warmer (>1,280°C) mantle that can flow slowly when acted on by force. The asthenosphere can undergo convection, like water in a pot, but the lithosphere cannot.

Continental lithosphere and oceanic lithosphere differ markedly in their thickness. On average, continental lithosphere has a thickness of 150 km, whereas old oceanic lithosphere has a thickness of about 100 km. (For reasons discussed later in this chapter, new oceanic lithosphere at a mid-ocean ridge is only 7 to 10 km thick.) Recall that the crustal part of continental lithosphere ranges from 25 to 70 km thick and consists of relatively *low-density* felsic and intermediate rock (see p. 26). In contrast, the crustal part of oceanic lithosphere is only 7 to 10 km thick and consists of relatively *high-density* mafic rock. Because of these differences, the surface of continental lithosphere "floats" at a higher elevation than does the surface of oceanic lithosphere.

The lithosphere forms the Earth's relatively rigid shell. But unlike the shell of a hen's egg, the lithosphere shell contains a number of major "breaks," which separate the lithosphere into distinct pieces. As noted earlier, we call the pieces **lithosphere plates,** or simply plates. We call the breaks **plate boundaries.** Geoscientists distinguish twelve major plates and several microplates. Some plates have familiar names (such as the North American Plate and the African Plate), while some do not (such as the Cocos Plate and the Juan de Fuca Plate).

As illustrated in ▶Figure 2.29, some plate boundaries follow continental margins, the boundary between a continent and an ocean, while others do not. For this reason, we distinguish between **active margins,** which are plate boundaries, and **passive margins,** which are not plate boundaries. Along passive margins, continental crust is thinner than in continental interiors (▶Fig. 2.30), and the upper part breaks into wedge-shaped slices. Thick (10–15 km) accumulations of sediment cover this thinned crust. The surface of this sediment layer is a broad, shallow (less than 500 m deep) region called the continental shelf, home to the major fisheries of the world. Note that some plates consist entirely of oceanic lithosphere or entirely of continental lithosphere, whereas some plates consist of both.

The Basic Premise of Plate Tectonics

We can now restate plate tectonics theory concisely as follows. The Earth's lithosphere is divided into plates that

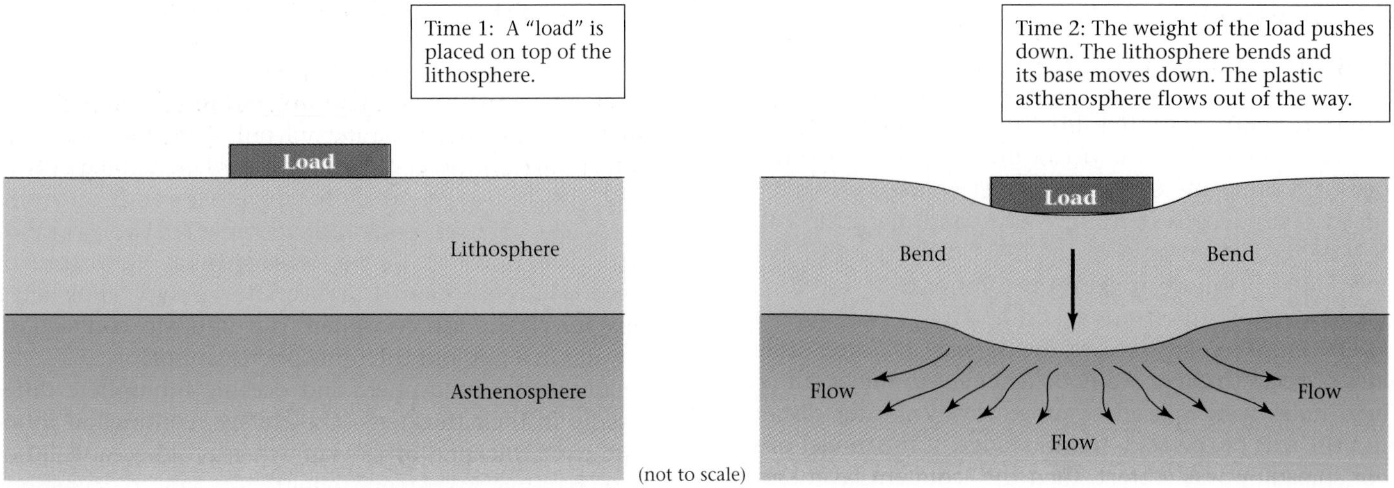

Time 1: A "load" is placed on top of the lithosphere.

Time 2: The weight of the load pushes down. The lithosphere bends and its base moves down. The plastic asthenosphere flows out of the way.

Load

Lithosphere

Asthenosphere

Load

Bend   Bend

Flow   Flow

Flow

(not to scale)

FIGURE 2.28 Lithosphere bends when a load is placed on it, whereas asthenosphere flows.

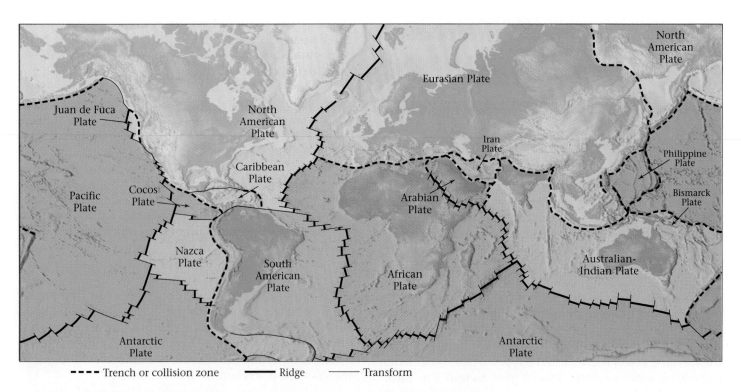

- - - Trench or collision zone   ——— Ridge   ——— Transform

FIGURE 2.29 The major plates making up the lithosphere. Note that some plates are all ocean floor, while some contain both continents and oceans. Thus, some plate boundaries lie along continental margins (coasts), while others do not. For example, the eastern border of South America is not a plate boundary, but the western edge is.

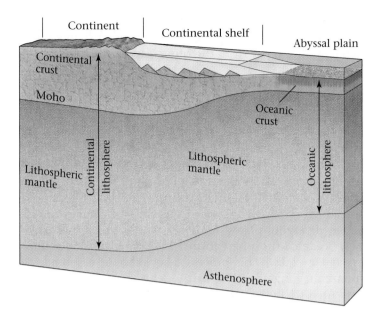

FIGURE 2.30 In this block diagram of a passive margin, note that the continental crust thins along the boundary (see Section 2.13, "Continental Rifting"). The sediment pile that accumulates over this thinned crust underlies the continental shelf.

move relative to each other and relative to the underlying asthenosphere. Plate movement occurs at rates of about 1 to 15 cm per year. As a plate moves, its internal area remains largely rigid and intact, but rock along plate boundaries undergoes deformation (cracking, sliding, bending, stretching, and squashing) as the plate grinds or scrapes against its neighbors or pulls away from its neighbors. As plates move, so do the continents that form part of the plates, resulting in continental drift. Because of plate tectonics, the map of Earth's surface constantly changes.

Identifying Plate Boundaries

How do we recognize the location of a plate boundary? The answer becomes clear from looking at a map showing the locations of earthquakes (▶Fig. 2.31). Recall from Chapter 1 that earthquakes are vibrations caused by shock waves that are generated where rock breaks and suddenly shears (slides) along a fault (a fracture on which sliding occurs). The hypocenter (or focus) of the earthquake is the spot where the fault begins to slip, and the epicenter marks the point on the surface of the Earth directly above the focus. Earthquake epicenters do not speckle the Earth's surface

FIGURE 2.31 The locations of most earthquakes fall in distinct bands, or belts. These earthquake belts define the positions of the plate boundaries. Compare this map with the plate boundaries on Figure 2.29.

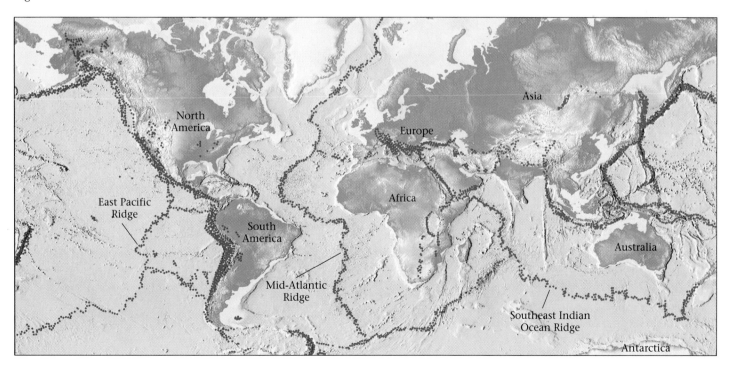

randomly, like buckshot on a target. Rather, the majority occur in relatively narrow, distinct belts. These earthquake belts define the position of plate boundaries, because the fracturing and slipping that occur along plate boundaries as plates move generate earthquakes. Plate interiors, regions away from the plate boundaries, remain relatively earthquake free because they are stronger and do not accommodate much movement. While earthquakes serve as the most definitive indicator of a plate boundary, other prominent geologic features also develop along plate boundaries, as we will see by the end of this chapter.

Geologists define three types of plate boundaries, based simply on the relative motions of the plates on either side of the boundary (▶Fig. 2.32a–c). A boundary at which two plates move apart from each other is a **divergent boundary.** A boundary at which two plates move toward each other so that one plate sinks beneath the other is a **convergent boundary.** And a boundary at

which one plate slips along the side of another plate is a **transform boundary.** Each type of boundary looks and behaves differently from the others, as we will now see.

2.9 DIVERGENT PLATE BOUNDARIES AND SEA-FLOOR SPREADING

At divergent boundaries, or spreading boundaries, two oceanic plates move apart by the process of sea-floor spreading. Note that an open space does not develop between diverging plates. Rather, as the plates move apart, new oceanic lithosphere forms along the divergent boundary (▶Fig. 2.33). This process takes place at a submarine mountain range called a **mid-ocean ridge** (such as the Mid-Atlantic Ridge, the East Pacific Rise, or the Southeast

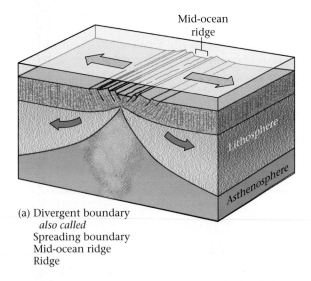

(a) Divergent boundary
also called
Spreading boundary
Mid-ocean ridge
Ridge

FIGURE 2.32 Geologists recognize three types of plate boundary based on the nature of relative movement at the boundary. (a) At a divergent boundary (its other names are listed below), two oceanic plates move away from each other. The lithosphere thickens with increasing distance from the ridge. (b) At a convergent boundary, one oceanic plate bends and sinks into the mantle beneath another plate. (c) At a transform boundary, two plates slide past one another along a vertical fault surface.

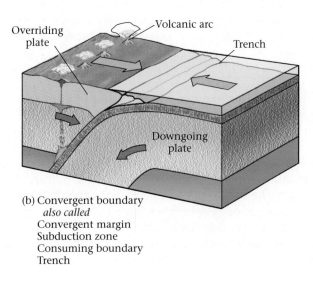

(b) Convergent boundary
also called
Convergent margin
Subduction zone
Consuming boundary
Trench

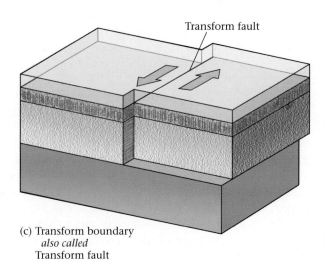

(c) Transform boundary
also called
Transform fault

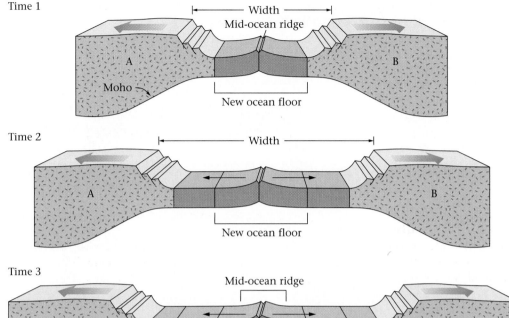

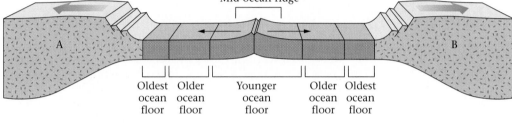

FIGURE 2.33 These sketches depict successive stages in sea-floor spreading along a divergent boundary (mid-ocean ridge); only the crust is shown. The top figure depicts an early stage of the process, after the mid-ocean ridge formed but before the ocean grew very wide. With time, as seen in the next two figures, the ocean gets wider and continent A drifts way from continent B. Note that the youngest ocean crust lies closest to the ridge.

Indian Ocean Ridge; Fig. 2.31) that rises 2 km above the adjacent abyssal plains of the ocean. Thus, geologists also commonly call a divergent boundary a mid-ocean ridge, or simply a ridge.

Characteristics of a Mid-Ocean Ridge

To characterize a divergent boundary more completely, let's look at one mid-ocean ridge in more detail (▶Fig. 2.34a). The Mid-Atlantic Ridge extends from the waters between northern Greenland and northern Scandinavia southward across the equator to the latitude opposite the southern tip of South America. Geologists have mapped segments of the Mid-Atlantic Ridge in detail, using sonar from ships and from research submarines. They have found that the formation of new sea floor takes place only along the axis (centerline) of the ridge. The axis lies at water depths of about 2 to 2.5 km.

The sea floor slopes away from the ridge axis, reaching the depth of the abyssal plain (4 to 5 km) at a distance of about 500 to 800 km from the ridge axis. Roughly speaking, the Mid-Atlantic Ridge is symmetrical—its eastern half looks like a mirror image of its western half. The ridge consists, along its length, of short segments (tens to hundreds of km long) linked by breaks called transform faults, which we will discuss later.

The Formation of Oceanic Crust at a Mid-Ocean Ridge

As noted above, sea-floor spreading does not create an open space between diverging plates. Rather, as each increment of spreading occurs, new sea floor develops in the space. How does this happen?

As sea-floor spreading takes place, hot asthenosphere (the soft, flowable part of the mantle) rises beneath the ridge (▶Fig. 2.34b). As this asthenosphere rises, it begins to melt, producing molten rock, or magma. Magma has a lower density than solid rock, so it behaves buoyantly and rises. It eventually accumulates in the crust below the ridge axis, filling a magma chamber. Some of the magma solidifies along the side of the chamber to make the coarse-grained, mafic igneous rock called gabbro. Some of the magma rises still higher to fill vertical cracks, where it solidifies and forms wall-like sheets, or dikes, of basalt (a fine-grained mafic igneous rock). Finally, some magma rises all the way to the surface of the sea floor at the ridge axis and spills out of small submarine volcanoes. The resulting lava cools to form a layer of basalt. The basalt occurs in small blobs called pillows. We can't easily see the submarine volcanoes because they occur at depths of more than 2 km beneath sea level, but they have been observed by the research submarine *Alvin*. *Alvin* has also detected chimneys spewing hot, mineralized water that rose

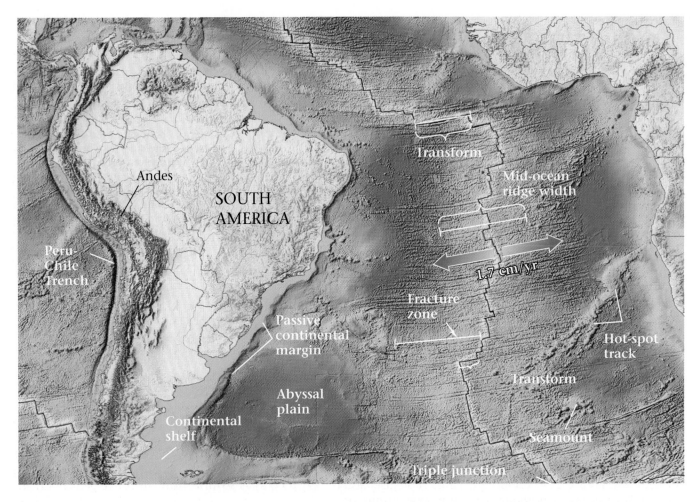

(a)

FIGURE 2.34 (a) A map showing the bathymetry of the Mid-Atlantic Ridge in the south Atlantic Ocean. The lighter colors are shallower depths. The map also shows the trench along the west coast of South America. (b) New lithosphere forms at a mid-ocean ridge. Rising hot asthenosphere partly melts underneath the ridge axis. The molten rock, magma, rises to fill a magma chamber in the crust. Some of the magma solidifies along the side of the chamber, to make coarse-grained mafic rock called gabbro. Some magma rises still farther to fill cracks, solidifying into basalt that forms wall-like sheets of rock called dikes. Finally, some magma rises all the way to the surface of the sea floor at the ridge axis. This magma, now called lava, spills out and forms a layer of pillow basalt. As sea-floor spreading continues, the oceanic crust breaks along faults. Also, as a plate moves away from a ridge axis and cools, the lithospheric mantle thickens.

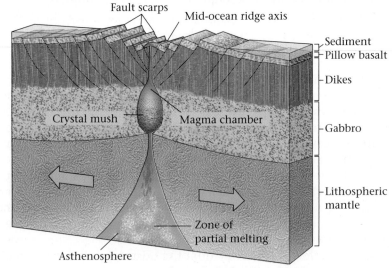

(b)

through cracks in the sea floor, after being heated by magma below the surface. These chimneys are called **black smokers** because the water they emit looks like dark smoke; the color comes from a suspension of tiny mineral grains that precipitate in the water as the water cools (▶Fig. 2.35).

As soon as it forms, new oceanic crust moves away from the ridge axis, and as this happens, more magma rises from below, so still more crust forms. In other words, like a vast, continuously moving conveyor belt, magma from the mantle rises to the Earth's surface at the ridge, solidifies to

FIGURE 2.35 A column of superhot water gushing from a vent (known as a "black smoker") along the mid-ocean ridge. The water has been heated by magma (molten rock) just below the surface. The cloud of "smoke" actually consists of tiny mineral grains; the elements making these minerals had been dissolved in the hot water, but when the hot water mixes with cold water of the sea, they precipitate. Many exotic species of life, such as giant worms, live around these vents.

form oceanic crust, and then moves laterally away from the ridge. Because all sea floor forms at mid-ocean ridges, the youngest sea floor occurs on either side of the ridge axis, and sea floor becomes progressively older away from the ridge (see Fig. 2.33). In the Atlantic Ocean, the oldest sea floor lies adjacent to the passive continental margins on either side of the ocean (▶Fig. 2.36). The oldest ocean floor on our planet occurs in the western Pacific Ocean; this crust formed 200 million years ago.

The tension (stretching force) applied to newly formed solid crust as spreading takes place breaks this new crust, resulting in the formation of faults. Slip on the faults causes divergent-boundary earthquakes and creates numerous cliffs, or scarps, that parallel the ridge axis.

The Formation of Lithospheric Mantle at a Mid-Ocean Ridge

So far, we've seen how oceanic crust forms at mid-ocean ridges. What about the formation of the mantle part of the oceanic lithosphere? Remember that this part consists of the cooler uppermost area of the mantle, in which temperatures are less than about 1,280°C. At the ridge axis, such temperatures occur almost at the base of the crust, because of the presence of rising hot asthenosphere and hot magma, so the lithospheric mantle beneath the ridge axis effectively doesn't exist. But as the newly formed oceanic crust moves

away from the ridge axis, the crust and the uppermost mantle directly beneath it gradually cool as they lose heat to the ocean above. As soon as mantle rock cools below 1,280°C, it becomes, by definition, part of the lithosphere.

As oceanic lithosphere continues to move away from the ridge axis, it continues to cool, so the lithospheric mantle, and therefore the oceanic lithosphere as a whole, grows progressively thicker (▶Fig. 2.37a, b). This process doesn't change the thickness of the oceanic crust, for the crust formed entirely at the ridge axis. The rate at which cooling and thickening occur decreases with distance from the ridge axis. In fact, by the time the lithosphere is about 80 million years old it has just about reached its maximum thickness. As lithosphere thickens and gets cooler and denser, it sinks into the asthenosphere, as a ship sinks slightly when taking on ballast. Thus, the ocean is deeper over older ocean floor than over younger ocean floor.

2.10 CONVERGENT PLATE BOUNDARIES AND SUBDUCTION

At convergent plate boundaries, or convergent margins, two plates, at least one of which is oceanic, move toward one another. But rather than butting each other like angry rams, one oceanic plate bends and sinks down into the asthenosphere beneath the other plate. Geologists refer to the sinking process as **subduction,** so convergent boundaries are also known as subduction zones. Because subduction at a convergent boundary consumes old ocean lithosphere and thus closes (or "consumes") oceanic basins, geologists also refer to convergent boundaries as consuming boundaries, and because they are delineated by deep-ocean trenches, they are sometimes simply called **trenches** (Fig. 2.34a). The amount of oceanic plate consumption worldwide, averaged over time, equals the amount of sea-floor spreading worldwide, so the surface area of the Earth remains constant through time.

Subduction occurs for a simple reason: oceanic lithosphere, once it has aged at least 10 million years, is denser than asthenosphere and thus can sink through the asthenosphere. When it lies flat on the surface of the asthenosphere, oceanic lithosphere doesn't sink because the resistance of the asthenosphere to flow is too great. However, once the end of the convergent plate bends down and slips into the mantle, it begins to sink like an anchor falling to the bottom of a lake (▶Fig. 2.38a, b). As the lithosphere sinks, asthenosphere flows out of its way, just as water flows out of the way of an anchor. Even though it is relatively soft and plastic the asthenosphere resists flow, so oceanic lithosphere can sink only very slowly, at a rate of less than about 15 cm per year.

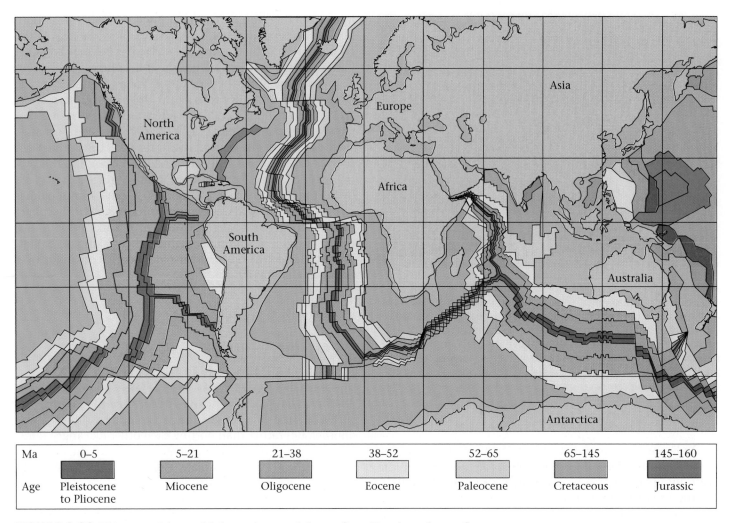

Ma	0–5	5–21	21–38	38–52	52–65	65–145	145–160
Age	Pleistocene to Pliocene	Miocene	Oligocene	Eocene	Paleocene	Cretaceous	Jurassic

FIGURE 2.36 This map of the world shows the age of the sea floor. Note how the sea floor grows older with increasing distance from the ridge axis. (Ma = million years ago)

FIGURE 2.37 (a) As sea floor ages, the dense lithospheric mantle thickens. (b) Like the ballast of a ship, older (thicker) lithosphere sinks deeper into the mantle. This is why the ocean above older ocean lithosphere is deeper than the ocean above younger ocean lithosphere.

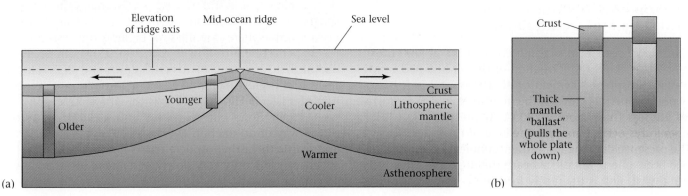

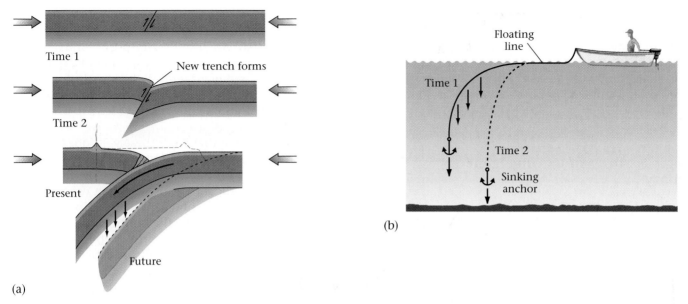

(a)

(b)

FIGURE 2.38 (a) The concept of subduction. A plate bends, and one piece pushes over the other. Oceanic lithosphere is denser than the underlying asthenosphere, but when it lies flat on the surface of the asthenosphere, it can't sink because the resistance of the asthenosphere to flow is too great. However, once the end of the plate is pushed into the mantle, the lithosphere begins to sink. (b) The process of sinking is like an anchor pulling a floating anchor line down. As a consequence, the bend in the plate (or in the anchor line) progressively moves with time.

Note that the downgoing plate, the plate that has been subducted, *must* be composed of oceanic lithosphere. The overriding plate, which does not sink, can consist of either oceanic or continental lithosphere. Continental crust cannot be subducted because it is too buoyant; the low-density rocks of continental crust act like a life preserver keeping the continent afloat. If continental crust moves into a convergent margin, subduction stops. Because of subduction, all ocean floor on the planet is less than about 200 million years old. Because continental crust cannot subduct, some continental crust has persisted at the surface of the Earth for over 3.8 billion years.

Earthquakes and the Fate of Subducted Plates

At convergent plate boundaries, the downgoing plate grinds along the base of the overriding plate, a process that generates large earthquakes. These earthquakes occur fairly close to the Earth's surface, so some of them trigger massive destruction in coastal cities. But earthquakes also happen in downgoing plates at greater depths, deep below the overriding plate. In fact, geologists have detected earthquakes within downgoing plates to a depth of 660 km. The belt of earthquakes in a downgoing plate is called a Wadati-Benioff zone, after its two discoverers (▶Fig. 2.39).

FIGURE 2.39 A Wadati-Benioff zone is a band of earthquakes that occur in subducted oceanic lithosphere. The discovery of these earthquakes led to the proposal of subduction.

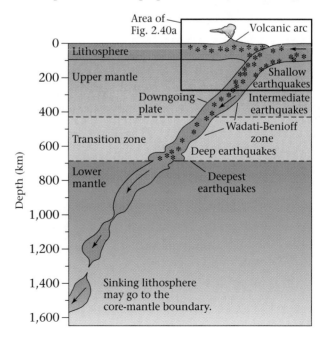

At depths greater than 660 km, conditions leading to earthquakes evidently do not occur. Recent evidence, however, indicates that some downgoing plates do continue to sink *below* a depth of 660 km—they just do so without generating earthquakes. In fact, studies suggest that the lower mantle may be a graveyard for old subducted plates.

Geologic Features of a Convergent Boundary

To become familiar with the various geologic features that occur along a convergent plate boundary, let's look at an example, the boundary between the western coast of the South American Plate and the eastern edge of the Nazca Plate (a portion of the Pacific Ocean floor). A deep-ocean trench, the Peru-Chile Trench, delineates this boundary (see Fig. 2.34a). Such trenches form where the plate bends as it starts to sink into the asthenosphere.

In the Peru-Chile Trench, as the downgoing plate slides under the overriding plate, sediment (clay and plankton) that had settled on the surface of the downgoing plate, as well as sand that fell into the trench from the shores of South America, gets scraped up and incorporated in a wedge-shaped mass known as an accretionary prism (▶Fig. 2.40a). An accretionary prism forms in basically the same way as a pile of snow in front of a plow, and like the snow, the sediment tends to be squashed and contorted during the formation of the prism (▶Fig. 2.40b).

A chain of volcanoes known as a **volcanic arc** develops behind the accretionary prism. As we will see in Chapter 4, the magma that feeds these volcanoes forms at or just above the surface of the downgoing plate when the plate reaches a depth of about 150 km below the Earth's surface. If the volcanic arc forms where an oceanic plate subducts beneath continental lithosphere, the resulting chain of volcanoes grows on the continent and forms a continental volcanic arc. In some cases, the plates compress together, causing a belt of faults to form (Fig. 2.40a). If, however, the volcanic arc forms where one oceanic plate subducts beneath another oceanic plate, the resulting volcanoes form a chain of islands known as a volcanic island arc. A marginal sea, or back-arc basin, the small ocean basin between an island arc and the continent, forms either in cases where subduction happens to begin offshore, trapping ocean lithosphere behind the arc, or where stretching of the lithosphere behind the arc leads

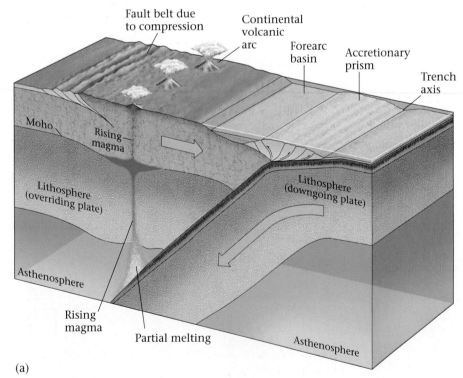

(a)

FIGURE 2.40 (a) This model shows the geometry of subduction along an active continental margin. The trench axis (lowest part of the trench) roughly defines the plate boundary. Numerous faults form in the accretionary prism, which is composed of material scraped off the sea floor. Behind the prism lies a basin (a forearc basin) of trapped sediment. A volcanic arc is created from magma that forms at or just above the surface of the downgoing plate. Here, the plate subducts beneath continental lithosphere, so the chain of volcanoes is called a continental arc. Faulting occurs on the backside of the arc. The Andes in South America and the Cascades in the United States are examples of such continental arcs. (b) The action of a bulldozer pushing snow or soil is similar to the development of an accretionary prism.

(b)

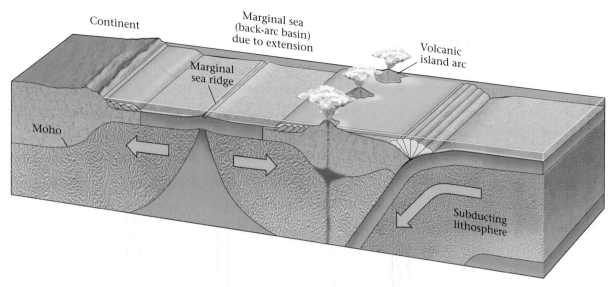

FIGURE 2.41 Subduction along an island arc. Here, the volcanoes build on the sea floor. Behind some island arcs, a marginal sea forms. This sea resembles a small ocean basin, with a spreading ridge that is created when the plate behind the arc moves away from the arc.

to the formation of a small spreading ridge between the arc and the continent (▶Fig. 2.41).

2.11 TRANSFORM PLATE BOUNDARIES

When researchers began to explore the bathymetry of mid-ocean ridges in detail, they discovered that mid-ocean ridges are not smooth, uninterrupted lines, but rather consist of short segments that appear to be offset from each other (see Fig. 2.34). Narrow belts of broken and irregular sea floor lie roughly at right angles to the ridge segments, intersect the ends of the segments, and extend beyond the ends of the segments. These belts came to be known as fracture zones. Originally, researchers incorrectly assumed that the entire length of each fracture zone was a fault, and that movement (slip) on a fracture zone had displaced segments of the mid-ocean ridge sideways, relative to each other. In other words, they imagined that a mid-ocean ridge initiated as a continuous, fence-like line that only later was broken up by faulting. But when information about the distribution of earthquakes along mid-ocean ridges became available, it was clear that this model could not be correct. Earthquakes, and therefore active fault movement, occur *only* on the segment of a fracture zone that lies between two ridge segments (▶Fig. 2.42a). The portions of fracture zones that extend beyond the edges of ridge segments, out into the abyssal plain, are not active.

The distribution of movement along fracture zones remained a mystery until a Canadian researcher, J. Tuzo Wilson, began to think about fracture zones in the context of the sea-flooring spreading concept. Wilson proposed that fracture zones formed *at the same time* as the ridge axis itself, and thus the ridge consisted of separate segments to start with. These segments were linked (not offset) by fracture zones. With this idea in mind, he drew a sketch map showing two ridge-axis segments linked by a fracture zone, and he drew arrows to indicate the direction that ocean crust was moving, relative to the ridge axis, as a result of sea-floor spreading (▶Fig. 2.42b, c). Look at Wilson's arrows. Clearly, the movement direction on the active portion of the fracture zone must be opposite to the movement direction that researchers originally thought occurred on the structure. Further, in Wilson's model, slip occurs only along the segment of the fracture zone between the two ridge segments (see Fig. 2.42c).

Wilson introduced the term transform fault for the actively slipping segment of a fracture zone between two ridge segments, and he pointed out that transform faults are a third type of plate boundary. Geologists now also call them transform boundaries, or simply transforms. At a transform boundary, one plate slides sideways past another, but no new plate forms and no old plate is consumed. Transform boundaries are, therefore, defined by a vertical fault on which the slip direction parallels the Earth's surface (see Fig. 2.42a).

So far we've discussed only transforms along mid-ocean ridges. Not all transforms link ridge segments. Some, such as the Alpine Fault of New Zealand, link trenches, while

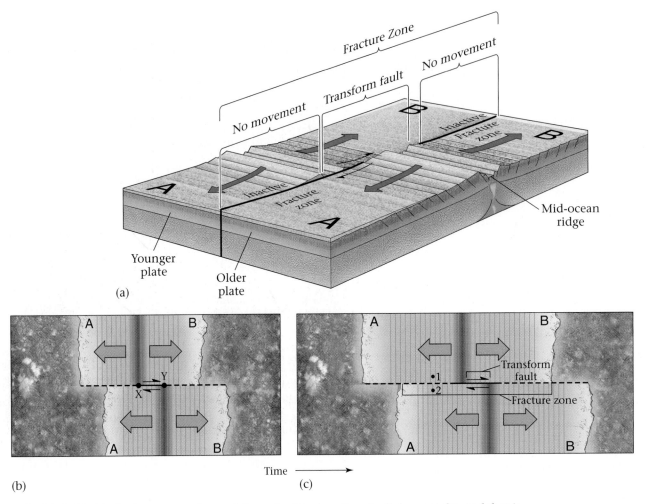

FIGURE 2.42 (a) The fracture zone beyond the ends of the transform fault does not slip, and thus is not a plate boundary. It does, however, mark the boundary between portions of the plate that are different in age. (b) In Wilson's interpretation, the ridge initiates at the same time as the transform fault, and thus was never continuous. Note that the way in which the fault slips (along the fracture zone between points X and Y) makes sense if sea-floor spreading takes place. Only the portion between X and Y is active—this is the transform fault. The dashed lines represent inactive portions of the fracture zone. (c) Even though the ocean grows, the transform fault can stay the same length. Note that the age of the sea floor on one side of the fracture zone is not the same as that of the sea floor on the other side. Point 1 on plate A is younger than point 2 because it lies closer to the ridge axis.

others link a trench to a ridge segment. Further, not all transform faults occur in oceanic lithosphere; a few cut across continental lithosphere. The San Andreas Fault, for example, which cuts across California, defines part of the plate boundary between the North American Plate and the Pacific Plate—the portion of California that lies to the west of the fault (including Los Angeles) is part of the Pacific Plate, while the portion that lies to the east of the fault is part of the North American Plate (▶Fig. 2.43a, b). On average, the Pacific Plate moves about 6 cm north, relative to North America, every year. If this motion continues, Los Angeles will become a suburb of Anchorage, Alaska, in about 100 million years.

2.12 SPECIAL LOCATIONS IN THE PLATE MOSAIC

Triple Junctions

So far, we've focused attention on boundaries—divergent (mid-ocean ridge), convergent (trench), and transform—between *two* plates. But there are several places where *three* plate boundaries intersect at a point. Geologists refer to these points as **triple junctions.** We name triple junctions after the types of boundaries that intersect. For example, the triple junction formed where the Southwest Indian

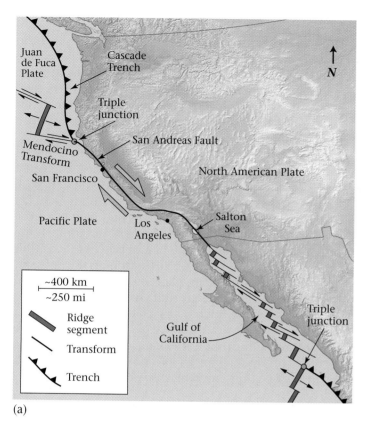

FIGURE 2.43 (a) The San Andreas Fault is a transform plate boundary between the Pacific Plate to the west and the North American Plate to the east. At its southeastern end, the San Andreas connects to spreading ridge segments in the Gulf of California. (b) The San Andreas Fault where it cuts across a dry landscape. The fault trace is the narrow valley running the length of the photo. The land has been pushed up slightly, along the fault; streams have cut small side canyons into this uplifted land.

Ocean Ridge intersects two arms of the Mid–Indian Ocean Ridge (this is the triple junction of the African, Antarctic, and Australian Plates) is a ridge-ridge-ridge triple junction (▶Fig. 2.44a). The triple junction north of San Francisco is a trench-transform-transform triple junction (▶Fig. 2.44b).

Hot Spots

Most subaerial (above sea level) volcanoes are situated in the volcanic arcs that border trenches. Small volcanoes also lie along mid-ocean ridges, but ocean water hides most of them. The volcanoes of volcanic arcs and mid-ocean ridges are plate-boundary volcanoes, formed as a consequence of movement along the boundary. Not all volcanoes on Earth are plate-boundary volcanoes, however. For example, the big island of Hawaii, a large volcano, lies in the middle of the Pacific Plate, and Yellowstone National Park, site of recent volcanic activity, lies in the interior of the North

American plate. Worldwide, geoscientists have identified about one hundred volcanoes that exist as isolated points and are not a consequence of movement at a plate boundary. These are called hot-spot volcanoes, or simply **hot spots** (▶Fig. 2.45). Most hot spots are located in the interiors of plates, away from the boundaries, but a few lie on mid-ocean ridges.

What causes hot-spot volcanoes? After studying examples in the interiors of ocean plates, J. Tuzo Wilson noted that an erupting hot-spot volcano formed an island at the end of a chain of now-dead (no-longer erupting) volcanic islands and seamounts. A volcano along a convergent plate boundary, in contrast, is one of many in a chain that are all active at about the same time. Wilson noted further that all the hot-spot volcanoes in the Pacific lie at the southeastern end of a northwest-southeast-lying chain of dead volcanoes (▶Fig. 2.46). With this image in mind, Wilson suggested in 1963 that a hot-spot volcano develops over a heat source in

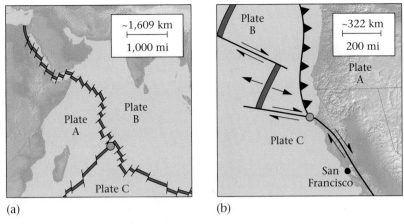

(a)   (b)

FIGURE 2.44 (a) A ridge-ridge-ridge triple junction (at the orange dot) in the western Indian Ocean. (b) A trench-transform-transform triple junction along the Pacific coast of the United States.

the mantle that is fixed relative to the moving plate; an active volcano represents the location of the heat source. The chain of seamounts and inactive (dead) volcanic islands linked to the hot-spot volcano indicates locations on the plate that were once over the source but have since moved off. Subsequently, the heat source came to be associated with a **mantle plume,** a column of very hot rock rising up through the mantle.

Let's look more closely at how Wilson's model works. Most mantle plumes are thought to originate just above the core-mantle boundary, where heat from the Earth's core warms the base of the mantle. (Recent evidence, however, suggests that some may originate in the upper mantle.) Because it expands as it heats, the hot rock above the core-mantle boundary becomes less dense than overlying mantle and begins to stream upward in a column-shaped mass, at the rate of a few centimeters per year. When the hot rock reaches the base of the lithosphere, it partially melts, for reasons discussed in Chapter 4. The magma formed by melting then rises through the lithosphere and erupts at a volcano on the Earth's surface (▶Fig. 2.47a). This location is the hot spot. All the while, the plate on which the volcano grows continues to shift, so eventually the volcano moves off the plume and dies, or goes extinct. Meanwhile, a new volcano grows over the plume. If the plume lasts for a long time in the same place, it will spawn a long chain of dead (extinct) volcanoes, a hot-spot track; again, at any one time, only the volcano over the plume can be active.

The Hawaiian chain provides a clear example of the volcanism associated with a hot spot. Volcanic eruptions occur today only on the big island of Hawaii (▶Fig.2.47b). Other

FIGURE 2.45 The dots indicate the locations of selected hot-spot volcanoes. The tails indicate hot-spot tracks. The most recent volcano (dot) is at one end of this track. Some of these are extinct, indicating that the plume no longer exists. Some hot spots are fairly recent and do not have tracks. Dashed tracks are places where track was broken by sea-floor spreading.

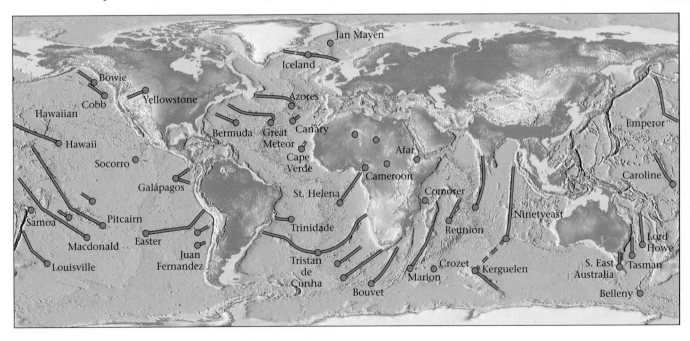

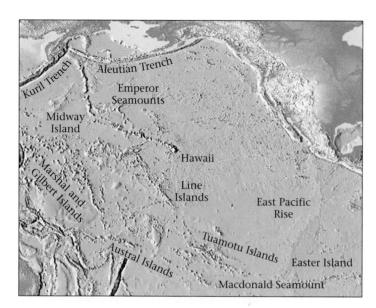

FIGURE 2.46 Bathymetric map showing hot-spot tracks in the Pacific Ocean. Note that the chains have a 40° bend in them, resulting from a change in the direction of motion of the Pacific Plate about 40 million years ago.

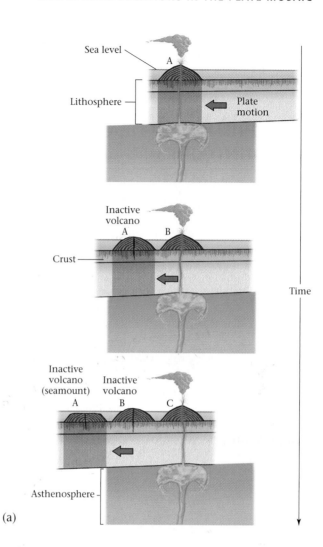

(a)

islands to the northwest are remnants of dead volcanoes. The oldest of these is Kauai. To the northwest of Kauai, still older volcanic remnants are found, and these continue in a straight line past Midway Island (the site of a major battle during World War II). To the northwest of Midway Island, most of the chain of volcanic remnants no longer pokes above sea level, and so we refer to them as the Emperor seamount chain.

Some hot spots lie within continents. For example, several have been active in the interior of Africa, and one now underlies Yellowstone National Park. The famous geysers (natural steam and hot-water fountains) of Yellowstone exist because hot magma, formed above the Yellowstone hot spot, lies not far below the surface of the park. A few hot spots lie on mid-ocean ridges. Where this happens, a volcanic island protrudes above sea level, because the hot spot produces far more magma than does a normal mid-ocean ridge. Iceland, for example, formed where a hot spot underlies the Mid-

FIGURE 2.47 (a) A mantle plume causes a hot spot to form at the base of a plate, leading to the growth of a volcano on the surface of the plate. As the plate moves, the volcano is carried off the hot spot; it then dies (becomes extinct), and a new volcano forms above the hot spot. As the process continues, a chain of extinct volcanoes develops, with the oldest one farthest from the hot spot. The extinct volcanoes gradually sink below sea level and become seamounts. Only the volcano above the hot spot erupts. The chain of islands is a hot-spot track. (b) The plume that forms a hot spot arises from the base of the mantle. Here, we see the plume that underlies the Hawaiian Islands.

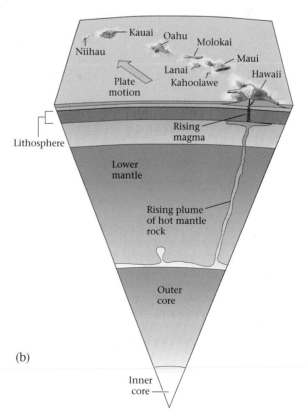

(b)

Atlantic Ridge. The extra volcanism of the hot spot built up the island of Iceland so that it rises almost 3 km above other places on the Mid-Atlantic Ridge.

2.13 THE BIRTH AND DEATH OF PLATE BOUNDARIES

The configuration of plates and plate boundaries visible on our planet today has not existed for all of geologic history, and will not exist indefinitely into the future. Because of plate motion, oceanic plates form and are later consumed, while continents merge and later split apart. How does a new divergent boundary come into existence, and how does a convergent boundary cease to exist? Most new divergent boundaries form when a continent splits and separates into two continents. We call this process **rifting.** A convergent boundary ceases to exist when a piece of buoyant lithosphere, such as a continent or an island arc, moves into the subduction zone. We call this process **collision.**

Continental Rifting

A **continental rift** is a linear belt in which continental lithosphere undergoes rifting, or pulls apart (▶Fig. 2.48). The lithosphere stretches horizontally, so it thins vertically, much like a piece of taffy you pull between your fingers. Near the surface of the continent, where the crust is cold and brittle, stretching causes rock to break and faults to develop. As a consequence of faulting, blocks of rock slide down, leading to the formation of a low area that gradually becomes buried by sediment. Lower in the crust, and even deeper in the lithospheric mantle, rock is warmer and softer, so stretching takes place in a plastic manner without breaking the rock. The whole region that stretches is the rift.

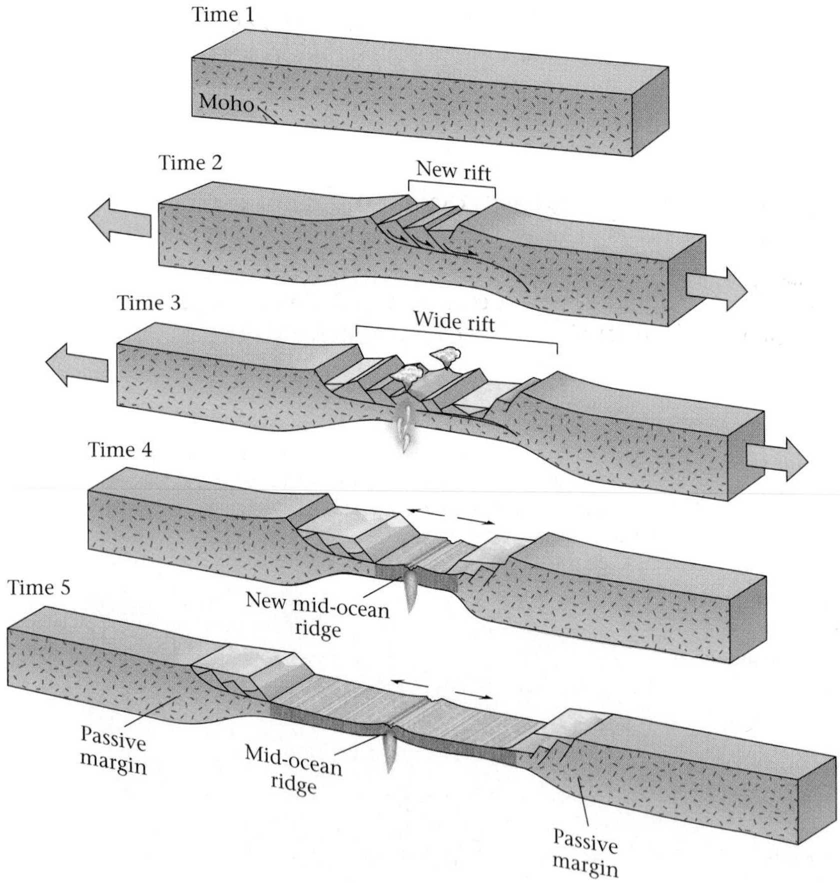

As continental lithosphere thins, hot asthenosphere rises beneath the rift and partly melts. This molten rock erupts at volcanoes along the rift. If rifting continues for a long enough time, the continent breaks in two, a new mid-ocean ridge forms, and sea-floor spreading begins. The relict of the rift evolves into a passive margin (see Fig. 2.30). In some cases, however, rifting stops before the continent splits in two. Then, the rift remains as a permanent scar in the crust, defined by a belt of faults, volcanic rocks, and a thick layer of sediment.

Perhaps the most spectacular example of a rift today occurs in eastern Africa; geoscientists aptly refer to it as the East African Rift (▶Fig. 2.49). To astronauts in orbit, the rift looks like a giant gash in the crust. On the ground, it consists of a deep trough bordered on both sides by high cliffs formed by faulting. Along the length of the rift, several major volcanoes smoke and fume; these include the snow-crested Mt. Kilimanjaro, towering over 6 km above the savanna. At its north end, the rift joins the Red Sea ridge and the Gulf of Aden ridge at a triple junction. Another major rift, known as the Basin and Range Province, breaks up the landscape of the western United States between Salt

FIGURE 2.48 When continental lithosphere stretches during continental rifting, the upper part of the crust breaks up into a series of faults. The lower part of the crust and the lithospheric mantle stretches more like soft plastic. The region that has stretched is the rift. With continued stretching, the crust becomes much thinner, and the asthenosphere that rises beneath the rift partly melts. As a consequence, volcanoes form in the rift. Eventually, the continent breaks in two, and a new mid-ocean ridge forms. With time, an ocean develops. The relicts of the stretched and broken crust of the rift underlie the thick sediment wedge of the passive margins. In this figure, we do not show the lithosphere mantle.

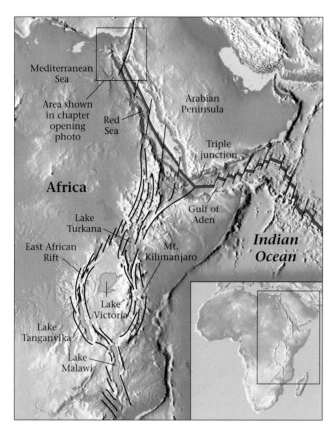

FIGURE 2.49 If the East African Rift were to continue growing, part of Africa would break off, forming a continental fragment. Note that the East African Rift intersects the Red Sea and the Gulf of Aden at a triple junction. The Red Sea and Gulf of Aden started as rifts but are now narrow oceans, bisected by new mid-ocean ridges. The inset shows the map area.

Lake City, Utah, and Reno, Nevada (▶Fig. 2.50). Here, movement on numerous faults tilted blocks of crust to form narrow mountain ranges, while sediment that eroded from the blocks filled the adjacent basins (the low areas between the ranges).

Collision

India was once a small, separate continent that lay far to the south of Asia. But subduction consumed the ocean between India and Asia, and India moved northward, finally slamming into the southern margin of Asia 40 to 50 million years ago. Continental crust, unlike oceanic crust, is too buoyant to subduct. So when India collided with Asia, the attached oceanic plate broke off and sank down into the deep mantle. But India pushed hard into Asia, squeezing the rocks and sediment that once lay between the two continents into the 8-km-high welt that we now know as

the Himalayan Mountains. During this process, not only did the surface of the Earth rise, but the crust became thicker. The crust beneath a collisional mountain range can be up to 60–70 km thick, about twice the thickness of normal continental crust.

Geoscientists refer to the process during which two buoyant pieces of lithosphere converge and squeeze together as **collision** (▶Fig. 2.51a, b). Some collisions involve two continents; some involve continents and an island arc. When a collision is complete, the convergent plate boundary that once existed between the two colliding pieces ceases to exist. Collisions yield some of the most spectacular mountains on the planet, such as the Himalaya and the Alps. They also yielded major mountain ranges in the past, which subsequently eroded away so that today we see only their relicts. For example, the Appalachian Mountains in the eastern United States were formed as a consequence of three collisions. After the last one, a collision between Africa and North America around 280 million years ago, North America became part of the Pangaea supercontinent. This supercontinent later rifted apart.

2.14 WHAT DRIVES PLATE MOTION?

When geoscientists first proposed plate tectonics, they thought the process occurred simply because convective flow in the asthenosphere actively dragged plates along,

FIGURE 2.50 The Basin and Range Province or Rift is the broad region between Reno, Nevada and Salt Lake City, Utah. Note the narrow north-south trending mountain ranges, the tips of fault-bounded blocks. The large arrows indicate the direction of stretching.

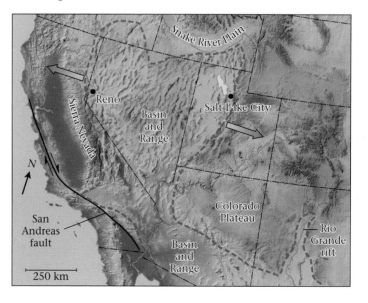

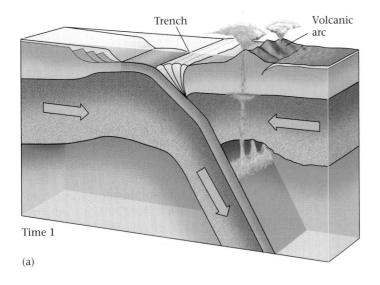

Time 1

(a)

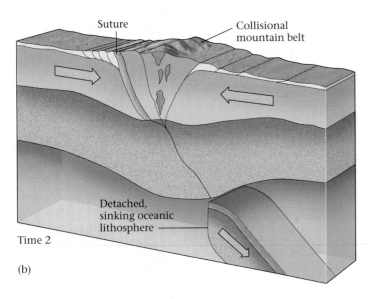

Time 2

(b)

FIGURE 2.51 (a) Before a continental collision takes place, subduction consumes an oceanic plate until it collides with another plate. Here, a passive continental margin collides with a continental volcanic arc. (b) After the collision, the oceanic plate detaches and sinks into the mantle. Rock caught in the collision zone gets broken, bent, and squashed, and forms a mountain range. Slivers of oceanic crust may be trapped along the boundary, or suture, between what once had been two continents. As the crust shortens horizontally, it thickens vertically.

as if the plates were rafts on a flowing river. Thus, early images depicting plate motion showed simple convection cells—elliptical flow paths of convecting asthenosphere—beneath mid-ocean ridges. At first glance, this hypothesis looked pretty good, but on closer examination, it failed. Among other reasons, it is impossible to draw a global

arrangement of convection cells that can explain the complex geometry of plate boundaries on Earth. Gradually, geoscientists came to the conclusion that convective flow within the asthenosphere occurs, but is not related in a simple way to the detailed geometry of plate motion. In other words, hot asthenosphere does rise in some places and sink in others, because of temperature and perhaps compositional contrasts in the mantle, and plate motion overall is a manifestation of this convection—but the local directions of this convective flow do not necessarily define the local directions of plate motion. Today, geoscientists favor the hypothesis that two forces—ridge-push force and slab-pull force—strongly influence the motion of individual plates.

Ridge-push force develops because mid-ocean ridges lie at a higher elevation than the adjacent abyssal plains of the ocean (▶Fig. 2.52a). To understand ridge-push force, imagine you have a glass containing a layer of water over a layer of honey. By tilting the glass momentarily and then returning it to its upright position, you can create a temporary slope in the boundary between these substances. While the boundary has this slope, gravity causes the elevated honey to push against the glass adjacent to the side where the honey surface lies at lower elevation. If the glass were to disappear suddenly, this force would push the honey out over the table. The geometry of a mid-ocean ridge resembles this situation. The surface of the sea floor is higher along a mid-ocean ridge axis than in adjacent abyssal plains. Thus, the surface of the sea floor overall slopes away from the ridge axis. Gravity causes the elevated lithosphere at the ridge axis to push on the lithosphere that lies farther from the axis (much as the tilted honey layer pushes on the side of the glass), making it move away. As lithosphere moves away from the ridge axis, new hot asthenosphere rises to fill the gap; it then moves away, cools, and itself becomes lithosphere. Note that the upward movement of asthenosphere beneath a mid-ocean ridge is a *consequence* of sea-floor spreading, not the cause.

Slab-pull force, the force that downgoing plates apply to oceanic lithosphere at a convergent margin, arises simply because lithosphere that was formed more than 10 million years ago is denser than asthenosphere, so it can sink into the asthenosphere (▶Fig. 2.52b). Thus, once an oceanic plate starts to sink, it gradually pulls the rest of the plate along behind it, like an anchor pulling down the anchor line. This "pull" is the slab-pull force.

Now let's summarize our discussion of forces that drive plate motions. Plates move away from ridges—in other words, sea-floor spreading occurs—in response to the ridge-push force, and as this happens, new asthenosphere rises to fill the space that opens between the plates. Old, cool oceanic lithosphere sinks down into the asthenosphere, creating a slab-pull force that tows the rest of the plate along with it. But ridge push and slab pull are

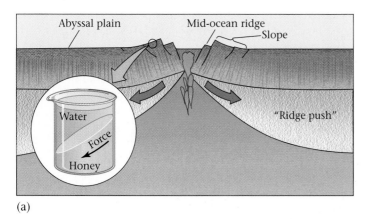

(a)

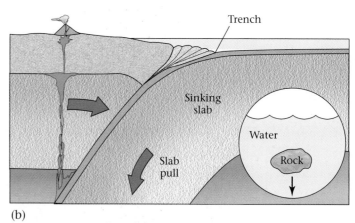

(b)

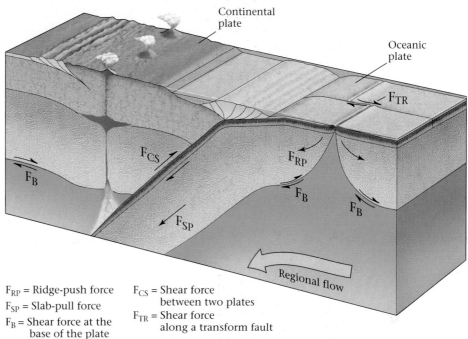

FIGURE 2.52 (a) A simplified profile (not to scale) of a mid-ocean ridge. Note that along the flanks of the ridge, the sea floor slopes. The elevation of the ridge causes an outward ridge-push force that drives the lithosphere plate away from the ridge. A similar situation exists in a glass containing honey and water. If the boundary between the honey and water tilts, the honey exerts an outward force at its base. (b) In this cross section illustrating slab-pull force, the oceanic plate is denser than the asthenosphere, so it sinks into the asthenosphere like a stone into water, only much more slowly. (c) In addition to ridge push and slab pull, the plates feel shear forces along their base. Their movement may also be resisted by shear along transform faults or at the base of a continent at a subduction zone.

F_{RP} = Ridge-push force
F_{SP} = Slab-pull force
F_B = Shear force at the base of the plate

F_{CS} = Shear force between two plates
F_{TR} = Shear force along a transform fault

(c)

not the only forces acting on the plate. The asthenosphere does convect and the flow of the asthenosphere probably exerts a force on the base of the plate, just as flowing water exerts a force on the bottom of a boat tied at a dock. If this force, or shear, happens to be in the same direction that the plate is already moving, it can speed up the plate motion, but if the shear is in the opposite direction, it might slow the plate down. Also, where one plate grinds against another, as occurs along a transform fault or at the base of an overriding plate at a convergent margin, friction (the force that resists sliding on a surface) may slow the plate down (▶Fig. 2.52c).

2.15 THE VELOCITY OF PLATE MOTIONS

How fast do plates move? It depends on your frame of reference. To illustrate this concept, imagine two cars speeding in the same direction down the highway. From the viewpoint of a tree along the side of the road, car A zips by at 100 km an hour, while car B moves at 80 km an hour. But relative to car B, car A moves at only 20 km an hour. Likewise, geologists use two different frames of reference for describing plate velocity (velocity = distance/time). If we describe the movement of plate A with respect to plate B, then we are talking about **relative plate velocity.** But if we describe the movement of both plates relative to a fixed

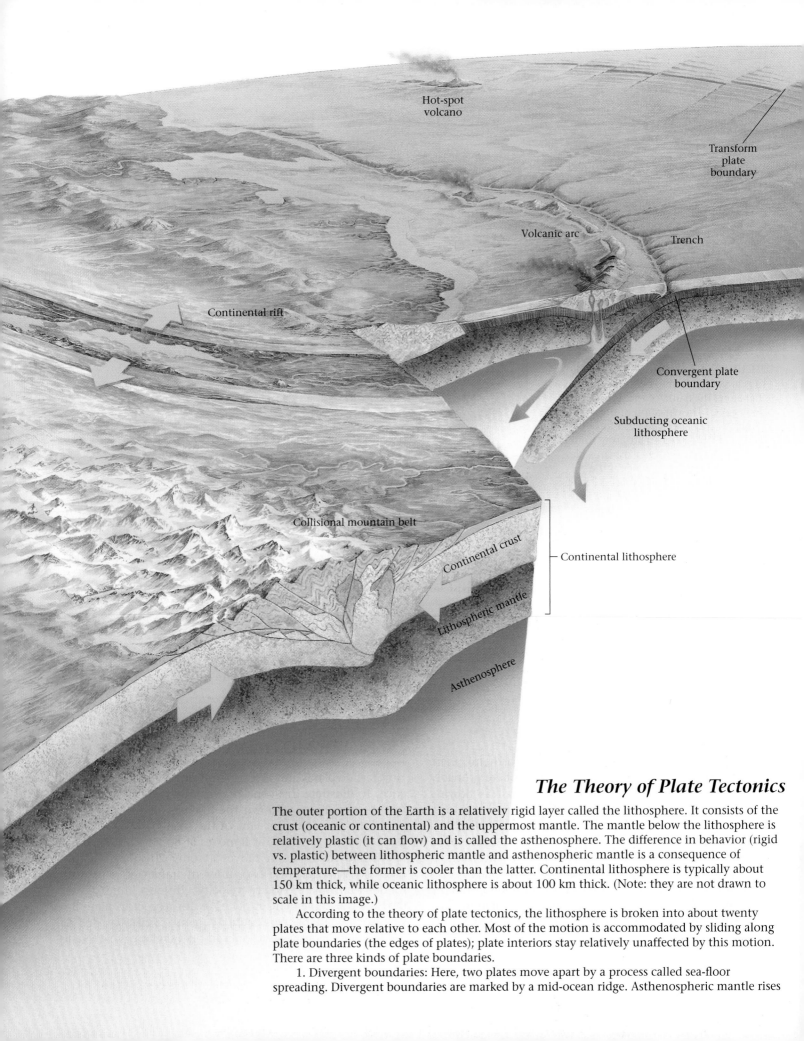

Hot-spot
volcano

Transform
plate
boundary

Volcanic arc

Trench

Continental rift

Convergent plate
boundary

Subducting oceanic
lithosphere

Collisional mountain belt

Continental crust

Continental lithosphere

Lithospheric mantle

Asthenosphere

The Theory of Plate Tectonics

The outer portion of the Earth is a relatively rigid layer called the lithosphere. It consists of the crust (oceanic or continental) and the uppermost mantle. The mantle below the lithosphere is relatively plastic (it can flow) and is called the asthenosphere. The difference in behavior (rigid vs. plastic) between lithospheric mantle and asthenospheric mantle is a consequence of temperature—the former is cooler than the latter. Continental lithosphere is typically about 150 km thick, while oceanic lithosphere is about 100 km thick. (Note: they are not drawn to scale in this image.)

According to the theory of plate tectonics, the lithosphere is broken into about twenty plates that move relative to each other. Most of the motion is accommodated by sliding along plate boundaries (the edges of plates); plate interiors stay relatively unaffected by this motion. There are three kinds of plate boundaries.

1. Divergent boundaries: Here, two plates move apart by a process called sea-floor spreading. Divergent boundaries are marked by a mid-ocean ridge. Asthenospheric mantle rises

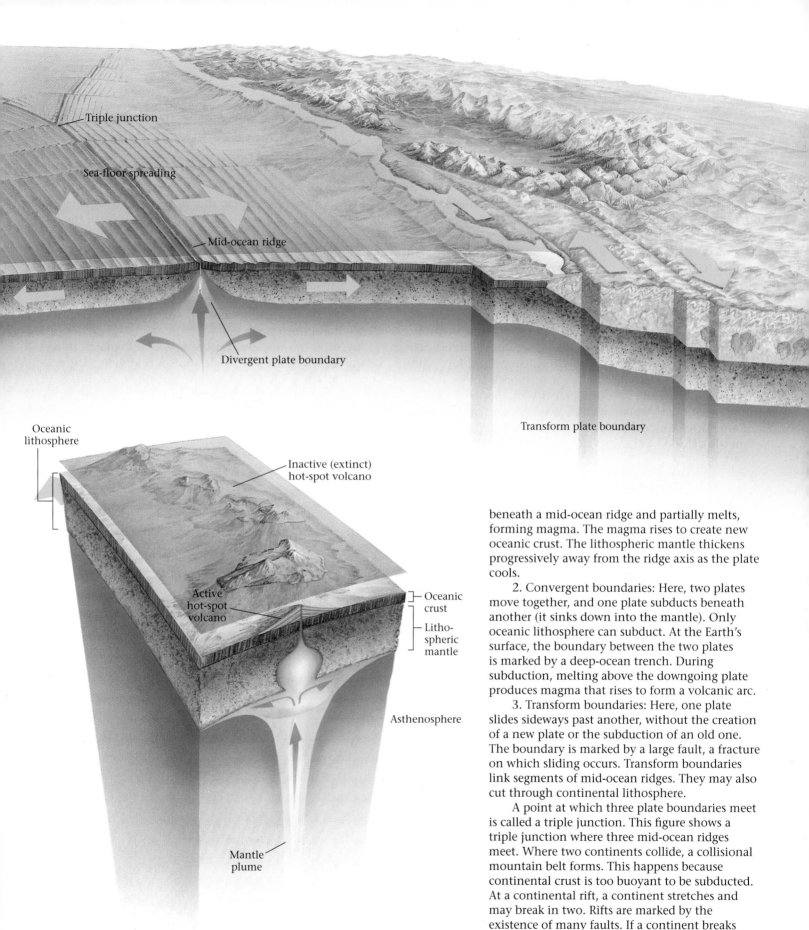

Triple junction

Sea-floor spreading

Mid-ocean ridge

Divergent plate boundary

Transform plate boundary

Oceanic
lithosphere

Inactive (extinct)
hot-spot volcano

Oceanic
crust

Litho-
spheric
mantle

Active
hot-spot
volcano

Asthenosphere

Mantle
plume

beneath a mid-ocean ridge and partially melts, forming magma. The magma rises to create new oceanic crust. The lithospheric mantle thickens progressively away from the ridge axis as the plate cools.

2. Convergent boundaries: Here, two plates move together, and one plate subducts beneath another (it sinks down into the mantle). Only oceanic lithosphere can subduct. At the Earth's surface, the boundary between the two plates is marked by a deep-ocean trench. During subduction, melting above the downgoing plate produces magma that rises to form a volcanic arc.

3. Transform boundaries: Here, one plate slides sideways past another, without the creation of a new plate or the subduction of an old one. The boundary is marked by a large fault, a fracture on which sliding occurs. Transform boundaries link segments of mid-ocean ridges. They may also cut through continental lithosphere.

A point at which three plate boundaries meet is called a triple junction. This figure shows a triple junction where three mid-ocean ridges meet. Where two continents collide, a collisional mountain belt forms. This happens because continental crust is too buoyant to be subducted. At a continental rift, a continent stretches and may break in two. Rifts are marked by the existence of many faults. If a continent breaks apart, a new mid-ocean ridge develops.

Hot-spot volcanoes form above plumes of hot mantle rock that rise from near the core-mantle boundary. As a plate drifts over a hot spot, it leaves a chain of extinct volcanoes.

point in the mantle, then we are speaking of **absolute plate velocity.**

We've already seen one method of determining relative plate motions. Geoscientists measure the distance of a known magnetic anomaly from the axis of a mid-ocean ridge, and they calculate the velocity of a plate relative to the ridge axis by applying this equation: plate velocity = distance from the anomaly to the ridge axis divided by the age of anomaly. The velocity of the plate on one side of the ridge relative to the plate on the other is twice this value. We've also seen a way to estimate absolute plate motions. If we *assume* that the position of a mantle plume does not change much for a long time, then the track of hot-spot volcanoes on the plate moving over the plume provides a record of the plate's absolute velocity and indicates the direction of movement (▶Fig. 2.53). (In reality, plumes are not completely fixed; geologists must use other, more complex methods to calculate absolute plate motions.) The Hawaiian Emperor seamount chain, for example, defines the absolute velocity of the Pacific Plate. Note that the Hawaiian chain runs northwest, whereas the Emperor chain curves north-northwest (Fig. 2.46). Radiometric dates of volcanic rocks from the bend indicate that

they formed about 43 million years ago. Thus, the direction in which the Pacific Plate moved changed significantly at this time.

Working from the calculations described above, geologists have determined that relative plate motions on Earth today occur at rates of about 1 to 15 cm per year. But these rates, though small, can yield large displacements given the immensity of geologic time; in a million years, a plate can move 100 km. Can we detect such slow rates? Until the last decade, the answer was no. Now the answer is yes. Satellites now orbiting the Earth provide us with the **global positioning system** (GPS). Sailors and pilots use GPS receivers to navigate, automobile drivers use them to find their destinations, and geologists use them to monitor plate motions. If we calculate carefully enough, we can detect displacements of millimeters per year.

In other words, we can now see the plates move! Taking into account many data sources that define the motion of plates, geologists have greatly refined the image of continental drift that Wegener tried so hard to prove nearly a century ago. We can now see how the map of our planet's surface has evolved during the past 400 million years (▶Fig. 2.54), and even before.

FIGURE 2.53 Relative plate velocities: the blue arrows show the rate and direction at which the plate on one side of the boundary is moving with respect to the plate on the other side. Outward-pointing arrows indicate spreading (divergent boundaries), inward-pointing arrows indicate subduction (convergent boundaries), and parallel arrows show transform motion. The length of an arrow represents the velocity. Absolute plate velocities: the red arrows show the velocity of the plates with respect to a fixed point in the mantle.

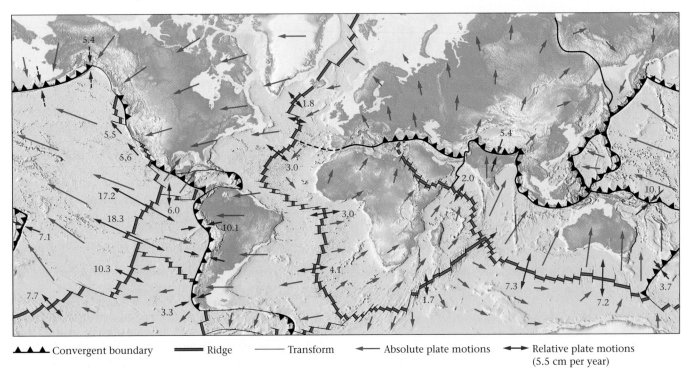

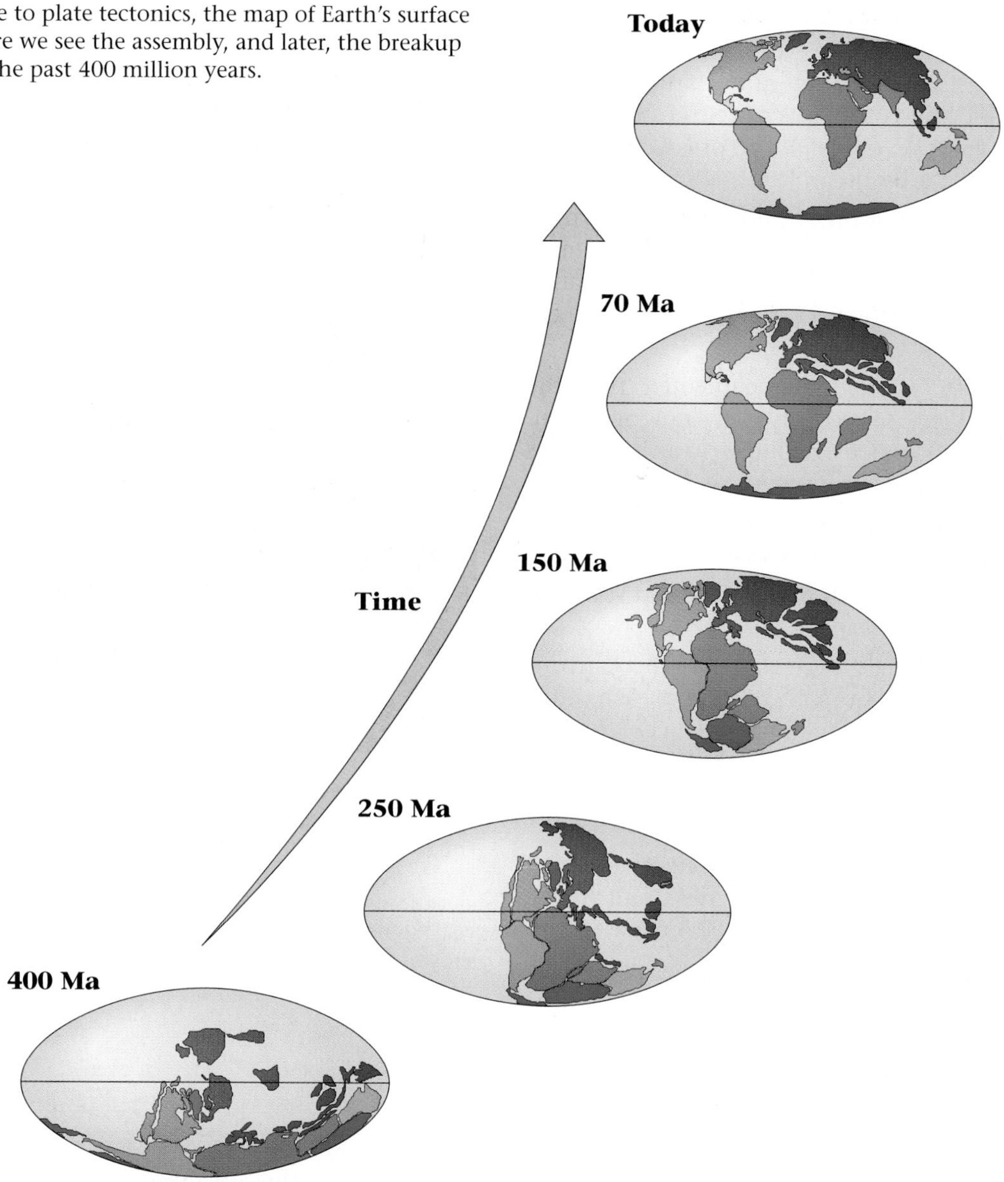

FIGURE 2.54 Due to plate tectonics, the map of Earth's surface slowly changes. Here we see the assembly, and later, the breakup of Pangaea during the past 400 million years.

Today

70 Ma

150 Ma

Time

250 Ma

400 Ma

CHAPTER SUMMARY

• Alfred Wegener proposed that continents had once been joined together to form a single huge supercontinent (Pangaea) and had subsequently drifted apart. This idea is the continental drift hypothesis.

• Wegener drew from several different sources of data to support his hypothesis: (1) coastlines on opposite sides of the ocean match up; (2) the distribution of late Paleozoic glaciers can be explained if the glaciers made up a polar ice cap over the southern end of Pangaea; (3) the distribution of late Paleozoic equatorial climatic belts is compatible with the concept of Pangaea; (4) the distribution of fossil species suggests the existence of a supercontinent; (5) distinctive rock assemblages that are now on opposite sides of the ocean were adjacent on Pangaea.

• Despite all the observations that supported continental drift, most geologists did not initially accept the idea because no one could explain *how* continents could move.

• Rocks retain a record of the Earth's magnetic field that existed at the time the rocks formed. This record is called paleomagnetism. By measuring paleomagnetism in successively older rocks, geologists found that the apparent position of the Earth's magnetic pole relative to the rocks changes through time. Successive positions of the pole define an apparent polar-wander path.

• Apparent polar-wander paths are different for different continents. This observation can be explained by continental drift: continents move with respect to each other, while the Earth's magnetic poles remain roughly fixed.

• Around 1960, Harry Hess proposed the hypothesis of sea-floor spreading. According to this hypothesis, new sea floor forms at mid-ocean ridges, above a band of upwelling mantle, then spreads symmetrically away from the ridge axis. As a consequence, an ocean can get progressively wider with time, and the continents on either side of the ocean basins drift apart. Eventually, the ocean floor sinks back into the mantle at deep-ocean trenches.

• Magnetometer surveys of the sea floor revealed marine magnetic anomalies. Positive anomalies, where the magnetic field strength is greater than expected, and negative anomalies, where the magnetic field strength is less than expected, are arranged in alternating stripes.

• During the 1950s, geologists documented that the Earth's magnetic field reverses polarity every now and then. The record of reversals, dated by radiometric techniques, is called the magnetic-reversal chronology.

• The proof of sea-floor spreading came from the interpretation of marine magnetic anomalies.

• Drilling of the sea floor confirmed its age and served as another proof of sea-floor spreading.

• The lithosphere, the rigid outer layer of the Earth, is broken into discrete plates that move relative to each other. Plates consist of the crust and the uppermost (cooler) mantle. Lithosphere plates effectively float on the underlying soft asthenosphere. Continental drift and sea-floor spreading are manifestations of plate movement.

• Most plate interactions occur along plate boundaries; the interiors of plates remain relatively rigid and intact. Earthquakes delineate the position of plate boundaries.

• There are three types of plate boundaries—divergent, convergent, and transform—distinguished from each other by the movement the plate on one side of the boundary makes relative to the plate on the other side.

• Divergent boundaries are marked by mid-ocean ridges. At divergent boundaries, sea-floor spreading takes place, a process that forms new oceanic lithosphere.

• Convergent boundaries are marked by deep-ocean trenches and volcanic arcs. At convergent boundaries, oceanic lithosphere of the downgoing plate is subducted beneath an overriding plate.

• Subducted lithosphere sinks back into the mantle. Its position can be tracked down to a depth of about 670 km by a belt of earthquakes known as a Wadati-Benioff zone.

• Transform boundaries, also called transform faults, are marked by large faults at which one plate slides sideways past another. No new plate forms and no old plate is consumed at a transform boundary.

• Triple junctions are points where three plate boundaries intersect.

• Hot spots are places where a plume of hot mantle rock rises and causes anomalous volcanism at an isolated volcano. As a plate moves over the hot spot, the volcano

THE VIEW FROM SPACE The Alps formed during the Cenozoic as a consequence of the collision between a "microplate" (including Italy) and the Eurasian plate. Such continental collisions build complex orogens. Because of the high elevation of its peaks, much of the Alps remains snow covered all year. Some Alpine valleys still contain glaciers. This image also shows the gentle curve of the Apennines on the Italian peninsula.

moves off the hot spot and dies, and a new volcano forms over the hot spot.

• A large continent can split into two smaller ones by the process of rifting. During rifting, continental lithosphere stretches and thins. If it finally breaks apart, a new mid-ocean ridge forms and sea-floor spreading begins.

• Convergent plate boundaries cease to exist when a buoyant piece of crust (a continent or an island arc) moves into the subduction zone. When that happens, collision occurs.

• Ridge-push force and slab-pull force contribute to driving plate motions. Plates move at rates of about 1–15 cm per year. Modern satellite measurements can detect these motions.

KEY TERMS

absolute plate velocity (p. 74)
abyssal plain (p. 45)
active margin (p. 53)

apparent polar-wander path (p. 43)
asthenosphere (p. 53)

black smoker (p. 58)
collision (pp. 68, 69)
continental drift (p. 35)
continental rift (p. 68)
convergent boundary (p. 56)
dipole (p. 41)
divergent boundary (p. 56)
global positioning system (p. 74)
hot spot (p. 65)
lithosphere (p. 53)
lithosphere plate (p. 53)
magnetic anomaly (p. 48)
magnetic pole (p. 40)
magnetic reversal (p. 49)
mantle plume (p. 66)
marine magnetic anomaly (p. 49)

mid-ocean ridge (pp. 45, 56)
paleomagnetism (p. 39)
paleopole (p. 43)
Pangaea (p. 35)
passive margin (p. 53)
plate boundary (p. 53)
plate tectonics (p. 36)
polarity (p. 40)
relative plate velocity (p. 71)
rifting (p. 68)
sea-floor spreading (pp. 36, 48)
subduction (pp. 36, 59)
transform boundary (p. 56)
trench (pp. 45, 59)
triple junction (p. 64)
volcanic arc (pp. 45, 62)

REVIEW QUESTIONS

1. What was Wegener's continental drift hypothesis? What was his evidence?
2. Was it possible for a dinosaur to walk from New York to Paris when Pangaea existed? Explain your answer.
3. How could paleomagnetic inclination be used to determine the ancient latitude of a continent?
4. How do apparent polar-wander paths show that the continents, rather than the poles, have moved?
5. Describe the basic characteristics of mid-ocean ridges, deep-ocean trenches, and seamount chains.
6. Describe the hypothesis of sea-floor spreading.
7. How did the observations of heat flow and seismicity support the hypothesis of sea-floor spreading.
8. What is a marine magnetic anomaly? How is it detected?
9. Describe the pattern of marine magnetic anomalies across a mid-ocean ridge. How is this pattern explained?
10. Did drilling into the sea floor contribute further proof of sea-floor spreading? If so, how?
11. What are the characteristics of a lithosphere plate?
12. How does oceanic lithosphere differ from continental lithosphere in thickness, composition, and density?
13. What are the basic premises of plate tectonics?
14. How do we identify a plate boundary?
15. Describe the three types of plate boundaries.
16. How does crust form along a mid-ocean ridge?
17. Why is subduction necessary on a nonexpanding Earth with spreading ridges?
18. Describe the major features of a convergent boundary.
19. Why are transform plate boundaries required on an Earth with spreading and subducting plate boundaries?
20. What is a triple junction?
21. How is a hot-spot track produced, and how can hot-spot tracks be used to track the past motions of a plate?
22. Describe the characteristics of a contintental rift, and give examples of where this process is occurring today.
23. Describe the process of continental collision, and give examples of where this process has occurred.
24. Discuss the major forces that move lithosphere plates.
25. Explain the difference between relative plate velocity and absolute plate velocity.

SUGGESTED READING

Allègre, C. 1988. *The Behavior of the Earth: Continental and Seafloor Mobility.* Cambridge, Mass.: Harvard University Press.

Butler, R. F. 1992. *Paleomagnetism: Magnetic Domains to Geologic Terranes.* Boston: Blackwell.

Condie, K. C. 2001. *Mantle Plumes & Their Record in Earth History.* Cambridge, UK: Cambridge University Press.

Condie, K. C. 2005. *Earth as an Evolving Planetary System.* Burlington, Mass.: Academic Press.

Cox, A., and R. B. Hart. 1986. *Plate Tectonics: How It Works.* Palo Alto, Calif.: Blackwell.

Glen, W. 1982. *The Road to Jaramillo: Critical Years of the Revolution in Earth Sciences.* Palo Alto, Calif.: Stanford University Press.

Kearey, P., and F. J. Vine. 1996. *Global Tectonics,* 2nd ed. Cambridge, Mass.: Blackwell.

LeGrand, H. E. 1988. *Drifting Continents and Shifting Theories.* Cambridge, UK: Cambridge University Press.

McFadden, P. L., and M. W. McElhinny. 2000. *Paleomagnetism: Continents and Oceans,* 2nd ed. San Diego: Academic Press.

McPhee, J. A. 1998. *Annals of the Former World.* New York: Farrar, Straus, and Giroux.

Moores, E. M., and R. J. Twiss. 1995. *Tectonics.* New York: Freeman.

Oreskes, N., ed. 2003. *Plate Tectonics: An Insider's History of the Modern Theory of the Earth.* Boulder: Westview Press.

Sullivan, W. 1991. *Continents in Motion: The New Earth Debate,* 2nd ed. New York: American Institute of Physics.

CHAPTER 3

Patterns in Nature: Minerals

This photo is real, not a computer collage! We're seeing the world's largest known mineral crystals jutting from the walls of a cave in Chihuahua, Mexico. The crystals are of the mineral gypsum; they formed by precipitation from a water solution.

3.1 INTRODUCTION

Zabargad Island rises barren and brown above the Red Sea, about 70 km off the coast of southern Egypt. Nothing grows on Zabargad, except for scruffy grass and a few shrubs, so no one lives there now. But in ancient times many workers toiled on this 5-square-km patch of desert, gradually chipping their way into the side of its highest hill. They hoped to find glassy green, pea-sized pieces of peridot, a prized gem. Carefully polished peridots were worn as jewelry by ancient Egyptians and may have been buried with them when they died. Eventually, some of the gems appeared in Europe, where jewelers set them into crowns and scepters (▶Fig. 3.1). These peridots now glitter behind glass cases in museums, millennia after first being pried free from the Earth, and perhaps 10 million years after first being formed by the bonding together of still more ancient atoms.

Peridot is one of about 4,000 minerals that have been identified on Earth, so far, and fascinate collectors and geologists alike. Mineralogists, people who specialize in the study of minerals, discover fifty to one hundred new minerals every year. Each different mineral has a name. Some names come from Latin, Greek, German, or English words describing a certain characteristic (for example, albite comes from the Latin word for

FIGURE 3.1 A royal crown containing a variety of valuable jewels. The large gemstone near the base of the crown is a green peridot.

note the basic scheme that geologists use to classify minerals. This chapter assumes that you understand the fundamental concepts of matter and energy, especially the nature of atoms, molecules, and chemical bonds. If you are rusty on these topics, please review the Appendix. We summarize basic terms from chemistry in ▶Box 3.1, for convenience.

3.2 WHAT IS A MINERAL?

To geologists, a **mineral** is a homogeneous, naturally occurring, solid substance with a definable chemical composition and an internal structure characterized by an orderly

FIGURE 3.2 (a) Museum specimen of malachite, a bright-green mineral containing copper (its formula is $Cu_2[CO_3][OH]_2$). Malachite is a mineral mined to produce copper, but because of its beauty, it is also used for jewelry. (b) Copper wire made by the processing of malachite and other copper-containing minerals.

(a)

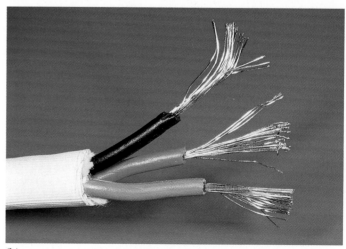

(b)

white, orthoclase comes from the German words meaning splits at right angles, and olivine is olive-colored); some honor a person (sillimanite was named for Benjamin Silliman, a famous nineteenth-century mineralogist); some indicate the place where the mineral was first recognized (illite was first identified in rocks from Illinois); and some reflect a particular element in the mineral (chromite contains chromium). Several minerals have more than one name—for example, peridot is the gem-quality version of a common mineral named olivine. Although the vast majority of mineral types are rare, forming only under special conditions, many are quite common and are found in a variety of rock types at Earth's surface.

Why study minerals? Without exaggeration, we can say that minerals are the building blocks of our planet. To a geologist, almost any study of Earth materials depends on an understanding of minerals, for minerals make up the rocks and sediments that make up the Earth and its landscapes. But minerals are also important from a practical standpoint. Industrial minerals serve as the raw materials for manufacturing chemicals, concrete, and wallboard. Ore minerals are the source of valuable metals like copper and gold and provide energy resources like uranium (▶Fig. 3.2a, b). Certain forms of minerals—gems—delight the eye as jewelry. Unfortunately, not all minerals are beneficial; some pose environmental hazards. No wonder **mineralogy,** the study of minerals, fascinates professionals and amateurs alike.

In this chapter, we begin with the geologic definition of a mineral, then look at how minerals form and at the main characteristics that enable us to identify them. Finally, we

arrangement of atoms in a crystalline structure. Almost all minerals are inorganic. Let's pull apart this mouthful of a definition and examine what its components actually mean.

- *Homogeneous:* Homogeneous materials are the same through and through—they cannot be physically broken into simpler components. When you smash a mineral specimen with a hammer, you get many tiny fragments of the same mineral.

- *Naturally occurring:* True minerals form by geological processes, not by the activity of humans. In recent decades, laboratory scientists have learned how to manufacture materials that are essentially identical to naturally occurring minerals. For example, some companies routinely manufacture diamonds by squeezing carbon under very high pressure. We sometimes call such materials "synthetic minerals," or "synthetic compounds" to distinguish them from true minerals.

- *Solid:* A solid is a form of matter that can maintain its shape indefinitely, and thus will not conform to the shape of its container. Liquids (like oil or water) and gases (like air) are not minerals.

- *Definable chemical composition:* This simply means that it is possible to write a chemical formula for a mineral (see Box 3.1). For example, the minerals diamond and graphite have the formula C because they consist entirely of carbon. Quartz has the formula SiO_2—it contains the elements silicon and oxygen in the proportion of one silicon atom for every two oxygen atoms. Some formulas are more complicated: for example, the formula for biotite is $K(Mg,Fe)_3(AlSi_3O_{10})(OH)_2$. Although many minerals, like quartz, have only one composition, others have compositions that can vary slightly.

- *Orderly arrangement of atoms:* The atoms that make up a mineral are not distributed randomly and cannot move

---

BOX 3.1
SCIENCE TOOLBOX

Some Basic Definitions from Chemistry

To describe minerals, we need to use several terms from chemistry (for a more in-depth discussion, see the Appendix). To avoid confusion, terms are listed in an order that permits each successive term to utilize previous terms.

- **Element:** a pure substance that cannot be separated into other elements.

- **Atom:** the smallest piece of an element that retains the characteristics of the element. An atom consists of a nucleus surrounded by a cloud of orbiting electrons; the nucleus is made up of protons and neutrons (except in hydrogen, whose nucleus contains only one proton and no neutrons). Electrons have a negative charge, protons have a positive charge, and neutrons have a neutral charge. An atom that has the same number of electrons as protons is said to be "neutral," in that it does not have an overall electrical charge.

- **Atomic number:** the number of protons in an atom of an element.

- **Atomic weight:** approximately the number of protons plus neutrons in an atom of an element.

- **Ion:** an atom that is not neutral. An ion that has an excess negative charge (because it has more electrons than protons) is an anion, whereas an ion that has an excess positive charge (because it has more protons than electrons) is a cation. We indicate the charge with a superscript. For example, Cl^- has a single excess electron; Fe^{2+} is missing two electrons.

- **Chemical bond:** an attractive force that holds two or more atoms together. For example, *covalent bonds* form when atoms share electrons. *Ionic bonds* form when a cation and anion (ions with opposite charges) get close together and attract each other. In materials with metallic bonds, some of the electrons can move freely.

- **Molecule:** two or more atoms bonded together. The atoms may be of the same element or of different elements.

- **Compound:** a pure substance that can be subdivided into two or more elements. The smallest piece of a compound that retains the characteristics of the compound is a molecule.

- **Chemical:** a general name used for a pure substance (either an element or a compound).

- **Chemical formula:** a shorthand recipe that itemizes the various elements in a chemical and specifies their relative proportions. For example, the formula for water, H_2O, indicates that water consists of molecules in which two hydrogens bond to one oxygen.

- **Chemical reaction:** a process that involves the breaking or forming of chemical bonds. Chemical reactions can break molecules apart or create new molecules and/or isolated atoms.

- **Mixture:** a combination of two or more elements or compounds that can be separated without a chemical reaction. For example, a cereal composed of bran flakes and raisins is a mixture—you can separate the raisins from the flakes without destroying either.

- **Solution:** a type of material in which one chemical (the solute) dissolves (becomes completely incorporated) in another (the solvent). In solutions, a solute may separate into ions during the process. For example, when salt (NaCl) dissolves in water, it separates into sodium (Na^+) and chloride (Cl^-) ions. In a solution, atoms or molecules of the solvent surround atoms, ions, or molecules of the solute.

- **Precipitate:** (noun) a compound that forms when ions in liquid solution join together to create a solid that settles out of the solution; (verb) the process of forming solid grains by separation and settling from a solution. For example, when saltwater evaporates, solid salt crystals precipitate and settle to the bottom of the remaining water.

around easily. Rather, they are fixed in a specific pattern that repeats itself over a very large region, relative to the size of atoms. A material in which atoms are fixed in an orderly pattern is called a crystalline solid. Mineralogists refer to the pattern itself (the imaginary framework representing the arrangement of atoms) as a **crystal lattice** (▶Fig. 3.3a, b).

• *Generally inorganic:* Most, but not all, minerals are inorganic, by which we mean that most do not contain organic chemicals. An organic chemical consists of carbon bonded to hydrogen, and in some cases varying amounts of oxygen, nitrogen, and other elements. Despite the name "organic," not all organic materials form in nature—plastic, for example, consists of organic chemicals. In this context, materials such as plastic, plant debris, sugar, fat, and protein are not minerals. Note that the presence of carbon alone does not make a mineral organic. For example, diamond and graphite are minerals composed of pure carbon, and calcite is a mineral composed of carbon bonded to oxygen and calcium—neither is organic.

With these definitions in mind, we can make an important distinction between a mineral and glass. Both minerals and glasses are solids, in that they can retain their shape indefinitely (see Appendix). But a mineral is crystalline, while glass is not. Whereas atoms, ions, or molecules in a mineral are ordered into a crystal lattice, like soldiers standing in formation, those in a glass are arranged in a semichaotic way, like a crowd of people at a party, in small clusters or chains that are neither oriented in the same way nor spaced at regular intervals (▶Fig. 3.3c, d).

If you ever need to figure out whether a substance is a mineral or not, just check it against the criteria listed above. Is motor oil a mineral? No—it's a liquid. Is table salt a mineral? Yes—it's a solid crystalline compound with the formula NaCl.

3.3 BEAUTY IN PATTERNS: CRYSTALS AND THEIR STRUCTURE

What Is a Crystal?

The word *crystal* brings to mind sparkling chandeliers, elegant wine goblets, and shiny jewels. But, as is the case with the word *mineral,* geologists have a more precise definition. A **crystal** is a single, continuous (that is, uninterrupted) piece of a crystalline solid, typically bounded by flat surfaces called **crystal faces** that grow naturally as the mineral forms. The word comes from the Greek *krystallos,* meaning ice. Many crystals have beautiful shapes that look like they belong in the pages of a geometry book. The angle between two adja-

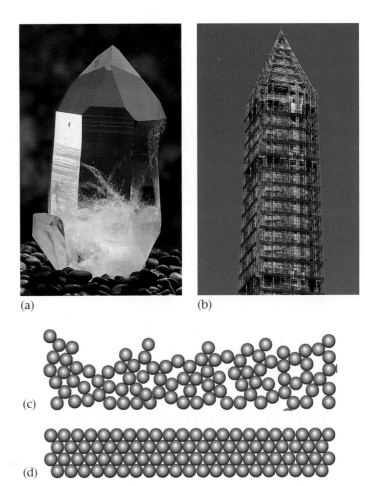

(a)   (b)

(c)

(d)

FIGURE 3.3 (a) Internally, this quartz crystal contains an orderly arrangement of atoms. (b) This gridwork of scaffolding surrounding the Washington Monument in Washington, D.C., provides an analogy for the fixed arrangements of atoms in a mineral. (c) Disordered atoms, as occur in glass, do not define a regular pattern. (d) Ordered atoms like these are found in a mineral.

cent crystal faces of one specimen is identical to the angle between the corresponding faces of another specimen. For example, a perfectly formed quartz crystal looks like an obelisk (▶Fig. 3.4a). The angle between the faces of the columnar part of a quartz crystal is exactly 120°. This rule holds regardless of whether the whole crystal is big or small and regardless of whether all of the faces are the same size (▶Fig. 3.4b). Crystals come in a great variety of shapes including cubes, trapezoids, pyramids, octahedrons, hexagonal columns, blades, needles, columns, and obelisks (▶Fig. 3.5a–h).

Because crystals have a regular geometric form, people have always considered crystals to be special, perhaps even a source of magical powers. For example, shamans of some cultures relied on talismans or amulets made of crystals, which supposedly brought power to their wearer or warded off evil spirits. In recent years, it has become fashionable for people to wear crystals around their necks with the thought that

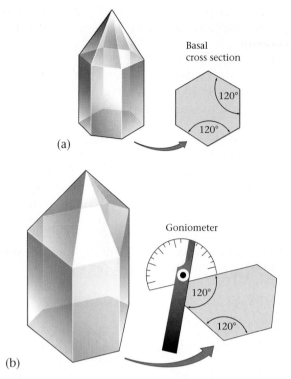

FIGURE 3.4 For a given mineral, the angle between two adjacent crystal faces in one specimen is the same as the angle between corresponding faces in another. (a) A small crystal of quartz whose vertical crystal faces happen to have the same size. The intersection between crystal faces makes an angle of 120°, as shown by the cross-section slice through the crystal. (b) A large crystal of quartz whose vertical crystal faces do not all have the same size. Even though the dimensions of the faces differ from one another, the angle between the faces is still 120°, as measured by a goniometer (an instrument that measures angles).

crystals can somehow channel the "life force" of the universe into a person's soul. Scientists have concluded, however, that crystals have no demonstrable effect on health or mood. For millennia, crystals have inspired awe because of the way they sparkle, but such behavior is simply a consequence of how crystal structures interact with light.

What's Inside a Crystal?

The scientific question of why crystals display the shapes and properties that they do has been a focus of study for over three centuries. In 1912 researchers discovered the key to this mystery. They showed that an X-ray beam passing through a crystal undergoes diffraction, meaning that it splits into tiny beams whose waves interfere with one another. The interference creates a pattern of spots on a projection screen (▶Fig. 3.6). Researchers knew

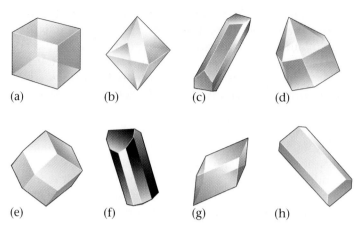

FIGURE 3.5 Crystals come in all kinds of shapes. Some are double pyramids, some are cubes, and some have blade shapes. Some crystals terminate at a point, and some terminate in a chisel-like wedge. (a) Halite, (b) diamond, (c) staurolite, (d) quartz, (e) garnet, (f) stibnite, (g) calcite, and (h) kyanite.

that diffraction occurs when electromagnetic radiation (light or X-rays) passes through a material composed of regularly spaced obstacles if the distance between the obstacles is comparable to the wavelength of the radiation. Thus, the observation of X-ray diffraction in crystals showed that crystals have an orderly internal arrangement of atoms or ions.

What do the insides of a mineral actually look like? We can picture atoms in minerals as tiny balls packed together tightly and held in place by chemical bonds. The way in which atoms are packed defines the **crystal structure** of the mineral.

To illustrate crystal structures, we look at a few examples. Halite (rock salt) consists of oppositely charged ions

FIGURE 3.6 A single X-ray beam undergoes diffraction when it passes through a crystal. As a consequence, it appears to divide into numerous beams that project as dots on a screen. X-rays interact this way with a crystal because of the crystal's orderly arrangement of atoms.

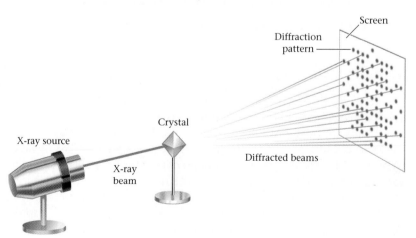

(atoms that have gained or lost electrons) that stick together because opposite charges attract. In halite, six chloride (Cl^-) ions surround each sodium (Na^+) ion, producing an overall arrangement of atoms that defines the shape of a cube (▶Fig. 3.7a, b). Diamond, by contrast, is a mineral made entirely of carbon. In diamond, each atom bonds to four neighbors arranged in the form a tetrahedron, so some naturally formed diamond crystals have the shape of a double tetrahedron (▶Fig. 3.7c, d). Graphite, another mineral composed entirely of carbon, behaves very differently from diamond. In contrast to diamond, graphite is so soft that it can be used as the "lead" in a pencil; as you move a pencil across paper, tiny flakes of graphite peel off the pencil point and adhere to the paper. This behavior occurs because the carbon atoms in graphite are not arranged in tetrahedra, but rather occur in sheets (▶Fig. 3.7e, f). The sheets are bonded to each other by weak bonds and thus can separate from each other easily.

Two different minerals (such as diamond and graphite) that have the same composition but different crystal structures are **polymorphs.**

The orderly arrangement of atoms inside a crystal—its crystal structure—provides one of nature's most spectacular examples of a pattern. The pattern on a sheet of wallpaper may be defined by the regular spacing of, say, clumps of flowers. Similarly, the pattern in a crystal is defined by the regular spacing of atoms (▶Fig. 3.8a, b). If the crystal contains more than one type of atom, the atoms alternate in a regular way. The orderly arrangement controls the outward shape, or morphology, of crystals. For example, if the atoms in a mineral are packed into the shape of a cube, a crystal of the mineral will have faces that intersect at 90° angles.

The pattern of atoms or ions in a mineral displays **symmetry,** meaning that the shape of one part of a mineral is a

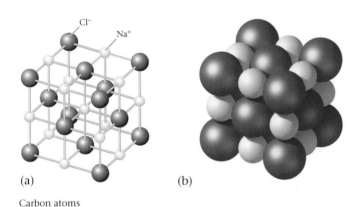

(a)   (b)

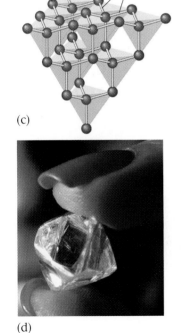

(c)

(d)

FIGURE 3.7 Minerals are composed of atoms that stick together by chemical bonds. (a) A ball-and-stick model of halite, composed of ionically bonded ions of sodium and chloride. The sticks represent bonds, and the balls represent atoms. (b) A ball model of halite. The different sizes of balls represent the relative sizes of ions. Note that the smaller sodium ions fit in between the larger chloride ions. (c) A ball-and-stick model of diamond, composed of covalently bonded atoms of carbon arranged in a tetrahedron. These bonds are very strong. (d) A photograph of a diamond crystal. (e) A ball-and-stick model of graphite. Note that the carbon atoms are arranged in sheets of hexagons; the sheets are held together by weak bonds. (f) A photograph of a graphite crystal.

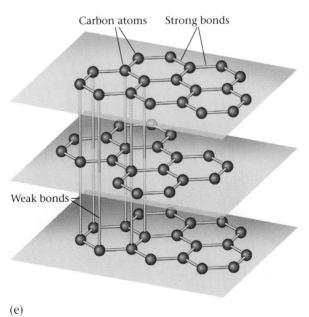

(e)

(f)

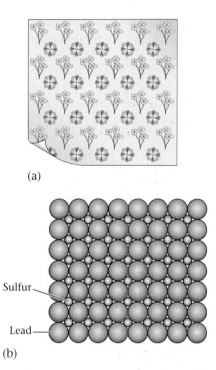

(a)

Sulfur

Lead

(b)

FIGURE 3.8 (a) Repetition of a flower motif in wallpaper illustrates a regular pattern. (b) On the face of a crystal of galena (a type of lead ore), lead and sulfur atoms pack together in a regular array.

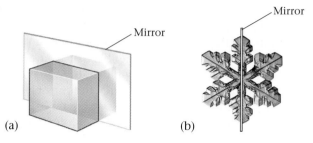

(a)   (b)

FIGURE 3.9 (a) Crystals have symmetry: one half of a halite crystal is a mirror image of the other half. (b) Snowflakes, crystals of ice, are symmetrical hexagons.

mirror image of the shape of another part. For example, if you were to cut a halite crystal in half and place one half against a mirror, the crystal would look whole again (▶Fig. 3.9a, b).

The Formation and Destruction of Minerals

New mineral crystals can form in five ways. First, they can form by the solidification of a melt, meaning the freezing of a liquid. For example, ice crystals, a type of mineral, are made by freezing water. Second, they can form by precipitation from a solution, meaning that atoms, molecules, or ions dissolved in water bond together and separate out of the water. Salt crystals, for example, develop when you evaporate saltwater. Third, they can form by solid-state diffusion, the movement of atoms or ions through a solid to arrange into a new crystal structure, a process that takes place very slowly. Fourth, minerals can form at interfaces between the physical and biological components of the Earth System by a process called biomineralization. This occurs when living organisms cause minerals to precipitate either within or on their bodies, or immediately adjacent to their bodies. For example, clams and other shelled organisms extract ions from water to produce mineral shells. Fifth, minerals can form directly from a vapor. This process, called fumerolic mineralization, typically occurs around volcanic vents or around geysers, for at such locations volcanic gases or steam enter the atmosphere and cool abruptly. Some of the bright yellow sulfur deposits found in volcanic regions form in this way.

The first step in forming a crystal is the chance formation of a seed, or an extremely small crystal (▶Fig. 3.10a–c).

FIGURE 3.10 (a) New crystals nucleate (begin to form) in a water solution. They grow inward from the walls of the container. (b) At a later time, the crystals have grown larger. (c) On a crystal face, atoms in the solution are attracted to the surface and latch on.

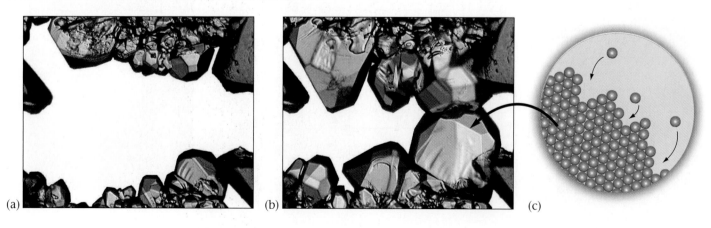

(a)   (b)   (c)

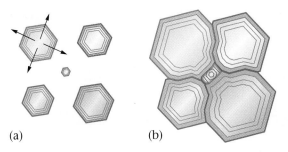

FIGURE 3.11 (a) Crystals grow outward from the central seed. (b) Crystals maintain their shape until they interfere with each other. When that happens, the crystal shapes can no longer be maintained.

Once the seed exists, other atoms in the surrounding material attach themselves to the face of the seed. As the crystal grows, crystal faces move outward from the center of the seed but maintain the same orientation. The youngest part of the crystal is always its outer edge (▶Fig. 3.11a, b).

In the case of crystals formed by the solidification of a melt, atoms begin to attach to the seed when the melt becomes so cool that thermal vibrations can no longer break apart the attraction between the seed and the atoms in the melt. Crystals formed by precipitation from a solution develop when the solution becomes saturated, meaning the number of dissolved atoms, ions, or molecules per unit volume of solution becomes so great that they can get close enough to each other to bond together. Sometimes crystals formed by precipitation from a solution grow from the walls of the solution's "container" (for example, a crack or pore in a rock). This process can form a spectacular **geode**, a mineral-lined cavity in rock formed when water solutions pass through the rock (▶Fig. 3.12a).

As crystals grow, they develop their particular crystal shape, based on the geometry of their internal structure. The shape is defined by the relative dimensions of the crystal (needlelike, sheetlike, etc.) and the angles between crystal faces. If a mineral's growth is uninhibited so that it displays well-formed crystal faces, then it is a euhedral crystal (▶Fig. 3.12b). Typically, however, the growth of minerals is restricted in one or more directions, because existing crystals act as obstacles. In such cases, minerals grow to fill the space that is available, and their shape is controlled by the shape of the surroundings. Minerals without well-formed crystal faces are anhedral grains (▶Fig. 3.13a, b).

A mineral can be destroyed by melting, dissolving, or some other chemical reaction. Melting involves heating a mineral to a temperature at which thermal vibration of the atoms or ions in the lattice break the chemical bonds holding them to the lattice. The atoms or ions then separate, either individually or in small groups, to move around again freely. Dissolution occurs when you immerse a mineral in a solvent, such as water. Atoms or ions then separate from the crystal face and are surrounded by solvent molecules. Chemical reactions can destroy a mineral when it comes in contact with reactive materials. For example, iron-bearing minerals react with air and water to form rust. The action of microbes in the environment can also destroy minerals. In effect, microbes "eat" certain minerals, using the energy stored in the chemical bonds that hold the atoms of the mineral together as their source of energy for metabolism.

3.4 HOW CAN YOU TELL ONE MINERAL FROM ANOTHER?

Amateur and professional geologists alike get a kick out of recognizing minerals. They'll be the show-offs in a museum who hover around the display case naming specimens without

FIGURE 3.12 (a) A geode, in which euhedral crystals of purple quartz (amethyst) grow from the wall into the center. (b) An enlargement of a euhedral crystal, showing that the surfaces are crystal faces.

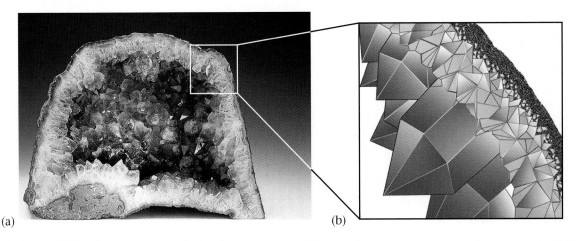

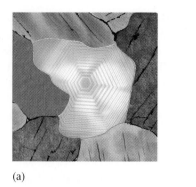

(a)   (b)

FIGURE 3.13 A crystal growing in a confined space is anhedral, meaning its surface is not composed of crystal faces. (a) A crystal stops growing when it meets the surfaces of other grains and continues growing to fill in gaps. (b) The resulting mineral grain, if it were to be separated from other grains, would have an anhedral shape.

FIGURE 3.14 The range of colors of quartz, displayed by different crystals: milky, clear, and rose quartz.

bothering to look at the labels. How do they do it? The trick lies in learning to recognize the basic physical properties (visual and material characteristics) that distinguish one mineral from another. Some physical properties, such as shape and color, can be seen from a distance. Others, such as hardness and magnetization, can be determined only by handling the specimen or by performing an identification test on it. Identification tests include scratching the mineral against another object, placing it near a magnet, weighing it, tasting it, or placing a drop of acid on it. Let's examine some of the physical properties most commonly used in mineral identification.

- *Color:* Color results from the way a mineral interacts with light. Sunlight contains the whole spectrum of colors; each color has a different wavelength. A mineral absorbs certain wavelengths, so the color you see when looking at a specimen represents the wavelengths the mineral does not absorb. Certain minerals always have the same color, but many show a range of colors. For example, quartz can be clear, white, purple, gray (smoky), red (rose), or just about anything in between (▶Fig. 3.14). Color variations in a mineral reflect the presence of tiny amounts of impurities.

- *Streak:* The **streak** of a mineral refers to the color of a powder produced by pulverizing the mineral. You can obtain a streak by scraping the mineral against an unglazed ceramic plate (▶Fig. 3.15). The color of a mineral powder tends to be less variable than the color of a whole crystal, and thus provides a fairly reliable clue to a mineral's identity. Calcite, for example, always yields a white streak even though pieces of calcite may be white, pink, or clear.

- *Luster:* **Luster** refers to the way a mineral surface scatters light. Geoscientists describe luster simply by comparing the

appearance of the mineral with the appearance of a familiar substance. For example, minerals that look like metal have a metallic luster, whereas those that do not have a non-metallic luster—the adjectives are self-explanatory (▶Fig. 3.16a, b). Terms used for types of nonmetallic luster include silky, glassy, satiny, resinous, pearly, or earthy.

- *Hardness:* **Hardness** is a measure of the relative ability of a mineral to resist scratching, and therefore represents the resistance of bonds in the crystal structure to being broken. The atoms or ions in crystals of a hard mineral are more strongly bonded than those in a soft mineral. Hard minerals can scratch soft minerals, but soft minerals cannot scratch hard ones. Diamond, the hardest mineral known, can scratch anything, which is why it is used to cut glass. In the early 1800s, a mineralogist named Friedrich Mohs listed some minerals in sequence of relative hardness; a mineral with a

FIGURE 3.15 A streak plate, showing the red streak of hematite.

(a)

(b)

FIGURE 3.16 (a) This specimen of pyrite looks like a piece of metal because of its shiny gleam; we call this metallic luster. (b) These specimens of feldspar have a nonmetallic luster. The white one on the left is plagioclase, and the pink one on the right is orthoclase (potassium feldspar, or "K-spar").

TABLE 3.1 Mohs Hardness Scale

Mohs #	Mineral or Substance
10	Diamond
9	Corundum (ruby)
8	Topaz
7	Quartz
6.5	*Steel file*
6	Orthoclase (K-feldspar)
5.5	*Steel knife; glass*
5	Apatite
4	Fluorite
3.5	*Copper penny*
3	Calcite
2.5	*Fingernail*
2	Gypsum
1	Talc

hardness of 5 can scratch all minerals with a hardness of 5 or less. This list, the **Mohs hardness scale,** helps in mineral identification. When you use the scale (▶Table 3.1), it might help to compare the hardness of a mineral with a common item such as your fingernail, a penny, or a glass plate. Note that not all of the minerals in Table 3.1 are common or familiar.

• *Specific gravity:* **Specific gravity** represents the density of a mineral, as specified by the ratio between the weight of a volume of the mineral and the weight of an equal volume of water at 4°C. For example, one cubic centimeter of quartz has a weight of 2.65 grams, whereas one cubic centimeter of water has a weight of 1.00 gram. Thus, the specific gravity of quartz is 2.65. In practice, you can develop a feel for specific gravity by hefting minerals in your hands. A piece of galena (lead ore) "feels" heavier than a similar-sized piece of quartz.

• *Crystal habit:* The **crystal habit** of a mineral refers to the shape of a single crystal with well-formed crystal faces, or to the character of an aggregate of many well-formed crystals that grew together as a group (▶Fig. 3.17). The habit depends on the internal arrangement of atoms in the crystal. A description of habit generally includes adjectives that define relative dimensions of the crystal and the geometric shape of the crystal. For example, crystals that are roughly the same length in all directions are called equant or blocky, those that are much longer in one dimension than in others are columnar or needle-like, those shaped like sheets of paper are platy, and those shaped like knife blades are bladed.

• *Fracture and cleavage:* Different minerals fracture (break) in different ways, depending on the internal arrangement of atoms. If a mineral breaks to form distinct planar surfaces that have a specific orientation in relation to the crystal structure, then we say that the mineral has **cleavage** and we refer to each surface as a cleavage plane (▶Fig. 3.18a–e). Cleavage forms in directions where the bonds holding atoms together in the crystal are the weakest. Some minerals have one direction of cleavage. For example, mica has very weak bonds in one direction but strong bonds in the other two directions. Thus, it easily splits into parallel sheets; the surface of each sheet is a cleavage plane. Other minerals have two or three directions of cleavage that intersect at a specific angle. For example, halite has three sets of cleavage planes that intersect at right angles, so

(a)

(b)

FIGURE 3.17 Crystal habit refers to the shape or character of the mineral. (a) Kyanite generally occurs in clusters of blades. (b) A spray of needlelike, or fibrous, crystals.

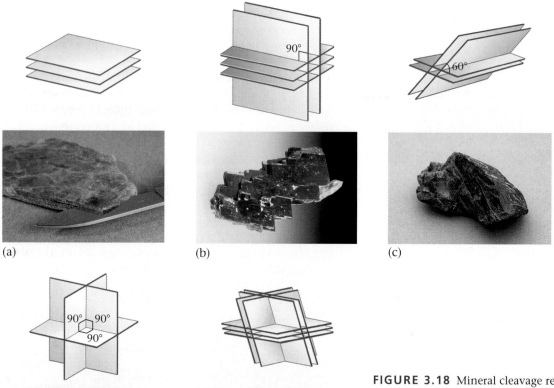

(a)

(b)

(c)

(d)

(e)

FIGURE 3.18 Mineral cleavage refers to the way a crystal breaks. Some crystals break in only one direction, some in two or three, and some in many. Others have no cleavage at all. (a) Muscovite has one direction of cleavage and splits into thin sheets. (b) Pyroxene has two directions of cleavage at right angles. (c) Amphibole has two directions of cleavage, where one direction makes an angle of 60° with respect to the other. (d) Halite has three directions of cleavage, all at right angles to each other. (e) Calcite has three directions of cleavage, one of which is inclined at an angle of less than 90°.

halite crystals break into little cubes. Materials that have no cleavage at all (because bonding is equally strong in all directions) break either by forming irregular fractures or by forming conchoidal fractures. **Conchoidal fractures** are smoothly curving, clamshell-shaped surfaces; they typically form in quartz and glass (►Fig. 3.19).

• *Special properties:* Some minerals have distinctive properties that readily distinguish them from other minerals. For example, calcite ($CaCO_3$) reacts with dilute hydrochloric acid (HCl) to produce carbon dioxide (CO_2) gas (►Fig. 3.20). Dolomite ($CaMg[CO_3]_2$) also reacts with acid, but not as strongly. Graphite makes a gray mark on paper (it's the "lead" in pencils), magnetite attracts a magnet, halite tastes salty, and plagioclase has striations (thin parallel corrugations or stripes) on cleavage planes.

FIGURE 3.20 Calcite reacts with hydrochloric acid to produce carbon dioxide gas.

3.5 ORGANIZING OUR KNOWLEDGE: MINERAL CLASSIFICATION

Close to 4,000 minerals are already known, and mineralogists discover 50 to 100 new minerals every year. Nevertheless, minerals can be separated into a small number of groups, or mineral classes. You may think, "Why bother?" Classification schemes are useful because they help organize information and streamline discussion. Biologists, for example, classify animals into groups based on how they feed their young and on the architecture of their skeletons, and botanists classify plants according to the way they reproduce

FIGURE 3.19 Minerals without cleavage break on random fractures. Some materials, such as quartz and glass, break on conchoidal fractures, which have a curving, scoop shape. The photo shows concoidal fracture surfaces on glass.

and by the shape of their leaves. In the case of minerals, a good means of classification eluded researchers until it became possible to determine the chemical makeup of minerals. A Swedish chemist, Baron Jöns Jacob Berzelius (1779–1848), analyzed minerals and noted chemical similarities among many of them. Berzelius, along with his students, then established that most minerals can be classified by specifying the principal anion (negative ion) within the mineral. We now take a look at these mineral groups, focusing especially on silicates, the class that constitutes most of the rock in the Earth.

The Mineral Classes

Mineralogists distinguish several principal classes of minerals, based on their chemical composition. Here are some of the major ones.

• *Silicates:* The fundamental component of most **silicates** in the Earth's crust is the SiO_4^{4-} anionic group. In the SiO_4^4 group, four oxygen atoms surround a single silicon atom. The oxygen atoms are arranged to define the corners of a tetrahedron, a pyramid-like shape with four triangular faces (►Fig. 3.21a–d). Mineralogists commonly refer to this anionic group as the **silicon-oxygen tetrahedron.** We can identify a huge variety of silicate minerals that differ from each other in the way the tetrahedra link and in the cations present in the mineral. We will learn more about silicates in the next section.

• *Oxides:* Oxides consist of metal cations bonded to oxygen anions. Typical oxide minerals include magnetite (Fe_3O_4) and hematite (Fe_2O_3).

• *Sulfides:* Sulfides consist of a metal cation bonded to a sulfide anion (S^{2-}). Examples include galena (PbS) and pyrite (FeS_2; Fig. 3.16a).

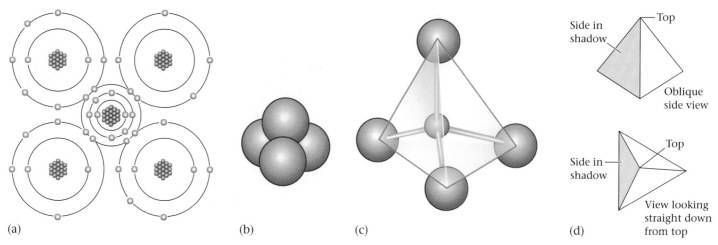

(a)   (b)   (c)   (d)

FIGURE 3.21 (a) The basic building block of a silicate mineral is the SiO_4^{4-} anionic group, also known as the silicon-oxygen tetrahedron. In this compound, the oxygens occupy the corners of a tetrahedron, which can be portrayed in different ways. (b) In this ball model, the small silicon atom is hidden, surrounded by larger oxygen atoms. (c) A tetrahedron has been superimposed on this ball-and-stick model. (d) Simple geometric sketches of quadrilaterals.

- *Sulfates:* Sulfates consist of a metal cation bonded to the SO_4^{2-} anionic group. Many sulfates form by precipitation out of water at or near the Earth's surface. An example is gypsum ($CaSO_4 \cdot 2H_2O$).

- *Halides:* The anion in a halide is a halogen ion (such as chloride [Cl^-] or fluoride [F^-]), an element from the second column from the right in the periodic table (see Appendix). Halite, or rock salt (NaCl; Fig. 3.18d), and fluorite (CaF_2), a source of fluoride, are common examples.

- *Carbonates:* In **carbonates,** the molecule CO_3^{2-} serves as the anionic group. Elements such as calcium or magnesium bond to this group. The two most common carbonates are calcite ($CaCO_3$; Fig. 3.18e) and dolomite ($CaMg[CO_3]_2$).

- *Native metals:* Native metals consist of pure masses of a single metal. The metal atoms are bonded by metallic bonds. Copper and gold, for example, may occur as native metals. A gold nugget is a mass of native gold that has been broken out of a rock.

Silicates: The Major Rock-Forming Minerals

Silicate minerals make up over 95% of the continental crust. Rocks of the oceanic crust and the Earth's mantle consist almost entirely of silicate minerals. Thus, silicate minerals are the most common minerals on Earth.

Most silicate minerals in the crust consist of combinations of a fundamental building block, the silicon-oxygen tetrahedron. Silica tetrahedra can link together, forming larger molecules, by sharing oxygen atoms. Silicate minerals are divided into seven groups, of which five are described below. Groups are distinguished from each other on the basis of how silicon-oxygen tetrahedra in the mineral link together. The number of links determines how many oxygen atoms are shared between tetrahedra and, therefore, the ratio of Si to O in the mineral.

- *Independent tetrahedra:* In this group, the tetrahedra are independent and do not share any oxygen atoms (▶Fig. 3.22a). The attraction between the tetrahedra and positive ions (metal cations, such as iron or magnesium) holds such minerals. This group includes olivine, a glassy green mineral that usually occurs in small, sugarlike crystals. Garnet is also a member of this group.

- *Single chains:* In a single-chain silicate, the tetrahedra link to form a chain by sharing two oxygen atoms (▶Fig. 3.22b). The most common of the many different types of single-chain silicates are pyroxenes, a group of black or dark-green minerals that occur in elongate crystals with two cleavage directions at 90° to each other (Fig. 3.18b).

- *Double chains:* In a double-chain silicate, the tetrahedra link to form a double chain by sharing two or three oxygen atoms (▶Fig. 3.22c). Amphiboles are the most common type; these typically are black or dark-brown elongate crystals, with two cleavage directions. You can distinguish amphiboles from pyroxenes because the cleavage planes of amphiboles lie at about 60° to each other (Fig. 3.18c).

- *Sheet silicates:* The tetrahedra in this group share three oxygen atoms and therefore link to form two-dimensional sheets (▶Fig. 3.22d). Other ions and, in some cases, water molecules fit between the sheets in some sheet silicates.

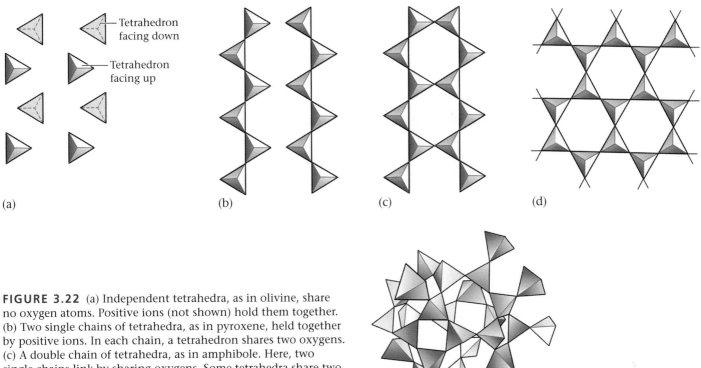

FIGURE 3.22 (a) Independent tetrahedra, as in olivine, share no oxygen atoms. Positive ions (not shown) hold them together. (b) Two single chains of tetrahedra, as in pyroxene, held together by positive ions. In each chain, a tetrahedron shares two oxygens. (c) A double chain of tetrahedra, as in amphibole. Here, two single chains link by sharing oxygens. Some tetrahedra share two oxygens, some share three. (d) A sheet of tetrahedra, as in mica. Each tetrahedron shares three oxygens. (e) A 3-D network (framework) of tetrahedra. Note that within the framework, each tetrahedron shares all four oxygens with its neighbors.

Because of their structure, sheet silicates have a single strong cleavage in one direction, and they occur in books of very thin sheets. In this group we find micas (see Fig. 3.18a). Clay minerals are also sheet silicates. Clays have a crystal structure similar to that of mica, but clays occur only in extremely tiny flakes.

• *Framework silicates:* In a framework silicate, each tetrahedron shares all four oxygen atoms with its neighbors, forming a 3-D structure (▶Fig. 3.22e). Examples include feldspar and quartz. The two most common feldspars are plagioclase, which tends to be white, gray, or blue; and orthoclase (also called potassium feldspar, or K-feldspar), which tends to be pink (see Fig. 3.16b).

3.6 SOMETHING PRECIOUS—GEMS!

Mystery and romance follow famous gems. Consider the stone now known as the Hope Diamond, recognized by name the world over. No one knows who first dug it out of the ground (▶Box 3.2). Was it mined in the 1600s, or was it stolen off an ancient religious monument? What we do know is that in the 1600s, a French trader named Jean Baptiste Tavernier obtained a large (112.5 carats, where 1 carat ≈ 200 milligrams), rare blue diamond in India, perhaps from a Hindu

statue, and carried it back to France. King Louis XIV bought the diamond and had it fashioned into a jewel of 68 carats. This jewel vanished in 1762 during a burglary. Perhaps it was lost forever—perhaps not. In 1830, a 44.5-carat blue diamond mysteriously appeared on the jewel market for sale. Henry Hope, a British banker, purchased the stone, which then became known as the Hope Diamond (▶Fig. 3.23). It changed

FIGURE 3.23 The Hope Diamond, now on display in the Smithsonian Institution, Washington, D.C.

BOX 3.2
THE REST OF THE STORY

Where Do Diamonds Come From?

As we saw earlier, diamond consists of the element carbon. Accumulations of carbon develop in a variety of ways: soot (pure carbon) results from burning plants at the surface of the Earth; coal (which consists mostly of carbon) forms from the remains of plants buried to depths of up to 15 km; and graphite develops from coal or other organic matter buried to still greater depths (15–70 km) in the crust during mountain building. Experiments demonstrate that the temperatures and pressures needed to form diamond are so extreme that, in nature, they generally occur only at depths of around 150 km below the Earth's surface—that is, in the mantle. Under these conditions, the carbon atoms that were arranged in hexagonal sheets in graphite rearrange to form the much stronger and more compact structure of diamond. (Of note, engineers can duplicate these conditions in the laboratory; corporations manufacture several tons of diamonds a year.)

How does carbon get down into the mantle, where it transforms into diamond? Geologists speculate that the process of subduction provides the means. Carbon-containing rocks and sediments in oceanic lithosphere plates at the Earth's surface can be carried into the mantle at a convergent plate boundary. This carbon transforms into diamond, some of which becomes trapped in the lithospheric mantle beneath continents. But if diamonds form in the mantle, then how do they return to the surface? One possibility is that the process of rifting cracks the continental crust and causes a small part of the underlying lithospheric mantle to melt. Magma generated during this process rises to the surface, bringing the diamonds with it. Near the surface, the magma cools and solidifies to form a special kind of igneous rock called kimberlite (named for Kimberley, South Africa, where it was first found). Diamonds brought up with the magma are frozen into the kimberlite. Kimberlite magma contains a lot of dissolved gas and thus froths to the surface very rapidly. Kimberlite rock commonly occurs in carrot-shaped bodies 50–200 m across and at least 1 km deep that are called kimberlite pipes (▶Fig. 3.24a).

Controversial measurements suggest that many of the diamonds that sparkle on engagement rings today were created when subduction carried carbon into the mantle 3.2 billion years ago. The diamonds sat at depths of 150 km in the Earth until two rifting events, one of which took place in the late Precambrian and the other during the late Mesozoic, released them to the surface, like genies out of a bottle. The Mesozoic rifting event led to the breakup of Pangaea.

In places where diamonds occur in solid kimberlite, they can be obtained only by digging up the kimberlite and crushing it, to separate out the diamonds (▶Fig. 3.24b). But nature can also break diamonds free from the Earth. In places where kimberlite has been exposed at the ground surface for a long time, the rock chemically reacts with water and air (a process called weathering; see Chapter 5). These reactions cause most minerals in kimberlite to disintegrate, creating sediment that washes away in rivers. Diamonds are so strong that they remain as solid grains in river gravel. Thus, many diamonds have been obtained simply by separating them from recent or ancient river gravel.

Diamond-bearing kimberlite pipes occur in many places around the world, particularly where very old continental lithosphere exists. Southern and central Africa, Siberia, northwestern Canada, India, Brazil, Borneo, Australia, and the U.S. Rocky Mountains all have pipes. Rivers and glaciers, however, have transported diamond-bearing sediments great distances from their original sources. In fact, diamonds have even been found in farm fields of the midwestern United States. Not all natural diamonds are valuable: value depends on color and clarity. Diamonds that contain imperfections (cracks, or specks of other material), or are dark gray in color, won't be used for jewelry. These stones, called industrial diamonds, are used instead as abrasives, for diamond powder is so hard (10 on the Mohs hardness scale) that it can be used to grind away any other substance.

Gem-quality diamonds come in a range of sizes. Jewelers measure diamond size in carats, where one "carat" equals 200 milligrams (0.2 grams)—one ounce equals 142 carats. (Note that a carat measures gemstone weight, whereas a "karat" specifies the purity of gold. Pure gold is 24 karat, whereas 18-karat gold is an alloy containing 18 parts of gold and 6 parts of other metals.) The largest diamond ever found, a stone called the Cullinan Diamond, was discovered in South Africa in 1905. It weighed 3,106 carats (621 grams) before being cut into nine large gems (the largest weighing 516 carats) as well as many smaller ones. By comparison, the diamond on a typical engagement ring weighs less than 1 carat. Diamonds are rare, but not as rare as their price suggests. A worldwide consortium of diamond producers stockpile the stones so as not to flood the market and drive the price down.

FIGURE 3.24 (a) A mine in a kimberlite pipe. (b) A raw diamond still imbedded in kimberlite.

(a)   (b)

hands several times until 1958, when the famous New York jeweler Harry Winston donated it to the Smithsonian Institution in Washington, D.C., where it now sits behind bulletproof glass in a heavily guarded display. Legend has it that whoever owns the stone suffers great misfortune.

What makes stones such as the Hope Diamond so special that people risk life and fortune to obtain them? What is the difference between a gemstone, a gem, and any other mineral? A gemstone is a mineral that has special value because it is rare and people consider it beautiful. A **gem** is a cut and finished stone ready to be set in jewelry. Jewelers distinguish between precious stones (such as diamond, ruby, sapphire, emerald), which are particularly rare and expensive, and semiprecious stones (such as topaz, tourmaline, aquamarine, garnet), which are less rare and less expensive. The category of semiprecious stones also includes opaque or translucent minerals such as lapis, malachite (Fig. 3.2a), and opal.

In everyday language, pearls and amber may also be considered gemstones. Unlike diamonds and garnets, which form inorganically in rocks, pearls form in living oysters when the oyster extracts calcium and carbonate ions from water and precipitates them around an impurity, such as a sand grain, embedded in its body. Thus, pearls are a result of biomineralization. Most pearls used in jewelry today are "cultured" pearls, made by artificially introducing round sand grains into oysters in order to stimulate pearl production. Amber is also formed by organic processes—it consists of fossilized tree sap. But because amber consists of organic compounds that are not arranged in a crystal structure, it does not meet the definition of a mineral.

In some cases, gemstones are merely pretty and rare versions of more common minerals. For example, ruby is a special version of the common mineral corundum, and emerald is a special version of the common mineral beryl (▶Fig. 3.25). As for the beauty of a gemstone, this quality lies basically in its color and, in the case of transparent gems, its *"fire"*—the way the stone bends and internally reflects the light passing through it, and disperses the light into a spectrum. Fire makes a diamond sparkle more than a similarly cut piece of glass.

Gemstones form in many ways. Some solidify from a melt along with other minerals of igneous rock, some form by diffusion in a metamorphic rock, some precipitate out of a water solution in cracks, and some are a consequence of the chemical interaction of rock with water near the Earth's surface. Many gems come from pegmatites, particularly coarse-grained igneous rocks formed by the solidification of steamy melt.

Most gems when used in jewelry are "cut" stones. The smooth **facets** on a gem are ground and polished surfaces made with a faceting machine (▶Fig. 3.26a). Facets are not the natural crystal faces of the mineral, nor are they cleavage planes, though gem cutters sometimes make the facets

FIGURE 3.25 Beryl crystals in rock.

parallel to cleavage directions and will try to break a large gemstone into smaller pieces by splitting it on a cleavage plane. A faceting machine consists of a doping arm, a device that holds a stone in a specific orientation; and a lap, a rotating disk covered with a wet paste of grinding powder and water. The gem cutter fixes a gemstone to the end of the doping arm and positions the arm so that it holds the stone against the moving lap. The movement of the lap grinds a facet. When the facet is complete, the gem cutter rotates the arm by a specific angle, lowers the stone, and grinds another facet. The geometry of the facets defines the cut of the stone. Different cuts have names, such as "brilliant," "French," "star," "pear," and "kite." Grinding facets is a lot of work—a typical engagement-ring diamond with a brilliant cut has fifty-seven facets (▶Fig. 3.26b)!

FIGURE 3.26 The shiny faces on gems in jewelry are made by a faceting machine. (a) In this faceting machine, the gem is held against the face of the spinning lap. (b) Top and side views show the many facets of a brilliant-cut diamond, and names for different parts of the stone.

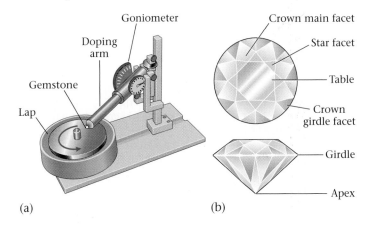

(a)

(b)

CHAPTER SUMMARY

• Minerals are homogeneous, naturally occurring, solid substances with a definable chemical composition and an internal structure characterized by an orderly arrangement of atoms, ions, or molecules in a lattice. Most minerals are inorganic.

• In the crystalline lattice of minerals, atoms occur in a specific pattern—one of nature's finest examples of ordering.

• Minerals can form by the solidification of a melt, precipitation from a water solution, diffusion through a solid, the metabolism of organisms, and precipitation from a gas.

• Close to 4,000 different types of minerals are known, each with a name and distinctive physical properties (color, streak, luster, hardness, specific gravity, crystal form, crystal habit, and cleavage).

• The unique physical properties of a mineral reflect its chemical composition and crystal structure. By observing these physical properties, you can identify minerals.

• The most convenient way for classifying minerals is to group them according to their chemical composition. Mineral classes include: silicates, oxides, sulfides, sulfates, halides, carbonates, and native metals.

• The silicate minerals are the most common on Earth. The silicon-oxygen tetrahedron, a silicon atom surrounded by four oxygen atoms, is the fundamental building block of silicate minerals.

• Groups of silicate minerals are distinguished from each other by the ways in which the silicon-oxygen tetrahedra that constitute them are linked.

• Gems are minerals known for their beauty and rarity. The facets on cut stones used in jewelry are made by grinding and polishing the stone with a faceting machine.

KEY TERMS

carbonates (p. 90)
cleavage (p. 87)
conchoidal fracture (p. 89)
crystal (p. 81)
crystal face (p. 81)
crystal habit (p. 87)
crystal lattice (p. 81)
crystal structure (p. 82)
facet (p. 93)
gem (p. 93)
geode (p. 85)
hardness (p. 86)
luster (p. 86)

mineral (p. 79)
mineralogy (p. 79)
Mohs hardness scale (p. 87)
polymorphs (p. 83)
silicates (p. 89)
silicon-oxygen tetrahedron (p. 89)
specific gravity (p. 87)
streak (p. 86)
symmetry (p. 83)

REVIEW QUESTIONS

1. What is a mineral, as geologists understand the term? How is this definition different from the everyday usage of the word?

2. Why is glass not a mineral?

3. Salt is a mineral, but the plastic making up an inexpensive pen is not. Why not?

4. In what way does the arrangement of atoms in a mineral define a pattern? How can X-rays be used to study these patterns?

5. Describe the several ways that mineral crystals can form.

6. Why do some minerals occur as beautiful euhedral crystals, whereas others occur as anhedral grains?

7. List and define the principal physical properties used to identify a mineral.

8. How can you determine the hardness of a mineral? What is the Mohs hardness scale?

9. How do you distinguish cleavage surfaces from crystal faces on a mineral? How does each type of surface form?

10. What is the prime characteristic that geologists use to separate minerals into classes?

11. On what basis do mineralogists organize silicate minerals into distinct groups?

12. What is the relationship between the way in which silicon-oxygen tetrahedra bond in micas and the characteristic cleavage of micas?

13. How do sulfate minerals differ from sulfide minerals?

14. Why are some minerals considered gems? How do you make the facets on a gem?

SUGGESTED READING

Campbell, G. 2002. *Blood Diamonds: Tracing the Deadly Path of the World's Most Precious Stones.* Boulder: Westview Press.

Ciprianni, C., A. Borelli, and K. Lyman, eds. 1986. *Simon and Schuster's Guide to Gems and Precious Stones.* New York: Simon & Schuster.

Deer, W. A., J. Zussman, and R. A. Howie. 1996. *An Introduction to the Rock-Forming Minerals,* 2nd ed. Boston: Addison-Wesley.

Hibbard, J. J. 2001. *Mineralogy: A Geologist's Point of View.* New York: McGraw-Hill.

Klein, C., C. S. Hurlbut, and J. D. Dana. 2001. *The Manual of Mineral Science,* 22nd ed. New York: John Wiley & Sons.

Krajick, K. 2001. *Barren Lands: An Epic Search for Diamonds in the North American Arctic.* New York: Henry Holt and Co.

Kurlansky, M. 2003. *Salt: A World History.* New York: Penguin USA.

Perkins, D. 2001. *Mineralogy,* 2nd ed. Upper Saddle River, N.J.: Pearson Education.

Rock Groups

It took years of back-breaking labor for nineteenth-century workers to chisel and chip ledges and tunnels through the hard rock of the Sierra Nevada, in their quest to run a rail line across this rugged range. In the process, the workers became very familiar with the nature of rock.

A.1 INTRODUCTION

During the 1849 gold rush in the Sierra Nevada Mountains of California, only a few lucky individuals actually became rich. The rest of the "forty-niners" either slunk home in debt or took up less glamorous jobs in new towns such as San Francisco. These towns grew rapidly, and soon the American West Coast was demanding large quantities of manufactured goods from East Coast factories. Making the goods was no problem, but getting them to California meant either a stormy voyage around the southern tip of South America or a trek with stubborn mule teams through the deserts of Nevada or Utah. The time was ripe to build a railroad linking the East

and West coasts of North America, and, with much fanfare, the Central Pacific line decided to punch one right through the peaks of the Sierras. In 1863, while the Civil War raged elsewhere in the United States, the company transported six thousand Chinese laborers across the Pacific in the squalor of unventilated cargo holds and set them to work chipping ledges and blasting tunnels. Foremen measured progress in terms of feet per day—if they were lucky. Along the way, untold numbers of laborers died of frostbite, exhaustion, mistimed blasts, landslides, or avalanches.

Through their efforts, the railroad laborers certainly gained an intimate knowledge of how rock feels and behaves—it's solid, heavy, and hard! They also found that some rocks seemed to break easily into layers but others did not, and some rocks were dark colored while others were light colored. They realized, like anyone who looks closely at rock exposures, that rocks are not just gray, featureless masses, but rather come in a great variety of colors and textures. Why are there so many distinct types of rocks? The answer is simple: there are many different ways in which rocks can form, and many different materials out of which rocks can be made. Because of the relationship between rock type and the process of formation, *rocks provide a historical record of geologic events and give insight into interactions among components of the Earth System.* The next few chapters are devoted to a discussion of rocks and a description of how rocks form; this interlude is a general introduction. Here, we learn what the term *rock* means to geologists, what rocks are made of, and how to distinguish the three principal groups of rocks. We also look at how geologists study rocks.

A.2 WHAT IS ROCK?

To geologists, **rock** is a coherent, naturally occurring solid, consisting of an aggregate of minerals or, less commonly, a mass of glass. Now let's take this definition apart.

- *Coherent:* A rock holds together, and it must be broken to separate into pieces. As a result of its coherence, rock can form cliffs or can be carved into sculptures. A pile of unattached grains does not constitute a rock.

- *Naturally occurring:* Geologists consider only naturally occurring materials to be rocks, so manufactured materials such as concrete and brick do not qualify.

- *An aggregate of minerals or a mass of glass:* The vast majority of rocks consist of an aggregate (a collection) of many mineral grains, or crystals, stuck or grown together. (A "grain" is any fragment or piece of mineral, rock, or glass. A "crystal" is a piece of a mineral that grew into its present shape.) Technically, a single

mineral crystal is simply a "mineral specimen," not a rock, even if it is meters long. Some rocks contain only one kind of mineral, whereas others contain several different kinds. A few of the rock types that form at volcanoes (see Chapter 4) consist of glass, which may occur either as a homogeneous mass or as an accumulation of tiny shards (flakes).

What holds rock together? Grains in nonglassy rock stick together to form a coherent mass either because they are bonded by natural **cement,** mineral material that precipitates from water and fills the space between grains (▶Fig. A.1a–c), or because they interlock with one another like pieces in a jigsaw puzzle (▶Fig. A.2a–d). Rocks whose grains are stuck together by cement are called **clastic;** rocks whose crystals interlock with one another are called **crystalline.** Glassy rocks hold together either because they originate as a continuous mass (that is, they have no separate grains) or because glassy grains welded together while still hot.

FIGURE A.1 (a) Hand specimens of sandstone. (b) A magnified image of the sandstone shows that it consists of round white sand grains, surrounded by cement. (c) This exploded image emphasizes how the cement surrounds the sand grains. We depict the cement using two different colors because the cement probably formed at different times from different solutions.

(a)

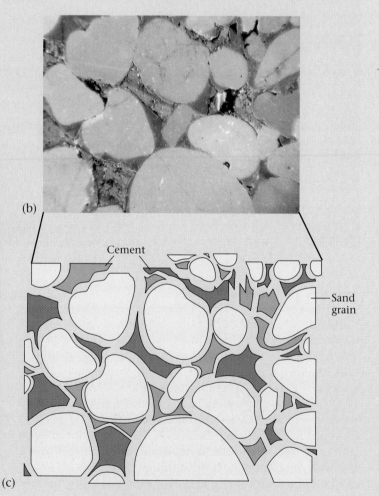

(b)

(c)

Cement

Sand grain

FIGURE A.2 (a) A hand specimen of granite, a rock formed when melt cools underground. (b) A photomicrograph (a photo taken through a microscope) shows the texture of granite is different from that of sandstone. In granite, the grains interlock with one another, like pieces of a jigsaw puzzle. (c) An artist's sketch emphasizes the irregular shapes of grains and how they interlock. (d) This exploded image highlights the grain shapes.

A.3 ROCK OCCURRENCES

At the surface of the Earth, rock occurs either as broken chunks (pebbles, cobbles, or boulders; see Chapter 5) that have moved from their point of origin by falling down a slope or by being transported in ice, water, or wind; or as **bedrock,** which is still attached to the Earth's crust. Geologists refer to an exposure of bedrock as an **outcrop.** An outcrop may appear as a rounded knob out in a field, as a ledge along a cliff or ridge, on the face of a stream cut (where running water cut down into bedrock), or along human-made road cuts and excavations (▶Fig. A.3a–c).

To people who live in cities or forests or on farmland, outcrops of bedrock may be unfamiliar, since they may be covered by vegetation, sand, mud, gravel, soil, water, asphalt, concrete, or buildings. Outcrops are particularly rare in regions such as the midwestern United States, where, during the past million years, ice-age glaciers melted and left behind thick deposits of debris (see Chapter 18).

A.4 THE BASIS OF ROCK CLASSIFICATION

People have developed classification schemes for just about every group of materials on the planet—for insects, trees, airplanes, books, and so on. Why? Classification schemes help us organize information and remember significant details about materials or objects, and they help us recognize similarities and differences among them.

Beginning in the late seventeenth century, geologists struggled to develop a sensible way to classify rocks, for they realized, as did miners from centuries past, that not all rocks are the same. One of the earliest classification schemes divided rocks into three groups—primary, secondary, and tertiary—on the basis of a perception (later proved incorrect) that the groups had formed in succession.

By the end of the eighteenth century, many geologists had realized that a "genetic scheme" for classifying rocks—a scheme that focuses on the origin (genesis) of rocks—would be the best approach for classifying rocks, and this is the

(a)

(b)

(c)

FIGURE A.3 (a) Outcrops (natural rock exposures) in a field along the coast of Scotland. (b) A stream cut, meaning an outcrop that forms when a stream's flow removes overlying soil and vegetation, in Brazil. Note that dense forest covers most of the adjacent hills, obscuring outcrops that may be exposed there. (c) Road cuts, such as this one along a highway near Kingston, New York, are made by setting off dynamite placed at the bottom of drill holes. Note that the layers of rock exposed in this road cut are curved or "folded" (see Chapter 9).

approach that we continue to use today. Using this approach, geologists recognize three basic groups: (1) **igneous rocks,** which form by the freezing (solidification) of molten rock, or melt (▶Fig. A.4a); (2) **sedimentary rocks,** which form either by the cementing together of fragments (grains) broken off preexisting rocks or by the precipitation of mineral crystals out of water solutions at or near the Earth's surface (▶Fig. A.4b); and (3) **metamorphic rocks,** which form when pre-existing rocks change into new rocks in response to a change

in pressure and temperature conditions, and/or as a result of squashing, stretching, or shear (▶Fig. A.4c). Metamorphic change occurs in the "solid state," which means that it does not require melting. Each of the three groups contains many different individual rock types, distinguished from one another by physical characteristics such as

• *grain size:* The dimensions of individual grains in a rock may be measured in millimeters or centimeters. Some

(a)

(b)

(c)

FIGURE A.4 Examples of the major rock groups. (a) Lava (molten rock) erupts at a volcano, then freezes to form basalt, an igneous rock. (b) Sand, deposited on a beach, eventually becomes buried to form layers of sandstone, a sedimentary rock, such as those exposed in the cliffs behind the beach. (c) When preexisting rocks become deeply buried during mountain building, the increase in temperature and pressure transforms them into metamorphic rocks. The lichen-covered outcrop in the foreground was once buried deep beneath a mountain range, but was later exposed when glaciers and rivers stripped off overlying rocks.

grains are so small that they can't be seen without a microscope, whereas others are as big as a fist or larger. Some grains are **equant,** meaning that they have the same dimensions in all directions; some are **inequant,** meaning that the dimensions are not the same in all directions (▶Fig. A.5a, b). In some rocks, all the grains are the same size, whereas other rocks contain a variety of different-sized grains.

- *composition:* A rock is a mass of chemicals. The term "rock composition" refers to the proportions of different chemicals making up the rock. The proportion of chemicals, in turn, reflects the proportion of different minerals constituting the rock.

- *texture:* This term refers to the arrangement of grains in a rock, that is, the way grains connect to one another and whether or not inequant grains are aligned parallel to each

other. The concept of rock texture will become easier to grasp as we look at different examples of rocks in the following chapters.

- *layering:* Some rock bodies appear to contain distinct layering, defined either by bands of different compositions or textures, or by the alignment of inequant grains so that they trend parallel to each other. Different types of layering occur in different kinds of rocks. For example, the layering in sedimentary rocks is called **bedding,** whereas the layering in metamorphic rocks is called **metamorphic foliation** (▶Fig. A.6a, b).

Each individual rock type has a name. Names come from a variety of sources. Some come from the dominant component making up the rock, some from the region where the rock was first discovered or is particularly abundant, some

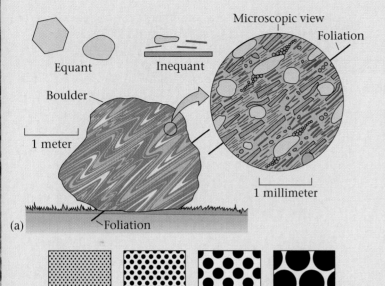

FIGURE A.5 (a) A rock is an aggregate of mineral grains. At high magnification, we can see that the rock consists of both equant and inequant grains of various sizes. In this example, the inequant grains are aligned parallel to one another. (b) Using this comparison chart, we can measure the size of the grains, in millimeters.

from a root word of Latin origin, and some from a traditional name used by people in an area where the rock is found. All told, there are hundreds of different rock names, though in this book we introduce only about thirty.

A.5 STUDYING ROCK

Outcrop Observations

The study of rocks begins by examining a rock in an outcrop. If the outcrop is big enough, such an examination will reveal relationships between the rock you're interested in and the rocks around it, and will allow you to detect layering. Geologists carefully record observations about an outcrop, then break off a **hand specimen,** a fist-sized piece, which they can examine more closely with a hand lens (magnifying glass). Observation with a hand lens enables geologists to identify sand-sized or larger mineral grains, and may enable them to describe the texture of the rock.

Thin-Section Study

Geologists often must examine rock composition and texture in minute detail in order to identify a rock and develop a hypothesis for how it formed. To do this, they take a specimen back to the lab, make a very thin slice (about 3/100 mm thick, the thickness of a human hair), and mount it on a glass slide. They study such a **thin section** (▶Fig. A.7a–c) with a petrographic microscope ("petro" comes from the Greek word for "rock"). A petrographic microscope differs from an ordinary microscope in that it illuminates the thin section with transmitted polarized light. This means that the illuminating light beam first passes through a special filter that makes all the light waves in the beam vibrate in the same plane, and then the light passes up through the thin section; an observer can look through the thin section as if it were a window.

FIGURE A.6 Layering in rocks. (a) Bedding in a sedimentary rock, in this case defined by alternating coarser and finer layers exposed in a cliff along the coast of Oregon. (b) Foliation in a metamorphic rock, in this case defined by alternating light and dark layers.

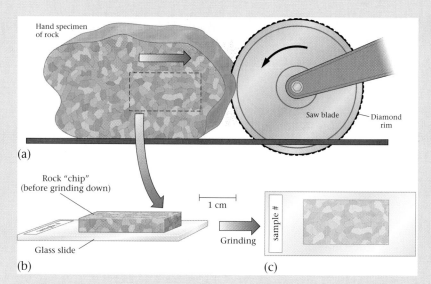

(a)

(b)   (c)

FIGURE A.7 (a) To prepare a thin section, a geologist cuts a brick-shaped chip out of rock using a rock saw (a rotating circular blade with a diamond-studded rim). (b) The chip is glued to a glass slide and then ground down until it's paper thin. (c) The thin section is then labeled and ready to examine.

When illuminated with transmitted polarized light, each type of mineral grain displays a unique suite of colors. The specific color the observer sees depends on both the identity of the grain and its orientation with respect to the waves of polarized light, for a crystal interferes with polarized light and allows only certain wavelengths to pass through.

The brilliant colors and strange shapes in a thin section rival the beauty of an abstract painting or stained glass. By examining a thin section with a petrographic microscope, geologists can identify most of the minerals constituting the rock and can describe the way in which the grains connect to each other (▶Fig. A.8a). A photograph taken through a petrographic microscope is called a **photomicrograph.**

High-Tech Analytical Equipment

Beginning in the 1950s, high-tech electronic instruments became available that enabled geologists to examine rocks on an even finer scale than is possible with a petrographic microscope. Modern research laboratories typically boast instruments such as electron microprobes, which can focus a beam of electrons on a small part of a grain to create a signal that defines the chemical composition of the mineral (▶Fig. A.8b); mass spectrometers, which analyze the proportions of atoms with different atomic weights contained in a rock; and X-ray diffractometers, which identify minerals by looking at the way X-ray beams pass through crystals in a rock. Such instruments, in conjunction with optical examination, can provide petrologists with highly detailed characterizations of rocks, which in turn help them understand how the rocks formed and where the rocks came from. This information enables geologists to use the study of rocks as a basis for deciphering Earth history.

FIGURE A.8 (a) A rock photographed through a petrographic microscope. The colors are caused by the interaction of polarized light with the crystals. The long dimension of this photo is 2 mm. (b) An electron microprobe uses a beam of electrons to analyze the chemical composition of minerals.

(a)

(b)

CHAPTER 4

Up from the Inferno: Magma and Igneous Rocks

A river of molten rock (lava) weaves across a stark terrain of already solidified igneous rock. The volcano, the vent from which the lava has spilled out onto Earth's surface from the interior, rises in the distance.

4.1 INTRODUCTION

Every now and then, an incandescent liquid—molten rock, or melt—fountains from a crater (pit) or crack on the big island of Hawaii. Hawaii is a **volcano,** a vent at which melt from inside the Earth spews onto the planet's surface. The transfer of melt from inside the Earth onto its surface is a volcanic eruption. Some of the melt, called **magma** while underground and **lava** once it has reached the Earth's surface, pools around the vent, while the rest flows down the mountainside as a viscous (syrupy) red-yellow stream called a lava flow. Near its source, the flow moves swiftly, cascading over escarpments at speeds of up to 60 km per hour (▶Fig. 4.1a). At the base of the mountain, the lava flow slows but advances nonetheless, engulfing roads, houses, or vegetation in its path. As the flow cools, it slows down. Its surface darkens and crusts over, occasionally breaking to reveal the hot, sticky mass that remains within (▶Fig. 4.1b). Finally, the flow stops moving entirely, and within days or weeks the once red-hot melt has become a hard, black solid through and through (▶Fig. 4.1c). A new **igneous rock,** made by the freezing of a melt, has formed. Considering the fiery heat of the melt from which igneous rocks develop, the name "igneous"— from the Latin *ignis,* meaning "fire"—

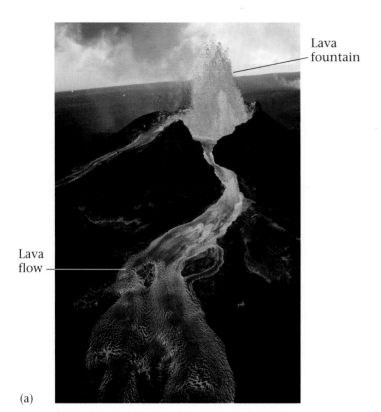

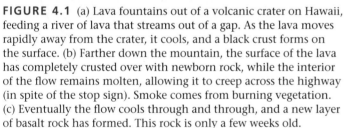

Lava
fountain

Lava
flow

(a)

(b)

(c)

FIGURE 4.1 (a) Lava fountains out of a volcanic crater on Hawaii, feeding a river of lava that streams out of a gap. As the lava moves rapidly away from the crater, it cools, and a black crust forms on the surface. (b) Farther down the mountain, the surface of the lava has completely crusted over with newborn rock, while the interior of the flow remains molten, allowing it to creep across the highway (in spite of the stop sign). Smoke comes from burning vegetation. (c) Eventually the flow cools through and through, and a new layer of basalt rock has formed. This rock is only a few weeks old.

makes sense. Igneous rocks are very common on Earth. They make up all of the oceanic crust and much of the continental crust.

It may seem strange to speak of freezing in the context of forming rock, when most people think of freezing as the transformation of liquid water to solid ice as the temperature drops below 0°C (32°F). Nevertheless, the freezing of liquid melt to form solid igneous rock represents the same phenomenon, except that igneous rocks freeze at high temperatures—between 650°C and 1,100°C. (To put such temperatures in perspective, remember that home ovens attain a maximum temperature of only 260°C.)

Clearly, if the Earth were not hot inside, igneous processes would not be possible. Where does our planet's internal heat come from? Much of the heat is left over from events that occurred during and relatively soon after Earth's formation. Specifically, intense bombardment by meteorites, as the Earth grew from a planetesimal into a planet, generated heat, for each time a meteorite collided with the

Earth, the object's kinetic energy (energy of motion) transformed into heat energy. You may have experienced this phenomenon if you've touched the warm head of a nail after banging on it with a hammer. Later, when the growing Earth underwent differentiation, iron sank to its center to form the core, a process that produced more heat because the sinking iron sheared against its surroundings as it sank. In the terminology of physics, gravitational potential energy transformed into heat energy during core formation (see Appendix). Shortly after core formation, a Mars-sized planet collided with the young Earth, blasting free the debris that then coalesced to form the Moon. This event heated the Earth again, because an immense amount of kinetic energy transformed into heat energy. In fact, collisions and core differentiation made the early Earth so hot that it was partially molten, and its surface was a sea of lava.

Ever since the heat-producing catastrophes of its early days, the Earth has been slowly cooling as heat radiates into space. Eventually, the sea of lava cooled and solidified into igneous rock, the first rocks to form on Earth. However, our planet would have cooled faster had it not been for the presence of radioactive elements, primarily in the crust. The decay of a radioactive atom produces a tiny amount of heat. But the cumulative effect of radioactive decay in the crust has been sufficient to slow the cooling of the whole planet. Earth remains very hot inside, with temperatures at the base of the lithosphere reaching almost 1,300°C, and temperatures at the center exceeding 4,700°C.

As we noted above, igneous rocks began to form fairly early in Earth history. But they clearly continue to form even today, as we've emphasized by our description of lava eruption in Hawaii. Thus, though the crust and mantle are mostly solid, production of melt still takes place at special locations inside the Earth.

Although some igneous rocks solidify at the surface during volcanic eruptions such as those on Hawaii, a vastly greater volume results from solidification of magma underground, out of sight. Rock made by the freezing of magma underground, after it has pushed its way (intruded) into preexisting rock of the crust, is **intrusive igneous rock,** and rock that forms by the freezing of lava above ground, after it spills out (extrudes) onto the surface of the Earth and comes into contact with the atmosphere or ocean, is

extrusive igneous rock (▶Fig. 4.2). Extrusive igneous rock includes both solid lava flows, formed when streams or mounds of lava solidify on the surface of the Earth, and deposits of **pyroclastic debris** (from the Greek word *pyro,* meaning "fire"). Such debris includes **volcanic ash,** consisting of fine particles of glass that form when a spray of lava erupts into the air and freezes instantly. Some ash billows up several kilometers into the sky above a volcano, and eventually drifts down in a snowlike ash fall. But some ash rushes down the side of the volcano in a scalding avalanche called an ash flow. Pyroclastic debris also includes larger fragments formed when clots of lava freeze in the air, or when the force of the eruption blasts apart preexisting rock of a volcano. We'll provide further detail about pyroclastic debris in Chapter 7.

A great variety of igneous rocks exist on Earth. To understand why and how these rocks form, and why there are so many different kinds, we now discuss why magma forms, why it rises, why it sometimes erupts as lava, and how it freezes in intrusive and extrusive environments. We then look at the scheme that geologists use to classify igneous rocks.

4.2 THE FORMATION OF MAGMA

As mentioned previously, the popular image that the solid crust of the Earth floats on a sea of molten rock is not correct. In reality, magma forms only in special places where preexisting solid rock melts. Below, we describe conditions that lead to melting.

Melting as a result of a decrease in pressure (decompression). The Earth is hot inside. We can portray the variation in temperature with depth on a graph by a curving line, the **geotherm.** Beneath typical oceanic crust, temperatures comparable to those of lava (650°–1,100°C) generally occur in the upper mantle (▶Fig. 4.3). But even though the upper mantle is very hot, its rock stays solid because it is also under high pressure from the weight of overlying rock. Pressure at great depth, simplistically, prevents atoms from breaking free of mineral crystals.

Because pressure prevents melting, a decrease in pressure can permit melting. Specifically, if the pressure affecting hot mantle rock decreases while the temperature remains unchanged, a magma forms. This kind of melting, called decompression melting, occurs where hot mantle rock rises to shallower depths in the Earth. (▶Fig. 4.4a–c).

Melting as a result of the addition of volatiles. Magma also forms at locations where chemicals called volatiles mix with hot mantle rock. Volatiles are elements or molecules, such as water (H_2O) and carbon dioxide (CO_2), that evaporate

FIGURE 4.2 In the intrusive realm, magma solidifies (freezes) to form intrusive igneous rocks. In the extrusive realm, lava and ash erupt and then solidify or accumulate, respectively, to form extrusive igneous rocks.

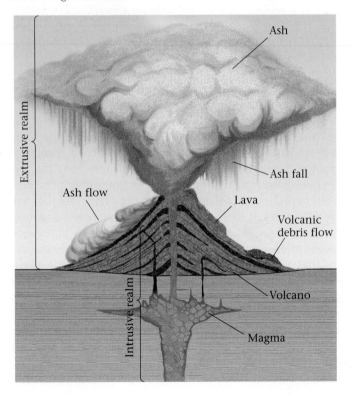

Ash

Extrusive realm

Ash fall

Ash flow

Lava

Volcanic debris flow

Intrusive realm

Volcano

Magma

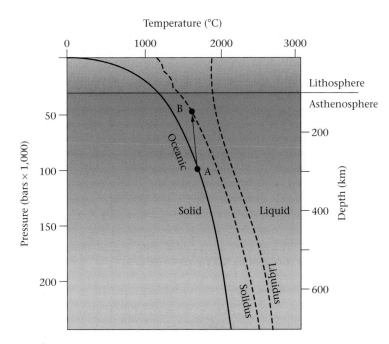

Temperature (°C)

FIGURE 4.3 The graph plots the Earth's geotherm (solid line), which specifies the temperature at various depths below oceanic lithosphere, as well as the solidus and liquidus (dashed lines) for peridotite, the ultramafic rock that makes up the mantle. The solidus represents conditions of pressure and temperature at which a rock *begins to melt,* whereas the liquidus represents the conditions of pressure and temperature at which the *last solid disappears.* The region of the graph between the liquidus and solidus represents conditions under which there can be a mixture of solid and melt. A rock that starts at pressure and temperature conditions indicated by point A, and then rises to point B, undergoes a decrease in pressure without much change in temperature. When it reaches the conditions indicated by point B, it begins to melt. This process is called decompression melting.

easily and can exist in gaseous forms at the Earth's surface. When volatiles mix with hot rock, they help break chemical bonds so if you add volatiles to a solid, hot dry rock, the rock begins to melt (▶Fig. 4.4d). In effect, adding volatiles decreases a rock's melting temperature. Melting due to addition of volatiles is sometimes called flux melting.

Melting as a result of heat transfer from rising magma. When magma from the mantle rises up into the crust, it brings heat with it. This heat raises the temperature of the surrounding crustal rock and, in some cases, the rise in temperature may be sufficient to melt part of the crustal rock (Fig. 4.4e). To picture the process, imagine injecting hot fudge into ice cream; the fudge transfers heat to the ice cream, raises its temperature, and causes it to melt. We call such melting heat-transfer melting, because it results from the transfer of heat from a hotter material to a cooler one.

4.3 WHAT IS MAGMA MADE OF?

All magmas contain silicon and oxygen, which bond to form the silicon-oxygen tetrahedron. But magmas also contain varying proportions of other elements such as aluminum (Al), calcium (Ca), sodium (Na), potassium (K), iron (Fe), and magnesium (Mg). Because magma is a liquid, its atoms do not lie in an orderly crystalline lattice but are grouped instead in clusters or short chains, relatively free to move with respect to one another.

"Dry" magmas contain no volatiles. "Wet" magmas, in contrast, include up to 15% dissolved volatiles such as water, carbon dioxide, nitrogen (N_2), hydrogen (H_2), and sulfur dioxide (SO_2). These volatiles come out of the Earth at volcanoes in the form of gas. Usually water constitutes about half of the gas erupting at a volcano. Thus, magma not only contains the elements that constitute solid minerals in rocks, but can also contain molecules that become water or air.

The Major Types of Magma

Geologists distinguish among the four major types of magmas by specifying the relative proportions of oxides. Common oxides include silica (SiO_2), iron oxide (FeO or Fe_2O_3), and magnesium oxide (MgO). The four magma types are as follows. **Felsic magma** (or silicic magma) contains about 66% to 76% silica. **Intermediate magma** contains about 52% to 66% silica. The name "intermediate" indicates that these magmas have a composition between that of felsic magma and mafic magma. **Mafic magma** contains about 45% to 52% silica. Mafic magma gets its name because it has a relatively high proportion of MgO and FeO or Fe_2O_3. The "ma" in the word "mafic" stands for "magnesium," and the "-fic" stands for "iron" (from the Latin *ferric*). **Ultramafic magma** contains only 38% to 45% silica.

Why are there so many different kinds of magma? Several factors control magma composition.

Source rock composition. The composition of a melt reflects the composition of the solid from which it was derived. Not all magmas form from the same source rock, so not all magmas have the same composition.

Partial melting. Under the temperature and pressure conditions that occur in the Earth, only about 2% to 30% of an original rock melts to produce magma at a given location—the temperature at sites of magma production simply never gets high enough to melt the entire original rock. **Partial melting** refers to the process by which only part of an original rock melts to produce magma (▶Fig. 4.5a). As it forms, magma migrates away from the site of melting, leaving behind the solid that can't melt. Since felsic minerals tend to melt at lower temperatures, magmas formed by partial melting tend to be more felsic than the original rock

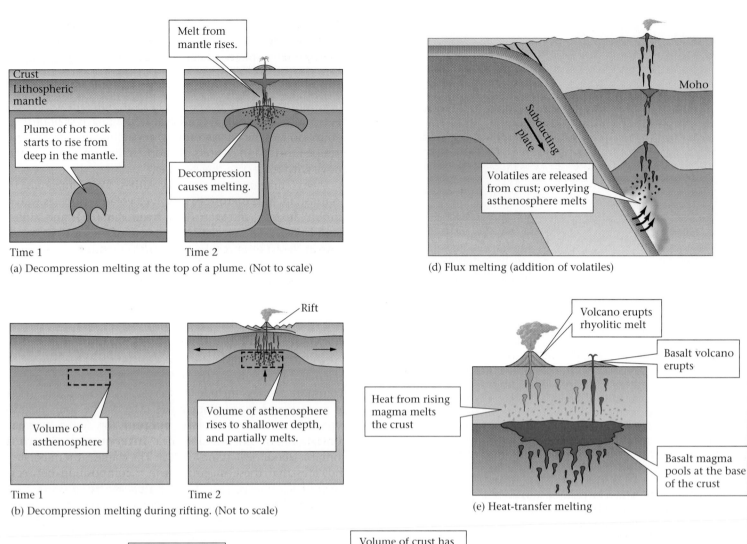

Melt from mantle rises.

Crust
Lithospheric mantle

Plume of hot rock starts to rise from deep in the mantle.

Decompression causes melting.

Time 1   Time 2

(a) Decompression melting at the top of a plume. (Not to scale)

Moho

Subducting plate

Volatiles are released from crust; overlying asthenosphere melts

(d) Flux melting (addition of volatiles)

Rift

Volume of asthenosphere

Volume of asthenosphere rises to shallower depth, and partially melts.

Time 1   Time 2

(b) Decompression melting during rifting. (Not to scale)

Volcano erupts rhyolitic melt

Basalt volcano erupts

Heat from rising magma melts the crust

Basalt magma pools at the base of the crust

(e) Heat-transfer melting

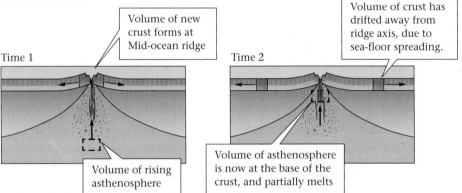

Volume of new crust forms at Mid-ocean ridge

Volume of crust has drifted away from ridge axis, due to sea-floor spreading.

Time 1

Time 2

Volume of rising asthenosphere

Volume of asthenosphere is now at the base of the crust, and partially melts

(c) Decompression melting at a mid-ocean ridge.

FIGURE 4.4 The main causes of melting and magma formation in the Earth. (a) Decompression melting happens when a mantle plume brings rock from deep in the mantle up to the base of the lithosphere. (b) Decompression melting happens when rifting thins the lithosphere, causing the underlying asthenosphere to rise. (c) Decompression melting happens when asthenosphere rises beneath the axis of a mid-ocean ridge, as sea-floor spreading occurs. (d) Flux melting occurs when volatiles escape from subducted oceanic crust into the overlying asthenosphere. (e) Heat-transfer melting occurs when heat rises into the base of the crust from basaltic magma that has pooled at the Moho or has intruded into the lower crust. The heat causes the lower crust to partially melt, producing felsic magma that rises.

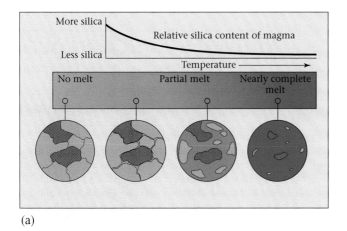

(a)

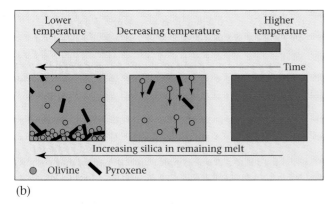

(b)

FIGURE 4.5 (a) The concept of partial melting. Rock does not all melt at once; at lower temperatures, only part of the rock melts. The first melt tends to be more felsic (silicic) than the later-formed melt, as the graph shows. When the rock first starts to melt, the molten rock forms films around still-solid grains. Grains that melt at lower temperatures melt first, whereas grains that melt at higher temperatures remain. With further melting, a crystal "mush" develops, with relict solid grains surrounded by melt. (b) The concept of fractional crystallization. The highest-melting-temperature (mafic) minerals begin to crystallize. Some of these early-formed minerals sink to the bottom of the magma body. Elements incorporated in these minerals, therefore, are extracted from the melt. The remaining melt becomes more silicic.

from which they were derived—for example, partial melting of an ultramafic rock produces a mafic magma. Variations in the degree of partial melting affect the magma composition.

Assimilation. As magma sits in a magma chamber before completely solidifying, it may incorporate chemicals derived from the wall rocks of the chamber. This process of contamination or **assimilation** takes place when wall rocks fall into the magma and then partially melt, or when heat from the magma partially melts the walls of the chamber.

Magma mixing. Different magmas formed in different locations from different sources may enter a magma chamber.

When this happens, the originally distinct magmas mix to create a new, different magma. For example, thoroughly mixing a felsic magma with a mafic magma in equal proportions produces an intermediate magma.

Fractional crystallization. Not all minerals freeze at the same temperature. For example, when a magma starts to cool, mafic minerals begin to form first. These minerals are denser than magma, so they start to sink to the floor of the magma chamber. As this happens, some react with the remaining melt. Mafic minerals are rich in iron and magnesium, so they take iron and magnesium out of the magma when they settle. As a result the remaining magma becomes more felsic (▶Fig. 4.5b). Because of this process of **fractional crystallization,** the remaining magma becomes progressively more felsic as it cools. (See ▶Box 4.1 to learn more about the sequence in which minerals form.)

4.4 THE MOVEMENT OF MAGMA AND LAVA

Forces Driving the Rise of Magma

If magma stayed put once it formed, new igneous rocks would not develop in or on the crust. But it doesn't stay put; magma tends to move upward, away from where it formed—in some cases, it reaches the Earth's surface and erupts at a volcano. This movement is a key component of the Earth System, because it transfers material from deeper parts of the Earth upwards, and provides the raw material for new rocks and for the atmosphere and ocean. But why does magma rise?

First, magma rises because it is less dense than surrounding rock. Thus, buoyancy drives magma upward just as it drives a wooden block up through water. Second, magma rises because the weight of overlying rock creates pressure at depth that literally squeezes magma upward. The same process happens when you step into a puddle barefoot and mud squeezes up between your toes.

Melt Viscosity

Viscosity, or resistance to flow, affects the speed with which magmas or lavas move. Magmas with low viscosity flow more easily than those with high viscosity, just as water flows more easily than molasses. Magma viscosity depends on temperature, volatile content, and silica content. Hotter magma is less viscous than cooler magma, just as hot tar is less viscous than cool tar, because thermal energy breaks bonds and allows atoms to move more easily. Similarly, magmas or lavas containing more volatiles are less viscous than dry (volatile-free) magmas, because the volatile atoms

BOX 4.1
THE REST OF THE STORY

Bowen's Reaction Series

In the 1920s, Norman L. Bowen, a geologist at the Carnegie Institution in Washington, D.C., and later at the University of Chicago, began a series of laboratory experiments designed to determine the sequence in which silicate minerals crystallize from a melt. Bowen first melted powdered mafic igneous rock in a sealed crucible by raising its temperature to about 1,280°C. Then he cooled the melt just enough to cause part of it to solidify, and he "quenched" the remaining melt by submerging it quickly in mercury. Quenching, which means undergoing a sudden cooling to form a solid, caused any remaining liquid to turn into glass, a material without a crystalline structure. The early-formed crystals were trapped in the glass. Bowen identified mineral crystals formed before quenching by examining thin sections with a microscope. Then he analyzed the composition of the remaining glass.

After repeated experiments at different temperatures, Bowen found that as new crystals form during crystallization, they extract certain chemicals preferentially from the melt. Thus, the chemical composition of the remaining melt progressively changes as the melt cools. Further, once formed, crystals continue to react with the remaining melt. Bowen described the specific sequence of mineral-producing reactions that take place in a cooling, initially mafic, magma. This sequence is now called **Bowen's reaction series** in his honor.

In a cooling melt, olivine and calcium-rich plagioclase form first. The Ca-plagioclase reacts with the melt to form more plagioclase, but the newer plagioclase contains more sodium (Na). Meanwhile, some olivine crystals react with the remaining melt to produce pyroxene, which may encase (i.e., surround) olivine crystals or even replace them. However, some of the olivine and Ca-plagioclase crystals settle out of the melt, taking iron, magnesium, and calcium atoms with them. By this process, the remaining melt becomes enriched with silica. As the melt continues to cool, plagioclase continues to form, with later-formed plagioclase having progressively more Na than earlier-formed plagioclase. Pyroxene crystals react with melt to form amphibole, and then amphibole reacts with the remaining melt to form biotite. All the while, crystals continue to settle out, so the remaining melt continues to become more felsic. At temperatures of between 650°C and 850°C, only about 10% melt remains, and this melt has a high silica content. At this stage, the final melt freezes, yielding felsic minerals such as quartz, K-feldspar, and muscovite.

On the basis of his observations, Bowen realized that there are two tracks to the reaction series. The "discontinuous reaction series" refers to the sequence olivine, pyroxene, amphibole, biotite, K-feldspar/muscovite/quartz: each step yields a different class of silicate mineral. The "continuous reaction series" refers to the progressive change from calcium-rich to sodium-rich plagioclase: the steps yield different versions of the same mineral (▶Fig. 4.6). In the discontinuous reaction series, the first mineral to form is composed of isolated silicon-oxygen tetrahedra; the second contains single chains of tetrahedra; the third, double chains of tetrahedra; the fourth, 2-D sheets of tetrahedra; and the last, 3-D network silicates.

It's important to note that not all minerals listed in the series appear in all igneous rock. For example, a mafic magma may completely crystallize before felsic minerals such as quartz or K-feldspar have a chance to form.

Bowen's studies provided a remarkable demonstration of how laboratory experiments can help us understand processes that take place in locations (such as a deep magma chamber) that no human can visit directly.

FIGURE 4.6 Bowen's reaction series. Minerals that crystallize at higher temperatures are at the top of the series. The left side of the diagram shows the discontinuous reaction series, consisting of a succession of different minerals. The right side shows the continuous reaction series, consisting of progressively changing plagioclase compositions. Rocks formed from minerals at the top of the series are mafic; rocks made from minerals at the bottom of the series are felsic. The short arrows indicate reactions.

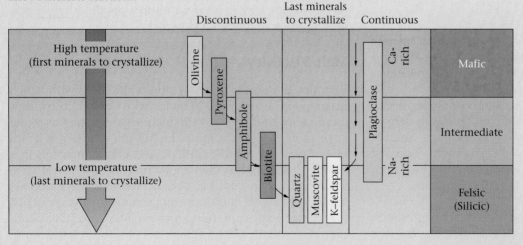

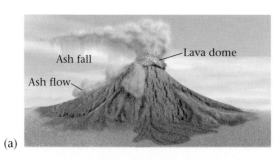

(a)

(b)   Not to scale

FIGURE 4.7 The behavior of erupting lava reflects its viscosity. (a) Viscous lava (silica-rich lava) forms a blob or moundlike dome at the volcano's vent. (b) Nonviscous lava (hot mafic lava) spreads out in a thin flow and can travel far from the vent; it can also fountain.

also tend to break apart bonds and may accumulate to form bubbles. Mafic magmas are less viscous than are felsic magmas, because silicon-oxygen tetrahedra tend to link together in the magma to create long chains that can't move past each other easily. Thus, hotter mafic lavas have relatively low viscosity and flow in thin sheets over wide regions, but cooler felsic lavas are highly viscous and clump at the volcanic vent (▶Fig. 4.7a, b).

4.5 EXTRUSIVE VERSUS INTRUSIVE ENVIRONMENTS

Recall that igneous rocks form in two environments. If lava erupts at the Earth's surface and freezes in contact with the atmosphere or the ocean, then the rock it forms is called extrusive igneous rock. The term implies that the melt was extruded from (it flowed or exploded out of) a volcanic vent. In contrast, if magma freezes underground, the rock it forms is called intrusive igneous rock, implying that the magma intruded into preexisting rock of the crust.

Extrusive Igneous Settings

Not all volcanic eruptions are the same, so not all extrusive rocks are the same. Some volcanoes erupt streams of low-viscosity lava that run down the flanks of the volcano and

then spread over the countryside. When this lava freezes, it forms a sheet of igneous rock, also known as a lava flow. In contrast, some volcanoes erupt viscous masses of lava that pile into domes, and still others erupt explosively, sending clouds of volcanic ash and debris skyward, and avalanches of ash that tumble down the sides of the volcano (▶Fig. 4.8).

Which type of eruption occurs depends largely on a magma's composition and volatile content. As noted above, mafic lavas tend to have low viscosity and spread in broad, thin flows. Volatile-rich felsic lavas tend to erupt explosively and form thick ash deposits. We discuss the products of extrusive settings in more detail in Chapter 7.

Intrusive Igneous Settings

Magma rises and intrudes into preexisting rock by slowly percolating upward between grains or by forcing open cracks. The magma that doesn't make it to the surface freezes solid underground in contact with preexisting rock and becomes intrusive igneous rock. Geologists commonly

FIGURE 4.8 Two styles of ash eruption have happened at this volcano. Some ash rises high in the sky, and some cascades down the flank of the volcano.

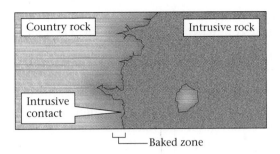

FIGURE 4.9 An intrusive contact, showing the baked zone, blocks of country rock surrounded by intrusive rock, and fingers of the intrusion protruding into the country rock.

refer to the preexisting rock into which magma intrudes as country rock, or wall rock, and the boundary between wall rock and an intrusive igneous rock as an intrusive contact (▶Fig. 4.9). If the wall rock was cold to begin with, then heat from the intrusion "bakes" and alters it in a narrow band along an intrusive contact (see Chapter 6).

Geologists distinguish among different types of intrusions on the basis of their shape. Tabular intrusions, or sheet intrusions, are planar and are of roughly uniform thickness. Most tabular intrusions are in the range of centimeters to tens of meters thick, and tens of meters to tens of kilometers long. In places where tabular intrusions cut across rock that does not have layering, a nearly vertical, wall-like tabular intrusion is a **dike,** whereas a nearly horizontal, tabletop-shaped tabular intrusion is a **sill.** In places where tabular intrusions enter rock that has layering (bedding or foliation), dikes are intrusions that cut across layering, whereas sills are intrusions that are parallel to layering (▶Figs. 4.10a, b; 4.11a–e). Some intrusions start to inject between layers but then dome upward, creating a blister-shaped intrusion known as a **laccolith** (▶Fig. 4.12a).

Plutons are irregular or blob-shaped intrusions that range in size from tens of meters across to tens of kilometers across (▶Figs. 4.12a–c). The intrusion of numerous plutons in a region creates a vast composite body that may be several hundred kilometers long and over 100 km wide; such immense masses of igneous rock are called **batholiths.** The rock making up the Sierra Nevada mountains of California is a batholith formed from plutons that intruded between 145 and 80 million years ago (▶Fig. 4.13a–c).

Where does the space for intrusions come from? Dikes form in regions where the crust is being stretched (such as in a rift). Thus, as the magma that makes a dike forces its way up into a crack (sometimes causing the crack to form in the first place), the crust stretches sideways (▶Fig. 4.14a). Intrusion of sills occurs near the surface of the Earth, so the pressure of the magma effectively pushes up the rock above the sill, leading to uplift of the Earth's surface (▶Fig. 4.14b).

How do plutons form? Some geologists propose that a pluton is a frozen "diapir," meaning a light-bulb-shaped blob of magma that pierced overlying rock and pushed it aside as it rose (▶Fig. 4.14c). Another explanation involves **stoping,** a process during which magma assimilates wall rock, and blocks of wall rock break off and sink into the magma (▶Fig. 4.14d). (If a stoped block does not melt entirely, but rather becomes surrounded by new igneous rock, it is a **xenolith,** after the Greek word *xeno,* meaning "foreign"; ▶Fig. 4.14e; see also Fig. 4.9.) More recently, geologists have proposed that plutons form by injection of several superimposed dikes or sills, which coalesce to become a single larger intrusion.

If intrusive igneous rocks form beneath the Earth's surface, why can we see them exposed today? The answer comes from studying the dynamic activity of the Earth.

FIGURE 4.10 (a) Dikes and sills appear as vertical or horizontal bands, respectively, on the face of an outcrop. (b) If we were to strip away the surrounding rock, dikes would look like walls, and sills would look like tabletops.

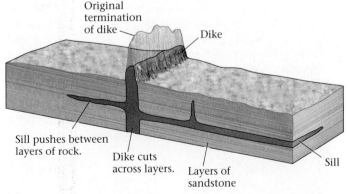

(a)

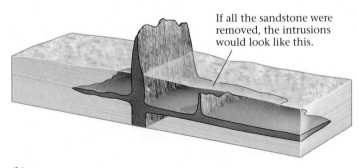

(b)

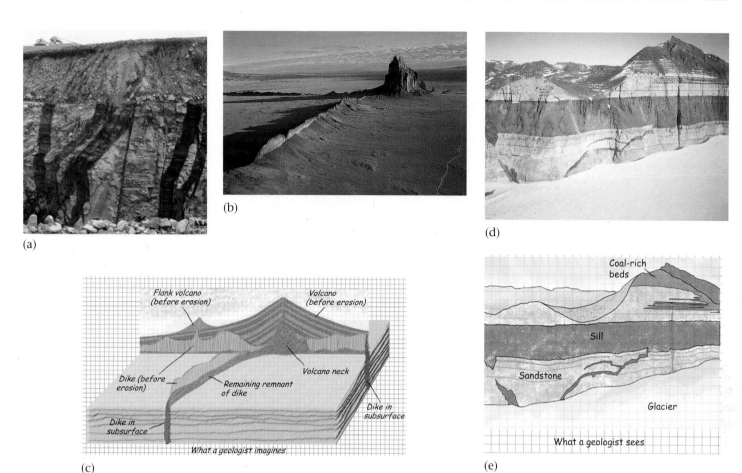

FIGURE 4.11 (a) Basalt dikes (black stripes) cutting granite in an Oklahoma quarry. A large layer of gravel overlies the bedrock, and rock-crushing equipment stands on top of the gravel. The quarry wall is about 50 m (150 feet) high. (b) At this ancient volcano called Shiprock, in New Mexico, ash and lava flows have eroded away, leaving a "volcanic neck" (the solid igneous rock that cooled in a magma chamber within the volcano). Large dikes radiate outward from the center, like spokes of a wheel. The softer rocks that once surrounded the dikes have eroded away, leaving wall-like remnants of the dikes exposed. (c) Shiprock was once in the interior of a volcano or below a volcano. The dikes radiated from this neck and cut across preexisting rock layers. (d) This dark sill, exposed on a cliff in Antarctica, consists of basalt; the white rock consists of sandstone, while the peak of the cliff consists of coal-rich beds. (e) This geologist's sketch shows the cliff face as viewed face on.

Over long periods of geologic time, mountain building, driven by plate interactions, slowly uplifts huge masses of rock. Moving water, wind, and ice eventually strip away great thicknesses of overlying rock and expose the intrusive rock that has formed below.

4.6 TRANSFORMING MAGMA INTO ROCK

What makes magma freeze? In nature, two phenomena lead to the formation of solid igneous rock from a magma. Magma may freeze if the volatiles dissolved within it bubble out. But more commonly, magma freezes simply when it cools below its freezing temperature and crystals start to grow. Because temperatures decrease toward the Earth's surface, magma automatically enters a cooler environment when it rises. The cooler environment may be cool country rock if the magma intrudes underground, or it may be the atmosphere or the ocean if the magma extrudes as lava at the Earth's surface.

The time it takes for a magma to cool depends on how fast it is able to transfer heat into its surroundings. To see why, think about the process of cooling coffee. If you pour hot coffee into a thermos bottle and seal it, the coffee stays hot for hours; because of insulation, the coffee in the thermos loses heat to the air outside only very slowly. Like the thermos bottle, surrounding rock acts as an insulator in that

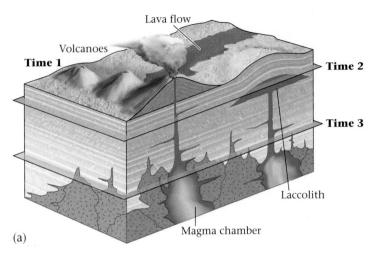

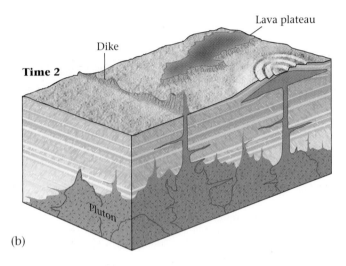

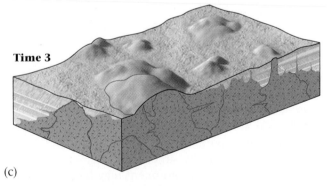

(a)

(b)

(c)

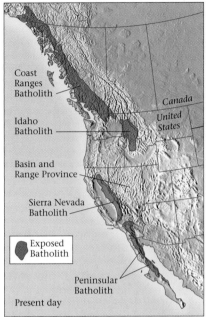

(a)

(b)

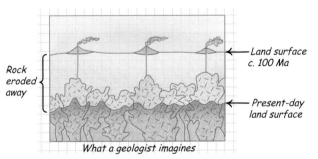

(c)

FIGURE 4.12 (a) While a volcano remains active, a magma chamber exists underground. Dikes, sills, and laccoliths intrude preexisting rock, and lava and ash erupt at the surface. Here, we see that the active magma chamber, the one currently erupting, is only the most recent of many in this area. Earlier ones solidified before the current erruption. Each mass is a pluton. A composite of many plutons is a batholith. (b) Later, the bulbous magma chamber freezes into a pluton. The soft parts of the volcano erode, leaving wall-like dikes and column-like volcanic necks. Hard lava flows create resistant plateaus. (c) With still more erosion, volcanic rocks and shallow intrusions disappear, and we see plutonic intrusive rocks.

FIGURE 4.13 (a) The batholiths of western North America today. (b) The Sierra Nevada Batholith as exposed today. The rounded, light-colored hills are all composed of granite-like intrusive igneous rock. (c) A geologist's sketch illustrates that today's land surface once lay several kilometers beneath a chain of volcanoes.

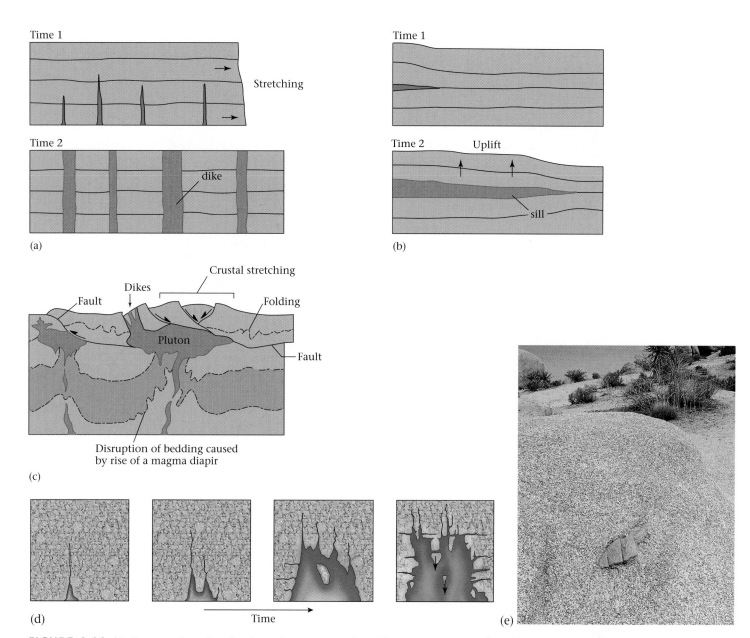

FIGURE 4.14 (a) Cross sections showing how the crust stretches sideways to accommodate dike intrusion. (b) Cross sections showing how intrusion of a sill may raise the surface of the Earth. (c) Ways in which crust accommodates emplacement of a pluton. (d) A magma stoping into country rock, gradually breaking off and digesting blocks as it moves. (e) Xenolith in a granite outcrop in the Mojave Desert.

it transports heat away from a magma very slowly, so magma underground (in an intrusive environment) cools slowly. In contrast, if you spill coffee on a table, it cools quickly because it loses heat to the cold air. Similarly, lava that erupts at the ground surface cools quickly because it is initially surrounded by air or water, which conduct heat away quickly.

Three factors control the cooling time of magma that intrudes below the surface.

• *The depth of intrusion:* Magma intruded deep in the crust, where it is surrounded by warm country rock, cool more slowly than shallow intrusions, surrounded by cold country rock.

Mantle plume and
a hot-spot volcano

Subduction yields
a volcanic arc.

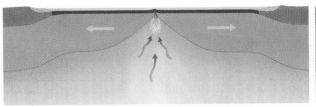

Melting occurs beneath
a mid-ocean ridge.

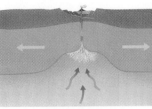

Melting occurs beneath
a continental rift.

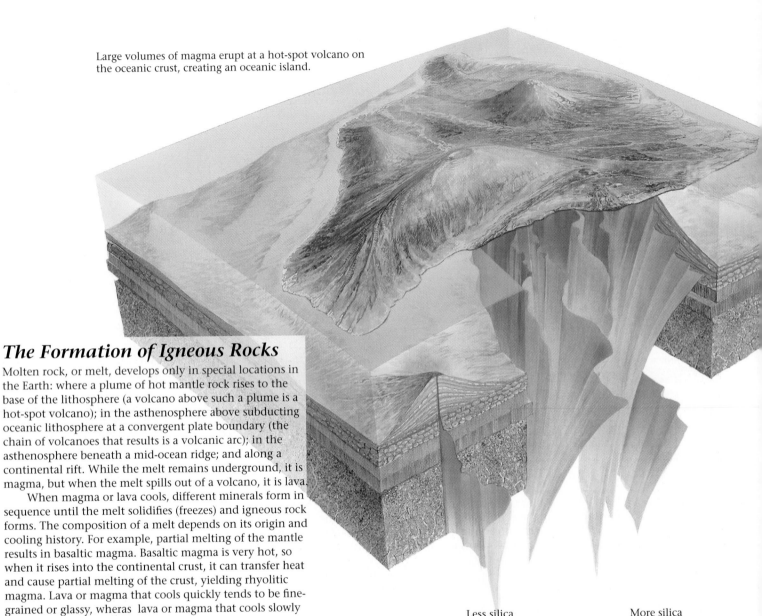

Large volumes of magma erupt at a hot-spot volcano on
the oceanic crust, creating an oceanic island.

The Formation of Igneous Rocks

Molten rock, or melt, develops only in special locations in
the Earth: where a plume of hot mantle rock rises to the
base of the lithosphere (a volcano above such a plume is a
hot-spot volcano); in the asthenosphere above subducting
oceanic lithosphere at a convergent plate boundary (the
chain of volcanoes that results is a volcanic arc); in the
asthenosphere beneath a mid-ocean ridge; and along a
continental rift. While the melt remains underground, it is
magma, but when the melt spills out of a volcano, it is lava.

When magma or lava cools, different minerals form in
sequence until the melt solidifies (freezes) and igneous rock
forms. The composition of a melt depends on its origin and
cooling history. For example, partial melting of the mantle
results in basaltic magma. Basaltic magma is very hot, so
when it rises into the continental crust, it can transfer heat
and cause partial melting of the crust, yielding rhyolitic
magma. Lava or magma that cools quickly tends to be fine-
grained or glassy, wheras lava or magma that cools slowly
tends to be coarse-grained.

Igneous rock that forms by the solidification of
magma underground is intrusive rock. Bloblike intrusions
are plutons; sheetlike intrusions are dikes if they cut across
preexisting layers, and sills if they intrude parallel to
preexisting layers. In rock containing no preexisting
layers, dikes are vertical and sills are horizontal. Intrusions
that are shaped like a blister are called laccoliths.

Volcanoes erupt both lava flows and pyroclastic debris
(ash and other fragmental material ejected explosively).
Igneous rock that forms by the extrusion of lava or the
accumulation of pyroclastic debris is extrusive rock.

Less silica
(basalt)

More silica
(granite)

Granite forms from the cooling of a felsic melt, such as in a
continental pluton, whereas basalt results from the cooling of
a mafic melt, as at an oceanic hot-spot volcano.

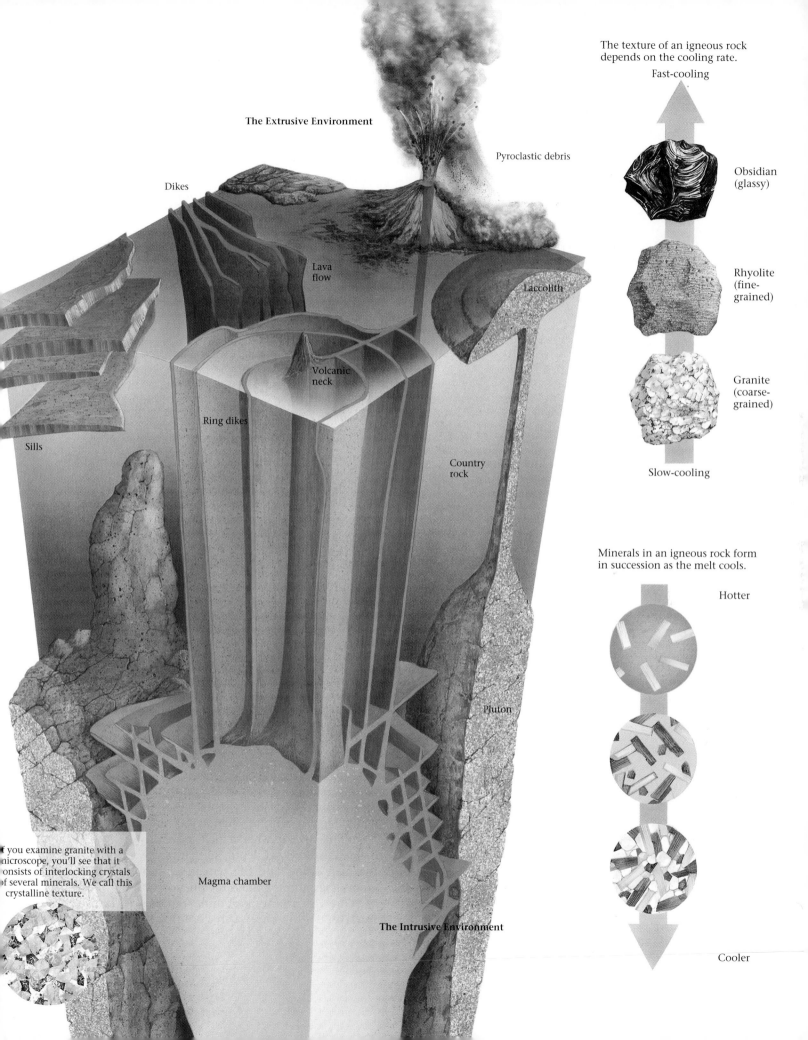

The Extrusive Environment

Dikes

Pyroclastic debris

Lava flow

Laccolith

Volcanic neck

Ring dikes

Country rock

Sills

Pluton

Magma chamber

The Intrusive Environment

If you examine granite with a microscope, you'll see that it consists of interlocking crystals of several minerals. We call this crystalline texture.

The texture of an igneous rock depends on the cooling rate.

Fast-cooling

Obsidian (glassy)

Rhyolite (fine-grained)

Granite (coarse-grained)

Slow-cooling

Minerals in an igneous rock form in succession as the melt cools.

Hotter

Cooler

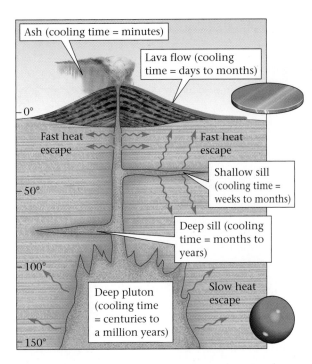

FIGURE 4.15 The cooling time of an intrusion decreases with greater depth (because country rock is hotter at greater depth). The cooling time also depends on the shape of the intrusion (a thin sheet cools faster than a sphere of the same volume) and on the size of the intrusion (a small intrusion cools faster than a large one). Thus, ash (formed when droplets of lava come in contact with air) cools fastest, followed by a thin sheet of lava, a shallow sill, a deep sill, and a deep pluton.

- *The shape and size of a magma body:* Heat escapes from an intrusion at the intrusion's surface, so the greater the surface area for a given volume of intrusion, the faster it cools. Thus, a pluton cools more slowly than a tabular intrusion with the same volume (because a tabular intrusion has a greater surface area across which heat can be lost), and a large pluton cools more slowly than a small pluton of the same shape. Similarly, droplets of lava cool faster than a lava flow, and a thin flow of lava cools more quickly than a thick flow (►Fig. 4.15).

- *The presence of circulating groundwater:* Water passing through magma absorbs and carries away heat, much like the coolant that flows around an automobile engine.

4.7 IGNEOUS ROCK TEXTURES

When talking about an igneous rock, a description of texture tells us whether the rock consists of glass, crystals, or fragments and may also indicate the size of the crystals or fragments.

- *Glassy texture:* Rocks made of a solid mass of glass, or of tiny crystals surrounded by glass are **glassy igneous rocks** (►Fig. 4.16a).

- *Interlocking texture:* Rocks that consist of mineral crystals that intergrow when the melt solidifies, and thus fit together like pieces of a jigsaw puzzle, have an interlocking texture (►Fig. 4.16b). Rocks with an interlocking texture are **crystalline igneous rocks.** The interlocking of crystals in these rocks occurs because once some grains have developed, they interfere with the growth of later-formed grains. The last grains to form end up filling irregular spaces between preexisting grains.

  Geologists distinguish subcategories of crystalline igneous rocks according to the size of the crystals. Coarse-grained (phaneritic) rocks have crystals large enough to be identified with the naked eye. Fine-grained (aphanitic) rocks have crystals too small to be identified with the naked eye. Porphyritic rocks have larger crystals surrounded by a mass of fine crystals. In a porphyritic rock, the larger crystals are called phenocrysts, while the mass of finer crystals is called groundmass.

- *Fragmental texture:* Rocks with a fragmental texture, consisting of igneous fragments that are packed together, welded together, or cemented together after having solidified, are **fragmental igneous rocks.**

Cooling time plays an important role in determining texture. The presence of glass indicates that cooling happened so quickly that the atoms within a lava didn't have time to arrange into crystal lattices. Crystalline rocks form when a melt cools more slowly. In crystalline rocks, grain size depends on cooling time. A melt that cools rapidly, but not rapidly enough to make glass, forms fine-grained (aphanitic) rock, because many crystal seeds form but none has time to grow large (►Fig. 4.16c). A melt that cools very slowly forms a coarse-grained (phaneritic) rock, because relatively few seeds form and each crystal has time to grow large (►Fig. 4.16d).

Because of the relationship between cooling time and texture, lava flows, dikes, and sills tend to be composed of fine-grained igneous rock. In contrast, plutons tend to be composed of coarse-grained rock. Plutons that intrude into hot country rock at great depth cool very slowly and thus tend to have larger crystals (that is, are coarser grained) than plutons that intrude into cool country rock at shallow depth, where they cool relatively rapidly. Porphyritic rocks (rocks in which larger grains, or phenocrysts, are embedded on finer groundmass) form when a melt cools in two stages. First, the melt cools slowly at depth, so that phenocrysts form. Then, the melt erupts and the remainder cools quickly, so that groundmass forms around the phenocrysts.

One important crystalline igneous rock type, **pegmatite,** doesn't necessarily cool slowly. Pegmatite, a very

FIGURE 4.16 (a) Thin section of a glassy, fine-grained volcanic rock. The light-colored grains are crystals, and the dark matrix (the region between the crystals) is glass. (b) This thin section of granite shows relatively large interlocking crystals. (c) Basalt, a fine-grained igneous rock; (d) granite; (e) obsidian; (f) pumice; (g) gabbro; (h) rhyolitic welded tuff.

coarse-grained igneous rock, contains crystals up to tens of centimeters across and occurs in tabular intrusions called pegmatite dikes. A variety of gemstones grow in pegmatites. Because pegmatite occurs in dikes, which generally cool quickly, the coarseness of the rock may seem surprising. Researchers have shown that pegmatites are coarse because they form from *water-rich* melts in which atoms can move around so rapidly that large crystals can grow very quickly.

4.8 CLASSIFYING IGNEOUS ROCKS

Because melts can have a variety of compositions and can freeze to form igneous rocks in many different environments above and below the surface of the Earth, we observe a wide spectrum of igneous rock types (▶Fig. 4.16c–h). We classify these according to their texture and composition. Studying a rock's texture tells us about the rate at which it cooled and therefore the environment in which it formed, whereas studying its composition tells us about the original source of the magma and the way in which the magma evolved before finally solidifying.

Glassy Igneous Rocks

Glassy texture develops more commonly in felsic igneous rocks because the presence of a high concentration of silica inhibits the easy growth of crystals. But basaltic and intermediate lavas can form glass if they cool rapidly enough. In some cases, a rapidly cooling lava freezes while it still contains a high concentration of gas bubbles—these bubbles remain as open holes known as **vesicles.** Geologists distinguish among several different kinds of glassy rocks.

- **Obsidian** is a mass of solid, felsic glass. It tends to be black or brown (Fig. 4.16e). Because it breaks conchoidally, sharp-edged pieces split off its surface when you hit a sample with a hammer. Pre-industrial people worldwide used such pieces for arrowheads, scrapers, and knife blades.

- **Pumice** is a glassy felsic volcanic rock that contains abundant vesicles, giving it the appearance of a sponge (Fig. 4.16f). Pumice forms by the quick cooling of frothy lava that resembles the foam head in a glass of beer. In some cases, pumice contains so many air-filled pores that it can actually float on water, like styrofoam. Ground-up pumice makes the grainy abrasive that blue-jean manufacturers use to "stonewash" jeans.

- **Scoria** is a glassy mafic volcanic rock that contains abundant vesicles (more than about 30%). Generally, the bubbles in scoria are bigger than those in pumice, and the rock, overall, looks darker.

Crystalline Igneous Rocks

The scheme for classifying the principal types of crystalline igneous rocks is quite simple. The different compositional classes are distinguished on the basis of silica content—ultramafic, mafic, intermediate, or felsic—whereas the different textural classes are distinguished according to whether or not the grains are coarse or fine. The chart in ▶Figure 4.17 gives the texture and composition of the most commonly used rock names. As a rough guide, the color of an igneous rock reflects its composition: mafic rocks tend to be black or dark gray, intermediate rocks tend to be lighter gray or greenish gray, and felsic rocks tend to be light tan to pink or maroon.

As an example note that, according to Figure 4.17a, rhyolite and granite have the same chemical composition but differ in grain size. Which of these two rocks will develop from a melt of felsic composition depends on the cooling rate. A felsic lava that solidifies quickly at the Earth's surface or in a thin dike or sill turns into fine-grained rhyolite, but the same magma, if solidifying slowly at depth in a pluton, turns into coarse-grained granite. A similar situation holds for mafic lavas—a mafic lava that cools quickly in a lava flow forms basalt, but a mafic magma that cools slowly forms gabbro (see Fig. 4.16g).

Fragmental Igneous Rocks

Geologists distinguish among different kinds of fragmental igneous rocks according to the size of the fragments and the way in which the fragments stick together. Fragmental rocks form when a flow shatters into pieces during its movement and then the pieces weld together; when a fountain of lava sends droplets of lava into the air, forming bombs or cinders that accumulate around the vent; or when ash, crystals, and preexisting volcanic rock blast from a volcano during an eruption. Fragmental rocks composed of debris that has been blasted out of a volcano or thrown out in a fountain are commonly called **pyroclastic rocks** (from the Greek *pyro* for "fire," and from the word *clastic,* which signifies that the rocks are composed of grains that were already solid when they stuck together). Because of the way in which they form, the fragments in pyroclastic rocks can be glassy.

- **Tuff** is a fine-grained pyroclastic igneous rock composed of volcanic ash and/or fragments of pumice. It may also include fragments of preexisting rock, such as rhyolite, making up a volcano. Tuff that settles from air is called air-fall tuff, and tuff formed by the welding together of hot shards is called welded tuff (see Fig. 4.16h).

- **Volcanic breccia** consists of larger fragments of volcanic debris, either that fall through the air and accumulate, or that form when a lava flow breaks into pieces.

- **Hyaloclasite** is formed from lava that erupts under water or ice and cools so quickly that it shatters into glassy fragments that then weld or cement together.

In this chapter, we've focused on the diversity of igneous rocks, and on why and where they form. We've seen that extrusive rocks erupt at volcanoes. In Chapter 7, we will examine volcanic eruptions in greater detail and will discuss the geologic settings at which igneous activity takes place.

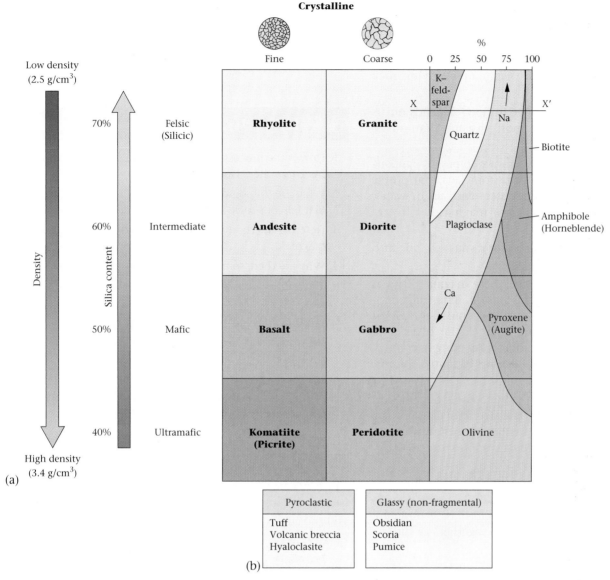

FIGURE 4.17 (a) Crystalline (nonglassy) igneous rocks are distinguished from each other by their grain size and composition. The right side of the chart shows the relative proportions of different minerals in the different rock types. To read this chart, draw a horizontal line next to a rock name; the minerals that the line crosses are the minerals found in that rock. (For example, granite [line X–X'] includes K-feldspar, quartz, plagioclase, amphibole, and biotite.) (b) The principal types of glassy and fragmental igneous rocks can be separated into several categories.

CHAPTER SUMMARY

• Magma is liquid rock (melt) under the Earth's surface. Lava is melt that has erupted from a volcano at the Earth's surface.

• Magma forms when hot rock in the Earth partially melts. This process only occurs under certain circumstances—when the pressure decreases (decompression), when volatiles (such as water or carbon dioxide) are added to hot rock, and when heat is transferred into the crust by magma rising from the mantle into the crust.

• Magma occurs in a range of compositions: felsic (silicic), intermediate, mafic, and ultramafic. The composition of magma reflects the original composition of the rock from which the magma formed and the way the magma evolves during assimilation and fractional crystallization.

- During partial melting, only part of the source rock melts to create magma. Magma tends to be more felsic than the rock from which it was extracted.

- Magma rises from depth because of its buoyancy and because pressure caused by the weight of overlying rock squeezes magma upward.

- Magma viscosity (its resistance to flow) depends on its composition. Felsic magma is more viscous than mafic magma.

- Geologists distinguish between two types of igneous rocks. Extrusive igneous rocks form from lava that erupts out of a volcano and freezes in contact with air or water. Intrusive igneous rocks develop from magma that freezes inside the Earth.

- Lava may solidify to form flows, or it may explode into the air to form ash.

- Intrusive igneous rocks form when magma intrudes into preexisting rock below Earth's surface. Blob-shaped intrusions are called plutons. Sheetlike, or tabular, intrusions that cut across layering in country rock are dikes, and sheetlike intrusions that form parallel to layering in country rock are sills. Huge intrusions, made up of many plutons, are known as batholiths.

- The rate at which intrusive magma cools depends on the depth at which it intrudes, the size and shape of the magma body, and whether circulating groundwater is present. The cooling time influences the texture of an igneous rock.

- Crystalline (nonglassy) igneous rocks are classified according to texture and composition. Glassy igneous rocks are classified according to texture (a solid mass is obsidian, while ash that has cemented or welded together is a tuff).

KEY TERMS

assimilation (p. 107)
batholith (p. 110)
Bowen's reaction series (p. 108)
crystalline igneous rock (p. 116)
dike (p. 110)
extrusive igneous rock (p. 104)
felsic magma (p. 105)
fractional crystallization (p. 107)
fragmental igneous rock (p. 116)
geotherm (p. 104)
glassy igneous rock (p. 116)
hyaloclasite (p. 118)
igneous rock (p. 102)

intermediate magma (p. 105)
intrusive igneous rock (p. 104)
laccolith (p. 110)
lava (p. 102)
mafic magma (p. 105)
magma (p. 102)
obsidian (p. 118)
partial melting (p. 105)
pegmatite (p. 116)
pluton (p. 110)
pumice (p. 118)
pyroclastic debris (p. 104)
pyroclastic rock (p. 118)
scoria (p. 118)
sill (p. 110)

stoping (p. 110)
tuff (p. 118)
ultramafic magma (p. 105)
vesicle (p. 118)

volcanic ash (p. 104)
volcanic breccia (p. 118)
volcano (p. 102)
xenolith (p. 110)

REVIEW QUESTIONS

1. How is the process of freezing magma similar to that of freezing water? How is it different?

2. What are the sources of heat in the Earth? How did the first igneous rocks on the planet form?

3. Describe the three processes that lead to the formation of magmas.

4. Why are there so many different types of magmas?

5. Why do magmas rise from depth to the surface of the Earth?

6. What factors control the viscosity of a melt?

7. What factors control the cooling time of a magma within the crust?

8. How does grain size reflect the cooling time of a magma?

9. What does the mixture of grain sizes in a porphyritic igneous rock indicate about its cooling history?

SUGGESTED READING

Best, M. G., and E. H. Christiansen. 2001. *Igneous Petrology.* 2nd ed. Oxford, UK: Blackwell Science.

Faure, G. 2000. *Origin of Igneous Rocks: The Isotopic Evidence.* New York: Springer-Verlag.

LeMaitre, R. E., ed. 2002. *Igneous Rocks: A Classification and Glossary of Terms,* 2nd ed. Cambridge, UK: Cambridge University Press.

Leyrit, H., C. Montenat, and P. Bordet, eds. 2000. *Volcaniclastic Rocks, from Magmas to Sediments.* London: Taylor and Francis.

Middlemost, E. A. K. 1997. *Magmas, Rocks and Planetary Development: A Survey of Magma/Igneous Rock Systems.* Boston: Addison/Wesley.

Philpotts, A. R. 2003. *Petrography of Igneous and Metamorphic Rocks.* Long Grove, Ill: Waveland Press.

Winter, J. D. 2001. *Introduction to Igneous and Metamorphic Petrology.* Upper Saddle River, N.J.: Prentice-Hall.

Young, D. A. 2003. *Mind over Magma.* Princeton, N.J.: Princeton University Press.

A Surface Veneer: Sediments, Soils, and Sedimentary Rocks

5.1 INTRODUCTION

In the 1950s, the government of Egypt decided to build the Aswan High Dam to trap water of the Nile River in a huge reservoir before the water could reach the Mediterranean Sea. To identify a good site for the dam's foundation, geologists drilled holes into the ground to find the depth to bedrock. They discovered that the present-day Nile River flows on the surface of a 1.5-km-thick layer of gravel, sand, and mud that fills what was once a canyon as large as the Grand Canyon (▶Fig. 5.1). How could the river once have carved a canyon 1.5 km deep, and why did this canyon later fill with sediment?

The origin of the pre-Nile canyon remained a mystery until the summer of 1970, when geologists had the opportunity to drill holes into the floor of the Mediterranean Sea to find out what lay beneath. They expected the sea floor to be covered with shells of plankton (tiny floating organisms) that had settled out of the water, or with clay that rivers had carried to the sea. To their surprise, however, they found that, in addition to clay and plankton shells, a 2-km-thick layer of halite and gypsum lies beneath the floor of the Mediterranean Sea. These minerals form when seawater dries up, allowing the salt in seawater to precip- itate. The researchers realized that to

These cliffs, near Bryce Canyon (Utah), expose beds of sedimentary rock deposited in lakes, and by streams, over 40 million years ago. Present-day erosion has produced aprons of debris at the base of the cliffs.

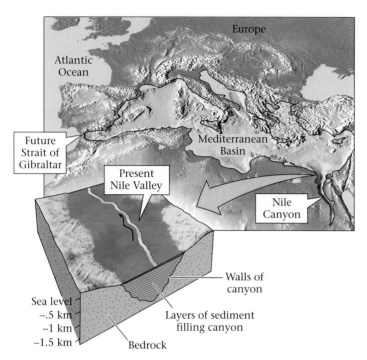

FIGURE 5.1 The present Nile River overlies a deep canyon, now buried by layers of sand and mud. The bottom of the canyon lies 1.5 km below sea level.

yield a layer that is 2 km thick, the entire Mediterranean would have had to dry up completely several times, with the sea refilling after each drying event. This discovery solved the mystery of the pre-Nile canyon. When the Mediterranean Sea dried up, the Nile River could dig into its substrate and cut a canyon. And when the sea refilled with water, this canyon flooded and filled with sand and gravel.

Why did the Mediterranean Sea dry up? Only 10% of the water in the Mediterranean comes from rivers, and since the sea lies in a hot, dry region, ten times that amount of water evaporates from its surface each year. Thus, most of the water filling the Mediterranean enters through the Strait of Gibraltar from the Atlantic Ocean—if this flow stops, the Mediterranean Sea evaporates. About 6 million years ago, the northward-drifting African Plate collided with the European Plate, forming a natural dam separating the Mediterranean from the Atlantic. When global sea level dropped, water stopped flowing in from the Atlantic and the Mediterranean evaporated. All the salt that had been dissolved in its water accumulated as a solid deposit of halite and gypsum on the floor of the resulting basin. When sea level rose, a gigantic flood rushed from the Atlantic into the Mediterranean, filling the basin again. This process repeated many times. About 5.5 million years ago, the Mediterranean rose to its present level and gravel, sand, and mud carried by the Nile River filled the Nile canyon.

Geologists refer to the kinds of deposits just described—sand, mud, gravel, halite and gypsum layers, shell fragments—as sediment. **Sediment,** broadly defined, consists of loose fragments of rocks or minerals broken off bedrock, mineral crystals that precipitate directly out of water, and shells or shell fragments. Important aspects of Earth history, including the amazing story of the Mediterranean Sea, come from studying sediments—not only those that remain "unconsolidated" (loose and not connected), but also those that have been bound together into sedimentary rock. Formally defined, **sedimentary rock** is rock that forms at or near the surface of the Earth by the cementing together of loose grains broken off pre-existing rock, by the precipitation of mineral aggregates from water solutions, or by the cementing together of shells and shell fragments. Layers of sediment and sedimentary rock are like the pages of a book, recording tales of ancient events and ancient environments on the ever-changing face of the Earth.

Sediments and sedimentary rocks only occur in the upper part of the crust—in effect, they form a surface veneer, or "cover," on older igneous and metamorphic rocks, which make up the "basement" of the crust (▶Fig. 5.2). This veneer ranges in thickness from nonexistent, in places where igneous and metamorphic rocks crop out at the Earth's surface, to several kilometers. Sedimentary rocks are a uniquely important rock type, both because they contain a record of Earth history and because they contain the bulk of the Earth's energy resources. Further, some sediments transform into soil, essential for life. Let's now look at how sediments, soils, and sedimentary rocks form, and what these materials can tell us about the Earth System.

5.2 WEATHERING: THE FORMATION OF SEDIMENT

The Mountains Crumble

Weathering refers to the processes that break up and corrode solid rock, eventually transforming it into sediment. We can say that rock that has not undergone weathering is unweathered or "fresh" (▶Fig. 5.3). All mountains and other features on the Earth's surface sooner or later crumble away because of weathering. Just as a plumber can unclog a drain by using physical force (with a plumber's snake) or by causing a chemical reaction (with a dose of liquid drain opener), nature can attack rocks in two ways, so geologists distinguish between two types of weathering: physical weathering and chemical weathering.

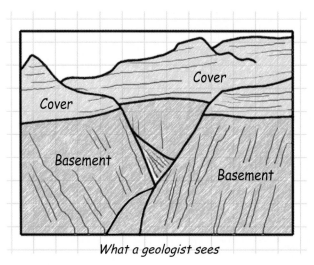

What a geologist sees

FIGURE 5.2 Near the bottom of the Grand Canyon, we can see the boundary between the sedimentary veneer, or cover (here, a succession of horizontal layers), and the older basement (here, the steep cliff of dark metamorphic rock). The Colorado River flows along the floor of the canyon. A geologist's sketch emphasizes the contact, or boundary, between cover and basement.

Physical Weathering

Physical weathering, sometimes referred to instead as mechanical weathering, breaks intact rock into unconnected grains or chunks, collectively called debris or detritus. Each size range of grains has a name (measurements are grain diameters):

- *boulders* more than 256 millimeters (mm)
- *cobbles* between 64 mm and 256 mm
- *pebbles* between 2 mm and 64 mm
- *sand* between 1/16 mm and 2 mm
- *silt* between 1/256 mm and 1/16 mm
- *mud* less than 1/256 mm

Many different phenomena contribute to physical weathering.

FIGURE 5.3 This outcrop shows the contrast between "fresh" and weathered granite. The rock below the notebook is fresh—the outcrop face is a fairly smooth fracture. The rock above the notebook is weathered—the outcrop face is crumbly, breaking into grains that have fallen and collected on the ledge.

Weathered granite

Unweathered (fresh) granite

Jointing. Rocks buried deep in the Earth's crust endure enormous pressure due to the weight of overlying rock or overburden. Rocks at depth are also warmer than rocks nearer the surface because of Earth's geothermal gradient (see Chapter 4). Over long periods of time, moving water, air, and ice at the Earth's surface grind away and remove overburden, so rock formerly at depth rises closer to the Earth's surface. As a result, the pressure squeezing this rock decreases, and the rock becomes cooler. A change in pressure and temperature causes rock to change shape slightly. Such changes cause hard rock to break into pieces. Natural cracks that form in rocks due to removal of overburden or due to cooling (and for other reasons as well; see Chapter 9) are known as **joints.**

Almost all rock outcrops contain joints; some joints are fairly planar, some curving, and some irregular. Joints can break rock into large or small rectangular blocks, onionlike sheets, irregular chunks, or pillarlike columns. For example, large granite plutons split into onionlike sheets along joints that lie parallel to the mountain face; this process is called exfoliation (▶Fig. 5.4a). Sedimentary rock layers tend to

(a)

(b)

FIGURE 5.4 (a) Exfoliation joints in the Sierra Nevada. (b) Vertical joints in sedimentary rock (Brazil). (c) Talus has accumulated at the base of these cliffs near Mt. Snowdon in Wales.

—Joint

(c)

break into rectangular blocks (▶Fig. 5.4b). Regardless of their orientation, the formation of joints turns formerly intact bedrock into loose blocks. Eventually, these blocks fall from the outcrop at which they formed. After a while, they may collect in an apron of talus, rock rubble at the base of a slope (▶Fig. 5.4c).

Frost wedging. Freezing water bursts pipes and shatters bottles because water expands when it freezes and pushes the walls of the container apart. The same phenomenon happens in rock. When the water trapped in a joint freezes, it forces the joint open and may cause the joint to grow. Such frost wedging helps break blocks free from intact bedrock (▶Fig. 5.5a).

Root wedging. Have you ever noticed how the roots of an old tree can break up a sidewalk? Even though the wood of roots doesn't seem very strong, as roots expand they apply pressure to their surroundings. Tree roots that grow into joints can push joints open in a process known as root wedging (▶Fig. 5.5b). Even the roots of small plants, fungi, and lichen get into the act, by splitting open small cracks and pores.

Salt wedging. In arid climates, dissolved salt in groundwater precipitates and grows as crystals in open pore spaces in rocks. This process, called salt wedging, pushes apart the surrounding grains and so weakens the rock that when exposed to wind and rain, the rock disintegrates into separate grains. The same phenomenon happens in coastal areas, where salt spray percolates into surface rock and then dries (▶Fig. 5.5c).

Thermal expansion. When the heat of an intense forest fire bakes a rock, the outer layer of the rock expands. On cooling, the layer contracts. This change creates forces in the rock sufficient to make the outer part of the rock break off in sheetlike pieces.

Animal attack. Animal life also contributes to physical weathering: burrowing creatures, from earthworms to gophers, push open cracks and move rock fragments. And in the past century, humans have become perhaps the most energetic agent of physical weathering on the planet. When we excavate quarries, foundations, mines, or roadbeds by digging and blasting, we shatter and displace rock that might otherwise have remained intact for millions of years more.

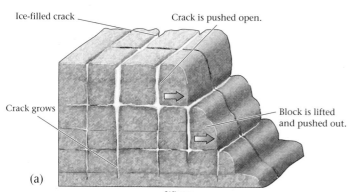

(b)

(c)

FIGURE 5.5 Examples of processes contributing to physical weathering. (a) During the summer, cracks are closed. During the winter, water in the cracks freezes and forces rocks apart. Ice can even lift blocks up. (b) The roots of this old pine tree near the rim of the Grand Canyon, in Arizona, grew in a joint. Eventually, the growth of the tree has pushed the joint open. (c) These gravestones, near the ruin of a medieval abbey on the seacoast near Whitby, England, absorbed salt from the sea spray. Salt wedging has resulted in honeycomb-like weathering.

Chemical Weathering

Up to now we've taken the plumber's-snake approach to breaking up rock; now let's look at the liquid-drain-opener approach. **Chemical weathering** refers to the chemical reactions that alter or destroy minerals when rock comes in contact with water solutions or air. Common reactions involved in chemical weathering include the following.

• *Dissolution.* Chemical weathering during which minerals dissolve into water is called dissolution. Dissolution primarily affects salts and carbonate minerals, but even quartz dissolves slightly (▶Fig. 5.6a–c; see Chapter 16).

• *Hydrolysis.* During hydrolysis, water chemically reacts with minerals and breaks them down (*lysis* means "loosen" in Greek) to form other minerals. For example, hydrolysis reactions in feldspar produce clay.

• *Oxidation.* Chemists refer to a reaction during which an element loses electrons as an oxidation reaction because

commonly such a loss takes place when elements combine with oxygen. The oxidation, or rusting, of iron, serves as an example. Oxidation reactions in rocks transform iron-bearing minerals (such as biotite and pyrite) into a rusty-brown mixture of various iron-oxide and iron-hydroxide minerals.

• *Hydration.* Hydration, the absorption of water into the crystal structure of minerals, causes some minerals, such as certain types of clay, to expand. Such expansion weakens rock.

Not all minerals undergo chemical weathering at the same rates. Some weather in a matter of months or years, whereas others remain unweathered for millions of years. The difference depends partly on crystal structure and partly on chemical composition. For example, when a granite (which contains quartz, mica, and feldspar) undergoes chemical weathering, everything but quartz transforms to clay. That's why beaches typically consist of quartz sand;

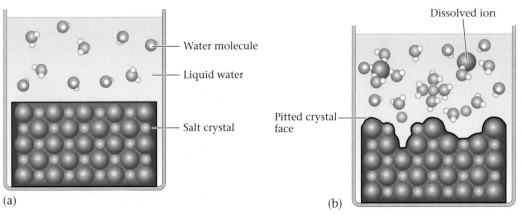

FIGURE 5.6 Weathering by dissolution. (a) A salt crystal consists of ions that can be attracted by polar water molecules. (b) Eventually, water molecules pluck sodium and chlorine ions off the face of the crystal, surround them, and carry them away. (c) Dissolution enlarges joints on the surface of a limestone outcrop and dissolves away sharp edges. In this example from Ireland, wildflowers find a home in the troughs that have formed by dissolution.

quartz is the only mineral left after the other minerals have turned to clay and washed away.

Until fairly recently, geoscientists tended to think of chemical weathering as a strictly inorganic chemical reaction, occurring entirely independently of life forms. But we now realize that organisms play a major role in the chemical-weathering process. For example, the roots of plants, fungi, and lichens secrete organic acids that help dissolve minerals in rocks; these organisms extract nutrients from the minerals. Microbes, such as bacteria, are amazing in that they literally eat minerals for lunch. Bacteria pluck off molecules from minerals and use the energy from the molecules' chemical bonds to supply their own life force.

Physical and Chemical Weathering Working Together

So far we've looked at the processes of chemical and physical weathering separately, but in the real world they happen together, aiding one another in disintegrating rock to form sediment.

Physical weathering speeds up chemical weathering. To understand why, keep in mind that chemical-weathering reactions take place at the surface of a material. Thus, the overall rate at which chemical weathering occurs depends on the ratio of surface area to volume—the greater the surface area, the faster the volume as a whole can chemically weather. When jointing (physical weathering) breaks a large block of rock into smaller pieces, the surface area increases, so chemical weathering happens faster (▶Fig. 5.7).

Similarly, chemical weathering speeds up physical weathering, because chemical weathering dissolves away grains or cements that hold a rock together, transforms hard minerals (like feldspar) into soft minerals (like clay), and causes minerals to absorb water and expand. These phenomena make rock weaker, so it can disintegrate more easily.

Notably, both kinds of weathering happen faster at edges, and even faster at the corners of broken blocks. This is because weathering attacks a flat face from only one direction, an edge from two directions, and a corner from three directions. Thus, with time, edges of blocks become blunt and corners become rounded (▶Fig. 5.8a). In rocks such as granite, which do not contain layering that can affect weathering rates, rectangular blocks transform into a spheroidal shape (▶Fig. 5.8b).

Not all rock types weather at the same rate. When different rocks in an outcrop undergo weathering at different rates, we say that the outcrop has undergone differential weathering. Because of differential weathering, cliffs composed of a variety of rock layers take on a stair-step or sawtooth shape (▶Fig. 5.8c). You can easily see the consequences of differential weathering if you walk through a graveyard. The inscriptions on some headstones are sharp and clear, whereas those on other stones have become blunted or have even disappeared (▶Fig. 5.9a, b). That's because the minerals in these different stones have different resistances to weathering. Granite, an igneous rock with a high quartz content, retains inscriptions the longest. But marble, a metamorphic rock composed of calcite, dissolves away relatively rapidly in acidic rain.

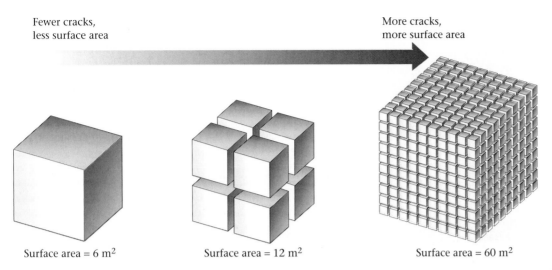

Fewer cracks, less surface area

More cracks, more surface area

Surface area = 6 m²   Surface area = 12 m²   Surface area = 60 m²

FIGURE 5.7 The surface area per unit volume of a block increases every time you break the block into more pieces. For example, a coherent 1 m³ block has a surface area of 6 m². Divide the block into eight pieces, and the surface area increases to 12 m², but the volume stays the same. The rate of chemical weathering increases as the surface area increases, because the weathering reactions occur at the surface. (To picture this, think about how fast granular sugar dissolves as compared with a solid cube of sugar.)

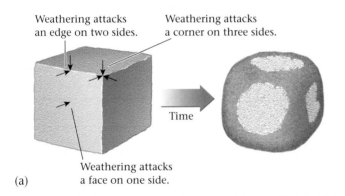

Weathering attacks an edge on two sides.

Weathering attacks a corner on three sides.

Time

Weathering attacks a face on one side.

(a)

(b)

(c)

FIGURE 5.8 (a) Weathering attacks more vigorously at edges and most vigorously at corners, resulting in a rounded block. (b) Spheroidal blocks of granite in Joshua Tree National Monument, California. (c) Sawtooth shape of an outcrop of weathered sedimentary rock in New Mexico. Resistant layers stick out relative to weak layers.

FIGURE 5.9 (a) Inscriptions in a granite headstone remain sharp for a long time. This example dates from 1856. (b) Inscriptions in a marble headstone weather away fairly rapidly. This example, from the same cemetery, dates from 1872.

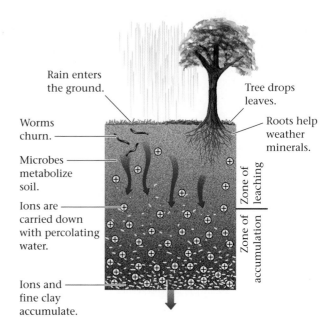

FIGURE 5.10 During the formation of soil, the downward percolation of water creates a zone of leaching and a zone of accumulation. The percolating water carries ions and clay downward. Soil formation also involves the metabolism of microbes and fungi and the addition of organic matter at the surface and underground.

5.3 SOIL

If you've ever had the chance to dig in a garden, you've seen firsthand that material in which flowers grow looks and feels different from beach sand or from potter's clay. We call material in a garden dirt or, more technically, soil. **Soil** consists of rock or sediment that has been modified by physical and chemical interaction with organic material and rainwater, over time, to produce a substrate that can support the growth of plants. Soil is one of our planet's most valuable resources, for without it there could be no agriculture, forestry, ranching, or home gardening.

Three processes taking place at or just below the surface of the Earth contribute to soil formation. First, chemical and physical weathering produces loose debris, new mineral grains (such as clay), and ions in solution. Second, rainwater percolates through the debris and carries dissolved ions and clay flakes downward. The region in which this downward transport occurs is the **zone of leaching,** because leaching means "extracting and absorbing." Farther down, new mineral crystals precipitate directly out of the water or form by reaction of the water with debris, and the water leaves behind its load of fine clay. The region in which new minerals and clay collect is the **zone of accumulation** (▶Fig. 5.10). Third, microbes, fungi, plants, and animals interact with sediment, by producing acids that weather grains, by absorbing nutrient atoms (such as K, Ca, Mg), and by leaving behind waste and remains. Plant roots and burrowing animals (insects, worms, and gophers) churn and break up the soil, and microbes metabolize mineral grains and release chemicals.

Because different soil-forming processes operate at different depths, soils typically develop distinct zones, known as **soil horizons,** arranged in a vertical sequence called a **soil profile** (▶Fig. 5.11). Let's look at an idealized soil profile, from top to bottom, using a soil formed in a temperate forest as our example. The highest horizon is the O-horizon (the prefix stands for organic), so called because it consists almost entirely of organic matter and contains barely any mineral matter. Below the O-horizon we find the A-horizon, in which humus has decayed further and has mixed with mineral grains (clay, silt, and sand). Water percolating through the A-horizon causes chemical weathering reactions to occur and produces ions in solution and new clay materials. The downward-moving water eventually carries soluble chemicals and fine clay deeper into the subsurface. The A-horizon constitutes dark gray to blackish-brown topsoil, the fertile portion of soil that farmers till for planting crops. In some places, the A-horizon grades downward into the E-horizon, a soil level that has undergone substantial leaching but has not yet mixed with organic material. Ions and clay leached and transported down from above accumulate in the B-horizon, or subsoil. As a result, new minerals form, and clay fills open spaces. Note, from our description, that the O-, A-, and E-horizons make up the zone of leaching, whereas the B-horizon makes up the zone of accumulation.

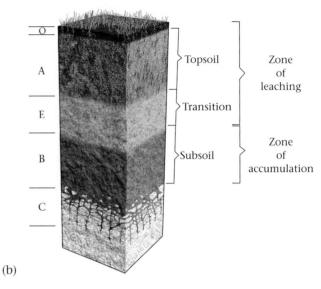

FIGURE 5.11 The process of soil formation results in distinctive soil profiles. (a) In this soil exposed on a cliff face, the dark layer (horizon) at the top is the organic-rich layer. Because of the redistribution of elements, the different horizons have different colors. (b) This schematic column shows idealized soil horizons.

Finally, at the base of a soil profile we find the C-horizon, which consists of material derived from the substrate that's been chemically weathered and broken apart, but has not yet undergone leaching or accumulation. The C-horizon grades downward into unweathered bedrock, or into unweathered sediment.

As farmers, foresters, and ranchers well know, the soil in one locality can differ greatly from the soil in another, in terms of composition, thickness, and texture. And crops that grow well in one type of soil may wither and die in another. Such diversity exists because the makeup of a soil depends on several soil-forming factors.

- *Climate:* Large amounts of rainfall and warm temperatures accelerate chemical weathering and cause most of the soluble elements to be leached. In regions with small amounts of rainfall and cooler temperatures, soils take a long time to develop and can retain unweathered minerals and soluble components. Climate seems to be the single most important factor in determining the nature of soils that develop.

- *Substrate composition:* Some soils form on basalt, some on granite, some on volcanic ash, some on recently deposited quartz silt. These different substrates consist of different materials, so the soils formed on them end up with different chemical compositions.

- *Slope steepness:* A thick soil can accumulate under land that lies flat. But on a steep slope, regolith may wash away before it can evolve into a soil. Thus, all other factors being equal, soil thickness increases as the slope angle decreases.

- *Wetness:* Depending on the details of local topography and on the depth below the surface at which groundwater occurs, some soil is wetter than other soil in the same region. Wet soils tend to contain more organic material than do dry soils.

- *Time:* Because soil formation is an evolutionary process, a young soil tends to be thinner and less developed than an old soil. The rate of soil formation varies greatly with environment.

- *Vegetation type:* Different kinds of plants extract or add different nutrients and quantities of organic matter to a soil. Also, some plants have deeper root systems than others and help prevent soil from washing away. Thus, the type of plants growing affects soil composition and thickness.

Soil scientists worldwide have struggled mightily to develop a rational scheme for classifying soils. Not all schemes utilize the same criteria, and even today there is not worldwide agreement on which works best. An older scheme, which worked reasonably well in the United States, divided soils into categories primarily on the basis of the elements that accumulated in the B-horizon (▶Fig. 5.12a–c). In this scheme, pedalfer soil forms in temperate climates from a substrate that contains aluminum (*al*) and iron (*fer*); they have well-defined horizons, including an O-horizon and an organic-rich A-horizon (▶Fig. 5.12a). Pedocal soils form in arid climates and tend to be thin. Such soils do not have an O-horizon, and their A-horizon contains unweathered minerals, rock fragments, and a relatively high concentration of

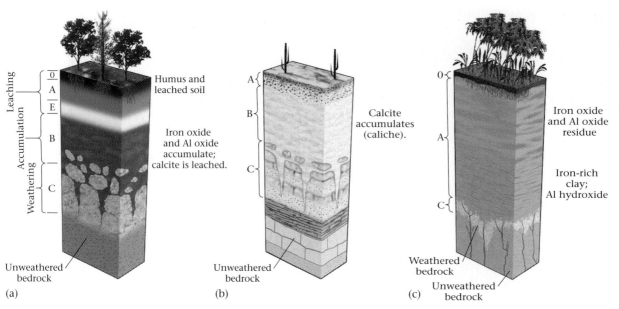

FIGURE 5.12 (a) In a pedalfer soil (alfisol) formed in a temperate climate, an O-horizon forms on top, and because of the moderate amount of rainfall, materials leached from the A-horizon can accumulate in the B-horizon. In this example, the C-horizon happens to consist of weathered granite. (b) Because of low rainfall, a thin pedocal soil (aridisol) in a desert has only a thin A-horizon. Soluble minerals, specifically calcite, that would be washed away in a temperate climate can accumulate in the B-horizon, creating caliche. In this example, the C-horizon consists of weathered limestone. (c) In a thick tropical laterite (oxisol), so much water percolates down from the heavy rainfall that all reactive minerals dissolve or break down and get carried away. This leaves only a residue of iron oxide and/or aluminum (Al) oxide, which are both very stable. There is no real zone of accumulation, but at depth, iron-rich clays collect. Here, the C-horizon consists of weathered metamorphic rock.

soluble minerals such as calcite, but very little organic matter. Calcite in a pedocal soil accumulates in the B-horizon and may cement the soil together, creating a solid mass sometimes called caliche or calcrete (►Fig. 5.12b). Because evaporation rates in desert regions are high, water sometimes moves by capillary action upward through pedocal soil, contributing to accumulation of salt and calcite in horizons near the ground surface.

Laterite soil forms in tropical regions where abundant rainfall drenches the land during the rainy season, and the soil dries during the dry season. Because so much percolating water passes down through the soil, just about all mineral components get leached out of the soil until only insoluble iron and/or aluminum oxide remain. In fact, so much water passes through the soil that accumulation cannot take place, even at depth, so this soil has no B-horizon (►Fig. 5.12c). During the dry season, capillary action brings water upward, contributing to the supply of oxide minerals. Iron oxide gives laterite a brick-red color.

In recent years, soil scientists have developed more complex and comprehensive classification schemes. One of these, the U.S. Comprehensive Soil Classification System, which distinguishes among twelve orders of soil, is based

on both physical characteristics and environment of formation (►Table 5.1). Different orders typically form in different environments. For example, an aridisol (≈ pedocal) is a soil that forms in arid climates, contains very little organic matter, and commonly contains caliche, whereas an alfisol (≈ pedalfer) develops in moist forests, has well-developed horizons, and contains abundant nutrients. Canadians use a different scheme focusing only on soils that develop north of the 40th parallel.

As we noted earlier, climate plays a major role in controlling the development of soil. Climate, in turn, depends in part on latitude and elevation. Thus, we can correlate the character and thickness of a soil with latitude and elevation.

As we have seen, soils take time to form, so soils capable of supporting crops or forests should be considered a natural resource worthy of protection. However, agriculture, overgrazing, and clear-cutting have led to the destruction of soil. Crops rapidly remove nutrients from soil, so if they are not replaced, the soil will not contain sufficient nutrients to maintain plant life. When the natural plant cover disappears, the surface of the soil becomes exposed to wind and water. Actions such as the impact of falling raindrops or the rasping of a plow break up the soil at the surface, with the

TABLE 5.1 Soil Orders (U.S. Comprehensive Soil Classification System)

Alfisol	Gray/brown, has subsurface clay accumulation and abundant plant nutrients. Forms in humid forests.
Andisol	Forms in volcanic ash.
Aridisol	Low in organic matter, has carbonate horizons. Forms in arid environments.
Entisol	Has no horizons. Formed very recently.
Gelisol	Underlain with permanently frozen ground.
Histosol	Very rich in organic debris. Forms in swamps and marshes.
Inceptisol	Moist, has poorly developed horizons. Formed recently.
Mollisol	Soft, black, and rich in nutrients. Forms in subhumid to subarid grasslands.
Oxisol	Very weathered, rich in aluminum and iron oxide, low in plant nutrients. Forms in tropical regions.
Spodosol	Acidic, low in plant nutrients, ashy, has accumulations of iron and aluminum. Forms in humid forests.
Ultisol	Very mature, strongly weathered soils, low in plant nutrients.
Vertisol	Clay-rich soils capable of swelling when wet, and shrinking and cracking when dry.

result that it can wash away in water or blow away as dust. When this happens, **soil erosion,** the removal of soil by running water or by wind, takes place (▶Fig. 5.13). In some cases, almost six tons of soil may be lost from an acre of land per year. Human activities increase rates of soil erosion by 10 to 100 times, so that it far exceeds the rate of soil formation. Droughts exacerbate the situation. For example, during the 1930s a succession of droughts killed off so much vegetation in the American plains that wind stripped the land of soil and caused devastating dust storms. Large numbers of people were forced to migrate away from the Dust Bowl of Oklahoma and adjacent areas.

The consequences of rainforest destruction on soil are particularly profound. In an established rainforest, lush growth provides sufficient organic debris so that trees can grow. But if the forest is logged, or cleared for agriculture, the humus rapidly disappears leaving laterite that contains few nutrients. Crop plants consume whatever nutrients there are so rapidly that the soil becomes infertile after only a year or two, useless for agriculture and unsuitable for re-growth of rainforest trees.

FIGURE 5.13 In this image, we can see that the lack of natural plant coverage has led to severe soil erosion by wind. Similar conditions produced the "Dust Bowl" of the 1930s.

5.4 CLASSES OF SEDIMENTARY ROCKS

So far, we've learned that weathering attacks bedrock and breaks it down to form dissolved ions and loose sediment grains. What happens next? The products of weathering may become components of soil, as we have seen, or they can become buried and transformed into sedimentary rock. Sedimentary rock develops from a variety of materials in a variety of environments. Thus, there are many different kinds of sedimentary rock. Geologists divide sedimentary rocks into four major classes, based on their mode of origin. (1) **Clastic sedimentary rock** consists of cemented-together solid fragments and grains derived from preexisting rocks (clastic comes from the Greek *klastos*, meaning "broken"). (2) **Biochemical sedimentary rock** consists of the shells of organisms. (3) **Organic sedimentary rock** consists of carbon-rich relicts of plants. And (4) **chemical sedimentary rock** is made up of minerals that precipitate directly from water solutions. Let's look at these classes in more detail.

Clastic Sedimentary Rocks

Formation. Nine hundred years ago, a thriving community of Native Americans inhabited the high plateau of Mesa Verde, Colorado. In the hollows beneath huge overhanging ledges, they built multistory stone-block buildings that have survived to this day. Clearly, the blocks are solid and durable—they are, after all, rock. But if you were to rub your thumb along one, it would feel gritty, and small grains of quartz would break free and roll under your thumb, for the block consists of quartz sand grains cemented together. Geologists call such rock a **sandstone.**

Sandstone is an example of clastic, or detrital, sedimentary rock, rock created from solid grains (**clasts,** or detritus) stuck together to form a solid mass. The grains can consist of individual minerals (such as grains of quartz or flakes of clay) or fragments of rock (such as pebbles of granite). The loose grains of sediment transform into clastic sedimentary rock by the following five steps (▶Fig. 5.14a, b).

- *Weathering:* Detritus forms by disintegration of bedrock in response to physical and chemical weathering.

- *Erosion:* **Erosion** refers to the combination of processes that separate rock or regolith from its substrate and carry it away.

- *Transportation:* Wind, water, or ice move sediment. The ability of a medium to carry sediment depends on its viscosity and velocity. Solid ice can carry sediment of any size, regardless of how slowly the ice moves. Very fast-moving, turbulent water can transport coarse fragments (cobbles and boulders), moderately fast-moving water can carry only sand and gravel, and slowly moving water carries only silt and mud. Strong winds can move sand and dust, but gentle breezes carry only dust.

- *Deposition:* **Deposition** is the process by which sediment settles out of the transporting medium. Sediment settles out of wind or moving water when these fluids slow, because as the velocity decreases, the fluid no longer has the ability to move sediment.

- *Lithification:* Geologists refer to the transformation of loose sediment into solid rock as **lithification.** The lithification of clastic sediment involves two steps. First, when the sediment has been buried, pressure caused by the weight of overlying material squeezes out water and air that had been trapped between clasts, and clasts compact together tightly. Compacted sediment may then be bound together to make coherent sedimentary rock by the process of **cementation.** Cement consists of minerals (commonly quartz or calcite) that precipitate from groundwater, and fill the spaces between clasts. Effectively, cement acts like glue and holds detritus together.

FIGURE 5.14 (a) The basic steps during the development of a sedimentary rock: weathering → erosion → transportation → deposition → lithification. (b) As sediment moves from its source to the site of deposition, it becomes finer grained.

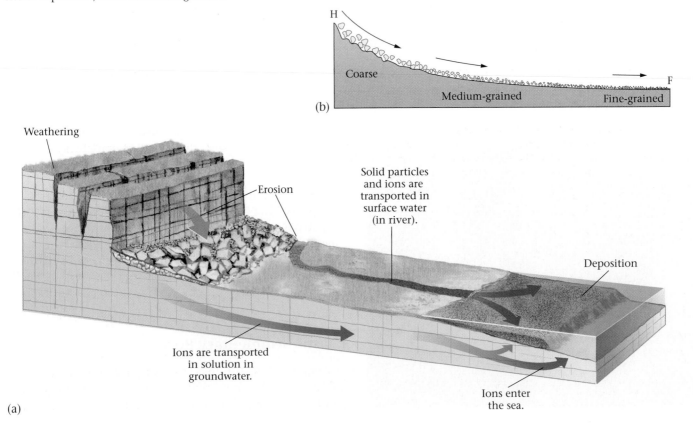

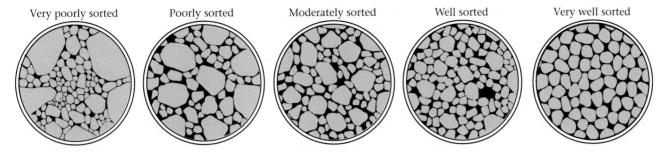

Very poorly sorted   Poorly sorted   Moderately sorted   Well sorted   Very well sorted

FIGURE 5.15 A poorly sorted sediment has a great variety of different clast sizes, whereas in a well-sorted sediment, all the clasts are the same size.

Describing and classifying clastic sedimentary rocks. Say that you pick up a clastic sedimentary rock and want to describe it sufficiently so that from your words alone, another person can picture the rock. What characteristics should you mention? Geologists find the following characteristics most useful:

- *Clast size.* This refers to the length of diameter of clasts making up a rock. Names used for clast size, listed in order from coarsest to finest, are: boulder, cobble, pebble, sand, silt, and clay.

- *Clast composition.* This refers to the makeup of clasts in sedimentary rock. Clasts may be composed of individual minerals or of rock fragments.

- *Angularity and sphericity.* Angularity indicates the degree to which clasts have smooth or angular corners and edges. Sphericity, in contrast, refers to the degree to which a clast is equidimensional, or resembles a sphere.

- *Sorting.* **Sorting** of clasts indicates the degree to which the clasts in a rock are all the same size or include a variety of sizes (▶Fig. 5.15). Well-sorted sediment consists entirely of sediment of the same size, whereas poorly sorted sediment contains a mixture of more than one grain size.

- *Character of cement.* Not all clastic sedimentary rocks have the same kind of cement. In some, the cement consists predominantly of silica (quartz), whereas in others, it consists predominantly of calcite.

With the above characteristics in mind, we can distinguish among several common types of clastic sedimentary rocks, listed in ▶Table 5.2 (▶Fig. 5.16a–l). This table provides common rock names. Specialists sometimes use other, more precise names based on more complex classification schemes.

Characteristics of a sedimentary rock provide clues to the source of the sediment, and to the environment of deposition. To see how, let's follow the fate of rock fragments as they gradually move from a cliff face in the mountains via a river to the seashore (Fig. 5.14). Different kinds of sediment develop along the route. Each of these types of sediment, if buried and lithified, would yield a different kind of sedimentary rock.

TABLE 5.2 Classifying Clastic Sedimentary Rocks

Clast Size*	Clast Character	Rock Name (Alternate Name)
Coarse to very coarse	Rounded pebbles and cobbles	**Conglomerate**
	Angular clasts	**Breccia**
	Large clasts in muddy matrix	**Diamictite**
Medium to coarse	Sand-sized grains	**Sandstone**
	• quartz grains only	• **quartz sandstone (quartz arenite)**
	• quartz and feldspar sand	• **arkose**
	• sand-sized rock fragments	• **lithic sandstone**
	• sand and rock fragments in a clay-rich matrix	• **wacke** (informally called **graywacke**)
Fine	Silt-sized clasts	**Siltstone**
Very fine	Clay and/or very fine silt	**Shale** (if it breaks into platy sheets)
		Mudstone (if it doesn't break into platy sheets)

*For precise diameters, see p. 123.

FIGURE 5.16 Examples of sediments and sedimentary rocks. (a) A sedimentary breccia. (b) Recently deposited stream gravel. (c) A conglomerate made of stream gravel that was later cemented together. (d) Hand samples of arkose. The whitish fragments consist of feldspar. (e) A photomicrograph (photograph of a thin section) showing quartz grains in a sandstone. The field of view is 3 mm. (f) Wind-deposited sand accumulations in Australia. (g) A thick sandstone bed forms an overhang that protects ancient Native American dwellings on Mesa Verde, Colorado. (h) A hand specimen of sandstone. (i) Mud along a dirt road in central Australia makes a slippery obstacle for drivers. (j) Thin shale beds beneath a sandstone bed in Pennsylvania. Note the coin for scale. (k) A photograph taken with a scanning electron microscope shows flakes of clay that are less than 2 microns across. (l) A photomicrograph of graywacke.

To start, imagine that some large blocks of granite tumble off a cliff and slam into other blocks already at the bottom. The impact shatters the blocks, producing a pile of angular fragments with sharp edges. If these fragments were to be cemented together, the resulting rock would be **breccia** (▶Fig. 5.16a). Later, a storm causes the fragments (clasts) to slide downslope into a turbulent river. In the river, clasts bang into each other and into the bed of the stream, a process that shatters them into still smaller pieces and breaks off their sharp edges. Angular clasts gradually become rounded clasts. When the storm abates and the river water slows, pebbles and cobbles stop moving and form a mound or bar of gravel. Burial and lithification of these rounded clasts would produce **conglomerate** (▶Fig. 5.16b, c).

If the gravel stays put for a long time, it undergoes chemical weathering. As a consequence, cobbles and pebbles break apart into individual mineral grains, eventually producing a mixture of quartz, feldspar, and clay. If another storm causes the river to rise and flow faster, these sediments start to wash downstream. Clay is so fine that it may remain suspended in the water and be carried far downstream. Sand, however, may drop out along the stream bed or stream banks when the flow slows between storms. In the resulting sandbars, not far from the source, we find a mixture of quartz and some feldspar grains—this sediment, if buried and lithified, would become **arkose** (▶Fig. 5.16d). Over time, feldspar grains in sand continue to weather into clay so that gradually, during successive events that wash the sediment farther downstream, the sand loses feldspar and ends up being composed almost entirely of durable quartz grains. This sediment, when buried and lithified, becomes quartz sandstone (▶Fig. 5.16e–h). Some of the sand may make it to the sea, where waves carry it to beaches. Meanwhile, silt and clay may accumulate in the flat areas bordering the stream (regions called floodplains; see Chapter 14) that become inundated only during floods, or in a wedge of sediment, called a delta, that accumulates in the sea at the mouth of the river. Some of the silt and mud may be collected in lagoons or mud flats along the shore. The silt, when lithified, becomes **siltstone,** and the mud, when lithified, becomes **shale** or **mudstone** (▶Fig. 5.16i–k).

Two of the rock types listed in Table 5.2 do not appear in the above narrative, because they don't form in the depositional settings just described. Diamictite forms either from debris flows (slurries consisting of mud mixed with larger clasts) both on land and under water, or in glacial settings where ice can transport clasts of all sizes. Wacke typically forms from the deposits of submarine avalanches (▶Fig. 5.16l).

You may have noted from this narrative that as sediment moves downstream, grains overall become smaller and rounder; grains composed of minerals that are susceptible to chemical weathering progressively break down and

disappear; and sediment becomes better sorted. Geologists use the term sediment maturity to refer to the degree to which a sediment has evolved from being just a crushed-up version of the original rock to a sediment that has lost its easily weathered components and has become well sorted and rounded.

Biochemical and Organic Sedimentary Rocks: Byproducts of Life

The Earth System involves interactions between living organisms and the physical planet. Numerous organisms have developed the ability to extract dissolved ions from seawater to make solid shells. Some organisms construct their shells out of calcium (Ca^{2+}) and carbonate (CO_3^{2-}) ions, which they merge to make the minerals calcite ($CaCO_3$) or its polymorph, aragonite, whereas other organisms make their shells out of dissolved silica (SiO_2). When the organisms die, the solid material in their shells becomes incorporated in biochemical sedimentary rock. Since carbonate shells are much more common than silica shells, most biochemical rocks are carbonates. Plants, algae, bacteria, and plankton also yield materials that can be incorporated in sedimentary rocks. The rocks formed from this material contain organic chemicals and pure carbon derived from organic chemicals. Such rocks are called organic sedimentary rocks. Geologists recognize several different types of biochemical and organic sedimentary rocks, which we now describe.

Limestone (biochemical). A snorkeler gliding above the Great Barrier Reef of Australia sees an incredibly diverse community of coral and algae, around which creatures such as clams, oysters, snails (gastropods), and lampshells (brachiopods) live, and above which plankton float (▶Fig. 5.17a). Though they all look so different from each other, many of these organisms share an important characteristic: they make solid shells of calcite (or its polymorph, aragonite). When the organisms die, their skeletons may stay in place, as is the case with reef builders such as coral; settle out of the water like snowflakes; or be moved by currents or waves to another location, where they eventually settle out. During transport, shells may break up into small fragments. Rocks formed from the calcite or aragonite skeletons of organisms are the biochemical version of **limestone,** a type of carbonate rock (▶Fig. 5.17b, c). There are many different kinds of limestones, distinguished by the material from which they formed. Three common types are: fossiliferous limestone, consisting of identifiable shells and shell fragments; micrite, consisting of carbonate mud; and chalk, consisting of plankton shells.

Typically, ancient limestone is a massive light-gray to dark-bluish-gray rock that breaks into chunky blocks—it doesn't look much like a pile of shell fragments. That's because several processes take place that can change the texture of the rock over time. For example, organisms living

(a)

(b)

(c)

FIGURE 5.17 (a) In this modern coral reef, corals produce shells of calcite or aragonite. If buried and preserved, these become limestone. (b) A quarry face in Vermont shows the typical gray color of limestone; this rock is over 400 million years old. The thinly laminated layers are carbonate mud; the white mounds are relicts of small reefs. (c) This specimen of fossiliferous limestone consists entirely of small fossil shells and shell fragments. Not all fossiliferous limestones contain such a high proportion of fossils.

in the depositional environment burrow into recently formed or deposited shells, break them up, and may even convert some to carbonate mud. Later, water passing through the rock precipitates new cement, and also dissolves some carbonate grains and causes new ones to grow.

Chert (biochemical). If you walk beneath the northern end of the Golden Gate Bridge in Marin County, California, north of San Francisco, you will find outcrops of reddish, almost porcelainlike rock occurring in 3- to 15-cm-thick layers, one on top of another in a sequence with a total thickness of hundreds of meters (▶Fig. 5.18a). Hit it with a hammer, and the rock cracks, almost like glass, creating smooth, spoon-shaped (conchoidal) fractures. Geologists call this rock biochemical chert; it's made from crypto-crystalline quartz (*crypto* is Greek for "hidden"), meaning quartz grains that are too small to be seen without the extreme magnification of an electron microscope. The chert beneath the Golden Gate Bridge formed from the shells of plankton (particularly microscopic animals called radiolaria and diatoms). The shells accumulate on the sea floor as a silica-rich ooze. Gradually, after burial, the shells dissolve, forming a silica-rich solution. Chert then precipitates from this solution.

Organic (carbonaceous) rocks: coal and oil shale. The Industrial Revolution of the nineteenth century, which transformed the world's economy from an agricultural to an industrial base, depended on power provided by steam engines. After decimating forests to provide fuel for these engines, industrialists turned to coal for fuel. Coal is a black, combustible rock consisting of over 50% carbon, and so differs markedly from the other sedimentary rocks discussed so far—the carbon of coal occurs as pure carbon or in organic chemicals, not in minerals. Still, we consider coal a sedimentary rock because it is made up of detritus deposited in layers (▶Fig. 5.18b). We'll look more at coal formation in Chapter 12. Here, we simply need to know that the carbon and the organic chemicals making coal come from the remains of plant material that died and accumulated on the floor of a forest or swamp. Coal forms when the remains become deeply buried. Then heat and pressure at depth can compact the plant material and drive off volatiles (hydrogen, water, carbon dioxide, ammonia), leaving a concentration of carbon.

Chemical sedimentary rocks. The colorful terraces, or mounds, around the vents of hot-water springs; the immense layers of salt that underlie the floor of the Mediterranean Sea; the smooth, sharp point of an ancient arrowhead—these materials all have something in common. They all consist of rock formed primarily by the precipitation of minerals out of water solutions. We call such rocks chemical sedimentary rocks. They typically have a crystalline texture, partly formed during the original precipitation and partly when, at a later time, new crystals grow at the expense of old ones.

Evaporites: the products of saltwater evaporation. In 1965, two daredevil drivers in jet-powered cars battled to be the first to set the land-speed record of 600 mph. On November 7,

(a) (b)

FIGURE 5.18 (a) This bedded chert, which crops out near the northern foundation of the Golden Gate Bridge in Marin County, California, north of San Francisco, developed on the deep sea floor by the deposition of plankton that secrete silica shells. The bends in the layers, called folds (see Chapter 9), formed when the layers were squeezed and wrinkled as they were scraped off the sea floor. (b) Coal is deposited in layers just like other kinds of sedimentary rocks. Here, we see a coal seam (a miner's term for a coal layer) between layers of sandstone and shale.

Art Arfons, in the *Green Monster,* peaked at 576.127 mph, but eight days later Craig Breedlove, driving the *Spirit of America,* reached 600.601 mph. Traveling at such speeds, a driver must maintain an absolutely straight line; any turn will catapult the vehicle out of control, because its tires simply can't grip the ground. Thus, high-speed trials take place on extremely long and flat racecourses. Not many places can provide such conditions—the Bonneville Salt Flats, near the Great Salt Lake of central Utah, do. The salt flats formed by the evaporation of an ancient salt lake. Under the heat of the Sun, the water turned to gas and drifted up into the atmosphere. The salt that had been dissolved in the water stayed behind.

Such salt precipitation occurs wherever there is saturated saltwater—along desert lakes with no outlet and along margins of restricted seas. For thick deposits of salt to form, large volumes of water must evaporate (▶Fig. 5.19a–d).

Because salt deposits form as a consequence of evaporation, geologists refer to them as **evaporites.** The specific type of salt constituting an evaporite depends on the amount of evaporation. When 80% of the water evaporates, gypsum forms; and when 90% of the water evaporates, halite precipitates.

Travertine (chemical limestone). **Travertine** is a rock composed of crystalline calcium carbonate (calcite and/or aragonite) formed by chemical precipitation from groundwater that has seeped out at the ground surface (in hot- or cold-water springs) or on the walls of caves. What causes this precipitation? It happens, in part, when the groundwater degasses, meaning that some of the carbon dioxide that had been dissolved in the groundwater bubbles out of solution. Dissolved carbon dioxide makes water better able to

dissolve carbonate, so removal of carbon dioxide decreases the ability of the water to hold dissolved carbonate. Precipitation also occurs when water evaporates and leaves behind dissolved ions, thereby increasing the concentration of carbonate. Various kinds of microbes live in the environments in which travertine accumulates, so biologic activity may also contribute to the precipitation process. Travertine produced at springs forms terraces and mounds that are meters or even hundreds of meters thick (▶Fig. 5.20). In cave settings, travertine builds up beautiful and complex growth forms called speleothems.

Dolostone: replacing calcite with dolomite. **Dolostone** differs from limestone in that it contains the mineral dolomite ($CaMg[CO_3]_2$). Where does the magnesium come from? Most dolostone forms by a chemical reaction between solid calcite and magnesium-bearing groundwater. Much of the dolostone you may find in an outcrop actually originated as limestone but later changed as dolomite replaced calcite. This change may take place beneath lagoons along a shore soon after the limestone formed, or a long time later, after the limestone has been buried deeply.

Chert (replacement). A tribe of Native Americans, the Onondaga, once lived off the land in eastern New York State. Here, outcrops of limestone contain layers of a black chert (▶Fig. 5.21). Because of the way it breaks, artisans could fashion sharp-edged tools (arrowheads and scrapers) from this chert, so the Onondaga collected it for their own tool-making industry and for use in trade with other people. Unlike the deep-sea (biochemical) chert described earlier, the chert collected by the Onondaga formed when cryptocrystalline quartz gradually replaced calcite crystals within

FIGURE 5.19 The process of forming evaporite deposits. (a) In lakes with no outlet, the tiny amount of salt brought in by fresh water streams stays behind as the water evaporates. Along the margins of the lake, salts precipitate. If the whole lake evaporates, a flat surface of salt forms. (b) White evaporite precipitated on dark brown sediment after evaporation of a temporary lake in Death Valley, California. The white on top of the mountain in the distance is snow. (c) Salt precipitation can also occur along the margins of a restricted marine basin, if saltwater evaporates faster than it can be resupplied. The entire restricted sea may dry up if it is cut off from the ocean. (d) Thick layers of salt accumulate in a rift and later are buried deeply. The salt then recrystallizes. Here, thick salt layers are being mined.

(b)

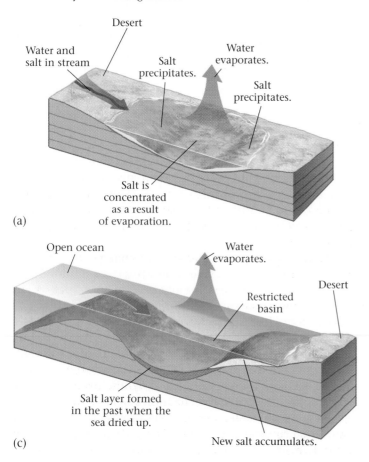

(a)

(c)

(d)

a body of limestone long after the limestone was deposited; geologists call such material replacement chert. Chert comes in many colors (black, white, red, brown, green, gray), depending on the impurities it contains. Petrified wood is chert that forms when silica-rich sediment, such as ash from a volcanic eruption, buries a forest. The silica dissolves in groundwater that then passes into the wood. Dissolved silica precipitates as cryptocrystalline quartz within wood, gradually replacing the wood's cellulose. The chert deposit retains the shape of the wood and even its growth rings. Some chert, known as agate, precipitates in concentric rings inside hollows in a rock.

5.5 SEDIMENTARY STRUCTURES

In the stark outcrop shown in Figure 5.8c, note the distinct lines across its face. In 3-D, we see that these lines are the traces of individual surfaces that separate the rock into sheets. In fact, sedimentary rocks in general contain distinctive layering. The layers themselves may have a characteristic internal arrangement of grains or distinctive markings on their surface. We use the term **sedimentary structure** for the layering of sedimentary rocks, surface features on layers formed during deposition, and the arrangement of grains within layers. Here, we examine some of the more important types.

(a)

(b)

FIGURE 5.20 (a) A travertine buildup at Mammoth Hot Springs in Yellowstone National Park. Note the terraces. (b) A travertine buildup on the wall of a cave in Puerto Rico.

FIGURE 5.21 Replacement chert occurring as nodules in a limestone. Chert forms the black band in these tilted layers.

carry only silt, which collects on the river bed. During a flood, the river carries sand and pebbles, so a layer of sandy gravel forms over the silt layer. Then, when the flooding stops, more silt buries the gravel. If these sediments become lithified and exposed for you to see, they appear as alternating beds of siltstone and sandy conglomerate (▶Fig. 5.22a–d).

During geologic time, long-term changes in a depositional environment can take place. Thus, a sequence of beds may differ markedly from sequences of beds above or below. If a sequence of strata is distinctive enough to be traced across a fairly large region, geologists call it a **stratigraphic formation,** or simply a formation (▶Fig. 5.23). For example, a region may contain a succession of alternating sandstone and shale beds deposited by rivers, overlain by beds of marine limestone deposited later when the region was submerged by the sea. A stratigrapher might identify the sequence of sandstone and shale beds as one formation and the sequence of limestone beds as another. Formations are often named after the locality where they were first found and studied.

Bedding and Stratification

Geologists have a jargon for discussing sedimentary layers. A single layer of sediment or sedimentary rock with a recognizable top and bottom is called a **bed;** the boundary between two beds is a bedding plane; several beds together constitute strata; and the overall arrangement of sediment into a sequence of beds is bedding, or stratification.

Why does bedding form? To find the answer, we need to think about how sediment is deposited. Changes in the climate, water depth, or the sediment source control the type of sediment deposited at a location at a given time. For example, on a normal day a slow-moving river may

Ripples, Dunes, and Cross Bedding: A Consequence of Deposition in a Current

Many clastic sedimentary rocks accumulate in moving fluids (wind, rivers, or waves). The movement of the fluids creates fascinating sedimentary structures at the interface between the sediment and the fluid. These structures are called bedforms. The bedforms that develop at a given location reflect factors such as the velocity of the flow and the size of the clasts. Though there are many types of bedforms, we'll focus on only two—ripples and dunes. The growth of

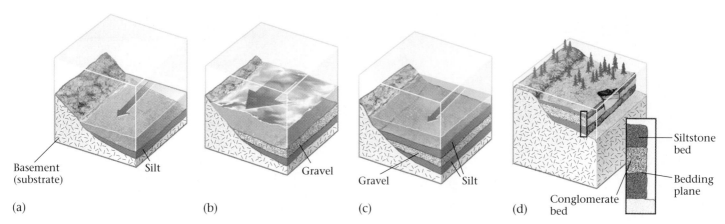

Basement
(substrate)   Silt   Gravel   Gravel   Silt   Conglomerate   Siltstone
                                                                                                         bed   bed

                                                                                                                                      Bedding
                                                                                                                                      plane

(a)   (b)   (c)   (d)

FIGURE 5.22 Bedding forms as a result of changes in the environment. (a) During normal river flow, a layer of silt is deposited. (b) During a flood, turbulent water brings in a layer of gravel. (c) When the river returns to normal, another layer of silt is deposited. (d) Later, after lithification, uplift, and exposure, a geologist sees these layers as beds on an outcrop.

FIGURE 5.23 A particularly thick bed, a sequence of beds of the same composition, or a sequence of beds of alternating rock types can be called a stratigraphic formation, if the sequence is distinctive enough to be traced across the countryside. In this photo of the Grand Canyon, we can see five formations. Formations that consist primarily of one rock type may take the rock-type name (for example, Kaibab Limestone), but formations containing more than one rock type may just be called a formation. The Supai Group is a group because it consists of several related formations, which are too thin to show here. Formations and groups are examples of stratigraphic units. Note that each formation consists of many beds, and that beds vary greatly in thickness. The boundaries between units are called contacts.

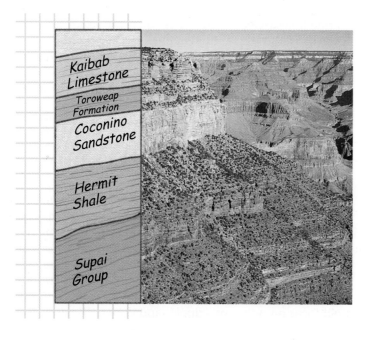

Kaibab
Limestone

Toroweap
Formation

Coconino
Sandstone

Hermit
Shale

Supai
Group

both produces cross bedding, a special type of lamination within beds.

Ripples (or **ripple marks**) are relatively small, generally no more than a few centimeters high, elongated ridges that form on a bed surface at right angles to the direction of current flow. The crest of a symmetric ripple is a sharp ridge, whereas the trough between adjacent ridges is a smooth, concave-up curve. You can find ripples on modern beaches and preserved on bedding planes of ancient rocks (▶Fig. 5.24a, b). Dunes look like ripples, only they are much larger. For example, dunes on the bed of a stream may be tens of centimeters high, and wind-formed dunes formed in deserts may be tens to over 100 meters high.

If you slice into a ripple or dune and examine it in cross section, you will find distinct internal laminations that are inclined at an angle to the boundary of the main sedimentary layer. Such laminations are called **cross beds.** Cross bedding forms directly as a consequence of the evolution of ripples or dunes. To see how, imagine a current of air or water moving uniformly in one direction (▶Fig. 5.25a). The current erodes and picks up clasts from the upstream part of the bedform (because here the fluid moves quickly) and deposits them on the downstream or leeward part of the bedform (because here the fluid moves more slowly). The process repeats as more sediment builds up on the leeward side of the bedform and then slips down, so with time the leeward side of the bedform builds in the downstream direction. The curving surface of the slip face establishes the shape of the cross beds. With time, a new cross-bedded layer builds out over a preexisting one. The boundary between two successive layers is called the main bedding, and the internal curving surfaces within the layer constitute the cross bedding (▶Fig. 5.25b).

FIGURE 5.24 (a) Modern ripples forming in the sand on a beach. (b) Ancient ripples preserved in a 1.5 billion-year-old layer of quartzite. This outcrop occurs in Wisconsin.

Turbidity Currents and Graded Beds

Sediment deposited on a submarine slope might not stay in place forever. For example, an earthquake or storm might disturb this sediment and cause it to slip downslope. If the sediment is loose enough, it mixes with water to create a murky, turbulent cloud. This cloud is denser than clear water, and thus flows downslope like an underwater avalanche (▶Fig. 5.26). We call this moving submarine suspension of sediment a turbidity current. Eventually, in deeper water where the slope becomes gentler, or the turbidity current spreads out, the turbidity current slows. When this happens, the sediment that it has carried starts to settle out. Larger grains sink faster through a fluid than do finer grains, so the coarsest sediment settles out first. Progressively finer grains accumulate on top, with the finest sediment (clay) settling out last. This process forms a graded bed—that is, a layer of sediment in which grain size varies from coarse at the bottom to fine at the top.

FIGURE 5.25 (a) Cross beds form as sand blows up the windward side of a dune and then accumulates on the slip face. (b) At a later time, we see that dunes migrate, and eventually bury the layers below. (c) On this cliff face of sandstone in Zion National Park, we see remnants of ancient sand dunes. Cross beds indicate the wind direction during deposition.

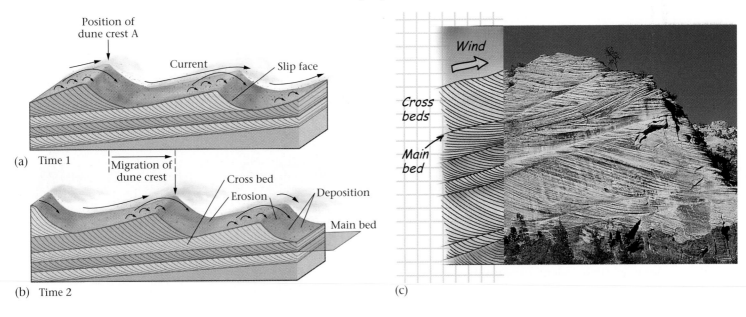

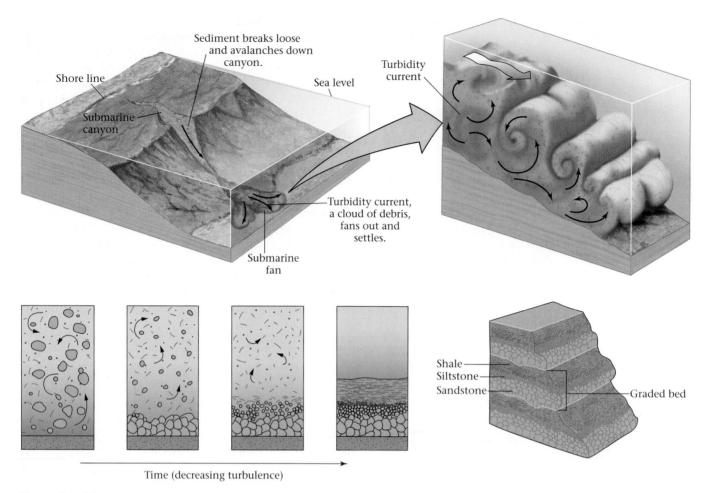

Time (decreasing turbulence)

FIGURE 5.26 An earthquake or storm triggers an underwater avalanche (turbidity current), which mixes sediment of different sizes together. When the current slows, the larger grains settle faster, producing a graded bed.

Bed-Surface Markings

A number of features appear on the surface of a bed as a consequence of events that happen during deposition or soon after, while the sediment layer remains soft. These bed-surface markings include the following.

- *Mud cracks:* If a mud layer dries up after deposition, it cracks into roughly hexagonal plates that typically curl up at their edges. We refer to the openings between the plates as mud cracks. Later, these fill with sediment and can be preserved (▶Fig. 5.27).

- *Scour marks:* As currents flow over a sediment surface, they may scour out small troughs called scour marks parallel to the current flow. These indentations can be buried and preserved.

FIGURE 5.27 Mud cracks formed when red mud dried up. Note how the edges of the mud plates curl up.

• *Fossils:* Fossils are relics of past life. Some fossils are shell imprints or footprints on a bedding surface (see Interlude D).

The Value of Studying Sedimentary Structures

Sedimentary structures are not just a curiosity, but rather serve as key clues that help geologists understand the environment in which clastic sedimentary beds were deposited. For example, the presence of ripple marks and cross bedding indicates that layers were deposited in a current. The presence of mud cracks indicates that the sediment layer was exposed to the air on occasion. Graded beds indicate deposition by turbidity currents. And fossil types can tell us whether sediment was deposited along a river or in the sea, for different species of organisms live in different environments. In the next section of this chapter, we examine these environments in greater detail.

5.6 SEDIMENTARY ENVIRONMENTS

Geologists refer to the conditions in which sediment was deposited as the sedimentary environment (or depositional environment). Examples include beach, glacial, and river environments. To identify these environments, geologists, like detectives, look for such clues as grain size, composition, sorting, and roundness of clasts, which can tell us how far the sediment has traveled from its source and whether it was deposited from the wind, from a fast-moving current, or from a stagnant body of water. Clues such as fossil content and sedimentary structures can help us decide whether the sediments were deposited subaerially, just off the coast, or in the deep sea.

Now let's look at some examples of different sedimentary environments, and the sediments deposited in them, by imagining that we are taking a journey from the mountains to the sea, examining sediments as we go. We begin with terrestrial sedimentary environments, those formed on dry land, and end with marine sedimentary environments, those formed along coasts and under the waters of the ocean. (See two-page art on pp. 148–149.) Of note, sediments deposited in terrestrial environments may oxidize (rust) when undergoing lithification in oxygen-bearing water, or if in contact with air. If this happens, the sedimentary beds develop a reddish color, and can be called redbeds.

Terrestrial (Nonmarine) Sedimentary Environments

Glacial environments. We begin high in the mountains, where it's so cold that more snow collects in the winter than melts away, so glaciers—rivers of ice—develop and slowly

flow downslope. Because ice is a solid, it can move sediment of any size. So as a glacier moves down a valley in the mountains, it carries along *all* the sediment that falls on its surface from adjacent cliffs or gets plucked from the ground at its base. At the end of the glacier, where the ice finally melts away, it drops its load and makes a pile of glacial till (►Fig. 5.28a). Till is unsorted and unstratified—it contains clasts ranging from clay size to boulder size all mixed together, with large clasts distributed through a matrix of silt and clay.

Mountain stream environments. As we walk down beyond the end of the glacier, we enter a realm where turbulent streams rush downslope in mountain valleys. This fast-moving water has the power to carry large clasts; in fact, during floods, boulders and cobbles tumble down the stream bed. Between floods, when water flow slows, the largest clasts settle out to form gravel and boulder beds, while the stream carries finer sediments like sand and mud away (►Fig. 5.28b). Sedimentary deposits of a mountain stream would, therefore, include conglomerate.

Alluvial fan environments. Our journey now takes us to the mountain front, where the fast-moving stream empties onto a plain. In arid regions, where there is not enough water for the stream to flow continuously, the stream deposits its load of sediment right at the mountain front, creating a large, wedge-shaped apron called an alluvial fan (►Fig. 5.28c). Deposition takes place here because when the stream pours from a canyon mouth and spreads out over a broader region, friction with the ground causes the water to slow down, and slow-moving water does not have the power to move coarse sediment. Because we are still so close to the mountains—the source of the sediment—the sand still contains feldspar grains, for these have not yet broken up and have not yet weathered into clay. Alluvial-fan sediments, when later buried and transformed into sedimentary rock, become arkose and conglomerate.

Sand-dune environments. In deserts, relatively few plants grow, so the ground lies exposed to the wind. The strongest winds can transport sand. As a result, large sand dunes of well-sorted sand may accumulate. Thus, thick layers of well-sorted sandstone, in which we see large cross beds, are relics of desert sand-dune environments (Fig. 5.25b).

Lake environments. From the dry regions, we continue our journey into a temperate realm, where water remains at the surface throughout the year. Some of this water collects in lakes, in which relatively quiet water can't move coarse sediment; any coarse sediment brought into the lake by a stream settles out at the stream's outlet. Only fine clay makes it out into the center of the lake, where it settles to form mud on the lake bed. Thus, lake sediments typically consist of finely laminated shale (►Fig. 5.29a).

(a)   (b)   (c)

FIGURE 5.28 (a) Glacial till deposited at the end of a melting glacier in New Zealand. (b) Coarse boulders deposited by a flooding mountain stream in California. (c) An alluvial fan in Death Valley, California. Note the road for scale.

River environments. The lake drains into a stream that carries water onward toward the sea. As we follow the stream, it merges with tributaries to become a large river, winding back and forth across a plain. Rivers transport sand, silt, and mud. The coarser sediments tumble along the bed in the river's channel, while the finer sediments drift along, suspended in the water (▶Fig. 5.29b). This fine sediment settles out along the banks of the river, or on the floodplain, the flat region on either side of the river that is covered with water only during floods. On the floodplain, mud layers dry out between floods, leading to the formation of mud cracks. Thus, river sediments lithify to form sandstone, siltstone, and shale. Typically, channels of coarser sediment are surrounded by layers of fine-grained floodplain deposits; in cross section, the channel has a lenslike shape (▶Fig. 5.29c, d).

Marine Sedimentary Environments

Marine delta deposits. After following the river downstream for a long distance, we reach its mouth, where it empties into the sea. Here, the river builds a delta of sediment out into the sea. Deltas were so named because the map shape of some deltas (such as the Nile Delta of Egypt) resembles the Greek letter *delta* (Δ), as we discuss further in Chapter 14. River water stops flowing when it enters the sea, so sediment settles out.

In 1885, an American geologist named G. K. Gilbert studied small deltas that formed where small streams (carrying gravel, sand, and silt) emptied into standing water. He

showed that such deltas contain three components (▶Fig. 5.30a): nearly horizontal topset beds composed of gravel, sloping foreset beds of gravel and sand (deposited on the sloping face of the delta), and nearly horizontal silty bottomset beds, formed at depth on the floor of the water body. Large river deltas are more complex and provide many different sedimentary environments (▶Fig. 5.30b). Also, sea-level changes may cause the position of the different environments to move with time. Deposits of a small delta contain topset, foreset, and bottomset beds. Those of a large, ocean-margin delta can be identified in the stratigraphic record as thick sequences in which deeper-water (offshore) sediments of a given age grade progressively into fluvial sediments in a shoreward direction.

Coastal beach sands. Now we leave the delta and wander along the coast. Oceanic currents transport sand along the coastline. The sand washes back and forth in the surf, so it becomes well sorted (waves winnow out mud and silt) and well rounded, and because of the back-and-forth movement of ocean water over the sand, the sand surface may become rippled. Thus, if you find well-sorted, medium-grained sandstone, perhaps with ripple marks, you may be looking at the remnants of a beach environment.

Shallow-marine clastic deposits. From the beach, we proceed offshore. In deeper water, where wave energy does not stir the sea floor, finer sediment accumulates. Because the water here

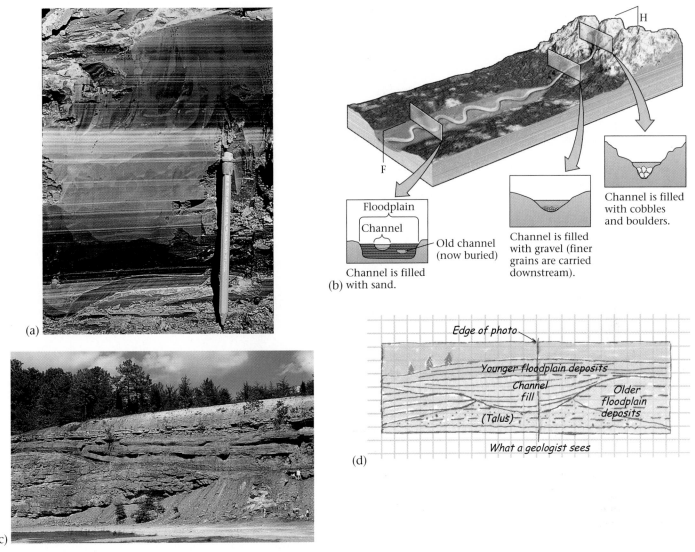

FIGURE 5.29 (a) Finely laminated lake-bed shales in Grenoble, France. (b) The character of river sediment varies with distance from the source. In the steep channel, the turbulent river can carry boulders and cobbles. As the river slows, it can carry only sand and gravel. And as the river winds across the floodplain, it carries sand, silt, and mud. Coarser sediment accumulates in the river channel, the finer sediment on the floodplain. (c) This exposure shows the lenslike shape of an ancient gravel-filled river channel in cross section. (d) A geologist's sketch portrays the channel shape.

may be only meters to a few tens of meters deep, geologists refer to this depositional setting as a shallow-marine environment. Clastic sediments that accumulate in this environment tend to be fine-grained, well-sorted, well-rounded silt, and they are inhabited by a great variety of organisms such as mollusks and worms. Thus, if you see smooth beds of silt-stone and mudstone containing marine fossils, you may be looking at shallow marine clastic deposits.

Shallow-water carbonate environments. In shallow-marine settings far from the mouth of a river, where relatively little clastic sediment (sand and mud) enters the water, warm, clear, nutrient-rich water hosts an abundance of organisms. Their shells, which consist of carbonate minerals make up most of the sediment that accumulates (▶Fig. 5.31a). The nature of carbonate sediment depends on the water depth. Beaches collect sand composed of shell fragments; lagoons (protected bodies of quiet water) are sites where carbonate mud accumulates; and reefs consist of coral and coral debris. Farther off-shore of a reef, we can find a sloping apron of reef fragments (▶Fig. 5.31b). Shallow-water carbonate environments transform into sequences of limestone.

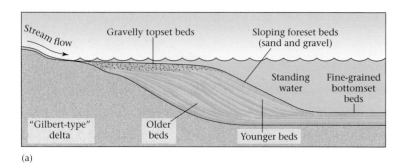

(a)

FIGURE 5.30 Sedimentary deposits of deltas. (a) A simple "Gilbert-type" delta formed where a small stream carrying gravel, sand, and silt enters a standing water body. The delta contains topset, foreset, and bottomset beds. (b) A large river delta is very complex and doesn't fit the simple Gilbert-type model. The great variety of local depositional environments in a delta setting are labeled. Note that, as time passes, the delta builds seaward.

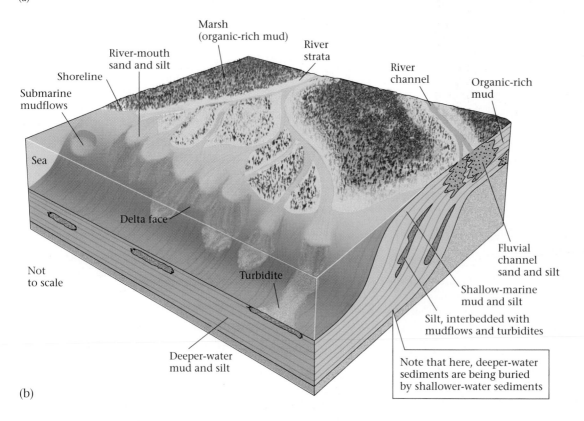

(b)

FIGURE 5.31 (a) A coral reef and adjacent lagoon surrounding an island in the South Pacific. (b) The different carbonate environments associated with a reef.

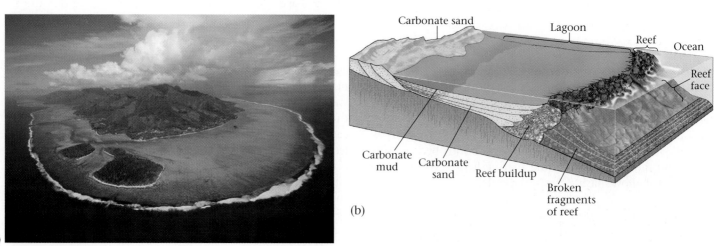

(a)

(b)

FIGURE 5.32 (a) These plankton shells, which make up deep-marine sediment, are so small that they could pass through the eye of a needle. (b) The chalk cliffs of Dover, England. These were originally deposited on the sea floor and later uplifted.

Deep-marine deposits. We conclude our journey by sailing offshore. Along the transition between coastal regions and the deep ocean, turbidity currents deposit graded beds (Fig. 5.26). Farther offshore, in the deep-ocean realm, only fine clay and plankton provide a source for sediment. The clay eventually settles out onto the deep sea floor, forming deposits of finely laminated mudstones, and plankton shells settle to form chalk (from calcite shells; ►Fig. 5.32a, b) or chert (from silica shells). Thus, deposits of mudstone, chalk, or bedded chert indicate a deep-marine origin.

5.7 SEDIMENTARY BASINS

The sedimentary veneer on the Earth's surface varies greatly in thickness. If you stand in central Siberia or south-central Canada, you will find yourself on igneous and metamorphic basement rocks that are over a billion years old—there are no sedimentary rocks anywhere in sight. Yet if you stand along the southern coast of Texas, you would have to drill through over 10 km of sedimentary beds before reaching igneous and metamorphic basement. Thick accumulations of sediment form only in regions where the surface of the Earth's lithosphere sinks, as sediment collects. Geologists use the term **subsidence** to refer to the sinking of lithosphere, and the term **sedimentary basin** for the sediment-filled depression. In what geologic settings do sedimentary basins form? An understanding of plate tectonics theory provides some answers.

Types of Sedimentary Basins in the Context of Plate Tectonics Theory

Geologists distinguish among different kinds of sedimentary basins on the basis of the region of a lithosphere plate in which they formed. Let's consider a few examples.

- *Rift basins:* These form in continental rifts, regions where the lithosphere has been stretched. As the rift grows, slip on faults drops blocks of crust down, creating low areas bordered by narrow mountain ridges. Also during rifting, warm asthenosphere rises beneath the rift and heats up the thin lithosphere. When rifting ceases, the rifted lithosphere then cools, thickens, and becomes denser. This heavier lithosphere sinks down, causing more subsidence, just as the deck of a tanker ship drops to a lower elevation when the ship fills with ballast.

- *Passive-margin basins:* These form along the edges of continents that are *not* plate boundaries. They are underlain by stretched lithosphere, the remnants of a rift whose evolution ultimately led to the formation of a mid-ocean ridge. Passive-margin basins form because subsidence of stretched lithosphere continues long after rifting ceases and sea-floor spreading begins. They fill with sediment carried to the sea by rivers and with carbonate rocks formed in coastal reefs. Sediment in a passive-margin basin can reach an astounding thickness of 15 to 20 km.

- *Intracontinental basins:* These develop in the interiors of continents, initially because of subsidence over an unsuccessful rift. They may continue to subside even hundreds of millions of years after they first formed, for reasons that are not well understood. Illinois and Michigan are each underlain with an intracontinental basin in which up to 7 km of sediment has accumulated.

- *Foreland basins:* These form on the continent side of a mountain belt because as the mountain belt grows, large slices of rock are pushed up and onto the surface of the continent. Such movement takes place by slip along large faults. The weight of these slices pushes down on the surface of the lithosphere, creating a wedge-shaped

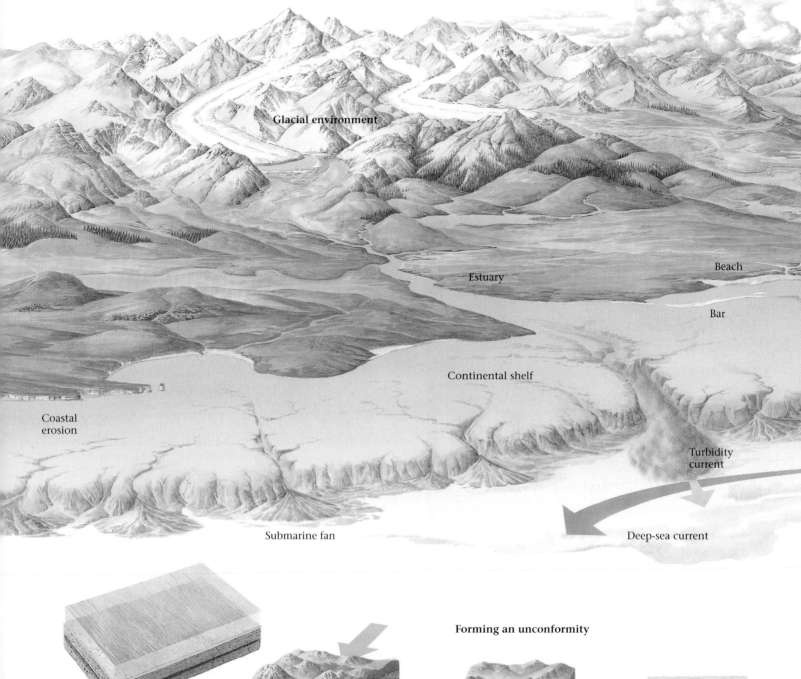

Glacial environment

Estuary

Beach

Bar

Continental shelf

Coastal erosion

Turbidity current

Submarine fan

Deep-sea current

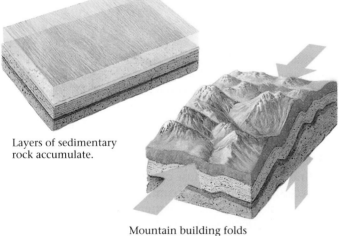

Layers of sedimentary rock accumulate.

Mountain building folds the rock layers.

Forming an unconformity

The mountains are eroded; the folded layers are submerged.

New sedimentary layers accumulate.

The Formation of Sedimentary Rocks

Categories of sedimentary rocks include clastic sedimentary rocks, chemical sedimentary rocks (formed from the precipitation of minerals out of water), and biochemical sedimentary rocks (formed from the shells of organisms). Clastic sedimentary rocks develop when grains (clasts) break off preexisting rock by weathering and erosion and are transported to a new location by

wind, water, or ice; the grains are deposited to create sediment layers, which are then cemented together. We distinguish among types of clastic sedimentary rocks on the basis of grain size.

The character of a sedimentary rock depends on the composition of the sediment and on the environment in which it accumulated. For example, glaciers carry sediment of all sizes, so

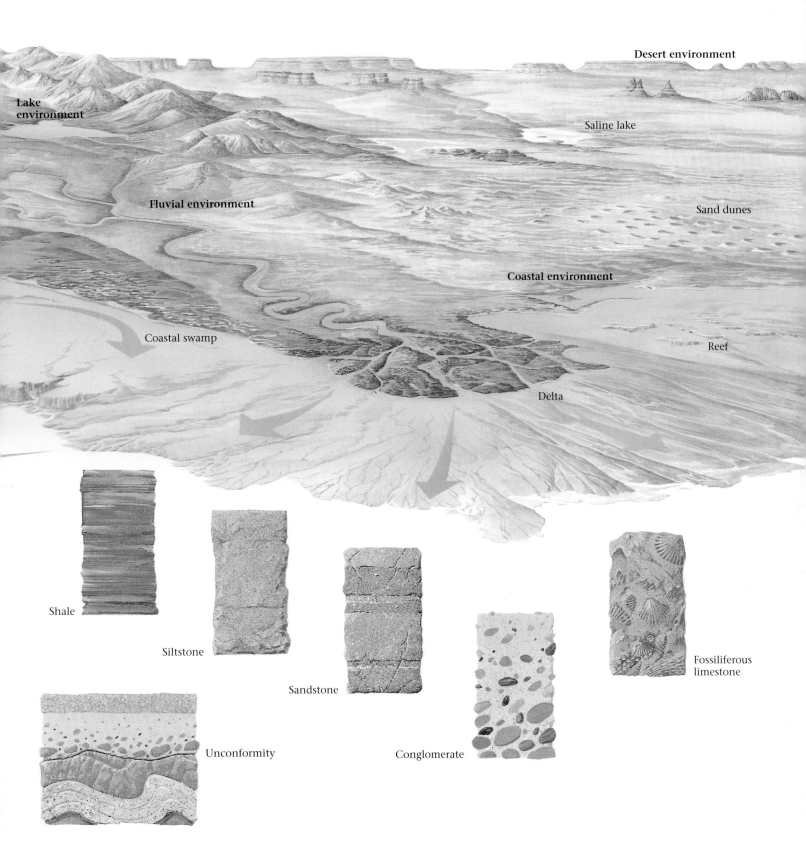

Desert environment

Lake environment

Saline lake

Fluvial environment

Sand dunes

Coastal environment

Coastal swamp

Reef

Delta

Shale

Siltstone

Sandstone

Fossiliferous limestone

Unconformity

Conglomerate

they leave deposits of poorly sorted (different-sized) till; streams deposit coarser grains in their channels and finer ones on floodplains; a river slows down at its mouth and deposits an immense pile of silt in a delta. Fossiliferous limestone develops on coral reefs. In desert environments, sand accumulates into dunes and evaporates precipitate in saline lakes. Offshore, submarine canyons channel avalanches of sediment, or turbidity currents, out to the deep-sea floor.

Sedimentary rocks tell the history of the Earth. For example, the layering, or bedding, of sedimentary rocks is initially horizontal. So where we see layers bent or folded, we can conclude that the layers were deformed during mountain building. Where horizontal layers overlie folded layers, we have an unconformity: for a time, sediment was not deposited, and/or older rocks were eroded away.

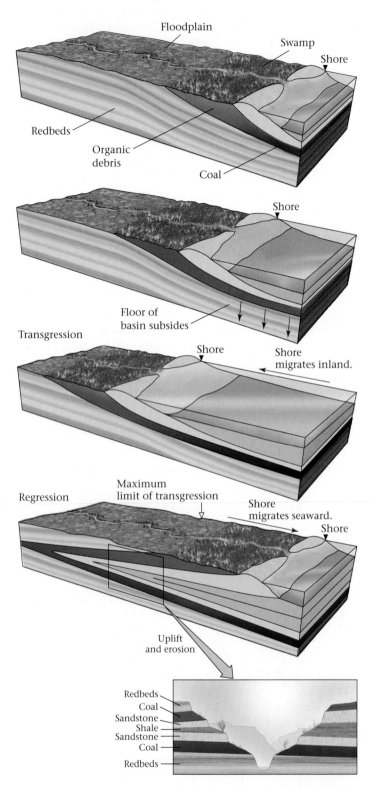

FIGURE 5.33 The concept of transgression and regression. As sea level rises and the shore migrates inland, coastal sedimentary environments overlap terrestrial environments. Eventually, deeper-water environments overlap shallower ones. Thus, a regionally extensive layer does not all form at the same time. During regression, sea level falls and the shore moves seaward.

depression adjacent to the mountain range that fills with sediment eroded from the range.

Transgression and Regression

Sea-level changes control the succession of sediments that we see in a sedimentary basin. At times during Earth history, sea level has risen by as much as a couple hundred meters, creating shallow seas that covered the interiors of continents. At other times sea level has fallen by a couple hundred meters, exposing even the continental shelves to air. Sea-level changes may be due to a number of factors, including climate changes, which control the amount of ice stored in polar ice caps, and changes in the volume of ocean basins, due to changes in the rate of sea-floor spreading. When sea level rises, the coast migrates inland. We call this process **transgression.** As the coast migrates, it gets progressively buried by deeper-water sediment. When sea level falls, the coast migrates seaward. We call this process **regression** (▶Fig. 5.33).

5.8 DIAGENESIS

Earlier in this chapter we discussed the process of lithification, by which sediment hardens into rock. Lithification is an aspect of a broader phenomenon called diagenesis. Geologists use the term **diagenesis** for all the physical, chemical, and biological processes that transform sediment into sedimentary rock and that alter characteristics of sedimentary rock after the rock has formed.

In the depositional environment, diagenesis includes bioturbation, growth of minerals in pore spaces and around grains, and replacement of existing crystals with new crystals. In buried sediment, diagenesis involves the compaction of sediment and growth of cement that leads to complete lithification. In fully lithified sedimentary rocks, diagenesis continues both as a result of chemical reactions between the rock and groundwater passing through the rock, and as a result of increases in temperature and pressure. These reactions may dissolve existing cement and/or form new cement, and may grow new minerals.

As temperature and pressure increase still deeper in the subsurface, the changes that take place in rocks become more profound. At sufficiently high temperature and pressure, a whole new assemblage of minerals forms, and/or mineral grains become aligned parallel to each other. Geologists consider such changes to be examples of metamorphism. The transition between diagenesis and metamorphism in sedimentary rocks is gradational. Most geologists consider changes that take place in rocks at temperatures of below about 150°C to be clearly diagenetic reactions, and those that occur in rocks at temperatures above about 300°C to be clearly metamorphic reactions. In the next chapter, we enter the true realm of metamorphism.

THE VIEW FROM SPACE A great variety of depositional environments occur on Earth. In these environments. sediments accumulate and may be buried deeply enough to transform into sedimentary rock. The type of sediment and the form of a deposit is an indication of the depositional environment. Here, we see an alluvial fan in the Taklimakan Desert of China, formed when an intermittent stream spreads out over a flat basin and divides into many small channels within and along which sediment accumulates.

CHAPTER SUMMARY

- Sediment consists of mineral grains and rock fragments derived from preexisting rock, mineral crystals that precipitate directly out of water, and shells.

- Rocks at the surface of the Earth undergo physical and chemical weathering. During physical weathering, intact rock breaks into pieces. During chemical weathering, rocks react with water and air to produce new minerals and ions in solution.

- Soil forms by the activities of organisms, by downward-percolating rainwater, and by the mixing in of organic matter. Distinct horizons can be identified in soil. The type of soil that forms depends on factors such as climate and source material.

- Geologists recognize four major classes of sedimentary rocks. Clastic (detrital) rocks form from cemented-together mineral grains and rock fragments that were first produced by weathering, then were transported, deposited, and lithified. Biochemical rocks develop from the shells of organisms. Organic rocks consist of plant debris or of altered plankton remains. Chemical rocks, such as evaporites, precipitate directly from water.

- Sedimentary structures include features such as bedding, cross bedding, graded bedding, ripple marks, dunes, and mud cracks. Their presence provides clues to depositional settings.

- Glaciers, mountain streams, mountain fronts, deserts, lakes, rivers, deltas, beaches, shallow seas, and deep seas each accumulate a different assemblage of sedimentary strata. Thus, by studying sedimentary rocks, we can reconstruct the characteristics of past environments.

- Thick piles of sedimentary rocks accumulate in sedimentary basins, regions where the lithosphere sinks, producing a depression at the Earth's surface.

- Sea level changes with time. Transgressions occur when sea level rises and the coastline migrates inland. Regressions occur when sea level falls and the coastline migrates seaward.

- Diagenesis involves processes leading to lithification and processes that alter sedimentary rock once it has formed.

KEY TERMS

arkose (p. 135)
bed (p. 139)
biochemical sedimentary rock (p. 131)
breccia (p. 135)
cementation (p. 132)
chemical sedimentary rock (p. 131)
chemical weathering (p. 125)
clastic (detrital) sedimentary rock (p. 131)
clasts (p. 132)
conglomerate (p. 135)
cross bed (p. 140)
deposition (p. 132)
diagenesis (p. 150)
dolostone (p. 137)
erosion (p. 132)
evaporites (p. 137)
joint (p. 123)
limestone (p. 135)
lithification (p. 132)
mudstone (p. 135)
organic sedimentary rock (p. 131)

physical (mechanical) weathering (p. 123)
regression (p. 150)
ripple mark (p. 140)
sandstone (p. 132)
sediment (p. 122)
sedimentary basin (p. 147)
sedimentary rock (p. 122)
sedimentary structure (p. 138)
shale (p. 135)
siltstone (p. 135)
soil (p. 128)
soil erosion (p. 131)
soil horizon (p. 128)
soil profile (p. 128)
sorting (p. 133)
stratigraphic formation (p. 139)
subsidence (p. 147)
transgression (p. 150)
travertine (p. 137)
weathering (p. 122)
zone of accumulation (p. 128)
zone of leaching (p. 128)

REVIEW QUESTIONS

1. Explain the circumstances that allowed the Mediterranean Sea to dry up.

2. How does physical weathering differ from chemical weathering?

3. Describe the processes that produce joints in rocks.

4. Feldspars are among the most common minerals in igneous rocks, but they are relatively rare in sediments. Why are they more susceptible to weathering and what common sedimentary minerals are produced by weathering of feldspar?

5. Describe the different horizons in a typical soil profile.

6. What factors determine the nature of soils in different regions?

7. Describe how a clastic sedimentary rock forms from its unweathered parent rock.

8. Clastic and chemical sedimentary rocks both consist of material that has been transported. How are they different?

9. Explain how biochemical sedimentary rocks form.

10. Describe how grain size and shape, sorting, sphericity, and angularity change as sediments move downstream.

11. Describe the two different kinds of chert. How are they similar? How are they different?

12. What kinds of conditions must exist for evaporite deposition to take place?

13. What minerals precipitate out of seawater first? next? last? What does this suggest when geologists find huge volumes of pure gypsum in the Earth's crust?

14. How does dolostone differ from limestone, and how does it form?

15. Describe how cross beds form. How can you read the current direction from cross beds?

16. Describe how a turbidity current forms and moves. How does it produce graded bedding?

17. Compare the deposits of an alluvial fan with those of a typical river environment and with those of a deep-marine deposit.

18. Why don't sediments accumulate everywhere? What types of tectonic conditions are required to create basins?

19. What kinds of changes may take place during diagenesis?

SUGGESTED READING

Boggs, S., Jr. 1998. *Petrology of Sedimentary Rocks.* Englewood Cliffs, N.J.: Prentice-Hall.

Boggs, S., Jr. 2000. *Principles of Sedimentology and Stratigraphy,* 3rd ed. Upper Saddle River, N.J.: Pearson Education.

Brady, N. C., and R. R. Weil. 2001. *The Nature and Properties of Soils,* 13th ed. Upper Saddle River, N.J.: Prentice-Hall.

Einsele, G. 2000. *Sedimentary Basins: Evolution, Facies, and Sediment Budget.* New York: Springer-Verlag.

Harpstead, M. I., et al. 2001. *Soil Science Simplified,* 4th ed. Ames: Iowa State University Press.

Leeder, M. R. 1999. *Sedimentology and Sedimentary Basins: From Turbulence to Tectonics.* Oxford, UK: Blackwell.

Prothero, D. R., and F. L. Schwab. 1998. *Sedimentary Geology.* New York: Freeman.

Reading, H. G., ed. 1996. *Sedimentary Environments: Processes, Facies, and Stratigraphy.* Oxford, UK: Blackwell.

Tucker, M. E. 2001. *Sedimentary Petrology,* 3rd ed. Oxford, UK: Blackwell.

Metamorphism: A Process of Change

Nothing in the world lasts, save eternal change.
— HONORAT DE BUEIL (1589–1650)

6.1 INTRODUCTION

Cool winds sweep across Scotland for much of the year. In this blustery climate, vegetation has a hard time taking hold, so the landscape provides countless outcrops of barren rock. During the latter half of the eighteenth century, a gentleman farmer and doctor named James Hutton became fascinated with the Earth and examined these outcrops, hoping to learn how rock formed. Hutton found that many features in the outcrops resembled the products of present-day sediment deposition or of volcanic activity, and soon he came to an understanding of how sedimentary and igneous rock forms. But Hutton also found rock that contained minerals and textures quite different from those in sedimentary and igneous samples. He described this puzzling rock as "a mass of matter which had evidently formed originally in the ordinary manner . . . but which is now extremely distorted in its structure . . . and variously changed in its composition."

The rock that so puzzled Hutton is now known as metamorphic rock, from the Greek words *meta*, meaning "beyond" or "change," and *morphe*, meaning "form." In modern terms, a **metamorphic rock** is a rock that forms from a preexisting rock, or **protolith,** when the protolith undergoes

An outcrop of Precambrian metamorphic rock, exposed in the Wasatch Mountains of Utah. The layering, or foliation, formed during metamorphism and then later was bent into an S-shape.

mineralogical and textural changes in response to modification of its physical or chemical environment. This means that during **metamorphism,** the process of forming metamorphic rock, new minerals may grow at the expense of old ones, and/or the shape, size, and arrangement of grains in the rock may change. These changes occur when the protolith is subjected to heat, pressure, differential stress (a push, pull, or shear), and/or hydrothermal fluids (hot-water solutions). Because metamorphic rocks do not form from melts, we say that *metamorphism is a solid-state process.*

Hutton did more than just note the existence of metamorphic rock—he also tried to understand why metamorphism takes place. Because he found metamorphic rocks adjacent to igneous intrusions, he concluded that metamorphism can take place when heat from an intrusion "cooks" the rock into which it intrudes. And because he realized that metamorphic rocks can also occur over broad regions, in the absence of intrusions, he speculated that metamorphism also takes place when rock becomes deeply buried.

From Hutton's day to the present, geologists have undertaken field studies, laboratory experiments, and theoretical calculations to flesh out the story of metamorphism. In this chapter, we present the results of their work. We begin by explaining the causes of metamorphism. Then, we describe the basis for classifying metamorphic rocks. We conclude by discussing the geologic settings in which these rocks form. As you will see, Hutton's speculations on the origin of metamorphic rock were basically correct, but they represent only part of the story—the rest of the story could not take shape until the theory of plate tectonics came along. Hutton's name appears elsewhere in this book, because he proposed so many of the fundamental principles of geology that geologists commonly refer to him as the "father of geology."

6.2 WHAT IS A METAMORPHIC ROCK?

If someone were to put a rock on a table in front of you, how would you know that it is metamorphic? First, metamorphic rocks can have a metamorphic texture, meaning that the grains in the rock have grown in place and interlock. Second, metamorphic rocks can possess metamorphic minerals, new minerals that grow only under metamorphic temperatures and pressures. In fact, metamorphism can produce a group of minerals which together make up a metamorphic mineral assemblage. And third, metamorphic rocks can have metamorphic foliation, defined by the parallel alignment of platy minerals (such as mica) and/or the presence of alternating light-colored and dark-colored layers. Development of these characteristics can make a meta-

morphic rock as different from its protolith as a butterfly is from a caterpillar. For example, metamorphism of red shale can yield a metamorphic rock consisting of aligned mica flakes and brilliant garnet crystals (▶Fig. 6.1a, b), and metamorphism of fossiliferous limestone can yield a metamorphic rock consisting of large interlocking crystals of calcite (▶Fig. 6.1c, d).

The formation of metamorphic textures and minerals takes place very slowly—it may take thousands to millions of years—and it involves several processes, which sometimes occur alone and sometimes together. The most common processes are:

- *Recrystallization,* which changes the shape and size of grains without changing the identity of the mineral constituting the grains (▶Fig. 6.2a).

- *Phase change,* which transforms one mineral into another mineral with the same composition but a different crystal structure. At an atomic scale, phase change involves the rearrangement of atoms.

- *Metamorphic reaction,* or *neocrystallization* (from the Greek *neos,* for "new"), which results in the growth of new mineral crystals that differ from those of the protolith (▶Fig. 6.2b). We can think of neocrystallization as being the result of one or more chemical reactions that effectively digest or decompose minerals of the protolith to produce new minerals of the metamorphic rock. For this to take place, atoms migrate, or diffuse, through solid crystals, a very slow process, and/or dissolve and reprecipitate at grain boundaries, sometimes with the aid of hydrothermal fluids.

- *Pressure solution,* which happens when a rock is squeezed more strongly in one direction than in others at relatively low pressures and temperatures, in the presence of water. Mineral grains dissolve where their surfaces are pressed against other grains, producing ions that migrate through the water to precipitate elsewhere (▶Fig. 6.2c).

- *Plastic deformation,* which happens at elevated temperatures; under these conditions some minerals behave like soft plastic in that if they are squeezed or stretched, they change shape and become flattened or elongate (▶Fig. 6.2d).

6.3 CAUSES OF METAMORPHISM

Caterpillars undergo metamorph*osis* because of hormonal changes in their bodies. Rocks undergo metamorph*ism* when they are subjected to heat, pressure, differential stress, and/or hydrothermal fluids. Let's now consider the details of how these agents of metamorphism operate.

FIGURE 6.1 (a) Hand specimen of red shale, consisting of clay flakes, quartz, and iron oxide (hematite). (b) Hand specimen of metamorphic rock (gneiss) containing biotite, quartz, feldspar, and bright purple garnets. A rock similar to the shale could have been the protolith of this gneiss. This hand specimen is about 10 cm wide. (c) This photomicrograph of a Devonian limestone shows that the rock consists of small fossil shells and shell fragments that have been cemented together. The circular grain is part of a crinoid. The field of view is about 3 mm. (d) This photomicrograph of marble shows how new crystals of metamorphic calcite grew to form an interlocking texture. This photo was taken with polarized light; the color and darkness of an individual grain depends on its orientation with respect to the light waves. The field of view is 5 mm.

Metamorphism Due to Heating

When you heat cake batter, the batter transforms into a new material—cake. Similarly, when you heat a rock, its ingredients transform into a new material—metamorphic rock. Why? Think about what happens to atoms in a mineral grain as the grain warms. Heat causes the atoms to vibrate rapidly, stretching and bending chemical bonds that lock atoms to their neighbors. If bonds stretch too far and break, atoms detach from their original neighbors, move slightly, and form new bonds with other atoms. Repetition of this process leads to rearrangement of atoms within grains, or to migration of atoms into and out of grains. As a consequence, recrystallization and/or neocrystallization take place, enabling a metamorphic mineral assemblage to grow in solid rock.

Metamorphism takes place at temperatures between those at which diagenesis occurs and those that cause melting. Roughly speaking, this means that most metamorphic rocks you find in outcrops on continents formed at temperatures of between 200°C and 850°C.

Protolith   Metamorphic rock

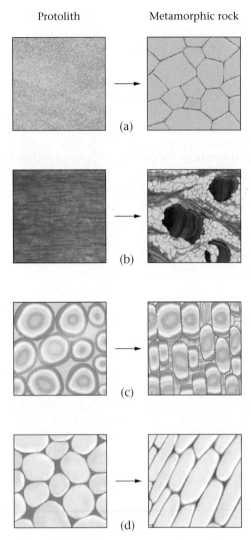

(a)

(b)

(c)

(d)

FIGURE 6.2 Examples of protoliths, and metamorphic rocks derived from them, illustrating different mechanisms of metamorphism. Each box represents a sketch of a thin section, with the field of view about 1 mm wide. (a) A protolith of siltstone recrystallizes to form metamorphic rock (quartzite) made of larger quartz crystals of the same mineral. (b) Metamorphic reactions (neocrystallization) in a protolith of silty shale will form a rock made up of quartz, mica, large garnets, and other minerals. (c) A protolith of oolitic limestone (an "oolite" is a tiny snowball-like sphere of calcite with internal concentric rings) undergoes pressure solution so that grains have dissolved on two sides. (d) A protolith of quartz sandstone deforms plastically to produce a metamorphic rock in which the quartz grains have been flattened into pancakes.

Metamorphism Due to Pressure

As you swim underwater in a swimming pool, water squeezes against you equally from all sides—in other words, your body feels pressure. Pressure can cause a material to collapse inward. For example, if you pull an air-filled balloon down to a depth of 10 m in a lake, the balloon becomes significantly smaller. Pressure can have the same effect on minerals. Near the Earth's surface, minerals with relatively open crystal structures are stable. However, if you subject these minerals to extreme pressure, denser minerals tend to form. Such transformations involve phase changes and/or neocrystallization.

Changing Pressure and Temperature Together

So far, we've considered changes in pressure and temperature as separate phenomena. But in the Earth, pressure and temperature change together with increasing depth. For example, at a depth of 8 km, temperature in the crust reaches about 200°C and pressure reaches about 2.3 kbar. If a rock slowly becomes buried to a depth of 20 km, as can happen during mountain building, temperature in the rock increases to 500°C, and pressure to 5.5 kbar. Experiments and calculations show that the stability of certain minerals and mineral assemblages depend on *both* pressure and temperature, so a metamorphic rock formed at 8 km does not contain the same minerals as one formed at 20 km.

We can illustrate the relation of mineral stability to pressure and temperature by studying the behavior of Al_2SiO_5 (aluminum silicate) as portrayed on a phase diagram, a graph with temperature indicated by one axis and pressure indicated by the other (▶Fig. 6.3). Al_2SiO_5 can exist as three different minerals: kyanite, andalusite, and sillimanite. Each of these minerals forms only under a specific range of temperatures and pressures, indicated by an area called a stability field, on the phase diagram.

FIGURE 6.3 We depict the stability fields for three metamorphic minerals (kyanite, andalusite, and sillimanite) that are polymorphs of Al_2SiO_5 (aluminum silicate) on a phase diagram. At a pressure of 2 kbar and a temperature of 450°C (point X), andalusite is stable. At 5 kbar and 450°C (point Y), kyanite is stable. At 5 kbar and 650°C (point Z), sillimanite is stable.

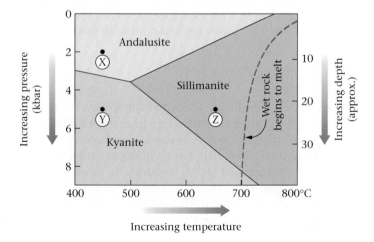

Differential Stress and Development of Preferred Mineral Orientation

Imagine that you have just built a house of cards and, being in a destructive mood, you step on it. The structure collapses because the downward push you apply with your foot exceeds the push provided by air in other directions. If we squeeze or stretch a material unequally from different sides, we subject the material to differential stress. In other words, under conditions of differential stress, the push or pull in one direction differs in magnitude from the push or pull in another direction (▶Fig. 6.4a). We distinguish two kinds of differential stress:

- *Normal stress:* Normal stress pushes or pulls perpendicular to a surface. We call a push compression and a pull tension. Compression flattens a material (▶Fig. 6.4b), whereas tension stretches a material.

- *Shear stress:* Shear stress, or shear, moves one part of a material sideways, relative to another. If, for example, you place a deck of cards on a table, then set your hand on top of the deck and move your hand parallel to the table, you shear the deck (▶Fig. 6.4c).

When rocks are subjected to differential stress at elevated temperatures and pressures, they can change shape without breaking. For example, a piece of rock undergoing compression or shearing during metamorphism may slowly become flatter. As it changes shape, the internal texture of a rock also changes, typically resulting in the development of **preferred mineral orientation.** By this, we mean that platy (pancake-shaped) grains lie parallel to one another, and elongate (cigar-shaped) grains align in the same direction. Both platy and elongate grains are inequant grains, meaning that the length of a grain is not the same in all directions; in contrast, equant grains have roughly the same dimensions in all directions (▶Fig. 6.5a). The preferred orientation of inequant minerals in a rock gives the rock metamorphic foliation (▶Fig. 6.5b).

FIGURE 6.4 The concept of differential stress. (a) Before you step on it, a house of cards feels only air pressure, equal from all sides. As you step on the cards, they feel a differential stress because the vertical push by your foot (large arrows) is greater than the horizontal push by air (small arrows), so the house flattens. (b) A normal stress (in this case, compression) applied to a ball of dough flattens the ball into a pancake. (c) A shear stress smears out a pack of cards parallel to the table.

FIGURE 6.5 Some basic shapes in nature. (a) Equant grains have roughly the same dimensions in all directions. Inequant grains can be either elongate (cigar-shaped) or platy (pancake shaped). (b) A drawing of a thin section, showing alignment of inequant mineral grains, producing a preferred mineral orientation.

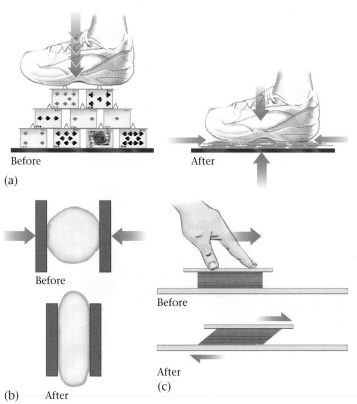

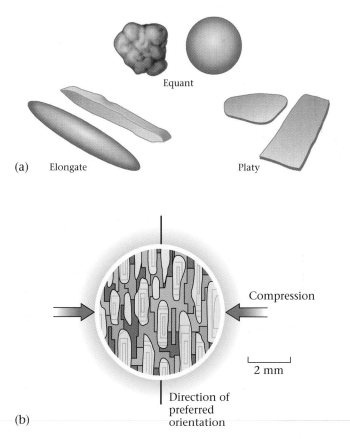

The Role of Hydrothermal Fluids

Metamorphic reactions usually take place in the presence of hydrothermal fluids, because water occurs throughout the crust. We initially defined hydrothermal fluids simply as "hot-water solutions." In fact, they actually can include hot water, steam, and so-called supercritical fluid. A supercritical fluid forms under high temperatures and pressures, and has characteristics of both liquid and gas. Hydrothermal fluids are *chemically active* in that they react chemically with rock. For example, hydrothermal fluids can dissolve certain minerals. As a consequence, such fluids do not consist of pure water, but rather are solutions.

Hydrothermal fluid plays many roles during metamorphism. First, fluids accelerate metamorphic reactions, for atoms involved in the reactions can migrate faster through a fluid than they can through a solid. Second, fluids provide water that can be absorbed during metamorphic reactions. Third, fluids passing through a rock may pick up some dissolved ions and drop off others (just as a bus on its route through a city picks up some passengers and drops off others) and thus can change the overall chemical composition of a rock during metamorphism. The process by which a rock's chemical composition changes because of a reaction with hydrothermal fluids is called **metasomatism.** Dissolved ions carried away by hydrothermal fluids either react with rocks elsewhere, reach the Earth's surface and wash away, or precipitate underground to form veins, mineral-filled cracks that cut across preexisting rock.

6.4 TYPES OF METAMORPHIC ROCKS

Coming up with a way to classify and name the great variety of metamorphic rocks on Earth hasn't been easy. After decades of debate, geologists find it most convenient to divide metamorphic rocks into two fundamental classes: foliated rocks and nonfoliated rocks. Each class contains several rock types. We distinguish nonfoliated rocks from each other primarily by their component minerals, whereas foliated rocks are distinguished partly by their component minerals and partly by the nature of their foliation.

Foliated Metamorphic Rocks

To understand this class of rocks, we first need to discuss the nature of **foliation** in more detail. The word comes from the Latin *folium*, for leaf. Geologists use foliation to refer to the repetition of planar surfaces or layers in a metamorphic rock (▶Fig. 6.6a). Some layers are indeed as thin as a leaf, or thinner, but some may be over a meter thick. Foliation can give metamorphic rocks a striped or streaked appearance in an outcrop, and/or give them the ability to split into thin sheets. A foliated metamorphic rock has foliation because it contains inequant mineral crystals that are aligned parallel to one another, defining preferred mineral orientation, and/or because the rock has alternating dark-colored and light-colored layers.

Foliated metamorphic rocks can be distinguished from one another according to their composition, their grain

FIGURE 6.6 (a) A block of rock with slaty cleavage splits along cleavage planes into thin sheets. Originally, the slate was shale and had sedimentary bedding. If you look carefully, you may find hints of the bedding, indicated by sandier layers, in the slate. Note that the bedding plane and cleavage plane, in this example, are perpendicular to each other. (b) Slate easily splits into thin sheets that can be used as shingles on roofs. Here, an old-style shingle maker in Wales plies his trade.

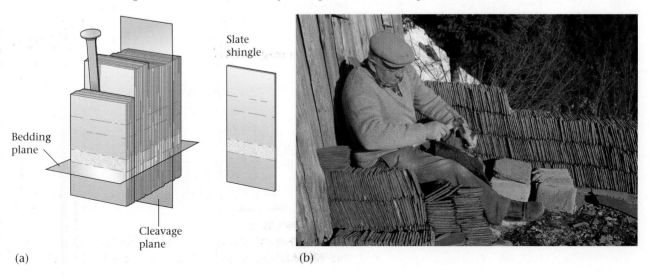

Slate shingle

Bedding plane

Cleavage plane

(a)

(b)

size, and the nature of their foliation. The most common types include:

- *Slate:* **Slate,** the finest-grained foliated metamorphic rock, forms by the metamorphism of shale (a sedimentary rock consisting of clay) under relatively low pressures and temperatures. The foliation, or slaty cleavage, in slate results from the development of a strong preferred orientation of clay and chlorite grains, for these grains are shaped like extremely tiny sheets of paper (Fig. 6.6a). Slate tends to split on slaty cleavage planes into thin, impermeable sheets that can serve as roofing material (▶Fig. 6.6b). Slaty cleavage forms in response to differential stress. Typically, cleavage planes form perpendicular to the direction of compression. For example, end-on compression of a sequence of horizontal shale beds produces vertical slaty cleavage (▶Fig. 6.7a, b).

- *Phyllite:* Phyllite is a fine-grained metamorphic rock with a foliation caused by the preferred orientation of very fine-grained white mica and, in some cases, chlorite. The word comes from the Greek word *phyllon*, leaf. The parallelism of translucent fine-grained mica gives phyllite a silky sheen known as phyllitic luster (▶Fig. 6.8a). Phyllite forms by the metamorphism of slate at a temperature high enough to cause neocrystallization; metamorphic reactions produce a new assemblage of minerals (fine-grained mica and chlorite) out of clay.

- *Metaconglomerate:* Under the metamorphic conditions that transform shale to slate or to phyllite, a protolith of conglomerate also undergoes changes and becomes metaconglomerate. Specifically, pressure solution and plastic deformation allow pebbles or cobbles to flatten and

become pancake shaped and/or elongate. The alignment of inequant clasts defines a foliation (▶Fig. 6.8b).

- *Schist:* **Schist** is a medium- to coarse-grained metamorphic rock that possesses a type of foliation, called schistosity, defined by the preferred orientation of large mica (that is, muscovite and/or biotite) flakes (▶Fig. 6.8c). Schist forms at a higher temperature than phyllite.

- *Gneiss:* **Gneiss** is a compositionally layered metamorphic rock, typically composed of alternating dark-colored and light-colored layers or lenses that range in thickness from millimeters to meters. This compositional layering, or gneissic banding, gives gneiss a striped appearance (▶Fig. 6.9a, b).

  How does the banding in gneiss form? Banding in some examples of gneiss evolved directly from the original bedding in a rock. For example, metamorphism of a protolith consisting of alternating beds of sandstone and shale produces a gneiss consisting of alternating beds of quartzite and mica schist. It is more common for gneissic banding to form when the protolith undergoes an extreme amount of shearing under conditions in which the rock can flow like soft plastic. Such flow stretches, folds, and smears out any kind of preexisting compositional contrasts in the rock and transforms them into aligned sheets. Finally, banding in some gneisses develops by an incompletely understood process called metamorphic differentiation. During this process, chemical reactions segregate different minerals into different layers.

- *Migmatite:* At very high temperatures, or if hydrothermal fluids enter the rock and lower its melting temperature, gneiss begins to partially melt, producing magma that is enriched in silica. In some cases, this melt does not move very far before freezing to form a light-colored (felsic) igneous rock. Lenses of this new igneous rock are surrounded by bands of relict gneiss, which consists of minerals left behind when the felsic melt seeped out; the relict gneiss tends to be dark-colored (mafic). The resulting mixture of igneous and relict metamorphic rock is called a **migmatite** (▶Fig. 6.10).

Nonfoliated Metamorphic Rocks

A nonfoliated metamorphic rock contains minerals that recrystallized, or new minerals that grew, during metamorphism, but it has no foliation. The lack of foliation may mean that metamorphism occurred in the absence of differential stress, or simply that most of the new crystals are equant. We list below some of the rock types that commonly can occur without foliation.

- *Hornfels:* Rock that undergoes metamorphism because of heating in the absence of differential stress becomes

FIGURE 6.7 (a) The end-on compression of a bed will create slaty cleavage at an angle perpendicular to the bedding. (b) Commonly, the rock folds (bends into curves) at the same time cleavage forms. Cleavage tends to be parallel to the axial plane of the fold, the imaginary plane that, simplistically, divides the fold in half (see Chapter 9). The dashed lines indicate the original shape of the rock body that was deformed.

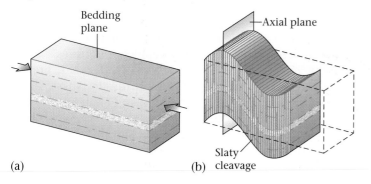

Bedding plane

Axial plane

(a)   (b) Slaty cleavage

(a)

(b)

(c)

FIGURE 6.8 (a) The sheen in this phyllite comes from the reflection of light off the tiny mica flakes that constitute the rock. The accordionlike crinkles (crenulation) on the surface are due to compression. (b) In a metaconglomerate, the pebbles and cobbles have pancakelike shapes and align with each other to define a foliation. The larger clasts in this photo are about 20 cm (5 inches) long. (c) A hand specimen of schist, with coarse mica crystals.

hornfels. The specific mineral assemblage in a hornfels depends on the composition of the protolith and on the temperature and pressure of metamorphism.

- *Quartzite:* **Quartzite** forms by the metamorphism of quartz sandstone. During metamorphism, preexisting quartz grains recrystallize, creating new, generally larger grains. In the process, the distinction between cement and grains disappears, open pore space disappears, and the grains become interlocking. In fact, recrystallization makes the grains of a quartzite weld together into a tight mosaic, so that when quartzite cracks, the fracture cuts across grain boundaries—in contrast, fractures in sandstone curve *around* grains. Quartzite looks glassier than sandstone and does not have the grainy, sandpaperlike surface characteristic of sandstone (▶Fig. 6.11a, b). Depending on the impurities contained in the quartz, quartzite can vary in color from white to gray, purple, or green.

- *Marble:* The metamorphism of limestone yields **marble.** During the formation of marble, calcite composing the protolith recrystallizes; as a consequence, fossil shells, pore space, and the distinction between grains and cement disappear. Thus, marble typically consists of a fairly uniform mass of interlocking calcite crystals.

  Sculptors love to work with marble because the rock is relatively soft and has a uniform texture that gives it the cohesiveness and homogeneity needed to fashion large, smooth, highly detailed sculptures. Marble comes in a variety of colors—white, pink, green, and black—depending on the impurities it contains. Michelangelo, one of the great Italian Renaissance artists, sought large, unbroken blocks of creamy white marble from the quarries in the Italian Alps for his masterpieces (▶Fig. 6.12a). This marble began as a pure limestone, formed from the shells of tiny marine

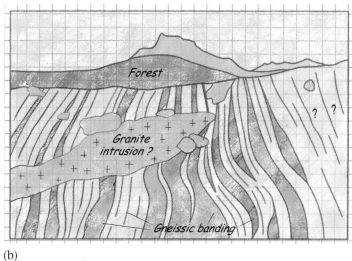

(a) (b)

FIGURE 6.9 (a) A glacially eroded surface exposing gneiss contains alternating bands of light-colored and dark-colored minerals. (b) A geologist's interpretation of the outcrop emphasizes the banding. A small intrusion of granite cuts across the banding. The loose boulders were dropped by the glacier when it melted.

organisms, and was metamorphosed to marble during the mountain-building event that produced the Alps.

Like quartzite, marble at the scale of grains commonly has no foliation, for it contains mostly equant grains. But impurities, such as iron oxide or graphite, may create beautiful color banding in marble, giving it a foliation that makes it a prized decorative stone (▶Fig. 6.12b).

FIGURE 6.10 An outcrop of migmatite in northern Michigan contains both light-colored (felsic) igneous rock and dark-colored (mafic) metamorphic rock. The mixture of the two rock types makes migmatite resemble marble cake. Migmatite forms under conditions that cause a metamorphic rock to undergo partial melting to produce a small amount of felsic magma.

6.5 THE INTENSITY OF METAMORPHISM

Metamorphic Grade

Not all metamorphism takes place under the same physical conditions. For example, rocks carried to a great depth beneath a mountain range undergo more intense metamorphism than do rocks closer to the surface. Geologists use the term **metamorphic grade,** in a somewhat informal way, to indicate the "intensity of metamorphism," meaning the amount or degree of metamorphic change (▶Fig. 6.13). (To provide a more complete indication of the intensity of metamorphism, geologists use the concept of "metamorphic facies"; see ▶Box 6.1.) Classification of metamorphic grade depends primarily on temperature, because temperature plays the dominant role in determining the extent of recrystallization and neocrystallization during metamorphism. Metamorphic rocks that form under relatively low temperatures (between about 200°C and 320°C) are low-grade rocks, and metamorphic rocks that form under relatively high temperatures (over 600°C) are high-grade rocks. Intermediate-grade rocks form under temperatures between these two extremes. Different grades of metamorphism yield different metamorphic mineral assemblages.

Geologists refer to metamorphism that occurs while temperature and pressure progressively *increases* as prograde metamorphism. During prograde metamorphism, recrystallization and neocrystallization produce coarser grains and

(a)

(b)

FIGURE 6.11 (a) A sandstone protolith, with a grainy surface. (b) A quartzite with a glassy surface. Both (a) and (b) consist predominantly of quartz, but they have different textures.

(a)

(b)

FIGURE 6.12 (a) The marble in this unfinished sculpture by Michelangelo is fairly soft and easy to carve, but it does not crumble. (b) Various impurities (e.g., iron-oxide minerals) create the color banding of decorative marble, as illustrated by the beautiful bands in this seventeenth-century column.

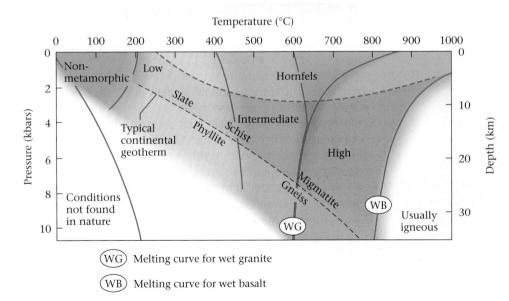

(WG) Melting curve for wet granite

(WB) Melting curve for wet basalt

FIGURE 6.13 A graph showing the approximate conditions of various grades. At low temperatures, only diagenesis takes place. At progressively higher temperatures, a rock passes from low to intermediate to high grade. Hornfels forms under conditions of relatively low pressure, where heat is provided by an adjacent magma intrusion. Slate, phyllite, schist, and gneiss form at progressively increasing depth in the crust, under conditions where temperature and pressure increase together. WG = conditions under which wet rock of granitic composition begins to melt. If this happens, migmatite can form (region with red stipple). If the rock does not contain water, melting doesn't begin until even higher temperatures; at that point, high-grade metamorphic rocks form. WB = conditions under which wet basalt begins to melt.

Metamorphic Facies

BOX 6.1
THE REST OF THE STORY

In the early years of the twentieth century, geologists working in Scandinavia, where erosion by glaciers has left beautiful, nearly un-weathered exposures, came to realize that metamorphic rocks, in general, do not consist of a hodgepodge of minerals formed at different times and in different places, but rather consist of a distinct assemblage of minerals that grew in association with each other at a certain pressure and temperature. It seemed that the mineral assemblage present in a rock more or less represented a condition of "chemical equilibrium," meaning that the chemicals making up the rock had organized into a group of mineral grains that were—to anthropomorphize a bit—comfortable with each other and their surroundings, and thus did not feel the need to change further. These geologists also realized that the specific mineral assemblage in a rock depends on pressure and temperature conditions, *and* on the composition of the protolith.

The above discovery led the geologists to propose the concept of metamorphic facies. A **metamorphic facies** is a set of metamorphic mineral assemblages indicative of a certain range of pressure and temperature. Each specific assemblage in a facies reflects a specific protolith composition. According to this definition, a given metamorphic facies includes several different kinds of rocks that differ from each other in terms of composition (i.e., mineral content)—but all the rocks

of a given facies formed under roughly the same temperature and pressure conditions. Geologists recognize several major facies, of which the major ones are zeolite, hornfels, greenschist, amphibolite, blueschist, eclogite, and granulite. The names of the different facies are based on a distinctive feature or mineral found in some of the rocks of the facies.

We can represent the approximate conditions under which metamorphic facies formed by using a pressure-temperature graph (▶Fig. 6.14). Each area on the graph, labeled with a facies name, represents the approximate range of temperatures and pressures in which mineral assemblages characteristic of that particular facies form. For example, a rock subjected to the pressure and temperature at Point A (4.5 kbar and 400°C) develops a mineral assemblage characteristic of the greenschist facies. As the graph implies, the boundaries between facies cannot be precisely defined, and there are likely broad transitions between facies. We can also portray the geothermal gradients of different crustal regions on the graph. Beneath mountain ranges, for example, the geothermal gradient passes through the zeolite, greenschist, amphibolite, and granulite facies. In contrast, in the accretionary prism that forms at a subduction zone, temperature increases slowly with increasing depth, so blueschist assemblages can form.

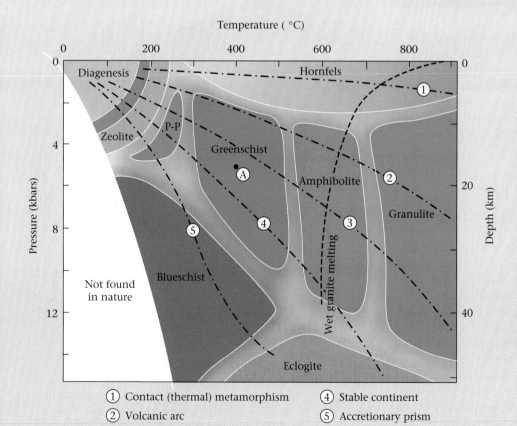

① Contact (thermal) metamorphism
② Volcanic arc
③ Collisional mountain belt
④ Stable continent
⑤ Accretionary prism

FIGURE 6.14 The common metamorphic facies. The boundaries between the facies are depicted as wide bands because they are gradational and approximate. Note that some amphibolite-facies rocks and all granulite-facies rocks form at pressure-temperature conditions to the right of the melting curve for wet granite. Thus, such metamorphic rocks develop only if the protolith is dry. The dotted lines indicate approximations of various geothermal gradients found on Earth. One of the facies depicted on the graph is not mentioned in the text: specifically, P-P = "prehnite-pumpellyite" facies, named for two unusual metamorphic minerals.

new mineral assemblages that are stable at higher temperatures and pressures (▶Fig. 6.15). As grade increases, metamorphic reactions release water, so high-grade rocks tend to be drier than low-grade rocks.

Metamorphism that takes place when temperatures and pressures progressively decrease is known as retrograde metamorphism. For retrograde metamorphism to occur, water must be added back to the rock. Thus, retrograde metamorphism does not happen unless hydrothermal fluids enter the rock. In fact, under cold and dry conditions, retrograde metamorphic reactions cannot proceed—it is for this reason that high-grade rocks formed early during Earth history have survived so that we see them exposed at the surface of the Earth today.

Index Minerals and Metamorphic Zones

Geologists have discovered that the presence of certain minerals, known as index minerals, in a rock indicates the approximate metamorphic grade of the rock. Geologists can determine the locations in a region where a particular index mineral first appears. The line on a map along which an index mineral first appears is called an isograd (from the Greek *iso*, meaning "equal"). All points along an isograd have approximately the same metamorphic grade. **Metamorphic zones** are the regions between two isograds; zones are named after an index mineral that was not present in the previous, lower-grade zone.

To compare rocks of different grades, you could take a hike from central New York State eastward into central Massachusetts in the eastern United States. Your path starts in a region where rocks were not metamorphosed, and it takes you into the internal part of the Appalachian Mountain belt, where rocks were intensely metamorphosed. As a consequence, you cross several metamorphic zones (▶Fig. 6.16).

6.6 METAMORPHISM IN THE CONTEXT OF PLATE TECTONICS

So far, we've discussed the nature of changes that occur during metamorphism, the agents of metamorphism (heat, pressure, differential stress, and hydrothermal fluids), the rock types that form as a result of metamorphism, and the concepts of metamorphic grade and metamorphic facies. With this background, let's now examine the geologic settings on Earth where metamorphism takes place, as viewed from the perspective of plate tectonics theory (see two-page art, pp. 168–169). Because of the wide range of possible metamorphic environments, metamorphism occurs at a wide range of conditions in the Earth. You will see that the conditions under which metamorphism occurs are not the same in all geologic settings. That's because the geothermal gradient

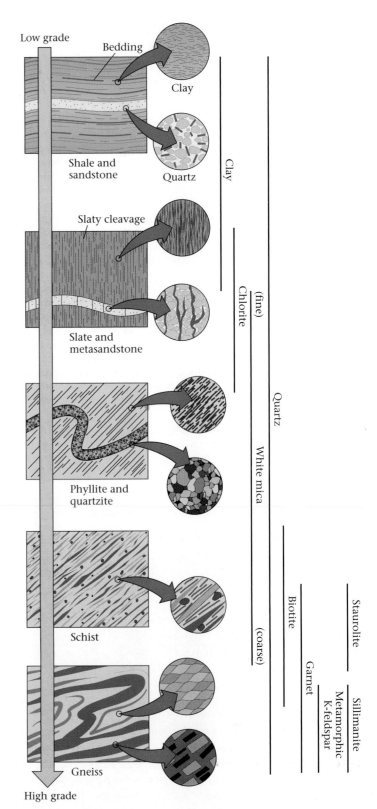

FIGURE 6.15 When shale progressively metamorphoses from low grade to high grade, it first becomes slate, then phyllite, then schist, then gneiss. In many cases, gneiss and schist can form under the same conditions. The side graph shows the stability range of various minerals.

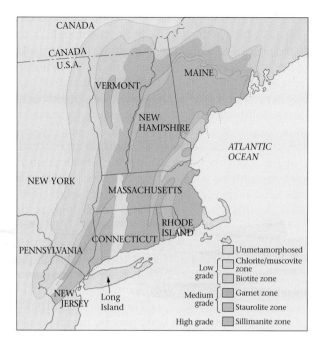

FIGURE 6.16 Simplified metamorphic zones as portrayed on a map of New England (U.S.A.). Isograds, defined by the first appearance of an index mineral, separate the zones.

(indicating the relation between temperature and depth), the extent to which rocks endure differential stress during metamorphism, and the extent to which rocks interact with hydrothermal fluids all depend on the geologic environment.

Thermal or Contact Metamorphism

Imagine a hot magma that rises from great depth beneath the Earth's surface and intrudes into cooler country rock at a shallow depth. Heat flows from the magma into the country rock, for heat always flows from hotter to colder materials. As a consequence, the magma cools and solidifies while the country rock heats up. In addition, hydrothermal fluids circulate through both the intrusion and the country rock. As a consequence of the heat and hydrothermal fluids, the country rock undergoes metamorphism, with the highest-grade rocks forming immediately adjacent to the pluton, where the temperatures were highest, and progressively lower-grade rocks forming farther away. The distinct belt of metamorphic rock that forms around an igneous intrusion is called a **metamorphic aureole** or contact aureole (▶Fig. 6.17). The width of an aureole depends on the size and shape of the intrusion, and on the amount of hydrothermal circulation—larger intrusions create wider aureoles.

The local metamorphism caused by igneous intrusion can be called either **thermal metamorphism** (see ▶Box 6.2), to emphasize that it develops in response to heat without a change in pressure and without differential stress, or

contact metamorphism, to emphasize that it develops adjacent to an intrusion. Because this metamorphism takes place without application of differential stress, aureoles contain hornfels, a nonfoliated metamorphic rock.

Contact metamorphism occurs anywhere that the intrusion of plutons occurs. According to plate tectonics theory, plutons intrude into the crust at convergent plate boundaries, in rifts, and during the mountain building that takes place when continents collide.

Burial Metamorphism

As sediment gets buried in a subsiding sedimentary basin, the pressure increases due to the weight of overburden, and the temperature increases due to the geothermal gradient. In the upper few kilometers, the temperatures and pressures are low enough that the changes taking place in the sediment can be considered to be manifestations of diagenesis. But at depths greater than about 8 to 15 km, depending on the geothermal gradient, temperatures may be great enough for metamorphic reactions to begin, and low-grade nonfoliated metamorphic rocks form. Metamorphism due only to the consequences of very deep burial is called **burial metamorphism.**

FIGURE 6.17 In a metamorphic aureole bordering an igneous intrusion, the highest-grade thermally metamorphosed rocks directly border the intrusion. The grade decreases away from the pluton. The gradation is analogous to the gradation from clay to pottery to porcelain, obtained by firing clay in an oven.

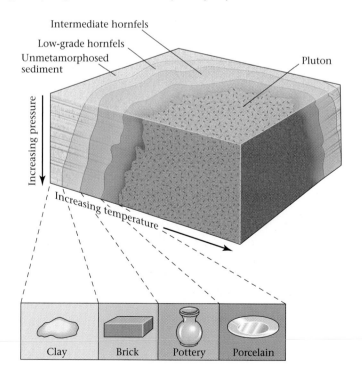

Dynamic Metamorphism

Faults are surfaces on which one piece of crust slides, or shears, past another. Near the Earth's surface (in the upper 10–15 km) this movement can fracture rock, breaking it into angular fragments or even crushing it to a powder. But at greater depths, rock is so warm that it behaves like soft plastic as shear along the fault takes place. During this process, the minerals in the rock recrystallize. We call this process dynamic metamorphism, because it occurs as a consequence of shearing alone, under metamorphic conditions, without requiring a change in temperature or pressure. The resulting rock, called a mylonite, has a foliation that roughly parallels the fault (▶Fig. 6.18a–c). Dynamic metamorphism takes place anywhere that faulting occurs at depth in the crust. Thus, mylonites can be found at all plate boundaries.

Dynamothermal or Regional Metamorphism

During the development of large mountain ranges, in response to either convergent-margin tectonics or continental collision, large slices of continental crust slip up and over other portions of the crust. As a consequence, rock that was once near the Earth's surface along the margin of a continent (▶Fig. 6.19a) ends up at great depth beneath the mountain range (▶Fig. 6.19b). In this new environment, three changes happen: (1) the protolith heats up because of the geothermal gradient and because of igneous activity; (2) the protolith endures greater pressure because of the weight of overburden; and (3) the protolith undergoes compression and shearing because of the differential stress. As a result of these changes, the protolith transforms into foliated metamorphic rock. The type of foliated rock that forms depends on the grade of metamorphism—slate forms at shallower depths, whereas schist and gneiss form at greater depths. Since the metamorphism we've just described involves not only heat but also compression and shearing, we can call it **dynamothermal metamorphism.** Typically, such metamorphism affects a large region, so geologists also call it **regional metamorphism.** Erosion eventually removes the mountains, exposing a belt of metamorphic rock that once lay at depth. Such belts may be hundreds of kilometers wide and thousands of kilometers long.

Hydrothermal Metamorphism

Hot magma rises beneath the axis of mid-ocean ridges, so when cold seawater sinks through cracks down into the oceanic crust along ridges, it heats up and transforms into

FIGURE 6.18 Dynamic metamorphism along a fault zone. (a) Note the band of sheared rock on either side of the slip surface. (b) The microscopic texture of rock outside the shear zone differs from that of the rock inside. (c) The block formed in (a) must have developed at a depth where metamorphic conditions exist, so that mylonite forms; otherwise, it would break up during movement.

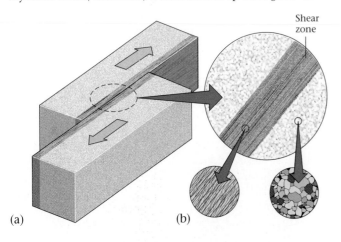

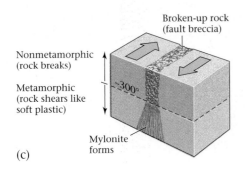

FIGURE 6.19 (a) Metamorphism occurs where there is plutonic activity along a convergent boundary. Some metamorphism may be thermal, but because of compression and shearing along convergent boundaries, some may be dynamothermal. (b) The sedimentary rock that lay at the top of a passive margin (point A) gets carried to great depth in a continental collision that leads to mountain building. As a result, it undergoes dynamothermal metamorphism. A broad region beneath a collision will lie in the field of metamorphism.

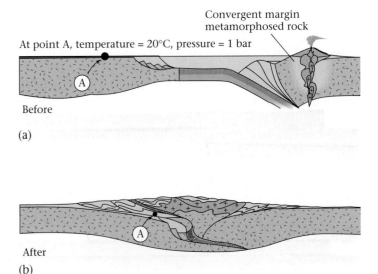

hydrothermal fluid. This fluid then rises through the crust, near the ridge, causing hydrothermal metamorphism of ocean-floor basalt (▶Fig. 6.20). Eventually, the fluid escapes through vents back into the sea; these vents are called black smokers (see Chapter 2).

Metamorphism in Subduction Zones

Blueschist is a relatively rare rock that contains an unusual blue-colored amphibole called glaucophane. Laboratory experiments indicate that glaucophane requires very high pressure but relatively low temperature to form. Such conditions do not develop in continental crust—usually, at the high pressure needed to produce glaucophane, temperature in continental crust is also high (see Box 6.1). So to figure out where blueschist forms, we must determine where high pressure can develop at relatively low temperature.

Plate tectonics theory provides the answer to this puzzle. Researchers found that blueschist occurs only in the accretionary prisms that form at subduction zones (see art, pp. 168–169). They realized that because prisms grow to be over 20 km thick, rock at the base of the prism feels high pressure (due to the weight of overburden). But because the subducted oceanic lithosphere beneath the prism remains cool, temperatures at the base of the prism remain relatively low. Under these conditions, glaucophane forms. Because of shear between the subducting plate and the overriding plate, blueschist develops a foliation.

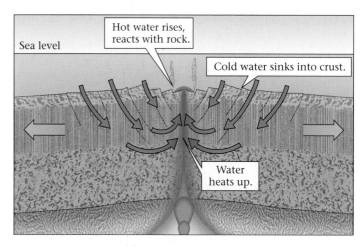

FIGURE 6.20 The circulation of hydrothermal fluids in response to igneous activity along mid-ocean ridges causes metamorphism of basalt in the oceanic crust.

Shock Metamorphism

When large meteorites slam into the Earth, a vast amount of kinetic energy instantly transforms into heat, and a pulse of extreme compression (a shock wave) propagates into the Earth. The heat may be sufficient to melt or even vaporize rock at the impact site, and the extreme compression of the shock wave causes the crystal structure of quartz

Pottery Making—An Analog for Thermal Metamorphism

BOX 6.2
THE HUMAN ANGLE

A crumbling brick in the wall of an adobe house, an earthenware pot, a stoneware bowl, and a translucent porcelain teacup may all be formed from the same lump of soft clay, scooped from the surface of the Earth and shaped by human hands. This pliable and slimy muck is a mixture of very fine clay minerals and quartz grains formed during the chemical weathering of rock and water. Fine potter's clay for making white china contains a particular clay mineral called kaolinite ($Al_2Si_2O_5[OH]_4 \cdot 2H_2O$), named after the locality in China (called Kauling, meaning "high ridge") where it was originally discovered.

People in arid climates make adobe bricks by forming damp clay into blocks, which they then dry in the sun. Such bricks can be used for construction *only* in arid climates, because if it rains heavily, the bricks will rehydrate and turn back into sticky muck—drying clay in the sun does not change the structure of the clay minerals.

To make a more durable material, brick makers place clay blocks in a kiln and bake ("fire") them at high temperatures. This process

makes the bricks hard and impervious to water. Potters use the same process to make jugs. In fact, fired clay jugs that were used for storing wine and olive oil have been found intact in sunken Greek and Phoenician ships that have rested on the floor of the Mediterranean Sea for thousands of years! Clearly, the firing of a clay pot fundamentally and permanently changes clay in a way that makes it physically different (see Fig. 6.17). In other words, firing causes a thermal metamorphic change in the mineral assemblage that composes pottery. The extent of the transformation depends on the kiln temperature, just as the grade of metamorphic rock depends on temperature. Potters usually fire earthenware at about 1,100°C and stoneware (which is harder than a knife or fork) at about 1,250°C. To produce porcelain—fine china—the clay must partially melt at even higher temperatures. Just as it begins to melt, the potter cools it quickly ("quenches" it). Quenching the melt creates glass, which gives porcelain its translucent, vitreous (glassy) appearance.

Environments of Metamorphism

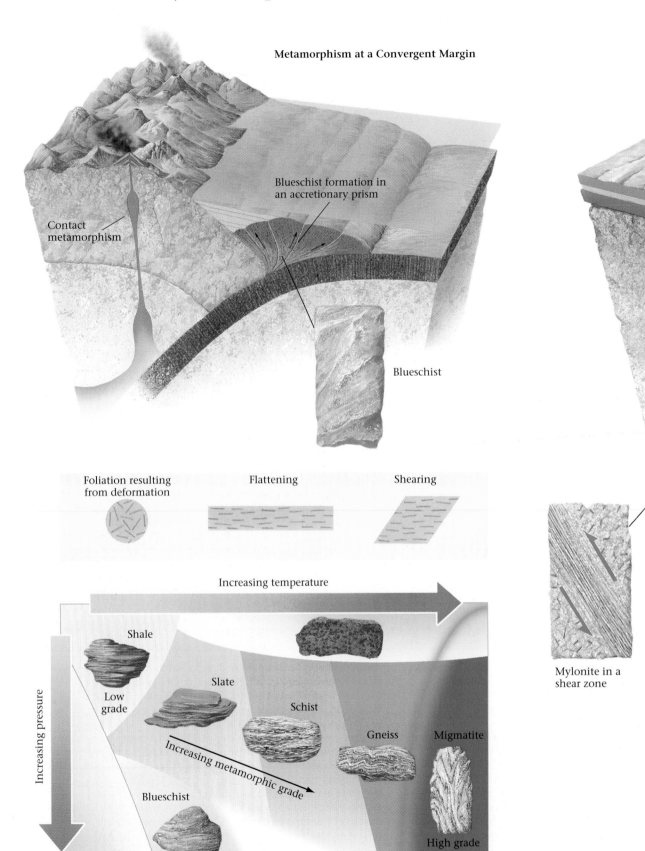

Metamorphism at a Convergent Margin

Contact metamorphism

Blueschist formation in an accretionary prism

Blueschist

Foliation resulting from deformation

Flattening

Shearing

Increasing temperature

Increasing pressure

Shale

Low grade

Slate

Schist

Gneiss

Migmatite

Increasing metamorphic grade

Blueschist

High grade

Mylonite in a shear zone

Hornfels formation

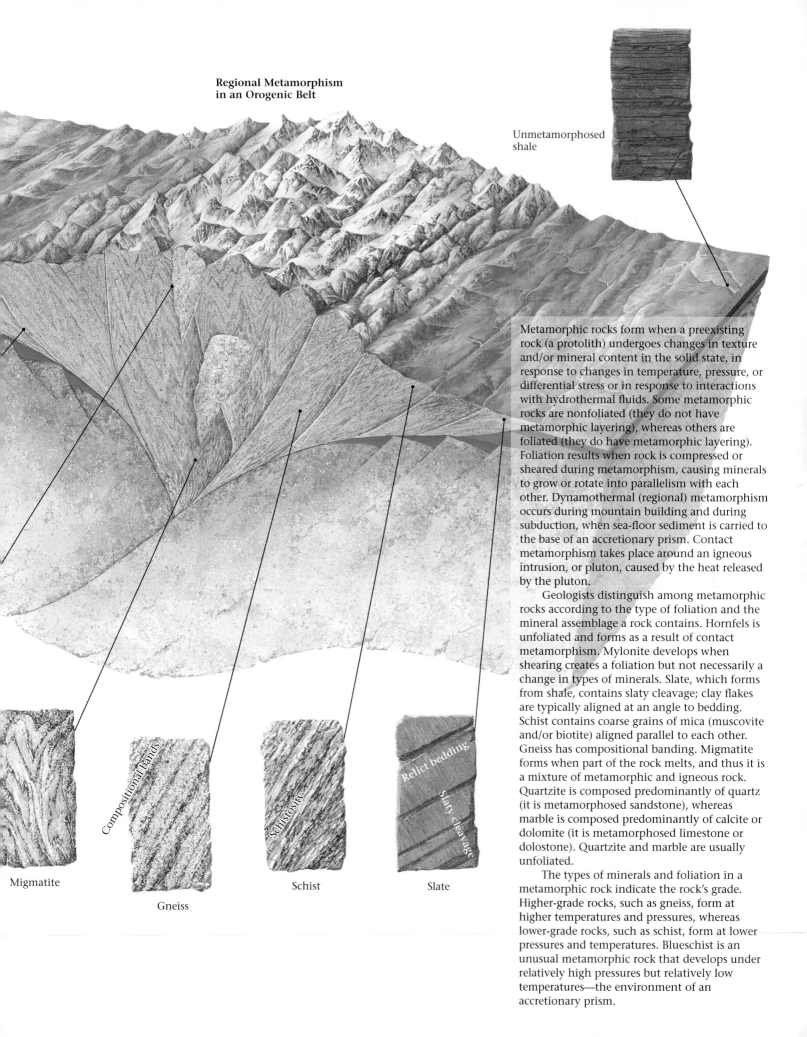

Regional Metamorphism in an Orogenic Belt

Unmetamorphosed shale

Metamorphic rocks form when a preexisting rock (a protolith) undergoes changes in texture and/or mineral content in the solid state, in response to changes in temperature, pressure, or differential stress or in response to interactions with hydrothermal fluids. Some metamorphic rocks are nonfoliated (they do not have metamorphic layering), whereas others are foliated (they do have metamorphic layering). Foliation results when rock is compressed or sheared during metamorphism, causing minerals to grow or rotate into parallelism with each other. Dynamothermal (regional) metamorphism occurs during mountain building and during subduction, when sea-floor sediment is carried to the base of an accretionary prism. Contact metamorphism takes place around an igneous intrusion, or pluton, caused by the heat released by the pluton.

Geologists distinguish among metamorphic rocks according to the type of foliation and the mineral assemblage a rock contains. Hornfels is unfoliated and forms as a result of contact metamorphism. Mylonite develops when shearing creates a foliation but not necessarily a change in types of minerals. Slate, which forms from shale, contains slaty cleavage; clay flakes are typically aligned at an angle to bedding. Schist contains coarse grains of mica (muscovite and/or biotite) aligned parallel to each other. Gneiss has compositional banding. Migmatite forms when part of the rock melts, and thus it is a mixture of metamorphic and igneous rock. Quartzite is composed predominantly of quartz (it is metamorphosed sandstone), whereas marble is composed predominantly of calcite or dolomite (it is metamorphosed limestone or dolostone). Quartzite and marble are usually unfoliated.

The types of minerals and foliation in a metamorphic rock indicate the rock's grade. Higher-grade rocks, such as gneiss, form at higher temperatures and pressures, whereas lower-grade rocks, such as schist, form at lower pressures and temperatures. Blueschist is an unusual metamorphic rock that develops under relatively high pressures but relatively low temperatures—the environment of an accretionary prism.

Migmatite

Compositional bands

Gneiss

Schistosity

Schist

Relict bedding

Slaty cleavage

Slate

grains in rocks below the impact site to suddenly undergo a phase change to a more compact mineral called coesite. Such changes are called shock metamorphism, to emphasize their relationship to a large impact.

6.7 BRINGING METAMORPHIC ROCKS BACK TO THE EARTH'S SURFACE

When you stand on an outcrop of metamorphic rock, you are standing on material that once lay many kilometers beneath the surface of the Earth. In areas of regional metamorphism, high-grade rocks rose, generally from a greater depth than did low-grade rocks—some high-grade rocks were once tens of kilometers below the surface. How does metamorphic rock return to the Earth's surface? Geologists refer to the overall process by which deeply buried rocks end up back at the surface as **exhumation.**

Exhumation results from several processes in the Earth System that happen simultaneously. Let's look at the spe-

cific processes that contribute to bringing high-grade metamorphic rocks from below a collisional mountain range back to the surface. First, as two continents progressively push together, the rock caught between them squeezes upward, or is uplifted, much like a ball of dough pressed in a vise (▶Fig. 6.21a); the upward movement takes place by slip on faults and by plastic like flow of rock. Second, as the mountain range grows, the crust at depth beneath it warms up and becomes softer. Eventually, the range starts to collapse under its own weight, much like a block of soft cheese placed in the hot sun, a process called extensional collapse (▶Fig. 6.21b). As a result of this collapse, the upper crust spreads out laterally. This movement stretches the upper part of crust in the horizontal direction and causes it to become thinner in the vertical direction. As the upper part of the crust becomes thinner, the deeper crust ends up closer to the surface. Third, erosion takes place at the surface (▶Fig 6.21c); weathering, landslides, river flow, and glacial flow together play the role of a giant rasp, stripping away rock at the surface and exposing rock that was once below the surface.

FIGURE 6.21 Three geologic phenomena together contribute to exhumation in a collisional mountain belt. Because of exhumation, the vertical distance between a point at depth in the belt decreases with time. (a) Collision squeezes rock in the mountain belt upward, like dough pressed in a vise. (b) The crust beneath the mountain range becomes warm and weak, so the mountain belt collapses, like a block of soft cheese placed in the hot sun. (c) Throughout the history of the mountain belt, erosion grinds rock off the surface and removes it, much like a giant rasp.

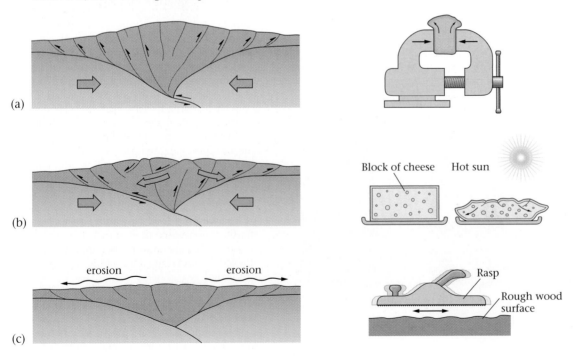

6.8 WHERE DO YOU FIND METAMORPHIC ROCKS?

If you want to study metamorphic rocks, you can start by taking a hike in a collisional or convergent mountain range (see Fig. 6.16). During the formation of such ranges, rocks undergo both contact metamorphism and regional metamorphism; because of exhumation, these rocks now lie exposed as towering cliffs of gneiss and schist. Where ancient mountain ranges once existed, we still find belts of metamorphic rocks cropping out at the ground surface even though the high peaks of the range have long since eroded away.

Huge expanses of metamorphic rock crop out in continental shields. Specifically, a **shield** is that part of a craton (the old, relatively stable part of a continent) where extensive areas of Precambrian rock crop out at the ground sur-face, because overlying younger rock has eroded away (►Fig. 6.22a). Geologists refer to the shield of North America as the Canadian Shield because it encompasses about half the land area of Canada (►Fig. 6.22b). Large shields also occur in South America, northern Europe, Africa, India, and Siberia. Rocks in shields were metamorphosed during a succession of Precambrian mountain-building events that were responsible for the growth of the continent in the first place—the oldest rocks on Earth occur in shields.

In the continental platform portion of a craton, a veneer of Paleozoic and Mesozoic sedimentary rocks covers most of the Precambrian metamorphic rocks. Here you can see these metamorphic rocks only where they were uplifted and exposed by erosion in younger mountain belts, or where rivers have cut down deeply enough to expose basement (►Fig. 6.22c).

FIGURE 6.22 (a) The distribution of shield areas (exposed Precambrian metamorphic and igneous rock) on the Earth. *(continued)*

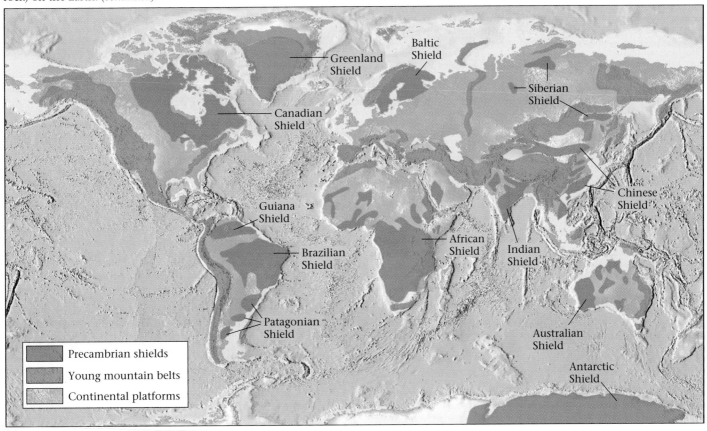

(a)

(b)

(c)

FIGURE 6.22 *(continued)* (b) The Canadian Shield as viewed from the air. Many lakes have formed on the low-lying landscape. (c) The walls of the Black Canyon of the Gunnison River in Colorado display high-grade metamorphic rocks. The stripes are pegmatite dikes that intruded the dark rock.

CHAPTER SUMMARY

• Metamorphism refers to changes in a rock that result in the formation of a metamorphic mineral assemblage, and/or metamorphic foliation, in response to change in temperature and/or pressure, to the application of differential stress, and to interaction with hydrothermal fluids.

• Metamorphism involves recrystallization, metamorphic reactions (neocrystallization), phase changes, pressure solution, and/or plastic deformation. If hot-water solutions bring in or remove elements, we say that metasomatism has occurred.

• Metamorphic foliation can be defined either by preferred mineral orientation (aligned inequant crystals) or by compositional banding. Preferred mineral orientation develops where differential stress causes the compression and shearing of a rock so that its inequant grains align parallel with each other.

• Geologists separate metamorphic rocks into two classes, foliated rocks and nonfoliated rocks, depending on whether the rocks contain foliation.

• The class of foliated rocks includes slate, metaconglomerate, phyllite, schist, amphibolite, and gneiss. The class of nonfoliated rocks includes hornfels, quartzite, and marble. Migmatite, a mixture of igneous and metamorphic rock, forms under conditions where partial melting begins.

• Rocks formed under relatively low temperatures are known as low-grade rocks, whereas those formed under high temperatures are known as high-grade rocks. Intermediate-grade rocks develop between these two extremes. Different metamorphic mineral assemblages form at different grades.

• Geologists track the distribution of different grades of rock by looking for index minerals. Isograds indicate the location at which index minerals first appear. A metamorphic zone is the region between two isograds.

• A metamorphic facies is a group of metamorphic mineral assemblages that develop under a specified range of temperature and pressure conditions.

• Thermal metamorphism (also called contact metamorphism) occurs in an aureole surrounding an igneous intrusion. Dynamically metamorphosed rocks form along faults, where rocks undergo plastic shearing. Burial metamorphism occurs at depth in a sedimentary basin. Dynamothermal metamorphism (also called regional metamorphism) results when rocks undergo heating and shearing during mountain building. Shock metamorphism happens during the impact of a meteorite. Hydrothermal metamorphism takes place due to the circulation of hot water in oceanic crust at mid-ocean ridges.

• We find extensive areas of metamorphic rocks in mountain ranges. Blueschist forms in accretionary prisms. Vast regions of continents known as shields expose Precambrian metamorphic rocks.

KEY TERMS

burial metamorphism (p. 165)
contact metamorphism (p. 165)
dynamothermal (regional)
    metamorphism (p. 166)
exhumation (p. 170)
foliation (p. 158)
gneiss (p. 159)
hornfels (p. 160)
marble (p. 160)
metamorphic aureole (p. 165)
metamorphic facies (p. 163)
metamorphic grade (p. 161)
metamorphic rock (p. 153)

metamorphic zone (p. 164)
metamorphism (p. 154)
metasomatism (p. 158)
migmatite (p. 159)
preferred mineral orientation
    (p. 157)
protolith (p. 153)
quartzite (p. 160)
schist (p. 159)
shield (p. 171)
slate (p. 159)
thermal metamorphism
    (p. 165)

REVIEW QUESTIONS

1. How are metamorphic rocks different from igneous and sedimentary rocks?

2. What two features characterize most metamorphic rocks?

3. What phenomena cause metamorphism?

4. What is metamorphic foliation, and how does it form?

5. How is a slate different from a phyllite? How is a phyllite different from a schist? How is a schist different from a gneiss?

6. Why is hornfels nonfoliated?

7. What is a metamorphic grade, and how can it be determined? How does grade differ from facies?

8. How does prograde metamorphism differ from retrograde metamorphism?

9. Describe the geologic settings where thermal, dynamic, and dynamothermal metamorphism take place.

10. Why does metamorphism happen at the site of meteor impacts and along mid-ocean ridges?

11. How does plate tectonics explain the combination of low-temperature but high-pressure minerals found in a blueschist?

12. Where would you go if you wanted to find exposed metamorphic rocks, and how did such rocks return to the surface of the Earth after being at depth in the crust?

THE VIEW FROM SPACE The Terra satellite took this image of a portion of the Canadian Shield in southeastern Canada, and the Adirondack Mountains of northeastern New York. The black areas are water (including Lake Ontario and the St. Lawrence River), the grayish and reddish portions of the map are areas underlain metamorphic rocks, whereas the tan areas are underlain by sedimentary rocks.

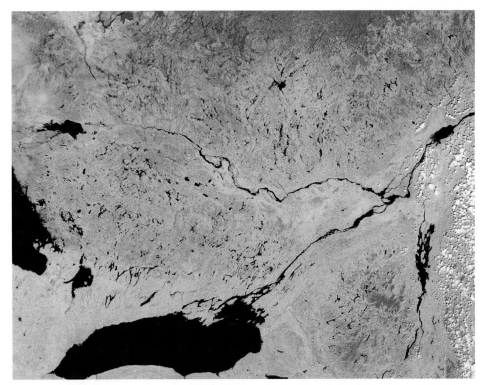

0 ⊢————⊣ 100 km

The Rock Cycle

The three rock types of the Earth System. (a) Igneous rock, here formed by the cooling of lava at a volcano. This lava solidified on top of a Hawaiian road. (b) Layers of sedimentary rock, here mesas in Monument Valley. (c) Metamorphic rock, here exposed in a Brazilian mountain belt. Over time, materials composing one rock type may be incorporated in another—this transfer of materials is called the rock cycle.

(a) (b) (c)

B.1 INTRODUCTION

"Stable as a rock." This familiar expression implies that a rock is permanent, unchanging over time—it isn't. In the time frame of Earth history, a span of over 4.5 billion years, atoms making up one rock type may be rearranged or moved elsewhere, eventually becoming part of another rock type. Later, the atoms may move again to form a third rock type, and so on. Geologists call the progressive transformation of Earth materials from one rock type to another the **rock cycle** (▶Fig. B.1), one of many examples of cycles acting in or on the Earth. (James Hutton, the eighteenth-century Scottish geologist, was the first person to visualize and describe the rock cycle.) We focus on the rock cycle here because it illustrates the relationships among the three rock types described in the previous three chapters.

There are many paths around or through the rock cycle. For example, igneous rock may weather and erode to produce sediment, which lithifies to form sedimentary rock. The new sedimentary rock may become buried so deeply that it transforms into metamorphic rock, which then could partially melt and produce magma. This magma later solidifies to form new igneous rock. We can symbolize this path as igneous → sedimentary → metamorphic → igneous. But alternatively, the metamorphic rock could be uplifted and eroded to form new sediment and then new sedimentary rock without melting, taking a shortcut path through the cycle that we can symbolize as igneous → sedimentary → metamorphic → sedimentary. Likewise, the igneous rock could be metamorphosed directly, without first turning to sediment. This metamorphic rock could be eroded to produce sediment that becomes sedimentary rock, defining another shortcut path:

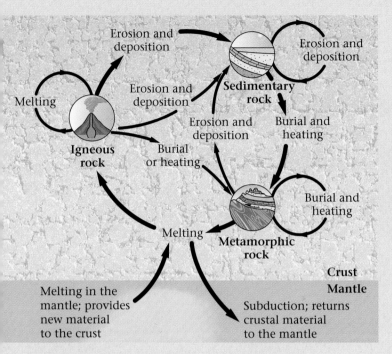

FIGURE B.1 The stages of the rock cycle, showing various alternative pathways.

igneous → metamorphic → sedimentary. To get a clearer sense of how the rock cycle works, we'll look at one example.

B.2 A CASE STUDY OF THE ROCK CYCLE

Material can enter the rock cycle when basaltic magma rises from the mantle. Suppose the magma erupts and forms basalt (an igneous rock) at a continental hot-spot volcano (▶Fig. B.2a). Interaction with wind, rain, and vegetation gradually weathers the basalt, physically breaking it into smaller fragments and chemically altering it to yield clay. Water washes the newly formed clay away and transports it downstream—if you've ever seen a brown-colored river, you've seen clay travelling to a site of deposition. Eventually the river reaches the sea, where the water slows down and the clay settles out.

Let's imagine, for this example, that the clay settles out along the margin of continent X and forms a deposit of mud. Gradually, through time, the mud becomes buried and the clay flakes pack tightly together. A new sedimentary rock, shale, forms. The shale resides 6 km below the continental shelf for millions of years, until the adjacent oceanic plate subducts and a neighboring continent, Y, collides with X. The edge of the encroaching con-

tinent pushes over the shale and buries it very deeply. As the mountains grow, the shale that had once been 6 km below the surface ends up 20 km below the surface, and under the pressure and temperature conditions present at this depth, it metamorphoses into schist (▶Fig. B.2b).

The story's not over. Once mountain building stops, erosion grinds away the mountain range, and exhumation brings some of the schist to the ground surface. This schist erodes to form sediment, which is carried off and deposited elsewhere to form new sedimentary rock. But other schist remains preserved below the surface. Eventually, continental rifting takes place at the site of the former mountain range, and the crust containing the schist begins to split apart. When this happens, heat brought up from rising magma causes some of the schist to melt partially and a new felsic magma forms. This felsic magma rises to the surface of the crust and freezes into rhyolite, a new igneous rock (▶Fig. B.2c). In terms of the rock cycle, we're back at the beginning, having once again made igneous rock (▶Fig. B.2d).

B.3 RATES OF MOVEMENT THROUGH THE ROCK CYCLE

We saw that not all atoms pass through the rock cycle in the same way. Similarly, not all atoms pass through the rock cycle at the same rate, and for that reason we find rocks of many different ages at the surface of the Earth. Some rocks remain in one form for less than a few million years, while others stay unchanged for most of Earth history. For example, rocks exposed on Precambrian shields have remained unchanged for billions of years. In contrast, a rock with an Appalachian Mountain address has passed through stages of the rock cycle many times in the past few billion years, because the eastern margin of North America has been subjected to multiple events of basin formation, mountain building, and rifting since the shield to the west developed.

Do the atoms in continental rocks ever return to the mantle? Yes. Some sediment that erodes off a continent ends up in deep-ocean trenches, and some of this gets carried back into the mantle by subduction. In fact, recent research suggests that metamorphic and igneous rocks at the base of the continental crust may be removed and transported down into the mantle at subduction zones.

Our tour of the rock cycle has focused on continental rocks. What about the oceans? Oceanic crust consists of igneous rock (basalt and gabbro) overlain by sediment. Because a layer of water blankets the crust, erosion does not affect it, so oceanic crustal rock does not follow the path into the sedimentary loop of the rock cycle. But sooner or later, oceanic crust subducts. When this happens, the rock

TIME 1

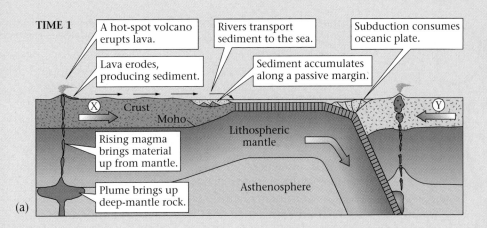

A hot-spot volcano erupts lava.

Rivers transport sediment to the sea.

Subduction consumes oceanic plate.

Lava erodes, producing sediment.

Sediment accumulates along a passive margin.

Crust

Moho

Lithospheric mantle

Rising magma brings material up from mantle.

Asthenosphere

Plume brings up deep-mantle rock.

(a)

TIME 2

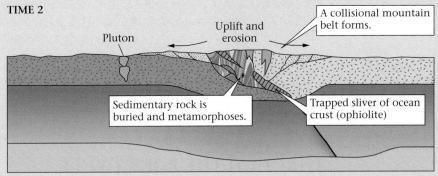

Pluton

Uplift and erosion

A collisional mountain belt forms.

Sedimentary rock is buried and metamorphoses.

Trapped sliver of ocean crust (ophiolite)

(b) ▨ Metamorphic rock ☐ Sediment eroded from mountains

TIME 3

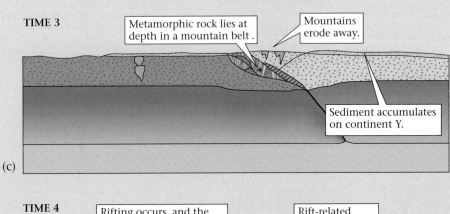

Metamorphic rock lies at depth in a mountain belt .

Mountains erode away.

Sediment accumulates on continent Y.

(c)

TIME 4

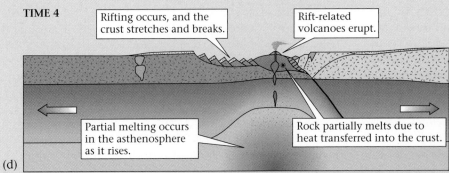

Rifting occurs, and the crust stretches and breaks.

Rift-related volcanoes erupt.

Partial melting occurs in the asthenosphere as it rises.

Rock partially melts due to heat transferred into the crust.

(d)

FIGURE B.2 (a) At the beginning of the rock cycle (time 1), atoms, originally making up peridotite in the mantle, rise in a mantle plume. The peridotite partially melts at the base of the lithosphere, and the atoms become part of a basaltic magma that rises through the lithosphere of continent X and erupts at a volcano. At this time, the atoms become part of the igneous rock in a lava flow. Weathering breaks the lava down, and rivers carry the resulting clay to a passive-margin basin. After burial, the atoms in the clay become part of a shale—a sedimentary rock. Note that the ocean floor to the east of the passive-margin basin subducts beneath continent Y. (b) At time 2, continents X and Y collide, and the shale ends up deep beneath the resulting mountain range (at the dot). Now the atoms become part of a schist—a metamorphic rock. (c) At time 3, the mountain range erodes away and the schist rises but does not reach the surface. (d) At time 4, rifting begins to split the continents apart, and igneous activity occurs again. At this time, the atoms of the schist become part of a new melt, which eventually freezes to form rhyolite, another igneous rock.

of the crust first undergoes metamorphism, for as it sinks, it feels progressively higher temperatures and pressures. And eventually, a little of the rock may melt and become new magma, which then rises at a volcanic arc.

B.4 WHAT DRIVES THE ROCK CYCLE IN THE EARTH SYSTEM?

The rock cycle occurs because the Earth is a dynamic planet. The planet's internal heat and gravitational field drive plate movements. Plate interactions cause the uplift of mountain ranges, a process that exposes rock to weathering, erosion, and sediment production. Plate interactions also generate the geologic settings in which metamorphism occurs, where rock melts to provide magma, and where sedimentary basins develop.

At the surface of the Earth, the gases initially released by volcanism collect to form the ocean and atmosphere. Heat (from the Sun) and gravity drive convection in the atmosphere and oceans, leading to wind, rain, ice, and currents—the agents of weathering and erosion. Weathering and erosion grind away at the surface of the Earth and send material into the sedimentary loop of the cycle. In the Earth System, life plays a key role by adding corrosive oxygen to the atmosphere, and by directly contributing to weathering. In sum, external energy (solar heat), internal energy (Earth's internal heat), gravity, and life all play a role in driving the rock cycle by keeping the mantle, crust, atmosphere, and oceans in constant motion.

Rock-Forming Environments and the Rock Cycle

Rocks form in many different environments. Igneous rocks develop where melt rises from depth and cools. Intrusive igneous rocks form where magma cools underground; extrusive igneous rocks form where lava and ash erupt at the surface.

Weathering and erosion break up existing rock and produce sediment. Different kinds of sediments develop in different places, reflecting both the composition of the source and the setting in which the sediment accumulates. We distinguish among sediment that collects in alluvial fans, desert dunes, river channels and floodplains, deltas, coral reefs, coastlines, the continental shelf, the deep sea, and the toe of a glacier. When this sediment eventually gets buried and undergoes lithification, new sedimentary rocks form.

Drainage networks collect surface water that can transport sediment to the ocean.

Sand dunes form from grains carried by the wind.

In a desert environment, rock weathers and fragments. Debris falls in landslides.

Flash floods carry sediment out of canyons to form an alluvial fan.

Volcanic eruptions emit lava and ash, which form new igneous rock at Earth's surface.

Sedimentary rocks make a cover on the surface of continents.

The crust and lithospheric mantle stretch and thin in a rift.

Magma rises from the mantle. Heat from this magma causes contact metamorphism.

Deep levels of continents consist of ancient metamorphic and igneous rocks. This is the basement of the continents.

Continental margins slowly sink and are buried by new sediment.

Partial melting occurs in the asthenosphere to produce new magma.

km

0

10

20

30

40

50

60

70

80

90

100

Glaciers erode rock and can transport sediment of all sizes.

In a region of continental collision, rocks that were near the surface are deeply buried and metamorphosed.

In humid climates, thick soils develop.

Magma that cools and solidifies underground forms igneous intrusions.

Along coastal plains, rivers meander. Sediment collects in the channel and floodplain.

Where a river enters the sea, sediment settles out to form a delta.

Reefs grow from calcite-secreting organisms. These will eventually turn into limestone.

Many different kinds of sediment accumulate along coastlines, building out a continental shelf.

Underwater avalanches carry a cloud of sediment that settles to form a submarine fan.

Fine clay and plankton shells settle on the oceanic crust.

The oceanic crust consists of igneous rocks formed at a mid-ocean ridge.

Under certain conditions, preexisting rocks can undergo change in the solid state—metamorphism—which produces metamorphic rocks. Contact metamorphism is due to heat released by an intrusion of magma. Regional metamorphism occurs where tectonic processes cause rocks from the surface to be buried very deeply.

Because the Earth is dynamic, environments change through time. Tectonic processes cause new igneous rocks to form. When exposed at the surface, these rocks weather to make sediment. The slow sinking of some regions creates sedimentary basins in which sediment accumulates and new sedimentary rocks form. Later, these rocks may be buried deeply and metamorphosed. Uplift as a result of mountain building exposes the rocks to the surface, where they may once again be transformed into sediment. This progressive transformation is called the rock cycle.

The Wrath of Vulcan: Volcanic Eruptions

Glowing waves rise and flow, burning all life on their way, and freeze into black, crusty rock which adds to the height of the mountain and builds the land, thereby adding another day to the geologic past. . . . I became a geologist forever, by seeing with my own eyes: the Earth is alive!
—HANS CLOOS (1886–1951), on seeing the eruption of Mt. Vesuvius (Italy)

During a volcanic eruption, molten rock rises from inside the Earth and either flows onto the surface as lava or blasts into the air as ash. This 1989–1990 eruption of Redoubt Volcano, Alaska, produced so much ash that it billowed into the stratosphere. A jumbo jet flying into the cloud lost all power, for the ash fouled the engines. Only after the plane had plummeted 2.5 km (8,000 feet) could the pilots restart the engines and bring the plane safely in for a landing.

7.1 INTRODUCTION

Every few hundred years, one of the hills on Vulcano, an island in the Mediterranean Sea off the western coast of Italy, rumbles and spews out molten rock, glassy cinders, and dense "smoke" (actually a mixture of various gases, fine ash, and very tiny liquid droplets). Ancient Romans thought that such eruptions happened when Vulcan, the god of fire, fueled his forges beneath the island to manufacture weapons for the other gods. Geologic study suggests, instead, that eruptions take place when hot magma, formed by melting inside the Earth, rises through the crust and emerges at the surface. No one believes the myth anymore, but the island's name evolved into the English word **volcano,** which geologists use to designate either an erupting vent through which molten rock reaches the Earth's surface or a mountain built from the products of eruption.

On the main peninsula of Italy, not far from Vulcano, another volcano, Mt. Vesuvius, towers over the nearby Bay of Naples. Two thousand years ago, a prosperous Roman resort

(a)

What a geologist imagines

(b)

(c)

FIGURE 7.1 (a) Pompeii, once buried by 6 m of volcanic debris from Mt. Vesuvius, was excavated by archaeologists in the late nineteenth century. Vesuvius rises in the distance. (b) What a geologist imagines: When Mt. Vesuvius erupted in 79 C.E., it was probably larger, as depicted in this sketch. (c) A plaster cast of an unfortunate inhabitant of Pompeii, found buried by ash in the corner of a room, where the person had crouched for protection. The flesh rotted away, leaving only an open hole that could be filled by plaster.

and trading town of 20,000 inhabitants sprawled along the shore, 5 km south of Vesuvius (▶Fig. 7.1a). This town was called Pompeii. In 79 C.E., earthquakes signaled the mountain's awakening. At 1:00 P.M. on August 24, a dark mottled cloud boiled up above Mt. Vesuvius's summit to a height of 27 km. As lightning sparked in its crown, the cloud drifted over Pompeii, turning day into night. Blocks and pellets of rock fell like hail, while fine ash and choking fumes enveloped the town (▶Fig. 7.1b). Frantic people rushed to escape, but for many it was too late. As the growing weight of volcanic debris began to crush buildings, an avalanche of hot ash swept over Pompeii, and by the next day the town had vanished beneath a 6-m-thick gray-black blanket. This covering protected the ruins of Pompeii so well that when archaeologists excavated the town 1,800 years later, they found an amazingly complete record of Roman daily life. During their

work, archaeologists discovered open spaces in the debris. Out of curiosity, they filled the spaces with plaster, and then dug away the surrounding ash. The spaces turned out to be fossil casts of Pompeii's unfortunate inhabitants; their bodies forever twisted in agony or huddled in despair (▶Fig. 7.1c).

Clearly, volcanoes are unpredictable and dangerous. Volcanic activity can build a towering, snow-crested mountain or can blast one apart. It can provide the fertile soil that enables a civilization to thrive, or it can snuff out a civilization in a matter of minutes. Because of the diversity of volcanic activity and its consequences, this chapter sets out ambitious goals. We first review the products of volcanic eruptions and the basic characteristics of volcanoes. Then we look again at the different kinds of volcanic eruptions on Earth. Volcanoes are *not* randomly distributed around the globe—their positions reflect the locations of plate boundaries, rifts, and hot spots. Finally, we examine the hazards posed by volcanoes, efforts by geoscientists to predict eruptions and help minimize the damage they cause, and the possible influence of eruptions on climate and civilization.

7.2 THE PRODUCTS OF VOLCANIC ERUPTIONS

The drama of a volcanic eruption transfers materials from inside the Earth to our planet's surface. Products of an eruption come in three forms—lava flows, pyroclastic debris, and gas (see art on pp. 188–189).

Lava Flows

Sometimes it races down the side of a volcano like a fast-moving, incandescent stream, sometimes it builds into a rubble-covered mound at a volcano's summit, and sometimes, it oozes like a sticky but scalding paste. Clearly, not all **lava** (molten rock that has extruded onto the Earth's surface) behaves in the same way when it rises out of a volcano. Therefore, not all **lava flows** (moving masses of molten lava, or sheets of rock formed when lava solidifies) look or behave the same. Why? The character of a lava flow primarily reflects its viscosity, or resistance to flow, and not all lavas have the same viscosity. Differences in viscosity depend on a variety of factors including chemical composition, temperature, gas content, and crystal content.

To illustrate the different ways in which lava behaves, we now examine flows of different compositions (▶Fig. 7.2a–c). Geologists give names to different lava compositions by specifying the silica content (SiO_2) relative to the sum of the iron oxide (FeO) and magnesium oxide (MgO) content. Lavas high in silica are called silicic, felsic, or rhyolitic; lavas with an intermediate silica content are called intermediate or andesitic; and lavas low in silica are called mafic or basaltic. Silica tends to link in long molecules whose presence increases lava viscosity. Thus, the greater the silica content, the stickier the lava (see Chapter 4).

Basaltic lava flows. Basaltic (mafic) lava has very low viscosity when it first emerges from a volcano because it contains relatively little silica and is hot. Thus, on the steep slopes near the summit of a volcano, it flows very quickly (▶Fig. 7.3a), sometimes at speeds of over 30 km/hour. The lava slows down to less-than-walking pace after it has traveled several kilometers and has started to cool. Most flows measure less than 10 km long, but the ends of some flows are as much as 500 km from the source.

How can lava travel so far? Although all the lava in a flow moves when it first emerges, rapid cooling causes the surface

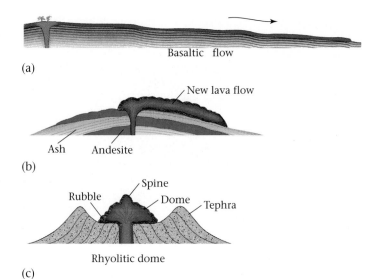

FIGURE 7.2 The character of a lava flow depends on the viscosity of the lava. Hotter lavas are less viscous than cooler lavas, and mafic lavas are less viscous than silicic lavas. (a) A basaltic lava flow is very fluid-like and can travel a great distance, forming a thin sheet. (b) An andesitic flow is too viscous to travel far, and it tends to break up as it flows. (c) Rhyolitic lava is so viscous that it piles up at the vent in the shape of a dome. In some cases, a towerlike spine pushes up from the center of the dome.

of the flow to crust over after the flow has moved a few kilometers from the source. The solid crust serves as insulation, allowing the hot interior of the flow to remain liquid and continue to flow. An insulated, tunnel-like conduit within a flow, through which lava moves, is a lava tube. In some cases, lava tubes drain and eventually become empty tunnels. Once a lava flow has covered the land and freezes, it becomes a solid blanket of rock.

The surface texture of a basaltic lava flow when it finally freezes reflects the timing of freezing relative to its movement. Flows with warm, pasty surfaces wrinkle into smooth, glassy, ropelike ridges; geologists have adopted the Hawaiian word **pahoehoe** (pronounced pa-hoy-hoy) for such flows (▶Fig. 7.3b). If the surface layer of the lava freezes and then breaks up due to the continued movement of lava underneath, it becomes a jumble of sharp, angular fragments, creating a rubbly flow also called by its Hawaiian name, **a'a'** (pronounced ah-ah) (▶Fig. 7.3c). Footpaths made by people living in basaltic volcanic regions follow the smooth surface of pahoehoe flows rather than the rough, foot-slashing surface of a'a' flows.

During the final stages of cooling, lava flows contract and may fracture into roughly hexagonal columns. This type of fracturing is called **columnar jointing** (▶Fig. 7.3d).

Basaltic flows that erupt under water look different from those that erupt on land, because the lava cools so much more quickly. In fact, water removes heat thirty times faster than air does (that's why you soon get hypothermia when immersed in cold water). Because of rapid cooling, submarine

(a)

(b)

(c)

(d)

FIGURE 7.3 Characteristics of basaltic (mafic) lava flows. (a) A fast-moving 1984 lava flow just after emerging from a vent on Mauna Loa Volcano, Hawaii. The flow is about 20 m wide. (b) A weathered pahoehoe lava flow at Craters of the Moon National Monument, in Idaho, displays the characteristic smooth, ropy surface. (c) An a'a' flow in Arizona has a rough and rubbly surface. (d) Lava flows contract when they cool, and some crack to form columnar joints. These columns crop out in Yellowstone Park.

basaltic lava forms a glass-encrusted blob, or pillow, on freezing (►Fig. 7.4). The rind of a pillow momentarily stops the flow's advance, but within minutes the pressure of the lava squeezing into the pillow breaks the rind, and a new blob of lava squirts out, and then itself freezes into a pillow.

Andesitic and rhyolitic lava flows. Because of its higher silica content and thus its greater viscosity, andesitic lava cannot flow as easily as basaltic lava. When erupted, andesitic lava first forms a large mound above the vent. This mound then advances slowly down the volcano's flank at only about 1–5 m a day, in a lumpy flow with a bulbous snout. Typically, andesitic flows are less than 10 km long. Because the lava moves so slowly, the outside of the flow has time to solidify; so as it moves, the surface breaks up into angular blocks, and the whole flow looks like a jumble of rubble.

Rhyolitic lava is the most viscous of all lavas because it is the most silicic and the coolest. Therefore, it tends to accumulate either in a domelike mass, called a lava dome, above the vent or in short and bulbous flows rarely more

FIGURE 7.4 Lava pillows form as a consequence of sub-aqueous basalt eruption on the floor of the South Pacific Ocean. The pillow in the foreground is approximately 1 m across.

than 1–2 km long. Sometimes rhyolitic lava freezes while still in the vent, and then pushes upward as a columnlike spire or spine up to 100 m above the vent. Rhyolitic flows, where they do form, have broken and blocky surfaces.

Pyroclastic Debris

Not all of the material that erupts at a volcano ends up as part of a lava flow. In some cases, bubble-filled lava begins to solidify in the vent of a volcano, forming scoria or pumice (see Chapter 4) that may eventually shatter and be ejected from the vent when an eruption begins. During basaltic eruptions, lava may spout from a vent in dramatic fountains, producing clots of lava that cool and become solid or nearly solid clasts by the time they land (▶Fig. 7.5a). During andesitic or rhyolitic eruptions, powerful explosions spray lava into the air, forming droplets that instantly freeze into **volcanic ash,** composed of tiny glass shards. The blast of a volcanic explosion can also shatter the preexisting solid rock that makes up the volcano itself, forming fragments of various sizes (▶Fig. 7.5b). Geologists refer to all fragmental material (glass shards, pumice or scoria fragments, and bits of preexisting rock) erupted from a volcano as **pyroclastic debris** (from the Latin word *pyro,* meaning "fire"), and they categorize the debris into several different types.

The finest pyroclastic debris, volcanic ash, (▶Fig. 7.5c) consists of powder-sized glass shards and pulverized rock generated during explosions. Pea- to plum-sized fragments are lapilli (from the Latin for "little stones"). If lapilli form from low-viscosity lava, they may develop streamlined form while flying through the air, yielding teardrop-shaped

glassy beads known as "Pelé's tears," after the Hawaiian goddess of volcanoes. (Some lapilli form when wet ash in an eruptive cloud sticks together to form ash balls.) When low-viscosity droplets rise from a pool of lava at the volcano's vent, they trail thin strands of lava behind them, and these strands freeze into filaments of brown glass known as Pelé's hair. Coarser pyroclastic debris includes blocks and bombs, which are apple- to refrigerator-sized fragments (▶Fig. 7.5c). Specifically, blocks are chunks of preexisting igneous rock torn from the walls of the vent, whereas bombs form when large lava blobs enter the air in a molten state and then solidify. Bombs typically become streamlined as they fall.

Unconsolidated deposits of pyroclastic grains, regardless of size, that have been erupted from a volcano constitute **tephra.** Ash, or ash mixed with lapilli, becomes **tuff** when lithified (transformed into coherent rock). Geologists refer to tuff formed from debris that settles from the air, like falling snow, as air-fall tuff. Yet not all tuff starts as an air-fall deposit. Some tuff forms from fast-moving turbulent avalanches of hot ash and lapilli that rushed down the flank of the volcano (▶Fig. 7.6a). Such a glowing avalanche is called a **pyroclastic flow** (or *nuée ardente,* French for "glowing cloud"). A sheet of tuff formed from a pyroclastic flow is an ignimbrite. Thick ignimbrite sheets are so hot, immediately after deposition, that soft ash at the base of the sheet—squeezed by the weight of overlying debris—fuses together to produce hard, welded tuff.

A devastating pyroclastic flow erupted from Mt. Pelée, a volcano on the otherwise quiet tropical West Indies island of Martinique. In April 1902, a small eruption shed fine white air-fall ash over the port town of St. Pierre, at the foot of the volcano. The air began to reek of sulfur, so inhabitants walked around with handkerchiefs covering their noses. Officials did not fully comprehend the threat and did not order an evacuation. Like a cork, frozen lava blocked the volcano's vent, but the pressure in the gas-rich magma beneath continued to build. On the morning of May 8, the cork popped, and like the froth that streams down the side of a champagne bottle, a pyroclastic flow swept down Pelée's flank. The cloud of burning ash, blocks, gas, and debris, at a temperature of 200°–450°C, rode a cushion of air and may have reached a velocity of over 300 km per hour before it slammed into St. Pierre two minutes later. One breath of the super-hot ash meant instant death, and within moments 28,000 people lay dead of asphyxiation or incineration. Buildings toppled into a chaos of rubble and twisted metal, stockpiles of rum barrels exploded and sent flaming liquor into the streets, and ships at anchor in the harbor capsized. Only two people survived; one was a prisoner who, though burned, was protected from the brunt of the cataclysm by the stout walls of his underground cell.

Similar eruptions began in the late 1990s on the island of Montserrat, another volcano in the eastern Caribbean (▶Fig. 7.6b, c). Numerous pyroclastic flows buried once-lush

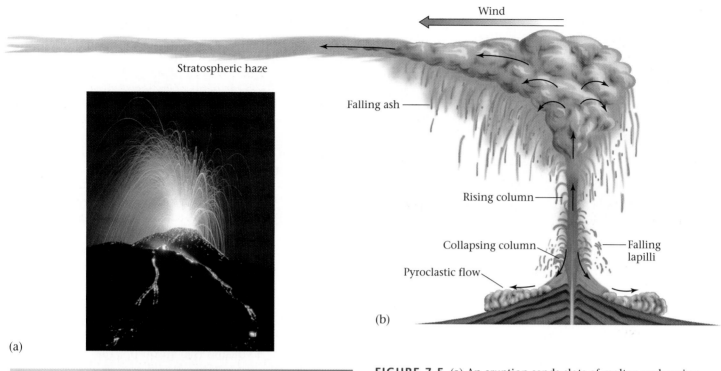

FIGURE 7.5 (a) An eruption sends clots of molten rock arcing into the sky. (b) Silicic (felsic) and intermediate volcanoes erupt large quantities of pyroclastic debris. Some ash may rise all the way to the stratosphere, while some falls back to earth, growing progressively finer farther away from the volcano. Ash may also cascade down the side of a volcano as a pyroclastic flow (also called a nuée ardente). Pyroclastic flows may develop when the upward force in the rising column diminishes, and the column collapses. (c) The rim of a volcano (crater is to the left) built from an accumulation of lapilli (smaller fragments) and bombs (larger pieces).

fields and forests and destroyed the towns, but fortunately this occurred after the island's 12,000 inhabitants had been evacuated from the danger zone.

Tephra accumulations on a volcano are quite weak. Thus, many tephra deposits slide downslope before they lithify. If tephra mixes with at least 10–25% water, it becomes a mobile mass called a volcanic debris flow. Particularly fluid, fast-moving volcanic debris flows—essentially slurries composed of a mixture of ash-rich debris and water—are **lahars.** A lahar can rush down valleys at speeds of up to 50 km/hour and can travel tens of kilometers. When it finally slows, a lahar settles and gradually drains, yielding a layer of debris composed of pebbles and cobbles of volcanic rock suspended in ashy mud (▶Fig. 7.7).

Volcanic Gas

Most magma contains dissolved gases, including water, carbon dioxide, sulfur dioxide, and hydrogen sulfide (H_2O, CO_2, SO_2, H_2S). In fact, up to 9% of a magma may consist of gaseous components. Generally, lavas with more silica contain a greater proportion of gas. Volcanic gases come out of solution when the magma approaches the Earth's surface and pressure decreases, just as bubbles come out of solution in a soda or in champagne when you pop the bottle top off. In low-viscosity magma, gas bubbles can rise faster than the magma moves, and thus most reach the surface of the magma and enter the atmosphere before the lava does. The last bubbles to form, however, freeze into the lava to create holes

(a)

(b)

FIGURE 7.6 (a) A pyroclastic flow rushes down the side of a volcano in Japan. (b) Some of the ash erupted from Montserrat blanketed the town of Plymouth.

called **vesicles** (▶Fig. 7.8). In high-viscosity magmas, the gas has trouble escaping because bubbles can't push through the sticky lava. When this happens, explosive pressures build inside or beneath the volcano.

7.3 THE ARCHITECTURE AND SHAPE OF VOLCANOES

As we saw in Chapter 4, melting in the upper mantle and lower crust produces magma, which rises into the upper crust. Typically, this magma accumulates underground in a **magma chamber,** an open space or a zone of highly fractured rock that can contain a large quantity of magma.

Some of the magma freezes in the magma chamber and transforms into intrusive igneous rock, but some rises through an opening or conduit to the Earth's surface and erupts to form a volcano (see art on pp. 188–189). In some volcanoes, the conduit has the shape of a vertical pipe, while in others the conduit is a crack called a **fissure** (▶Fig. 7.9a–c). Initially, a fissure may erupt a curtain of lava.

With time, the solid products of eruption (lava and/or pyroclastic debris) accumulate around a conduit to form a mound or cone. At the top of the mound, a circular depression called a **crater** (shaped like a bowl, up to 500 m across and 200 m deep) develops, either during eruption as material accumulates around the summit vent or just after eruption as the summit collapses into the drained conduit. Eruptions that happen in the summit crater are summit eruptions. In some

FIGURE 7.7 A lahar flowed down a river on the flanks of Mt. St. Helens following the volcano's 1980 eruption, destroying homes and property along the banks of the river.

FIGURE 7.8 Vesicles are the holes made by gas bubbles trapped in a freezing lava. This boulder of basalt, from Sunset Crater National Monument, in Arizona, contains vesicles of various sizes.

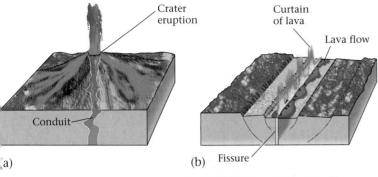

FIGURE 7.9 (a) Some volcanoes erupt out of a circular vent above a tube-shaped conduit. (b) Other volcanoes erupt out of a long crack, called a fissure, and produce a curtain of lava. (c) A "curtain of fire" formed as lava erupts from a fissure.

volcanoes, a secondary conduit or fissure breaks through along the sides, or flanks, of the volcano, causing a flank eruption (▶Fig. 7.10a).

After major eruptions, the center of the volcano may collapse into the large, drained magma chamber below, creating a **caldera,** a big circular depression (up to thousands of meters across and up to several hundred meters deep) with steep walls and a fairly flat floor (▶Fig. 7.10a–d). Some calderas form when a volcano explodes, blasting so much material into the air that only a portion of the volcano remains. Note that calderas differ from craters in terms of size, shape, and mode of formation.

Geologists distinguish among three different shapes of subaerial volcano. **Shield volcanoes,** so named because

FIGURE 7.10 (a) The plumbing beneath a volcano can be complex. A central vent may lie directly above the magma chamber, but some of the lava may erupt at flank vents. (b) During an eruption, the magma chamber beneath a volcano is inflated with magma. (c) If the eruption drains the magma chamber, the volcano collapses downward to form a circular depression called a caldera. Later eruptions may build a resurgent dome within the caldera. (d) A moderate-sized caldera has formed at the summit of Mt. Kilimanjaro, in the East African Rift.

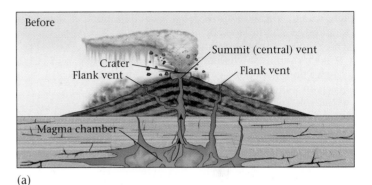

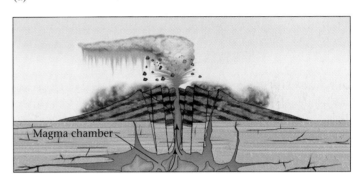

Volcano

Volcanic eruptions are a sight to behold, and in some cases a hazard to fear. Beneath a volcano, magma formed in the upper mantle or the lower crust rises to fill a magma chamber near the Earth's surface. When the pressure in this magma chamber becomes great enough, magma forces upward through a conduit, or crack, to the ground surface, and erupts.

Once molten rock has erupted at the surface, it is lava. Some lava spills down the side of the volcano to make a lava flow. Lava flows eventually cool, forming solid rock. In some cases, lava spatters or fountains out of the volcanic vent in little blobs or drops that cool quickly in the air to create fragmental igneous rock called tephra, or cinders. Larger blobs ejected by a volcano become volcanic bombs, which attain a streamlined shape as they fall. Cinders may accumulate in a cone-shaped pile called a cinder cone.

Styles of eruption

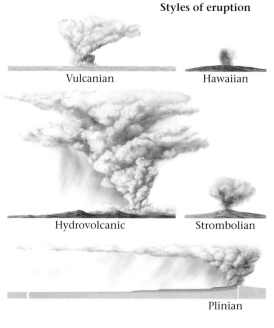

Vulcanian

Hawaiian

Hydrovolcanic

Strombolian

Plinian

Volcano starts to erupt.

Full magma chamber

Ash and debris

Main explosive eruption

Magma chamber empties.

Newly formed caldera

Collapsed blocks

Empty magma chamber

New volcanic cone grows.

Lake fills caldera.

Caldera formation, such as occurred at Crater Lake, Oregon

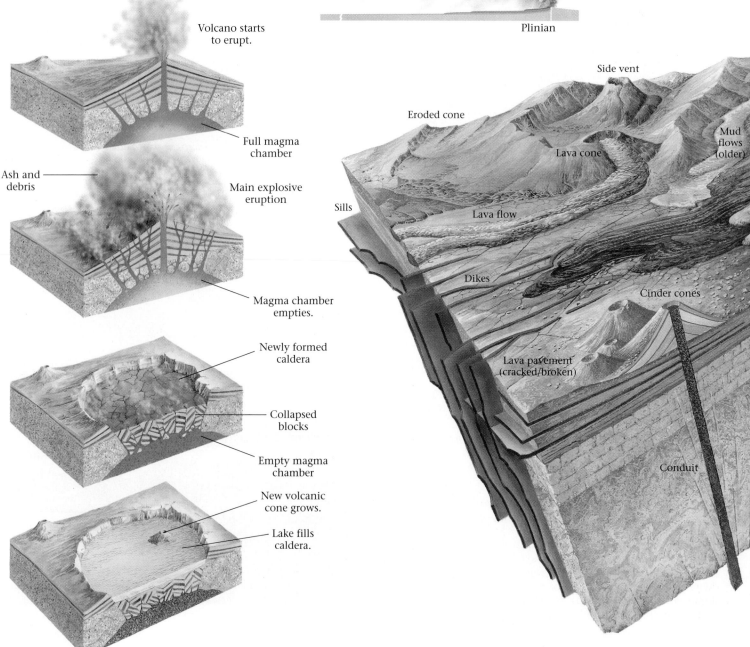

Side vent

Eroded cone

Mud flows (older)

Lava cone

Sills

Lava flow

Dikes

Cinder cones

Lava pavement (cracked/broken)

Conduit

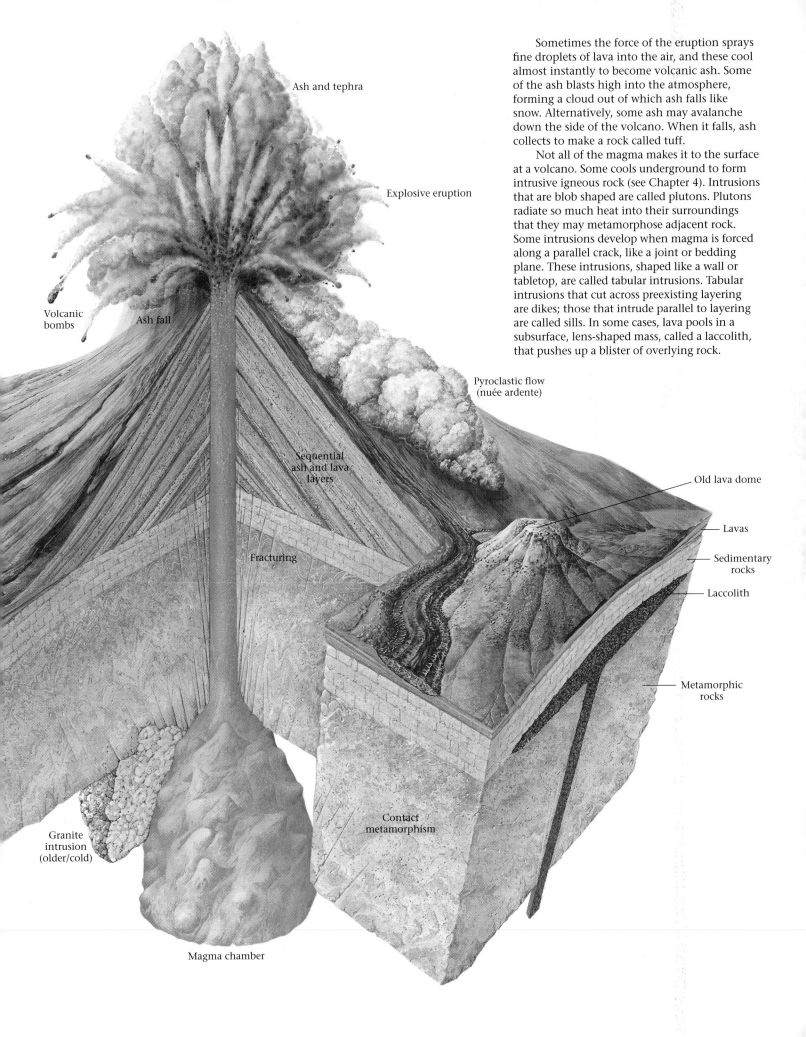

Ash and tephra

Explosive eruption

Volcanic
bombs

Ash fall

Pyroclastic flow
(nuée ardente)

Sequential
ash and lava
layers

Old lava dome

Lavas

Sedimentary
rocks

Fracturing

Laccolith

Granite
intrusion
(older/cold)

Metamorphic
rocks

Contact
metamorphism

Magma chamber

Sometimes the force of the eruption sprays
fine droplets of lava into the air, and these cool
almost instantly to become volcanic ash. Some
of the ash blasts high into the atmosphere,
forming a cloud out of which ash falls like
snow. Alternatively, some ash may avalanche
down the side of the volcano. When it falls, ash
collects to make a rock called tuff.

Not all of the magma makes it to the surface
at a volcano. Some cools underground to form
intrusive igneous rock (see Chapter 4). Intrusions
that are blob shaped are called plutons. Plutons
radiate so much heat into their surroundings
that they may metamorphose adjacent rock.
Some intrusions develop when magma is forced
along a parallel crack, like a joint or bedding
plane. These intrusions, shaped like a wall or
tabletop, are called tabular intrusions. Tabular
intrusions that cut across preexisting layering
are dikes; those that intrude parallel to layering
are called sills. In some cases, lava pools in a
subsurface, lens-shaped mass, called a laccolith,
that pushes up a blister of overlying rock.

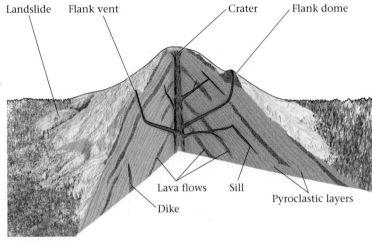

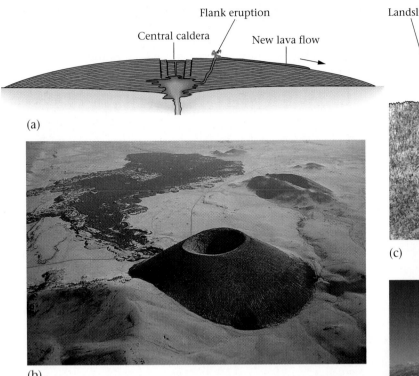

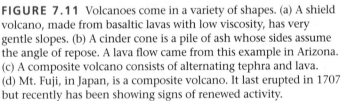

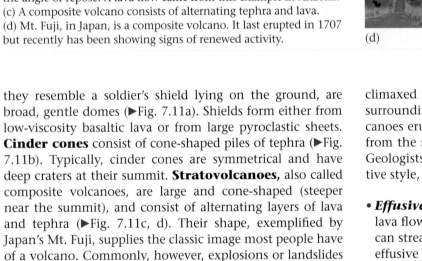

FIGURE 7.11 Volcanoes come in a variety of shapes. (a) A shield volcano, made from basaltic lavas with low viscosity, has very gentle slopes. (b) A cinder cone is a pile of ash whose sides assume the angle of repose. A lava flow came from this example in Arizona. (c) A composite volcano consists of alternating tephra and lava. (d) Mt. Fuji, in Japan, is a composite volcano. It last erupted in 1707 but recently has been showing signs of renewed activity.

they resemble a soldier's shield lying on the ground, are broad, gentle domes (▶Fig. 7.11a). Shields form either from low-viscosity basaltic lava or from large pyroclastic sheets. **Cinder cones** consist of cone-shaped piles of tephra (▶Fig. 7.11b). Typically, cinder cones are symmetrical and have deep craters at their summit. **Stratovolcanoes,** also called composite volcanoes, are large and cone-shaped (steeper near the summit), and consist of alternating layers of lava and tephra (▶Fig. 7.11c, d). Their shape, exemplified by Japan's Mt. Fuji, supplies the classic image most people have of a volcano. Commonly, however, explosions or landslides destroy the classic shape.

7.4 ERUPTIVE STYLES: WILL IT FLOW, OR WILL IT BLOW?

The 1990 eruption of Kilauea in Hawaii produced lakes and rivers of lava that cascaded down the volcano's flanks. In contrast, the 1980 eruption of Mt. St. Helens in Washington climaxed with a tremendous explosion that blanketed the surrounding countryside with tephra. Clearly, different volcanoes erupt in different ways. In fact, successive eruptions from the same volcano may even differ from one another. Geologists refer to the character of an eruption as the eruptive style, and make the following distinctions.

- *Effusive eruptions:* These eruptions produce mainly lava flows. Most yield low-viscosity basaltic lavas, which can stream tens to hundreds of kilometers. During effusive eruptions, lava may pool in lakes around the vent, or spray up in fountains.

- *Explosive eruptions:* These eruptions produce clouds and avalanches of pyroclastic debris (▶Box 7.1). Pyroclastic eruptions happen when gas expands in the rising magma but cannot escape. Eventually, the pressure becomes so great that it blasts the lava, along with previously solidified volcanic rock, out of the volcano. In some cases, an explosive eruption blasts the volcano apart and leaves behind a large caldera. Such explosions, awesome in their power and catastrophic in their consequences, eject cubic kilometers of igneous particles upward at initial speeds of up to 90 m per second. The resulting plume of debris resembles the mushroom cloud above a nuclear explosion. Coarse-grained ash and lapilli settle from the cloud close to the volcano, whereas finer ash settles farther away.

Note that the type of volcano (shield, cinder cone, or composite) depends on its eruptive style. Volcanoes that have only effusive eruptions become shield volcanoes, those that generate small pyroclastic eruptions due to fountaining lava yield cinder cones, and those that alternate between effusive and large pyroclastic eruptions become composite volcanoes. Large explosions yield calderas and blanket the surrounding countryside with ash. If the ash is hot enough, as happens in a glowing avalanche, the ash welds to form sheets of ignimbrite. Why are there such contrasts in eruptive style, and therefore in volcano shape? Eruptive style depends on the viscosity and gas content of the magma in the volcano. These characteristics, in turn, depend on the composition and temperature of the magma and on the environment (subaerial or submarine) in which the eruption occurs.

Traditionally, geologists have classified volcanoes according to their eruptive style, each style named after a well-known example (Hawaiian, Vulcanian, etc.) as described in books focused on volcanoes (see art on pp. 188–189). Next, we focus on relating eruptive styles to the geologic setting in which the volcano forms, in the context of plate tectonics theory (▶Fig. 7.12).

FIGURE 7.12 A map showing the distribution of volcanoes around the world, and the basic geologic settings in which volcanoes form, in the context of plate-tectonics theory.

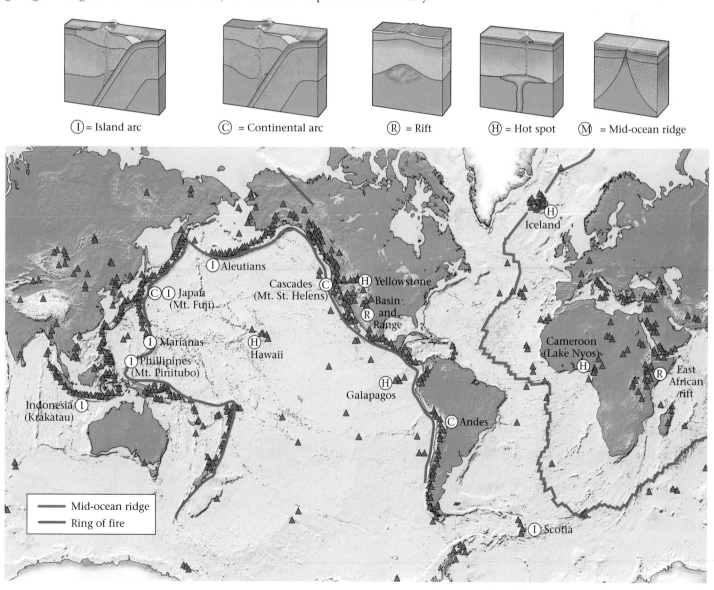

BOX 7.1
THE HUMAN ANGLE

Volcanic Explosions to Remember

Explosions of volcanic-arc volcanoes generate enduring images of destruction. Let's look at two notable historic cases (see ▶Fig. 7.13).

Mt. St. Helens, a snow-crested stratovolcano in the Cascades of the northwestern United States, had not erupted since 1857. However, geologic evidence suggested that the mountain had a violent past, punctuated by many explosive eruptions. On March 20, 1980, an earthquake announced that the volcano was awakening once again. A week later, a crater 80 m in diameter burst open at the summit and began emitting gas and pyroclastic debris. Geologists who set up monitoring stations to observe the volcano noted that its north side was beginning to bulge markedly, suggesting that the volcano was filling with magma and that the magma was making the volcano expand like a balloon. Their concern that an eruption was imminent led local authorities to evacuate people in the area.

The climactic eruption came suddenly. At 8:32 A.M. on May 18, the geologist David Johnston, monitoring the volcano from a distance of 10 km, shouted over his two-way radio, "Vancouver, Vancouver, this is it!" An earthquake had triggered a huge landslide that caused 3 cubic km of the volcano's weakened north side to slide away. The sudden landslide released pressure on the magma in the volcano, causing a sudden and violent expansion of gases that blasted through the side of the volcano (▶Fig. 7.14a–c). Rock, steam, and ash screamed north at the speed of sound and flattened a forest and everything in it over an area of 600 square km (▶Fig. 7.14d, e). Tragically, Johnston, along with sixty others, vanished forever. Water-saturated ash formed viscous slurries, or lahars, that flooded river valleys, carrying away everything in their path. Seconds after the sideways blast, a vertical column carried about 540 million tons of ash (about 1 cubic km) 25 km into the sky, where the jet stream carried it away so that it was able to circle the globe. In towns near the volcano, a blizzard of ash choked roads and buried fields. Measurable quantities of ash settled over an area of 60,000 square km. When the eruption was over, the once conelike peak of Mt. St. Helens had disappeared—the summit now lay 440 m lower, and the once snow-covered mountain was a gray mound with a large gouge in one side. The volcano came alive again in 2004, but did not explode.

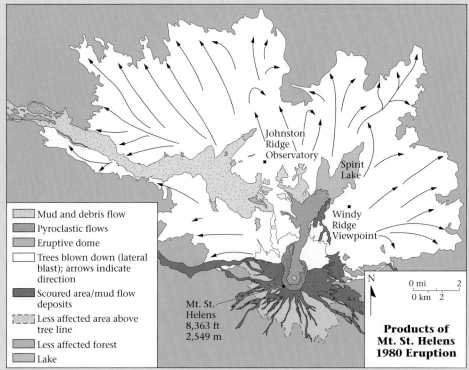

FIGURE 7.13 (a) The graph shows the relative amounts of pyroclastic debris (in cubic km) ejected during major historic eruptions of the past two centuries. Notice that the 1815 Tambora eruption was over five times larger than the 1883 Krakatau eruption, which in turn was over five times larger than the 1980 Mt. St. Helens eruption. (b) Profile of Krakatau, before and after the eruption. Note that a new resurgent dome (Anak Krakatau) has formed. (c) A map illustrating the dimensions of the region destroyed by the 1980 eruption of Mt. St. Helens. The arrows indicate the blast direction.

An even greater explosion happened in 1883 (Fig. 7.13b). Krakatau, a volcano in the sea between Indonesia and Sumatra, where the Indian Ocean floor subducts beneath Southeast Asia, had grown to become a 9-km-long island rising 800 m (2,600 feet) above the sea. On May 20, the island began to erupt with a series of large explosions, yielding ash that settled as far as 500 km away. Smaller explosions continued through June and July, and steam and ash rose from the island, forming a huge black cloud that rained ash into the surrounding straits. Ships sailing by couldn't see where they were going, and their crews had to shovel ash off the decks.

Krakatau's demise came at 10 A.M. on August 27, perhaps when the volcano cracked and the magma chamber flooded with seawater.

The resulting blast, five thousand times greater than the Hiroshima atomic bomb explosion, could be heard as far as 4,800 km away, and subaudible sound waves traveled around the globe seven times. Giant waves pushed out by the explosion slammed into coastal towns, killing over 36,000 people. Near the volcano, a layer of ash up to 40 m thick accumulated. When the air finally cleared, Krakatau was gone, replaced by a submarine caldera some 300 m deep. All told, the eruption shot 20 cubic km of rock into the sky. Some ash reached elevations of 27 km. Because of this ash, people around the world could view spectacular sunsets during the next several years.

FIGURE 7.14 The eruption of Mt. St. Helens, 1980. (a) Before the eruption, the magma chamber is a zone of broken rock inside the volcano. (b) The magma chamber fills, and the side of the volcano bulges outward. (c) The weakened north flank suddenly slipped, releasing the pressure on the magma chamber. The sudden decrease in pressure caused dissolved gases in the magma to expand and blast laterally out of the volcano. (d) The eruptive cloud. (e) The neighboring forest, flattened by a blast of rock, steam, and ash.

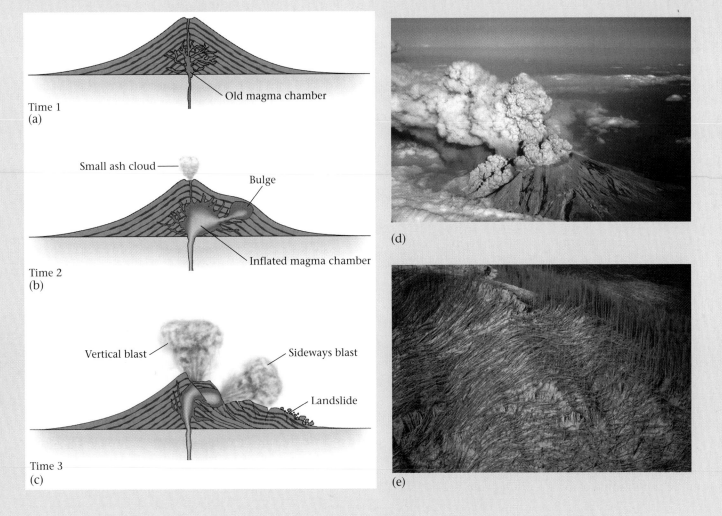

7.5 PLATE BOUNDARY AND RIFT ERUPTIONS

Mid-Ocean Ridges

Products of mid-ocean ridge volcanism cover 70% of our planet's surface. We don't generally see this volcanic activity, however, because the ocean hides most of it beneath a blanket of water. Mid-ocean ridge volcanoes, which develop along fissures parallel to the ridge axis, are not all continuously active. Each one turns on and off in a time scale measured in tens to hundreds of years. They erupt basalt, which, because it cools so quickly underwater, forms pillow-lava mounds. Water that heats up as it circulates through the crust near the magma chamber bursts out of hydrothermal (hot-water) vents along these mounds (see Chapter 2). These vents are black smokers.

Convergent Boundaries

Most of the subaerial volcanoes on Earth lie along convergent plate boundaries (subduction zones). The volcanoes form when volatile compounds such as water and carbon dioxide released from the subducting plate rise into the overlying hot mantle, causing melting and producing magma, which then rises through the lithosphere and erupts. Some of these volcanoes start out as submarine volcanoes and later grow into volcanic island arcs (such as the Aleutians of Alaska), while others grow on continental crust, building continental volcanic arcs (such as the Cascade volcanic chain of Washington and Oregon). Typically, individual volcanoes in volcanic arcs lie about 50–100 km apart. Subduction zones border over 60% of the Pacific Ocean, creating a 20,000-km-long chain of volcanoes known as the Ring of Fire.

Many different kinds of magma form at volcanic arcs. As a result, these volcanoes sometimes have effusive eruptions and sometimes pyroclastic eruptions—and occasionally they explode. Such eruptions yield composite volcanoes, such as the elegant symmetrical cone of Mt. Fuji (see Fig. 7.11d), and the blasted-apart hulk of Mt. St. Helens (see Fig. 7.14d).

Continental Rifts

The rifting of continental crust yields a wide array of different types of volcanoes, because (as in the case of continental hot spots) the magma that feeds these volcanoes comes both from the partial melting of the mantle and from the partial melting of the crust. Rifts host basaltic fissure eruptions, in which curtains of lava fountain up or linear chains of cinder cones develop. They also host explosive rhyolitic volcanoes, and in some places even stratovolcanoes.

7.6 HOT-SPOT ERUPTIONS

Hot spots are points on the Earth's surface where volcanism does not appear to be the direct consequence of standard plate-boundary processes (sea-floor spreading, subduction, or rifting). Some hot spots, such as the Hawaiian hot spot and the Yellowstone hot spot, have been so named because they do not occur along plate boundaries. Others, such as the Iceland hot spot, have been so named because they are sites of unusually abundant magma production. All told, about 50 hot spots exist on Earth today. But even after decades of study, the origin of hot spots remains unclear. Geologists still debate three possible models of genesis: (1) Hot spots lie above fingerlike plumes of hot mantle that flow up from the core-mantle boundary; (2) Hot spots lie above shallow plumes that originate in the upper mantle; (3) Hot spots form where a crack propagates into the lithosphere, like a zipper unzipping. Let's now look at a few examples of hot-spot volcanoes.

Oceanic Hot-Spot Volcanoes (Hawaii)

When a hot-spot volcano first forms on oceanic lithosphere, basaltic magma erupts at the surface of the sea floor. At first, such submarine eruptions yield an irregular mound of pillow lava. With time, the volcano grows up above the sea surface and becomes an island. When the volcano emerges from the sea, the basalt lava that erupts no longer freezes so quickly, and thus flows as a thin sheet over a great distance. Thousands of thin basalt flows pile up, layer upon layer, to build a broad, dome-shaped shield volcano with gentle slopes (▶Fig. 7.15). As the volcano grows, portions of it can't resist the pull of gravity and slip seaward, creating large submarine slumps.

Continental Hot-Spot Volcanoes (Yellowstone National Park)

Yellowstone Park lies at the northeast end of a track of volcanism, marked by a string of calderas. The oldest of these calderas, at the southwest end of the track, erupted 16 million years ago (▶Fig. 7.16a, b). Recent and ongoing activity beneath Yellowstone National Park has yielded fascinating landforms, volcanic rock deposits, and geysers. Eruptions at the Yellowstone hot spot differ from those on Hawaii in an important way: unlike Hawaii, the Yellowstone hot spot erupts both basaltic lava and rhyolitic pyroclastic debris. This happens because basaltic magma rising from the asthenosphere heats up and partially melts the continental crust. Rhyolitic eruptions produce thick tuffs that now crop out as yellow and red rocks in the canyon of the Yellowstone River (▶Fig. 7.16c).

About 627,000 years ago, an immense pyroclastic flow, as well as a cloud of ash, blasted out of the Yellowstone region. Close to the eruption, ignimbrites up to tens of

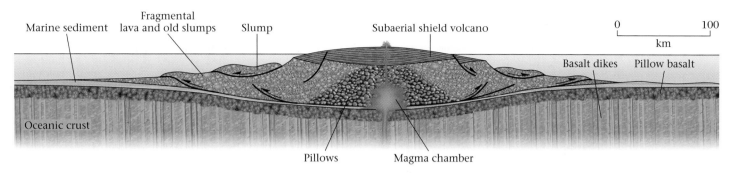

FIGURE 7.15 The inside of an oceanic hot-spot volcano is a mound of pillow basalt built on the surface of the oceanic crust. When the mound emerges above sea level, a shield volcano forms on top. Volcanic debris accumulates along the margin of the volcano. The weak material occasionally slumps seaward on sliding surfaces (indicated with arrows).

FIGURE 7.16 (a) Volcanic rocks from hot spots formed in two places in the western United States. First, the Columbia River basalt plateau erupted about 17 million years ago (m.y. = million years), when the plume rose beneath a rift. Then, perhaps, a new plume or a different part of the same plume rose beneath northern Nevada, and/or the lithosphere started splitting. Successive eruptions yielded the calderas along the Snake River Plain; the numbers indicate the age of the calderas (millions of years). The hot spot now lies underneath Yellowstone National Park. (b) Numerous flows of basalt piled one on top of the other in the Snake River Plain. The Snake River has cut a canyon through these basalts. (c) The "yellow stone" of Yellowstone National Park consists of felsic tuffs. The Yellowstone River has been able to cut a deep canyon through these tuffs, because they are relatively soft.

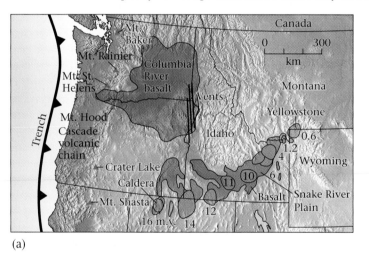

(a)

(b)

(c)

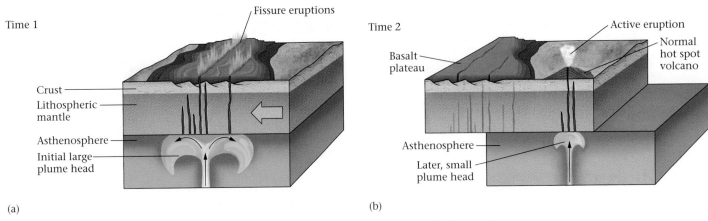

FIGURE 7.17 (a) When the lithosphere cracks and rifts above the bulbous head of a plume, huge amounts of magma rise and erupt through fissures, producing sheets of basalt that pile up to form a plateau. (b) Later, the bulbous plume head no longer exists, leaving only a narrower plume stalk. The lithosphere has moved relative to the plume, so a track of hot-spot volcanoes begin to form.

meters thick formed, and ash from the giant cloud sifted down over the United States as far east as the Mississippi River. The eruption, 1,000 times more powerful than that of Mt. St. Helens, left a huge caldera, almost 100 km across, which now dominates the landscape of Yellowstone. Magma remains beneath Yellowstone today, causing geyser activity.

Flood-Basalt Eruptions

When a plume first rises, it has a large bulbous head in which there is a huge amount of partially molten rock. If the crust above the plume stretches and rifts, voluminous amounts of lava erupt along fissures. The low-viscosity lava spreads out in sheets over vast areas. Geologists refer to these sheets as **flood basalt.** Over time, eruption of basalts builds a broad plateau (▶Fig. 7.17a, b). Geologists refer to a broad area covered by flood basalt as a large igneous province (LIP). About 15 million years ago, rifting above a plume created the region that now constitutes the Columbia River Plateau of Washington and Oregon (Fig. 7.16a). Gradually, layer upon layer erupted, creating a pile of basalt covering a region of 220,000 square km. Even larger flood-basalt provinces have formed elsewhere in the world, notably the Deccan Plateau of India (▶Fig. 7.18), the Paraná Basin of Brazil, and the Karroo Plateau of South Africa.

Iceland—a Hot Spot on a Ridge

Iceland is one of the few places on Earth where mid-ocean ridge volcanism has built a mound of basalt that protrudes above the sea. The island formed where a hot spot lies beneath the Mid-Atlantic Ridge—the presence of this

hot spot (probably due to an underlying mantle plume) means that far more magma erupted here than beneath other places along the ridge. Because Iceland straddles a divergent plate boundary, it is being stretched apart, with

FIGURE 7.18 The flood basalts of western India, known as the "Deccan traps," are exposed in a canyon near the village of Ajanta. Between about 100 B.C.E. and 700 C.E., Buddhists carved a series of monasteries and meeting halls into the solid basalt. These are decorated by huge statues, carved in place, as well as spectacular frescoes, painted on cow-dung plaster.

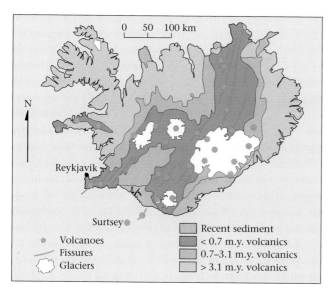

FIGURE 7.19 Iceland consists of volcanic rocks that erupted from a hot spot along the Mid-Atlantic Ridge. Because the island straddles a divergent boundary, it gradually stretches, leading to the formation of faults. The central part of the island is an irregular northeast-trending rift, where we find the youngest rocks of the island.

faults forming as a consequence. Indeed, the central part of the island is a narrow rift, in which the youngest volcanic rocks of the island have erupted (▶Fig. 7.19). This rift *is* the trace of the Mid-Atlantic Ridge. Faulting cracks the crust and so provides a conduit to a magma chamber. Thus, eruptions on Iceland tend to be fissure eruptions, yielding either curtains of lava that are many kilometers long or linear chains of small cinder cones (see Fig. 7.9b).

7.7 VOLCANOES IN THE LANDSCAPE

Why do volcanoes look the way they do? First of all, the shape of a volcano depends on whether it has been erupting recently or ceased erupting long ago. The shape (shield, stratovolcano, or cinder cone) of an erupting volcano depends primarily on the eruptive style, because at an erupting volcano, the process of construction happens faster than the process of erosion. For example, in southern Mexico, the volcano Paricutín began to spatter out of a cornfield on February 20, 1943. Its eruption continued for nine years, and by the end, 2 cubic km of tephra had piled up into a cone almost half a kilometer high.

Once a volcano stops erupting, erosion attacks. The rate at which erosion destroys a volcano depends on whether it's composed of pyroclastic debris or lava. Cinder cones and ash piles can wash away quickly. For example, in the summer of 1831, a cinder cone grew 60 m above the surface of the Mediterranean Sea. As soon as the island appeared, Italy, Britain, and Spain laid claim to it, and shortly the island had at least seven different names. But the volcano stopped erupting, and within six months it was gone, fortunately before a battle for its ownership had begun. In contrast, composite or shield volcanoes, which have been armor plated by lava flows, can withstand the attack of water and ice for quite some time.

In the end, however, erosion wins out, and you can tell an old volcano that has not erupted for a long time from a volcano that has erupted recently by the extent to which river or glacial valleys have been carved into its flanks. In some cases, the softer exterior of a volcano completely erodes away, leaving behind the plug of harder frozen magma that once lay just beneath the volcano, as well as the network of dikes that radiate from this plug (▶Fig. 7.20a–c). You can see good examples of such landforms at Shiprock, New Mexico (Fig. 4.11b), and at Devil's Tower, Wyoming (▶Fig. 7.21).

FIGURE 7.20 (a) The shape of an active volcano is defined by the surface of the most recent lava flow or ash fall. Little erosion affects the surface. (b) An inactive volcano that has been around long enough for the surface to be modified by erosion. In humid climates, these volcanoes have gullies carved into their flanks and may be partially covered with forest. (c) A long-dead (extinct) volcano has been so deeply eroded that only the neck of the volcano may remain.

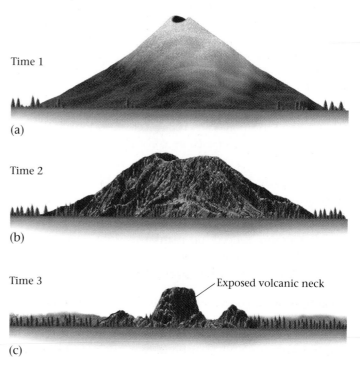

FIGURE 7.21 Devil's Tower, Wyoming, rises 260 m above the surrounding land surface. It formed when a mass of magma cooled beneath a volcano, about 40 million years ago. Huge columnar joints, 2.5 m wide at the base, developed when the magma cooled. Subsequently, erosion stripped away overlying softer tuff and flows and exposed the mass. In Native American legend, the ribbed surface of Devil's Tower represents the claw marks of a giant bear, trying to reach a woman who sought refuge on the Tower's summit.

7.8 BEWARE: VOLCANOES ARE HAZARDS!

Like earthquakes, volcanoes are natural hazards that have the potential to cause great destruction to humanity, in both the short term and the long term. According to one estimate, volcanic eruptions in the last two thousand years have caused about a quarter of a million deaths—much fewer than those caused by earthquakes, but nevertheless a sizable number. Considering the rapid expansion of cities, far more people live in dangerous proximity to volcanoes today than ever before, so if anything, the hazard posed by volcanoes has gotten worse—imagine if a Krakatau-like explosion were to occur next to a major city today. Let's now look at the different kinds of threats posed by volcanic eruptions.

Hazards Due to Eruptive Materials

Threat of lava flows. When you think of an eruption, perhaps the first threat that comes to mind is the lava that flows from a volcano. Indeed, on many occasions lava has overwhelmed towns. Basaltic lava from effusive eruptions is the greatest threat, because it can flow quickly and spread over a broad area. In Hawaii, recent lava flows have buried roads, housing developments, and cars (see photo (a) on p. 174). In one place, basalt almost completely submerged a parked (and empty) school bus (▶Fig. 7.22a). Usually people have time to get out of the way of such flows, but not necessarily with

their possessions. All they can do is watch helplessly from a distance as an advancing flow engulfs their homes (▶Fig. 7.22b). Before the lava even touches it, the building may burst into flames from the intense heat. The most disastrous lava flow in recent times came from the 2002 eruption of Mt. Nyiragongo in the Congo (▶Fig. 7.22c). Lava flows traveled almost 50 km and flooded the streets of Goma, encasing the streets with a 2-m-thick layer of basalt. The flows destroyed almost half the city and turned 300,000 people into refugees.

Threat of ash and lapilli. During a pyroclastic eruption, large quantities of ash erupt into the air, later to fall back to Earth. Close to a volcano, pumice and lapilli tumble out of the sky, smashing through or crushing roofs of nearby buildings, and can accumulate into a blanket up to several meters thick. Winds can carry fine ash over a broad region. In the Philippines, for example, a typhoon spread heavy air-fall ash from the 1991 eruption of Mt. Pinatubo so that it covered a 4,000-square-km area (▶Fig. 7.22d). Because of heavy rains, the ash became soggy and heavy, and it was particularly damaging to roofs. Ash buries crops, may spread toxic chemicals that poison the soil, and insidiously infiltrates machinery, causing moving parts to wear out.

Fine ash from an eruption can also present a hazard to airplanes. Like a sandblaster, the sharp, angular ash abrades turbine blades, greatly reducing engine efficiency. The ash, along with sulfuric acid formed from the volcanic gas, scores windows and damages the fuselage. Also, when heated inside a jet engine, the ash melts, creating a liquid that sprays around the turbine and freezes; the resulting glassy coating restricts the air flow and causes the engine to flame out. In 1982, a British Airways 747 flew through the ash cloud over a volcano on Java. Corrosion turned the windshield opaque, and ingested ash caused all four engines to fail. For thirteen minutes, the plane glided earthward, dropping from 11.5 km (37,000 feet) to 3.7 km (12,000 feet) above the black ocean below. As passengers assumed a brace position for ditching at sea, the pilots tried repeatedly to restart the engines. Suddenly, in the oxygen-rich air of lower elevations, the engines roared back to life. The plane swooped back into the sky and headed for an emergency landing in Jakarta, where, without functioning instruments and with an opaque windshield, the pilot brought the 263 passengers and crew back to the ground safely. To land, he had to squint out an open side window, with only his toes touching the controls.

Threat of pyroclastic flows. Pyroclastic flows race down the flanks of a volcano at speeds of 100–300 km per hour (▶Fig. 7.22e). The largest can travel tens to hundreds of kilometers. The volume of ash contained in such glowing avalanches is not necessarily great—St. Pierre on Martinique was covered only by a thin layer of dust after the pyroclastic flow from Mt. Pelée had passed (see p. 184)—but the cloud can be so hot and poisonous that it means instant death to

FIGURE 7.22 (a) This empty school bus was engulfed by a basalt flow in Hawaii. (b) When lava at over 1,000°C comes close to a house, the house erupts in flame. (c) Residents of Goma, in west-central Africa, walking over lava-filled streets after the 2002 eruption of a nearby volcano. (d) A blizzard of ash falling from the eruption of Mt. Pinatubo, in the Philippines, blankets a nearby town in ghostly white. (e) A pyroclastic flow rushes toward fleeing firefighters in Japan, during the eruption of Mt. Unzen. (f) A devastating lahar buried the town of Armero, Colombia.

anyone caught in its path, and because it moves so fast, the force of its impact can flatten buildings and forests (Fig. 7.6a, b).

Other Hazards Related to Eruptions

Threat of the blast. Most exploding volcanoes direct their fury upward. But some, like Mt. St. Helens, explode sideways. The forcefully ejected gas and ash, like the blast of a bomb, flattens everything in its path. In the case of Mt. St. Helens, the region around the volcano had been a beautiful pine forest; but after the eruption, the once-towering trees were stripped of bark and needles and lay scattered over the hill slopes like matchsticks (see Fig. 7.14e).

Threat of landslides. Eruptions commonly trigger large landslides along a volcano's flanks. The debris, composed of ash and solidified lava that erupted earlier, can move quite fast (250 km per hour) and far. During the eruption of Mt. St. Helens, 8 billion tons of debris took off down the mountainside, careered over a 360-m-high ridge, and tumbled down a river valley, until the last of it finally came to rest over 20 km from the volcano.

Threat of lahars. When volcanic ash and other debris mix with water, the result is a slurry that resembles freshly mixed concrete. A flow of this slurry, known as a lahar, can move downslope at speeds of over 50 km per hour. Because lahars are denser and more viscous than water, they pack more force than flowing water and can literally carry away everything in their path. The lahars of Mt. St. Helens traveled more than 40 km from the volcano, following existing drainages. When they had passed, they left a gray and barren wake of mud, boulders, broken bridges, and crumpled houses, as if a giant knife had scraped across the landscape. Widespread lahars also swept down the flanks of Mt. Pinatubo in 1991, the water provided by typhoonal and monsoonal rains.

Lahars may develop in regions where snow and ice cover an erupting volcano, for the eruption melts the snow and ice, thereby creating an instant supply of water. Perhaps the most destructive lahar of recent times accompanied the eruption of the snow-crested Nevado del Ruiz in Colombia on the night of November 13, 1985. The lahar surged down a valley of the Rio Lagunillas like a 40-m-high wave, hitting the sleeping town of Armero, 60 km from the volcano. Other pulses of lahar followed. When they had passed, 90% of the buildings in the town were gone, replaced by a 5-m-thick layer of mud (▶Fig. 7.22f), which now entombs the bodies of 25,000 people.

Threat of earthquakes. Earthquakes accompany almost all major volcanic eruptions, for the movement of magma breaks rocks underground. Such earthquakes may trigger landslides on the volcano's flanks and can cause buildings to collapse and dams to rupture, even before the eruption itself begins.

Threat of tsunamis (giant waves). Where explosive eruptions occur in the sea, the blast and the underwater collapse of a caldera generate huge sea waves, or tsunamis, tens of meters high. Most of the 36,000 deaths attributed to the 1883 eruption of Krakatau were due not to ash or lava, but rather to tsunamis that slammed into nearby coastal towns (see Box 7.1).

Threat of gas. We have already seen that volcanoes erupt not only solid material, but also large quantities of gases such as water vapor, carbon dioxide, sulfur dioxide, and hydrogen sulfide. Usually the gas eruption accompanies the lava and ash eruption, with the gas contributing only a minor part of the calamity. But occasionally gas erupts alone and snuffs out life in its path without causing any other damage. Such an event occurred in 1986 near Lake Nyos in Cameroon, western Africa.

Lake Nyos is a small but deep lake filling the crater of an active hot-spot volcano in Cameroon. Though only 1 km across, the lake reaches a depth of over 200 m. Because of its depth, the cool bottom water of the lake does not mix with warm surface water, and for many years the bottom water remains separate from the surface water. During this time, carbon dioxide gas slowly bubbles out of cracks in the floor of the crater and dissolves in the cool bottom water. Apparently, by August 21, 1986, the bottom water had become supersaturated in carbon dioxide. On that day, perhaps triggered by a landslide or wind, the lake "burped" and expelled a forceful froth of CO_2 bubbles. Because it is denser than air, this invisible gas flowed down the flank of the volcano and spread out over the countryside for about 23 km before dispersing. Although not toxic, carbon dioxide cannot provide oxygen for metabolism or oxidation. When the gas cloud engulfed the village of Nyos, it quietly put out the cooking fires and suffocated the sleeping inhabitants, most of whom died where they lay. The next morning, the landscape looked exactly as it had the day before, except for the lifeless bodies of 1,742 people and about 6,000 head of cattle (▶Fig. 7.23).

Taken together, the various damaging phenomena accompanying a volcanic eruption can devastate a society. For example, archeological research suggests that the Minoan culture, which thrived in the eastern Mediterranean during the Bronze Age (beginning 3000 B.C.E.), disappeared during the century following explosive eruptions of the Santorini volcano in 1645 B.C.E. All that remains of Santorini today is a huge caldera whose rim projects above sea level as the Greek island of Thera. Ash clouds, tsunami, and earthquakes generated by Santorini may have so disrupted their daily life

FIGURE 7.23 Cattle near Lake Nyos, Cameroon, fell where they stood, victims of a cloud of carbon dioxide.

that the Minoan people moved elsewhere, leaving behind their elaborately decorated palaces.

7.9 PROTECTION FROM VULCAN'S WRATH

Active, Dormant, and Extinct Volcanoes

Geologists refer to volcanoes that are erupting, have erupted recently, or are likely to erupt soon as **active volcanoes** and distinguish them from **dormant volcanoes,** which have not erupted for hundreds to thousands of years but do have the potential to erupt again in the future. Volcanoes that were active in the past but have shut off entirely and will never erupt in the future are called **extinct volcanoes.** As examples, geologists consider Hawaii's Kilauea to be active, for it currently erupts and has erupted frequently during recorded history. In contrast, Mt. Rainier in the Cascades last erupted centuries to millennia ago, but since subduction continues along the western edge of Oregon and Washington, the volcano could erupt in the future, and so it is considered dormant. Devil's Tower, in Wyoming, is the remnant of a volcano that was active millions of years ago but is now extinct, for the geologic cause for volcanism in the area no longer exists.

Predicting Eruptions

Little can be done to predict an eruption at a given active or dormant volcano beyond a few months or years, except to define the recurrence interval, the average time between eruptions. But short-term (weeks to months) predictions of impending volcanic activity, unlike short-term predictions of earthquakes, *are* actually feasible. Some volcanoes send out distinct warning signals announcing that an eruption may take place very soon, for as magma squeezes into the magma chamber, it causes a number of changes that geologists can measure.

- *Earthquake activity:* Movement of magma generates vibrations in the Earth. When magma flows into a volcano, rocks surrounding the magma chamber crack, and blocks slip with respect to each other. Such cracking and shifting also causes earthquakes. Thus, in the days or weeks preceding an earthquake, the region between 1 and 7 km beneath a volcano becomes seismically active.

- *Changes in heat flow:* The presence of hot magma increases the local heat flow, the amount of heat passing through rock. In some cases, the increase in the heat flow melts snow or ice on the volcano, triggering floods and lahars even before an eruption occurs.

- *Changes in shape:* As magma fills the magma chamber inside a volcano, it pushes outward and can cause the surface of the volcano to bulge; the same effect happens when you blow into a balloon.

- *Increases in gas emission and steam:* Even though magma remains below the surface, gases bubbling out of the magma, or steam formed by the heating of groundwater by the volcano, percolate upward through cracks in the Earth and rise from the volcanic vent. So an increase in the volume of gas emission, or of new hot springs, indicates that magma has entered the ground below.

Because geologists can determine when magma has moved into the magma chamber of a volcano, government agencies now send monitoring teams to a volcano at the first sign of activity. These teams set up instruments to record earthquakes, measure the heat flow, determine changes in the volcano's shape, and analyze emissions.

Controlling Volcanic Hazards

Danger assessment maps. Let's say that a given volcano has the potential to erupt in the near future. What can we do to prevent the loss of life and property? Since we can't prevent the eruption, the first and most effective precaution is to define the regions that can be directly affected by the eruption—to compile a volcanic danger-assessment map (▶Fig. 7.24). These maps delineate areas that lie in the path of potential lava flows, lahars, debris flows, or pyroclastic flows.

River valleys initiating on the flanks of a volcano are particularly dangerous areas, because lahars may flow down them. Before the 1991 eruption of Mt. Pinatubo in the Philippines, geologists had defined areas potentially in the path of pyroclastic flows and had predicted which river

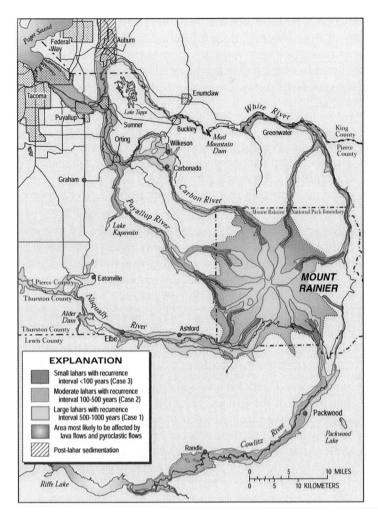

FIGURE 7.24 A danger-assessment map for the Mt. Rainier area in Washington (courtesy of the U.S. Geological Survey). The different colors on the map indicate different kinds of hazards. Note that lahars can travel long distances down river valleys—some may threaten the city of Tacoma.

valleys were likely hosts for lahars. Although the predicted pyroclastic-flow paths proved to be accurate, the region actually affected by lahars was much greater. Nevertheless, many lives were saved by evacuating people in areas thought to be under threat.

Evacuation. Unfortunately, because of the uncertainty of prediction, the decision about whether or not to evacuate is a hard one. In the case of Mt. St. Helens, hundreds of lives were saved in 1980 by timely evacuation, but in the case of Mt. Pelée in 1902, thousands of lives were lost because warning signs were ignored.

Diverting flows. In traditional cultures, people believed that gods or goddesses controlled volcanic eruptions, so when a volcano rumbled, they provided offerings to ap-

pease the deity. Sometimes, people have used direct force to change the direction of a flow or even to stop it. For example, during a 1669 eruption of Mt. Etna, a lively volcano on the Italian island of Sicily, basaltic lava formed a glowing orange river that began to spill down the side of the mountain. When the flow approached the town of Catania, 16 km from the summit, fifty townspeople protected by wet cowhides boldly hacked through the chilled side of the flow to create an opening through which the lava could exit. They hoped thereby to cut off the supply of lava feeding the end of the flow, near their homes. Their strategy worked, and the flow began to ooze through the new hole in its side. But unfortunately, the diverted flow began to move toward the neighboring town of Paterno. Five hundred men of Paterno then chased away the Catanians so that the hole would not be kept open, and eventually the flow swallowed part of Catania.

More recently, people have used high explosives to blast breaches in the flanks of flows, and have built dams and channels to divert flows. Major efforts to divert flows from a 1983 eruption of Mt. Etna, and again in 1992, were successful. Inhabitants of Iceland used a particularly creative approach in 1973 to stop a flow before it overran a town: they sprayed cold seawater onto the flow to freeze it in its tracks (►Fig. 7.25). The flow did stop short of the town, but whether this was a consequence of the cold shower it received remains unknown.

7.10 VOLCANOES AND CLIMATE

In 1783, Benjamin Franklin was living in Europe, serving as the American ambassador to France. The summer of that year seemed to be unusually cool and hazy. Franklin,

FIGURE 7.25 As a basalt flow encroached on this town in Iceland, firefighters used forty-three pumps to dump over 6 million cubic m of seawater on the lava to freeze it and stop the flow.

who was an accomplished scientist as well as a statesman, couldn't resist seeking an explanation for this phenomenon, and soon learned that in June of 1783, a huge volcanic eruption had taken place in Iceland. He wondered if the "smoke" from the eruption had prevented sunlight from reaching the Earth, thus causing the cooler temperatures. Franklin reported this idea at a meeting, and by doing so, may well have been the first scientist ever to suggest a link between eruptions and climate.

Franklin's idea seemed to be confirmed in 1815, when Mt. Tambora in Indonesia exploded. Tambora's explosion ejected over 100 cubic km of ash and pumice into the air. Ten thousand people were killed by the eruption and the associated tsunami. Another 82,000 died of starvation. The sky became so hazy that stars dimmed by a full magnitude. Temperatures dipped so low in the Northern Hemisphere that 1816 became known as "the year without a summer." The unusual weather of that year inspired writers and artists. For example, Byron's 1816 poem "Darkness" contains the gloomy lines "The bright Sun was extinguish'd, and the stars / Did wander darkling in the eternal space . . . Morn came and went—and came, and brought no day."

Geoscientists have witnessed other examples of eruption-triggered coolness more recently. In the months following the 1883 eruption of Krakatau and the 1991 eruption of Pinatubo, global temperatures noticeably dipped. To study the effect of volcanic activity on climate even further in the past, geologists have studied ice from the glaciers of Greenland and Antarctica. Glacial ice has layers, each of which represents the snow that fell in a single year. Some layers contain concentrations of sulfuric acid, formed when SO_2 from volcanic gas dissolves in the water from which snow forms. These layers indicate years in which major eruptions occurred. Years in which ice contains acid correspond to years during which the thinness of tree rings elsewhere in the world indicates a cool growing season.

How can a volcanic eruption create these cooling effects? When a large explosive eruption takes place, fine ash and aerosols (tiny liquid droplets) enter the stratosphere. It takes only about two weeks for the ash and aerosols to circle the planet. They stay suspended in the stratosphere for many months to years, because they are above the weather and do not get washed away by rainfall. The haze they produce causes cooler average temperatures because it absorbs incoming visible solar radiation during the day but does not absorb the infrared radiation that rises from the Earth's surface at night.

7.11 VOLCANOES ON OTHER PLANETS

We conclude this chapter by looking beyond the Earth, for our planet is not the only one in the solar system to have hosted volcanic eruptions. We can see the effects of volcanic activity on our nearest neighbor, the Moon, just by looking up on a clear night. The broad darker areas of the Moon, the maria (singular mare) (after the Latin word for "sea"), consist of flood basalts that erupted over 3 billion years ago (▶Fig. 7.26). Geologists propose that the flood basalts formed when huge meteors collided with the Moon, blasting out giant craters. Crater formation decreased the pressure in the Moon's mantle so that it underwent partial melting. Fluid basaltic magma rose to the surface and filled the craters.

On Venus, about 22,000 volcanic edifices have been identified. Some of these even have caldera structures at their crests. Though no volcanoes currently erupt on Mars, the planet's surface displays a record of a spectacular volcanic past. The largest known mountain in the solar system, Olympus Mons (▶Fig. 7.27a), is an extinct shield volcano on Mars. The base of Olympus Mons is 600 km across, and its peak rises 25 km above the surrounding plains.

Active volcanism currently occurs on Io, one of the many moons of Jupiter. Cameras in the *Galileo* spacecraft have recorded huge volcanoes on Io in the act of spraying plumes of sulfur gas into space (▶Fig. 7.27b) and have tracked immense, moving lava flows. Different colors of erupted material make the surface of this moon resemble a pizza. Researchers have proposed that the volcanic activity is due to tidal power: the gravitational pull exerted by Jupiter and by other moons alternately stretches and then squeezes Io, creating sufficient friction to keep Io's mantle hot. Geologists have also recently found evidence of active volcanism on Titan, a moon of Saturn. During this activity, icy fluids (perhaps liquid methane and/or ammonia) erupted.

FIGURE 7.26 The Mare of the Moon, the broad dark areas, are composed of flood basalts. This image is a composite of three images.

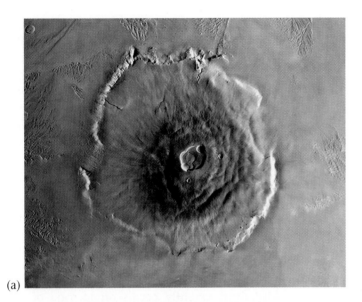

(a)

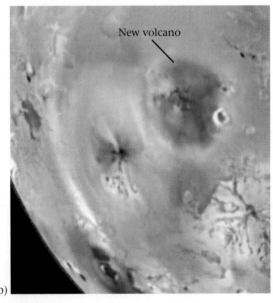

New volcano

(b)

FIGURE 7.27 (a) Satellites orbiting Mars have provided this digital image of Olympus Mons, an immense shield volcano. Notice the caldera at the summit. (b) A satellite image caught a volcano on Io, one of the moons of Jupiter, in the act of erupting. The bluish bubble is a cloud of erupting gas.

CHAPTER SUMMARY

•   Volcanoes are vents at which molten rock (lava), pyroclastic debris, gas, and aerosols erupt at the Earth's surface. A hill or mountain created from the products of an eruption is also called a volcano.

•   The characteristics of a lava flow depend on its viscosity, which in turn depends on its temperature and composition.

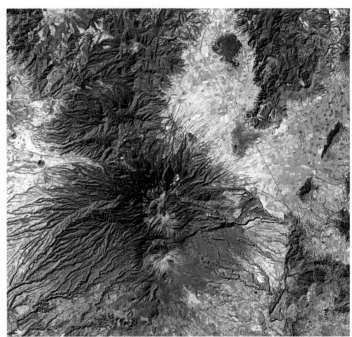

THE VIEW FROM SPACE Subduction of the Pacific Ocean floor beneath Mexico causes partial melting in the Earth's mantle. The melting yields magma which rises into the crust of Mexico. Some magma makes it to the ground surface and erupts, forming a mound of ash and lava called a volcano. This example, Colima Volcano, has two eruptive centers. The northern one is older. A radial drainage network has formed on the flanks of the volcanoes.

•   Basaltic lavas can flow great distances. Pahoehoe flows have smooth, ropy surfaces, whereas a'a' flows have rough, rubbly surfaces. Andesitic and rhyolitic lava flows tend to pile into mounds at the vent.

•   Pyroclastic debris includes powder-sized ash, marble-sized lapilli, and apple- to refrigerator-sized blocks and bombs. Some falls from the air, whereas some forms glowing avalanches that rush down the side of the volcano.

•   Eruptions may occur at a volcano's summit or from fissures on its flanks. The summit of an erupting volcano may collapse to form a bowl-shaped depression called a caldera.

•   A volcano's shape depends on the type of eruption. Shield volcanoes are broad, gentle domes. Cinder cones are steep-sided, symmetrical hills composed of tephra. Composite volcanoes can become quite large and consist of alternating layers of pyroclastic debris and lava.

•   The type of eruption depends on several factors, including the lava's viscosity and gas content. Effusive eruptions produce only flows of lava, whereas explosive eruptions produce clouds and flows of pyroclastic debris.

- Different kinds of volcanoes form in different geologic settings that can be understood in the context of plate tectonics theory.

- Volcanic eruptions pose many hazards: lava flows overrun roads and towns, ash falls blanket the landscape, pyroclastic flows incinerate towns and fields, landslides and lahars bury the land surface, earthquakes topple structures and rupture dams, tsunamis wash away coastal towns, and invisible gases suffocate nearby people and animals.

- Eruptions can be predicted through changes in heat flow, changes in shape of the volcano, earthquake activity, and the emission of gas and steam.

- We can minimize the consequences of an eruption by avoiding construction in danger zones and by drawing up evacuation plans. In a few cases, it may be possible to divert flows.

- Immense mare of flood basalts cover portions of the Moon. The largest known volcano in the solar system, Olympus Mons, towers over the surface of Mars. Satellites have documented evidence for eruptions on Io, a moon of Jupiter, and on Titan, a moon of Saturn.

KEY TERMS

a'a' (p. 182)	lava (p. 182)
active volcano (p. 201)	lava flow (p. 182)
caldera (p. 187)	magma chamber (p. 186)
cinder cone (p. 190)	pahoehoe (p. 182)
columnar jointing (p. 182)	pyroclastic debris (p. 184)
crater (p. 186)	pyroclastic flow (p. 184)
dormant volcano (p. 201)	shield volcano (p. 187)
effusive eruption (p. 190)	stratovolcano (p. 190)
explosive eruption (p. 190)	tephra (p. 184)
extinct volcano (p. 201)	tuff (p. 184)
fissure (p. 186)	vesicles (p. 186)
flood basalt (p. 196)	volcanic ash (p. 184)
lahar (p. 185)	volcano (p. 180)

REVIEW QUESTIONS

1. Describe the three different kinds of material that can erupt from a volcano.

2. Describe the differences between a pyroclastic flow and a lahar.

3. Describe the differences among shield volcanoes, stratovolcanoes, and cinder cones. How are these differences explained by the composition of their lavas and other factors?

4. Why do some volcanic eruptions consist mostly of lava flows, while others are explosive and do not produce flows?

5. Describe the activity in the mantle that leads to hot-spot eruptions.

6. How do continental-rift eruptions form flood basalts?

7. Contrast an island volcanic arc with a continental volcanic arc. What is a hot-spot volcano?

8. Identify some of the major volcanic hazards, and explain how they develop.

9. How do geologists predict volcanic eruptions?

10. Explain how steps can be taken to protect people from the effects of eruptions.

SUGGESTED READING

Cattermole, P. 1996. *Planetary Volcanism,* 2nd ed. Chichester, England: John Wiley & Sons.

Chester, D. 1993. *Volcanoes and Society.* London: Edward Arnold.

De Boer, J. Z., and D. T. Sanders. 2001. *Volcanoes in Human History: The Far-Reaching Effects of Major Eruptions.* Princeton: Princeton University Press.

Decker, R. W., and B. B. Decker. 1997. *Volcanoes.* New York: W. H. Freeman.

Fisher, R. V., G. Heiken, and J. B. Hulen. 1997. *Volcanoes: Crucibles of Change.* Princeton, N. J.: Princeton University Press.

Francis, P. 1993. *Volcanoes: A Planetary Perspective.* Oxford, UK: Clarendon Press.

Schminck, H. U. 2004. *Volcanism.* Berlin: Springer.

Sigurdsson, H., et al., eds. 2000. *Encyclopedia of Volcanoes.* San Diego: Academic Press.

CHAPTER 8

A Violent Pulse: Earthquakes

8.1 INTRODUCTION

An oblique air view, generated by a computer, of the San Francisco region. The red lines are faults on which earthquakes have occurred. The complex topography of this region results from fault movement.

As the morning of January 17, 1994, approached, residents of Northridge, a suburb near Los Angeles, slept peacefully in anticipation of the Martin Luther King Day holiday. But beneath the quiet landscape, a disaster was in the making. For many years, the imperceptibly slow movement of the Pacific Plate relative to the North American Plate had been bending the rocks making up the California crust. But like a stick that you flex with your hands, rock can bend only so far before it snaps (▶Fig. 8.1a, b). Under California, the "snap" happened at 4:31 A.M., 10 km down. It sent pulses of energy racing through the crust at an average speed of 11,000 km (7,000 miles) per hour, ten times the speed of sound.

When the energy pulses, or shocks, reached the Earth's surface, the ground bucked up and down and swayed from side to side. Sleepers bounced off their beds, homes slipped off their foundations, and freeway bridges disconnected from their supports. As more and more shocks arrived, walls swayed and toppled, roofs collapsed, and rail lines buckled (▶Fig. 8.2). Early risers brewing coffee in their kitchens tumbled to the floor, under attack by dishes and cans catapulting out of cupboards. Trains left their tracks, and steep hill slopes bordering the coast gave way,

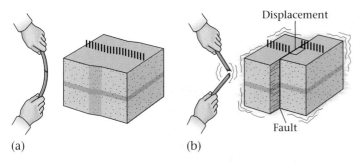

(a)   (b)

FIGURE 8.1 Most earthquakes happen when rock in the ground first bends slightly and then suddenly snaps and breaks, as does a stick you flex in your hands. (a) Before an earthquake, the crust bends (the amount of bending is greatly exaggerated here). (b) When the crust breaks, sliding suddenly occurs on a fault, generating vibrations.

FIGURE 8.2 In the 1994 Northridge, California, earthquake, this building facade tore free of its supports and collapsed.

dumping heaps of rock, mud, and broken houses onto the beach below. Ruptured gas lines fed fires that had been ignited by water heaters and sparking wires in the rubble. Then, forty seconds after it started, the motion stopped, and the shouts and sirens of rescuers replaced the crash and clatter of breaking masonry and glass. A major **earthquake**—an episode of ground shaking—had occurred.

Earthquakes have affected the Earth since the solid crust first formed. Most are a consequence of lithosphere-plate movement; they punctuate each step in the growth of mountains, the drift of continents, and the opening and closing of ocean basins. And, perhaps of more relevance to us, earthquakes have afflicted human civilization since the construction of the first village, and they have directly caused the deaths of over 3.5 million people during the past two millennia. Ground shaking, giant waves, landslides, and fires during earthquakes turn cities to rubble. The destruction caused by some earthquakes may even have changed the course of civilization.

Earthquakes are a fact of life on planet Earth: almost 1 million detectable earthquakes happen every year. Fortunately, most cause no damage or casualties, because they are too small or they occur in unpopulated areas. But a few hundred earthquakes per year rattle the ground sufficiently to damage buildings and injure their occupants, and every five to twenty years, on average, a great earthquake triggers a horrific calamity. What geologic phenomena generate earthquakes? Why do earthquakes take place where they do? How do they cause damage? Can we predict when earthquakes will happen, or even prevent them from happening? These questions have puzzled seismologists (from *seismos*, Greek for "shock" or "earthquake"), geoscientists who study earthquakes, for decades. In this chapter, we seek some of the answers, answers that can help those living in earthquake-prone regions to cope.

8.2 CAUSES OF EARTHQUAKES

Ancient cultures offered a variety of explanations for **seismicity** (earthquake activity), most of which involved the action or mood of a giant animal or god. Scientific study suggests that seismicity can occur for several reasons, including

- the sudden formation of a new **fault** (a fracture on which sliding occurs),
- sudden slip on an existing fault,
- a sudden change in the arrangement of atoms in the minerals of rock,
- movement of magma in a volcano,
- the explosion of a volcano,
- giant landslides,
- a meteorite impact, or
- underground nuclear-bomb tests.

As we learned in Chapter 1, the place within the Earth where rock ruptures and slips, or the place where an explosion occurs, is the **hypocenter** (or focus) of the earthquake. Energy radiates from the hypocenter. The point on the surface of the Earth that lies directly above the hypocenter is the **epicenter** (▶Fig. 8.3).

The formation and movement of faults cause the vast majority of destructive earthquakes, so typically the hypocenter of an earthquake lies on a fault plane (the surface of the fault). Thus, we'll focus our investigation on how faults develop and why their movement generates earthquakes.

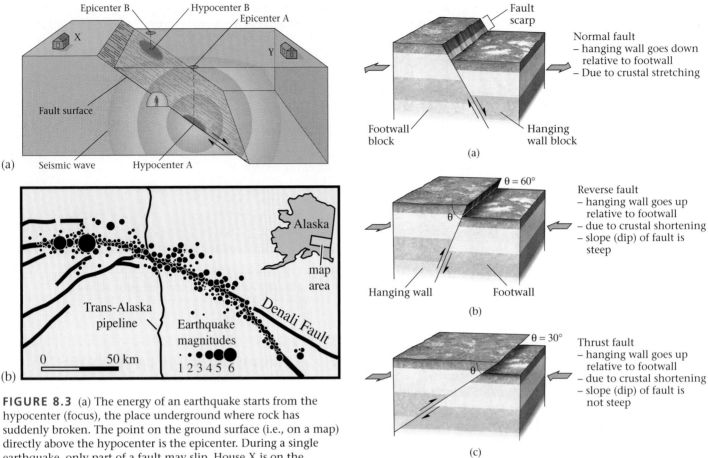

FIGURE 8.3 (a) The energy of an earthquake starts from the hypocenter (focus), the place underground where rock has suddenly broken. The point on the ground surface (i.e., on a map) directly above the hypocenter is the epicenter. During a single earthquake, only part of a fault may slip. House X is on the footwall, and house Y is on the hanging wall. The miner excavating a tunnel along the fault has the hanging wall over his head and the footwall under his feet. (b) A simplified map of the Denali National Park region, Alaska, shows the trace of the Denali fault and epicenters of earthquake events along the fault during a period in late 2002. The diameter of a circle represents the size (magnitude) of the shock it represents, as we discuss later.

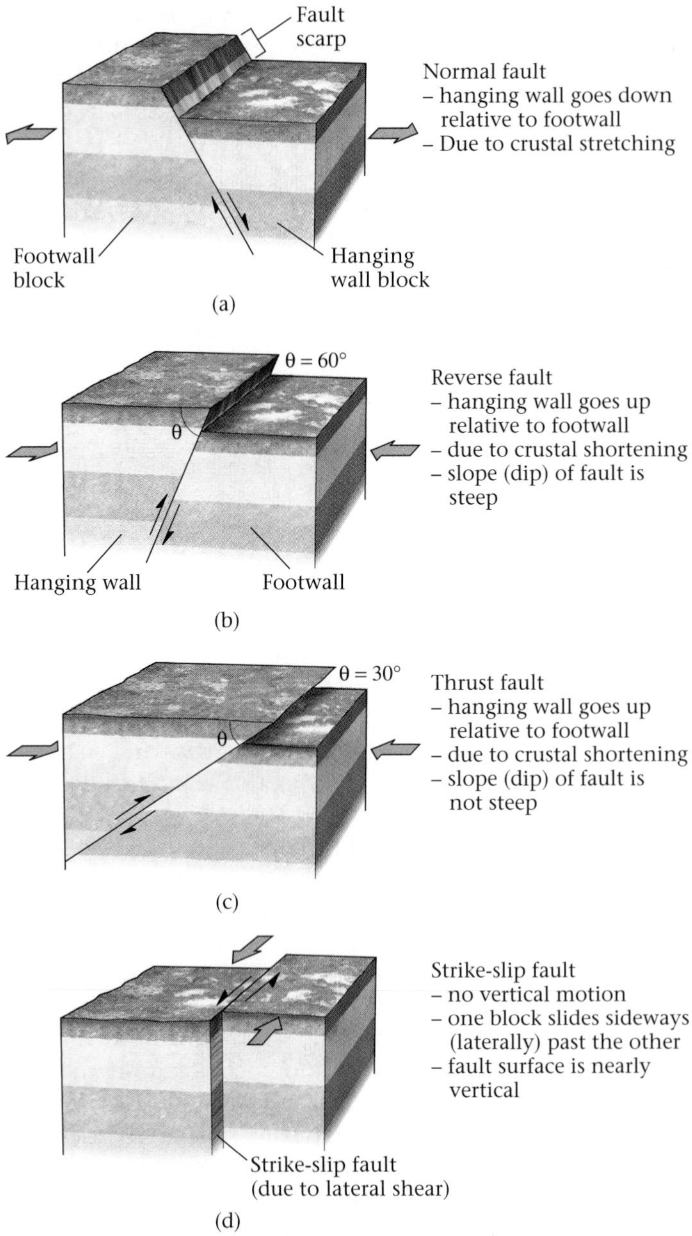

How Faulting Causes Earthquakes

Faults are fractures on which slip or sliding occurs (see Chapter 9). They can be pictured simplistically as planes that cut through the crust. Nineteenth-century miners who encountered faults in mine tunnels referred to the rock mass above a sloping fault plane as the hanging wall, because it hung over their heads, and the rock mass below the fault plane as the footwall, because it lay beneath their feet (see Fig. 8.3). The miners described the direction in which rock masses slipped on a fault by specifying the direction that the hanging wall moved in relation to the footwall, and we still use these terms today. When the hanging wall slips down the slope of the fault, it's a normal fault. When the hanging wall slips up the slope, it's a reverse fault if steep and a thrust fault if shallowly sloping (▶Fig. 8.4a–c). Strike-slip faults

FIGURE 8.4 The basic types of faults. Note that faults are distinguished from each other by the nature of the slip. (a) Normal fault, (b) Reverse fault, (c) Thrust fault, (d) Strike-slip fault.

have near-vertical planes on which slip occurs parallel to an imaginary horizontal line, called a strike line, on the fault plane—no up or down motion takes place here (▶Fig. 8.4d). In Chapter 9, we will introduce other types of faults.

Normal faults form in response to stretching or extension of the crust. Reverse or thrust faults develop in response to squeezing (compression) and shortening of the crust, and strike-slip faults form where one block of crust slides past another laterally. By measuring the distance between the two ends of a distinctive sedimentary bed or igneous dike that's been offset by a fault, geologists define the **displacement,** the amount of slip, on the fault (▶Fig. 8.5a, b).

Faults are found almost everywhere—but don't panic! Not all of them are likely to be the source of earthquakes. Faults that have moved recently or are likely to move in the near future are called active faults (and if they generate earthquakes, news media sometimes refer to them as earthquake faults); faults that last moved in the distant past and probably won't move again in the near future (but are still recognizably faults because of the displacement across them) are called inactive faults. Some faults have been inactive for billions of years.

Not all faults intersect the ground surface, but some do. Geologists refer to the intersection between a fault and the ground surface as the fault trace, or fault line (Fig. 8.5b). In places where an active normal or reverse fault intersects the ground, movement on the fault displaces the ground surface and generates a small step called a **fault scarp** (Fig. 8.4a).

What is the relationship between faulting and earthquakes? Earthquakes can happen either when rock breaks and a new fault forms, or when a preexisting fault suddenly slips again. Let's look more closely at these two causes.

• *Earthquakes due to fault formation:* Imagine that you grip each side of a brick-shaped block of rock with a clamp. Apply an upward push on one of the clamps and a downward push on the other. By doing so, you have applied a stress to the rock (see Chapter 6). At first, the rock bends slightly but doesn't break (▶Fig. 8.6a). In fact, if you were to stop applying stress at this stage, the rock would return to its original shape. Geologists refer to such a phenomenon as elastic behavior—the same phenomenon happens when a rubber band returns to its original shape or a bent stick straightens out after you let go. Now repeat the experiment, but bend the rock even more. If you bend the rock far enough, a number of small cracks or breaks start to form (▶Fig. 8.6b). Eventually the cracks connect to one another to form a fracture that cuts across the entire block of rock. The instant that such a throughgoing fracture forms, the block breaks in two. The rock on one side suddenly slides past the rock on the other side, and any elastic bending that had built up is released so the rock straightens out (▶Fig. 8.6c). Because sliding occurs, the fracture at this stage has become a fault. Such fault formation and slip, like the snap of a stick, generates vibrations—an earthquake.

A fault can't slip forever, for friction eventually slows and stops the movement, just as friction slows and stops a book sliding across a table. Friction, defined simply as the force that resists sliding on a surface, exists because all surfaces in the real world are slightly rough. Thus, as one surface slides along another, bumps protruding from one surface snag on the bumps of the opposing surface, like little anchors, resisting movement.

FIGURE 8.5 (a) This wooden fence was built across the San Andreas Fault. During the 1906 San Francisco earthquake, slip on the fault broke and offset the fence; the displacement of the fence indicates that the fault is strike-slip, as we see no evidence of up or down motion. The rancher quickly connected the two ends of the fence so no cattle could escape. (b) The amount the fence was offset indicates the displacement on the fault.

(a)

(b)   *What a geologist sees*

Elastic bending

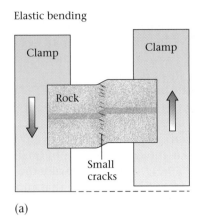

(a)

Cracking

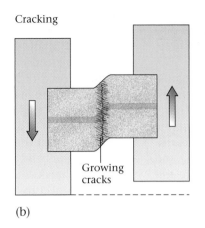

(b)

Rupture and sliding

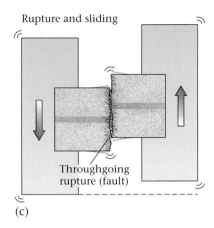

(c)

FIGURE 8.6 The stages in the development of a fault can be illustrated by breaking a rock block gripped at each end by a clamp. (a) As one clamp moves relative to the other (indicated by arrows), the rock begins to bend elastically and develop strain. (b) If bending continues, the rock begins to crack, and then the cracks grow and start to connect. (c) When the cracks connect sufficiently to make a throughgoing fracture, the rock ruptures into two pieces. The instant one piece slides past another, the rupture becomes a fault.

• *Earthquakes due to slip on a preexisting fault:* Once a fault comes into being, it can act like a scar in the lithosphere, in that it remains weaker than surrounding, intact lithosphere. Nevertheless, because of friction, a fault won't slip again until the applied stress becomes great enough. Put another way, stress will build in the lithosphere until the stress can overcome frictional resistance to sliding on a fault. Then the fault slips again. When slipping takes place, the bumps or protrusions on the fault surface break off or plow into the opposing wall—this process generates earthquake vibrations. Note that after each slip event, no earthquakes happen on the fault until stress builds again. Geologists refer to such alternation between stress buildup, when no slip occurs, and slip events (earthquakes) as **stick-slip behavior.**

The breaking of rock that occurs either when a new fault initiates, or when protrusions on an existing fault break during sliding, generates the energy of an earthquake. The concept that earthquakes happen in a region of lithosphere because stresses build up, causing rock to bend elastically until slip on a fault occurs, at which time elastic bending decreases, has come to be known as the **elastic-rebound theory.** The relative movement of one plate with respect to another, along a plate boundary, generates the stress that drives the process.

The series of shocks generated by the sudden slip on a fault and the subsequent vibrations create the shaking we feel as an earthquake. The bigger the amount of slip and the greater the amount of rock that moves, the greater the size of the vibrations and, therefore, the larger the earthquake.

A major earthquake along a fault may be preceded by smaller ones, called **foreshocks,** which possibly result

from the development of the smaller cracks that will eventually link up to form a major rupture. Smaller earthquakes that follow a major earthquake, called **aftershocks,** may occur for days to several weeks. The largest aftershock tends to be ten times smaller than the main shock; most are much smaller. Aftershocks happen because the movement of rock during the main earthquake creates new stresses, which may be large enough to reactivate small portions of the main fault or to activate small, nearby faults.

How much of a fault slips during an earthquake? The answer depends on the size of the earthquake: the larger the earthquake, the larger the slipped area and the greater the amount of slip. For example, the major earthquake that hit San Francisco, California, in 1906 ruptured a 430-km-long (measured parallel to the Earth's surface) by 15-km-high (measured perpendicular to the Earth's surface) segment of the San Andreas Fault. The fault slip (the amount of sliding along the fault) died out at the ends of the segment, but toward the middle it reached 7 m, in a strike-slip sense. (Note that the amount of slip on a fault during an earthquake varies along the length of the fault.) Slip on a thrust fault caused the 1964 Good Friday earthquake in southern Alaska: at depth in the Earth, slip reached a maximum of 12 m, and at the Earth's surface, the faulting uplifted the ground over a 500,000-square-km area by as much as 2 m. Smaller earthquakes, such as the one that hit Northridge in 1994, resulted in about 0.5 m slip on a break that was about 5 km long and 5 km wide. The smallest-felt earthquakes (which rattle the dishes but not much more) result from displacements measured in millimeters to centimeters.

Although the cumulative movement on a fault during a human life span may not amount to much, over geologic time the cumulative movement becomes significant. For

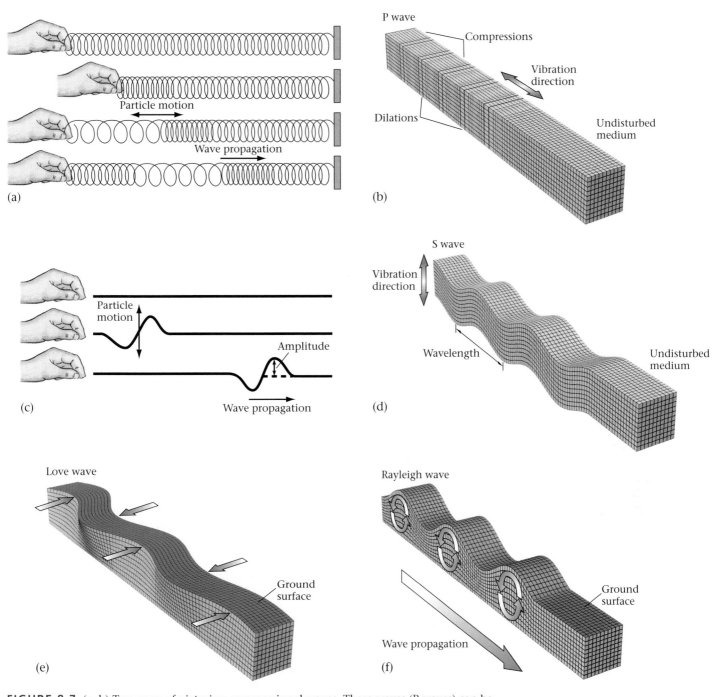

FIGURE 8.7 (a, b) Two ways of picturing compressional waves. These waves (P-waves) can be generated by pushing on the end of a spring. The pulse of energy compresses in sequence down the length of spring. Note that the back-and-forth motion of the coils occurs in the same direction that the wave travels. The wavelength of P-waves is defined by the distance between successive pulses of compression. (c, d) Two ways of picturing shear waves. These waves (S-waves) resemble the waves in a rope. Note that the back-and-forth motion occurs in a direction perpendicular to the direction the wave travels. The wavelength of S-waves is defined as the distance between successive peaks (or troughs). (e) L-waves cause the surface of the ground to shear sideways, like a moving snake. (f) R-waves make the surface of the ground go up and down. Underground, particles follow an elliptical orbit, as shown.

example, if earthquakes occurring on a reverse fault cause 1 cm of uplift over ten years, on average, the fault's movement will yield 1 km of uplift after 1 million years.

on a fault deep in the crust causes less damage than one caused by slip on a fault near the surface.

8.3 SEISMIC WAVES

How does the energy emitted at the hypocenter of an earthquake travel to the surface or even pass through the entire Earth? Like other kinds of energy, earthquake energy travels in the form of waves. We call these waves **seismic waves** (or earthquake waves). You feel them if you touch one end of a brick and tap the other end with a hammer—the energy of the hammer blows travels to your fingertip in the form of waves.

Seismologists distinguish among different types of seismic waves on the basis of where and how the waves move. **Body waves** pass through the interior of the Earth, whereas **surface waves** travel along the Earth's surface. Waves in which particles of material move back and forth parallel to the direction in which the wave itself moves are called **compressional waves.** As a compressional wave passes, the material first compresses (or squeezes together), then dilates (or expands). To see this kind of motion in action, push on the end of a spring and watch as the little pulse of compression moves along the length of the spring. Waves in which particles of material move back and forth perpendicular to the direction in which the wave itself moves are called **shear waves.** To see shear-wave motion, jerk the end of a rope up and down and watch how the up-and-down motion travels along the rope. With these concepts in mind, we can define four basic types of seismic waves (▶Fig. 8.7a–f):

- P-waves (P stands for "primary") are compressional body waves.

- S-waves (S stands for "secondary") are shear body waves.

- R-waves (R stands for Rayleigh, the name of a physicist) are surface waves that cause the ground to ripple up and down.

- L-waves (L stands for Love, the name of a seismologist) are surface waves that cause the ground to ripple back and forth, creating a snakelike movement.

P-waves travel the fastest and thus arrive first. S-waves travel more slowly, at about 60% of the speed of P-waves, so they arrive later. Surface waves are the slowest of all.

Friction absorbs energy as waves pass through a material, and waves bounce off layers and obstacles in the Earth, so the amount of energy carried by seismic waves decreases the farther they travel. People near the epicenter of a large earthquake may be thrown off their feet, but those 100 km away barely feel it. Similarly, an earthquake caused by slip

8.4 MEASURING AND LOCATING EARTHQUAKES

Most news reports about earthquakes provide information on the size and location of an earthquake. What does this information mean, and how do we obtain it? What's the difference between a great earthquake and a minor one? How do seismologists locate an epicenter? Understanding how a seismograph works and how to read the information it provides will allow you to answer these questions.

Seismographs and the Record of an Earthquake

When, in 1889, a German physicist realized that a pendulum in his lab had moved in reaction to a deadly earthquake that had occurred in Japan, his observation confirmed speculations that earthquake energy can pass through the planet. On reading of this discovery, other researchers saw a way to construct an instrument, called a **seismograph** (or seismometer), that can systematically record the ground motion from an earthquake happening anywhere on Earth. Seismologists now use two basic configurations of seismographs, one for measuring vertical (up-and-down) ground motion and the other for measuring horizontal (back-and-forth) ground motion (▶Fig. 8.8a, b). Typically, the instruments are placed on bedrock in sheltered areas, away from traffic and other urban noise (▶Fig. 8.8c).

A mechanical vertical-motion seismograph consists of a heavy weight (like a pendulum) suspended from a spring (▶Fig. 8.9a–c). The spring connects to a sturdy frame that has been bolted to the ground. A pen extends sideways from the weight and touches a vertical revolving cylinder of paper that has been connected to the seismograph frame. When an earthquake wave arrives and causes the ground surface to move up and down, it makes the seismograph frame also move up and down. The weight, however, because of its inertia (the tendency of an object at rest to remain at rest), remains fixed in space. As a consequence, the revolving paper roll moves up and down under the pen; the pen, therefore, traces out the waves representing the up-and-down movement. Note that if the paper cylinder did not revolve, the pen would move back and forth in the same place on the paper. A mechanical horizontal-motion seismograph works on the same principle, except that the paper cylinder is horizontal and the weight hangs from a wire. Sideways back-and-forth

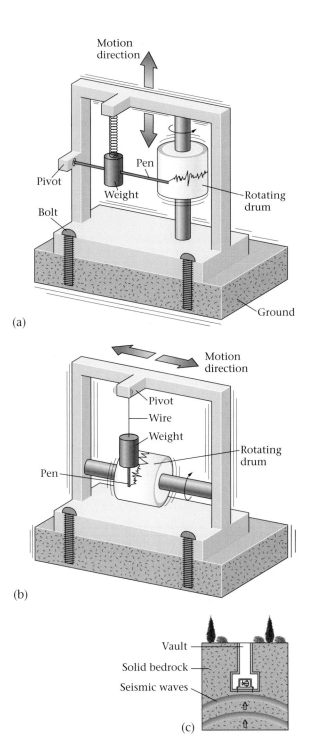

FIGURE 8.8 (a) A vertical-motion seismograph records up-and-down ground motion. (b) A horizontal-motion seismograph records back-and-forth ground motion. (c) Seismographs are bolted to the bedrock in a protected shelter or vault.

movement of this seismograph causes the pen to trace out waves (see Fig. 8.8b). In sum, the key to a seismograph is the presence of a weight that stays fixed in space while everything else moves around it.

The waves traced by the pen on a seismograph provides a record of the earthquake called a **seismogram** (▶Fig. 8.9d, e). In order to be able to determine when a particular earthquake wave arrives, the record also displays lines representing time. At first glance, a typical seismogram looks like a messy squiggle of lines, but to a seismologist it contains a wealth of information. The horizontal axis represents time, and the vertical axis represents the amplitude (the size) of the seismic waves. The instant at which an earthquake wave appears at a seismograph station is the arrival time of the wave. The first squiggles on the record represent P-waves, because P-waves travel the fastest. Next come the S-waves, and finally the surface waves (Rayleigh and Love). Typically, the surface waves have the largest amplitude and arrive over a relatively long period of time.

Seismologists the world over have agreed to certain standards for measuring earthquakes, and they calibrate their seismographs to a precise time signal so that they can compare seismograms from different parts of the world. Today, seismologists work with digital records produced by electronic seismographs. In these instruments, the simple weight has been replaced by a heavy cylindrical magnet surrounding a coil of wire. The coil connects to a solid frame anchored to the ground. During an earthquake, the magnet stays in place while the coil and the frame move. This movement generates an electrical current whose voltage indicates the amount of movement. Seismologists can now obtain data from thousands of stations constituting the worldwide seismic network.

Finding the Epicenter

How do we find the location of an earthquake's epicenter? The key to this problem comes from measuring the difference between the time that the P-wave arrives, and the time that the S-wave arrives, at a seismograph station. P-waves and S-waves pass through the interior of the Earth at different velocities. Thus, the farther the waves travel, the greater the distance between them. As an analogy, consider an automobile race. The first car travels at 100 mph and the second at 80 mph. At the starting gate (representing the earthquake's hypocenter), the cars line up next to one another, but a half hour into the race, the faster car (P-wave) is 10 miles ahead of the second car (S-wave), and an hour into the race, the faster car is 20 miles ahead of the second car. Similarly, the delay between P-wave and S-wave arrival times increases as the distance from the epicenter increases (▶Fig. 8.10a, b). We can represent this time delay on a graph with a line called a travel-time curve, which plots the time since the earthquake began on the vertical axis and the distance to the epicenter on the horizontal axis (▶Figure 8.10c).

To use a travel-time curve for determining the distance to an epicenter, start by measuring the time difference between

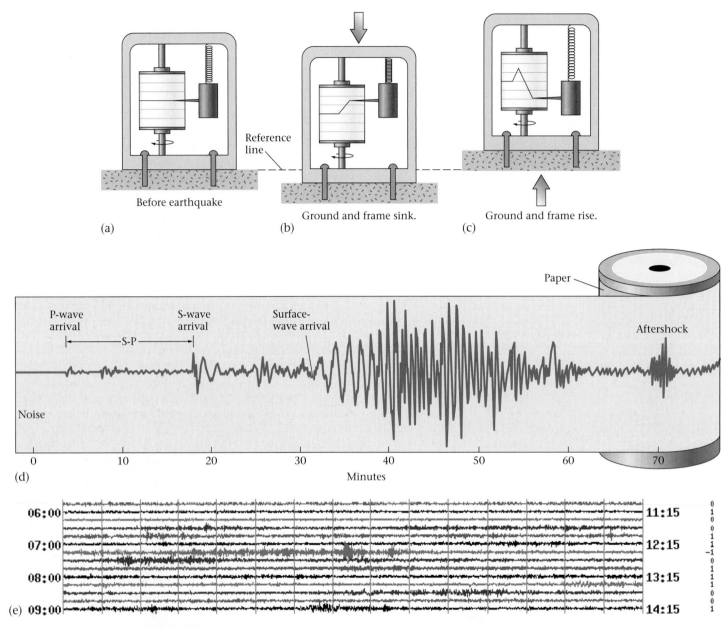

FIGURE 8.9 How a seismograph works (here, a vertical-motion seismograph). (a) Before an earthquake, the pen traces a straight line. (b) During an earthquake, when the ground and the frame of a seismograph go down, the weight stays in place because of inertia, so the pen rises relative to the paper roll. (c) When the ground and the seismograph frame rise, the pen goes down. (d) This close-up of the record (seismogram) for a single earthquake shows the signals generated by different kinds of seismic waves. (e) Digital seismic record from a seismograph station in Arkansas. The vertical lines represent minutes. Colors have no meaning, they just make the figure more readable. All of the earthquakes shown are small.

the P- and S-waves on your seismogram (Fig. 8.10c). Then draw a line segment on a piece of paper to represent this amount of time, at the scale used for the vertical axis of the graph. Move the line segment back and forth until one end lies on the P-wave curve and the other end lies on the

S-wave curve. Extend the line down to the horizontal axis, and simply read off the distance to the epicenter (Fig. 8.10c). You have now determined the distance at which the time difference between the two waves equals the time difference you observed.

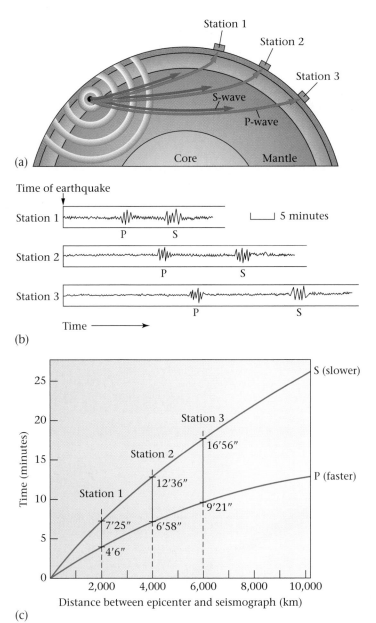

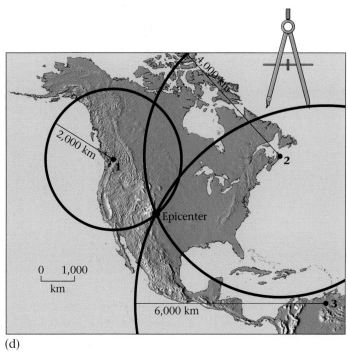

(d)

FIGURE 8.10 (a) Different seismic waves travel at different velocities. Thus, different waves arrive at seismograph stations at different times. After following curved paths through the Earth (see Interlude C), P-waves arrive first, then S-waves. (b) The greater the distance between the epicenter and the seismograph station, the greater the time delay between the P-wave and S-wave arrival times. In this example, station 1 is closest to the epicenter, and station 3 is farthest away from it. Note that the P-wave arrives later at station 3 than at station 1, and that the time interval between P- and S-wave arrivals is greater at station 3 than at station 1. Arrivals at station 2 are in between. (c) We can represent the contrasting arrival times of P-waves and S-waves on a travel-time curve. (d) If an earthquake epicenter lies 2,000 km from station 1, we draw a circle with a radius of 2,000 km around the station. Following the same procedure for stations 2 and 3, we can locate the epicenter: it lies at the intersection point of the three circles.

The analysis of one seismogram tells you only the distance between the epicenter and the seismograph station; it does not tell you in which direction the epicenter lies. To determine the map position of the epicenter, we use a method called triangulation, by plotting the distance from the epicenter to three stations. For example, say you know that the epicenter lies 2,000 km from station 1, 4,000 km from station 2, and 6,000 km from station 3. On a map, draw a circle around each station, such that the radius of the circle is the distance between the station and the epicenter, at the scale of the map. The epicenter lies at the intersection of the three circles, for this is the only point at which the epicenter has the appropriate measured distance from all three stations (▶Fig. 8.10d).

Defining the Size of an Earthquake

Some earthquakes shake the ground violently and cause extensive damage, whereas others can barely be felt. Seismologists have developed means to define size in a uniform way, so that they can make systematic comparisons among earthquakes. Seismologists use two distinct approaches for characterizing the relative sizes of earthquakes—the first yields a number called the intensity and the second yields a number called magnitude.

Mercalli intensity scale. In 1902, the Italian scientist Giuseppe Mercalli developed the first widely used scale for characterizing earthquake size. This scale, called the

Faulting in the Crust

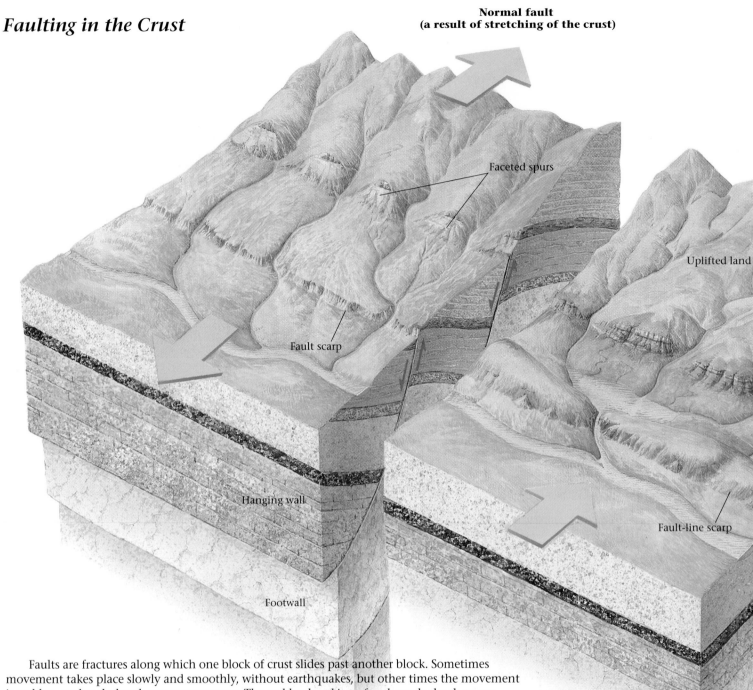

Normal fault
(a result of stretching of the crust)

Faceted spurs

Uplifted land

Fault scarp

Hanging wall

Fault-line scarp

Footwall

Faults are fractures along which one block of crust slides past another block. Sometimes movement takes place slowly and smoothly, without earthquakes, but other times the movement is sudden, and rocks break as a consequence. The sudden breaking of rock sends shock waves, called seismic waves, through the crust, creating vibrations at the Earth's surface—an earthquake.

Geologists recognize three types of faults. If the hanging-wall block (the rock above a fault plane) slides down the fault's slope relative to the footwall block (the rock below the fault plane), the fault is a normal fault. (Normal faults form where the crust is being stretched apart, as in a continental rift.) If the hanging-wall block is being pushed up the slope of the fault relative to the footwall block, then the fault is a reverse fault. (Reverse faults develop where the crust is being compressed or squashed, as in a collisional mountain belt.) If one block of rock slides past another and there is no up or down motion, the fault is a strike-slip fault. Strike-slip fault planes tend to be nearly vertical.

If a fault displaces the ground surface, it creates a ledge called a fault scarp. Sometimes we can identify the trace (or line) of a fault on the land surface because the rock of the hanging wall has a different resistance to erosion than the rock of the footwall; a ledge formed along this line due to erosion is a fault-line scarp. Where fault scarps cut a system of rivers and valleys, the ridges are truncated to make triangular facets. Strike-slip faults may offset ridges, streams, and orchards sideways. If there is a slight extension along the fault, the land surface sinks, and a sag pond develops.

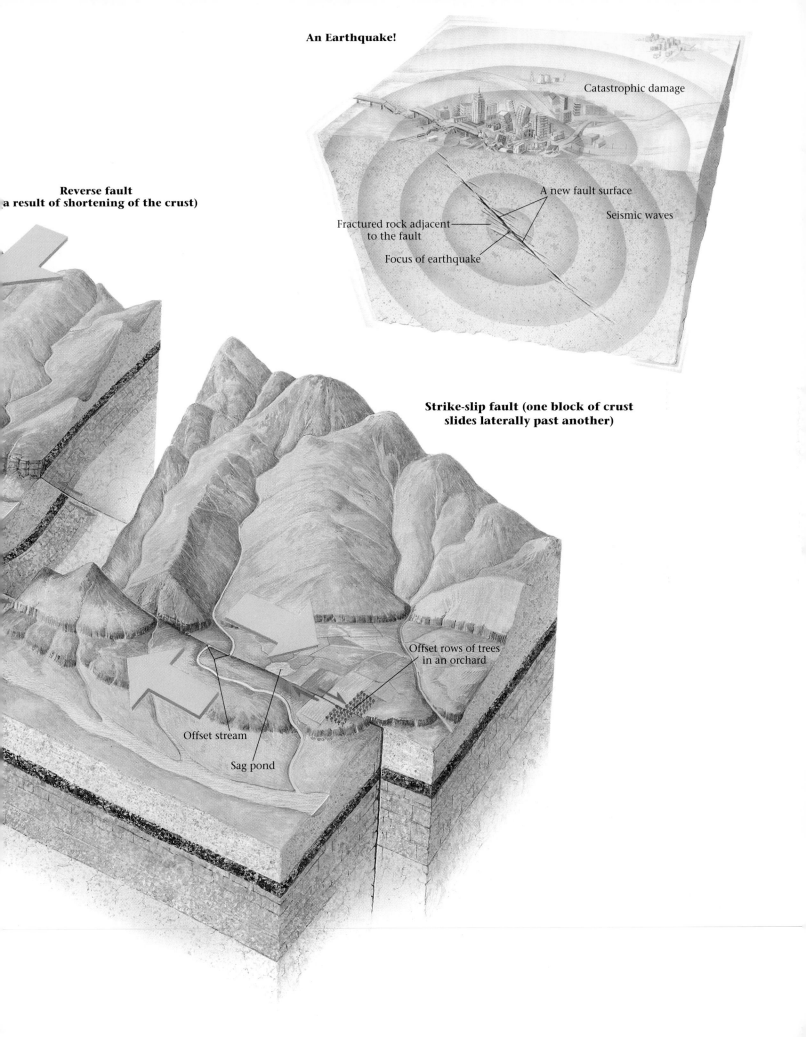

An Earthquake!

Catastrophic damage

**Reverse fault
(a result of shortening of the crust)**

A new fault surface

Seismic waves

Fractured rock adjacent
to the fault

Focus of earthquake

**Strike-slip fault (one block of crust
slides laterally past another)**

Offset rows of trees
in an orchard

Offset stream

Sag pond

TABLE 8.1 Mercalli Intensity Scale

M	Destructiveness (Perceptions of the Extent of Damage)
I	Detected only by seismic instruments; causes no damage.
II	Felt by a few stationary people, especially in upper floors of buildings; suspended objects, such as lamps, may swing.
III	Felt indoors; standing automobiles sway on their suspensions; it seems as though a heavy truck is passing.
IV	Shaking awakens some sleepers; dishes and windows rattle.
V	Most people awaken; some dishes and windows break, unstable objects tip over; trees and poles sway.
VI	Shaking frightens some people; plaster walls crack, heavy furniture moves slightly, and a few chimneys crack, but overall little damage occurs.
VII	Most people are frightened and run outside; a lot of plaster cracks, windows break, some chimneys topple, and unstable furniture overturns; poorly built buildings sustain considerable damage.
VIII	Many chimneys and factory smokestacks topple; heavy furniture overturns; substantial buildings sustain some damage, and poorly built buildings suffer severe damage.
IX	Frame buildings separate from their foundations; most buildings sustain damage, and some buildings collapse; the ground cracks, underground pipes break, and rails bend; some landslides occur.
X	Most masonry structures and some well-built wooden structures are destroyed; the ground severely cracks in places; many landslides occur along steep slopes; some bridges collapse; some sediment liquifies; concrete dams may crack; facades on many buildings collapse; railways and roads suffer severe damage.
XI	Few masonry buildings remain standing; many bridges collapse; broad fissures form in the ground; most pipelines break; severe liquefaction of sediment occurs; some dams collapse; facades on most buildings collapse or are severely damaged.
XII	Earthquake waves cause visible undulations of the ground surface; objects are thrown up off the ground; there is complete destruction of buildings and bridges of all types.

Mercalli intensity scale, defines the intensity of an earthquake by the amount of damage it causes—that is, by its destructiveness. We denote different Mercalli intensities (M) with Roman numerals, as shown in ▶Table 8.1. Note that the specification of the intensity of an earthquake depends on a subjective assessment of the damage, not on a particular measurement with an instrument. Note also that the Mercalli intensity varies with distance from the epicenter, because earthquake energy dies out as the waves travel farther through the Earth. Seismologists draw lines, called contours, around the epicenter to delimit zones in which the earthquake has a specific intensity (▶Fig. 8.11).

Earthquake magnitude scales. When you read a report of an earthquake disaster in the news, you will likely come across a phrase that reads something like, "An earthquake with a magnitude of 7.2 struck the city yesterday at noon." What does this mean? The **magnitude** of an earthquake is a number that indicates its relative size as determined by measuring the maximum amplitude of ground motion recorded by a seismograph. By "amplitude of ground motion," we mean the amount of up-and-down or sideways motion of the ground. The larger the ground motion, the greater the deflection of a seismograph pen or needle as it traces out a seismogram. For a calculation of magnitude, a seismologist accommodates for the distance between the epicenter and the seismograph, so magnitude does not depend on this

FIGURE 8.11 This map shows the contours of Mercalli intensity for the 1886 Charleston, South Carolina, earthquake. Note that near the epicenter, ground shaking reached M = X, and in New York City, ground shaking reached M = II to III.

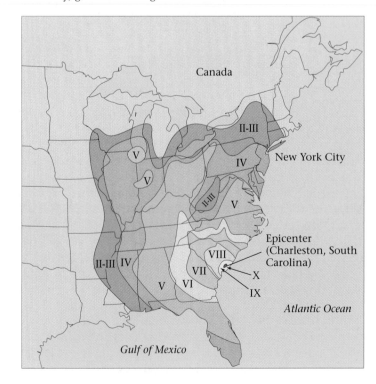

distance, and a calculation based on data from any seismograph anywhere in the world will yield the same result.

The American seismologist Charles Richter developed the concept of defining and measuring earthquake magnitude in 1935. The scale he proposed came to be known as the **Richter scale.** You can determine a Richter magnitude by measuring the amplitude of the largest deflection, on a seismograph, generated in response to seismic waves that have a period of one second, as recorded by a seismic station 100 km from the epicenter. The period for a set of earthquake waves is the time interval between the arrival of successive waves. At 100 km from the epicenter, the S-wave overlaps with the fastest surface waves, so the largest amplitude wave is probably a combination of the S-wave and a surface wave. Since most seismograph stations do not happen to lie exactly 100 km from the epicenter, seismologists use a chart to adjust for distance from the epicenter (▶Fig. 8.12).

Richter's scale became so widely used that news reports often include wording such as, "The earthquake registered a 7.2 on the Richter scale." These days, however, seismologists use several different magnitude scales, not just the Richter scale, because the original Richter scale actually works well only for shallow, nearby earthquakes (specifically earthquakes whose epicenters are less than 600 km from the seismograph, and whose hypocenters are less than 15 km below the surface). Because of the distance limitation, a number on the original Richter scale is now called a local magnitude (M_L). A scale based on measuring the amplitudes of certain R-waves is called a surface-wave magnitude (M_s), and a scale to describe the size of earthquakes deeper than 50 km based on measurements of certain body waves is a body-wave magnitude (m_b).

The M_L, m_b, and M_s scales have limitations—they cannot accurately define the sizes of very large earthquakes, because for an earthquake above a given size, the scales give roughly the same magnitude number regardless of how large the earthquake really is. Because of this problem, seismologists developed the moment magnitude (M_w) scale. To calculate the moment magnitude, it is necessary to measure the amplitude of a number of different seismic waves, determine the area of the slipped portion of the fault that moved, determine how much slip occurred, and define physical characteristics of the rock that broke during faulting. Moment magnitude numbers may be different from other magnitude numbers for the same earthquake. For example, the largest recorded earthquake in history hit Chile in 1960 and had an M_s of 8.5 but an M_w of 9.5.

What magnitude is given in modern reports of earthquakes in the newspaper? It depends on how soon the article appears after an earthquake has taken place. For early reports, seismologists report a preliminary magnitude, which is an M_L, m_b, or M_s, because these magnitudes can be calculated fairly quickly. Later on, after they have had the chance to collect the necessary data, seismologists report an M_w, which becomes the number now generally used for archival records.

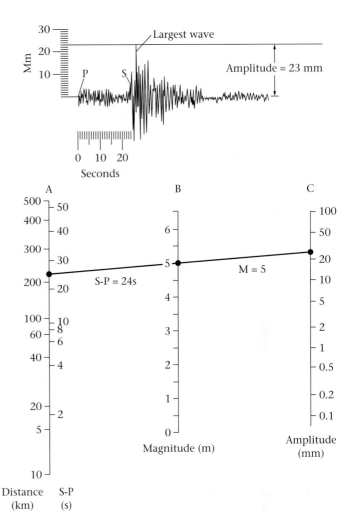

FIGURE 8.12 To calculate the Richter magnitude from a seismogram, first measure the S minus P time to determine the distance to the epicenter; then measure the height, or amplitude (in mm), of the largest wave recorded by the seismograph. Draw a line from the point on column A representing the S-minus-P time to the point on column C representing the wave amplitude, and read the Richter magnitude (m) off column B. Note that if the earthquake were much closer, then the same amplitude in the seismogram would yield a smaller-magnitude earthquake. We must take the distance to the epicenter into account because seismic waves grow smaller in amplitude with increasing distance from the epicenter.

All magnitude scales are logarithmic, meaning that an increase of one unit of magnitude represents a tenfold increase in the maximum ground motion. Thus, a magnitude 8 earthquake results in ground motion that is ten times greater than that of a magnitude 7 earthquake, and a thousand times greater than that of a magnitude 5 earthquake. To make life easier, seismologists use familiar adjectives to describe the size of an earthquake, as listed in ▶Table 8.2.

As we pointed out earlier, earthquakes release energy. Seismologists can calculate the energy release from equations

TABLE 8.2 Adjectives for Describing Earthquakes

Adjective	Magnitude	Approximate maximum intensity at epicenter	Effects
great	>8.0	X to XII	major to total destruction
major	7.0 to 7.9	IX to X	great damage
strong	6.0 to 6.9	VII to VIII	moderate to serious damage
moderate	5.0 to 5.9	VI to VII	slight to moderate damage
light	4.0 to 4.9	IV to V	felt by most; slight damage
minor	<3.9	III or smaller	felt by some; hardly any damage

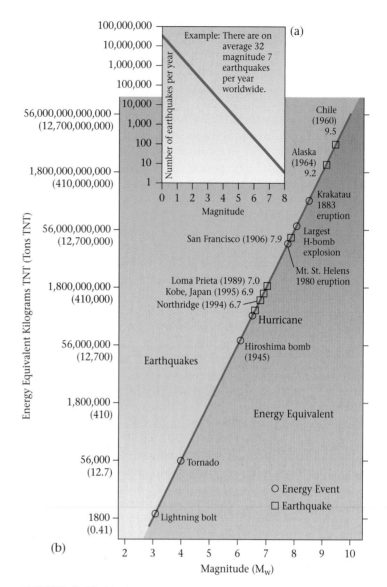

FIGURE 8.13 (a) This graph emphasizes that the energy released by an earthquake increases dramatically with magnitude. (b) This graph illustrates how the number of earthquakes of a given magnitude decreases with increasing magnitude. According to the graph, there are over 1 million $M_w = 1$ earthquakes per year, but only about 10,000 $M_w = 4$ earthquakes.

that relate moment magnitude to energy. Not all versions of this calculation yield the same result, so energy estimates must be taken as an approximation. According to some researchers, a magnitude 6 earthquake releases about as much energy as the atomic bomb that was dropped on Hiroshima in 1945, and the 1964 Alaska earthquake released significantly more energy than the largest hydrogen bomb ever detonated. Notably, an increase in magnitude by one integer represents approximately a thirty-two-fold increase in energy. Thus, a magnitude 8 earthquake releases about 1 million times more energy than a magnitude 4 earthquake (▶Fig. 8.13a). In fact, a single magnitude 8.9 earthquake releases as much energy as the entire *average* global annual release of seismic energy coming from all other earthquakes combined! Fortunately, such large earthquakes occur much less frequently than small earthquakes; there are about 100,000 magnitude 3 earthquakes every year, but a magnitude 8 earthquake happens only about once every three to five years (▶Fig. 8.13b).

8.5 WHERE AND WHY DO EARTHQUAKES OCCUR?

Earthquakes do not take place everywhere on the globe. By plotting the distribution of earthquake epicenters on a map, seismologists find that most, but not all, earthquakes occur in

fairly narrow **seismic belts,** or seismic zones. Most seismic belts define plate boundaries. Earthquakes within these belts are plate-boundary earthquakes. Ones that occur away from plate boundaries are intraplate earthquakes (the prefix *intra* means "within") (▶Fig. 8.14). Notably, 80% of the earthquake energy released on Earth comes from plate-boundary earthquakes in the belts surrounding the Pacific Ocean.

Similarly, earthquakes do not occur at random depths in the Earth. Seismologists distinguish three classes of earthquakes based on hypocenter (focus) depth: shallow-focus earthquakes occur in the top 20 km of the Earth,

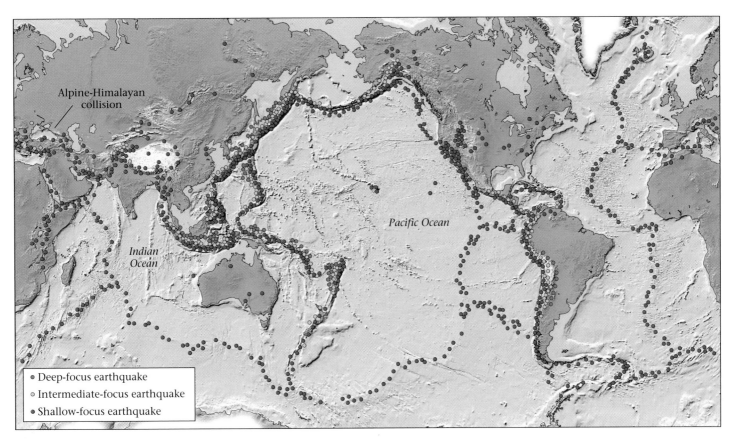

FIGURE 8.14 A simplified map of epicenters shows that most earthquakes occur in distinct belts that define plate boundaries, and most take place in the highly populated Pacific Rim region. Epicenters that do not lie along plate boundaries are the result of rifting or collision, or are in intraplate settings. Intermediate- and deep-focus earthquakes occur only along convergent plate boundaries, except for a few that happen in collisional zones.

intermediate-focus earthquakes take place between 20 and 300 km, and deep-focus earthquakes occur down to a depth of about 670 km. Earthquakes cannot happen deeper in the Earth, because rock at great depth is too ductile, so it flows instead of breaking when subjected to stress. In this section, we look at the characteristics of earthquakes in various geologic settings and learn why earthquakes take place where they do.

Earthquakes at Plate Boundaries

The majority of earthquakes happen at faults along plate boundaries; the relative motion between plates is accommodated primarily by slip on these faults. We find different kinds of faulting at different types of plate boundaries.

Divergent plate-boundary seismicity. At divergent plate boundaries (mid-ocean ridges), two oceanic plates form and move apart. Divergent boundaries are segmented, and spreading segments are linked by transform faults. There-fore, two kinds of faults develop at divergent boundaries. Along spreading segments, newly formed crust at or near the mid-ocean ridge stretches and ruptures, generating normal faults, whereas along the transform faults that link spreading segments, strike-slip faulting occurs (▶Fig. 8.15). Seismicity along mid-ocean ridges takes place at shallow depths.

Convergent-plate-boundary seismicity. Convergent plate boundaries are complicated regions at which several different kinds of earthquake take place. Shallow-focus earthquakes occur in both the subducting plate and the overriding plate. Specifically, as the downgoing plate begins to subduct, it bends and shears along the base of the overriding plate. Large thrust faults develop along the contact between the downgoing and overriding plates, and shear on these faults generates great earthquakes. In some cases, the push applied by the downgoing plate compresses the overriding plate and causes shallow-focus reverse faulting in the overriding plate.

In contrast to other types of plate boundaries, convergent plate boundaries also host intermediate- and deep-focus

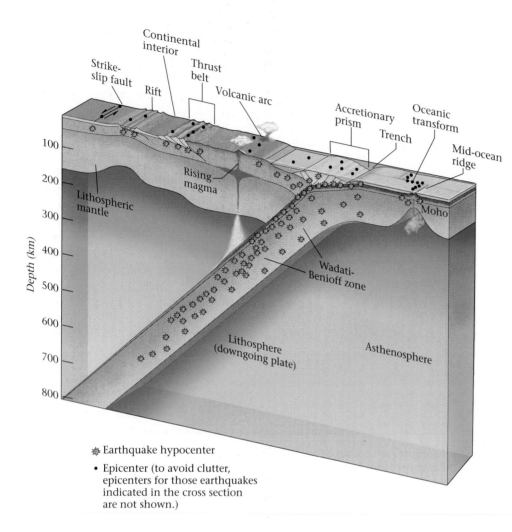

FIGURE 8.15 Earthquakes along oceanic divergent boundaries and transform boundaries are all shallow focus. Shallow-focus earthquakes also occur in the overriding plate of a convergent boundary. Earthquakes in the downgoing plate of a convergent boundary define the Wadati-Benioff zone and include intermediate- and deep-focus earthquakes.

earthquakes. These occur in the downgoing slab as it sinks into the mantle, defining the sloping band of seismicity called a **Wadati-Benioff zone,** after the seismologists who first recognized it (Fig. 8.15). Intermediate- and deep-focus earthquakes happen partly in response to stresses caused by shear between the downgoing plate and the mantle, and partly by the slab pull of the deeper part of the plate on the shallower part.

Considering the depth at which they occur, intermediate- and deep-focus earthquakes of a Wadati-Benioff zone present a problem: Shouldn't the Earth be too warm to undergo brittle deformation at such depths? Seismologists suggest that intermediate-focus earthquakes can happen because the downgoing plate takes a long time to heat up, and the rock in the plate's interior remains cool and brittle enough to break even down to about 300 km. Deep-focus earthquakes, however, remain a mystery. Some researchers suggest that they occur when minerals in the rock suddenly collapse as a result of great pressure, and the atoms re-arrange into new minerals that take up less space—the sudden change in volume would generate a shock.

The fate of subducting lithosphere below a depth of 670 km remains unclear. Recent evidence suggests that some plate material accumulates at 670 km, while some sinks still deeper into the mantle, but no longer generates earthquakes, because all minerals have completed the transition to more compact forms, and the rock is too warm to break in a brittle manner.

Earthquakes in southern Alaska, Japan, the western coast of South America, Mexico, and other places along the Pacific Rim are examples of convergent boundary earthquakes. One such event happened on Good Friday of 1964, near Anchorage, Alaska. This earthquake had a seismic-moment magnitude of 9.2, and its vibrations wrecked substantial parts of towns all along the coast (▶Fig. 8.16). In 1995, an earthquake with a magnitude (M_W) of 6.9 struck Kobe, Japan, causing immense devastation and killing over 5,000 people (▶Fig. 8.17a, b).

FIGURE 8.16 A portion of 4th Avenue in Anchorage, Alaska, collapsed during the 1964 earthquake, when the slope on which it was built gave way, and slid downhill.

Transform plate-boundary seismicity. At transform plate boundaries, where one plate slides past another without the production or consumption of oceanic lithosphere, most faulting results in strike-slip motion. The majority of trans-form faults in the world link segments of oceanic ridges (Fig. 8.15), but a few, such as the San Andreas Fault of California and the Alpine Fault of New Zealand, cut across continental lithosphere (▶Fig. 8.18). All transform-fault earthquakes have a shallow focus, so the larger ones on land can cause disaster.

As an example of a transform-fault earthquake, consider the slip of the San Andreas Fault near San Francisco in 1906 (▶Fig. 8.19a). In the wake of the gold rush, San Francisco was a booming city with broad streets and numerous large buildings. But it was built on the transform boundary along which the Pacific Plate moves north at an average of 6 cm per year relative to North America. Because of the stick-slip behavior of the fault, this movement doesn't occur smoothly but happens in little jerks, each of which causes an earthquake. At 5:12 A.M. on April 18, the fault slipped by as much as 6 m, and earthquake waves slammed into the city. Witnesses watched in horror as the street undulated like ocean waves. Buildings swayed and banged together, laundry lines stretched and snapped, church bells rang, and then towers, facades, and houses toppled. Judging from the damage, seismologists estimate that the largest shock would have registered a seismic moment magnitude of 7.9. Most building collapse took place downtown (▶Fig. 8.19b). Fire followed soon after, consuming huge areas of the city, for most buildings were made of wood. In the end, about five hundred people died, and a quarter of a million were left homeless.

FIGURE 8.17 (a) Simplified map of earthquake epicenters (black dots) and plate boundaries (subduction zones) in Japan. The size of the dots indicates magnitude. Note the location of Kobe. (b) A collapsed building after the Kobe earthquake.

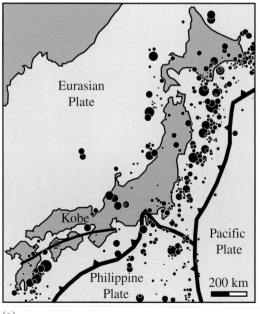

(a)　　　　　　　　　　　　　　　　(b)

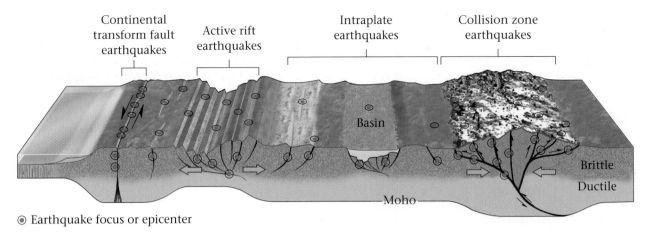

Continental transform fault earthquakes   Active rift earthquakes   Intraplate earthquakes   Collision zone earthquakes

Basin

Moho

Brittle
Ductile

⊚ Earthquake focus or epicenter

FIGURE 8.18 Earthquakes occur in continental transform faults (such as the San Andreas), in continental rifts (the East African Rift, the Basin and Range Province), in intraplate settings (usually by reactivating old faults), and in collision zones (mountain ranges). Continental earthquakes mostly happen in the brittle crust, at depths of less than about 15 km.

The San Francisco earthquake has not been the only one to strike along the San Andreas and nearby related faults. Over a dozen major earthquakes have happened on these faults during the past two centuries, including the 1857 magnitude 8.7 earthquake just east of Los Angeles, and the 1989 magnitude 7.1 Loma Prieta earthquake, which occurred 100 km south of San Francisco but nevertheless shut down a World Series game and caused the collapse of a double-decker freeway.

Earthquakes at Continental Rifts and Collision Zones

Continental rifts. The stretching of continental crust at continental rifts generates normal faults (see Fig. 8.18). Active rifts today include the East African Rift, the Basin and Range Province (mostly in Nevada, Utah, and Arizona), and the Rio Grande Rift (in New Mexico). In all these places, shallow-

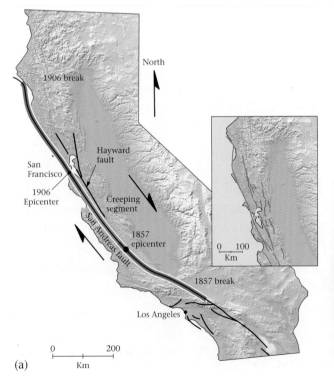

(a)

(b)

FIGURE 8.19 (a) Map showing the portion of the San Andreas Fault that ruptured during the 1906 earthquake. A large portion of the fault in southern California ruptured during the 1857 Fort Tejon earthquake ($M_w = 8.0$). (b) A street in San Francisco after the 1906 earthquake.

focus earthquakes occur, similar in nature to the earthquakes at mid-ocean ridges. But in contrast to mid-ocean ridges, these seismic zones can be located under populated areas, like Salt Lake City, and thus cause major damage.

Collision zones. Two continents collide when the oceanic lithosphere that once separated them has been completely subducted. Such collisions produce great mountain ranges such as the Alpine-Himalayan chain. Though a variety of earthquakes happen in collision zones, the most common result from movement on thrust faults (Fig. 8.18).

The complex collision between Africa and Europe generated the stress that triggered an earthquake near Lisbon, Portugal, on All Saints' Day (November 1), 1755. According to an eyewitness account,

> The first alarm was a rumbling noise that sounded like exceptionally heavy traffic. Then there was a brief pause and a devastating shock followed, lasting over two minutes, that brought down roofs, walls, facades of churches, palaces and houses and shops in a dreadful deafening roar of destruction. Close on this came a third trembling to complete the disaster, and then a dark cloud of suffocating dust settled fog-like on the ruins of the city.[1]

About fifteen minutes later, fires began to spread and consume the rubble, and then the waters of the harbor "rocked and rose menacingly, and then poured in three great towering waves over its banks."[1] Aftershocks lasted for days. Innumerable artworks by great Renaissance masters were destroyed, along with a library that housed all the records of Portuguese exploration. The effects of the Lisbon earthquake went beyond the immediate destruction. Voltaire (1694–1778), the great French writer, immortalized the earthquake in his novel *Candide*. How, pondered Voltaire, could this world be the best of all possible worlds if such a calamity could happen and ruin good people? The earthquake, and Voltaire's writing about it, brought to an end the "era of optimism" and ushered in the "era of realism" in Europe.

Stress resulting from a collision between the Indian subcontinent and Asia caused the catastrophic magnitude 7.6 earthquake of October, 2005, in the Kashmir region along the border of India and Pakistan. In this region of poorly constructed homes, ground shaking resulted in the collapse of whole towns. Associated landslides ripped out roads, denying access to potential rescuers. At least 86,000 people perished, and another 4 million were left homeless. Because of the high elevation and inaccessibility of the epicentral region, victims continued to perish long after the earthquake due to lack of food, shelter, and medical care.

Intraplate Earthquakes

Some earthquakes occur in the interiors of plates and are not associated with plate boundaries, active rifts, or collision zones (Fig. 8.18). These **intraplate earthquakes** account for only about 5% of the earthquake energy released in a year. Almost all have a shallow focus. Seismologists are still trying to understand the causes of intraplate earthquakes. Most favor the idea that force applied to the boundary of the plate can cause the interiors of plates to break at weak, preexisting fault zones, some of which first formed in the Precambrian.

Because the crust in an intraplate setting tends to be composed of stronger rock, overall, than the crust along a plate boundary, it tends to transmit seismic waves more efficiently than does plate-boundary crust. As an analogy, if you were to bang a hammer on a block of solid rock and on a block of styrofoam, you would feel the impact on the other side of the rock, but not on the other side of the styrofoam because the impact would be absorbed by the styrofoam. Thus, intraplate earthquakes can be felt over a very broad region.

In Europe, a number of intraplate earthquakes have been recorded. For example, an earthquake with a magnitude of 4.8 hit central England, near Birmingham, in 2002. In North America, intraplate earthquakes occur in the vicinity of New Madrid, Missouri; Charleston, South Carolina; eastern Tennessee; Montréal, Quebec; and the Adirondack Mountains, New York. Several large events have happened in these locations. For example, a magnitude 7.3 earthquake occurred near Charleston in 1886, ringing church bells up and down the coast and vibrating buildings as far away as Chicago (Fig. 8.11). In Charleston itself, over 90% of the buildings were damaged, and sixty people died. In the early nineteenth century, the region of New Madrid, which lies near the Mississippi River, was inhabited by a small population of Native Americans and an even smaller population of European descent (▶Fig. 8.20a, b). During the winter of 1811–1812, three magnitude 8 to 8.5 earthquakes struck the region. In the words of a settler:

> The Mississippi first seemed to recede from its banks, and its waters gathered up like a mountain, leaving for a moment many boats on the bare sand. . . . Then, rising 15 to 20 feet perpendicularly . . . the banks overflowed with a retrograde current rapid as a torrent. The boats were now torn from their moorings . . . the river took with it whole groves of young cottonwood trees.[2]

The earthquakes resulted from slip on faults in a Precambrian rift that underlies the Mississippi Valley.

---

[1]James R. Newman, "The Lisbon Earthquake," in Rhodes and Stone, eds., *Language of the Earth* (New York: Pergamon Press, 1981), p. 59.

[2]John Gribbin, *This Shaking Earth: Earthquakes, Volcanoes, and Their Impact on Our World* (New York: Putnam, 1978), p. 15.

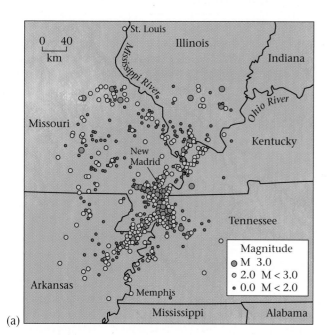

FIGURE 8.20 (a) The epicenters of recent small earthquakes in the New Madrid, Missouri, area, recorded by modern seismic instruments. The earthquakes, most of which are minor, appear to occur along two sets of ancient, weak faults that break in response to stress in the interior of North America. The New Madrid zone is not a plate boundary. (b) The New Madrid earthquake destroyed cabins and disrupted the Mississippi River.

8.6 DAMAGE FROM EARTHQUAKES

An area ravaged by a major earthquake is a heartbreaking sight. The terror and sorrow etched on the faces of survivors mirror the inconceivable destruction. This destruction comes as a result of many processes.

Ground Shaking and Displacement

An earthquake starts suddenly and may last from a few seconds to a few tens of seconds. Different kinds of earthquake waves cause different kinds of ground motion (▶Fig. 8.21a–d). If the motions are large enough, they may knock you down or even throw you in the air. In great earthquakes, the ground's movement can have an amplitude of as much as 1 m, but in moderate earthquakes, motions fall in the range of a few centimeters or less.

The nature and severity of the shaking at a given location depend on four factors: (1) the magnitude of the earthquake, because larger-magnitude events release more energy; (2) the distance from the hypocenter, because earthquake energy decreases as waves pass through the Earth; (3) the nature of the substrate at the location (that is, the character and thickness of different materials beneath the ground sur-

face); in places underlain by unconsolidated sediment or landfill, earthquake waves tend to be amplified and thus cause more ground motion than they would in hard, coherent bedrock; and (4) the frequency of the earthquake waves (where frequency equals the number of oscillations that pass a point in a specified interval of time).

If you're out in an open field during an earthquake, ground motion alone won't kill you—you may be knocked off your feet and bounced around a bit, but your body is too flexible to break. Buildings aren't so lucky (▶Fig. 8.22a–g). When earthquake waves pass, buildings sway, twist back and forth, or lurch up and down, depending on the type of wave motion. As a result, connectors between the frame and facade of a building may separate, so the facade crashes to the ground. The flexing of walls shatters windows and makes roofs collapse. Floors may rise up and slam down on the columns that support them. This motion can crush the columns, so some buildings collapse with their floors piling on top of one another like pancakes in a stack (▶Fig. 8.23a). The majority of earthquake-related deaths and injuries result when people are hit by flying debris or are crushed beneath falling walls or roofs. Aftershocks worsen the problem, because they may be the final straw that breaks already weakened buildings, trapping rescuers.

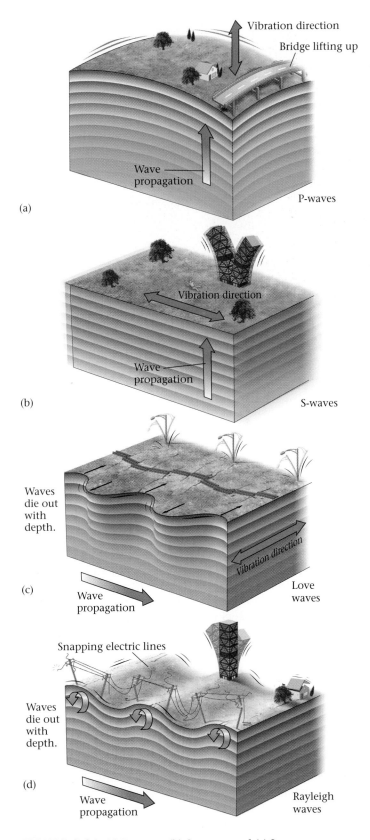

FIGURE 8.21 (a) P-waves, (b) S-waves, and (c) Love waves cause the ground to undulate laterally. (d) Rayleigh waves make the ground undulate in a rolling motion, like the sea surface.

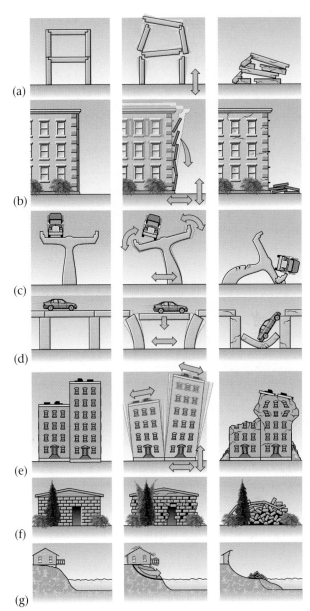

FIGURE 8.22 Shaking causes (a) a concrete-slab building (or bridge) to disconnect and collapse, (b) a building's facade to fall off, (c) a poorly supported bridge to topple, (d) a bridge span to disconnect and collapse, (e) neighboring buildings to collide and shatter (floors inside a tall building may collapse), (f) a concrete-block, brick, or adobe building to crack apart and collapse, (g) a steep cliff to collapse, carrying buildings with it.

Roads, rail lines, and bridges suffer the same fate as buildings: ground motion breaks and buckles them (▶Fig. 8.23b, c). In the case of bridges, the span may rise off its supports, then slam down so the bridge collapses, or the span may slip off the supports. For unlucky drivers on the bridge, the road literally disappears in front of them.

(a)

(b)

(c)

FIGURE 8.23 (a) During a 1999 earthquake in Turkey, this concrete building collapsed when the supports gave way and piled onto each other like pancakes. (b) Reinforced concrete bridge supports were crushed during the 1994 Northridge earthquake. (c) An elevated bridge, balanced on a single row of columns, simply tipped over during the 1995 Kobe, Japan, earthquake.

If the ruptured portion of a fault intersects the ground surface, it displaces the ground surface. Strike-slip faults cause lateral displacements, whereas normal or reverse faults cause vertical displacements and fault scarps. If a building, fence, road, pipeline, or rail line straddles a fault trace, slippage on the fault can crack the structure and separate it into two pieces.

Landslides

The shaking of an earthquake can cause ground on steep slopes or ground underlain by weak sediment to give way. This movement results in a landslide, the tumbling and flow of soil and rock downslope. Landslides occur commonly along the coast of California, for movement on faults has rapidly uplifted this coastline in the past few million years, resulting in the development of steep cliffs. When earthquakes take place, the cliffs collapse, often carrying expensive homes down to the beach below (▶Fig. 8.24). Such events lead to the misperception that "Cali-

fornia will someday fall into the sea." Although small portions of the coastline do collapse, the state as a whole remains firmly attached to the continent, despite what Hollywood scriptwriters say.

Sediment Liquefaction

Places where the substrate consists of wet sediment can be particularly hazardous during an earthquake. In beds of wet sand or silt, ground shaking causes the sediment grains to try and settle together. But because the spaces (pores) between grains are filled with water, this can't happen; instead, the water pressure in the pores increases and pushes the grains apart, and the wet silt or sand becomes a fluid-like slurry. Ground shaking similarly transforms wet clay into a viscous liquid. The abrupt loss of strength of a wet sediment (either sand or clay) in response to ground shaking is a phenomenon called **liquefaction,** and it can cause major damage during an earthquake. Buildings whose foundations lie in liquefied material may sink or even tip over (▶Fig. 8.25).

FIGURE 8.24 A landslide along the California coast, triggered by the Northridge earthquake, carried part of a luxury home down to the beach below.

FIGURE 8.25 The foundation of this apartment complex in Hsingchung, Taiwan, liquefied during a 1999 earthquake, and the buildings tipped over, intact.

If liquefaction occurs in the sediment beneath a slope, the ground above may give way and slide down the slope. Such an event happened during the 1964 Alaska earthquake, when the sediment beneath a housing development in the coastal suburb of Turnagain Heights liquefied (▶Fig. 8.26a–c). The land broke into a series of blocks that slid seaward on the liquefied sediment, carrying with it dozens of houses that were transformed into a jumble of splintered wood and shattered windows.

In some cases, the liquefaction of sand layers below the ground surface makes the sand erupt through holes or cracks in overlying sediment, producing small mounds of sand called sand volcanoes or sand boils (▶Fig. 8.27a). Sometimes fountains of wet sand spurt 10 m into the sky during an earthquake. Liquefaction may also cause bedding in unconsolidated sequences of sediment to break up, and ground settling due to underlying liquefaction may cause large cracks to develop in the overlying sediments.

Fire

The shaking during an earthquake can make lamps, stoves, or candles with open flames tip over, and it may break wires or topple power lines, generating sparks. As a consequence, areas already turned to rubble, and even areas not so badly damaged, may be consumed by fire. Ruptured gas pipelines

and oil tanks feed the flames, sending columns of fire erupting skyward (▶Fig. 8.28). Firefighters might not even be able to reach the fires, because the doors to the firehouse won't open or rubble blocks the streets. Moreover, firefighters may find themselves without water, for ground shaking and landslides damage water lines.

If a fire gets a good start, it can become an unstoppable inferno. Most of the destruction of the 1906 San Francisco earthquake, in fact, resulted from fire. For three days, the blaze spread through the city until firefighters contained it by blasting a firebreak. By then, five hundred blocks of structures had turned to ash, causing twenty times as much financial loss as the shaking itself. When a large earthquake hit Tokyo in September 1923, fires set by cooking stoves spread quickly through the wood-and-paper buildings, creating an inferno that heated the air above the city. The hot air rose like a balloon, and when cool air rushed in, creating wind gusts of over 100 mph, the wind stoked the blaze and incinerated 120,000 people.

Tsunamis

The azure waters and palm-fringed islands of the Indian Ocean's east coast hide one of the most seismically active plate boundaries on Earth—the Sunda trench. Along this convergent boundary, the Indian Ocean floor subducts at about 6 cm per year, leading to slip on large thrust faults. Just before 8:00 A.M. on December 26, 2004, the crust above an 1300-km-long by 100-km-wide portion of one of these faults lurched westward by as much as 15 m. The rupture started at the

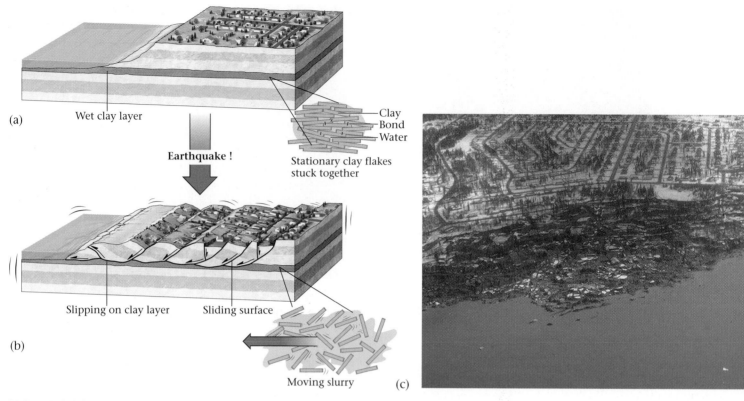

(a) Wet clay layer

Clay
Bond
Water

Stationary clay flakes
stuck together

Earthquake !

Slipping on clay layer   Sliding surface

(b)

Moving slurry   (c)

FIGURE 8.26 The Turnagain Heights (Alaska) disaster. (a) Before the earthquake, wet clay packed together in a subsurface layer of compacted but wet mud. (b) Ground shaking caused liquefaction of the wet clay layer in the sediment beneath a housing development. As a consequence, the land slumped and slid seaward on an array of sliding surfaces, carrying the houses with it. The ground broke into many slices. (c) The housing development after the slide.

FIGURE 8.27 (a) Ground shaking may cause wet sand buried below the surface to squirt up between cracks in overlying compacted mud. The result is a row of sand volcanoes. (b) Photo of ground cracking due to liquefaction of sediment underground.

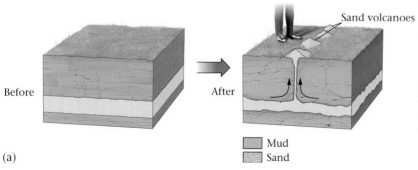

Before

After

Sand volcanoes

Mud
Sand

(a)   (b)

FIGURE 8.28 Broken gas mains erupt fountains of flame after the Northridge, California earthquake.

epicenter, and then propagated north at 2.8 km/s; thus, the rupturing process took nine minutes. This slip triggered a great earthquake $M_w = 9.3$) and pushed the sea floor up by tens of centimeters. The rise of the sea floor, in turn, shoved up the overlying water. Because the area that rose was so broad, the volume of displaced water was immense. As a consequence, tragedy of unimaginable extent was about to unfold.

Water from above the upthrust sea floor began moving outward from above the fault zone, a process that generated a series of giant waves, or **tsunami,** traveling at speeds of about 800 km/hour (500 mph)—almost the speed of a jet plane (▶Fig. 8.29a–f). Fifteen minutes later, the first tsunami

struck Banda Aceh, a city at the north end of the island of Sumatra, as it was waking to a beautiful, cloudless day. First, the sea receded much farther than anyone had ever seen, exposing large areas of reefs that normally remained submerged even at low tide. People walked out onto the exposed reefs in wonder. But then, with a rumble that grew to a roar, a wall of frothing water began to build in the distance and approach land. Puzzled bathers first watched, and then ran inland in panic when the threat became clear (▶Fig. 8.30a). As the tsunami approached shore, friction with the seafloor had slowed it to less than 30 km/hour, but it still moved faster than people can run. In places, the wave front reached heights of 15 to 30 m (45 to 100 feet) as it slammed into Banda Aceh (▶Fig. 8.30b).

The impact of the water ripped boats from their moorings, snapped trees, battered buildings into rubble, and tossed cars and trucks like toys. And the water just kept coming, eventually flooding low-lying land as far as about 7 km inland. It drenched forests and fields with salt water (deadly to plants) and buried fields and streets with up to a meter of sand and mud (▶Fig. 8.30c, d). Eventually the water slowed and then began to rush back to the shore, but at Banda Aceh, at least two more tsunamis struck before the first one had entirely receded, so the water remained high for some time. When the water level finally returned to normal, a jumble of flotsam, as well as the bodies of unfortunate victims, floated out to sea and drifted away forever.

Geologists refer to the tsunami that struck Banda Aceh as a near-field (or local) tsunami, because of its proximity to the earthquake. But the horror of Banda Aceh was merely a preamble to the devastation that would soon visit other

FIGURE 8.29 (a) Before a tsunami forms. (b) Sinking of the sea floor above a normal fault creates a void, and water rushes to fill it. Water mounds up over the fault. (c) A similar process happens in response to a reverse or thrust fault. (d) This time, the rising sea floor shoves up the water surface.

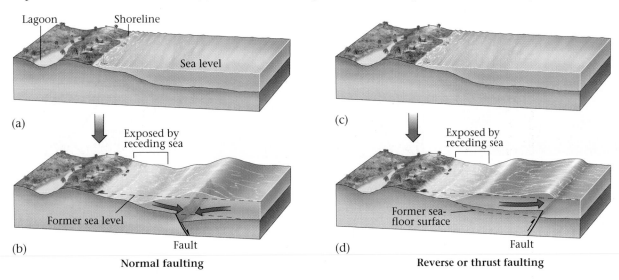

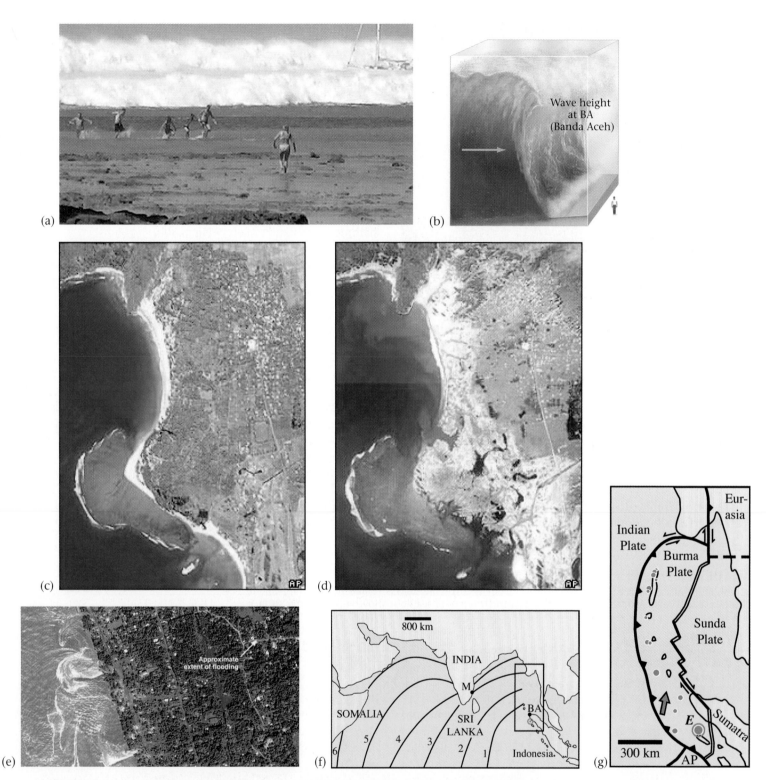

FIGURE 8.30 The Indian Ocean tsunami disaster of 2004. (a) A tourist took this photo of the tsunami approaching the shore of Sumatra. (b) At its highest, the wave front was many times the height of a person. (c, d) Satellite photos of the Indonesian province of Aceh before and after the tsunami struck. Note that the city was washed away, and that the beach vanished. (e) A satellite photo of the tsunami submerging land up to 500 m inland of the coastline of Sri Lanka; (f) A map of the Indian Ocean showing the position of the wave at various times after the earthquake. The enlargement of the boxed area (g) shows the configuration of plate boundaries near the epicenter.

stretches of Indian Ocean coast. Far-field (or distant) tsunamis crossed the ocean and struck Sri Lanka 2.5 hours after the earthquake, the coast of India half an hour later, and the coast of Africa, on the west side of the Indian Ocean, 5.5 hours after the earthquake (▶Fig. 8.30d, e). Coastal towns vanished, fishing fleets sank, and beach resorts collapsed into rubble. In the end, more than 220,000 people died that day.

Tsunami is a Japanese word that translates literally as "harbor wave," an apt name because tsunamis can be partially damaging to harbor towns. This name replaces the older term, "tidal wave." Tsunamis have no relationship to tides, and calling them tidal waves can be misleading. Tsunamis can be caused by undersea earthquakes but, as we will see in Chapter 13, they can also be set off by large undersea landslides.

Regardless of cause, tsunamis are very different from familiar, wind-driven storm waves. Large wind-driven waves can reach heights of 10 to 30 meters in the open ocean. But even such monsters have wavelengths of only tens of meters, and thus only affect the surface realm of the ocean. (Water beneath a wave moves down to a depth of about half a wavelength, so a submarine cruising at 100 m depth will move through calm water, even if a storm rages above.) In contrast, although a tsunami in deep water may cause a rise in sea level of at most only a few tens of centimeters—a ship crossing one wouldn't even notice—tsunamis have wavelengths of tens to hundreds of kilometers and thus affect the entire depth of the ocean. In simpler terms, we can think of the width of a tsunami, in map view, as being more than 100 times the width of a wind-driven wave. Because of this difference, a storm wave and a tsunami have very different effects when they strike the shore.

When a water wave approaches the shore, friction between the base of the wave and the sea floor slows the bottom of the wave, so the back of the wave catches up to the front, and the added volume of water builds the wave higher (▶Fig. 8.31a). The top of the wave may fall over the front of the wave and cause a breaker. In the case of a wind-driven wave, the breaker may be tall when it washes onto the beach, but the wave is so narrow that it doesn't contain much water. Thus, the wave makes it only part way up the beach before it runs out of water, friction slows it to a stop, and gravity causes the water to spill back seaward. In the case of a tsunami, the wave is so wide that, as friction slows the wave, it builds into a plateau that can be tens of meters high and many kilometers wide. Thus, when a tsunami reaches shore, it contains so much water that it crosses the beach and just keeps on going (▶Fig. 8.31b, c).

The 2004 Indian Ocean event remains etched in people's minds because of the immense death toll and the nonstop news coverage. But it is not unique. Tsunamis generated by the great Chilean earthquake of 1960 ($M_w = 9.5$) destroyed coastal towns of South America and crossed the

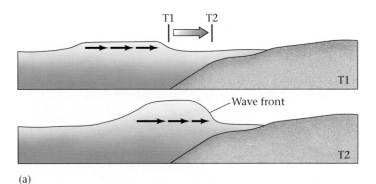

(a)

Wind-driven waves contain a small volume of water, and do not submerge higher areas.

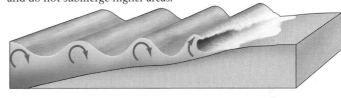

(b)

Tsunamis are so wide (measured perpendicular to shore) that, like a plateau of water, they submerge the land.

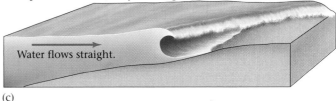

Water flows straight.

(c)

FIGURE 8.31 (a) As a tsunami approaches the shore, friction between water and the sea floor slows the wave, allowing the back part of the wave to catch up with the front part. As a consequence, the wave becomes much higher than it was in the open ocean as the front moves from its position at Time 1 to its position at Time 2. (b) Wind-driven waves may be quite high, but they are so narrow that they contain relatively little water, and they only go part-way up the beach until the water stops moving landward. The water in such a wave has a circular motion. (c) When a tsunami washes ashore, the water moves landward in a straight line. A tsunami is like a plateau of water—it contains so much water that the water keeps moving inland and submerges a broad area.

Pacific, causing a 10.7-m-high wall of water to strike Hawaii 15 hours later. When, 21 hours after the earthquake, the tsunami reached Japan, it flattened coastal villages and left 50,000 people homeless. The tsunami following the 1964 Good Friday earthquake in Alaska destroyed ports at Valdez and Kodiak (▶Fig. 8.32a). Typically, an earthquake generates several tsunamis that arrive on distant shores as much as an hour apart (▶Fig. 8.32b).

Because of the danger of tsunamis, predicting their arrival can save thousands of lives. A tsunami warning center in Hawaii keeps track of earthquakes around the Pacific and uses data relayed from tide gauges and sea-floor pressure gauges to

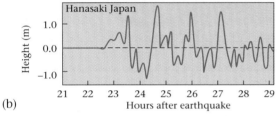

(a)

(b)

FIGURE 8.32 (a) In the aftermath of a tsunami that hit Kodiak, Alaska, after the 1964 earthquake, boats rest amid the rubble of a coastal town. (b) Tidal gauge at Hanasaki, Japan, twelve hours later.

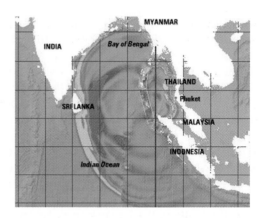

FIGURE 8.33 A computer model showing the 2004 Indian Ocean tsunami about 2 hours after the earthquake. The yellow areas have the greatest wave height.

determine whether a particular earthquake has generated a tsunami. If observers detect a tsunami, they flash warnings to authorities around the Pacific. In recent years, to help in the effort, researchers have used computer models that predict how tsunamis propagate (▶Fig. 8.33). Unfortunately, no warning system existed in the Indian Ocean, and even though Hawaiian observers detected the earthquake and realized that a tsunami was likely, they did not have the means to contact many local authorities. Even if they had, affected towns had no evacuation plans in place. A multinational team has begun work to remedy this situation.

Disease

Once the ground shaking and fires have stopped, disease may still threaten lives in an earthquake-damaged region. Earthquakes cut water and sewer lines, destroying clean water supplies and exposing the public to bacteria, and they cut transportation lines, preventing food and medicine from reaching the area. The severity of such problems depends on the ability of emergency services to cope.

8.7 PREDICTING THE "BIG ONE"

We have seen that large earthquakes occurring near population centers cause catastrophe. Needless to say, many lives could be saved if only it were possible to know exactly when and where an earthquake will happen, so people could evacuate dangerous buildings, turn off gas and electricity, and, with more warning, build stronger structures.

Can seismologists predict earthquakes? The answer depends on the time frame of the prediction. With our present understanding of the distribution of seismic zones and the frequency at which earthquakes occur, we can make long-term predictions (on the time scale of decades to centuries). For example, with some certainty, we *can* say that a major earthquake will rattle California during the next one hundred years, and that a major earthquake won't strike central Canada during the next ten years. But despite extensive research, seismologists *cannot* make accurate short-term predictions (on the time scale of hours to weeks). Thus, we cannot say, for example, that an earthquake will happen in Montreal at 2:43 P.M., January 17, 2010.

In this section, we look at the scientific basis of both long- and short-term predictions and consider the consequences of a prediction. Seismologists refer to studies leading to predictions as seismic-risk, or seismic-hazard assessment. On the basis of this work, they produce maps that assign levels of seismic risk to different regions.

Long-Term Predictions

When making a prediction, we use the word *probability,* because a prediction gives only the likelihood of an event. For example, a seismologist may say, "The probability of a major earthquake occurring in the next twenty years in this state is twenty percent." This sentence implies that there's a one-in-five chance that the earthquake will happen during a twenty-year period. Urban planners and civil engineers can use long-term predictions to help create building codes for a region—stronger buildings make sense for regions with greater seismic risk. They may also use predictions to determine whether or not to build vulnerable structures such as nuclear power plants, hospitals, or dams in potentially seismic areas. Seismologists base long-term earthquake predictions on two pieces of information: the identification of seismic zones and the **recurrence interval** (the average time between successive events).

To identify a seismic zone, seismologists produce a map showing the epicenters of earthquakes that have happened during a set period of time (say, thirty years). Clusters or belts of epicenters define the seismic zone. The basic premise of long-term earthquake prediction can be stated as follows: a region in which there have been many earthquakes in the past will likely experience more earthquakes in the near future. Seismic zones, therefore, are regions of greater seismic risk. This doesn't mean that a disastrous earthquake can't happen far from a seismic zone—they can and do—but the risk is less.

Epicenter maps can be produced using only data from the past fifty years or so, because before that time seismologists did not have enough seismographs to locate epicenters accurately. Fortunately, geologists can assess seismic risk by examining landforms for evidence of recent faulting. For example, the presence of a distinct fault scarp in a landscape indicates that faulting has happened so recently that erosion has not yet had time to grind away the evidence (▶Fig. 8.34a).

To determine the recurrence interval for large earthquakes in a seismic zone, seismologists must determine when large earthquakes happened there in the past. Since the historical record does not provide information far enough back in time, we must study geologic evidence for great earthquakes. For example, we can examine sedimentary strata near a fault to find layers of sand volcanoes and disrupted bedding in the stratigraphic record. Each layer, whose age can be determined by using radiocarbon dating of plant fragments, records the time of an earthquake. We can obtain additional information by dating offset soil horizons now buried beneath other sediment (▶Fig. 8.34b). With such information, seismologists are able to produce regional maps illustrating seismic risk (▶Fig. 8.35a–c).

Seismologists suspect that places called seismic gaps, where a known active fault has not slipped for a long time, may be particularly dangerous (▶Fig. 8.35c). In a seismic gap, either the fault must be moving nonseismically, or the stress must be building up to be released by a major earthquake in the future.

Short-Term Predictions

Short-term prediction, which could lead to such precautions as evacuating dangerous buildings, shutting off gas and electricity, and readying emergency services, is not reliable and

FIGURE 8.34 The effects of recent fault movement on the land surface. (a) A normal fault truncates ridges, creating triangular facets. (b) Here, earthquake events are represented by disrupted bedding (1), an offset ancient soil horizon, or paleosol (2), a layer of sand volcanoes (3), and a bent tree caused by the formation of a fault scarp (4). We can date these events from the wood fragments in the strata. If the disrupted bedding formed in 100 B.C.E., the offset paleosol in 800 C.E., the sand volcanoes in 1340, and the tilted tree in 1950, then the time gaps between earthquakes are 900, 540, and 610 years. The recurrence interval is 683 years (900 + 540 + 610 divided by 3), with a large degree of uncertainty.

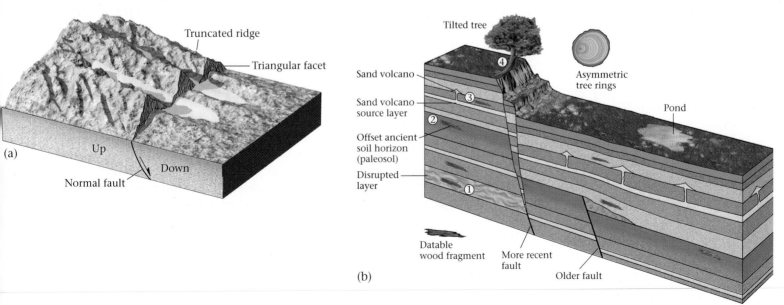

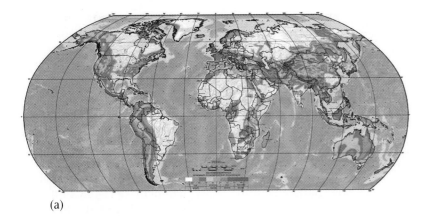

(a)

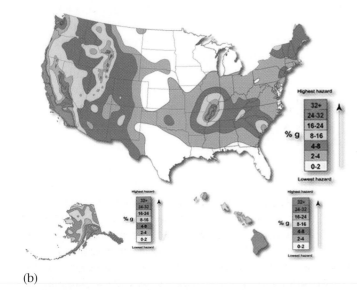

(b)

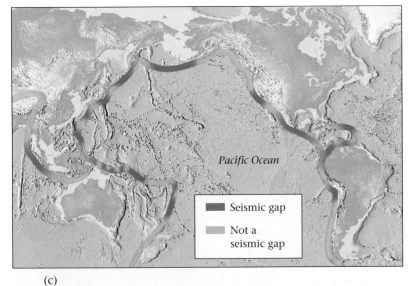

(c)

FIGURE 8.35 (a) A global seismic hazard map. The regions with darker (redder) colors have a greater probability of experiencing a large earthquake. (b) Map of seismic hazard in the United States. Darker (redder) colored areas are regions facing the greatest hazard. (c) Earthquakes may be more likely to occur in the seismic gaps around the Pacific in the near future.

may never be. Nevertheless, there are clues to imminent earthquakes, and seismologists have been working hard to understand them. The first clue comes from the detection of foreshocks. A swarm (a cluster of events during a short period of time) of foreshocks may indicate the cracking that precedes a major rupture along a fault zone. In this regard, foreshocks are analogous to the cracking noises you hear just before a tree limb breaks off and falls down. But foreshocks do not always occur, and even if they do, they usually can be identified only in hindsight, because they may be indistinguishable from other small earthquakes.

Another possible data source for short-term predictions comes from the precise laser surveying of the ground. Before an earthquake, a region of crust may undergo distortion, either in response to the buildup of elastic strain in the rock, or because of the development of small, open cracks that cause the crust to increase in volume. The land surface may bulge or sink, or straight lines on the ground may become bent. The detection of such movements by laser surveying can hint at an upcoming earthquake. More recently, researchers have been able to use satellite data to detect distortion of the land surface.

Recently, geologists have begun to use computer models of stress to predict where stress buildups may lead to earthquakes. Use of such stress-triggering models has provided substantial insight into seismic activity along the North Anatolian Fault, in Turkey (▶Fig. 8.36a). This fault is a large strike-slip fault along which Turkey slips westward. Earthquakes happen again and again along the fault—in fact, ruins of several ancient cities lie along the trace of the fault. Since 1939, 11 major earthquakes have occurred along the fault—each ruptured a different portion of the fault (▶Fig. 8.36b). Overall, the progression of faulting has been westward. By modeling the stress changes resulting from each earthquake, geologists have determined new places where stress accumulates and where earthquakes are likely.

Other changes that have been explored but have not yet proved to be precursors of earthquakes include the following: changes in the water level in wells; appearance of gases, such as radon or helium, in wells; changes in the electrical conductivity of rock underground; and unusual animal behavior (for instance, dogs howling). Believers in these proposed clues suggest that they all reflect the occurrence of cracking in the crust prior to an earthquake, but most investigators remain skeptical.

As long as short-term predictions remain questionable, emergency service planners must ask: "What if a prediction is wrong?" Should schools and offices be shut because of a prediction? Should millions be spent to evacuate people? Should a city be deserted,

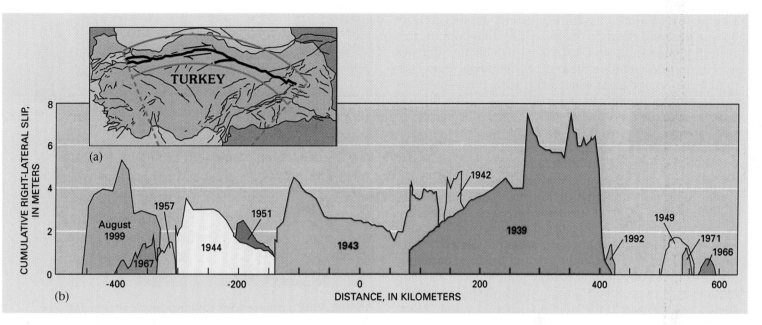

FIGURE 8.36 (a) A map of Turkey, showing the North Anatolian fault. (b) A graph representing regions that slipped during various earthquakes. The horizontal axis represents location along the fault, and the vertical axis represents the amount of slip.

allowing for the possibility of looters? Should the public be notified, or should only officials be notified, creating a potential for rumor? If the prediction proves wrong, can seismologists be sued? No one really knows the answers to these questions.

8.8 EARTHQUAKE ENGINEERING AND ZONING

The loss of human life from an earthquake of a given size varies widely. The loss depends on a number of factors, most notably the proximity of an epicenter to a population center, the depth of the hypocenter, the style of construction in the epicentral region, whether or not the earthquake occurred in a region of steep slopes or along the coast, whether building foundations are on solid bedrock or on weak substrate, whether the earthquake happened when people were outside or inside, and whether the government was able to provide emergency services promptly.

For example, a 1988 earthquake in Armenia was not much bigger than a 1971 earthquake in southern California, but it caused almost five hundred times as many deaths (24,000 versus 50). The difference in death toll reflects differences in the style and quality of construction and the characteristics of the substrate. The unreinforced concrete-slab buildings and masonry houses of Armenia collapsed, whereas the structures in California had, by and large, been erected according to building codes that take into account stresses caused by earthquakes. Most flexed and twisted but did not fall down and crush people. The terrible 1976 earthquake in T'ang-shan, China, killed over a quarter of a million people because the ground beneath the epicenter had been weakened by coal mining and collapsed, and because buildings were so poorly constructed. Mexico City's 1985 earthquake proved disastrous because the city lies over a sedimentary basin whose composition and bowl-like shape focused seismic energy, and caused buildings of a certain height to resonate (vibrate particularly strongly). During the 1989 Loma Prieta quake in California, portions of Route 880 in Oakland that were built on a weak substrate collapsed, whereas portions built on bedrock or gravel remained standing.

The differences in the destructiveness of earthquakes demonstrate that we can mitigate or diminish their consequences by taking sensible precautions. Clearly, earthquake engineering (the designing of buildings that can withstand shaking) and earthquake zoning (the determination of where land is stable and where it might collapse) can help save lives and property.

In regions prone to large earthquakes, buildings should be constructed so they are able to withstand vibrations without collapsing (▶Fig. 8.37a–c). They should be somewhat flexible so that ground motions can't crack them, and supports should be strong enough to maintain loads far in excess of the loads caused by the static (nonmoving) weight of the building. Bridge support columns should also be constructed with earthquakes in mind. Wrapping steel cables around the columns makes them many times stronger. Bolting the bridge spans to the top of a column prevents the spans from bouncing off. Concrete-block buildings, unreinforced concrete, and unreinforced brick buildings crack and tumble under conditions where wood-frame, steel-girder, or reinforced concrete buildings remain standing. Traditional heavy, brittle tile roofs shatter and bury the inhabitants inside, whereas sheet-metal or asphalt shingle roofs do not. Loose decorative stone and huge open-span roofs do not fare well when vibrated and should be avoided.

Similarly, developers should avoid construction on land underlain by weak mud that could liquefy. They should not build on top of, on, or at the base of steep escarpments because the escarpments could fail and produce landslides, and they should avoid locating large population centers downstream of dams (which could crack and collapse, causing a flood). And they should avoid constructing vulnerable buildings directly over active faults, because fault movement could crack and destroy the buildings. Cities in seismic zones need to draw up emergency plans to deal with disaster. Communication centers should be situated in safe localities, and strategies need to be implemented for providing supplies under circumstances where roads may be impassable.

Finally, communities and individuals should learn to protect themselves during an earthquake. In your home, keep emergency supplies, bolt bookshelves to walls, strap the water heater in place, install locking latches on cabinets, know how to shut off the gas and electricity, know how to find the exit, have a fire extinguisher handy, and know where to go to find family members. Schools and offices should have earthquake-preparedness drills (▶Fig. 8.38). As long as plates continue to move, earthquakes will continue to shake. But we can learn to live with them.

FIGURE 8.38 Japanese schoolchildren practice earthquake preparedness during an earthquake drill.

FIGURE 8.37 How to prevent damage and injury during an earthquake. (a) Wrapping a bridge's support columns in cable (preventing buckling of the columns) and bolting the span to the columns (preventing the span from separating from the columns) keeps the bridge from collapsing as easily. (b) Buildings will be stronger if they are wider at the base and if cross beams are added inside. (c) Placing buildings on rollers (or shock absorbers) will lessen the severity of the vibrations.

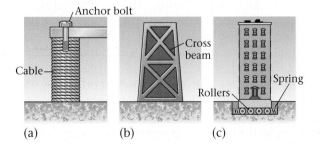

CHAPTER SUMMARY

- Earthquakes are episodes of ground shaking. Earthquake activity is called seismicity.

- Most earthquakes happen when rock breaks during faulting. A fault is a fracture on which sliding occurs. The place where rock breaks and earthquake energy is released is called the hypocenter (focus), and the point on the ground directly above the hypocenter is the epicenter.

- Active faults are faults on which movement is currently taking place. Inactive faults ceased being active long ago, but can still be recognized because of the displacement across them. Displacement on active faults that intersect the ground surface may yield a fault scarp.

- According to elastic-rebound theory, during fault formation, rock elastically bends, then cracks. Eventually, cracks link to form a throughgoing rupture on which sliding occurs. When this happens, the rock breaks and vibrates, and this generates an earthquake. Faults exhibit stick-slip behavior, in that they move in sudden increments.

- Earthquake energy travels in the form of seismic waves. Body waves, which pass through the interior of the Earth, include P-waves (compressional waves) and S-waves (shear waves). Surface waves, which pass along the surface of the Earth, include R-waves (Rayleigh waves) and L-waves (Love waves).

- We can detect earthquake waves by using a seismograph.

- Seismograms demonstrate that different earthquake waves arrive at different times, because they travel at different velocities. Using the difference between P-wave and S-wave arrival times, seismologists can pinpoint the epicenter location.

- The Mercalli intensity scale is based on documenting the damage caused by an earthquake. Magnitude scales, such as the Richter scale, are based on measuring the amount of ground motion, as indicated by traces of waves on a seismogram. The seismic-moment magnitude scale takes into account the amount of slip, the length and depth of the rupture, and the strength of the ruptured rock.

- A magnitude 8 earthquake yields about ten times as much ground motion as a magnitude 7 earthquake and releases about thirty-two times as much energy.

- Most earthquakes occur in seismic belts, or zones, of which the majority lie along plate boundaries. Intraplate earthquakes happen in the interior of plates. Different kinds of earthquakes happen at different kinds of plate boundaries.

- Earthquake damage results from ground shaking (which can topple buildings), landslides (set loose by vibration), sediment liquefaction (the transformation of compacted clay into a muddy slurry), fire, and tsunamis (giant waves).

- Seismologists predict that earthquakes are more likely in seismic zones than elsewhere, and can determine the recurrence interval (the average time between successive events) for great earthquakes. But it may never be possible to pinpoint the exact time and place at which an earthquake will take place.

- Earthquake hazards can be reduced with better construction practices and zoning, and by knowing what to do during an earthquake.

KEY TERMS

aftershock (p. 210)
body wave (p. 212)
compressional wave (p. 212)
displacement (p. 209)
earthquake (p. 207)
elastic-rebound theory (p. 210)
epicenter (p. 207)
fault (p. 207)
fault scarp (p. 209)
foreshock (p. 210)
hypocenter (focus) (p. 207)
intraplate earthquake (p. 225)
liquefaction (p. 228)
magnitude (p. 218)

Mercalli intensity scale (p. 218)
recurrence interval (p. 235)
Richter scale (p. 219)
seismic belt (zones) (p. 220)
seismicity (p. 207)
seismic wave (p. 212)
seismogram (p. 213)
seismograph (seismometer) (p. 212)
shear wave (p. 212)
stick-slip behavior (p. 210)
surface wave (p. 212)
tsunami (p. 231)
Wadati-Benioff zone (p. 222)

REVIEW QUESTIONS

1. Compare normal, reverse, and strike-slip faults.
2. Describe elastic rebound theory and the concept of stick-slip behavior.
3. Describe the motions of the four types of seismic waves. Which are body waves, and which are surface waves?
4. Explain how the vertical and horizontal components of an earthquake are detected on a seismograph.
5. Explain the contrasts among the different scales used to describe the size of an earthquake.
6. How does seismicity on mid-ocean ridges compare with seismicity at convergent or transform boundaries? Do all earthquakes occur at plate boundaries?
7. What is a Wadati-Benioff zone, and why was it important in understanding plate tectonics?
8. Describe the types of damage caused by earthquakes.
9. What is a tsunami, and why does it form?
10. Explain how liquefaction occurs in an earthquake, and how it can cause damage.
11. How are long-term and short-term earthquake predictions made? What is the basis for determining recurrence interval, and what does a recurrence interval mean?
12. What types of structure are most prone to collapse in an earthquake? What types are most resistant to collapse?

SUGGESTED READING

Bolt, B. A. 1999. *Earthquakes,* 4th ed. New York: Freeman.

Bryant, E. 2001. *Tsunami: The Underrated Hazard.* Cambridge, UK: Cambridge University Press.

Fradkin, P. L. 1999. *Magnitude 8.* Berkeley: University of California Press.

Geschwind, C. H. 2001. *California Earthquakes: Science, Risk and the Politics of Hazard Mitigation.* Baltimore: Johns Hopkins University Press.

Hough, S. E. 2002. *Earthshaking Science: What We Know (and Don't Know) about Earthquakes.* Princeton, N.J.: Princeton University Press.

Ritchie, D., and A. E. Gates. 2001. *Encyclopedia of Earthquakes and Volcanoes.* New York: Facts on File.

Stein, S., and M. Wysession. 2002. *An Introduction to Seismology, Earthquakes and Earth Structure.* Boston: Blackwell Science.

Yeats, R. S. 2001. *Living with Earthquakes in California: A Survivor's Guide.* Corvallis: Oregon State University Press.

Seeing Inside the Earth

C.1 INTRODUCTION

We live on the Earth's skin and can see light years into space just by looking up. But when we look down—that's another story! We can't use our eyes to look through rock, because it is opaque. So how do we learn about what's inside this planet? Tunneling and drilling aren't much help because they do little more than prick the surface. Fortunately, as discussed in Chapter 1, nineteenth-century geologists realized that measurements of the Earth's mass and shape provide indirect clues to the mystery of what's inside and, from these clues, they determined that the Earth is not homogeneous but rather consists of three concentric layers: a crust (of low density), a mantle (of intermediate density), and a core (of high density). Further study showed that the crust beneath continents differs from the crust beneath oceans. Continental crust consists of a variety of felsic, intermediate, and mafic rock, whereas oceanic crust consists almost entirely of mafic rock (▶Fig. C.1). However, the determination of the depths of the boundaries between Earth's layers and the division of the layers into sublayers with distinct properties could not be made until the twentieth century,

when studies of seismic waves became available. By studying the speed and direction in which seismic waves travel through the Earth, seismologists can effectively "see" details of the planet's internal layers.

In this interlude, we look at the behavior of seismic waves as they pass through our planet, and we learn how this behavior characterizes Earth's interior. We begin by reviewing a few key points about seismic waves, then move on to the phenomena of wave reflection and refraction. Finally, we witness the discoveries of the different layer boundaries in the Earth. This interlude, incorporating information about earthquakes and seismic waves provided in Chapter 8, completes the journey to the center of the Earth that we began in Chapter 1.

C.2 THE MOVEMENT OF SEISMIC WAVES THROUGH THE EARTH

Recall that a sudden rupture of intact rock or the frictional slip of rock on a fault produces seismic waves. These waves move outward from the point of rupture, the earthquake hypocenter, in all directions at once. The boundary between the rock through which a wave has passed and the rock through which it has not yet passed is called a **wave front.** A wave front expands outward from the earthquake focus like a growing bubble. We can represent a succession of waves in a drawing by a series of concentric wave fronts. The changing position of an imaginary point on a wave front as the front moves through rock is called a **seismic ray.** Note that seismic rays are perpendicular to wave fronts, so that each point on the wave front follows a slightly different ray (▶Fig. C.2). The time it takes for a wave to travel from the focus to a seismograph station along a given ray is the **travel time** along that ray.

The ability of a seismic wave to travel through a certain material, and the velocity at which it travels, depend on the character of the material. Factors such as density (mass per unit volume), rigidity (how stiff

FIGURE C.1 The nineteenth-century three-layer image of the Earth, showing the crust, mantle, and core. The inset shows the contrast between continental crust (felsic, intermediate, and mafic rock) and oceanic crust (mafic rock).

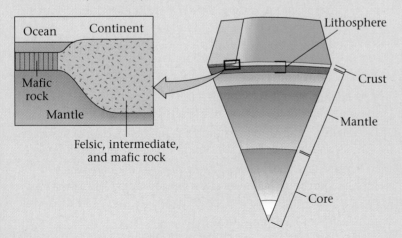

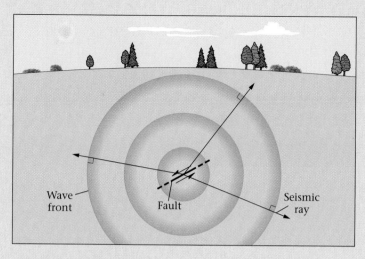

FIGURE C.2 An earthquake sends out waves in all directions. Seismic rays are perpendicular to the wave fronts.

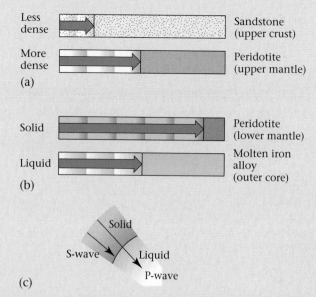

FIGURE C.3 (a) Seismic waves travel at different velocities in different rock types. For example, they travel faster in peridotite than in sandstone. (b) Seismic waves travel faster in solid peridotite than in a liquid such as molten iron alloy. (c) Both P-waves and S-waves can travel through a solid, but only P-waves can travel through a liquid.

or resistant to twisting a material is), and compressibility (how much a material's volume changes in response to squashing) all affect seismic-wave movement. Studies of seismic waves reveal the following:

- Seismic waves travel at different velocities in different rock types (▶Fig. C.3a). For example, P-waves travel at 8 km per second in peridotite (an ultramafic igneous rock), but at only 3.5 km per second in sandstone (a porous sedimentary rock). Therefore, waves accelerate or slow down if they pass from one rock into another.

- In general, seismic waves travel more slowly in a liquid than in a solid. For example, they travel more slowly in magma than in solid rock, and more slowly in molten iron alloy than in solid peridotite (▶Fig. C.3b).

- Both P-waves and S-waves can travel through a solid, but only P-waves can travel through a liquid (▶Fig. C.3c).

C.3 THE REFLECTION AND REFRACTION OF WAVE ENERGY

Shine a flashlight into a container of water so that the light ray hits the boundary (or interface) between water and air at an angle. Some of the ray bounces off the water surface and heads back up into the air, while some enters the water (▶Fig. C.4a). The light ray that enters the water bends at the air-water boundary, so that the angle between the ray and the boundary in the air is different from the angle between the ray and the boundary in the water. Physicists refer to the light ray that bounces off the air-water boundary and heads back into the air as the reflected ray, and the ray that bends

at the boundary as the refracted ray. The phenomenon of bouncing off is **reflection,** and the phenomenon of bending is **refraction.** Wave reflection and refraction take place at the interface between two materials, if the wave travels at different velocities through the two materials.

The amount and direction of refraction at a boundary depend on the contrast in wave velocity across the boundary and on the angle at which a wave hits the interface. As a rule, if waves enter a material through which they travel more slowly, the rays representing the waves bend away from the interface. If the waves enter a material through which they travel faster, the rays bend toward the interface. For example, the light ray in ▶Figure C.4b bends down when hitting the air-water boundary, because light travels more slowly in water. This relation makes sense if you picture a car driving from a paved surface diagonally onto a sandy beach—the wheel that rolls onto the sand first slows down relative to the wheel still on the pavement, causing the car to turn. If the ray were to pass from a material in which it travels slowly into one in which it travels more rapidly, it would bend up (▶Fig. C.4c).

Seismic energy also travels in the form of waves, so seismic waves reflect and/or refract when reaching the interface between two rock layers if the waves travel at different velocities in the two layers. For example, imagine a layer of sandstone overlying a layer of basalt. Seismic velocities in sandstone are slower than in basalt, so as seismic waves reach the boundary, some reflect while some refract.

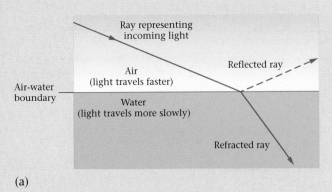
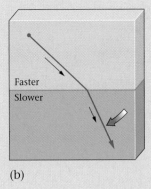
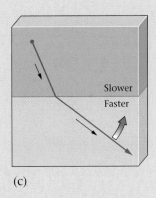

(a)

(b)

(c)

FIGURE C.4 (a) When it reaches the boundary between water and air, a ray of light partly reflects and partly refracts. The refracted ray bends down as it enters the water. (b) A ray that enters a slower medium bends away from the boundary (like light reaching water from air). (c) A ray that enters a faster medium bends toward the boundary.

C.4 DISCOVERING THE CRUST-MANTLE BOUNDARY

The concept that seismic waves refract at boundaries between different layers allowed geologists to define the depth of the core-mantle boundary. In 1909, Andrija Mohorovičić, a Croatian seismologist, noted that P-waves arriving at seismograph stations less than 200 km from the epicenter traveled at an average speed of 6 km per second, whereas P-waves arriving at seismographs more than 200 km from the epicenter traveled at an average speed of 8 km per second. To explain this observation, he suggested that P-waves reaching nearby seismographs followed a shallow path through the crust and traveled more slowly, but P-waves reaching distant seismographs followed a deeper path through the mantle and traveled more rapidly.

To understand Mohorovičić's proposal, examine ►Figure C.5a, which shows P-waves, depicted as rays, generated by an earthquake in the crust. Ray C, the shallower wave, travels through the crust directly to a seismograph. Ray M, the deeper wave, heads downward, refracts at the crust-mantle boundary, curves through the mantle, refracts again at the boundary, and then proceeds through the crust up to the seismograph. At stations less than 200 km from the epicenter, ray C arrives first, because it has a shorter distance to travel. But at stations more than 200 km from the epicenter (►Fig. C.5b), ray M arrives first, even though it has farther to go, because it travels faster for much of its length. Calculations based on this observation require the crust-mantle boundary beneath continents to be at a depth of about 35–40 km. As we learned in Chapter 1, this boundary is now called the **Moho,** in honor of Mohorovičić.

FIGURE C.5 (a) Seismic waves traveling only in the crust reach a nearby seismograph first, because they have a shorter distance to travel. Here, the waves following ray C arrive before those following ray M. (b) Seismic waves traveling for most of their path in the mantle reach a distant seismograph first, because they travel faster in the mantle than in the crust. Here, waves following ray M arrive before those following ray C. This observation led to the discovery of the Moho, the boundary surface between the crust and mantle.

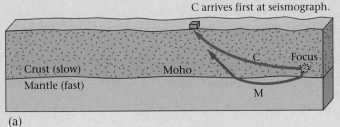
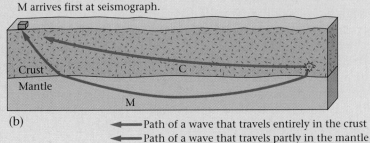

(a)

(b)

C.5 DEFINING THE STRUCTURE OF THE MANTLE

By studying travel times, seismologists have determined that seismic waves travel at different velocities at different depths in the mantle. Between about 100 and 200 km deep in the mantle beneath oceanic lithosphere, seismic velocities are slower than in the overlying lithospheric mantle (▶Fig. C.6). In this **low-velocity zone,** the prevailing temperature and pressure conditions cause peridotite to partially melt by up to 2%. The melt, a liquid, coats solid grains and fills voids between grains. Because seismic waves travel more slowly through liquids than through solids, the coatings of melt slow seismic waves down. In the context of plate tectonics theory, the low-velocity zone is the weak layer on which oceanic lithosphere plates move. Below the low-velocity zone, the mantle does not contain melt. Geologists do not find a well-developed low-velocity zone beneath continents.

Below about 200 km, seismic-wave velocities everywhere in the mantle increase with depth. Seismologists interpret this increase to mean that mantle peridotite becomes progressively less compressible, more rigid, and denser with depth. This proposal makes sense, considering that the weight of overlying rock increases with depth, and as pressure increases, the atoms making up rock squeeze together more tightly and bonds are not as free to move. Because of refraction, the progressive increase in seismic velocity with depth causes seismic rays to curve in the mantle. To understand the shape of a curved ray, look at ▶Figure C.7a, which represents a portion of the mantle by a series of imaginary layers, each permitting a slightly greater seismic-wave velocity than the layer above. Every time a seismic ray crosses the boundary between adjacent layers, it refracts a little toward the boundary. After the ray has crossed several layers, it has bent so much that it begins to head back up toward the top of the stack. Now if we replace the stack of distinct layers with a single layer in which velocity increases with depth at a constant rate, the wave follows a smoothly curving path (▶Fig. C.7b, c).

Between 410 km and 660 km deep (see Fig. C.6), seismic velocity increases in a series of abrupt steps, so the stack of layers in Figure C.7a is actually a somewhat realistic image. Experiments suggest that these **seismic-velocity discontinuities** occur at depths where phase changes occur (see Chapter 8) meaning that pressure causes atoms in minerals to rearrange and pack together more tightly, thereby changing the rock's density, compressibility, and rigidity. Researchers

FIGURE C.6 The velocity of P-waves in the mantle changes with depth. Note the low-velocity zone between 100 and 200 km, and the sudden jumps in velocity defining the transition zone between 410 and 660 km.

FIGURE C.7 (a) In a stack of layers in which seismic waves travel at different velocities (fastest in the lowest layer), a seismic ray gradually bends around and heads back to the surface. The curve consists of several distinct segments. (b) If the mantle's density increased gradually with depth, the ray would define a smooth curve. (c) Since the velocity of seismic waves increases with depth, wave fronts are oblong and seismic rays curve.

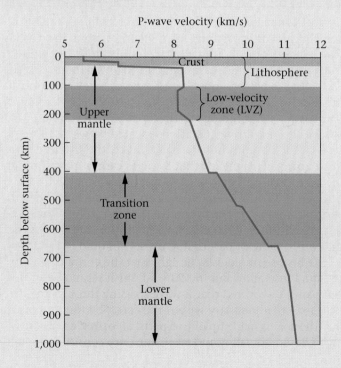

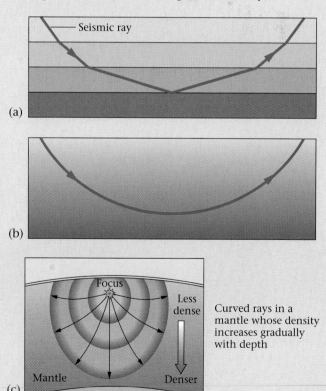

are still trying to determine if chemical changes also occur at the discontinuities. Because of these seismic-velocity discontinuities, as we learned in Chapter 1 but now can see more clearly, seismologists subdivide the mantle into the **upper mantle** (above 410 km), the **transition zone** (between 410 and 660 km), and the **lower mantle** (below 660 km) (see Fig. 1.21).

C.6 DISCOVERING THE CORE-MANTLE BOUNDARY

During the first decade of the twentieth century, seismologists installed seismographs at many stations around the world, expecting to be able to record waves produced by a large earthquake anywhere on Earth. In 1914, one of these seismologists, Beno Gutenberg, discovered that P-waves from an earthquake do not arrive at seismographs lying in a band between 103° and 143° from the earthquake epicenter,

FIGURE C.8 P-waves do not arrive in the interval between 103° and 143° from an earthquake's epicenter, defining the P-wave shadow zone. The wave that arrives at 103° passed just above the core-mantle boundary. The next wave bent down so far that it arrives at about 160°. And the wave that arrives at 143° bent only slightly as it passed through the core. The inset shows the shadow zone on a globe. Note that P-waves bend down at the core-mantle boundary because velocities are slower in the core than in the mantle (the core is less rigid).

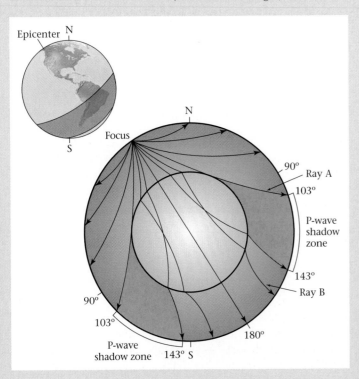

as measured along the circumference of the Earth. This band is now called the **P-wave shadow zone** (▶Fig. C.8). If the density of the Earth increased gradually with depth all the way to the center, the shadow zone would not exist, because rays passing into the interior would curve up and reach every point on the surface. Thus, the presence of a shadow zone means that deep in the Earth a major interface exists where seismic waves *abruptly* refract down (implying that the velocity of seismic waves suddenly decreases). This interface, now called the **core-mantle boundary,** lies at a depth of about 2,900 km.

To see why the P-wave shadow zone exists, follow the two seismic rays labeled A and B in Figure C.8. Ray A curves smoothly in the mantle (we are ignoring seismic-velocity discontinuities in the mantle) and passes just above the core-mantle boundary before returning to the surface. It reaches the surface 103° from the epicenter. In contrast, ray B just penetrates the boundary and refracts down into the core. Ray B then curves through the core and refracts again when it crosses back into the mantle. As a consequence, ray B intersects the surface at more than 143° from the epicenter.

The downward bending of seismic waves when they pass from the mantle down into the core indicates that seismic velocities in the core are slower than in the mantle. Thus, even though the core is deeper and denser than the mantle, the outer part of the core must be less rigid than the mantle.

C.7 DISCOVERING THE NATURE OF THE CORE

Geologists had inferred, based on measurements of the Earth's density and on the study of meteorites, that the Earth's core consists of iron alloy. But it was not until the modern study of how seismic waves pass through the interior of the Earth that it became possible to answer a key question: Is the core, or at least part of it, solid or liquid? A study of S-waves gave seismologists the answer. They found that S-waves do not arrive at stations located between 103° and 180° from the epicenter (a band called the **S-wave shadow zone**). This means that S-waves cannot pass through the core at all—otherwise, an S-wave headed straight down through the Earth would appear on the other side. Remember that S-waves can travel only through solids. Thus, the fact that S-waves do not pass through the core means that the core, or at least part of it, consists of liquid (▶Fig. C.9a).

At first, seismologists thought that the entire core might be liquid iron alloy. But in 1936, a Danish seismologist, Inge Lehmann, discovered that P-waves passing through the core reflected off a boundary within the core. She then proposed that the core is made up of two parts: an **outer core** consisting of liquid iron alloy and an **inner core** consisting of solid

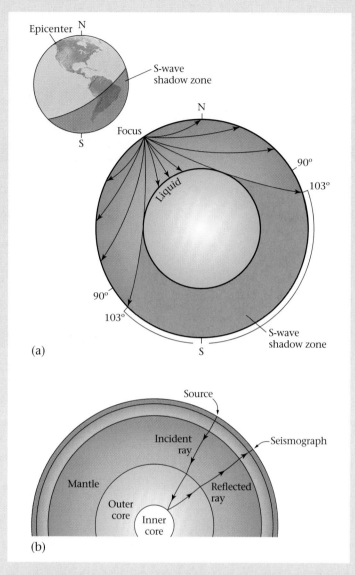

FIGURE C.9 (a) The S-wave shadow zone covers about a third of the globe and exists because shear waves cannot pass through the liquid outer core. (b) The solid inner core was detected when seismologists observed that some seismic waves generated by nuclear explosions reflected off a boundary within the core.

iron alloy. Lehmann's work defined the existence of the inner core but could not locate the depth at which the inner core/outer core interface occurs. This depth was located by measuring the exact time it took for seismic waves generated by nuclear explosions to penetrate the Earth, bounce off the inner core–outer core boundary, and return to the surface (▶Fig. C.9b). The measurements showed that the inner core–outer core boundary occurs at a depth of about 5,155 km. Recent studies of seismic waves passing through the inner core have shown that the inner core rotates slightly faster than the rest of the Earth—it makes an extra revolution about once every twenty-five years.

C.8 FINE-TUNING OUR IMAGE OF THE EARTH'S LAYERS

Seismic Tomography

In recent years, sophisticated computers have been able to use global seismic data to create a three-dimensional image of seismic-wave velocities within the Earth. This type of analysis, known as **seismic tomography,** resembles the method employed by a medical CAT-scan machine ("CAT" stands for "computer-aided tomography"). Tomography (the word comes from the Greek word for "slice") allows us to picture variations in seismic velocities visible on slices through the Earth. In seismic tomography studies, researchers compare travel times for seismic waves that follow different paths through the Earth, and they distinguish regions in which waves travel unexpectedly fast from regions where waves travel unexpectedly slowly. Overall, tomographic studies show that a layered image of the Earth like that depicted in Figure C.7a, though reasonable as a rough approximation, is not accurate in detail (▶Fig. C.10a, b). Rather, it appears that there can be different seismic velocities at a given depth in the Earth.

For example, seismologists have discovered distinct zones of lower-than-expected velocities and distinct zones of higher-than-expected velocities at the same depth in the mantle. These velocity contrasts indicate that there are blotchy variations in the nature of materials within the mantle. Seismologists suggest that these variations in turn reflect temperature variations—hotter rocks are less rigid than cooler rocks. The distribution of hotter and cooler zones in the mantle supports the theory that convection occurs in the mantle—the hotter zones are rising, and the cooler zones are sinking (▶Fig. C.11). Seismic-tomography studies can also detect remnants of subducted oceanic plates.

Seismic-Reflection Profiling

Seismic techniques are also letting us fine-tune our image of the crust. During the past half century, geologists have found that by using dynamite, by banging large weights against the Earth's surface, or by releasing bursts of compressed air into the ocean, they can create artificial seismic waves that propagate down into the Earth and reflect off the boundaries between different layers of rock in the crust. By recording the time at which these reflected waves return to the surface, geologists effectively create a cross-sectional view of the crust called a **seismic-reflection profile** (▶Fig. C.12a, b). This image defines the depths at which specific strata occur and reveals the presence of subsurface folds (bends in layers) and faults. Oil companies must obtain seismic-reflection profiles, despite their high cost, because they allow geologists to identify likely locations for oil and gas deposits underground.

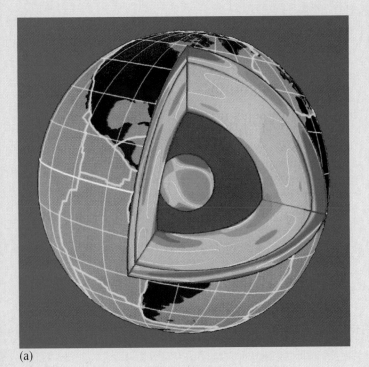

(a)

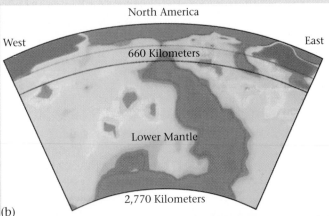

North America

West   East

660 Kilometers

Lower Mantle

2,770 Kilometers

(b)

FIGURE C.10 Tomographic images show regions of faster and slower velocities. Redder colors indicate regions where seismic waves travel more slowly than expected; bluer and purple colors are regions where seismic waves travel faster than expected (the mantle is cooler and more rigid). In the slowest regions, waves travel 1.5% more slowly; in the fastest regions, waves travel 1.5% faster than expected. (a) Tomographic image of the whole Earth. The brown region is the outer core, a liquid. (b) Closeup of the mantle beneath North America. The high-velocity area may be the remnant of a cold subducted slab.

In the past decade, computers have become so sophisticated that geologists can now produce three-dimensional seismic-reflection images of the crust. These provide so much detail that geologists can trace out a ribbon of sand representing the channel of an ancient stream even where the sand lies buried kilometers below the Earth's surface.

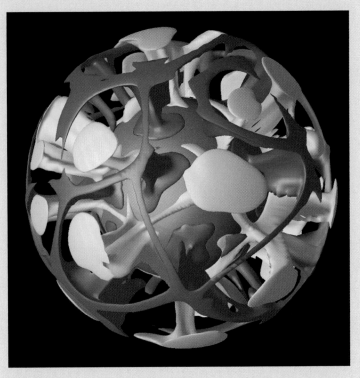

FIGURE C.11 This image is a computer-generated 3-D model depicting convective flow in the mantle. The upwelling regions, depicted in yellow, consist of rising hot mantle, and the downwelling regions, depicted in blue, consist of sinking cooler mantle. The red sphere inside is the surface of the core.

C.9 AN INTEGRATED VIEW OF THE EARTH: NEW DISCOVERIES

Through painstaking effort, seismologists have developed a graph, known as a **velocity-versus-depth curve,** that shows the depths at which seismic velocity suddenly changes and thus helped to establish the principal layers and sublayers in the Earth (▶Fig. C.13). This graph is an average for the whole Earth. Modern research now focuses on developing refined graphs for specific locations. In addition, the deep interior of the Earth has become one of the most active research areas of recent decades, as geologists have developed exotic new techniques to study the nature of materials under very high pressures and temperatures. Recently, a major new research initiative, called *EarthScope* began. This initiative involves placing hundreds of seismographs in a grid across swaths of the United States. Seismologists hope that data collected from these instruments will enable them to greatly refine images of the interior and test current models of it (▶Fig. C.14).

(a)

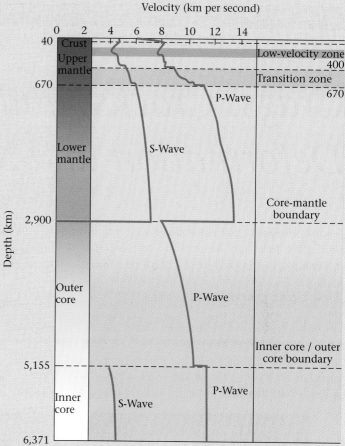

FIGURE C.13 The velocity-versus-depth profile for the Earth.

FIGURE C.14 The modern view of a complex and dynamic Earth interior. Note the convecting cells, the mantle plumes, and the subducted-plates.

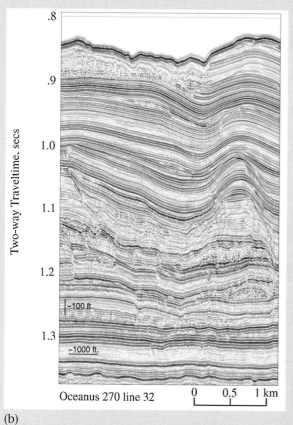

(b)

FIGURE C.12 (a) Trucks thumping on the ground to generate the signal needed for making a seismic-reflection profile. (b) A seismic-reflection profile. The colored stripes are distinct horizons in the stratigraphic sequence.

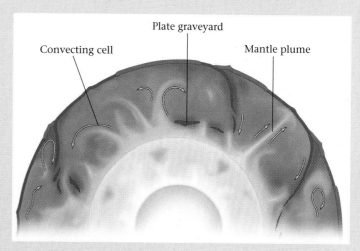

CHAPTER 9

Crags, Cracks, and Crumples: Crustal Deformation and Mountain Building

Innumerable peaks, black and sharp, rose grandly into the dark blue sky, their bases set in solid white, their sides streaked and splashed with snow, like ocean rocks with foam. . . . [Mountains] are nature's poems carved on tables of stone. . . . How quickly these old monuments excite and hold the imagination!
—JOHN MUIR, from *WILDERNESS ESSAYS*

Some of the world's most beautiful scenery can be found in mountainous regions. Here, gazing on Mt. Cook in the southern Alps of New Zealand, we see evidence of the many processes that contribute to the development of mountain scenery.

9.1 INTRODUCTION

Geographers call the peak of Mt. Everest "the top of the world," for this mountain, which lies in the Himalayas of south Asia, rises higher than any other on Earth (▶Fig. 9.1). The cluster of flags on Mt. Everest's summit flap at 8.85 km (29,029 feet) above sea level—almost at the cruising height of modern jets. No one can survive very long at the top, for the air there is too thin to breathe. In fact, even after spending weeks acclimating to high-altitude conditions at a base camp a couple of kilometers below the summit, most climbers still need to use bottled oxygen during their summit attempt. In 1953, the British explorer Sir Edmund Hillary and a Nepalese guide, Tenzing Norgay, became the first to reach the summit. By 1999, about 750 more people had also succeeded—but 150 died trying. So many climbs end in death because success depends not just on the skill of the climber, but also on the path of the jet stream, a 200-km-per-hour current of air that flows at high elevations. If the jet stream crosses the summit, it engulfs climbers in heat-robbing winds that can freeze a person's face, hands, and feet even if they're swaddled in high-tech clothing.

FIGURE 9.1 Mt. Everest (the large peak in the center) and the surrounding Himalayas, as viewed looking straight down from the space shuttle *Atlantis.*

Mountains draw nonclimbers as well, for everyone loves a vista of snow-crested peaks. To geologists, mountains provide one of the most obvious indications of dynamic activity on Earth. To make a mountain, Earth forces lift cubic kilometers of rock skyward against the pull of gravity. Erosion, over time, grinds away at a mountain to make sediment, and in the process sculpts awesome jagged topography.

The process of forming a mountain not only raises the surface of the crust, but also causes rocks to undergo **deformation,** a process by which rocks bend or break in response to compression, tension, or shearing. Deformation produces **geologic structures,** including **joints** (cracks), **faults** (fractures on which one body of rock slides past another), **folds** (bends or wrinkles), and **foliation** (layering resulting from the alignment of mineral grains or the development of compositional bands). Mountain building may also involve metamorphism and melting. In this chapter, we learn about geologic structures and other phenomena that happen during mountain building. We will also learn why mountains and structures form, in the context of plate tectonics theory.

9.2 MOUNTAIN BELTS AND THE CONCEPT OF OROGENY

With the exception of the large volcanoes formed over hot spots, mountains do not occur in isolation, but rather as part of linear ranges variously called **mountain belts** or **orogens** (from the Greek words *oros,* meaning "mountain," and *genesis,* meaning "formation"). Geographers define about a dozen major orogens and numerous smaller ones worldwide (▶Fig. 9.2). Some large orogens contain smaller ranges within.

A mountain-building event, or **orogeny,** has a limited lifetime. The process begins, lasts for tens of millions of years, and then ceases. After an orogeny ceases, erosion may eventually bevel the land surface almost back to sea level, sometimes in as little as 50 million years. Thus, the mountain ranges we see today are comparatively young; most of Earth's present mountainous topography didn't exist before the Cretaceous Period. But even long after erosion has eliminated its peaks, a belt of deformed (contorted or broken) and metamorphosed rocks remains to define the location of an ancient orogen.

9.3 ROCK DEFORMATION IN THE EARTH'S CRUST

Deformation and Strain

As noted above, orogeny causes deformation (bending, breaking, shortening, stretching, or shearing), which in turn yields geologic structures. To get a visual sense of deformation, let's compare a road cut along a highway in the central Great Plains of North America, a region that has not undergone orogeny, with a cliff in the Alps (▶Fig. 9.3a, b).

The road cut, which lies at an elevation of only about 100 m above sea level, exposes nearly horizontal beds of sandstone and shale—these beds have the same orientation that they had when first deposited. Notably, sand grains in sandstone beds of this outcrop have a nearly spherical shape (the same shape they had when deposited), and clay flakes in the shale lie roughly parallel to the bedding, because of compaction. Rock of this outcrop is undeformed, meaning that it contains no geologic structures other than a few joints.

In the Alpine cliff, exposed at an elevation of 3 km, rocks look very different. Here, we find layers of quartzite and slate (the metamorphic equivalent of sandstone and shale) in contorted beds whose shapes resemble the wrinkles in a rug that has been pushed across the floor. These wrinkles are folds. Quartz grains in the quartzite are not spheres, but resemble flattened eggs, and the clay flakes in slate are aligned parallel to each other and tilt at a steep angle to the bedding. In fact, the rock splits on planes called cleavage that parallel the flattened sand grains and clay flakes, and thus cut across the bedding at a steep angle. Finally, if we try tracing the quartzite and slate layers along the outcrop face, we find that they abruptly terminate at a sloping surface marked by broken-up rock. This surface is a fault. In this example, the quartzite and slate slid along the fault from where they first formed to get to their present location on top of marble layers.

Clearly, the beds in the Alpine cliff have been deformed, and as a result the cliff exposes a variety of geologic structures. Beds no longer have the same shape and position that

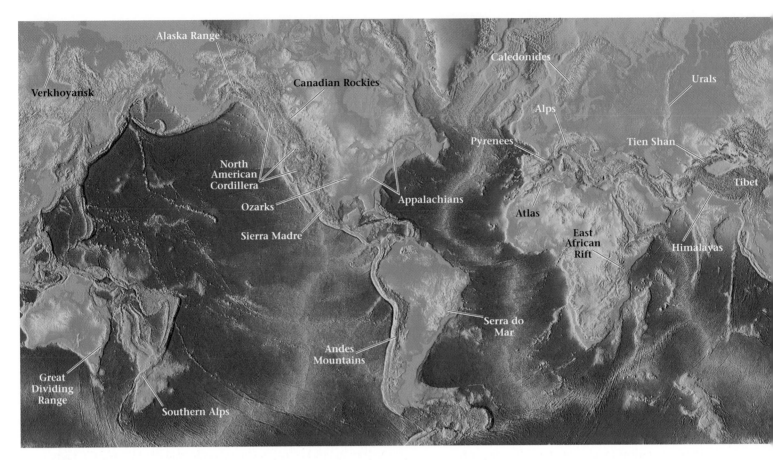

FIGURE 9.2 Digital map of world topography, showing the locations of major mountain ranges.

FIGURE 9.3 (a) This road cut exposes flat-lying beds of Paleozoic shale and sandstone along a highway, occurring in the interior region of North America. The region has *not* been involved in orogeny subsequent to the deposition of the beds. A few vertical joints cut the beds. Inset: An enlargement showing that the undeformed sandstone has spherical grains. (b) In this diagram of an Alpine mountain cliff, note the folded layers of quartzite and slate and the fault. Inset: Grains of sand in the quartzite have become flattened and are aligned parallel to each other. The slate has cleavage.

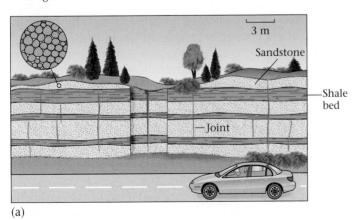

(a)

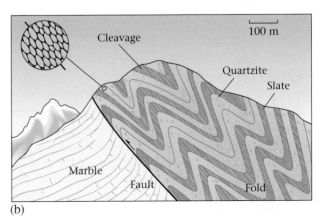

(b)

they had when first formed, and the shape and orientation of grains has changed. In sum, deformation includes one or more of the following (▶Fig. 9.4a–d): (1) a change in location (translation); (2) a change in orientation (rotation); and (3) a change in shape (distortion). Deformation can be fairly obvious when observed in an outcrop (▶Fig. 9.5a, b).

Geologists refer to the change in shape that deformation causes as **strain.** We distinguish among different kinds of strain according to how the rock changes shape. If a layer of rock becomes longer, it has undergone stretching, but if the layer becomes shorter, it has undergone shortening (▶Fig.

9.6a–c). If a change in shape involves the movement of one part of a rock body past another so that angles between features in the rock change, the result is called shear strain (▶Figs. 9.6d; 9.7).

Brittle vs. Ductile Deformation

Imagine that a plate tumbles off a table and lands on a hard floor—the plate breaks and shatters into pieces. Similarly, if you strike a glass window with a ball, the window cracks and may even shatter, and if you bend a dry stick, the stick fractures in two. All such phenomena—breaking, shattering, cracking, and fracturing—serve as familiar examples of **brittle deformation.** Now, imagine that you squeeze a ball of soft dough between a book and a table top—the dough flattens into a pancake. Similarly, if you bend a stick of chewing gum, it changes from a plane into a curve, and if you pull on the ends of a warm copper block, it stretches into a wire. During such **ductile deformation,** objects change shape without visibly breaking (▶Fig. 9.8a–d).

What actually happens within mineral grains during these two different kinds of deformation? Recall that the

FIGURE 9.4 The components of deformation. (a) A block of rock changes location when it moves from one place to another. (b) It changes orientation when it tilts or rotates around an axis. (c) It changes shape when its dimensions change, or when once-planar surfaces become curved. (d) Folds and faults represent deformation, because they involve changes in location (e.g., sliding has occurred on a fault), orientation (a layer has tilted to form a fold), and shape (the squares in the undeformed layer have become rectangles or parallelograms in the deformed layer).

(a)   Changing location

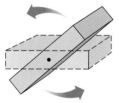

(b)   Changing orientation

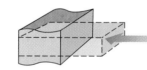

(c)   Changing shape

Before deformation

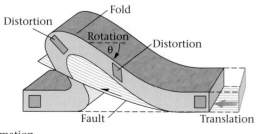

After deformation

(d)

FIGURE 9.5 (a) These flat-lying beds of sediment in Badlands National Monument, South Dakota, have not been deformed. (b) These folded layers of quartzite and dark schist, in Australia, have been deformed.

(a)

(b)

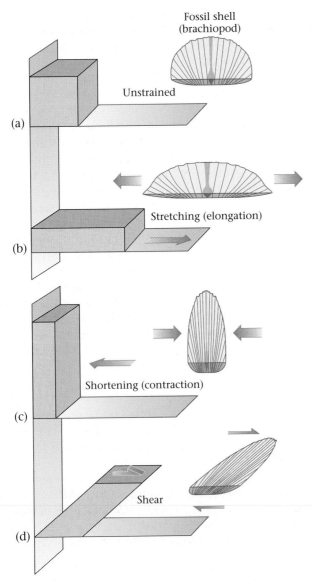

FIGURE 9.6 Different kinds of strain. (a) An unstrained cube and an unstrained fossil shell (brachiopod). (b) Horizontal stretching changes the cube into a brick whose long dimension parallels the direction of stretching, and it makes the brachiopod longer. (c) Horizontal shortening changes the cube into a brick whose long dimension lies perpendicular to the shortening direction, and it makes the brachiopod taller. (d) Shear strain tilts the cube over and transforms it into a parallelogram, and it changes the angular relationships in the brachiopod.

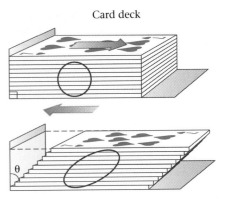

FIGURE 9.7 You can simulate shear strain by moving a deck of cards so that each card slides a little with respect to the one below. Note how a circle drawn on the side of the deck changes shape to become an ellipse, and that the angle between the bottom of the deck and the back side of the deck has changed from a right angle to an acute angle.

atoms that make up mineral grains are connected by chemical bonds. During brittle deformation, many of the bonds break and stay broken, leading to the formation of a permanent crack across which material no longer connects. During ductile deformation, simplistically, some bonds break but new ones quickly form. In this way, the atoms within grains rearrange, and the grains change shape without any cracks forming. (A textbook in structural geology provides more detail on the many complex processes that can contribute to ductile deformation.)

Why do rocks inside the Earth sometimes deform brittlely and sometimes ductilely? The behavior of a rock depends on:

- *Temperature:* Warm rocks tend to deform ductilely, whereas cold rocks tend to deform brittlely. To see this contrast, try an experiment with a candle. Chill a candle in a freezer, then press its middle against the edge of a table; the candle will brittlely snap in two. But if you first warm the candle in an oven, it will ductilely bend without breaking when pressed against the table.

- *Pressure:* Under great pressures deep in the Earth, rock behaves more ductilely than it does under low pressures near the surface. Pressure effectively prevents rock from separating into fragments.

- *Deformation rate:* A sudden change in shape causes brittle deformation, whereas a slow change in shape causes ductile deformation. For example, if you hit a thin marble bench with a hammer, it shatters, but if you leave the bench alone for a century, it gradually sags without breaking.

- *Composition:* Some rock types are softer than others; for example, halite (rock salt) deforms ductilely under conditions in which granite deforms brittlely.

Considering that pressure and temperature both increase with depth in the Earth, geologists find that in typical continental crust, rocks generally behave brittlely above about 10–15 km, but they generally behave ductilely below; we call this depth the brittle-ductile transition. Earthquakes in continental crust happen only above this depth because these earthquakes involve brittle breaking.

FIGURE 9.8 (a) Brittle deformation occurs when you drop a plate and it shatters. (b) Ductile deformation takes place when you squash a soft ball of dough beneath a book and the dough flattens into a pancake without breaking. (c) Cracks (joints) in an outcrop result from brittle deformation. (d) Folds, like these in the marble of a quarry wall, form without breaking a rock, and thus represent ductile deformation.

Stress—the Cause of Deformation

Up to this point, we've focused on picturing the consequences of deformation. Describing the causes of deformation is a bit more challenging in the context of an introductory geology book. In captions for displays about mountain building, museums and national parks typically dispense with the issue by using the phrase "The mountains were caused by forces deep within the Earth." But what does this mean? Isaac Newton defined force by noting that if you apply a force to an object, the object speeds up, slows down, or changes direction. Applying this concept to geology, we see that phenomena such as plate interactions (such as continent-continent collisions) apply forces to rock and thus cause rock to change location, orientation, or shape. In other words, the application of forces in the Earth indeed causes deformation.

Geologists, however, use the word *stress* instead of *force* when talking about the cause of deformation. We define the stress acting on a plane as the force applied *per unit area* of the plane. The need to distinguish between stress and force arises because the actual consequences of applying a force depend not just on the amount of force but also on the area over which the force acts. A simple pair of experiments shows why. Experiment 1: Stand on a single, empty aluminum can (▶Fig. 9.9a). All of your weight—a force—focuses entirely on the can, and the can crushes. Experiment 2: Place a board atop 100 cans and stand on the board (▶Fig. 9.9b). In this case, your weight is distributed across 100 cans, so the force acting on any one can is not enough to crush it. In both experiments, the force caused by the weight of your body was the same, but in experiment 1 the force was applied over a small area so a large stress developed, whereas in experiment 2 the same force was applied over a large area so only a small stress developed. How does this concept apply to geology? During mountain building, the force of one plate interacting with another is distributed across the area of contact between the two plates, so the deformation resulting at any specific location actually depends on the stress developed at that location, not on the total force involved in the plate interaction.

Different kinds of stress occur in rock bodies. **Compression** develops when a rock is squeezed, **tension** occurs when a rock is pulled apart, and **shear stress** develops when one side of a rock body moves sideways past the other

FIGURE 9.9 (a) When you stand on a single can, you apply enough force to the can to crush it, for the can feels a large stress. (b) When you stand on a board resting on 100 cans, you apply the same force to the board, but now it is spread out over 100 cans. Therefore, each can feels only a small stress and does not crumple.

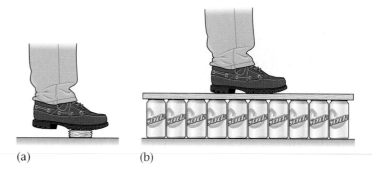

side (▶Fig. 9.11a–d). **Pressure** refers to a special stress condition in which the same push acts on all sides of an object.

Note that "stress" and "strain" have different meanings to geologists, even though we tend to use them interchangeably in everyday English: stress refers to the amount of force applied per unit area of a rock, whereas strain refers to the change in shape of a rock. Thus, stress *causes* strain. Specifically, compression causes shortening, tension leads to stretching, and shear stress produces shear strain. Pressure can cause an object to become smaller, but will not cause it

to change shape. With our knowledge of stress and strain, we can now look at the nature and origin of various classes of geologic structures.

9.4 JOINTS: NATURAL CRACKS IN ROCKS

If you look at the photographs of rock outcrops in this book, you'll notice thin black lines that cross the rock

---

Describing the Orientation of Structures

When discussing geologic structures, it's important to be able to communicate information about their orientation. For example, does a fault exposed in an outcrop at the edge of town continue beneath the nuclear power plant 3 km to the north, or does it go beneath the hospital 2 km to the east? If we knew the fault's orientation, we might be able to answer this question. To describe the orientation of a geologic structure, geologists picture the structure as a simple geometric shape, then specify the angles that the shape makes with respect to a horizontal plane (a flat surface parallel to sea level), a vertical plane (a flat surface perpendicular to sea level), and the north direction (a line of longitude).

Let's start by observing *planar* structures such as faults, beds, and joints. We call these structures planar because they resemble a geometric plane. A planar structure's orientation can be specified by its strike and dip. The **strike** is the angle between an imaginary horizontal line (the strike line) on the plane and the direction to true north (▶Fig. 9.10a, b). We measure the strike with a magnetic compass. The **dip** is the angle of the plane's slope—more precisely,

the angle between a horizontal plane and the dip line, an imaginary line parallel to the steepest slope on the plane, as measured in a vertical plane perpendicular to the strike. We measure the dip angle with a clinometer, a type of protractor that measures slope angles. A horizontal plane has a dip of 0°, and a vertical plane has a dip of 90°. We represent strike and dip on a geologic map using the symbol shown in Figure 9.10b.

A linear structure resembles a line rather than a plane; examples of linear structures include scratches or grooves on a rock surface. Geologists specify the orientation of linear structures by giving their plunge and bearing (▶Fig. 9.10c). The plunge is the angle between a line and horizontal, as measured with a clinometer, in the vertical plane that contains the line. A horizontal line has a plunge of 0°, and a vertical line has a plunge of 90°. The bearing is the compass heading of the line—more precisely, the angle between the projection of the line on the horizontal plane and the direction to true north.

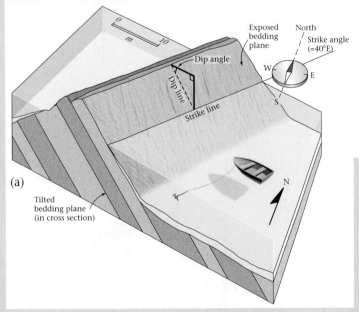

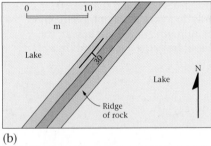

(b)

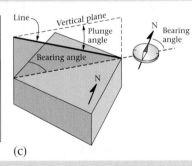

(c)

FIGURE 9.10 (a) We use strike and dip to measure the orientation of planar structures such as these tilted beds. The strike is the compass angle between the strike line and true north. The dip is the steepest slope on the plane. Note that the strike line and the dip line are perpendicular to one another. (b) On a map, the line segment represents the strike direction, and the tick on the segment represents the dip direction. The number indicates the dip angle as measured in degrees. (c) To specify the orientation of a line, we use plunge and bearing.

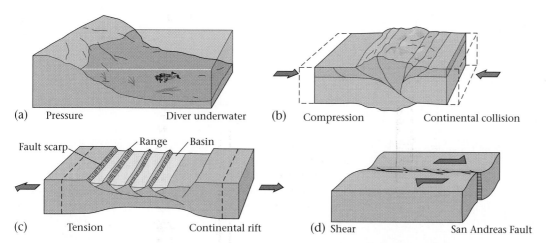

(a) Pressure Diver underwater (b) Compression Continental collision

(c) Tension Continental rift (d) Shear San Andreas Fault

FIGURE 9.11 We represent the direction and magnitude of stress acting on each face of an object by arrows; the lengths of the arrows represent the magnitude of the stress. (a) Pressure occurs when an object feels the same stress on all sides. A diver feels pressure when submerged. (b) Compression takes place when an object is squeezed. Compression occurs during growth of a collisional mountain range. (c) Tension develops when the opposite ends of an object are pulled in opposite directions. Tension occurs during growth of a continental rift. (d) Shear stress occurs when one surface of an object slides relative to the other surface (we depict the shear direction with half arrows). Shear stress parallel to the Earth's surface causes slip on the San Andreas Fault.

faces (see Fig. 9.8c). These lines represent traces of natural cracks along which the rock broke and separated into two pieces during brittle deformation. Geologists refer to such natural cracks as joints. Rock bodies do *not* slide past each other on joints. (Since joints are roughly planar structures, we define their orientation by their strike and dip; see ▶Box 9.1.)

Joints develop in response to tensional stress in brittle rock: a rock splits open because it has been pulled slightly apart. They may form for a variety of geologic reasons. For example, some joints form when a rock cools and contracts, because contraction makes one part of a rock pull away from the adjacent part. Others develop when rock layers formerly at depth undergo a decrease in pressure as overlying rock erodes away, and thus change shape slightly. Still others form when rock layers bend. Joints may control erosion in certain regions, for water can seep into joints and cause the rock bordering the joints to weather faster (▶Fig. 9.12a).

FIGURE 9.12 (a) Sandstone of Arches National Park, Utah, contains many uniformly oriented joints. Weathering along joints has produced narrow slotlike troughs along the joints. (b) The veins in this outcrop are composed of milky white quartz. They fill fractures in gray shale.

(a)

(b)

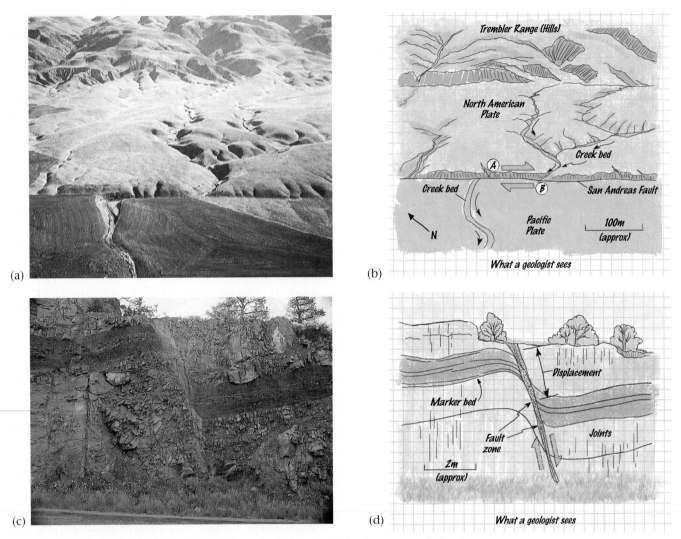

FIGURE 9.13 (a) An oblique air photo showing the San Andreas Fault displacing a creek flowing from the Tremblor Range (background) into the Carizzo Plain, California. (b) What a geologist sees in the previous photo. (c) A road cut in the Rocky Mountains of Colorado, showing a fault offsetting strata in cross section. Note that the fault is actually a band of broken rock about 50 cm wide. (d) What a geologist sees looking at the Rocky Mountain road cut.

If groundwater seeps through joints for a long period of time, minerals such as quartz or calcite may precipitate out of the groundwater and fill the joint. Such mineral-filled joints are called **veins** and look like white stripes cutting across a body of rock (▶Fig. 9.12b). Some veins contain small quantities of valuable metals, such as gold.

Geotechnical engineers, people who study the geologic setting of construction sites, pay close attention to jointing when recommending where to put roads, dams, and buildings. Water flows much more easily through joints than it does through solid rock, so it would be a bad investment to situate a water reservoir over rock with closely spaced joints—the water would leak down into the joints. Also, building a road on a steep cliff composed of jointed rock could be risky,

for joint-bounded blocks separate easily from bedrock, and the cliff might collapse.

9.5 FAULTS: FRACTURES ON WHICH SLIDING HAS OCCURRED

After the San Francisco earthquake of 1906, geologists found a rupture that ripped across the landscape near the city. Where this rupture crossed orchards, it offset rows of trees, and where it crossed a fence, it broke the fence in two; the western side of the fence moved northward by about 2 m (Fig. 8.5a). The rupture represents the trace of the San

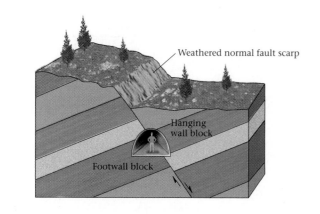

(a)

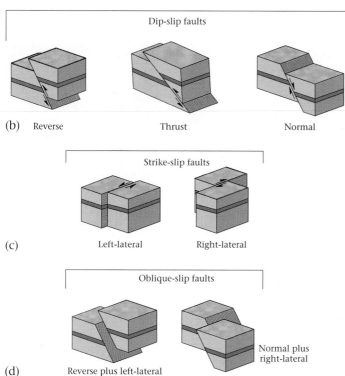

Dip-slip faults

(b) Reverse   Thrust   Normal

Strike-slip faults

(c) Left-lateral   Right-lateral

Oblique-slip faults

(d) Reverse plus left-lateral   Normal plus right-lateral

FIGURE 9.14 (a) A hanging-wall block and a footwall block, relative to a sloping fault surface. The weathered fault scarp is an exposure of the fault at the ground surface. (b) Three types of dip-slip faults, on which sliding parallels the dip line. (c) Two types of strike-slip faults, on which sliding parallels the strike line. (d) Two examples of oblique-slip faults, on which sliding takes place diagonally along the surface.

move. Others accommodate the sliding of rocks in the crust at depth and remain invisible at the surface unless they are later exposed by erosion (▶Fig. 9.13c, d).

Geologists study faults not only because the movement on some faults causes earthquakes, but also because they juxtapose bodies of rock that did not originally lie adjacent to each other and thus complicate the arrangement of rocks at the Earth's surface.

Fault Classification

Geologists have developed terminology to classify faults and describe movement on them. The fault plane can be vertical, horizontal, or at some angle in between. In the case of non-vertical faults (those that slope at an angle), we define the hanging-wall block as the rock above the fault plane, and the footwall block as the rock below the fault plane (▶Fig. 9.14a). If you stand in a tunnel along a fault plane, the hanging-wall block looms over your head, and the footwall block lies under your feet. We distinguish several types of faults.

• *Dip-slip vs. strike-slip vs. oblique-slip faults:* On dip-slip faults, sliding occurs up or down the slope of the fault (therefore, up or down the dip); on **strike-slip faults,** one block slides past another horizontally (therefore, parallel to the strike line); and on oblique-slip faults, sliding occurs diagonally on the fault plane (▶Fig. 9.14b–d).

• *Types of dip-slip faults:* We subdivide dip-slip faults into two kinds, depending on which way the hanging-wall block moves relative to the footwall block. On **thrust faults** and **reverse faults,** the hanging-wall block moves up the slope of the fault. Thrust faults differ from reverse faults only in terms of the fault-plane's slope (or dip)—thrust faults have a slope (or dip) of less than about 35°, whereas reverse faults have a slope of greater than 35°. On **normal faults,** the hanging-wall block moves down the slope of the fault.

• *Types of strike-slip faults:* Geologists distinguish between two types of strike-slip faults, based on the relative movement of one side of the fault with respect to the other. If you stand facing the fault, you can say it is a left-lateral strike-slip fault if the block on the far side slipped to your left, and that it is a right-lateral strike-slip fault if the block on the far side slipped to your right.

Recognizing Faults

How do you recognize a fault when you see one? The most obvious criterion is the occurrence of **displacement,** or offset, meaning the amount of movement across a fault plane (▶Fig. 9.15b). Displacement disrupts the layers in rocks, so that layers on one side of a fault are not continuous with layers on the other side.

Andreas Fault (▶Fig. 9.13a, b). As we have seen, a fault is a fracture on which sliding occurs. Slip events, or faulting, generate earthquakes. Faults, like joints, are planar structures, so we represent their orientation by strike and dip.

Faults riddle the Earth's crust. Some are currently active (sliding has been occurring on them in recent geologic time), but most are inactive (sliding on them ceased millions of years ago). Some faults, such as the San Andreas, intersect the ground surface and thus displace the ground when they

FIGURE 9.15 (a) A thrust fault, on which a distinct layer has been offset. (b) A geologist's sketch emphasizes the offset. Point B was originally adjacent to point A. (c) A fault scarp formed after an earthquake in Nevada on a normal fault. (d) Fault breccia consists of broken-up rock. (e) Slip lineations on a fault surface indicate direction of movement.

Faults may also leave their mark on the landscape. Those that intersect the ground surface while they are active can displace natural landscape features (such as stream valleys or glacial moraines; Fig. 9.13a) and human-made features (such as highways, fences, or rows of trees in orchards). Displacement on a dip-slip or oblique-slip fault will make a small step on the ground surface; this step is called a **fault scarp** (▶Fig. 9.15c). And because faults tend to break up rock, the fault may be preferentially eroded. If this happens, the fault trace (the line of intersection between the fault and the ground surface) will be marked by a linear valley.

Fault surfaces and their borders typically look different from bedding planes. For example, faulting under brittle conditions may crush or break adjacent rock. If this shattered rock consists of visible angular fragments, then it is called fault breccia (▶Fig. 9.15d), but if it consists of a fine powder, then it is called fault gouge. Some fault surfaces are polished and grooved by the movement of the hanging wall past the footwall. Polished fault surfaces are called slickensides, and linear grooves on fault surfaces are slip lineations (▶Fig. 9.15e). We specify the orientation of a slip lineation by giving its plunge and bearing (see Box 9.1).

9.6 FOLDS: CURVING ROCK LAYERS

Imagine a carpet lying flat on the floor. Push on one end of the carpet, and it will wrinkle or contort into a series of wavelike curves. Stresses developed during mountain building and other tectonic processes can similarly wrinkle or contort bedding and foliation (or other planar features) in rock. The result—a bend or curve displayed by the shape of a rock layer—is called a fold.

Not all folds look the same—some look like arches, some like troughs, and some have other shapes. To describe these shapes, we first label the parts of a fold (▶Fig. 9.16a).

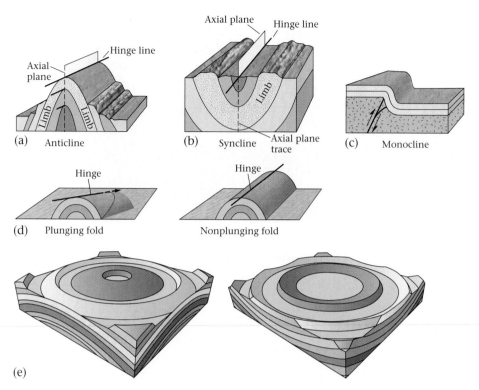

FIGURE 9.16 (a) An anticline (archlike fold) and (b) a syncline (troughlike fold), showing the hinge, limbs, and axial plane. (c) A monocline. Note how it resembles a stair step. (d) If the hinge of a fold is tilted, it's a plunging fold; if the hinge is horizontal, it's a nonplunging fold. (e) Domes (dome-shaped) and basins (bowl-shaped).

The hinge refers to the portion of the fold where curvature is greatest, and the limbs are the sides of the fold that show less curvature. The axial surface (axial plane, if non-curving) is an imaginary surface that encompasses the hinges of successive layers. With these terms in hand, we can now describe types of folds.

- *Anticlines, synclines, and monoclines:* Folds that have an archlike shape in which the limbs dip away from the hinge are called **anticlines,** whereas folds with a troughlike shape in which the limbs dip toward the hinge are called **synclines** (▶Fig. 9.16a, b). A monocline has the shape of a carpet draped over a stairstep (▶Fig. 9.16c).

- *Nonplunging and plunging folds:* If the hinge is horizontal, the fold is called a nonplunging fold, but if the hinge is tilted, the fold is called a plunging fold (▶Fig. 9.16d).

- *Domes and basins:* A fold with the shape of an overturned bowl is called a **dome,** whereas a fold shaped like a right-side-up bowl is called a **basin** (▶Fig. 9.16e). Some small domes and basins are found

at outcrops, but others measure hundreds of kilometers across.

Now see if you can use this terminology to identify the various folds shown in ▶Figure 9.17a–e.

You can recognize folds by the pattern of rock layers on the ground surface (▶Fig. 9.18a). For example, a nonplunging anticline involving sedimentary layers appears as a series of parallel stripes, with the oldest layer in the center and progressively younger layers away from the center; the stripes are symmetrically positioned around the hinge (▶Fig. 9.18b). In a nonplunging syncline involving sedimentary layers, the youngest layers crop out in the center and the oldest at the margins (Fig. 9.18b). Layers in a plunging fold have a U-shape on the ground surface (▶Fig. 9.18c, d). We can represent the hinge of the fold by a heavy line bordered by outward-pointing arrows for an anticline and inward-pointing arrows for a syncline. Domes and basins both show circular outcrop patterns that look like a bull's-eye—the oldest layer occurs in the center of a dome, while the youngest layer is located in the center of a basin.

FIGURE 9.17 (a) An anticline, exposed in a road cut near Kingston, New York. (b) A syncline, exposed in a road cut near Sideling Hill, in Maryland. (c) A train of folds exposed in sea cliffs in eastern Ireland. Note that the axial planes of these folds are not vertical. (d) What a geologist sees in the previous photo.

Folds develop in two principal ways. During formation of flexural folds, a stack of layers bends, and slip occurs between the layers. The same phenomenon happens when you bend a deck of cards—to accommodate the change in shape, the cards slide with respect to each other. Flow folds form when the rock, overall, is so soft that it behaves like weak plastic and slowly flows; these folds develop simply because different parts of the rock body flow at different rates.

Folds develop for a variety of reasons. Some layers wrinkle up, or buckle, in response to end-on compression (▶Fig. 9.19a). Others form where shear stress gradually moves one part of a layer up and over another part (▶Fig. 9.19b). Still others develop where rock layers move up and over bends in a fault and must curve to conform with the fault's shape (▶Fig. 9.19c). Finally, some folds form when a block of basement moves and bends the overlying sedimentary layers (▶Fig. 9.19d).

9.7 TECTONIC FOLIATION IN ROCKS

In an undeformed sandstone, the grains of quartz are roughly spherical, and in an undeformed shale, clay flakes press together into the plane of bedding so that shales tend to split parallel to the bedding. During ductile deformation, however, internal changes take place in a rock that gradually modify the original shape and arrangement of grains. For example, quartz grains may transform into cigar shapes, elongate ribbons, or tiny pancakes, and clay flakes may recrystallize or reorient so that they lie at an angle to the bedding. Overall, deformation can produce inequant grains and can cause them to align parallel to each other, thereby generating metamorphic foliation in the rock. We refer to layering created by the alignment of deformed and/or reoriented grains as tectonic foliation (▶Fig. 9.20a).

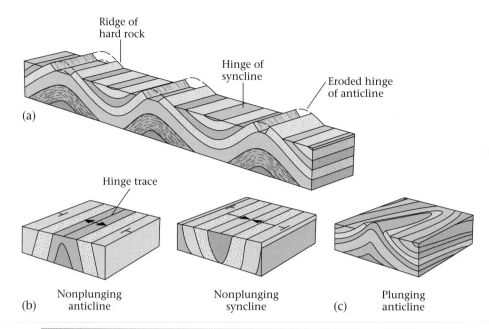

FIGURE 9.18 (a) After erosion, folds involving alternating strong and weak rock layers may control the positions of valleys and ridges on the land surface. Commonly, valleys form over synclines and in the hinge area of anticlines. (b) On the ground surface, the same layers appear on either side of the hinge of a nonplunging fold. (c) On the ground surface, the layers of a plunging anticline curve around the hinge. (d) Aerial photo of Sheep Mountain anticline, Wyoming, a plunging anticline. The strike and dip symbols indicate the attitude of bedding.

We introduced foliations, such as slaty cleavage, schistosity, and gneissic layering, in Chapter 6. Here we add to the story by noting that they form in response to flattening and shearing in ductilely deforming rocks—in other words, foliations indicate that the rock has developed a strain. For example, in rocks with slaty cleavage, the cleavage planes lie perpendicular to the direction of the shortening strain, so the cleavage may be parallel to the axial plane of folds (▶Fig. 9.20b). In schists and gneisses, the foliation commonly lies parallel to or at a slight angle to the direction of shear, because shear smears grains out into the plane of shearing (▶Fig. 9.20a–d).

9.8 UPLIFT AND THE FORMATION OF MOUNTAIN TOPOGRAPHY

Leonardo da Vinci, the Renaissance artist and scientist, enjoyed walking in the mountains, sketching rock ledges and examining the rocks he found there. In the process, he discovered marine shells (fossils) in limestone beds cropping out a kilometer above sea level, and suggested that the rock containing the fossils had risen from below sea level up to its present elevation. Contemporary geologists agree with da Vinci, and now refer to the process by which the surface of the Earth moves vertically from a

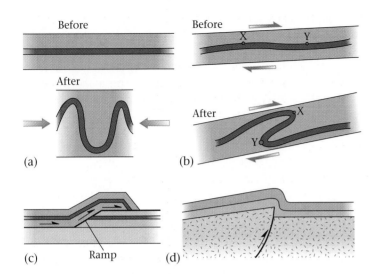

FIGURE 9.19 Different causes of folding. (a) If a layer becomes shortened along its length, it buckles (wrinkles up like a rug). (b) If a layer is sheared, it gradually bends over on itself to form a fold. (c) When layers move up and over step-shaped faults, they must bend into folds. (d) Faulting at depth may fold a layer closer to the ground surface. The folded layers drape over the uplifted fault block to form a monocline.

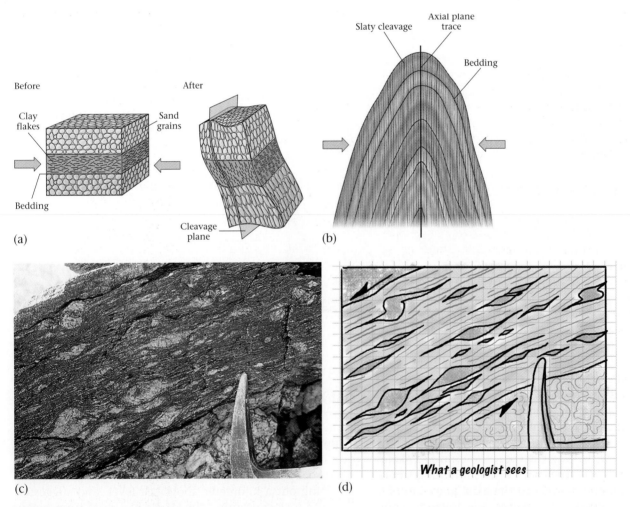

FIGURE 9.20 The development of tectonic foliation in rock. (a) Compression of a layer. Shortening occurs in one direction and lengthening in the other. Quartz grains flatten, and clay grains reorient. As a result, the rock develops cleavage. (b) Slaty cleavage oriented parallel to the axial plane of a fold and perpendicular to the direction of shortening. (c) Schistosity oriented at a low angle to the direction of shear. Note how large grains are all parallel to each other. (d) A geologist's sketch of the outcrop showing shear movement.

FIGURE 9.21 There are over 2 km of vertical relief between the base of Wyoming's Grand Teton Mountains and the peak. And the base already lies at a high elevation. The exposed rocks once lay 12 km or more beneath the surface of the Earth. Clearly, mountain building results in large vertical displacements of the surface of the crust.

lower to a higher elevation as uplift. Mountain building requires substantial uplift of the Earth's surface (►Fig. 9.21). For example, as noted earlier, Mt. Everest, the highest mountain on Earth, rises 8.85 km above sea level (about 0.06% of the Earth's diameter). In this section we look at why uplift occurs, how erosion carves rugged landscapes into uplifted crust, and why mountains can't get much higher than Mt. Everest.

Crustal Roots and Mountain Heights

In the mid-1800s, Sir George Everest undertook a survey of India, which at the time formed part of the British Empire. Everest, aside from contributing his name to Earth's highest mountain, discovered that the mass of the Himalaya Mountains was great enough to deflect the plumb bob (a lead weight at the end of a string) he was using to determine the vertical direction when setting up surveying equipment. But he was surprised to note that the amount of deflection caused by the mountains was actually *less* than it should have been, given their size. Why? A nineteenth-century British scientist, George Airy, keeping in mind that Earth's crustal rocks are less dense than its mantle rocks, came up with an explanation. Airy suggested that the crust of the Earth is thicker beneath the Himalayas than elsewhere, and thus that a low-density crustal root protrudes downward into the dense mantle beneath the range. A mountain range with a low-density crustal root has less mass overall than a mountain range underlain by dense mantle, and so it exerts less pull on a plumb bob.

Work in the twentieth century has confirmed that collisional and convergent mountain ranges do sit above crustal roots. Whereas typical continental crust has a thickness of about 35–40 km (measured from the surface to the Moho),

the crust beneath some mountain belts may reach a thickness of 50–70 km (about double its normal thickness) (►Fig. 9.22a). Mountain building in these belts shortens the crust horizontally and thickens it vertically. Crustal roots are important because without their buoyancy, mountain ranges would not be so high. To picture the importance of a crustal root in keeping some mountain ranges high, imagine that you place a set of wooden blocks in a bathtub. The thicker blocks float higher than the thinner blocks; in fact, the highest elevation occurs at the top of the thickest block (►Fig. 9.22b). The extra thickness of continental crust beneath collisional and convergent mountain ranges causes their high elevations.

Keep in mind that in the Earth System, it's not just the continental crust that floats on the mantle. Rather, it's the continental lithosphere, including the crust plus the lithospheric mantle, as a whole that floats on the asthenosphere. The soft, but solid, asthenosphere can slowly flow out of the way as the base of the lithosphere sinks. The crust acts like a

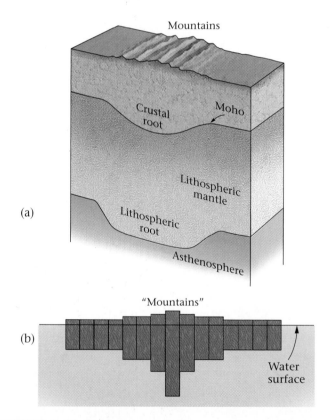

FIGURE 9.22 (a) Mountain belts have crustal roots, meaning that where the land surface rises to a higher elevation, the crust underneath is thicker. (b) In general, the lithosphere obeys Archimedes' principle of buoyancy. The surface of a thicker (longer) buoyant block rises to a higher elevation than the surface of a thinner (shorter) buoyant block. Also, the base of a thicker block extends down to greater depths.

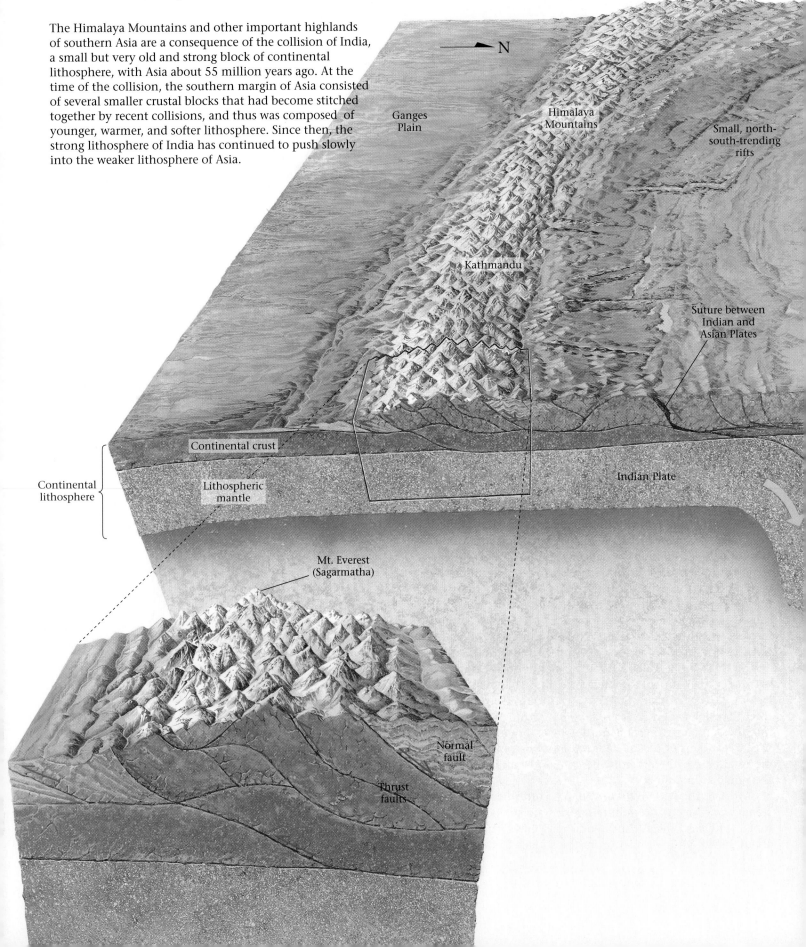

The Collision of India with Asia

The Himalaya Mountains and other important highlands of southern Asia are a consequence of the collision of India, a small but very old and strong block of continental lithosphere, with Asia about 55 million years ago. At the time of the collision, the southern margin of Asia consisted of several smaller crustal blocks that had become stitched together by recent collisions, and thus was composed of younger, warmer, and softer lithosphere. Since then, the strong lithosphere of India has continued to push slowly into the weaker lithosphere of Asia.

N

Ganges Plain

Himalaya Mountains

Small, north-south-trending rifts

Kathmandu

Suture between Indian and Asian Plates

Continental crust

Indian Plate

Continental lithosphere

Lithospheric mantle

Mt. Everest (Sagarmatha)

Normal fault

Thrust faults

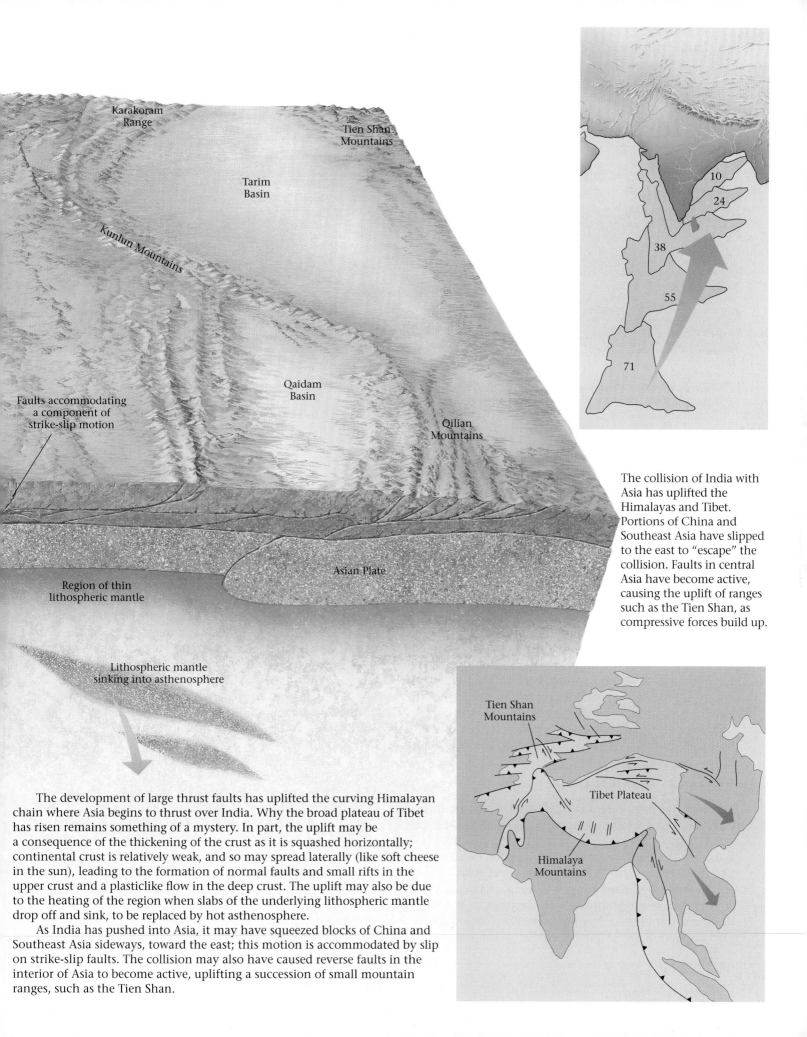

Karakoram Range

Tien Shan Mountains

Tarim Basin

Kunlun Mountains

Faults accommodating a component of strike-slip motion

Qaidam Basin

Qilian Mountains

Region of thin lithospheric mantle

Asian Plate

Lithospheric mantle sinking into asthenosphere

The collision of India with Asia has uplifted the Himalayas and Tibet. Portions of China and Southeast Asia have slipped to the east to "escape" the collision. Faults in central Asia have become active, causing the uplift of ranges such as the Tien Shan, as compressive forces build up.

The development of large thrust faults has uplifted the curving Himalayan chain where Asia begins to thrust over India. Why the broad plateau of Tibet has risen remains something of a mystery. In part, the uplift may be a consequence of the thickening of the crust as it is squashed horizontally; continental crust is relatively weak, and so may spread laterally (like soft cheese in the sun), leading to the formation of normal faults and small rifts in the upper crust and a plasticlike flow in the deep crust. The uplift may also be due to the heating of the region when slabs of the underlying lithospheric mantle drop off and sink, to be replaced by hot asthenosphere.

As India has pushed into Asia, it may have squeezed blocks of China and Southeast Asia sideways, toward the east; this motion is accommodated by slip on strike-slip faults. The collision may also have caused reverse faults in the interior of Asia to become active, uplifting a succession of small mountain ranges, such as the Tien Shan.

Tien Shan Mountains

Tibet Plateau

Himalaya Mountains

(a)

(b)

FIGURE 9.23 Evidence of erosion in mountain ranges. (a) Glacially carved peaks in Switzerland. (b) Deep valleys cut by rivers into the weathered bedrock of eastern Brazil.

buoy that holds the lithosphere up; the bigger the buoy (the thicker the crust), the higher the lithosphere floats. In some cases, the dense lithospheric mantle may eventually detach and drop off the base of the plate and cause the crust to rise a bit higher, like a balloon that has dropped some ballast.

Clearly, buoyancy forces play an important role in the process of uplifting mountain belts. The condition that exists when the buoyancy force pushing lithosphere up equals the gravitational force pulling lithosphere down is called **isostasy,** or isostatic equilibrium. In most places, isostatic equilibrium exists at the surface of the crust, so that the surface elevation of the crust reflects the level at which the lithosphere naturally floats. If a geologic event happens that changes the density or thickness of the lithosphere, then the surface of the crust slowly rises or falls to reestablish isostatic equilibrium, a process called isostatic compensation.

What Goes Up Must Come Down

The image we most often associate with a mountain range is that of rugged topography with spirelike peaks, knife-edge ridges, precipitous cliffs, and deep valleys. This type of landscape develops because of erosion by ice and water that, over time, sculpts uplifted land. Water (rain or snow) falling on uplifted land has gravitational potential energy. The water flows downslope, and in the process carries away debris that tumbled down steep slopes. This sediment-laden water also grinds away at the bedrock that it flows over.

The specific style of topography found within a mountain range depends on the climate. If conditions in the mountains become cool enough, glaciers form and, as they flow, carve pointed peaks and steep-sided valleys, as we will see in Chapter 18. But in many ranges, we can see glacially carved landscapes even though there are no glaciers today. Such landscapes formed during the last ice age and stopped forming when the ice melted away, less than 14,000 years ago (▶Fig. 9.23a). At lower elevations, or in warmer climates, mountain landscapes reflect the consequences of river erosion (▶Fig. 9.23b). Soil formation and vegetation growth may blunt escarpments, creating rounded hills.

Mountains much higher than Mt. Everest cannot exist on the Earth, for two reasons. First, erosion attacks mountains as soon as they rise, and in some cases it can tear mountains down as fast as uplift occurs. Second, rock does not have infinite strength. As mountains rise, the weight of overlying rock presses down on rock buried at depth in the crust. While this is happening, the rock buried at depth gradually grows warmer and softer because of heat rising from the Earth's interior. Eventually, the weight of the mountain range causes the warm, soft rock at depth to flow slowly. Effectively, the mountains begin to collapse under their own weight and spread laterally like soft cheese that has been left out in the summer sun. Geologists call this process orogenic collapse. Together, uplift, erosion, and orogenic collapse bring metamorphic and plutonic rocks that were once deep in the crust up toward the Earth's surface, a process called exhumation.

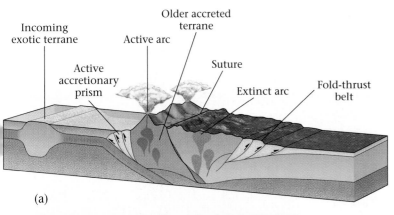

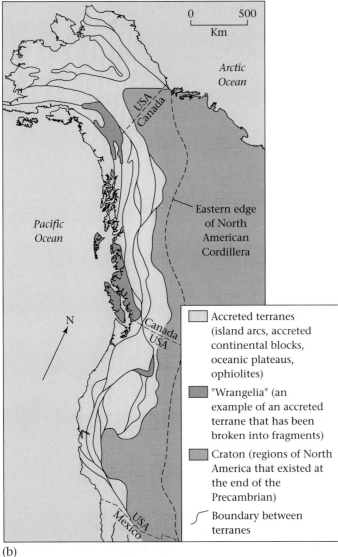

(a)

(b)

FIGURE 9.24 (a) In a convergent-margin orogen, volcanic arcs form, and compression may occur. Where this occurs, a large mountain range develops. Exotic terranes may collide with the convergent margin and accrete to the orogen. (b) Much of the western portion of the North American Cordillera consists of accreted terranes.

9.9 CAUSES OF MOUNTAIN BUILDING

Before plate tectonics theory became established, geologists were just plain confused about how mountains formed. Plate tectonics theory provided a new approach to interpreting mountains: as noted earlier, mountains form in response to convergent boundary deformation, continental collisions, and rifting. Below, we look at these different settings and the types of mountains and geologic structures that develop in each one.

Mountains Related to Convergence

Along the margins of continental convergent plate boundaries, where oceanic lithosphere subducts beneath a continent, a continental volcanic arc forms, and compression between the two plates causes a mountain range to rise (▶Fig. 9.24a). If plate movements push the continent tightly against the subduction zone, compression on the continent side of the volcanic arc generates a fold-thrust belt. In such belts, numerous thrust faults and associated folds develop, accommodating significant horizontal shortening of the crust. Convergence along the western coast of South America, for example, has generated a fold-thrust belt on the eastern side of the Andes. A similar belt formed on the eastern side of the North American Cordillera during the Mesozoic and Cenozoic.

During convergent-margin orogeny, offshore island volcanic arcs, oceanic plateaus, and small fragments of continental crust may drift into the convergent margin. These blocks are too buoyant to subduct, so they collide with the convergent margin and accrete, or attach, to the continent. Geologists refer to such blocks before they attach as exotic terranes and after they have attached as accreted terranes. The western half of the North American Cordillera consists largely of accreted terranes (▶Fig. 9.24b).

Mountains Related to Continental Collision

Once the oceanic lithosphere between two continents completely subducts, the continents themselves collide with each other. Continental collision results in the formation of large mountain ranges such as the present-day Himalayas or the Alps (▶Fig. 9.25) and the Paleozoic Appalachian Mountains. The final stage in the growth of the Appalachians happened when Africa and North America collided.

During collision, intense compression generates fold-thrust belts on the margins of the orogen (▶Fig. 9.26). In the interior of the orogen, where one continent overrides the edge of the other, high-grade metamorphism occurs, accompanied by formation of flow folds and tectonic foliation. During this process, the crust below the orogen thickens to as much as twice its normal thickness. Gradually, rocks

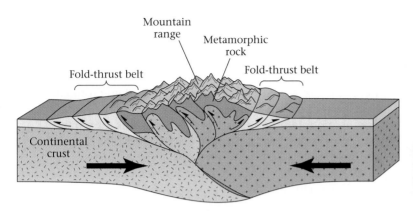

FIGURE 9.25 In a collisional orogen, two continents collide. The compression that results from the collision shortens and thickens the continental crust so that a large mountain range develops. Fold-thrust belts form along the margins of the orogen.

squeeze upward in the hanging walls of large thrust faults and later become exposed by exhumation. Finally, as noted earlier, rock at depth in the orogen heats up and becomes so weak that the mountain belt may collapse and spread out sideways. The broad Tibet Plateau may have formed in part when crust, thickened during the collision of India with Asia, spread to the northeast (see art on pp. 264–65).

Mountains Related to Continental Rifting

Continental rifts are places where continents are splitting in two. When rifts first form, uplift occurs, because as the lithosphere thins, hot asthenosphere rises, making the remaining lithosphere warmer and less dense. Because the lithosphere is less dense, it becomes more buoyant and thus rises to reestablish isostatic equilibrium.

As rifting continues, stretching causes normal faulting in the brittle crust above (▶Fig. 9.27). Movement on the normal faults drops down blocks of crust, producing deep, sediment-filled basins separated by narrow, elongate mountain ranges that contain tilted rocks. These ranges are sometimes called fault-block mountains. In addition, the rising asthenosphere beneath the rift partially melts, generating magmas that rise to form volcanoes within the rift. Today, the East African Rift clearly shows the configuration of rift-related mountains and volcanoes. And in North America, rifting yielded the broad Basin and Range Province of Utah, Nevada, and Arizona (▶Fig. 9.28).

9.10 CRATONS AND THE DEFORMATION WITHIN THEM

A **craton** consists of crust that has not been affected by orogeny for at least the last 1 billion years. Because orogeny

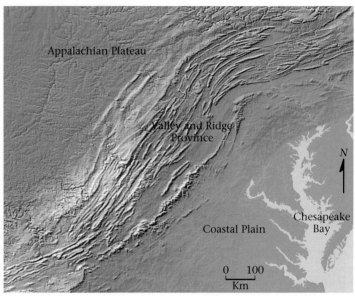

FIGURE 9.26 Relief map of the Valley and Ridge Province, Pennsylvania to Virginia. The ridges, which outline the shapes of plunging folds in the fold-thrust belt, are composed of resistant sandstone beds.

happened so long ago in cratons, their crust has become quite cool, and therefore relatively strong and stable. We can divide cratons into two provinces: shields, in which Precambrian metamorphic and igneous rocks crop out at the ground surface, and platforms, where a relatively thin layer of Phanerozoic sediment covers the Precambrian rocks (▶Fig. 9.29).

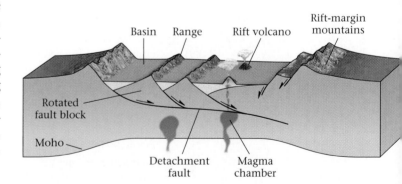

FIGURE 9.27 When the crust stretches in a continental rift, rift-related mountains form, as do normal faults. Displacement on the faults leads to the tilting of crustal blocks and the formation of basins. The basins fill with sediment eroded from the adjacent mountains. The exposed ends on the tilted blocks create long, narrow ranges. In the United States, the region containing such a structure is called the Basin and Range Province (located in Nevada, Utah, and Arizona).

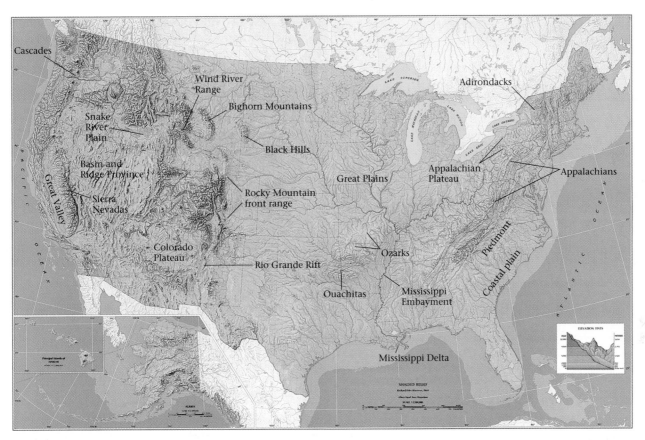

FIGURE 9.28 Reference map: topography of the United States.

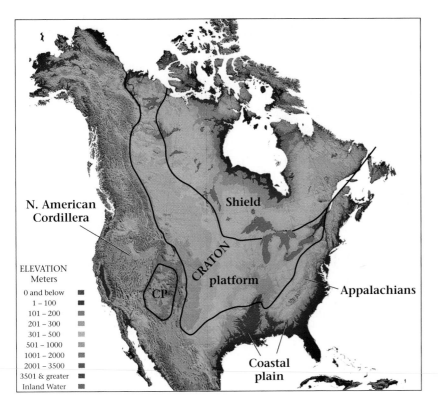

FIGURE 9.29 A digital elevation map of North America. The different colors indicate different elevations and define topographic provinces. The high, rugged area in the east is the Appalachian range and the one in the west is the North American Cordillera. The flat lands along the east and south coasts are the coastal plain, where the eroded roots of Paleozoic mountains were covered by sediment during the Cretaceous and early Cenozoic. The craton forms the interior of the continent. It can be subdivided into the shield, where Precambrian basement rock crops out, and the platform, where Paleozoic and Mesozoic sedimentary rock has buried the Precambrian basement.

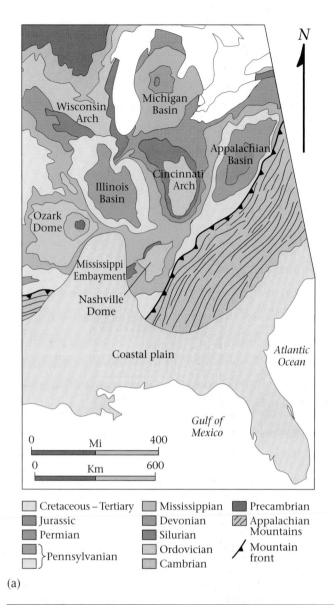

Cretaceous – Tertiary | Mississippian | Precambrian
Jurassic | Devonian | Appalachian Mountains
Permian | Silurian | Mountain front
Pennsylvanian | Ordovician
 | Cambrian

(a)

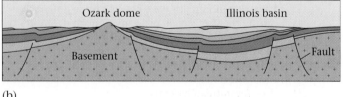

(b)

FIGURE 9.30 (a) Geologic map of the mid-continent region of the United States, showing the basins and domes and the faults that cut the region. (b) This cross section illustrates the geometry of the regional basins and domes. Note how layers of strata thin toward the crest of the Ozark Dome and thicken toward the center of the Illinois Basin. Some of the faults originated as normal faults and later moved again as reverse faults.

In shield areas, we find intensively deformed metamorphic rocks—abundant examples of flow folds and tectonic foliation. That's because the crust making the cratons was deformed during a succession of orogenies in the Precambrian. Recent studies of the Canadian Shield, which occupies much of the eastern two-thirds of Canada, for example, reveal the traces of Himalaya-like collision zones, Andean-like convergent boundaries, and East African–like rifts, all formed more than 1 billion years ago (some over 3 billion years ago). These orogens are so old that erosion has worn away the original topography, in the process exposing deep crustal rocks at the Earth's surface.

In the cratonic platform, strata do display deformation features, but in contrast to the deformation of orogens, cratonic-platform deformation is less intense. The cratonic platform of the U.S. Midwest region includes two classes of structures: regional basins and domes, and local zones of folds and faults.

Regional basins and regional domes are broad areas that gradually sank or rose, respectively (▶Fig. 9.30a, b). For example, in Missouri, strata arch across a broad dome, the Ozark dome, whose diameter is 300 km. Individual sedimentary layers thin toward the top of the dome, because less sediment accumulated on the dome than in adjacent basins. Erosion during more recent geologic history has produced the characteristic bull's-eye pattern of a dome, with the oldest rocks (Precambrian granite) exposed near the center. In the Illinois basin, strata appear to warp downward into a huge bowl that is also about 300 km across. Strata get thicker toward the center, indicating that the floor of the basin sank so there was more room for sediment to accumulate. The Illinois basin also has a bull's-eye shape, but here the youngest strata are exposed in the center.

Folds and faults are hard to find in the cratonic platform, because most do not cut the ground surface. But subsurface studies indicate that faults do occur at depth. Monoclines, step-shaped folds, develop over these faults; the folds formed as a block of basement pushes up. Most of these zones were likely active when major orogenies happened along the continental margin. This relation suggests that the orogenies produced enough stress in the craton to cause faults to move, but not enough to generate large mountains or to create foliation.

9.11 MEASURING MOUNTAIN BUILDING IN PROGRESS

Mountains are not just "old monuments," as John Muir mused. The rumblings of earthquakes and the eruptions of

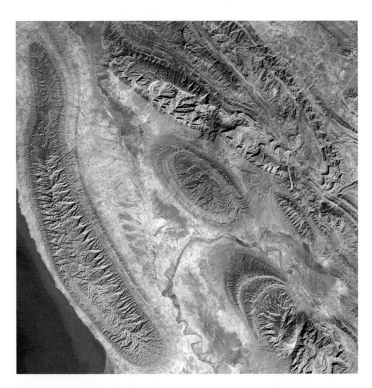

THE VIEW FROM SPACE Subduction of the Persian Gulf beneath the Asian continent, in the vicinity of Konari, Iran, produced doubly plunging folds. Dipping strata outline the ellipsoidal shape of the folds. The large one, near the coast, is about 10 km wide. Some of the folds are cored by salt. This is a tectonically active landscape, so deformation is warping the land surface.

volcanoes attest to present-day movements in mountains. Geologists can measure the rates of these movements through field studies and satellite technology. For example, geologists can determine where coastal areas have been rising relative to the sea level by locating ancient beaches that now lie high above the water. And they can tell where the land surface has risen relative to a river by identifying places where a river has recently carved a new valley down into sediments that it had previously deposited. In addition, geologists now can use the satellite **global positioning system (GPS)** to measure rates of uplift and horizontal shortening in orogens. With present technology, we can "see" the Andes shorten horizontally at a rate of a couple of centimeters per year, and we can "watch" as mountains along this convergent boundary rise by a couple of millimeters per year.

CHAPTER SUMMARY

- Mountains occur in linear ranges called mountain belts, orogenic belts, or orogens. An orogen forms during an orogeny, or mountain-building event. Orogenies, which last for millions of years, are a consequence of continental collision, subduction at a convergent plate boundary, or rifting.

- Mountain building causes rocks to bend, break, shorten, stretch, and shear. Because of such deformation, rocks change their location, orientation, and shape.

- During brittle deformation, rocks crack and break into two or more pieces. During ductile deformation, rocks change shape without breaking.

- Rocks undergo three kinds of stress: compression, tension, and shear. Strain refers to the way rocks change shape when subjected to a stress. Compression causes shortening, tension causes stretching, and shear stress leads to shear strain.

- Deformation results in the development of geologic structures.

- Joints are natural cracks in rock, formed in response to tension under brittle conditions. Veins develop when minerals precipitate out of water passing through joints.

- Faults are fractures on which there has been shearing. On normal faults, the hanging-wall block slides down the surface; on reverse faults, the hanging-wall block slides up the surface; on strike-slip faults, rock on one side of the fault slides horizontally past the other; and on oblique-slip faults, rock slides diagonally across the surface.

- Folds are curved layers of rock. Anticlines are archlike folds, synclines are troughlike, monoclines resemble the shape of a carpet draped over a stair step, basins are shaped like a bowl, and domes are shaped like an overturned bowl.

- Tectonic foliation forms when grains are flattened or rotated so that they align parallel with one another, or when new platy grains grow parallel to one another.

- Large mountain ranges are underlain by relatively buoyant roots. The height of such mountains is controlled by isostasy.

- Once uplifted, mountains are sculpted by the erosive forces of glaciers and rivers. Also, when the crust thickens during mountain building, the deep crust eventually becomes warm and weak and begins to flow, leading to orogenic collapse.

- Mountain belts formed by convergent margin tectonism may incorporate accreted terranes. Continental collision, which resulted in the Alps, Himalayas, and

Appalachians, generates metamorphic rocks and tectonic foliation. Fold-thrust belts form on the continental edge of collisional and convergent-margin orogens. Tilted blocks of crust in rifts become narrow, elongate mountain ranges called fault-block mountains.

• Cratons are the old, relatively stable parts of continental crust. They include shields, where Precambrian rocks are exposed at the surface, and platforms, where Precambrian rocks are buried by a thin layer of sedimentary rock. Broad regional domes and basins form in platform areas.

• With modern GPS technology, it is now possible to measure the slow uplift of mountains.

KEY TERMS

anticline (p. 259)
basin (p. 259)
brittle deformation (p. 251)
compression (p. 253)
craton (p. 268)
deformation (p. 249)
dip (p. 254)
displacement (p. 257)
dome (p. 259)
ductile deformation (p. 251)
fault (p. 249)
fault scarp (p. 258)
fold (p. 249)
foliation (p. 249)
geologic structures (p. 249)
global positioning system (GPS) (p. 271)

isostasy (isostatic equilibrium) (p. 266)
joint (p. 249)
mountain belt (p. 249)
normal fault (p. 257)
orogen (p. 249)
orogeny (p. 249)
pressure (p. 254)
reverse fault (p. 257)
shear stress (p. 253)
strain (p. 251)
strike (p. 254)
strike-slip fault (p. 257)
syncline (p. 259)
tension (p. 253)
thrust fault (p. 257)
veins (p. 256)

REVIEW QUESTIONS

1. What changes do rocks undergo during formation of an orogenic belt such as the Alps?

2. What is the difference between brittle and ductile deformation?

3. What factors determine whether a rock will behave in brittle or ductile fashion?

4. How are stress and strain different?

5. How is a fault different from a joint?

6. Compare the motion of normal, reverse, and strike-slip faults.

7. How do you recognize faults in the field?

8. Describe the differences among an anticline, a syncline, and a monocline.

9. Discuss the relationship between foliation and deformation.

10. Describe the principle of isostasy.

11. Discuss the processes by which mountain belts form in convergent margins, in continental collisions, and in continental rifts.

12. How are the structures of a craton different from those of an orogenic belt?

SUGGESTED READING

Condie, K. 2004. *Earth as an Evolving Planetary System.* New York: Academic Press.

Davis, G. H., and S. J. Reynolds. 1996. *Structural Geology of Rocks and Regions.* 2nd ed. New York: Wiley.

Hancock, P. L., ed. 1994. *Continental Deformation.* Oxford, England: Pergamon Press.

McPhee, J. 1998. *Annals of the Former World.* New York: Farrar, Straus and Giroux.

Moores, E. M., and R. J. Twiss. 1995. *Tectonics.* New York: Freeman.

van der Pluijm, B. A., and S. Marshak. 2004. *Earth Structure: An Introduction to Structural Geology and Tectonics.* 2nd ed. New York: W. W. Norton.

Memories of Past Life: Fossils and Evolution

D.1 THE DISCOVERY OF FOSSILS

If you pick up a piece of sedimentary rock, you may find shapes that look like shells, bones, leaves, or footprints (▶Fig. D.1). The origin of these shapes mystified early thinkers. Some thought that the shapes had simply grown underground, in solid rock. Today, geologists consider such **fossils** (from the Latin word *fossilis,* which means "dug up") to be remnants or traces of ancient living organisms now preserved in rock, and that they form when organisms become buried by sediment. This viewpoint, though first proposed by the Greek historian Herodotus in 450 B.C.E. and revived by Leonardo da Vinci in 1500 C.E., did not become widely accepted until the publication in 1669 of a book on the subject by a Danish physician named Nicholaus Steno (1638–1686). Steno noted the similarity between fossils known as "dragon's tongues" and the teeth of modern sharks, and concluded that fossil-containing rocks originated as loose sediment that had incorporated the remains of organisms, and that when the sediment hardened into rock, the organisms became part of the rock. The understanding of fossils increased greatly thanks to the efforts of a British scientist, Robert Hooke (1635–1703), who described and sketched fossils in detail and who realized that most represent extinct species, meaning species that lived in the past but no longer exist. During the following two centuries, geologists described thousands of fossils and established museum collections (▶Fig. D.2).

The nineteenth century saw **paleontology,** the study of fossils, ripen into a science. Work with fossils went beyond description alone when William Smith, a British engineer who supervised canal construction in England during the 1830s, noted that different fossils occur in different layers of strata within a sequence of sedimentary rocks. In fact, Smith realized that he could define a distinctive succession of fossils, and that a given species would be present only for a specific interval of strata. This discovery made it possible to use fossils as a basis for determining the age of one sedimentary

FIGURE D.2 A drawer of labeled fossils in a museum. Paleontologists from around the world study such collections to help identify unknown specimens.

FIGURE D.1 This bedding surface in limestone contains fossils of organisms that lived about 420 million years ago. These particular species no longer exist on Earth.

rock layer. Fossils, therefore, became an indispensable tool for studying geologic history and the evolution of life. In this interlude, we introduce fossils, an understanding of which serves as essential background for the next two chapters.

D.2 FOSSILIZATION

What Kinds of Rocks Contain Fossils?

Most fossils are found in sediments or sedimentary rocks, for they form when organisms die and become buried by sediment, or when organisms travel over or through sediment and leave their mark. Rocks formed from sediments deposited under anoxic (oxygen-free) conditions in quiet water (such as lake beds or lagoons) preserve particularly fine specimens. Rocks made from sediments deposited in high-energy environments, on the other hand—where strong currents tumble shells and bones and break them up—contain at best only small fragments of fossils mixed with other clastic grains.

Fossils can survive low grades of metamorphism, but not the recrystallization and new mineral growth, and in some cases shearing, that occur during intermediate- and high-grade metamorphism. Similarly, fossils do not occur in igneous rocks that crystallize directly from melt, for organisms can't live in molten rock, and if engulfed by molten rock, they will be incinerated. Fossils can occur in deposits formed from air-fall ash, for the ash settles just like sediment and can bury an organism or a footprint.

Forming a Fossil

Paleontologists, scientists who study fossils, refer to the process of forming a fossil as **fossilization.** To see how a typical fossil develops in sedimentary rock, let's follow the fate of an old dinosaur as it searches for food along a riverbank (▶Fig. D.3). On a scalding summer day, the hungry dinosaur, plodding through the muddy ground, succumbs to the heat and collapses dead into the mud. Over the coming days, scavengers strip the skeleton of meat and scatter the bones among the dinosaur footprints. But before the bones have time to weather away, the river floods and buries the bones, along with the footprints, under a layer of silt. More silt from succeeding floods buries the bones and prints still deeper in a chemically stable environment, so that the bones cannot be reworked by currents or disrupted by burrowing organisms. Later, sea level rises and a thick sequence of marine sediment buries the fluvial sediment.

Eventually, the sediment containing the bones turns to rock (siltstone). The footprints remain outlined by the contact between the siltstone and mud, while the bones reside within the siltstone. Minerals precipitating from groundwater passing through the siltstone gradually replace some of the chem-

icals constituting the bones, until the bones themselves have become like rock. The buried bones and footprints are now fossils. One hundred million years later, uplift and erosion expose the dinosaur's grave. Part of a fossil bone protrudes from a rock outcrop. A lucky paleontologist observes the fragment and starts excavating, gradually uncovering enough of the bones to permit reconstruction of the beast's skeleton. Further digging uncovers footprints. The dinosaur rises again, but this time in a museum. In recent years, bidding wars have made some fossil finds extremely valuable. For example, a skeleton of a *Tyrannosaurus rex,* a 67-million-year-old dinosaur found in South Dakota, sold at auction for $7.6 million in 1997. The specimen, named Sue after its discoverer, now stands in the Field Museum of Chicago.

Similar tales can be told for fossil seashells buried by sediment settling in the sea, for insects trapped in hardened tree sap (amber), and for mammoths drowned in the muck of a tar pit. In all cases, fossilization involves the burial and preservation of an organism or the trace (a footprint or burrow) of an organism. Once buried, the organism may be altered to varying degrees by pressure from overlying rock and chemical interaction with groundwater.

The Many Different Kinds of Fossils

Perhaps when you think of a fossil, you picture either a dinosaur bone or the imprint of a seashell in rock. In fact, paleontologists distinguish many different kinds of fossils, according to the specific way in which the organism was fossilized. Let's look at examples of these categories.

- *Frozen or dried body fossils:* In a few environments, whole bodies of organisms may be preserved. Most of these fossils are fairly young, by geologic standards—their ages can be measured in thousands, not millions, of years. Examples include woolly mammoths that became incorporated in the permafrost (permanently frozen ground) of Siberia and have stayed frozen since their death. (▶Fig. D.4a).

- *Body fossils preserved in amber or tar:* Insects landing on the bark of trees may become trapped in the sticky sap or resin the trees produce. This golden syrup envelops the insects and over time hardens into **amber,** the semiprecious "stone" used for jewelry. Amber can preserve insects, as well as other delicate organic material such as feathers, for 40 million years or more (▶Fig. D.4b).

  Tar similarly acts as a preservative. In isolated regions where oil has seeped to the surface, the more volatile components of the oil evaporate away and bacteria degrade what remains, leaving behind a sticky residue. At one such locality, the La Brea Tar Pits in Los Angeles, tar accumulated in a swampy area. While grazing or drinking at the swamp, or while attacking at the swamp, animals became mired in the tar and sank into it. Their bones have been remarkably well preserved for over 40,000 years.

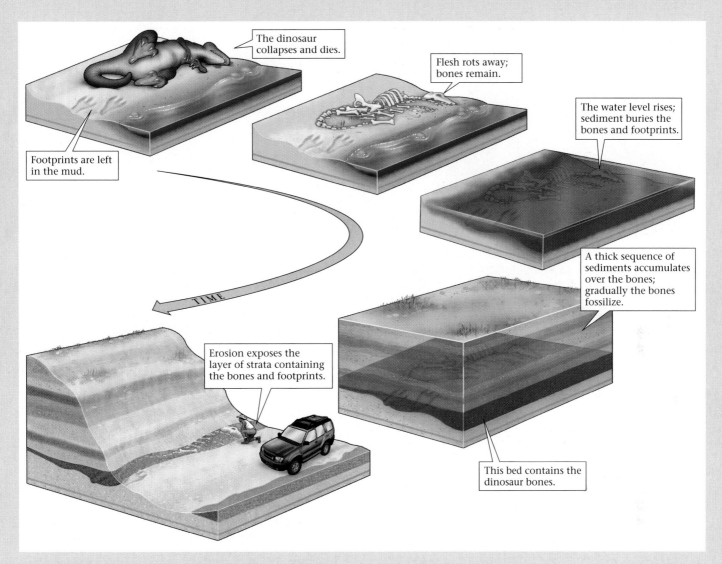

The dinosaur collapses and dies.

Flesh rots away; bones remain.

The water level rises; sediment buries the bones and footprints.

Footprints are left in the mud.

A thick sequence of sediments accumulates over the bones; gradually the bones fossilize.

TIME

Erosion exposes the layer of strata containing the bones and footprints.

This bed contains the dinosaur bones.

FIGURE D.3 How a dinosaur eventually becomes a fossil.

- *Preserved or replaced bones, teeth, and shells:* Bones (the internal skeletons of vertebrate animals) and shells (the external skeletons of invertebrate animals) consist of durable minerals, which may survive in rock. Some bone or shell minerals are not stable, and they recrystallize (▶Fig. D.4c). But even when this happens, the shape of the bone or shell remains in the rock.

- *Permineralized organisms:* **Permineralization** refers to the process by which minerals precipitate in porous material, such as wood or bone, from groundwater solutions that have seeped into the pores. **Petrified wood,** for example, forms by permineralization of wood, so that cell interiors are replaced with silica, causing the wood to become chert. In fact, the word *petrified* literally means "turned to stone." The more resistant cellulose of the wood transforms into an organic film that remains after permineralization, so

that the fine detail of the wood's cell structure can be seen in a petrified log (▶Fig. D.4d).

- *Molds and casts of bodies:* As sediment compacts around a shell, it conforms to the shape of the shell or body. If the shell or body later disappears, because of weathering and dissolution, a cavity called a mold remains (▶Fig. D.4e). (Sculptors use the same term to refer to the receptacle into which they pour bronze or plaster.) A mold preserves the delicate shape of the organism's surface; it looks like an indentation on a rock bed. The sediment that had filled the mold also preserves the organism's shape; this cast protrudes from the surface of the adjacent bed.

- *Carbonized impressions of bodies:* Impressions are simply flattened molds created when soft or semisoft organisms (leaves, insects, shell-less invertebrates, sponges, feathers, jellyfish) get pressed between layers of sediment. Chemical

FIGURE D.4 (a) A frozen mammoth, found in the permafrost (permanently frozen ground) of Siberia about 100,000 years after it died. It still had flesh and fur. (b) A piece of amber containing two fossil insects. (c) Fossil dinosaur bones exposed on a tilted bed of sandstone in Dinosaur National Monument, Utah. (d) Petrified wood from the Petrified National Forest, Arizona. Petrified wood is much harder than the surrounding tuff and thus remains after the tuff has eroded away. (e) Molds and casts of organisms. (f) The carbonized impression of fern fronds. (g) Dinosaur footprints in mudstone. (h) Worm burrows on a block of siltstone. (Lens cap for scale.)

reactions eventually remove the organic material that composed the organism, leaving only a thin film of carbon on the surface of the impression (▶Fig. D.4f).

- *Trace fossils:* These include footprints, feeding traces, burrows, and dung (coprolites) that organisms leave behind in sediment (▶Fig. D.4g, h).

- *Chemical fossils:* Living things consist of complex organic chemicals. When buried with sediment and subjected to diagenesis, some of these chemicals are destroyed, but some either remain intact or break down to form different, but still distinctive, chemicals. A distinctive chemical derived from an organism and preserved in rock is called a chemical fossil. (Such chemicals may also be called molecular fossils or biomarkers.)

Paleontologists also find it useful to distinguish among different fossils on the basis of their size. Macrofossils are fossils large enough to be seen with the naked eye. But some rocks and sediments also contain abundant microfossils, which can be seen only with a microscope or even an electron microscope. Microfossils include remnants of plankton, algae, bacteria, and pollen.

Fossil Preservation

Not all living organisms become fossils when they die. In fact, only a small percentage do, for it takes special circumstances—namely, one or more of the following four—to produce a fossil.

- *Death in an anoxic (oxygen-poor) environment:* A dead squirrel by the side of the road won't become a fossil. As time passes, birds, dogs, or other scavengers may come along and eat the carcass. And if that doesn't happen, maggots, bacteria, and fungi will infest the carcass and gradually digest it. Flesh that has not been eaten, or does not rot, reacts with oxygen in the atmosphere and transforms into carbon dioxide gas. The remaining skeleton weathers in air and turns to dust. Thus, before the dead squirrel can become incorporated in sediment, it has vanished. In order for fossilization to occur, a carcass must settle into an oxygen-poor environment, where oxidation reactions happen slowly, where scavenging organisms aren't as abundant, and where bacterial metabolism takes place very slowly. In such environments the organism won't rot away before it has a chance to be buried and preserved.

- *Rapid burial:* If an organism dies in a depositional environment where sediment accumulates rapidly, it may be buried before it has time to rot, oxidize, be eaten, be completely broken up, or be consumed by burrowing organisms. For example, if a storm suddenly buries an oyster bed with a thick layer of silt, the oysters die and become part of the sedimentary rock derived from the sediment.

- *The presence of hard parts:* Organisms without durable shells or skeletons, collectively called hard parts, commonly won't be fossilized, for soft flesh decays long before hard parts do under most depositional conditions. For this reason, paleontologists know much more about the fossil record of bivalves (a class of organisms, including clams and oysters, with strong shells) than they do about the fossil record of jellyfish (which have no shells) or spiders (which have very fragile shells).

- *Lack of metamorphism:* Metamorphism may destroy fossils, either by dissolving them away, or by causing recrystallization or neocrystallization sufficient to obscure the fossil's shape.

By carefully studying modern organisms, paleontologists have been able to provide rough estimates of the **preservation potential** of organisms, meaning the likelihood that an organism will be buried and eventually transformed into a fossil. For example, in a typical modern-day shallow-marine environment, such as the mud-and-sand sea floor close to a beach, about 30% of the organisms have sturdy shells and thus a high preservation potential, 40% have fragile shells and a low preservation potential, and the remaining 30% have no hard parts at all and are not likely to be fossilized except in special circumstances. Of the 30% with sturdy shells, though, few happen to die in a depositional setting where they actually *can* become fossilized.

Extraordinary Fossils: A Special Window to the Past

Though only hard parts survive in most fossilization environments, paleontologists have discovered a few special locations where rock contains relics of soft parts as well; such fossils are known as **extraordinary fossils.** We've already seen how extraordinary fossils such as insects and even feathers can be preserved in amber, and how complete skeletons have been found in tar pits. In some cases, such finds have yielded examples of ancient DNA. Extraordinary fossils may also be preserved in sediment that accumulates on the anoxic floor of lakes or lagoons or the deep ocean. Here, oxidation cannot occur, and flesh does not rot before burial. Carcasses of animals that settle into the mud gradually become fossils, but because they were buried before the destruction of their soft parts, fossil impressions of their soft parts surround the fossils of their bones.

A small quarry near Messel, in western Germany, for example, has revealed extraordinary fossils of 49-million-year-old mammals, birds, fish, and amphibians that died in a shallow-water lake (▶Fig. D.5a). Bird fossils from the quarry include the delicate imprints of feathers, bat fossils come complete with impressions of ears and wing flaps, and mammal fossils have an aura of carbonized fur. In southern Germany, exposures of the Solenhofen Limestone, an approximately 150-million-year-old rock derived from carbonate mud deposited in a stagnant lagoon, contain extraordinary fossils of six hundred species, including *Archaeopteryx,* one of the earliest birds (►Fig. D.5b). And exposures of the Burgess Shale in the Canadian Rockies of British Columbia have yielded a plentitude of fossils showing what shell-less invertebrates that inhabited the deep-sea floor about 510 million years ago looked like.

D.3 CLASSIFYING LIFE

The classification of fossils follows the same principles used for classifying living organisms. The principles of classification were first proposed in the eighteenth century by Carolus Linnaeus, a Swedish biologist. The study of how to classify organisms is now referred to as taxonomy. Linnaeus's scheme has a hierarchy of divisions. First, all life is divided into kingdoms, and each kingdom consists of one or more phyla. A phylum, in turn, consists of several classes; a class, of several orders; an order, of several families; a family, of several genera; and a genus, of one or more species. Kingdoms, then, are the broadest category and species the narrowest. Biologists now recognize six kingdoms (►Fig. D.6):

- *Archaea:* microorganisms typically found in extreme environments such as hot springs and in very acidic or very alkaline water;

- *Eubacteria:* "true bacteria," including the kind that cause infections;

- *Protista:* various unicellular and simple multicellular organisms; these include algae (such as diatoms) and forams, two of the major plankton types in the oceans;

- *Fungi:* mushrooms and yeast, etc.;

- *Plantae:* trees, grasses, and ferns, etc.; and

- *Animalia:* sponges, corals, snails, dinosaurs, ants, and people, etc.

Biologists have realized that Archaea are as different from Eubacteria as both are from the other kingdoms. Specifically, the last four kingdoms share enough common characteristics, at the cellular level, to be lumped together as a group called Eukarya. Thus, biologists now divide life into three "domains": Archaea, Eubacteria, and Eukarya. The cells of Eukaryotic organisms have distinct nuclei and internal membranes, whereas those of Eubacteria and Archaea do not. The latter two domains have so-called prokaryotic cells.

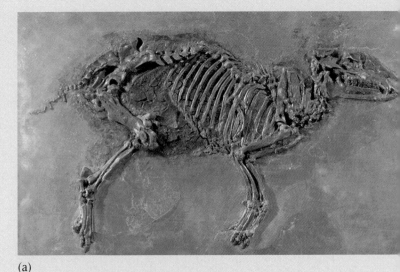

(a)

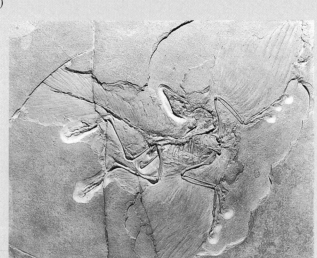

(b)

FIGURE D.5 Extraordinary fossils: (a) a mammal from Messel, Germany; (b) *Archaeopteryx* from the Solenhofen Limestone.

D.4 CLASSIFYING FOSSILS

Paleontologists traditionally distinguish different fossil species from each other according to **morphology** (the form or shape) alone. A fossil clam is a fossil clam because it looks like one. In some cases, the characteristics that distinguish one fossil species from another may be quite subtle (such as the number of ridges on the surface of a shell, or the relative length of different leg bones), but even beginners can distinguish the major groups of fossils from one another on sight.

There's nothing magical about classifying fossils. Major characteristics at the level of class or higher are fairly easy to distinguish just from the way they look. For instance, skeletons of birds are hard to mistake for skeletons of

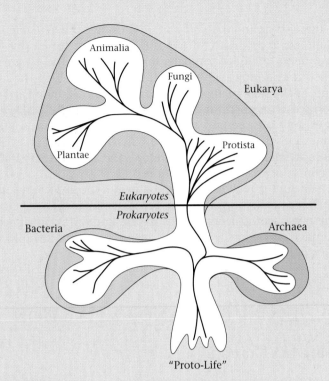

FIGURE D.6 The basic kingdoms of life on Earth. Archaea and Eubacteria are both prokaryotes, because, among other characteristics, they don't have nuclei. All other life forms are eukaryotes, because they consist of cells with nuclei.

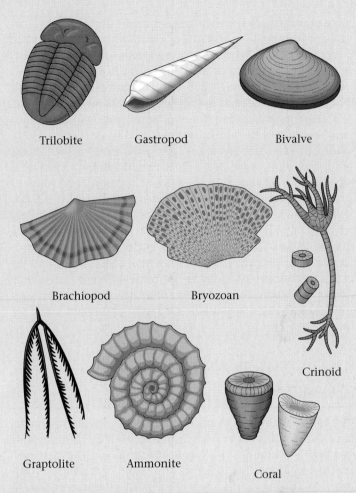

FIGURE D.7 Common types of invertebrate fossils. (Not to scale.)

mammals, and snail shells are hard to mistake for clam shells. But classification can also be difficult, especially if specimens are incomplete, and in many cases classification can be controversial.

▶Figure D.7 shows examples of some of the major types of invertebrate fossils. With this chart, you should be able to identify many of the fossils you'll find in a common bed of limestone. Particularly common invertebrate fossils include the following:

- *Trilobites:* These have a segmented shell that is divided lengthwise into three parts.

- *Gastropods* (snails): Most fossil specimens of gastropods have a shell that does not contain internal chambers.

- *Bivalves* (clams and oysters): These have a shell that can be divided into two similar halves.

- *Brachiopods* (lamp shells): The top and bottom parts of these shells have different shapes. Shells typically have ridges radiating out from the hinge.

- *Bryozoans:* These are colonial animals. Their fossils resemble a screenlike grid of cells.

- *Crinoids* (sea lilies): These organisms look like a flower but actually are animals. Their shells have a stalk consisting of numerous circular plates stacked on top of each other.

- *Graptolites:* These look like tiny carbon saw blades in a rock. They are remnants of colonial animals that floated in the sea.

- *Cephalopods:* These include ammonites, with a spiral shell, and nautiloids, with a straight shell. Their shells contain internal chambers and have ridged surfaces.

- *Corals:* These include colonial organisms that form distinctive mounds or columns.

D.5 THE FOSSIL RECORD

A Brief History of Life

As we will see in Chapter 10, the fossil record defines the long-term evolution of life on planet Earth. Archaea and bacteria fossils are found in rocks as old as about 3.7 billion years, indicating that these organisms are the earliest forms of life on Earth. Researchers once thought that archaea somehow developed by chemical reactions in concentrated "soups" that formed when seawater trapped in shallow pools evaporated. Growing evidence suggests, however, that they may instead have developed in warm groundwater beneath the Earth's surface or at hydrothermal vents on the sea floor.

For the first billion years or so of life history, archaea and bacteria were the only types of life on Earth. Then, about 2.5 Ga (billion years ago), organisms of the protist kingdom first appeared. Early multicellular organisms, shell-less invertebrates of the animal kingdom, and fungi of the plant kingdom came into existence perhaps 1.0–1.5 Ga. Within each kingdom, life radiated (divided) into different phyla, and within each phylum, life radiated into different classes. The great variety of shelly invertebrate classes appeared a little over 540 million years ago, and many organisms that persist today appeared at different times since then: first came invertebrates, then fish, land plants, amphibians, reptiles, and finally birds and mammals.

D.6 IS THE FOSSIL RECORD COMPLETE?

By some estimates, more than 250,000 species of fossils have been collected and identified to date, by thousands of geologists working on all continents during the past two centuries. These fossils define the framework of life evolution on planet Earth. But the record is not complete—known fossils cannot account for every intermediate step in the evolution of every organism. Considering that as many as 5 million species may be living on Earth today (not counting bacteria), over the billions of years that life has existed there may have been 5 billion to 50 billion species. Clearly, known fossils represent at most a tiny percentage of these species. Why is the record so incomplete?

First, despite all the fossil-collecting efforts of the past two centuries, paleontologists have not even come close to sampling every cubic centimeter of sedimentary rock exposed on Earth. Just as biologists have not yet identified every living species of insect, paleontologists have not yet identified every species of fossil. New species and even genera of fossils continue to be discovered every year.

Second, not all organisms are represented in the rock record, because not all organisms have a high preservation potential. As noted earlier, fossilization occurs only under special conditions, and thus only a minuscule fraction of the organisms that have lived on Earth become fossilized. There may be few, if any, fossils of a vast number of extinct species, so we have no way of knowing that they ever existed.

Finally, as we will learn in Chapter 10, the sequence of sedimentary strata that exists on Earth does not account for every minute of time since the formation of our planet.

Sediments accumulate only in environments where conditions are appropriate for deposition and not for erosion—sediments do not accumulate, for example, on the dry great plains or on mountain peaks, but do accumulate in the sea and in the floodplains and deltas of rivers. Because Earth's climate changes through time and because the sea level rises and falls, certain locations on continents are sometimes sites of deposition and sometimes aren't, and on occasion are sites of erosion. Therefore, the sequence of strata records only part of geologic time.

In sum, a rock sequence provides an incomplete record of Earth history, organisms have a low probability of being preserved, and paleontologists have found only a small percentage of the fossils preserved in rock. So the incompleteness of the fossil record should come as no surprise.

D.7 EVOLUTION AND EXTINCTION

Darwin's Grand Idea

As a young man in England in the early nineteenth century, Charles Darwin had been unable to settle on a career but had developed a strong interest in natural history. Therefore, he jumped at the opportunity to serve as a naturalist aboard H.M.S. *Beagle* on an around-the-world surveying cruise. During the five years of the cruise, from 1831 to 1836, Darwin made detailed observations of plants, animals, and geology in the field and amassed an immense specimen collection from South America, Australia, and Africa. Just before departing on the voyage, a friend gave him a copy of Charles Lyell's 1830 textbook *Principles of Geology*, which argued in favor of James Hutton's proposal that the Earth had a long history and that geologic time extended much further into the past than did human civilization. A visit to the Galápagos Islands, off the coast of Peru, was a turning point in Darwin's thinking. The naturalist was most impressed with the variability of Galápagos finches, and he marveled not only at the fact that different varieties of the bird occurred on different islands, but at how each variety had adapted to utilize a particular food supply. With Lyell's writings in mind, Darwin developed a hypothesis that the finches had begun as a single species but had branched into several different species when isolated on different islands. This proposal implied that a species could change, or undergo **evolution,** throughout long periods of time and that new species could appear.

The crux of Darwin's argument is simply this: because populations of organisms cannot grow exponentially forever, they must be limited by competition for scarce resources in the environment. In nature, only organisms capable of survival can pass on their characteristics to the next generation. In each new generation, some individuals have characteristics that make them more fit, whereas some have characteristics that make them less fit. The fitter organisms are more likely to survive and produce offspring. Thus, beneficial characteristics that they possess get passed on to the next generation. Darwin called this process **natural selection,** because it occurs on its own in nature. According to Darwin, when natural selection takes place over long periods of time—geologic time—it eventually produces new organisms that differ so significantly from their distant ancestors that the new organisms can be considered to constitute a new species. If environmental conditions change, or if competitors enter the environment, species that do not evolve and become better adapted to survive eventually die off.

Darwin's view of evolution has been successfully supported by many observations, and so far has not been definitively disproved by any observation or experiment. Also, it can be used to make testable predictions. Thus, scientists now refer to Darwin's idea as the **theory of evolution.** In the century and a half since Darwin published his work, the science of genetics has developed and has provided insight into *how* evolution works. With the discovery of DNA in 1953, biologists began to understand the molecular nature of mutations, and thus of evolution.

The theory of evolution provides a conceptual framework in which to understand paleontology. By studying fossils in sequences of strata, paleontologists are able to observe progressive changes in species through time, and can determine when some species die out and other species appear. But because of the incompleteness of the fossil record, questions remain as to the rates at which evolution takes place during the course of geologic time. Originally, it was assumed that evolution happened at a constant, slow rate—this concept is called **gradualism.** More recently, however, researchers have suggested that evolution takes place in fits and starts: evolution occurs very slowly for quite a while (the species are in equilibrium), and then during a relatively short period it takes place very rapidly. This concept is called **punctuated equilibrium.** Factors that could cause sudden pulses of evolution include (1) a sudden mass extinction event, during which many organisms disappear, leaving ecological niches open for new species to colonize; (2) a sudden change in the Earth's climate that puts stress on organisms—organisms that evolve to survive the new stress survive, whereas others become extinct; (3) the sudden formation of new environments, as may happen when rifting splits apart a continent and generates a new ocean with new coastlines; and (4) the isolation of a breeding population.

Extinction: When Species Vanish

Extinction occurs when the last members of a species die, so there are no parents to pass on their genetic traits to offspring. Some species become extinct as they evolve into new species, whereas others just vanish, leaving no hereditary offspring. These days, we take for granted that species become extinct, because a great number have, unfortunately, vanished from the Earth during human history. Before the 1770s, however, few geologists thought that extinction occurred; they thought that fossils that didn't resemble known species must have living relatives somewhere on Earth. Considering that large parts of the Earth remained unexplored, this idea wasn't so far fetched. But by the end of the eighteenth century, it became clear that numerous fossil organisms did not have modern-day counterparts. The bones of mastodons and woolly mammoths, for example, were too different from those of elephants to be of the same species, and the animals were too big to hide.

Twentieth-century studies concluded that many different phenomena can contribute to extinction. Some extinctions may happen suddenly, when all members of a species die off in a short time, whereas others may occur over longer periods, when the replacement rate of a population simply becomes lower than the mortality rate. By examining the number of species on Earth through time (in other words, by studying variations in the diversity of life, or biodiversity), paleontologists have found that the rate of extinction varies through time. Generally, the rate is fairly slow, but on occasion a **mass extinction event** occurs, during which a large number of species worldwide disappear. At least five major mass extinction events have happened during the past half billion years (▶Fig. D.8). These

FIGURE D.8 This graph illustrates the variation in diversity of life with time. Steep dips in the curve mean that the number of species on Earth suddenly decreased substantially. The largest dips represent major mass extinctions.

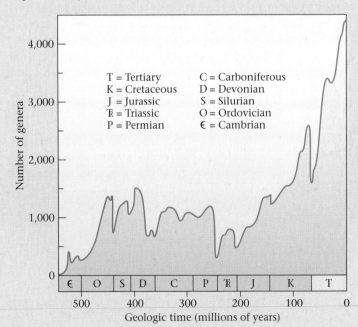

T = Tertiary   C = Carboniferous
K = Cretaceous   D = Devonian
J = Jurassic   S = Silurian
Ⱦ = Triassic   O = Ordovician
P = Permian   Ɛ = Cambrian

Number of genera

4,000

3,000

2,000

1,000

0

Ɛ O S D C P Ⱦ J K T

500   400   300   200   100   0

Geologic time (millions of years)

events define the boundaries between some of the major intervals into which geologists divide time. For example, a major extinction event marks the end of the Cretaceous Period, 65 million years ago. During this event, all dinosaur species (with the exception of their modified descendants, the birds) vanished, along with many marine invertebrate species. Some researchers have suggested that extinction events are periodic, but this idea remains controversial.

Following are some of the geologic factors that may cause extinction.

- *Global climate change:* At times, the Earth's mean temperature has been significantly colder than today's, whereas at other times it has been much hotter. Because of a change in climate, an individual species may lose its habitat, and if it cannot adapt to the new habitat or migrate to stay with its old one, the species will disappear.

- *Tectonic activity:* Tectonic activity causes mountain building, the gradual vertical movement of the crust over broad regions, changes in sea-floor spreading rates, and changes in the amount of volcanism. These phenomena can make sea level rise or fall, or can modify the distribution and area of habitats. Species that cannot adapt die off.

- *Asteroid or comet impact:* Many geologists have concluded that impacts of large meteorites or asteroids with the Earth have been catastrophic for life. An impact would send so much dust and debris into the atmosphere that it would blot out the Sun and plunge the Earth into darkness and cold. Such a change, though relatively short lived, could interrupt the food chain.

- *The appearance of a predator or competitor:* Some extinctions may happen simply because a new predator appears on the scene. Some researchers suggest that this phenomenon explains the mass extinction that occurred during the past 20,000 years, when a vast number of large mammal species vanished from North America. The timing of these extinctions appears to coincide with the appearance of the first humans (fierce predators) on the continent. If a more efficient competitor appears, the competitor steals an ecological niche from the weaker species, whose members can't obtain enough food and thus die out.

KEY TERMS

amber (p. 274)
evolution (p. 280)
extinction (p. 281)
extraordinary fossil (p. 277)
fossil (p. 273)
fossilization (p. 274)
gradualism (p. 281)
mass extinction event (p. 281)
morphology (p. 278)

natural selection (p. 281)
paleontology (p. 273)
permineralization (p. 275)
petrified wood (p. 275)
preservation potential (p. 277)
punctuated equilibrium (p. 281)
theory of evolution (p. 281)

SUGGESTED READING

Benton, M. J. 2003. *When Life Nearly Died: The Greatest Mass Extinction of All Time*. London: Thames & Hudson.

Cutle, A. 2003. *The Seashell on the Mountaintop: A Story of Science, Sainthood, and the Humble Genius Who Discovered a New History of the Earth*. New York: E. P. Dutton.

Knoll, A. H. 2003. *Life on a Young Planet: The First Three Billion Years of Evolution on Earth*. Princeton, N.J.: Princeton University Press.

Prothero, D. R. 1998. *Bringing Fossils to Life: An Introduction to Paleontology*. Boston: WCB/McGraw-Hill.

CHAPTER 10

Deep Time: How Old Is Old?

If the Eiffel Tower were now representing the world's age, the skin of paint on the pinnacle-knob at its summit would represent man's share of that age; and anybody would perceive that that skin was what the tower was built for. I reckon they would, I dunno.
—MARK TWAIN (1835–1910)

10.1 INTRODUCTION

In May of 1869, a one-armed Civil War veteran named John Wesley Powell set out with a team of nine geologists and scouts to explore the previously un-mapped expanse of the Grand Canyon, the greatest gorge on Earth. Though Powell and his companions battled fearsome rapids and the pangs of starvation, most managed to emerge from the mouth of the canyon three months later (▶Fig. 10.1). During their voyage, seemingly insurmountable walls of rock both imprisoned and amazed the explorers, and led them to pose important questions about the Earth and its history, questions that even casual tourists to the canyon ponder today: Did the Colorado River sculpt this marvel, and if so, how long did it take? When did the rocks mak-ing up the walls of the canyon form? Was there a time *before* the colorful layers accumulated? These questions pertain to **geologic time,** the span of time since Earth's formation. Powell realized that, like the pages in a book, the rock layers of the Grand Canyon contain some of the record of Earth's history.

In this chapter, we first learn how geologists developed the con-cept of geologic time and thus a frame of reference for describing the ages of rocks, fossils, structures, and landscapes. Then we look at the tools

How do geologists determine the age of rocks? This cliff face in Missouri shows two rock units. The sedimentary layers on the left were deposited on the rhyolite to the right. From this relationship, we can determine the relative age of the two units—the rhyolite is older. But to determine the age of the rhyolite in years—its numerical age—we must date it radiometrically.

FIGURE 10.1 Woodcut illustration of the "noonday rest in Marble Canyon," from J. W. Powell, *The Exploration of the Colorado River and Its Canyons* (1895). The explorers have just entered a quiet stretch of the river: "We pass many side canyons today that are dark, gloomy passages back into the heart of the rocks."

geologists use to determine the age of the Earth and its features. With the concept of geologic time in hand, a hike down a trail into the Grand Canyon becomes a trip into the distant past, into what authors call *deep time*. The geological discovery that our planet's history extends billions of years into the past changed humanity's perception of the Universe as profoundly as did the astronomical discovery that the limit of space extends billions of light years beyond the edge of our solar system.

10.2 THE CONCEPT OF GEOLOGIC TIME

Setting the Stage for Studying the Past

Before the dawn of modern science, most cultures assumed that geologic time was not much longer than human history, and that the Earth essentially formed as we see it today. This view was challenged by James Hutton (1726–1797), who based his conclusions on relationships that he observed in the rocky crags of his native land. Hutton lived during the

Age of Enlightenment, when philosophers like Voltaire, Kant, Hume, and Locke encouraged people to cast aside the constraints of dogma and think for themselves, and when the discovery of physical laws by Sir Isaac Newton made people look to natural, not supernatural, processes to explain the Universe. As Hutton wandered around Scotland, he came up with an idea that provides the foundation of geology and led to his title: "father of geology." This idea, the principle of **uniformitarianism,** states simply that physical processes we observe today also operated in the past and were responsible for the formation of the geologic features we see in outcrops; more concisely, *the present is the key to the past.* If this principle is correct, Hutton reasoned, the Earth must be much older than human history, for observed geologic processes work very slowly.

Hutton was not a particularly clear writer, and it took the efforts of subsequent geologists to clarify the implications of the principle of uniformitarianism and to publicize them. Once this had been accomplished, geologists around the world began to apply their growing understanding of geologic processes to define and interpret geologic events of the Earth's past.

Relative versus Numerical Age

Like historians, geologists want to establish both the sequence of events that created an array of geologic features (such as rocks, structures, and landscapes) and the exact dates on which the events happened. We specify the age of one feature with respect to another as its **relative age** and the age of a feature given in years as the **numerical age** (or, in older literature, the "absolute age"). Geologists developed ways of defining relative age long before they did so for numerical age, so we will look at relative-age determination first.

10.3 PHYSICAL PRINCIPLES FOR DEFINING RELATIVE AGE

Nineteenth-century geologists such as Charles Lyell recast the ideas of earlier researchers into formal, usable geological principles. These principles, defined below, continue to provide the basic framework within which geologists read the record of Earth history.

- *The principle of **uniformitarianism:*** As noted earlier, physical processes we observe operating today also operated in the past, at roughly comparable rates (▶Fig. 10.2a, b). Put simply, the present is the key to the past.

- *The principle of **superposition:*** In a sequence of sedimentary rock layers, each layer must be younger than the one below, for a layer of sediment cannot accumulate unless there is already a substrate on which it can collect. Thus, the layer at the bottom of a sequence is the oldest, and the layer at the top is the youngest (▶Fig. 10.2c).

- *The principle of **original horizontality:*** Sediments on Earth settle out of a fluid in a gravitational field. Typically, the surfaces on which sediments accumulate (such as a floodplain or the bed of a lake or sea) are fairly horizontal. If sediments were deposited on a steep slope, they would likely slide downslope before lithification, and so would not be preserved as sedimentary layers. Therefore, layers of sediment when originally deposited are fairly horizontal. With this principle in mind, we realize that when we see folds and tilted beds, we are seeing the consequences of deformation that postdates deposition (▶Fig. 10.2d).

- *The principle of original continuity:* Sediments generally accumulate in continuous sheets. If today you find a sedimentary layer cut by a canyon, then you can assume that the layer once spanned the canyon but was later eroded by the river that formed the canyon (▶Fig. 10.2e).

- *The principle of **cross-cutting relations:*** If one geologic feature cuts across another, the feature that has been cut is older. For example, if an igneous dike cuts across a sequence of sedimentary beds, the beds must be older than the dike (▶Fig. 10.2f). If a fault cuts across and displaces layers of sedimentary rock, then the fault must be younger than the layers.

- *The principle of inclusions:* If an igneous intrusion contains fragments of another rock, the fragments must be older than the intrusion (▶Fig. 10.2g). If a layer of sediment deposited on an igneous layer includes pebbles of the igneous rock, then the sedimentary layer must be younger (▶Fig. 10.2h). The fragments (xenoliths) in an igneous body and the pebbles in the sedimentary layer are inclusions, or pieces of one material incorporated in another. The rock containing the inclusion must be younger than the inclusion.

- *The principle of baked contacts:* An igneous intrusion "bakes" (metamorphoses) surrounding rocks. The rock that has been baked must be older than the intrusion (▶Fig. 10.2i).

Now let's use these principles to determine the relative ages of features shown in ▶Figure 10.3a. In so doing, we develop a geologic history of the region, defining the relative ages of events that took place there.

Let's start our analysis by looking at the sedimentary sequence. By the principle of superposition, we know that the oldest layer, the limestone labeled 1, occurs at the bottom and that progressively younger beds lie above the limestone. We can confirm that layer 1 predates layer 2 by applying the principle of inclusions, because layer 2 contains pebbles (inclusions) of layer 1. Because the same holds true for the other layers, the sedimentary beds from oldest to youngest are 1, 2, 3, 4, 5, 6, 7. (There are other rocks below layer 1, but we do not see them in our cross section.) Considering the principle of original horizontality, we conclude that the layers were folded sometime after deposition.

Now let's look at the relationships between the igneous rocks and the sedimentary rocks. The granite pluton cuts across the folded sedimentary rocks, so the intrusion of the pluton occurred after the deposition of the sedimentary beds and after they were folded. The layer of igneous rock that parallels the sedimentary beds could be either a sill that intruded between the sedimentary layers or a flow that spread out over the sandstone and solidified before the shale was deposited. From the principle of inclusions, we deduce that the layer is a sill, because it contains xenoliths of both the underlying sandstone and the overlying shale. Since the sill is folded, it intruded before folding took place. By applying the principle of baked contacts, we also can tell that the sill intruded before the pluton did, because the baked zones (metamorphic aureole) surrounding the pluton affected the sill. The dike cuts across both the pluton and the sill, as well as all the sedimentary layers, and thus formed later.

Finally, let's consider the fault and the land surface. Because the fault offsets the granite pluton and the sedimentary beds, according to the principle of cross-cutting relations the fault must be younger than those rocks. But the fault itself has been cut by the dike, and so must be older than the dike. The present land surface erodes all rock units and the fault, and thus must be younger. We can now propose the following geologic history for this region (▶Fig. 10.3b): (a) deposition of the sedimentary sequence, in order from layers 1 to 7; (b) intrusion of the sill; (c) folding of the sedimentary layers and the sill; (d) intrusion of the granite pluton; (e) faulting; (f) intrusion of the dike; (g) formation of the land surface.

10.4 ADDING FOSSILS TO THE STORY: FOSSIL SUCCESSION

As Britain entered the Industrial Revolution in the late eighteenth and early nineteenth centuries, new factories demanded coal to fire their steam engines. The government decided to build a network of canals to transport coal and iron, and hired an engineer named William Smith (1769–1839) to survey the excavations. These excavations provided fresh exposures of bedrock, which previously had been covered by vegetation. Smith learned to recognize distinctive layers of sedimentary rock and to identify the fossil assemblage (the group of fossil species)

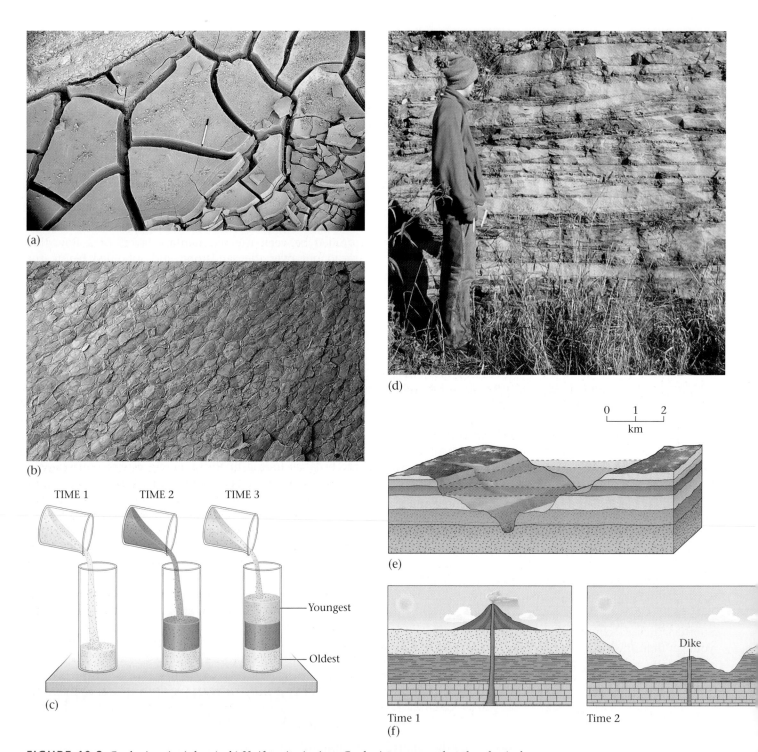

FIGURE 10.2 Geologic principles: (a, b) Uniformitarianism: Geologists assume that the physical processes that formed the mud cracks in a tidal flat of (a) also formed the mud cracks preserved on the surface of a Paleozoic siltstone bed shown in (b). (c) Superposition: If you fill a glass cylinder with different colors of sand, the oldest sand (white) must be on the bottom. (d) Original horizontality: When sediment is originally deposited, it forms a flat layer. If layers do not become deformed, they can remain horizontal for hundreds of millions of years, as illustrated by these beds of Paleozoic sandstone in Wisconsin. (e) Original continuity: We assume that the layers of sediment were originally continuous across the canyon before the canyon developed. (f) Cross-cutting relations: The sedimentary beds existed first; then they were cut by igneous magma rising to a volcano, so today we see a dike cutting across the bedding. Since the dike does the cutting, it must be younger.

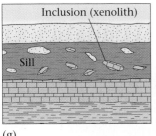 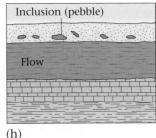 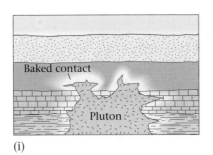

(g)   (h)   (i)

FIGURE 10.2 *(continued)* (g) Inclusions: In the left-hand example, a sill intrudes between a limestone layer and a sandstone one. The sill incorporates fragments (inclusions) of limestone and sandstone, so it must be younger. (h) In the right-hand example, the igneous rock is a lava flow that existed before the sandstone was deposited; the sandstone contains pebbles (inclusions) of lava. (i) Baked contacts: The intrusion of the pluton creates a metamorphic aureole (baked contact) in the surrounding rock, so the pluton must be younger.

that they contained. He also realized that a particular fossil species can be found only in a limited interval of strata, and not above or below this interval. Thus, once a fossil species disappears at a horizon in a sequence of strata, it never reappears higher in the sequence. In other words, extinction is forever. Smith's observation has been repeated at millions of locations around the world, and has been codified as the principle of **fossil succession.**

To see how this principle works, examine ►Figure 10.4a, which depicts a sequence of strata. Bed 1 at the base contains fossil species A, bed 2 contains fossil species A and B, bed 3 contains B and C, bed 4 contains C, and so on. From these data, we can define the range of specific fossils in the sequence, meaning the interval in the sequence in which the fossils occur. Note that the sequence contains a definable succession of fossils (A, B, C, D, E, F), that the range in which a particular species occurs may overlap with the range of other species, and that once a species vanishes, it does not reappear higher in the sequence.

Because of the principle of fossil succession, we can define the relative ages of strata by looking at fossils. For example, if we find a bed containing fossil A, we can say that the bed is older than a bed containing, say, fossil F (►Fig. 10.4b). Geologists have now identified and determined the relative ages of hundreds of thousands of fossil species.

10.5 UNCONFORMITIES: GAPS IN THE RECORD

James Hutton often strolled along the coast of Scotland because the shore cliffs provided good exposures of rock, stripped of soil and shrubbery. He was particularly puzzled

by an outcrop along the shore at Siccar Point. One sequence of rock exposed there consisted of alternating beds of gray sandstone and shale, but another consisted of red sandstone and conglomerate (►Fig. 10.5a, b). The beds of gray sandstone and shale were nearly vertical, whereas the beds of red sandstone and conglomerate were roughly horizontal. Further, the horizontal layers seemed to lie across the truncated ends of the vertical layers, like a handkerchief lying across a row of books. We can imagine that as Hutton examined this odd geometric relationship, the tide came in and deposited a new layer of sand on top of the rocky shore. With the principle of uniformitarianism in mind, Hutton suddenly realized the significance of what he saw. Clearly, the gray sandstone–shale sequence had been deposited, then tilted, and then truncated by erosion before the red sandstone–conglomerate beds were deposited.

Hutton deduced that the surface between the gray and red rock sequences represented a long interval of time during which new strata were not deposited at Siccar Point and older strata may have been eroded away. We now call such a surface, representing a period of nondeposition and possibly erosion, an **unconformity.** Essentially, an unconformity forms wherever the land surface does not receive and accumulate sediment. Geologists recognize three kinds of unconformity:

- *Angular unconformity:* Rocks below an angular unconformity were tilted or folded before the unconformity developed (►Fig. 10.6a). Thus, an angular unconformity cuts across the underlying layers; the layers below have a different orientation from the layers above. (We can see an angular unconformity in the outcrop at Siccar Point.) Angular unconformities form where rocks were tilted by either folding or faulting, before being exposed at the Earth's surface.

• *Nonconformity:* A nonconformity is a type of unconformity at which sedimentary rocks overlie intrusive igneous rocks and/or metamorphic rocks (▶Fig. 10.6b). The igneous or metamorphic rocks must have cooled, been uplifted, and been exposed by erosion to form the substrate on which new sedimentary rocks were deposited.

• *Disconformity:* Imagine that a sequence of sedimentary beds has been deposited beneath a shallow sea. Then sea level drops, exposing the recently deposited beds for some time. During this time, no new sediment accumulates, and some of the preexisting sediment gets eroded away. Later, sea level rises, and a new sequence of sediment accumulates over the old. The boundary between the two sequences is a **disconformity** (▶Fig. 10.6c). Even though the beds above and below the disconformity are parallel, the contact between them represents an interruption in deposition.

The succession of strata at a particular location provides a record of Earth history there. But because of unconformities, the record preserved in the rock layers is incomplete. It's as if geologic history is being chronicled by a tape recorder that turns on only intermittently—when it's on (times of deposition), the rock record accumulates, but when it's off (times of nondeposition and possibly erosion), an unconformity develops. Because of unconformities, no single location on Earth contains a complete record of Earth history.

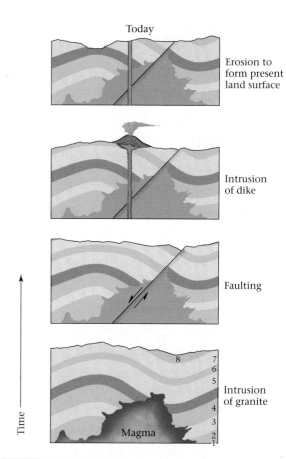

FIGURE 10.3 (a) Geologic principles allow us to interpret the sequence of events leading to the development of the features shown here. Beds 1–7 were deposited first. Intrusion of the sill came next, followed by folding, intrusion of the granite, faulting, intrusion of the dike, and erosion to yield the present land surface. (b) Sequence of geologic events leading to the geology shown in (a). Note that layer 8 has been eroded away completely.

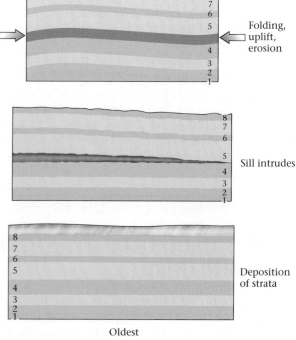

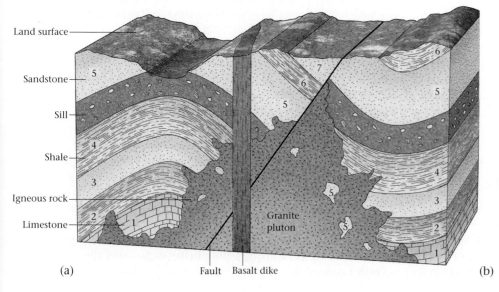

(a)

(b)

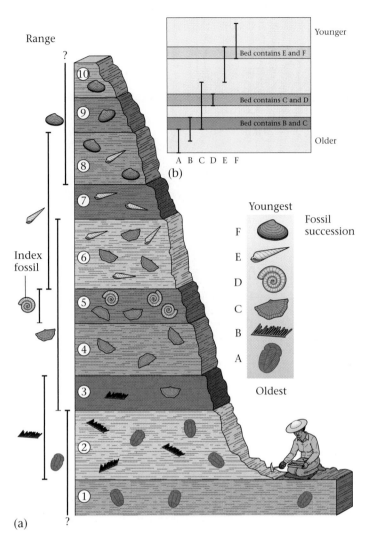

(a)

(b)

FIGURE 10.4 (a) The principle of fossil succession. Note that each species has only a limited range in a succession of strata, and ranges of different fossils may overlap. Widespread fossils with a short range are index fossils. Here, D is an index fossil. (b) Overlapping fossil ranges can be used to limit the relative age of a given bed and to determine the relative ages of beds. For example, a bed containing fossils E and F must be younger than a bed containing B and C. Note that a bed containing C alone could be older than or younger than a bed containing D alone.

FIGURE 10.5 (a) The Siccar Point unconformity in Scotland, on the coast about 60 km east of Edinburgh. (b) A geologic interpretation of the unconformity.

10.6 STRATIGRAPHIC FORMATIONS AND THEIR CORRELATION

Geologists summarize information about the sequence of strata at a location by drawing a **stratigraphic column.** Typically, we draw columns to scale, so that the relative thicknesses of layers portrayed on the column reflect the thicknesses of layers in the outcrop. Then, for ease of reference, geologists divide the sequence of strata represented on a column into **stratigraphic formations** ("formations," for short), a sequence of layers of a specific rock type or group of rock types that can be traced over a fairly broad region. The boundary surface between two formations is a type of geologic **contact.** (Typically, a formation has a specific geologic age. However, in cases where the formation was deposited during a transgression, a gradual migration of the coast as sea level rises, the age at one location may differ from the age at another.)

Let's see how the concept of a stratigraphic formation applies to the Grand Canyon. The walls of the canyon look striped, because they expose a variety of rock types that differ

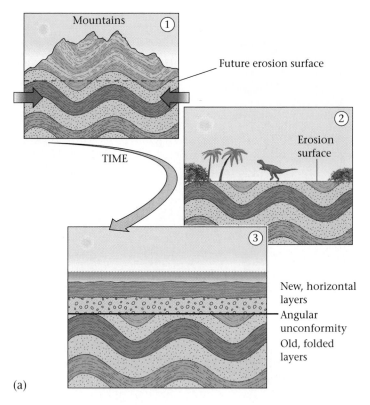

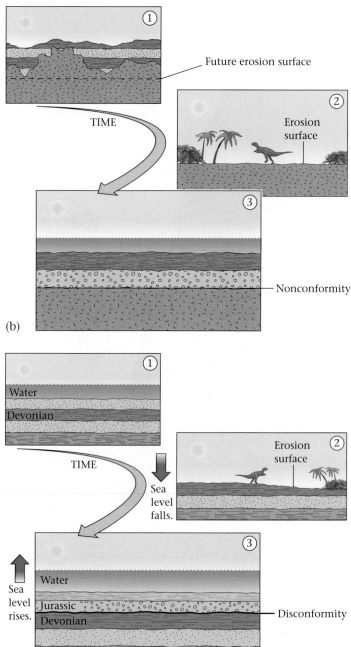

FIGURE 10.6 (a) The stages during the development of an angular unconformity: (1) mountains form and layers are folded; (2) erosion removes the mountains, creating an erosion surface; (3) sea level rises and new, horizontal layers of sediment are deposited. (b) The stages during the development of a nonconformity: (1) a pluton intrudes sedimentary rocks; (2) erosion removes all the sedimentary layers and cuts down into the crystalline rock, making an erosion surface; (3) sea level rises and new sedimentary layers are deposited above the erosion surface. (c) The stages during the development of a disconformity: (1) layers of sediment are deposited; (2) sea level drops and an erosion surface forms; (3) sea level rises and new sedimentary layers accumulate. Note that regardless of the details, an unconformity represents a surface of erosion and/or a period of nondeposition.

in color and in resistance to erosion. Geologists identify major contrasts and use them as a basis for dividing the strata into formations, each of which may consist of many beds (▶Fig. 10.7a, b, c). Note that some formations include a single rock type, whereas others include interlayered beds of two or more rock types. Also, note that not all formations have the same thickness. Typically, geologists name a formation after a locality where it was first identified. If the formation consists of only one rock type, we may incorporate that rock

type in the name (for example, Kaibab Limestone), but if the formation contains more than one rock type, we use the word "formation" in the name (such as Toroweap Formation). Note that in the formal name of a formation, all words are capitalized. Several related formations in a succession may be lumped together as a stratigraphic group.

While he was excavating canals in England, William Smith discovered that formations cropping out at one locality resembled formations cropping out at another, in that

(a)

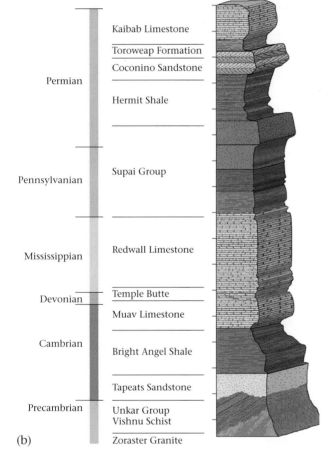

Permian
- Kaibab Limestone
- Toroweap Formation
- Coconino Sandstone
- Hermit Shale

Pennsylvanian
- Supai Group

Mississippian
- Redwall Limestone

Devonian
- Temple Butte

Cambrian
- Muav Limestone
- Bright Angel Shale
- Tapeats Sandstone

Precambrian
- Unkar Group
- Vishnu Schist
- Zoraster Granite

(b)

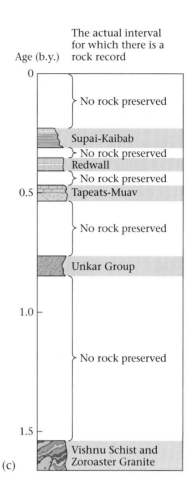

Age (b.y.)   The actual interval for which there is a rock record

0

No rock preserved

Supai-Kaibab

No rock preserved

Redwall

No rock preserved

Tapeats-Muav

No rock preserved

Unkar Group

No rock preserved

Vishnu Schist and Zoraster Granite

(c)

FIGURE 10.7 (a) The walls of the Grand Canyon expose several formations. (b) The sequence of strata can be represented on a stratigraphic column. The vertical scale gives relative thicknesses. The right-hand edge of the column represents resistance to erosion (e.g., Coconino Sandstone is more resistant than Hermit Shale). (c) Because of unconformities, the stack of strata exposed in the Grand Canyon represents only bits and pieces of geologic history. If the strata are projected on a numerical time scale, you can see that large intervals of time are not accounted for.

their beds looked similar and contained similar fossil assemblages. In other words, Smith was able to define the age relationship between the strata at one locality and the strata at another, a process called **correlation.**

How does correlation work? Typically, geologists correlate formations between *nearby* regions based on similarities in rock type. We call this method "lithologic correlation"

(►Fig. 10.8a–c). For example, the sequence of strata on the southern rim of the Grand Canyon clearly correlates with the sequence on the northern rim, because they contain the same rock types in the same order. Before deformation and/or erosion, the layers that correlate were continuous. In some cases, a sequence contains a key bed, or marker bed, a particularly unique layer that provides a

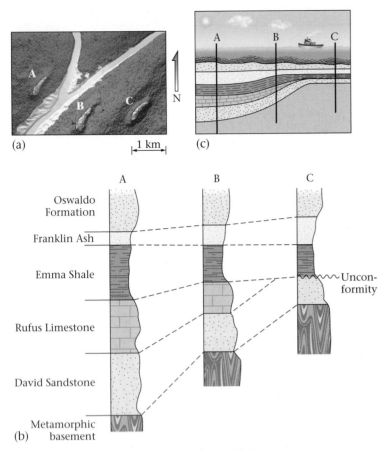

(a)

├─ 1 km ─┤

(c)

A   B   C

Oswaldo
Formation

Franklin Ash

Emma Shale

~~Uncon-
formity

Rufus Limestone

David Sandstone

Metamorphic
basement

(b)

FIGURE 10.8 The principle of lithologic correlation. (a) These three outcrops of rock are a couple of kilometers from each other. (b) The stratigraphic sections at each location are somewhat different, but the columns can be correlated with each other by matching rock types. Note that the Franklin Ash is a key bed, a distinctive layer that can be correlated with certainty. And the Rufus Limestone "pinches out" (thins and disappears along its length), so the contact between the Emma Shale and the David Sandstone in column C is an unconformity. (c) Geologists reconstruct a sedimentary basin using correlation. The eastward thinning of sedimentary layers suggests that the basin tapered to the east.

definitive basis for correlation. But to correlate units over *broad* areas, we must rely on fossils to define the relative ages of sedimentary units. We call this method "fossil correlation." Geologists must rely on fossil correlation for studies of broad areas because sources of sediments and depositional environments may change from one location to another. Thus, the beds deposited at one location during a given time interval may look quite different from the beds deposited at another location during the same time interval. But if fossils of the same relative age occur at both locations, we can say that the strata at the two locations correlate.

Now let's look at an example of fossil correlation by tracing the individual formations exposed in the Grand Canyon

into the mountains just north of Las Vegas, 150 km to the west (▶Fig. 10.9a, b). Near Las Vegas, we find a sequence of sedimentary rocks that includes a limestone formation called the Monte Cristo Limestone. The Monte Cristo Limestone contains fossils of the same age as occur in the Redwall Limestone of the Grand Canyon, but it is much thicker. Because the formations contain fossils of the same age, we conclude that they were deposited during the same time interval, and thus we say that they correlate with one another. Note also that not only are the units thicker in the Las Vegas area than in the Grand Canyon area, but there are more of them. This discovery indicates that during part of the time when thick sediments were deposited near Las Vegas, none accumulated near the Grand Canyon. Thus, the contact beneath the Grand Canyon's Redwall Limestone is an unconformity.

By correlating strata at many locations, William Smith realized that he could trace individual formations of strata over fairly broad regions. In 1815, he plotted the distribution of formations and created the first modern **geologic**

FIGURE 10.9 The principle of fossil correlation: (a) Because they both contain "P-age" fossils, we can say that the Redwall Limestone of the Grand Canyon correlates with the Monte Cristo Limestone near Las Vegas. But the fossils in the Muav Limestone, which lies directly beneath the Redwall at the Grand Canyon, contain "N-age" fossils and thus correlate with those of a unit 1,000 m below the Monte Cristo Limestone. The sequence of strata between the two limestones at Las Vegas is not represented by any rock at the Grand Canyon; thus, the Redwall-Muav contact is an unconformity. (b) A sedimentary basin thinned radically between Las Vegas and the Grand Canyon. Thus, the sequence of strata deposited at the Grand Canyon was thinner and less complete than the section deposited at Las Vegas.

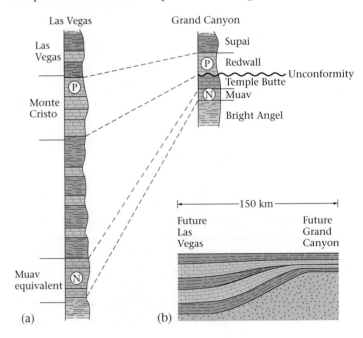

Las Vegas

Las
Vegas

Monte
Cristo

Muav
equivalent

Grand Canyon

Supai

Redwall

Temple Butte

Muav

Bright Angel

Unconformity

├─ 150 km ─┤

Future
Las
Vegas

Future
Grand
Canyon

(a)

(b)

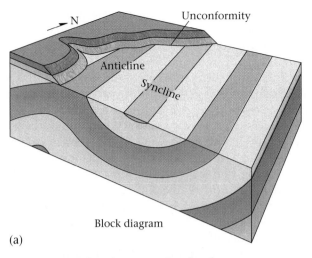

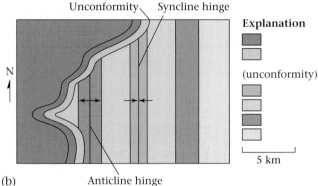

FIGURE 10.10 (a) The block diagram depicts an angular unconformity with horizontal strata above and folded strata below. (b) The geologic map shows what this ground surface would look like if viewed from above. On the map, the contacts have been projected onto a flat piece of paper.

map, which portrays the spatial distribution of rock units at the Earth's surface. As ▶Figure 10.10 shows, we can use the information provided by a geologic map to identify geologic structures.

10.7 THE GEOLOGIC COLUMN

As stated earlier, no one locality on Earth provides a complete record of our planet's history, because stratigraphic columns can contain unconformities. But by correlating rocks from locality to locality at millions of places around the world, geologists have pieced together a *composite* stratigraphic column, called the **geologic column,** that *represents* the entirety of Earth history (▶Fig. 10.11a, b). The column is divided into segments, each of which represents a specific interval of time. The largest subdivisions break Earth history into the Hadean,

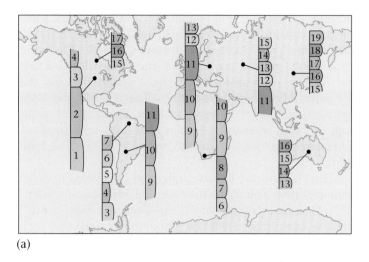

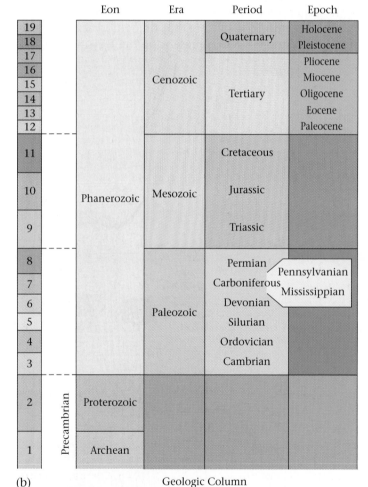

FIGURE 10.11 (a) The geologic column was constructed by determining the relative ages of stratigraphic columns from around the world. Each of these little columns represents the stratigraphy at a given location. (b) By correlation, the strata in the columns can be stacked in a sequence representing most of geologic time. This is the geologic column. Note that the column was first built without knowledge of numerical ages, so we depict the sequence of periods but do not indicate their relative duration here.

Archean, Proterozoic, and Phanerozoic **eons.** (The first three together constitute the **Precambrian.**) The suffix "-zoic" means "life," so "Phanerozoic" means "visible life," and "Proterozoic" means "earlier life." The earliest life, bacteria and archaea, appeared during the Archean Eon. In the Phanerozoic Eon, organisms with hard parts (shells and, later, skeletons) became widespread, so there are abundant fossils from this eon, whereas in Precambrian time, only small organisms with no shells existed, so Precambrian fossils are hard to find.

The Phanerozoic Eon is subdivided into **eras.** In order from oldest to youngest, they are the Paleozoic ("ancient life"), Mesozoic ("middle life"), and Cenozoic ("recent life") eras. We can further divide each era into **periods** and each period into **epochs.**

Where do the names of the periods come from? They refer either to localities where a fairly complete stratigraphic column representing that time interval was identified (for example, rocks representing the Devonian Period crop out near Devon, England) or to a characteristic of the time (rocks from the Carboniferous Period contain a lot of coal). The terminology was not set up in a planned fashion that would make it easy to learn. Instead, it grew haphazardly in the years between 1760 and 1845, as geologists began to refine their understanding of geologic history and fossil succession.

The succession of fossils preserved in strata of the geologic column defines the course of life's evolution throughout Earth history (▶Fig. 10.12). Simple bacteria appeared during the Archean Eon, but complex shell-less invertebrates did not evolve until the late Proterozoic. The appearance of invertebrates with shells defines the Precambrian-Cambrian boundary. At this time there was a sudden diversification in life, with many new genera appearing over a relatively short interval—this event is called the **Cambrian explosion.**

The first vertebrates, fish, appeared during the Ordovician Period. Before the Silurian Period, the land surface was barren

FIGURE 10.12 Life evolution in the context of the geologic column.

of multicellular life; in Silurian time, land plants spread over the landscape. Amphibians appeared during the Devonian. Though reptiles evolved during the Pennsylvanian Period, the first dinosaurs did not pound across the land until the Triassic. Dinosaurs continued to inhabit the Earth until their sudden extinction at the end of the Cretaceous Period. For this reason, geologists refer to the Mesozoic Era as the "Age of Dinosaurs." Small mammals also appeared during the Triassic Period, but the diversification (development of many different species) of mammals to fill a wide range of ecological niches did not happen until the beginning of the Cenozoic Era, so we now call the Cenozoic the "Age of Mammals." Birds also appeared during the Mesozoic (specifically, at the beginning of the Cretaceous Period), but underwent great diversification in the Cenozoic Era.

To conclude our discussion, let's see how the geologic column comes into play when correlating strata across a region. We return to the Colorado Plateau of Arizona and Utah, in the southwestern United States (▶Fig. 10.13 a–g); because of the lack of vegetation in this region, you can easily see bedrock exposures on the walls of cliffs and canyons, and some of these exposures are so beautiful that they have become national parks. The oldest sedimentary rock of the region crops out at the base of the Grand Canyon. This rock, the Unkar Group, rests unconformably on schist and granite, does not contain fossils, and underlies Cambrian sandstone. Thus, the Unkar Group is Precambrian. By the principle of superposition, the schist and granite below it must also be Precambrian. The walls of the Grand Canyon contain strata up through the Permian age—the light-colored rock that crops out at the top of the canyon is the Permian Kaibab Limestone. Rocks younger than the Kaibab no longer remain at the Grand Canyon, due to erosion, so to see younger strata, we must travel to Zion Canyon. Here, many layers of red sandstone and shale overlie the Kaibab Limestone. Fossils indicate that these strata were deposited in Triassic and Jurassic times. To see what lay above the strata at Zion, we must move farther west to Bryce Canyon or the cliffs of Cedar Breaks. The oldest strata exposed here correlate with the youngest strata at Zion. But at Bryce and Cedar Breaks, these rocks are overlain with Jurassic, Cretaceous, and Tertiary strata. Note that by using the names of the geologic column, geologists can efficiently define relative-age relationships among the strata of the different parks.

Walking through these parks is thus like walking through time—each rock layer gives an indication of the climate and topography of the region in the past (see art pp. 298–99). For example, when the Precambrian metamorphic and igneous rocks exposed in the inner gorge of the Grand Canyon first formed, the region was a high mountain range, perhaps as dramatic as the Himalayas today. When the fossiliferous beds of the Kaibab Limestone at the rim of the canyon first developed, the region was a Bahama-like carbonate reef and platform, bathed in a warm, shallow sea. And when the rocks making up the towering red cliffs of sandstone in Zion Canyon were deposited, the region was a Sahara-like desert, blanketed with huge sand dunes.

10.8 NUMERICAL AGE AND THE RADIOMETRIC CLOCK

Geologists since the days of Hutton could determine the relative ages of geologic events, but they had no way to specify numerical ages. Thus, they could not define a time line for Earth history or determine the duration of events. This situation changed with the discovery of radioactivity. Simply put, radioactive elements decay at a constant rate that can be measured in the lab and can be specified in years. In the 1950s, geologists first developed techniques for using measurements of radioactive elements to calculate the ages of rocks. We call the science of dating geologic events in years **radiometric dating** (or geochronology). In succeeding decades, techniques of radiometric dating have been vastly improved. We begin our discussion of this technique by first learning about radioactive decay.

Radioactive Decay and the Concept of a Half-Life

All atoms of a given element have the same number of protons in their nucleus—we call this number the atomic number (see Appendix). However, not all atoms have the same number of neutrons in their nucleus. Therefore, not all atoms of a given element have the same atomic weight— roughly the number of protons plus neutrons. Different versions of an element, called **isotopes** of the element, have the same atomic number but a different atomic weight. For example, all uranium atoms have 92 protons, but the uranium-238 isotope (abbreviated ^{238}U) has an atomic weight of 238 and thus has 146 neutrons, whereas the ^{235}U isotope has an atomic weight of 235 and thus has 143 neutrons.

Some isotopes of an element are stable, meaning that they last essentially forever. Radioactive isotopes are unstable: after a given time, they undergo a change called radioactive decay, which converts them into a different element. Radioactive decay can take place by a variety of reactions that change the atomic number of the nucleus and thus form a different element. In these reactions, the isotope that undergoes decay is the parent isotope, while the decay product is the daughter isotope. For example, rubidium-87 (^{87}Rb) decays to strontium-87 (^{87}Sr), potassium-40 (^{40}K) decays to argon-40 (^{40}Ar), and uranium-238 (^{238}U) decays to form lead-206 (^{206}Pb).

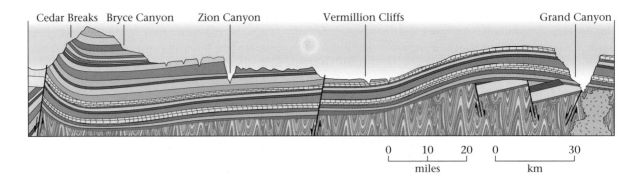

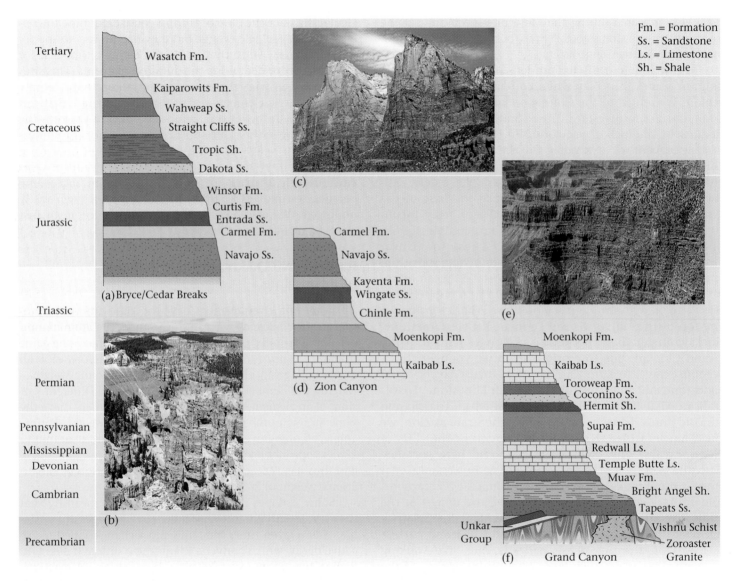

FIGURE 10.13 The correlation of strata among the various national parks of Arizona and Utah: (a, b) Bryce Canyon/Cedar Breaks, (c, d) Zion Canyon, (e, f) the Grand Canyon. The inset at the top shows a cross section of the region.

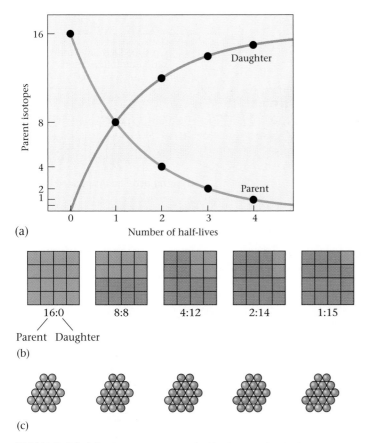

(a)

(b)

16:0   8:8   4:12   2:14   1:15

Parent Daughter

(c)

FIGURE 10.14 The concept of a half-life. (a) Graph showing the decrease in the number of parent isotopes as time passes; the curve is exponential, meaning that the rate of change in the parent-daughter ratio decreases with time. (b) The ratio of parent-to-daughter isotopes changes with the passage of each successive half-life. (c) A cluster of isotopes undergoing decay. There's no way to predict which parent will decay in a given time interval.

after a third half-life, 2 more parent isotopes have decayed, so the crystal contains 2 parent and 14 daughter isotopes. For a given decay reaction, the half-life is a constant.

Radiometric Dating Techniques

Like the ticktock of a clock, radioactive decay proceeds at a known rate and thus provides a basis for telling time. In other words, because an element's half-life is a constant, we can calculate the age of a mineral by measuring the ratio of parent to daughter isotopes in the mineral.

How do geologists actually obtain a radiometric date (or radiometric age)? First, we must find the right kind of elements to work with. Although there are many different pairs of parent and daughter isotopes among the known elements, only a few have long enough half-lives, and occur in sufficient abundance in minerals, to be useful for radiometric dating. Particularly useful elements are listed in ▶Table 10.1. Each radioactive element has its own half-life. Note that carbon dating is *not* used for dating rocks (see ▶Box 10.1 on p. 300).

Second, we must identify the right kind of minerals to work with. Not all minerals contain radioactive elements, but fortunately some common minerals do. For example, feldspar, mica, and hornblende contain potassium and rubidium; zircon contains uranium; and garnet contains samarium. Once we have identified appropriate minerals containing appropriate elements, we can set to work. Radiometric dating consists of the following steps.

- *Collecting the rocks:* Geologists collect unweathered rocks for dating, for the chemical reactions that happen during weathering may allow isotopes to leak out of minerals.

- *Separating the minerals:* The rocks are crushed, and the appropriate minerals are separated from the debris.

Physicists cannot specify how long an individual radioactive isotope will survive before it decays, but they *can* measure how long it takes for half of a group of parent isotopes to decay. This time is called the **half-life** of the isotope. ▶Figure 10.14 can help you visualize the concept of a half-life. Imagine a crystal containing 16 radioactive parent isotopes. (In real crystals, the number of atoms would be much larger.) After one half-life, 8 isotopes have decayed, so the crystal now contains 8 parent and 8 daughter isotopes. After a second half-life, 4 of the remaining parent isotopes have decayed, so the crystal contains 4 parent and 12 daughter isotopes. And

TABLE 10.1 Isotopes Used in the Radiometric Dating of Rocks

Parent → Daughter	Half-Life (years)	Minerals in Which the Isotopes Occur
$^{147}Sm \rightarrow {}^{143}Nd$	106 billion	Garnets, micas
$^{87}Rb \rightarrow {}^{87}Sr$	48.8 billion	Potassium-bearing minerals (mica, feldspar, hornblende)
$^{238}U \rightarrow {}^{206}Pb$	4.5 billion	Uranium-bearing minerals (zircon, uraninite)
$^{40}K \rightarrow {}^{40}Ar$	1.3 billion	Potassium-bearing minerals (mica, feldspar, hornblende)
$^{235}U \rightarrow {}^{207}Pb$	713 million	Uranium-bearing minerals (zircon, uraninite)

Sm = samarium, Nd = neodymium, Rb = rubidium, Sr = strontium, U = uranium, Pb = lead, K = potassium, Ar = argon

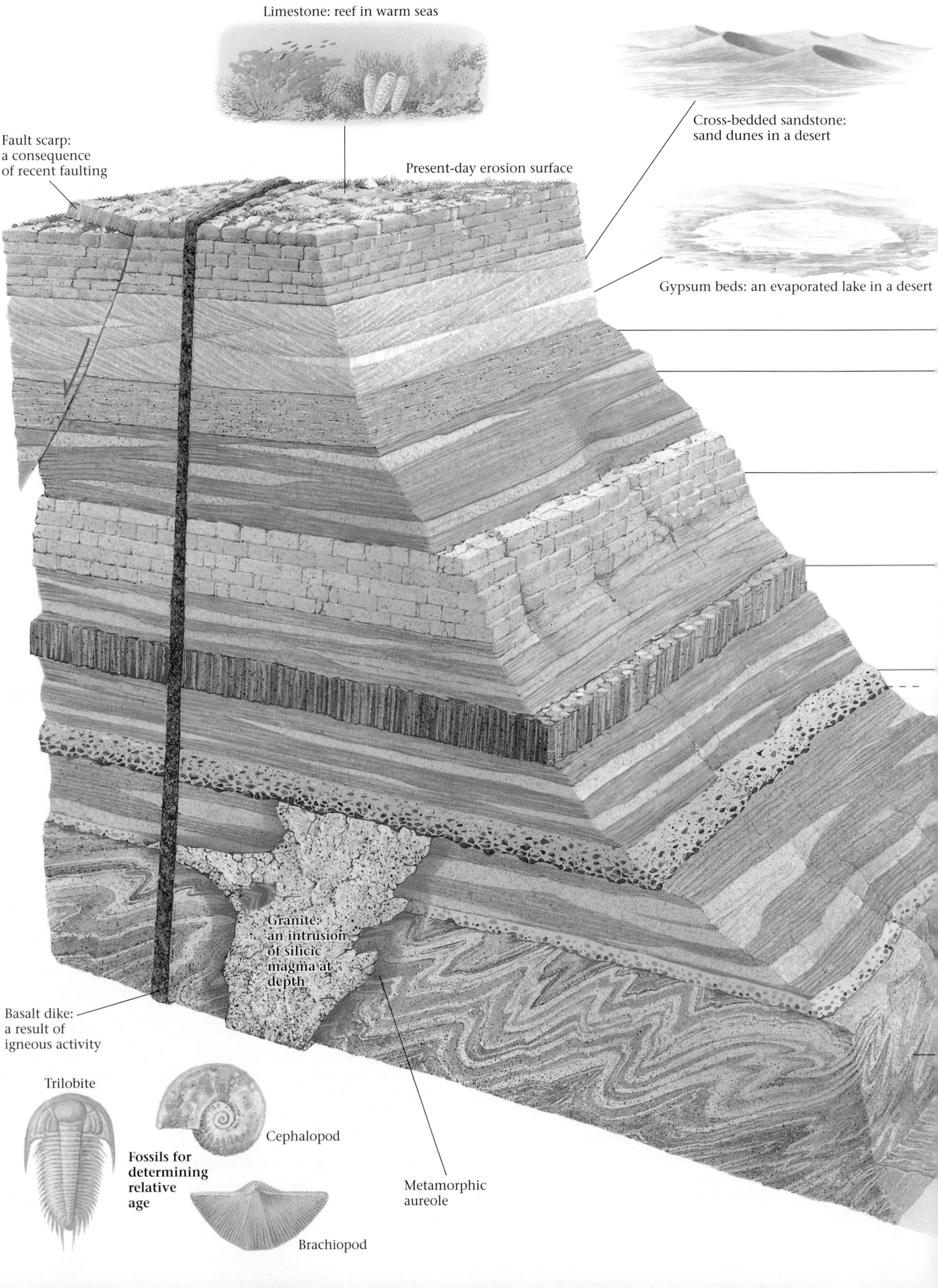

Limestone: reef in warm seas

Cross-bedded sandstone:
sand dunes in a desert

Gypsum beds: an evaporated lake in a desert

Fault scarp:
a consequence
of recent faulting

Present-day erosion surface

Basalt dike:
a result of
igneous activity

Granite:
an intrusion
of silicic
magma at
depth

Trilobite

Cephalopod

Fossils for
determining
relative
age

Brachiopod

Metamorphic
aureole

The Record in Rocks: Reconstructing Geologic History

Ignimbrite (welded tuff): an explosive volcanic eruption

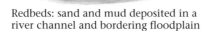

Limestone: reef in warm seas

Redbeds: sand and mud deposited in a river channel and bordering floodplain

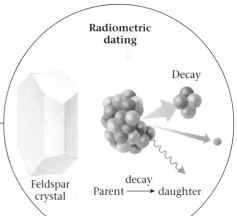

Radiometric dating

Decay

Feldspar crystal

decay

Parent ⟶ daughter

Basalt lava: flows from a volcano

Conglomerate: debris eroded from a cliff

Redbeds: sand and mud deposited by distributaries of a delta plain

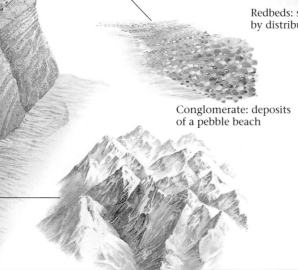

– – – Unconformity

Conglomerate: deposits of a pebble beach

Gneiss: metamorphism at depth beneath a mountain belt

When geologists examine a sequence of rocks exposed on a cliff, they see a record of Earth history that can be interpreted by applying the basic principles of geology, searching for fossils, and using radiometric dating. In this canyon, we see evidence for many geologic events. The layers of sediment (and the sedimentary structures they contain), the igneous intrusions, and the geologic structures tell us about past climates and past tectonic activity.

The insets show the way the region looked in the past, based on the record in the rocks. For example, the presence of gneiss at the base of the canyon indicates that at one time the region was a mountain belt, for the protoliths of the gneiss were buried deeply. Unconformities indicate that the region underwent uplift and erosion. Sedimentary successions record transgressions and regressions of the sea, igneous rocks are evidence of volcanic and intrusive activity, and faults indicate deformation.

Clearly, the land surface portrayed in this painting was sometimes a river floodplain or a delta (indicated by redbeds), sometimes a shallow sea (limestone), and sometimes a desert dune field (cross-bedded sandstones). And at several times in the past, volcanic activity occurred in the region. We can gain insight into the age of the sedimentary rocks by studying the fossils they contain, and the age of the igneous and metamorphic rocks by using radiometric dating methods.

• *Extracting parent and daughter isotopes:* To separate out the parent and daughter isotopes from minerals, geologists either dissolve the minerals in acid or evaporate portions of them with a laser (▶Fig. 10.15a).

• *Analyzing the parent-daughter ratio:* Geologists pass the dissolved or evaporated atoms through a mass spectrometer, a complex instrument that uses a very strong magnet to separate isotopes from one another according to their respective weights (▶Fig. 10.15b).

At the end of the laboratory process, geologists can define the ratio of parent to daughter isotopes in a mineral, and from this ratio calculate the age of the mineral. Needless to say, the description of the procedure here has been simplified—in reality, obtaining a radiometric date is time consuming and expensive and requires complex calculations.

What Does a Radiometric Date Mean?

At high temperatures, atoms in a crystal lattice vibrate so rapidly that chemical bonds can break and reattach relatively easily. As a consequence, parent and daughter isotopes escape from or move into crystals, so parent-daughter ratios are meaningless. Because radiometric dating is based

(a)

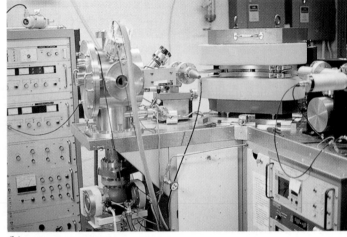

(b)

FIGURE 10.15 (a) A lab used in preparing samples for radiometric dating. The air must be exceedingly clean so that stray parent or daughter isotopes don't contaminate the samples. (b) The heart of this mass spectrograph, used for measuring isotope ratios, is a large magnet (the yellow coils).

BOX 10.1
THE REST OF THE STORY

Carbon-14 Dating

Many people who have heard of **carbon-14 (^{14}C) dating** assume that it can be used to define the numerical age of rocks. But this is not the case. Rather, ^{14}C dating tells us the ages of *organic material*—such as wood, cotton fibers, charcoal, flesh, bones, and shells—that contains carbon originally extracted from the atmosphere by photosynthesis in plants. ^{14}C, a radioactive isotope of carbon, forms naturally in the atmosphere when cosmic rays (charged particles from space) bombard atmospheric nitrogen-14 (^{14}N) atoms. When plants consume carbon dioxide during photosynthesis, or when animals consume plants, they ingest a tiny amount of ^{14}C along with ^{12}C, the more common isotope of carbon. After an organism dies and can no longer exchange carbon with the atmosphere, the ^{14}C in its body begins to decay back to ^{14}N

again. Thus, the ratio of ^{14}C to ^{12}C changes at a rate determined by the half-life of ^{14}C.

We can use ^{14}C dating to determine the age of prehistoric fire pits or of organic debris in sediment. ^{14}C has a short half-life—only 5,730 years. Thus, the method cannot be used to date anything older than about 70,000 years, for after that time essentially no ^{14}C remains in the material. But this range makes it a useful tool for geologists studying sediments of the last ice age and for archaeologists studying ancient cultures or prehistoric peoples. Since most rocks do not contain organic carbon, and are significantly older than 70,000 years, we cannot determine the age of rocks by using the ^{14}C dating method.

on the parent-daughter ratio, the "radiometric clock" starts only when crystals become cool enough for both parent and daughter isotopes to be locked into the lattice. The temperature below which isotopes are no longer free to move is called the blocking temperature of a mineral. When we specify a radiometric date for a rock, we are defining the time at which a specific mineral in the rock cooled below its blocking temperature.

With the concept of blocking temperature in mind, we can interpret the meaning of radiometric dates. In the case of igneous rocks, radiometric dating tells you when a magma or lava cooled to form a solid, cool igneous rock. In the case of metamorphic rocks, a radiometric date tells you when a rock cooled from the high temperature of metamorphism down to a low temperature.

Can we radiometrically date a sedimentary rock directly? No. If we date the minerals in a sedimentary rock, we determine only when the minerals making up the sedimentary rock first crystallized as part of an igneous or metamorphic rock, not the time when the minerals were deposited as sediment nor the time when the sediment lithified to form a sedimentary rock. For example, if we date the feldspar grains contained in a granite pebble in a conglomerate, we're dating the time the granite cooled below feldspar's blocking temperature, not the time the pebble was deposited by a stream. The age of mineral grains in sediment, however, can be useful. In recent years, geologists have undertaken studies to determine the ages of detrital (clastic) grains; by doing so, they can learn the age of the rocks in the region in which the sediment originated.

Other Methods of Determining Numerical Age

The rate of tree growth depends on the season. During the spring, trees grow rapidly and produce lighter, less dense wood, but during the winter, trees grow slowly or not at all, and produce darker, denser wood. Thus, wood contains recognizable annual growth rings (▶Fig. 10.16a). Such tree rings provide a basis for determining age. If you've ever wondered how old a tree that's just been cut down might be, just look at the stump and count the rings. Notably, by correlating clusters of distinctive rings in the older parts of living trees with comparable clusters of rings in dead logs, scientists can extend the tree-ring record back for many thousands of years, allowing geologists to track climate changes back into prehistory.

Seasonal changes also affect rates of such phenomena as shell growth, snow accumulation, clastic sediment deposition, chemical sediment precipitation, and production of organic material. Geologists have learned to use growth rings in shells, as well as rhythmic layering in sediments and in glacial ice (▶Fig. 10.16b), to date events numerically back through the past few hundred years of Earth history.

(a)

(b)

FIGURE 10.16 (a) Tree rings; (b) rhythmic layers in the ice of a glacier.

10.9 ADDING NUMERICAL AGES TO THE GEOLOGIC COLUMN

We have seen that radiometric dating can be used to date the time when igneous rocks formed and when metamorphic rocks metamorphosed, but not when sedimentary rocks were deposited. So how do we determine the numerical age of a sedimentary rock? We must answer this question if we want to add numerical ages to the geologic column—remember, the column was originally constructed by studying only the *relative* ages of fossil-bearing sedimentary rocks.

Geologists obtain dates for sedimentary rocks by studying cross-cutting relationships between sedimentary rocks and datable igneous or metamorphic rocks. For example, if we find a sedimentary rock layer deposited unconformably on a datable igneous rock, we know that the sedimentary rock must be younger than the igneous rock. If we find a datable igneous dike that cuts across beds of sedimentary rock, then the datable rock must be younger. And if a datable ash layer spread out over a layer of sediment and then was buried by another layer of sediment, the datable rock or ash must be younger than the underlying sediment and older than the overlying sediment (▶Fig. 10.17).

Geologists have searched the world for localities where they can recognize cross-cutting relations between datable igneous rocks and sedimentary rocks. By radiometrically dating the igneous rocks, they have been able to provide numerical ages for the boundaries between all the geological periods. For example, work from around the world shows that the Cretaceous Period actually began about 145 million years ago and ended 65 million years ago. So the

bed in Figure 10.17 was deposited during the middle part of the Cretaceous, not at the beginning or end.

The constant discovery of new data requires the numbers defining the boundaries of periods to change, which is why the term "numerical age" is preferred to "absolute age." In fact, around 1995, new research showed that the Cambrian-Precambrian boundary occurred at about 542 million years ago, in contrast to previous, less definitive studies that had placed the boundary at 570 million years ago. ▶Figure 10.18 shows the currently favored numerical ages of periods and eras in the geologic column. This dated column is commonly called the **geologic time scale.** Because of the numerical constraints provided by the geologic time scale, when geologists say that the first dinosaurs appeared during the Triassic Period, they mean that dinosaurs appeared after 251 million years ago.

10.10 THE AGE OF THE EARTH

During the eighteenth and nineteenth centuries, before the discovery of radiometric dating, scientists came up with a great variety of clever solutions to the question "How old is the Earth?"—all of which have since been proven wrong. Lord William Kelvin, a nineteenth-century physicist renowned for his discoveries in thermodynamics, made the most influential scientific estimate of the Earth's age of his time. Kelvin calculated how long it would take for a planet to cool down from a temperature as hot as the Sun's, and concluded that the Earth is about 24 million years old.

Kelvin's view contrasted with the idea being promoted by followers of Hutton, Lyell, and Darwin, who argued that if the concepts of uniformitarianism and evolution were correct, the Earth must be much older, because physical processes that shape the Earth and form its rocks and the process of natural selection that yields the diversity of species take a long time. Geologists and physicists debated the age issue for many years. The route to a solution came in 1896, when Henri Becquerel announced the discovery of radioactivity. Geologists immediately realized that the Earth's interior was producing some heat from the decay of radioactive material. This realization uncovered one of the flaws in Kelvin's argument: Kelvin had assumed that no new heat was produced after the planet first formed. Because radioactivity constantly generates new heat in the Earth, the planet has cooled down much more slowly than Kelvin had calculated. The discovery of radioactivity not only invalidated Kelvin's estimate of the Earth's age, it also led to the development of radiometric dating.

Since the 1950s, geologists have scoured the planet to identify its oldest rocks. Samples from several localities (Wyoming, Canada, Greenland, and China) have yielded

FIGURE 10.17 The Cretaceous sandstone bed was deposited unconformably on a 125-million-year-old granite pluton, so it must be younger than the granite. The 80-million-year-old dike cuts across the sandstone, so the sandstone must be older than the dike. Thus, this Cretaceous sandstone bed must be between 125 and 80 million years old. Similarly, the Paleocene bed was unconformably deposited over the dike and lies beneath a 50-million-year-old ash. Thus, the Paleocene bed must be between 80 and 50 million years old. Note that the data give only an age range.

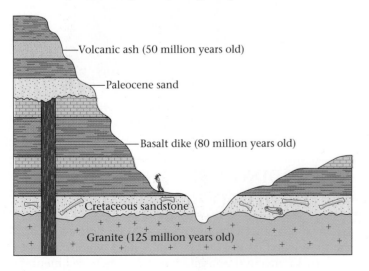

Volcanic ash (50 million years old)

Paleocene sand

Basalt dike (80 million years old)

Cretaceous sandstone

Granite (125 million years old)

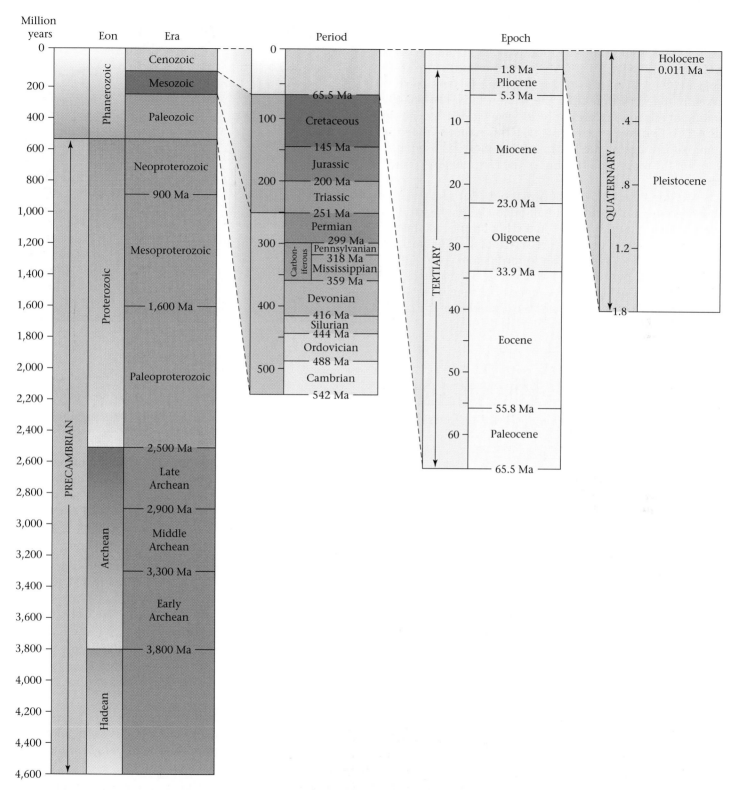

FIGURE 10.18 The geologic time scale assigns numerical ages to the intervals on the geologic column. Note that we have to change to a larger scale to portray the ages of intervals higher in the column. This time scale utilizes 2004 numbers favored by the International Commission on Stratigraphy. For further information, see the *GeoWhen Database* on the web, prepared by Robert A. Rohde, at http://www.stratigraphy.org/geowhen.

dates as old as 3.96 billion years. And sandstones found in Australia contain clastic grains that yielded dates of 4.1–4.2 billion years, indicating that rock as old as 4.2 billion years did once exist (▶Fig. 10.19). Models of the Earth's formation suggest that all objects in the solar system developed at roughly the same time from the same nebula. Radiometric dating of meteors and Moon rocks have yielded ages as old as 4.57 billion years; geologists take this number to be the approximate age of the Earth, leaving more than enough time for the rocks and life forms of the Earth to have formed and evolved.

We don't find 4.57 billion-year-old (Ga) rocks in the crust because during the first half-billion years of Earth history, rocks in the crust remained too hot for the radiometric clock to start (their temperature stayed above the blocking temperature), and/or crust that formed before about 4.0 Ga was destroyed by an intense meteorite bombardment at about 4.0 Ga. Geologists have named the time interval between the birth of the Earth and the formation of the oldest dated rock the Hadean Eon.

10.11 PICTURING GEOLOGIC TIME

The mind grows giddy gazing so far back into the abyss of time.
—JOHN PLAYFAIR (1747–1819), British geologist
who popularized the works of Hutton

The number 4.57 billion is so staggeringly large that we can't begin to comprehend it. If you lined up this many pennies in a row, they would make an 87,400-km-long line that would wrap around the Earth's equator more than twice. Notably, at the scale of our penny chain, human history is only about 100 city blocks long.

Another way to grasp the immensity of geologic time is to make a scale model, which we do by equating the entire 4.57 billion years to a single calendar year. On this scale, the oldest rocks preserved on Earth date from early February, and the first bacteria appear in the ocean on February 21. The first shelly invertebrates appear on October 25, and the first amphibians crawl out onto land on November 20. On December 7, the continents coalesce into the supercontinent of Pangaea. The first mammals and birds appear about December 15, along with the dinosaurs, and the Age of Dinosaurs ends on December 25. The last week of December represents the last 65 million years of Earth history, including the entire Age of Mammals. The first humanlike ancestor appears on December 31 at 3 P.M., and our species, *Homo sapiens,* shows up an hour before midnight. The last ice age ends a minute before midnight, and all of recorded human history takes place in the last 30 seconds. To put it another way, human history occupies the last 0.000001% of Earth history.

FIGURE 10.19 One of the oldest dated rocks on Earth, this 3.96-billion-year-old gneiss comes from the Northwest Territories of Canada.

CHAPTER SUMMARY

• The concept of geologic time, the span of time since the Earth's formation, developed when early geologists suggested that the Earth must be very old if geologic features were formed by the same natural processes we see today.

• Relative age specifies whether one geologic feature is older or younger than another; numerical age provides the age of a geologic feature in years.

• Using such principles as uniformitarianism, superposition, original horizontality, original continuity, cross-cutting relations, inclusions, and baked contacts, we can construct the geologic history of a region.

• The principle of fossil succession states that the assemblage of fossils in a sequence of strata changes from base to top. Once a fossil species becomes extinct, it never reappears.

• Strata are not deposited continuously at a location. An interval of nondeposition and/or erosion is called an unconformity. Geologists recognize three kinds: angular unconformity, nonconformity, and disconformity.

• A stratigraphic column shows the succession of formations in a region. A given succession of strata that can be traced over a fairly broad region is called a stratigraphic formation. The process of determining the relationship between strata at one location and strata at another is called correlation. A geologic map shows the distribution of formations.

- A composite stratigraphic column that represents the entirety of geologic time is called the geologic column. The geologic column's largest subdivisions, each of which represents a specific interval of time, are eons. Eons are subdivided into eras, eras into periods, and periods into epochs.

- The numerical age of rocks can be determined by radiometric dating. This is because radioactive elements decay at a constant rate known as a half-life.

- The radiometric date of a mineral specifies the time at which the mineral cooled below a certain temperature. We can use radiometric dating to determine when an igneous rock solidified and when a metamorphic rock cooled from high temperatures. To date sedimentary strata, we must examine cross-cutting relations with dated igneous or metamorphic rock.

- Other methods for dating materials include counting growth rings in trees and rhythmic layers in glaciers.

- From the radiometric dating of meteors and Moon rocks, geologists conclude that the Earth formed about 4.57 billion years ago. Our species, *Homo sapiens,* has been around for only 0.000001% of geologic time.

KEY TERMS

Cambrian explosion (p. 294)
carbon-14 (^{14}C) dating (p. 300)
contact (p. 289)
correlation (p. 291)
cross-cutting relation (p. 285)
disconformity (p. 288)
eon (p. 294)
epoch (p. 294)
era (p. 294)
formation (stratigraphic formation) (p. 289)
fossil succession (p. 287)
geologic column (p. 293)
geologic map (pp. 292–93)
geologic time (p. 283)
geologic time scale (p. 302)

half-life (p. 297)
isotope (p. 295)
numerical age (absolute age) (p. 284)
original horizonality (p. 285)
period (p. 294)
Precambrian (p. 294)
radiometric dating (geochronology) (p. 295)
relative age (p. 284)
stratigraphic column (p. 289)
stratigraphic formation (p. 289)
superposition (p. 284)
unconformity (p. 287)
uniformitarianism (p. 284)

REVIEW QUESTIONS

1. Compare numerical age and relative age.

2. Describe the principles that allow us to determine the relative ages of geologic events.

3. How does the principle of fossil succession allow us to determine the relative ages of strata?

4. How does an unconformity develop? Describe the differences among the three kinds of unconformities.

5. Describe two different methods of correlating rock units. How was correlation used to develop the geologic column? What is a stratigraphic formation?

6. What does the process of radioactive decay entail?

7. How do geologists obtain a radiometric date? What are some of the pitfalls in obtaining a reliable one?

8. Why can't we date sedimentary rocks directly?

9. How are growth rings and ice cores useful in determining the ages of geologic events?

10. What is the age of the oldest rocks on Earth? What is the age of the oldest rocks known? Why is there a difference?

SUGGESTED READING

Dalrymple, G. B. 1994. *The Age of the Earth.* Palo Alto, Calif.: Stanford University Press.

Gould, S. J. 1987. *Time's Arrow, Time's Cycle.* Cambridge, Mass.: Harvard University Press.

Gradstein, F. J., and A. Smith. 2005. *A Geologic Time Scale 2004.* Cambridge: Cambridge University Press.

Lyell, C., and J. A. Secord. 1998. *Principles of Geology* (abridged reprint of the original). New York: Penguin Classics.

Prothero, D. R., and R. H. Dott, Jr. 2003. *Evolution of the Earth.* 7th ed. New York: McGraw-Hill.

Repcheck, J. 2003. *The Man Who Found Time: James Hutton and the Discovery of Earth's Antiquity.* New York: Perseus Books Group.

Rudwick, M. J. S. 1985. *The Meaning of Fossils: Episodes in the History of Palaeontology.* Chicago: University of Chicago Press.

Winchester, S. 2002. *The Map That Changed the World: William Smith and the Birth of Modern Geology.* New York: Harper Perennial.

A Biography of Earth

I weigh my words well when I assert that the man who should know the true history of the bit of chalk which every carpenter carries about in his breeches pocket, though ignorant of all other history, is likely, if he will think his knowledge out to its ultimate results, to have a truer and therefore a better conception of this wonderful universe and of man's relation to it than the most learned student who [has] deep-read the records of humanity [but is] ignorant of those of nature.
—THOMAS HENRY HUXLEY, from ON A PIECE OF CHALK (1868)

A geologist at work in the field, here on the shore of Kangaroo Island, Australia. Field observations have allowed geologists to work out the history of Earth.

11.1 INTRODUCTION

In 1868, a well-known British scientist, Thomas Henry Huxley, presented a public lecture on geology to an audience in Norwich, England. Seeking a way to convey his fascination with Earth history to people with no previous geological knowledge, he focused his audience's attention on the piece of chalk he had been writing with (see epigraph above). And what a tale the chalk has to tell! Chalk, a type of limestone, consists of microscopic marine algae shells and shrimp feces. The specific chalk that Huxley held came from beds deposited in Cretaceous time (the name *Cretaceous,* in fact, derives from the Latin word for "chalk"). These beds now form the White Cliffs of Dover. Geologists in Huxley's day knew of similar chalk beds in outcrops throughout much of Europe, and had discovered that the chalk contains not only plankton shells but also fossils of bizarre swimming reptiles, fish, and invertebrates—species absent in the seas of today. Clearly, when the chalk was deposited, warm seas holding unfamiliar creatures covered some of what is dry land today.

Clues in his humble piece of chalk allowed Huxley to demonstrate to his audience that the geography and inhabitants of the Earth in the past differed markedly from those today, and thus that *the Earth has a history.* In the many decades since Huxley's lecture, field and laboratory studies worldwide have allowed geologists to develop an overall image of Earth's history. With every new geologic discovery, this image, once a blur, comes progressively more in focus. We now see a complex, evolving Earth System in which physical and biological components interact pervasively in ways that have transformed a formerly barren, crater-pocked surface into countless environments supporting a diversity of life. Fossil finds of the past two decades demonstrate that life began when the Earth was still quite young; life has affected surface and near-surface processes ever since. And as we pointed out near the beginning of this book, the map of our planet constantly changes as continents rift, drift, and collide, and as ocean basins open and close.

We will discuss the nature of overall global change further in Chapter 19. In this chapter, we offer a concise geological biography of our planet, from its birth 4.57 billion years ago to the present. We see how continents came into existence and have waltzed across the globe ever since. We also learn when mountain-building events took place and how Earth's climate and sea level have changed through time. And while these physical changes have taken place, life has evolved. To simplify the discussion, we use the following abbreviations: Ga (for "billion years ago"), Ma ("million years ago"), and Ka ("thousand years ago").

11.2 THE HADEAN EON: HELL ON EARTH?

A viscid pitch boiled in the fosse below
and coated all the bank with gluey mire.
I saw the pitch; but I saw nothing in it
except the enormous bubbles of its boiling
which swelled and sank, like breathing, through all the pit.
—DANTE, *THE INFERNO,* CANTO XXI

James Hutton, the eighteenth-century Scottish geologist who was the first to provide convincing evidence that the Earth was vastly older than human civilization, could not measure Earth's age directly, and indeed speculated that there may be "no vestige of a beginning." But radiometric dating studies conducted in recent decades have shown that it is possible, in fact, to assign a numerical age to our planet's formation. Specifically, dates obtained for a class of meteorites thought to be representative of the planetesimal cloud out of which the Earth formed consistently yield an age of 4.57 billion years. Geologists currently take this age to be the Earth's birth date. It appears that the Earth formed within 30 million years of the time when the Sun's nuclear furnace first fired up. But a clear record of Earth history, as recorded in continental crustal rocks, does not begin until about 3.80 Ga, for crustal rocks older than 3.80 billion years are exceedingly rare.

Geologists refer to the mysterious time interval between the birth of Earth and 3.80 Ga as the **Hadean Eon** (from the Greek *Hades,* the god of the underworld). A number of clues provide the basic framework of Hadean history.

The Hadean Eon began with the formation of the Earth by the accretion of planetesimals (see Chapter 1). As the planet grew, collection and compression of matter into a dense ball generated substantial heat, and each time another meteorite collided with the Earth, its kinetic energy transformed into more heat. Radioactive decay produced still more heat within the newborn planet.

Eventually, the Earth became hot enough to partially melt, and when this happened, by about 4.5 Ga, it underwent internal **differentiation**—gravity pulled molten iron down to the center of the Earth, where it accumulated to form the core. A mantle, composed of ultramafic rock, remained as a thick shell surrounding the core (see Chapter 2). Internal differentiation, which may have occurred relatively quickly, generated sufficient heat to make the Earth even hotter. Researchers suggest that soon after—or perhaps during—differentiation, a Mars-sized proto-planet collided with the Earth. The energy of this collision blasted away a significant fraction of Earth's mantle, which mixed with the shattered fragments of the colliding planet's mantle to form a ring of silicate-rock debris orbiting the Earth. Heat generated by the collision substantially melted Earth's remaining mantle, making it so weak that the iron core of the colliding body sank through it and merged with the Earth's core. Meanwhile, the ring of debris surrounding the Earth coalesced to form the Moon.

In the wake of differentiation and Moon formation, the Earth was so hot that its surface was mostly an ocean of seething magma, supplied by intense eruption of melts rising from the mantle—it resembled Dante's Inferno (▶Fig. 11.1). Here and there, rafts of solid rock formed temporarily on the surface of the magma ocean, but these eventually sank and remelted. This stage lasted at least until about 4.4 Ga. After that time, because of the decrease in radioactive heat generation, Earth *might* have become cool enough for solid rocks to form at its surface. The evidence for this statement comes from western Australia, where geologists have found 4.4 Ga grains of a durable mineral called zircon in sandstone beds. The zircons must have originally formed in igneous rock, but were eroded and later deposited in sedimentary beds. Existence of these grains suggests that at least some solid igneous rocks had formed by 4.4 Ga.

During the hot early stages of the Hadean Eon, rapid outgassing of the Earth's mantle took place. This means that volatile (gassy) elements or compounds originally incorporated in mantle minerals were released and erupted at the Earth's surface, along with lava. The gases accumulated to constitute an unbreathable atmosphere of water (H_2O), methane (CH_4), ammonia (NH_3), hydrogen (H_2), nitrogen (N_2), carbon dioxide (CO_2), sulfur dioxide (SO_2), and other

FIGURE 11.1 A painting of the early Hadean Earth. Note the magma ocean, with small crusts of solid basalt floating about. If the sky had been clear, streaks of meteors would have filled it, and the Moon would have appeared much larger than it does today, because it was closer. Probably, however, an observer on Earth could not have seen the sky, because the atmosphere contained so much water vapor and volcanic ash.

gases. Some researchers have speculated that gases from comets colliding with Earth during the Hadean may have contributed components of the early atmosphere. If the Hadean Earth's surface was sufficiently cool for an extensive solid crustal rock to form beginning at 4.4 Ga, then the first oceans may have accumulated soon thereafter, when water in the atmosphere condensed and fell as rain.

Studies of cratering on the Moon suggest that the rate of meteorite bombardment of Earth may have increased drastically between about 4.0 and 3.9 Ga. This late bombardment would have obliterated any crust or ocean that had formed up to that time. After this bombardment ceased, new crust formed, some of which has survived ever since. Geologists have found marine sediments in rock sequences as old as 3.8 billion years, indicating that liquid-water oceans existed at least by this time. Thus, if you could see the Earth as the Hadean Eon drew to a close, you would see barren, crater-pocked lands poking above oceans; both land and sea may have been partially covered by ice, and both lay beneath a CO_2- and SO_2-rich atmosphere.

11.3 THE ARCHEAN EON: THE BIRTH OF THE CONTINENTS AND THE APPEARANCE OF LIFE

The date that marks the boundary between the **Archean** (from the Greek word for "ancient") **Eon** and the Hadean Eon has been placed at about 3.8 Ga. Effectively, it marks the time

at which intense meteorite bombardment of Earth had ceased. Crustal rocks, including rocks that originated as marine sediments, have survived intact since 3.8 Ga.

Geologists still argue about whether plate tectonics operated in the early part of the Archean Eon in the same form as today. Some researchers picture an early Archean Earth with rapidly moving small plates, numerous volcanic island arcs, and abundant hot-spot volcanoes. Others propose that early Archean lithosphere was too warm and buoyant to subduct, and that plate tectonics could not have operated until the later part of the Archean; these authors argue that plume-related volcanism was the main source of new crust until the late Archean. Regardless of which model ultimately proves more accurate, it is clear that the Archean was a time of significant change in the map of the Earth. During that time the volume of continental crust increased significantly.

How did the continental crust come to be? According to one model, relatively buoyant (felsic and intermediate) crustal rocks formed both at subduction zones and at hot-spot volcanoes. Frequent collisions sutured volcanic arcs and hot-spot volcanoes together, creating progressively larger blocks called protocontinents (▶Fig. 11.2a, b). Some of these protocontinents developed rifts that filled with basalt. As time passed, protocontinents became cooler and stronger, and by 2.7 Ga the first cratons, long-lived blocks of

FIGURE 11.2 (a) During the Archean Eon, rocks that would eventually make up the continental crust began to form. At early subduction zones, volcanic island arcs, composed of relatively buoyant rocks, developed, and large shield volcanoes formed over hot spots. Larger crustal blocks underwent rifting, creating rift basins that filled with volcanic rocks, and the erosion of crustal blocks deposited graywacke on the sea floor. (b) Successive collisions brought all the buoyant fragments (labeled Ⓐ through Ⓓ) together to form a protocontinent. Melting at the base of the protocontinent may have caused the formation of plutons.

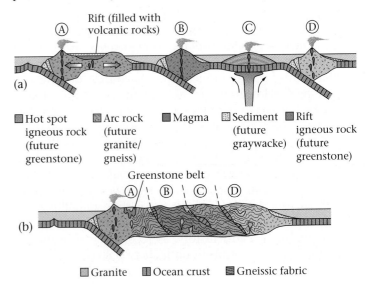

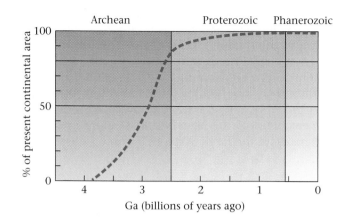

FIGURE 11.3 As time progressed, the area of the Earth covered by continental crust increased, though not all geologists agree about the rate of increase. This model shows growth beginning about 3.9 Ga and continuing rapidly until the end of the Archean Eon. The rate slowed substantially in the Proterozoic Eon.

durable continental crust, had developed. By the end of the Archean Eon, about 80% of continental crust had formed (▶Fig. 11.3). Volcanic activity during this time continued to supply gas to the atmosphere.

Clearly, the Archean Eon saw many firsts in Earth history—the first continents appeared during the Archean, and probably the first life. Geologists use three sources of evidence to identify early life:

- *Chemical (molecular) fossils, or biomarkers:* These are durable chemicals that are produced only by the metabolism of living organisms.

- *Isotopic signatures:* By analyzing the ratio of ^{12}C to ^{13}C in carbon-rich sediment geologists can determine if the sediment once contained the bodies of organisms, because organisms preferentially incorporate ^{12}C.

- *Fossil forms:* Given appropriate depositional conditions, fossils of bacteria or archaea cells can be preserved in rock.

The search for the earliest evidence of life continues to make headlines in the popular media. Most geologists currently conclude that life has existed on Earth since at least 3.5 Ga, for rocks of this age contain clear isotopic signatures of organisms. The oldest undisputed fossils of bacteria and archaea occur in 3.2 Ga rocks (▶Fig. 11.4a). Some rocks of this age contain **stromatolites,** distinctive mounds of sediment produced by mats of cyanobacteria. Stromatolites form because cyanobacteria secrete a mucouslike substance to which

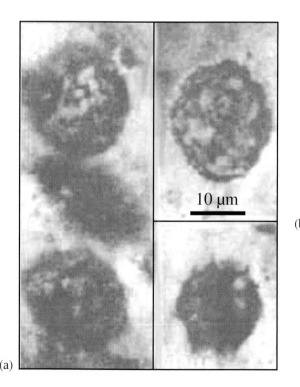

(a)

(b)

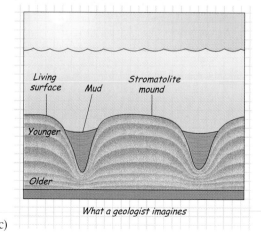

(c)

FIGURE 11.4 (a) Some of the first fossils, found in 3.2-million-year-old rocks, are spherical bacteria cells from South African chert. (b, c) This outcrop shows a cross section through a stromatolite deposit. Stromatolites are mounds of bacteria that grow upward with time, with living bacteria making up the mounds' surface.

sediment settling from water sticks. As the mat gets buried, new cyanobacteria colonize the top of the sediment, building a mound upward (▶Fig. 11.4b, c). Biomarkers in Archean sediments indicate that photosynthetic organisms appeared by 2.7 Ga.

What specific environment on the Archean Earth served as the cradle of life? Laboratory experiments conducted in the 1950s led many researchers to conclude that life began in warm pools of water, beneath a methane- and ammonia-rich atmosphere, under a stormy sky streaked by bolts of lightning. The only problem with this theory is that more recent evidence suggests that the early atmosphere consisted mostly of CO_2 and N_2, with relatively little methane and ammonia. More recent studies point instead to submarine hot-water vents, so-called black smokers, as the hosts of the first organisms. These vents emit clouds of ion-charged solutions from which sulfide minerals precipitate and build chimneys. The earliest life in the Archean Eon may well have been thermophilic (heat-loving) bacteria or archaea that dined on pyrite at dark depths in the ocean alongside these vents. Only when photosynthesis, the ability to extract energy from sunlight, evolved could organisms populate shallow marine environments.

As the Archean Eon came to a close, the first continents existed, and life had colonized not only the depths of the sea but also the shallow marine realm. Plate tectonics had commenced, continental drift was taking place, collisional mountain belts were forming, and erosion was occurring. The atmosphere was gradually accumulating oxygen, although probably this gas still accounted for only a very small percentage of the air. The stage was set for another major change in the Earth System.

11.4 THE PROTEROZOIC EON: TRANSITION TO THE MODERN WORLD

The **Proterozoic** (from the Greek words meaning "first life") **Eon** spans roughly 2 billion years, from about 2.5 billion years ago to the beginning of the Cambrian Period, 542 million years ago—thus, it encompasses almost half of Earth's history. The eon received its now misleading name before fossil organisms were discovered in Archean rock.

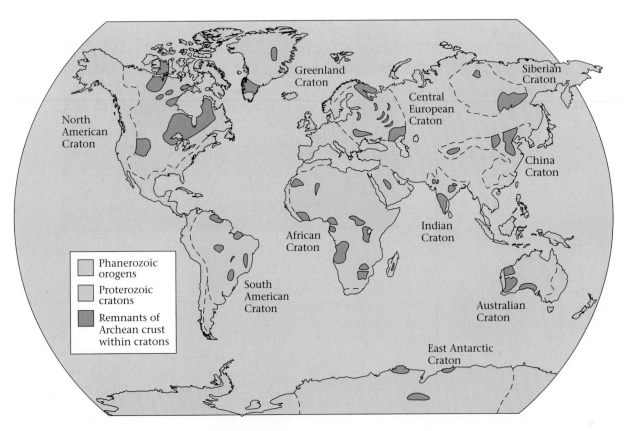

FIGURE 11.5 The distribution of Precambrian crust. The cratons formed in Proterozoic time, either from magmas rising from the mantle or by the metamorphism and deformation of existing Archean crust.

During Proterozoic time, Earth's surface environment changed from the unfamiliar world of possibly small, fast-moving plates, small continents, and an oxygen-poor atmosphere to the more familiar world of mostly large plates, large continents, and an oxygen-rich atmosphere.

First, let's look at changes to the continents. New continental crust continued to form during the Proterozoic Eon, but at progressively slower rates—in fact, by the middle of the eon, over 90% of the Earth's continental crust had formed. Also, collisions between the Archean continents, as well as the accretion of new volcanic island arcs and hotspot volcanoes, resulted in the assembly of large continents whose interiors were located far from orogenic belts. These continents cooled and strengthened to become cratons. Most of the large cratons that exist today had formed by about 1.8 Ga (▶Fig. 11.5).

Consider the geology of a large craton a little more closely, by examining the interior of North America. The Precambrian rocks constituting this craton crop out extensively in Canada. Geologists refer to this region as the Canadian Shield. A **shield** consists of a broad, low-lying region of exposed Precambrian rocks. (In the United States, Phanerozoic strata bury most of the Precambrian rocks, forming a province called a **cratonic platform,** or continental platform.) Here we see Precambrian rocks only where they have been exposed by erosion of the sedimentary cover. The Canadian Shield includes several Archean blocks that were sutured together along huge collisional belts. Each of these blocks and belts has been given a name (▶Fig. 11.6). The U.S. portion of the craton grew when a series of volcanic island arcs and continental slivers accreted (attached) to the southern margin of the Canadian Shield between 1.8 and 1.6 Ga. In the Midwest, huge granite plutons intruded much of this accreted region, and rhyolite covered it, between 1.5 and 1.3 Ga.

Successive collisions ultimately brought together most continental crust on Earth into a single supercontinent, named **Rodinia,** by around 1 Ga. The last major collision during the formation of Rodinia produced a large collisional orogen called the Grenville orogen. If you look at a popular (though not totally proven) reconstruction of Rodinia, you can identify the crustal provinces that would eventually become the familiar continents of today (▶Fig. 11.7a). In this reconstruction, the future Antarctica and Australia lay somewhere along the western coast of North America.

Several studies suggest that sometime between 800 and 600 Ma (million years ago), Rodinia "turned inside out," in that Antarctica, India, and Australia broke away from western North America and swung around and collided with the future South America, possibly forming a short-lived supercontinent that some geologists refer to as Pannotia (▶Fig. 11.7b).

Much of the transformation of the Earth's atmosphere from the oxygen-poor volcanic gas mix of the Archean Eon

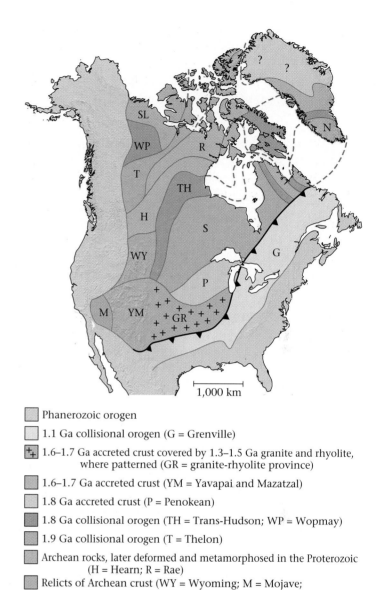

Phanerozoic orogen

1.1 Ga collisional orogen (G = Grenville)

1.6–1.7 Ga accreted crust covered by 1.3–1.5 Ga granite and rhyolite, where patterned (GR = granite-rhyolite province)

1.6–1.7 Ga accreted crust (YM = Yavapai and Mazatzal)

1.8 Ga accreted crust (P = Penokean)

1.8 Ga collisional orogen (TH = Trans-Hudson; WP = Wopmay)

1.9 Ga collisional orogen (T = Thelon)

Archean rocks, later deformed and metamorphosed in the Proterozoic (H = Hearn; R = Rae)

Relics of Archean crust (WY = Wyoming; M = Mojave; S = Superior; N = Nain; SL = Slave)

FIGURE 11.6 The North American craton consists of a collage of different fragments, stitched together during orogenies of Precambrian time. (The explanation names these fragments and orogens.) We distinguish between collisional orogens, where large blocks joined, and accretionary orogens, composed of volcanic island arcs, hot-spot volcanoes, and continental slivers plastered onto the margin of a larger continent at a convergent margin.

to the oxygen-rich mix we breathe today occurred during the Proterozoic Eon (▶Fig. 11.8; note that the present concentration of oxygen, 21%, was not attained until the Phanerozoic Eon). Part of the evidence for this statement comes from studying the distribution of banded iron formation (BIF). This is a special kind of sedimentary rock containing alternating layers of iron-oxide minerals and chert. Its deposition may have been triggered by an increase in the

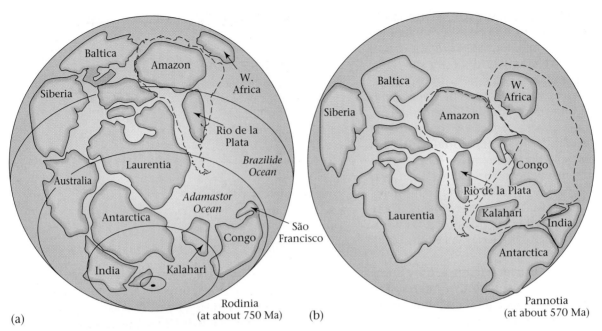

(a) Rodinia (at about 750 Ma)   (b) Pannotia (at about 570 Ma)

FIGURE 11.7 Supercontinents at the end of the Precambrian. (a) Rodinia formed around 1 Ga and lasted until about 700 Ma. (b) According to one model, by about 570 Ma, Antarctica, India, and Australia had broken off the western margin of Rodinia and had swung around and collided with the eastern margin of the future South America, to create a new, short-lived supercontinent called Pannotia. Pannotia broke apart at about 550 Ma.

dissolved oxygen content of sea water—the presence of oxygen could cause precipitation of iron that had been soluble in sea water in the absence of abundant oxygen. Formation of BIF may have absorbed oxygen and kept atmospheric concentrations of the gas low until about 1.8 Ga. After this time, oxygen concentrations in the atmosphere appear to have increased substantially.

This new oxygen-rich atmosphere had a profound effect on the Earth, for it permitted a great diversification of life and the eventual conquest of the land by living organisms. Life could become more complex because oxygen-dependent (aerobic) metabolism produces energy much more efficiently than does oxygen-free (anaerobic) metabolism—eating sulfide minerals may sustain bacteria, but it can't keep multicellular organisms alive! The addition of substantial oxygen to the atmosphere also set the stage for the eventual move of life onto the land, hundreds of millions of years after the Proterozoic. Oxygen provides the raw material from which ozone (O_3) forms—ozone protects the land surface from harmful ultraviolet rays.

So far, we've talked only of *prokaryotic* organisms, those which consist of single cells that do not contain a nucleus. Geologists have concluded, based on chemical fossils, that the first *eukaryotic* cells, the cells that make up more complex organisms, may have originated as early as 2.7 Ga. Controversial fossil forms of eukaryotic organisms have been found in rocks ranging from 2.1 to 1.2 Ga, but undeniable eukaryotic fossil organisms occur in 1.0 Ga rocks.

The last half-billion years of the Proterozoic Eon saw the remarkable transition from simple organisms into complex ones. Ciliate protozoans (single-celled organisms coated with fibers that give them mobility) appear at about 750 Ma. A great leap forward in complexity of organisms occurred during the next 150 million years, for sediments deposited perhaps as early as 620 Ma and certainly by 565 Ma contain

FIGURE 11.8 The graph shows the change in the oxygen content of Earth's atmosphere through time. The vertical axis represents the percentage of the atmosphere that consists of oxygen. Today, oxygen accounts for about 21% of the atmosphere, but in Archean time it constituted only about 0.000000001%.

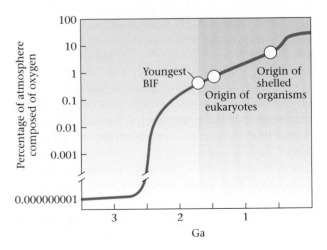

several types of multicellular organisms that together constitute the **Ediacaran fauna,** named for a region in southern Australia. Ediacaran species survived into the beginning of the Cambrian before becoming extinct. Though their nature is enigmatic, fossil forms suggest that some of these invertebrate (shell-less) organisms resembled jellyfish, while others resembled worms (▶Fig. 11.9).

Radical climate shifts occurred on Earth at the end of the Proterozoic Eon. Specifically, accumulations of glacial sediments occur in virtually all sequences of latest Proterozoic. What's strange about the occurrence of these sediments is that they can be found even in continental coastal regions that were located at the equator during these times. This observation implies that the *entire* planet was cold enough for glaciers to form at the end of the Proterozoic. Geologists still are debating the history of these global "ice ages" (see Chapter 18), but in one model, glaciers covered the land, and perhaps the entire ocean froze, creating what geologists refer to as **snowball Earth.** The shell of ice cut off the oceans from the atmosphere, causing oxygen levels in the sea to drop drastically, so many life forms died off. The icy sheath also prevented atmospheric CO_2 from dissolving in seawater, but it did not prevent volcanic activity from continuing to add CO_2 to the atmosphere. According to this model, Earth would have remained a snowball forever, were it not for volcanic CO_2, for CO_2 is a greenhouse gas, meaning that it traps heat in the atmosphere much as glass planes trap heat in a greenhouse (see Chapter 19). Thus, as the CO_2 concentration increased, Earth warmed up and eventually the glaciers rapidly melted. Life may have survived snowball Earth conditions only near submarine black smokers and near hot springs. When the ice vanished, life rapidly expanded into new environments, and new species, such as the Ediacaran fauna, could evolve.

FIGURE 11.9 *Dickinsonia,* a fossil of the Ediacaran fauna, an assemblage of soft-bodied organisms that appeared in late Proterozoic seas.

11.5 THE PHANEROZOIC EON: LIFE DIVERSIFIES, AND TODAY'S CONTINENTS FORM

The **Phanerozoic** (Greek for "visible life") **Eon** encompasses the last 542 million years of Earth history. Its name reflects the appearance of diverse organisms with hard shells or skeletons that became the well-preserved fossils you can easily find in rock outcrops. Geologists divide the Phanerozoic into three eras—the Paleozoic (Greek for "ancient life"), the Mesozoic ("middle life"), and the Cenozoic ("recent life")—to emphasize the changes in Earth's living population throughout the eon. The solid Earth also changed during this time. Continental blocks rearranged, and new supercontinents formed, only to break apart again. Numerous orogenies built mountain ranges, the most recent of which persist today. Shell-secreting organisms, burrowing organisms, and plants invaded the sedimentary environment and changed the style of deposition. Below, we look at the changes in the planet's map (its paleogeography) and in its life forms during the three eras of the Phanerozoic Eon.

11.6 THE PALEOZOIC ERA: FROM RODINIA TO PANGAEA

The Early Paleozoic Era (Cambrian-Ordovician Periods, 542–444 Ma)

Paleogeography. At the beginning of the Paleozoic Era, Pannotia broke up, yielding smaller continents including Laurentia (composed of North America and Greenland), Gondwana (South America, Africa, Antarctica, India, and Australia), Baltica (Europe), and Siberia (▶Fig. 11.10). Several passive-margin basins formed along the edges of new continents. In addition, sea level rose, so that vast areas of continents were flooded with shallow seas called **epicontinental seas.** For example, by the end of the Cambrian Period, the only dry land in Laurentia was an oval region centered on Hudson's Bay, so most of what is now the United States lay beneath water (▶Fig. 11.11). In many places, water depths in epicontinental seas reached only a few meters, creating a well-lit environment in which life abounded. Deposition in the seas yielded layers of fossiliferous sediment. Whereas a relatively thin layer of sediment formed in Laurentia's interior (the region that would become the continental platform), a much thicker layer accumulated in the passive-margin basins that fringed the continent. Sea level, however, did not stay high for the entire early Paleozoic Era; regressions and transgressions occurred, the former marked by unconformities and the latter by accumulations of sediment (▶Box 11.1).

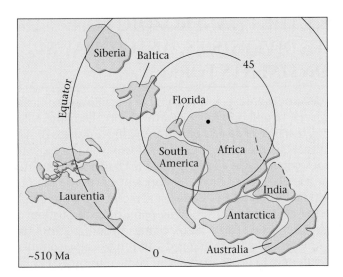

FIGURE 11.10 The distribution of continents in the Cambrian Period (about 510 Ma). Note that Gondwana lay near the equator, and Laurentia straddled the equator. The map shows how the Earth might have looked as viewed from the South Pole. Note that Florida had not yet connected to North America (part of Laurentia).

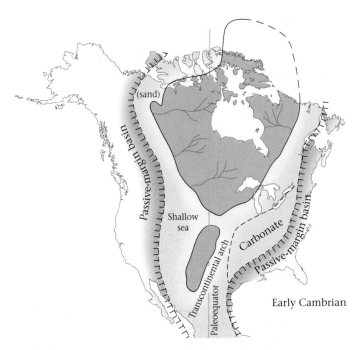

FIGURE 11.11 A paleogeographic map of North America, showing the regions of dry land and shallow sea in the early Cambrian Period. Note the edges of the passive margins, which define the borders of the continent. At this time, the regions between the edge of the passive margin and the coast did not yet exist. The transcontinental arch was a ridge of dry land.

The geologically peaceful world of the early Paleozoic Era in Laurentia abruptly came to a close in the Middle Ordovician Period, for at this time its eastern margin rammed into a volcanic island arc and other crustal fragments. The resulting collision, called the Taconic orogeny, deformed and metamorphosed strata of the former passive-margin basin and produced a mountain range.

Life evolution. Beginning with the Cambrian Period, life on Earth left a clear record of evolution, because so many organisms developed shells of durable minerals. The fossil record indicates that soon after the Cambrian began, life underwent remarkable diversification. This event, which paleontologists refer to as the **Cambrian explosion of life,** took about 20 million years. What caused this explosion? No one can say for sure, but considering that it occurred roughly at the time a supercontinent broke up, it may have had something to do with the new ecological niches and the isolation of populations that resulted when small continents formed and drifted apart.

The first animals to appear in the Cambrian Period had simple tube- or cone-shaped shells, but soon thereafter, the shells became more complex. Small fossils called conodonts, which resemble teeth, are found in Cambrian strata. Their presence suggests that creatures with jaws had appeared at this time, hinting that shells evolved as a means of protection against predators. By the end of the Early Cambrian, trilobites were grazing the sea floor. Trilobites shared the environment with mollusks, brachiopods, nautiloids, and echinoderms (▶Fig. 11.12; see Interlude D). Thus, a complex food chain arose, which included plankton, bottom feeders,

and at the top, giant predators. Mounds of sponges with mineral skeletons formed the first reefs, and graptolites (floating organisms that contained a serrated mineral band), coral, and gastropods appeared. The Ordovician saw the appearance of crinoids and the first vertebrate animals, jawless

FIGURE 11.12 A museum diorama illustrates what marine organisms living during the early Paleozoic may have looked like. Here, we see creatures such as trilobites and nautiloids.

Stratigraphic Sequences and Sea-Level Change

BOX 11.1
SCIENCE TOOLBOX

As we have seen in this chapter, there have been intervals in geologic history when the interior of continents were submerged beneath a shallow sea, and times when they were high and dry. This observation implies that sea level, relative to the surface of the continent, rises and falls through geologic time. When a continental surface is dry and exposed to the atmosphere, weathering and erosion grind away at previously deposited rock and, as a result, produce a continent-wide unconformity. In the early 1960s, an American stratigrapher named Larry Sloss introduced the name **stratigraphic sequence** to refer to the strata deposited on a continent during periods when the continent was submerged. Such sequences are bounded above and below by regional unconformities.

We can picture the deposition of an idealized sequence as follows. As the starting condition, much of the surface of a continent lies above sea level, and an unconformity develops. Then, as a transgression takes place, the margin of the continent floods first, then the interior. Thus, the preexisting unconformity gets progressively buried toward the interior, and the bottom layers of the newly deposited sequence tend to be older near the margin of the continent than toward the interior. The base of a sequence consists of terrestrial strata (river alluvium). As transgression progresses, this sediment is buried by nearshore sediment, then by deeper-water sediment (▶Fig. 11.13a). Eventually, nearly the entire width of the continent, with the exception of highlands and mountain belts, lies below sea level. When regression begins, the interior of the continent is exposed first and then the margins. So a new unconformity forms first in the interior and then later along the margins. Many shorter-duration transgressions and regressions may happen during a single long-duration rise or fall. Taken together, this information may provide insight into the history of the global rise and fall of sea level, though this interpretation remains controversial (▶Fig. 11.13b). Some sea-level changes may reflect changes in sea-floor-spreading rates and thus in the volume of mid-ocean ridges; some may reflect hot-spot activity; some may reflect variations in Earth's climate that caused the melting or formation of ice sheets; others may represent changes in areas of continents accompanying continental collisions; and still others may reflect the warping of continental surfaces as a result of mountain building.

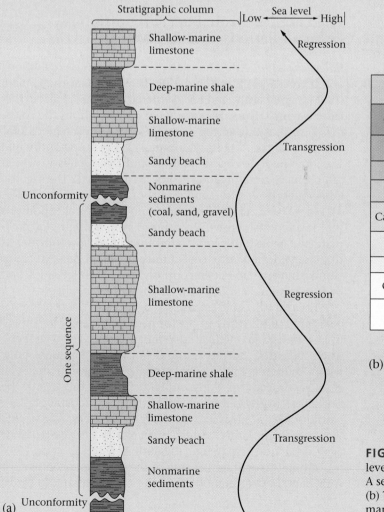

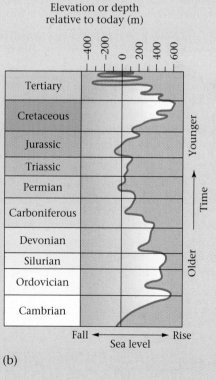

FIGURE 11.13 (a) Different types of strata are deposited as sea level rises and falls. Unconformities develop when sea level is low. A sequence is the interval between two major unconformities. (b) This graph, based on the study of stratigraphy in passive-margin basins, estimates the rise and fall of sea level through time.

fish. At the end of the Ordovician, mass extinction took place, perhaps because of the brief glaciation and associated sea-level lowering of the time.

The Middle Paleozoic Era (Silurian-Devonian Periods, 444–359 Ma)

Paleogeography. As the world entered the Silurian Period, the global climate warmed (leading to so-called **greenhouse conditions**), sea level rose, and the continents flooded once again. In some places, where water in the epicontinental seas was clear and could exchange with water from the oceans, huge reef complexes grew, forming a layer of fossiliferous limestone on the land.

More orogeny took place during the middle Paleozoic Era. For example, in eastern North America, the Appalachian region that had been affected by the Taconic orogeny underwent a collisional orogeny once again in the Devonian Period, this time with small continental masses that drifted in from the east (▶Fig. 11.14). Geologists refer to this event as the Acadian orogeny. The Acadian orogeny corresponds to the Caledonian orogeny of northern Europe.

Throughout the middle Paleozoic, the western margin of North America continued to be a passive-margin basin.

FIGURE 11.14 This paleogeographical map illustrates the movement of the Avalon microcontinent (including England and southern Ireland) toward its ultimate collision with the eastern margin of North America. During the process, the intervening Iapetus Ocean was subducted. The remnant of the Taconic orogen fringed the eastern margin of North America.

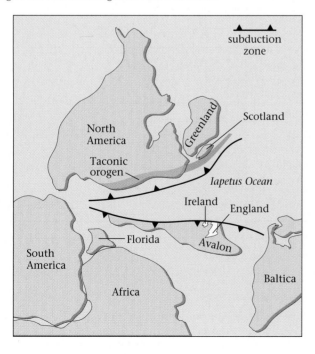

Finally, in the Late Devonian, the quiet environment of the west-coast passive margin ceased, possibly because of a collision with an island arc. This event, known as the Antler orogeny, pushed slices of deep-marine strata eastward, and up and over shallow-water strata. It was the first of many orogenies to affect the western margin of the continent (▶Fig. 11.15).

Life evolution. Life on Earth underwent radical changes in the middle Paleozoic Era. In the sea, new species of trilobites, eurypterids, gastropods, crinoids, and bivalves replaced species that had disappeared during the mass extinction at the end of the Ordovician Period. On land, vascular plants, with woody tissues and seeds and veins for transporting water and food, rooted for the first time. With the evolution of veins and wood, plants could grow much larger, and by the Late Devonian Period the land surface hosted swampy forests with tree-sized relatives of club mosses and ferns. Also at this time, spiders, scorpions, insects, and crustaceans came to exploit both dry-land and freshwater habitats, and jawed fish such as sharks and bony fish began to cruise the oceans. Finally, at the very end of the Devonian Period, the first amphibians crawled out onto land and breathed with lungs.

The Late Paleozoic Era (Carboniferous-Permian Periods, 359–251 Ma)

Paleogeography. After the peak of the greenhouse conditions and high sea levels in the middle Paleozoic Era, the climate cooled significantly, and Earth entered the late Paleozoic **icehouse conditions.** Seas gradually retreated from the continents, so that during the Carboniferous Period regions that had hosted the limestone-forming reefs of epicontinental seas now became coastal swamps and river deltas in which sand, shale, and organic debris accumulated. During the Carboniferous Period, Laurentia lay near the equator (▶Fig. 11.16a, b), so it retained tropical and semitropical conditions that favored lush growth in swamps even though global climate had cooled; this growth left thick piles of woody debris that transformed into coal after and by burial. Much of Gondwana and Siberia, in contrast, lay at high latitudes, and by the Permian Period they were covered by ice sheets.

The late Paleozoic Era also saw a succession of continental collisions, culminating in the formation of a single supercontinent, **Pangaea,** by the Mid-Permian Period. The largest collision occurred during Carboniferous and Permian time, when Laurentia joined with Gondwana, causing the Alleghenian orogeny of North America. During this event, the final stage in the development of the Appalachians, eastern North America squashed against northwestern Africa, and what is now the Gulf Coast region of North America squashed against the northern margin of South America.

Along the continental side of the Alleghenian orogen, a wide band of deformation called the Appalachian fold-thrust

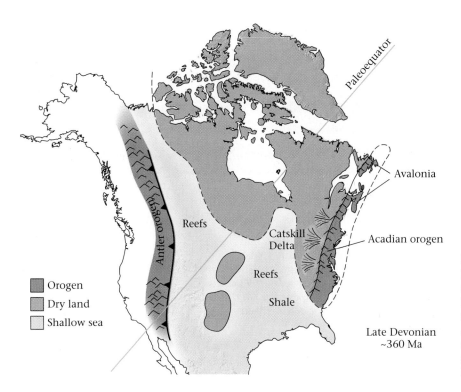

Orogen
Dry land
Shallow sea

Paleoequator

Antler orogen

Reefs

Catskill Delta

Reefs

Shale

Avalonia

Acadian orogen

Late Devonian
~360 Ma

FIGURE 11.15 This paleogeographical map of North America depicts the distribution of land and sea and the position of mountain belts in the late Devonian Period. Note that the Acadian orogeny occurred in the east and, somewhat later, the Antler orogeny in the west. The Acadian orogen shed a vast delta, called the Catskill Delta, of sand and conglomerate into the eastern interior of North America.

FIGURE 11.16 (a) This map shows the distribution of the continents of Pangaea as viewed from the South Pole (black dot). (b) This paleogeographical map shows the distribution of land and sea and the location of the Alleghenian and Hercynian orogens. Extensive coal swamps lay along the interior coast. The Ancestral Rockies (Late Paleozoic uplifts of the Rocky Mountain region) also developed at this time, as did smaller uplifts in the Midwest.

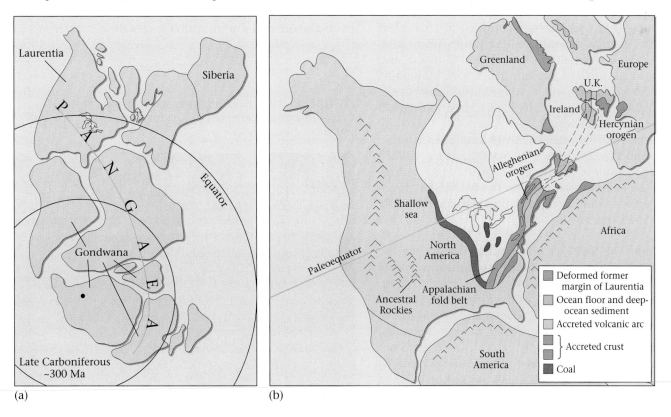

Laurentia

Siberia

PANGAEA

Equator

Gondwana

Late Carboniferous
~300 Ma

(a)

Greenland

Europe

U.K.

Ireland

Hercynian orogen

Alleghenian orogen

Shallow sea

Paleoequator

North America

Africa

Ancestral Rockies

Appalachian fold belt

South America

Deformed former margin of Laurentia
Ocean floor and deep-ocean sediment
Accreted volcanic arc
Accreted crust
Coal

(b)

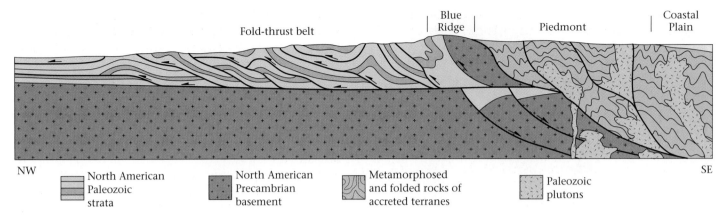

NW

| North American Paleozoic strata | North American Precambrian basement | Metamorphosed and folded rocks of accreted terranes | Paleozoic plutons |

SE

FIGURE 11.17 A cross section of the Alleghenian orogen. Note that the northwestern half is a thin-skinned fold-thrust belt, involving deformation of strata by thrust faults and folds above a sliding surface at the top of the Precambrian basement. The Blue Ridge exposes a slice of Precambrian North American basement. But in the metamorphic regions to the east, the crust consists of slices (accreted terranes) added on to North America during the Paleozoic Era. In the Piedmont, the accreted terranes are exposed, but in the Coastal Plain, these rocks have been buried by Cretaceous and Tertiary strata.

belt formed. In this province, you can see a distinctive style of deformation, called thin-skinned deformation, characterized by displacement on thrust faults in sedimentary strata above the Precambrian basement (▶Fig. 11.17). At depth, the thrust system merges with a near-horizontal sliding surface, or detachment. Movement on the faults produces folds in overlying beds.

Forces generated during the Alleghenian orogeny were so strong that faults in the continental crust clear across North America moved, creating uplifts (local high areas) and sediment-filled basins in the Midwest and even in the region of the present-day Rocky Mountains (Colorado, New Mexico, and Wyoming). Geologists refer to the late Paleozoic uplifts of the Rocky Mountain region as the **Ancestral Rockies.**

The assembly of Pangaea involved a number of other collisions around the world as well. Notably, Africa collided with southern Europe to form the Hercynian orogen. Also, a rift or small ocean in Russia closed to create the Ural Mountains, and parts of China along with other fragments of Asia attached to southern Siberia.

Life evolution. The fossil record indicates that during the late Paleozoic Era, plants and animals continued to evolve toward more familiar forms. In coal swamps, fixed-wing insects including huge dragonflies flew through a tangle of ferns, club mosses, and scouring rushes, and by the end of the Carboniferous Period insects such as the cockroach, with foldable wings, appeared (▶Fig. 11.18). Forests containing gymnosperms ("naked seed" plants such as conifers) and cycads (trees with a palmlike stalk peaked by a fan of fernlike fronds) became widespread in the Permian Period.

Amphibians and, later, reptiles populated the land. The appearance of reptiles marked the evolution of a radically new component in animal reproduction: eggs with protective covering. Such eggs permitted reptiles to reproduce without returning to the water, and thus allowed the group to populate previously uninhabitable environments. Early members of the reptile group included forerunners of mammals. The late Paleozoic Era came to a close with two major mass-extinction events, during which over 90% of marine species

FIGURE 11.18 A museum diaorama illustrating life in a Carboniferous coal swamp. The wingspan of the giant dragonfly was about 1 m (3 feet).

disappeared. Why these particular events occurred remains a controversy; but there is interesting evidence that the terminal Permian mass extinction occurred as a result of a huge meteor impact that, like the terminal Mesozoic event we will discuss shortly, disrupted the food chain.

11.7 THE MESOZOIC ERA: WHEN DINOSAURS RULED

The Early and Middle Mesozoic Era (Triassic–Jurassic Periods, 251–145 Ma)

Paleogeography. Pangaea, the supercontinent formed in the late Paleozoic Era, lasted for about 100 million years, until during the Late Triassic and Early Jurassic Periods, rifts formed and the supercontinent began to break up. By the end of the Jurassic Period, rifting had succeeded in forming the Mid-Atlantic Ridge, and the North Atlantic Ocean started to form (▶Fig. 11.19). At first, the Atlantic was narrow and shallow, and evaporation made its water so salty that thick evaporite deposits, which now underlie much of the Gulf Coast region, accumulated.

According to the record of sedimentary rocks, during the Triassic and Early Jurassic Earth had greenhouse conditions,

FIGURE 11.19 The breakup of Pangaea began in the Late Triassic Period, and by the Late Jurassic the North Atlantic had formed. Note that at this time, the South Atlantic remained closed, and southern Asia assembled out of a number of smaller continental fragments and volcanic island arcs.

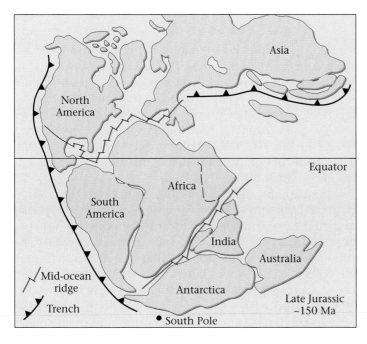

but during the Late Jurassic and Early Cretaceous, it had cool conditions. During the early Mesozoic, the interior of Pangaea remained a nonmarine environment in which red sandstones and shales, now exposed in the spectacular cliffs of Zion National Park, were deposited. By the Middle Jurassic Period, sea level began to rise. During the resulting transgression, a shallow sea submerged much of the Rocky Mountain region.

On the western margin of North America, convergent-margin tectonics became the order of the day. Beginning with Late Permian and continuing through Mesozoic time, subduction generated volcanic island arcs and caused them, along with microcontinents and hot-spot volcanoes, to collide with North America. Thus, North America grew in land area by the addition (accretion) of crustal fragments and island arcs—exotic terranes—on its western margin. From the end of the Jurassic through the Cretaceous Period, a major continental volcanic arc, the Sierran arc, formed along the western margin of North America; we'll learn more about this arc later.

Life evolution. During the early Mesozoic Era, a variety of new species of already established plant and animal groups appeared, filling the ecological niches left vacant by the Late Permian mass extinction. New creatures such as swimming reptiles appeared in the sea. Corals became the predominant reef builders, and have remained so ever since. On land, gymnosperms and reptiles diversified, and the Earth saw its first turtles and flying reptiles. More dramatic, at the end of the Triassic Period, the first true dinosaurs appeared. Dinosaurs differed from other reptiles in that their legs were positioned under their bodies rather than off to the sides, and they were possibly warm-blooded. By the end of the Jurassic Period, gigantic sauropod dinosaurs (which weighed up to 100 tons), along with other familiar ones such as stegosaurus, thundered across the landscape, and the first feathered birds took to the skies (▶Fig. 11.20). The earliest ancestors of mammals appeared at the end of the Triassic Period, in the form of small, ratlike creatures.

The Late Mesozoic Era (Cretaceous Period, 145–65 Ma)

Paleogeography. During the Cretaceous Period, the Earth's climate continued to shift to warmer, greenhouse conditions, and sea level rose significantly, reaching levels that had not been attained for the previous 200 million years. Great seaways flooded most of the continents, producing the thick deposits of chalk in Europe that Thomas Henry Huxley discussed as well as layers of limestone and sandstone in the western interior of North America (▶Fig. 11.21). In the latter part of the Cretaceous Period, a shark could have swum from the Gulf of Mexico to the Arctic Ocean.

FIGURE 11.20 During the Jurassic, giant dinosaurs inhabited the land.

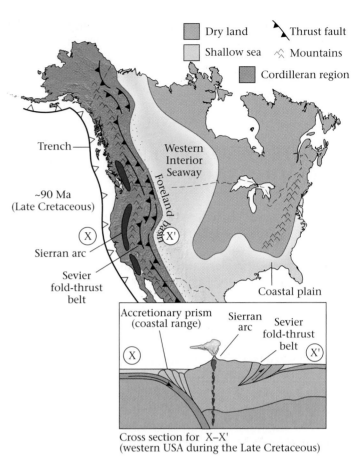

Cross section for X–X'
(western USA during the Late Cretaceous)

FIGURE 11.21 Paleogeographical map of North America in the Late Cretaceous Period, showing the distribution of land and sea and the convergent orogen of the Cordillera. Note the position of the Sierran arc and the Sevier fold-thrust belt (inset). An interior seaway stretched from the Gulf of Mexico to the Arctic Ocean.

The breakup of Pangaea continued through the Cretaceous Period, with the opening of the South Atlantic Ocean and the separation of South America and Africa from Antarctica and Australia. India broke away from Gondwana and headed rapidly northward toward Asia (▶Fig. 11.22). Along the continental margins of the newly formed Mesozoic oceans, new passive-margin basins developed that, like their predecessors of the early Paleozoic Era, filled with great thicknesses of sediments. For example, along the Gulf Coast, a wedge of sediment over 15 km thick has accumulated.

Along western North America, the Sierran arc, a large continental volcanic arc that had initiated at the end of the Jurassic Period, continued to be active. This arc resembled the one that currently lies along the Andes on the western edge of South America. Though the volcanoes of the arc have long since eroded away, we can see their roots in the form of the plutons that now constitute the granitic batholith of the Sierra Nevada. Compressional forces along the western North American convergent boundary activated large thrust faults east of the arc, an event geologists refer to as the Sevier orogeny. This orogeny produced a fold-thrust belt whose remnants you can see today in the Canadian Rockies and in western Wyoming.

At the end of the Cretaceous Period, continued compression along the convergent boundary of western North America caused large faults in the region of Wyoming, Colorado, eastern Utah, and northern Arizona to slip (▶Fig. 11.23a). In contrast to the faults of fold-thrust belts, these faults penetrated deep into the Precambrian rocks of the continent, and thus movement on them generated basement uplifts. During the formation of a basement uplift, overlying layers of earlier strata warped into large monoclines, folds whose shape resembles the drape of a carpet over a step (▶Fig. 11.23b, c). This event, which geologists call the **Laramide orogeny,** formed the structure of the present Rocky Mountains (▶Fig. 11.23a, d).

By mapping the sea floor, geologists have discovered that sea-floor spreading may have been particularly fast, and that huge submarine plateaus composed of hot-spot volcanic rocks formed during the Cretaceous Period. The presence of hot spots implies that the Cretaceous may have been a time when particularly large mantle plumes, called **superplumes,** reached the base of the lithosphere. Hot-spot and mid-ocean-ridge activity may have influenced the climate and sea level. Large mid-ocean ridges

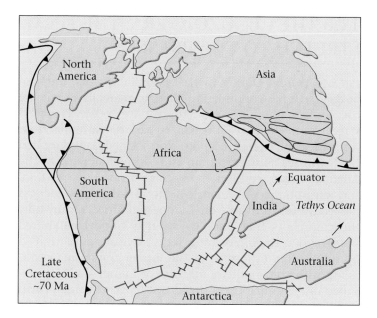

FIGURE 11.22 By the Late Cretaceous Period, the Atlantic Ocean had formed, and India was moving rapidly northward, ultimately colliding with Asia.

and large plateaus displaced seawater, and volcanic eruptions released huge quantities of carbon dioxide (CO_2) into the atmosphere. An increase in the concentration of CO_2 in the atmosphere leads to an increase in atmospheric temperature, for CO_2 in the air traps infrared radiation rising from the Earth and causes the air to warm up. (In this way, CO_2 acts as a greenhouse gas; see Chapter 19.) This increase would melt ice sheets, and thus contribute to the rise of sea level.

Life evolution. In the seas of the late Mesozoic world, modern fish appeared and became dominant. What set them apart from earlier fish were their short jaws, rounded scales, symmetrical tails, and specialized fins. Huge swimming reptiles and gigantic turtles (with shells up to 4 m across) preyed on the fish. On land, cycads largely vanished, and angiosperms (flowering plants), including hardwood trees, began to compete successfully with conifers for dominance of the forest. Dinosaurs reached their peak of success at this time, inhabiting almost all environments on Earth. Social herds of grazing dinosaurs roamed the plains, preyed on by the fearsome *Tyrannosaurus rex* (a Cretaceous, not a Jurassic, dinosaur, despite what Hollywood says!). Pterosaurs, with wingspreads of up to 11 m, soared overhead, and birds began to diversify. Mammals also diversified and developed larger brains and more specialized teeth, but they remained small and ratlike.

FIGURE 11.23 (a) During the Laramide orogeny, deformation shifted eastward, from the Sevier fold-thrust belt to the belt of Laramide uplifts. (b) The geometry of a basement uplift before erosion. Note that the fault penetrates the basement, and that the overlying sediment warps into a fold that resembles a carpet draped over a stair step. (c) After erosion, the Precambrian rocks in the core of the uplift crop out at the ground surface. (d) Digital elevation model of the Rocky Mountain Front. The view is looking west, over Denver.

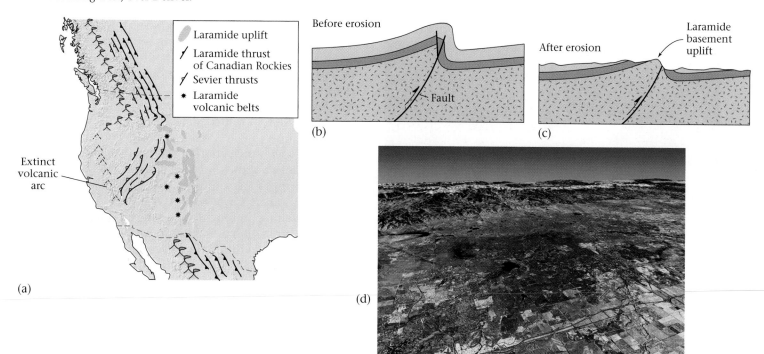

The "K-T boundary event." Geologists first recognized the K-T boundary (K stands for "Cretaceous" and T for "Tertiary") from eighteenth-century studies that identified an abrupt global change in fossil assemblages. Until the 1980s, most geologists assumed the faunal turnover took millions of years. But modern dating techniques indicate that this change happened almost instantaneously and that it signaled the sudden mass extinction of most species on Earth. The dinosaurs, which had ruled the planet for over 150 million years, simply vanished, along with 90% of some plankton species in the ocean and up to 75% of plant species. What kind of catastrophe could cause such a sudden and extensive mass extinction? From data collected in the 1970s and 1980s, most geologists have concluded that the Cretaceous Period came to a close, at least in part, as a result of the impact of a 10-km-wide meteorite at the site of the present-day Yucatán peninsula in Mexico, where a 100-km-wide by 16-km-deep scar called the Chicxulub crater lies buried by younger sediment (▶Fig. 11.24).

The impact caused so much destruction because it not only formed a crater, blasting huge quantities of debris into the sky, but probably also generated 2-km-high tsunamis that inundated the shores of continents and generated a blast of hot air that set forests on fire. The blast and the blaze together would have ejected so much debris into the atmosphere that for months there would have been perpetual night and winterlike cold. In addition, chemicals ejected into the air would have combined with water to produce acid rain. These conditions would cause photosynthesis to all but cease, and thus would break the food chain and probably trigger extinctions.

FIGURE 11.24 This painting illustrates the collision of a huge meteorite with the Earth at the end of the Cretaceous Period.

11.8 THE CENOZOIC ERA: THE FINAL STRETCH TO THE PRESENT

Paleogeography. During the last 65 million years, the map of the Earth has continued to change, gradually producing the configuration of continents we see today. The final stages of the Pangaea breakup separated Australia from Antarctica and Greenland from North America, and formed the North Sea between Britain and continental Europe. The Atlantic Ocean continued to grow because of sea-floor spreading on the Mid-Atlantic Ridge, and thus the Americas moved westward, away from Europe and Africa. Meanwhile, the continents that once constituted Gondwana drifted northward as the intervening Tethys Ocean was consumed by subduction. Collisions of the former Gondwana continents with the southern margins of Europe and Asia resulted in the formation of the largest orogenic belt on Earth today, the **Alpine-Himalayan chain** (▶Fig. 11.25). India and a series of intervening volcanic island arcs and microcontinents collided with Asia to form the Himalayas and the Tibetan Plateau to the north, while Africa along with some volcanic island arcs and microcontinents collided with Europe to produce the Alps in the west and the Zagros Mountains of Iran in the east.

As the Americas moved westward, convergent plate boundaries evolved along their western margins. In South America, convergent-boundary activity has built the Andes, which remains an active orogen to the present day. In North America, convergent-boundary activity continued without interruption until the Eocene Epoch, yielding, as we have seen, the Laramide orogen. Then, because of the rearrangement of plates off the western shore of North America, beginning about 40 Ma, a transform boundary replaced the convergent boundary in the western part of the continent (▶Fig. 11.26a–c). When this happened, volcanism and compression ceased in western North America, the San Andreas fault system formed along the coast of the United States, and the Queen Charlotte fault system developed off the coast of Canada. Along the San Andreas and Queen Charlotte faults today, the Pacific Plate moves northward with respect to North America at a rate of about 6 cm per year. In the western United States, convergent-boundary tectonics continues only in Washington, Oregon, and northern California where subduction of the Juan de Fuca Plate leads to the volcanism of the Cascade volcanic chain.

As convergent tectonics ceased in the western United States south of the Cascades, the region began to undergo rifting (stretching) in roughly an east-west direction. The result was the formation of the **Basin and Range Province**, a broad continental rift that has caused the region to stretch to twice its original width (▶Fig. 11.27). The Basin and Range gained its name from its topography—the province contains long, narrow mountain ranges separated from each

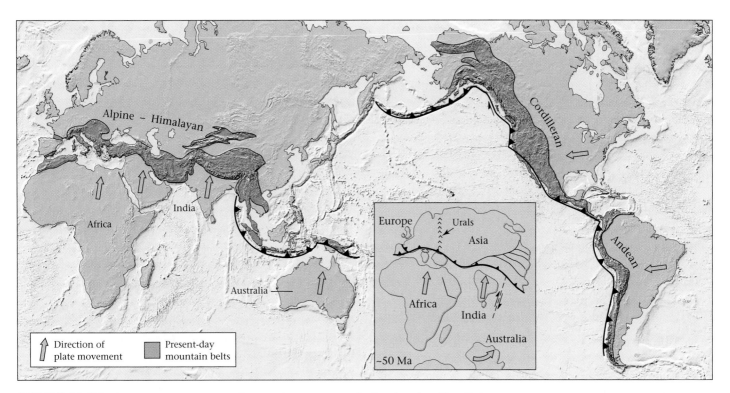

FIGURE 11.25 The two main continental orogenic systems on the Earth today. The Alpine-Himalayan system formed when pieces of Gondwana (Africa, India, and Australia) migrated north and collided with Asia (inset). The Cordilleran and Andean systems are the consequence of convergent-boundary tectonism along the eastern Pacific.

FIGURE 11.26 The western margin of North America changed from a convergent-plate boundary into a transform-plate boundary when the Farallon-Pacific Ridge was subducted. (a, b) The Farallon Plate was moving toward North America, while the Pacific Plate was moving parallel to the western margin of North America. (c) The Basin and Range Province opened as the San Andreas Fault developed. Subduction along the West Coast today only occurs where the Juan de Fuca Plate, a remnant of the Farallon Plate, continues to subduct.

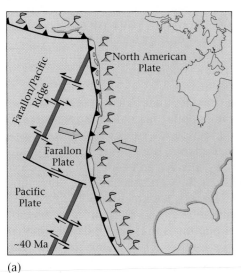

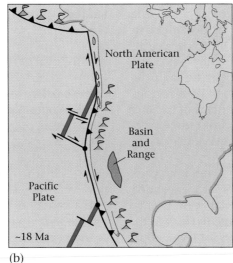

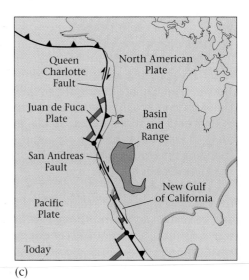

(a) (b) (c)

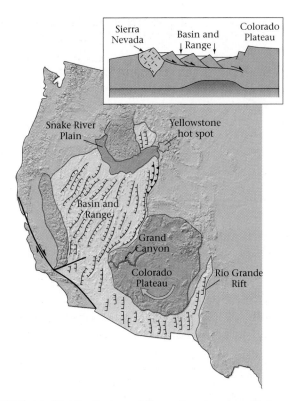

FIGURE 11.27 The Basin and Range Province is a rift (inset). The northern part has opened more than the southern part, causing the Sierran arc to swing westward and rotate. The Rio Grande Rift is a small rift that links to the Basin and Range. The Colorado Plateau is a block of craton bounded by the Rio Grande Rift to the east and the Basin and Range to the west.

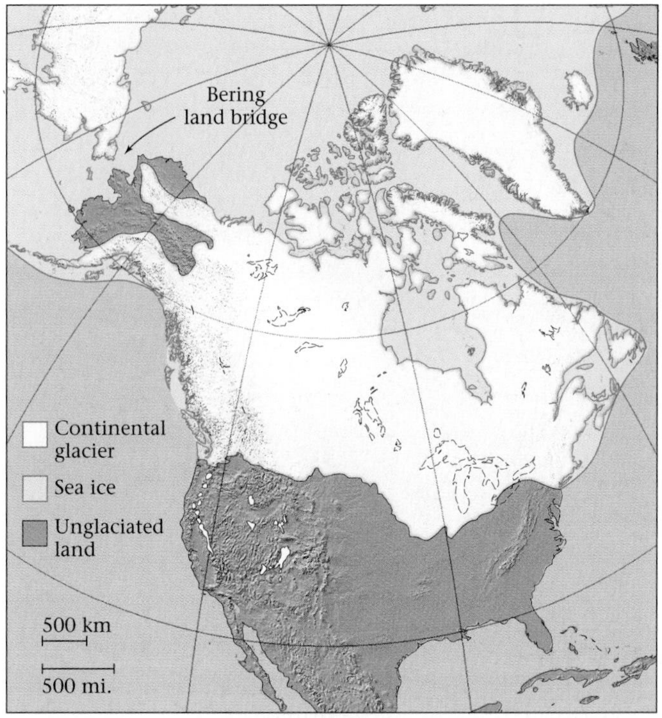

FIGURE 11.28 North America during the maximum advance of the Pleistocene ice sheet. Because sea level was so low (water was stored in the ice sheet on land), a land bridge formed across the Bering Strait, allowing people and animals to migrate from Asia to America.

other by flat, sediment-filled basins. This geometry reflects the normal faulting resulting from stretching: crust of the region was broken up by faults, and movement on the faults created elongate depressions.

Recall that in the Cretaceous Period, the world experienced greenhouse conditions and sea level rose so that extensive areas of continents were submerged. During the Cenozoic Era, however, the global climate rapidly shifted to icehouse conditions, and by the early Oligocene Epoch, Antarctic glaciers reappeared for the first time since the Triassic. The climate continued to grow colder through the Late Miocene Epoch, leading to the formation of grasslands in temperate climates. About 2.5 Ma, the Isthmus of Panama formed, separating the Atlantic completely from the Pacific, changing the configuration of oceanic currents, and allowing the Arctic Ocean to freeze over.

In the overall cold climate of the last 2 million years, the Quaternary Period, continental glaciers have expanded and retreated across northern continents at least twenty times, resulting in the **Pleistocene ice age** (▶Fig. 11.28). Each time the edge of the glacier advanced, sea level fell so much that the continental shelf became exposed to air, and

a land bridge formed across the Bering Strait, west of Alaska, providing migration routes for animals and people from Asia into North America; a partial land bridge also formed from southeast Asia to Australia, making human migration to Australia easier. Erosion and deposition by the glaciers created much of the landscape we see today in northern temperate regions. About 11,000 years ago, the climate warmed, and we entered the interglacial time interval we are still experiencing today (see Chapter 18). This most recent interval of time is the Holocene.

Life evolution. When the skies finally cleared in the wake of the K-T boundary catastrophe, plant life recovered, and soon forests of both angiosperms and gymnosperms reappeared. A new group of plants, the grasses, sprang up and began to dominate the plains in temperate and subtropical climates by the middle of the Cenozoic Era. The dinosaurs, however, were gone for good. Mammals rapidly diversified into a variety of forms to take their place. In fact, most of the modern groups of mammals that exist today originated at the beginning of the Cenozoic Era, giving this time the nickname "Age of Mammals." During the latter part of the era, remarkably huge mammals appeared (such as mammoths, giant beavers, giant bears, and giant sloths), but these became extinct during the past 10,000 years, perhaps because of hunting by humans.

It was during the Cenozoic that our own ancestors first appeared. Apelike primates diversified in the Miocene Epoch (about 20 Ma), and the first humanlike primate appeared at about 4 Ma, followed by the first members of the human genus, *Homo,* at about 2.4 Ma. Fossil evidence, primarily from Africa, indicates that *Homo erectus,* capable of making stone axes, appeared about 1.6 Ma, and the line leading to *Homo sapiens* (our species) diverged from *Homo neanderthalensis* (Neanderthal man) about 500,000 years ago. According to the fossil record, modern people appeared about 150,000 years ago. Thus, much of human evolution took place during the radically shifting climatic conditions of the Pleistocene Epoch.

We end our brief biography of Earth for now with the appearance of *Homo sapiens.* As summarized by the art on pages 326–27, this history has been shaped by complex plate interactions (including continental drift and collisions), sea-level changes, atmospheric changes, and life evolution. Clearly, the Earth System has changed significantly over time. We'll pick up the thread of this story in Chapter 19, where we discuss ideas of how the Earth System may change in the future.

CHAPTER SUMMARY

• Earth formed about 4.57 billion years ago. For part of the first 600 million years, the Hadean Eon, the planet was so hot that its surface was a magma ocean.

• The Archean Eon began about 3.8 Ga, when permanent continental crust formed. The crust assembled out of volcanic arcs and hot-spot volcanoes that were too buoyant to subduct. The atmosphere contained little oxygen, but the first life forms—bacteria and archaea—appeared.

• In the Proterozoic Eon, which began at 2.5 Ga, Archean cratons collided and sutured together along orogenic belts and large Proterozoic cratons. Photosynthesis by organisms added oxygen to the atmosphere. By the end of the Proterozoic, complex but shell-less marine invertebrates populated the planet. Most continental crust accumulated to form a supercontinent called Rodinia at about 1 Ga.

• At the beginning of the Paleozoic Era, rifting yielded several smaller continents. Sea level rose and fell a number of times, creating sequences of strata in continental interiors. Continents began to collide and coalesce again, leading to orogenies and, by the end of the era, another supercontinent, Pangaea. Early Paleozoic evolution produced many invertebrates with shells, and jawless fish. Land plants and

insects appeared in the middle Paleozoic. And by the end of the eon, there were land reptiles and gymnosperm trees.

• In the Mesozoic Era, Pangaea broke apart and the Atlantic Ocean formed. Convergent-boundary tectonics dominated along the western margin of North America. Dinosaurs appeared in Late Triassic time and became the dominant land animal through the Mesozoic Era. During the Cretaceous Period, sea level was very high, and the continents flooded. Angiosperms appeared at this time, along with modern fish. A huge mass extinction event, which wiped out the dinosaurs, occurred at the end of the Cretaceous Period, probably because of the impact of a large meteorite with the Earth.

• In the Cenozoic Era, continental fragments of Pangaea began to collide again. The collision of Africa and India with Asia and Europe formed the Alpine-Himalayan orogen. Convergent tectonics has persisted along the margin of South America, creating the Andes, but ceased in North America when the San Andreas Fault formed. Rifting in the western United States during the Cenozoic Era produced the Basin and Range Province. Various kinds of mammals filled niches left vacant by the dinosaurs, and the human genus, *Homo,* appeared and evolved throughout the radically shifting climate of the Pleistocene Epoch.

KEY TERMS

Alpine-Himalayan chain (p. 322)
Ancestral Rockies (p. 318)
Archean Eon (p. 308)
Basin and Range Province (p. 322)
Cambrian explosion (p. 314)
cratonic (continental) platform (p. 311)
differentiation (p. 307)
Ediacaran fauna (p. 313)
epicontinental sea (p. 313)
greenhouse conditions (p. 316)
Hadean Eon (p. 307)
icehouse conditions (p. 316)
Laramide orogeny (p. 320)
Pangaea (p. 316)
Phanerozoic Eon (p. 313)
Pleistocene ice age (p. 324)
Proterozoic Era (p. 310)
Rodinia (p. 311)
shield (p. 311)
snowball Earth (p. 313)
stratigraphic sequence (p. 315)
stromatolites (p. 309)
superplumes (p. 320)

REVIEW QUESTIONS

1. Why are there no rocks on Earth that yield radiometric dates older than 4 billion years?
2. Describe the condition of the crust, atmosphere, and oceans during the Hadean Eon.
3. How did the atmosphere and tectonic conditions change during the Proterozoic Eon?

(continued on page 328)

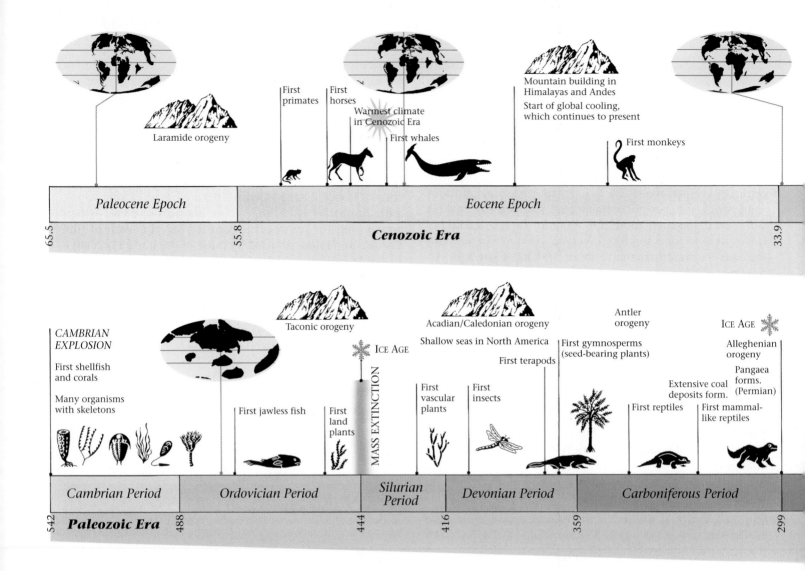

First primates
First horses

Laramide orogeny

Warmest climate
in Cenozoic Era

First whales

Mountain building in
Himalayas and Andes

Start of global cooling,
which continues to present

First monkeys

Paleocene Epoch	Eocene Epoch

65.5   55.8

Cenozoic Era

33.9

*CAMBRIAN
EXPLOSION*

First shellfish
and corals

Many organisms
with skeletons

First jawless fish

First
land
plants

Taconic orogeny

ICE AGE

MASS EXTINCTION

Acadian/Caledonian orogeny

Shallow seas in North America

First terapods

First
vascular
plants

First
insects

Antler
orogeny

First gymnosperms
(seed-bearing plants)

ICE AGE

Alleghenian
orogeny

Pangaea
forms.
(Permian)

Extensive coal
deposits form.

First reptiles

First mammal-
like reptiles

Cambrian Period	Ordovician Period	Silurian Period	Devonian Period	Carboniferous Period

542   **Paleozoic Era**   488   444   416   359   299

Formation of
the Earth
from planetesimals

Formation of
the Moon

Formation of
Earth's atmosphere

Oldest known rocks

Much of the Earth's surface is volcanic rock,
forming unstable regions of erupting lava.

Oldest known fossils: single-celled
organisms (prokaryotes) and
stromatolite-forming
cyanobacteria

Sizable continental areas begin to form.

Extensive
shallow seas
on the margin
of continents
and deposition
of banded-iron
formation

Hadean Eon	Archean Eon	

4,567   3,800   **Precambrian**   2,500

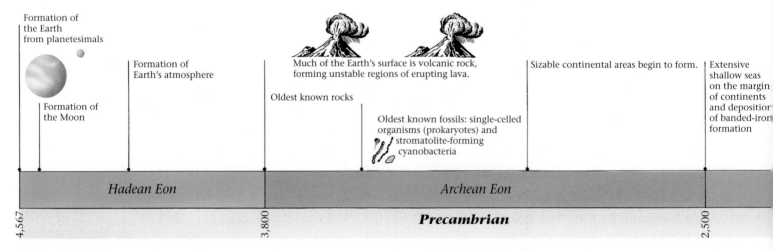

The Evolution of Earth

Earth has not had a static history; because of plate tectonics and its
consequences (continental drift, sea-floor spreading, volcanism,
and so on), the map of the Earth constantly changes. Distinct
mountain-building events, or orogenies, have taken place during
this process. The fossil record suggests that life first appeared within
the first few hundred million years of our planet's existence, soon
after a liquid-water ocean had accumulated, and like the planet
itself has constantly changed ever since. The progressive change of
the assemblage of species of life on Earth is called evolution.

The earliest life forms were microscopic. By the end of the
Precambrian, complex multicellular organisms had formed, and a
burst of evolution at the Precambrian-Cambrian boundary yielded a
diversity of invertebrates with shells. During the past half-billion years,
several new major groups of organisms have appeared, and countless

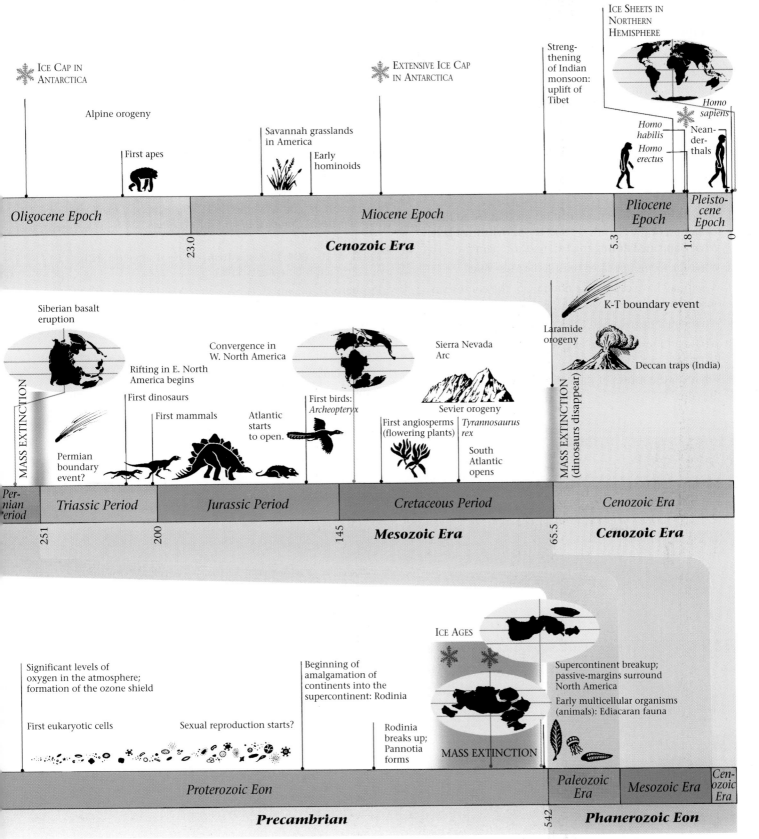

Cenozoic Era

Ice Cap in Antarctica

Alpine orogeny

First apes

Savannah grasslands in America

Early hominoids

Strengthening of Indian monsoon: uplift of Tibet

Extensive Ice Cap in Antarctica

Ice Sheets in Northern Hemisphere

Homo habilis
Homo erectus

Neanderthals

Homo sapiens

| Oligocene Epoch | Miocene Epoch | Pliocene Epoch | Pleistocene Epoch |

23.0 · Cenozoic Era · 5.3 · 1.8 · 0

Mesozoic Era

Siberian basalt eruption

MASS EXTINCTION

Permian boundary event?

Rifting in E. North America begins

First dinosaurs

First mammals

Convergence in W. North America

Atlantic starts to open.

First birds: Archeopteryx

Sierra Nevada Arc

Sevier orogeny

First angiosperms (flowering plants)

Tyrannosaurus rex

South Atlantic opens

Laramide orogeny

K-T boundary event

Deccan traps (India)

MASS EXTINCTION (dinosaurs disappear)

| Permian Period | Triassic Period | Jurassic Period | Cretaceous Period | Cenozoic Era |

251 · 200 · 145 · Mesozoic Era · 65.5 · Cenozoic Era

Precambrian / Phanerozoic Eon

Significant levels of oxygen in the atmosphere; formation of the ozone shield

First eukaryotic cells

Sexual reproduction starts?

Beginning of amalgamation of continents into the supercontinent: Rodinia

Rodinia breaks up; Pannotia forms

ICE AGES

MASS EXTINCTION

Supercontinent breakup; passive-margins surround North America

Early multicellular organisms (animals): Ediacaran fauna

| Proterozoic Eon | Paleozoic Era | Mesozoic Era | Cenozoic Era |

Precambrian · 542 · Phanerozoic Eon

Geological time (million years ago)

species have become extinct. Evidence suggests that evolution is not a continuous, gradual process, but occurs in pulses, separated by intervals of time during which the assemblage of species is fairly stable.

The last few hundred million years of Earth history have seen life leave the ocean and spread across the land. During the Mesozoic Era, dinosaurs roamed the Earth—then vanished abruptly 65 million years ago, perhaps as a result of a meteorite's colliding with the Earth. Since then, mammals have diversified into a great variety of species. And the last 100,000 years or so have witnessed the evolution of our own species, *Homo sapiens*. Considering the changes that people have brought about to the Earth System, the appearance of our species is clearly a major event in the history of the planet. The time scale is the 2004 time scale (International Committee on Stratigraphy).

4. What evidence do we have that the Earth nearly froze over twice during the Proterozoic Eon?

5. How did the Cambrian explosion of life change the nature of the living world?

6. How did the Alleghenian and Ancestral Rockies orogenies affect North America?

7. What are the major classes of organisms that appeared in the Paleozoic?

8. Describe the plate-tectonic conditions that led to the formation of the Sierran arc and the Sevier thrust belt. What happened during the Laramide orogeny?

9. What life forms appeared during the Mesozoic?

10. What may have caused the flooding of the continents during the Cretaceous Period?

11. What could have caused the K-T extinctions?

12. What continents formed as a result of the breakup of Pangaea?

13. What are the causes of the uplift of the Himalayas and the Alps?

SUGGESTED READING

Benton, M. J., 2004. *Vertebrate Palaeontology.* 3rd ed. Oxford, UK: Blackwell.

Cloud, P. 1988. *Oasis in Space: Earth History from the Beginning.* New York: W. W. Norton.

Condie, K. C. 2004. *Earth as an Evolving Planetary System.* New York: Academic Press.

Fastovsky, D. E., and D. B. Weishampel. 2005. *The Evolution and Extinction of the Dinosaurs.* 2nd ed. Cambridge, UK: Cambridge University Press.

Gould, S. J., et al. 1993. *The Book of Life.* New York: Norton.

Gradstein, F., J. Ogg, and A. Smith. 2004. *A Geologic Time Scale 2004.* Cambridge, UK: Cambridge University Press.

Hartmann, J., and R. Miller. 1991. *The History of the Earth: An Illustrated Chronicle of an Evolving Planet.* New York: Workman.

Knoll, A. H. 2003. *Life on a Young Planet.* Princeton, N.J.: Princeton University Press.

Prothero, D. R., and R. H. Dott, Jr. 2003. *Evolution of the Earth.* 7th ed. New York: McGraw-Hill.

Rodgers, J. J. W. 1994. *A History of the Earth.* Cambridge, UK: Cambridge University Press.

Schopf, J. W., ed. 2002. *Life's Origin: The Beginnings of Biological Evolution.* Los Angeles: University of California Press.

Stanley, S. M. 1999. *Earth System History.* New York: Freeman.

Wicander, R., and J. S. Monroe. 2003. *Historical Geology: Evolution of Earth and Life through Time.* 4th ed. Belmont, CA: Brooks Cole.

Windley, B. F. 1995. *The Evolving Continents.* 3rd ed. New York: Wiley.

CHAPTER 12

Riches in Rock: Energy and Mineral Resources

12.1 INTRODUCTION

In the extreme chill of a midwinter night, a pan of water freezes almost instantly. But the low temperature doesn't stop a wolf from stalking its prey—the wolf's legs carry it through the ice and snow, and its body radiates heat. To keep living in this way, the wolf must find materials from which to build its body, and from which to produce the energy that keeps its body warm and in motion. We refer to any materials that can be of use, either to make something or to provide energy, as a **resource.** In this context, **energy** is simply the capacity to do work, to cause something to happen, or to cause change in a system. The resources that a wolf seeks include mice, insects, berries, and water. From these resources, a wolf obtains the mix of chemicals that it needs to make its own bones and flesh. Energy comes from the breaking of chemical bonds in molecules of sugar, protein, and carbohydrates as the wolf metabolizes its food.

The earliest humans were hunter-gatherers and could survive on pretty much the same resources that a wolf needs. But when people discovered how fire could be used for cooking and heating, how weapons could facilitate hunting, how shelters made the winter nights more comfortable, and how harvests could be improved by

Humans utilize vast quantities of energy and mineral resources every day. Most of these resources come from geologic materials. Here, we see the wall of an immense open pit mine, dug to obtain copper-containing minerals disseminated through the rock.

using plows and fertilizers, their need for resources started to increase dramatically. In the twenty-first century, an average citizen of an industrialized country uses more than 100 times the amount of resources used by a hunter-gatherer.

Where do people obtain resources? Today, most come from the Earth, and are either geologic materials or the products of geologic processes. That is why we discuss resources in a geology book. In our discussion, we distinguish between two kinds of resources—**energy resources,** which can be used to produce heat and electricity or move machines, and **mineral resources,** which provide natural inorganic chemicals from which products can be manufactured. To understand where to find energy and mineral resources, why we run the risk of running out of some resources, and how the use of resources affects the environment, we must understand the geologic origin of resources and the settings where they can be found or disposed of. Searching for resources, and dealing with the consequences of their use, provides jobs for tens of thousands of geologists worldwide.

12.2 SOURCES OF ENERGY IN THE EARTH SYSTEM

When you get down to basics, only five fundamental sources provide energy on the Earth:

- *Energy from the Sun:* Solar radiation (or sunlight) produced by nuclear fusion in the Sun bathes the Earth's surface. We can convert solar energy directly into electricity or heat, though so far this has been done only on a small scale. Most of the solar energy that we use was first trapped by plants.

  Plants convert solar energy into biomass (such as wood) through the process of photosynthesis. **Photosynthesis** involves chemical reactions that, driven by sunlight, combine carbon dioxide and water to produce sugar. Plants then use the sugar to manufacture more complex chemicals. Burning releases the potential energy stored in biomass. During burning, organic molecules react with oxygen to produce carbon dioxide, water, and carbon (soot). The flames that you see in a wood fire consist of glowing gases released and heated by this reaction. Since biomass stores energy in a usable form, we can consider it to be a fuel. As we see later in this chapter, people can also convert biomass into other fuels, such as alcohol.

  Some of the plant and algae biomass that grows on the surface of the land, or in oceans or lakes, becomes food that feeds animal life. Some rots away and some gets buried along with sediment. Under appropriate conditions, buried plants, algae, and single-celled animals change into fuels underground. We refer to these fuels (which include oil, gas, and coal) as **fossil fuels,** to emphasize that they come from ancient organisms preserved in rock for geologic time. Burning fossil fuels releases energy.

- *Energy involving gravity:* Gravitational attraction of the Moon, and to a lesser extent of the Sun, causes ocean tides. The flow of water in and out of channels during tidal changes can drive water wheels or turbines that can run machinery or produce electricity. Gravity also causes rain and snow that fall at higher elevations to flow downhill to lower elevations, and causes wind to blow as cooler air sinks and warmer air rises—flowing rivers and wind can run machinery or produce electricity. In the case of river flow and wind, energy production also involves the Sun. Solar energy evaporates the water that enters the atmosphere and eventually falls as rain or snow. Solar energy also heats air, causing it to become buoyant and rise (like a hot-air balloon), creating the opportunity for cooler air to sink and generate wind.

- *Energy from chemical reactions:* Certain inorganic chemicals react with each other to produce light and heat. Dynamite explosions are extreme examples of such energy production. As we will see, hydrogen fuel cells produce energy from chemical reactions.

- *Energy from nuclear fission:* Atoms of radioactive elements split into smaller pieces, a process called nuclear fission (see Appendix). During fission, a tiny amount of mass transforms into a large amount of energy. This type of energy runs nuclear power plants and nuclear submarines.

- *Energy from Earth's internal heat:* Heat produced during the formation and early history of the Earth, along with continuing radioactive decay of elements in the crust, keeps the interior of the Earth warm. This internal heat may be utilized to produce **geothermal energy.**

During the course of civilization, the sources of energy that people use have changed (▶Fig. 12.1). Prior to the industrial age, direct burning of wood and other biomass provided most of humanity's energy needs. But by the second half of the nineteenth century, deforestation had nearly destroyed this resource, and energy needs had increased so dramatically that other fuels had to come into use. In the succeeding sections we discuss the geology of the various energy sources, beginning with oil and gas, the dominant fuels of the present day.

12.3 OIL AND GAS

What Are Oil and Gas?

For reasons of economics and convenience, industrialized societies today rely primarily on oil (petroleum) and natural gas for their energy needs. Oil and natural gas consist of **hydro-**

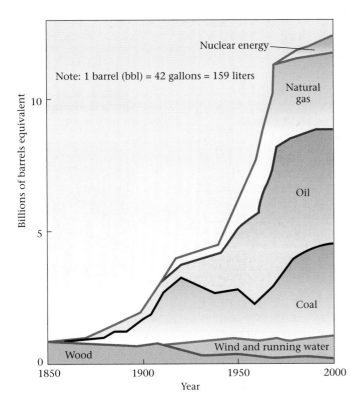

FIGURE 12.1 The graph demonstrates how energy needs have increased in the past 150 years, and how different energy resources have been used to fill those needs. Oil and natural gas together now account for more than half the world's energy usage.

carbons, chainlike or ringlike molecules made of carbon and hydrogen atoms. For example, bottled gas (propane) has the chemical formula C_3H_9. Chemists consider hydrocarbons to be a type of organic chemical, so named because similar carbon-based chemicals make up living organisms.

Some hydrocarbons are gaseous and invisible, some resemble a watery liquid, some appear syrupy, and some are solid. The viscosity (ability to flow) and the volatility (ability to evaporate) of a hydrocarbon product depend on the size of its molecules. Hydrocarbon products composed of short chains of molecules tend to be less viscous (they can flow more easily) and more volatile (they evaporate more easily) than products composed of long chains, because the long chains tend to tangle up with each other. Thus, short-chain molecules occur in gaseous form (natural gas) at room temperature, moderate-length-chain molecules occur in liquid form (gasoline and oil), and long-chain molecules occur in solid form (tar).

Where Do Oil and Gas Form?

Many people believe the false notion that hydrocarbons come from buried trees or the carcasses of dinosaurs. In fact, the primary sources of the organic chemicals in oil and gas are dead algae and plankton bodies. When algae and plankton die, they settle to the bottom of a lake or sea. Because their cells are so tiny, they can only accumulate in quiet-water environments in which clay also settles, so typically the cells mix with clay to create an organic-rich, muddy ooze. For this ooze to be preserved, it must be deposited in oxygen-poor water; otherwise, the organic chemicals in the ooze would react with oxygen or be eaten by bacteria, and thus would decompose quickly and disappear. But in some quiet-water environments (oceans, lagoons, or lakes), dead algae and plankton can get buried by still more sediment before being destroyed. Eventually, the muddy organic ooze lithifies and becomes black organic shale (in contrast to regular shale, which consists only of clay), which contains the raw materials from which hydrocarbons eventually form. Thus, we refer to organic shale as a **source rock.**

If organic shale is buried deeply enough (2 to 4 km), it becomes warmer, since temperature increases with depth in the Earth. Chemical reactions slowly transform the organic material in the shale into waxy molecules called **kerogen** (▶Fig. 12.2). Shale containing kerogen is called **oil shale.** If oil shale warms to temperatures of greater than about 90°C, the kerogen molecules break down to form oil and natural gas molecules. At temperatures over about 160°C, any remaining oil breaks down to form natural gas, and at temperatures over 250°C, organic matter transforms into graphite. Thus, oil itself forms only in a relatively narrow range of temperatures, called the oil window, which generally can exist only in the topmost 6 to 9 km of the crust.

12.4 MAKING AN OIL RESERVE

Oil and gas do not occur in all rocks at all locations. That's why the desire to control oil fields, regions that contain significant amounts of accessible oil underground, has played a role in sparking wars. The known supply of oil and gas held underground is a **hydrocarbon reserve;** if the reserve consists dominantly of oil, it is an oil reserve. Reserves are not randomly distributed around the Earth (▶Fig. 12.3). For example, countries bordering the Persian Gulf contain the world's largest oil reserves. Specifically, Saudi Arabia has reserves of 265 billion barrels. Iraq, Kuwait, Iran, and the United Arab Emirates together have another 373 billion barrels. (A barrel is 42 gallons or 159 liters.) All other countries in the world, combined, have 384 billion barrels. Of these, the United States, by far the largest consumer of oil, has reserves of only 22 billion barrels.

We now show how the development of a reserve requires a source rock, a reservoir rock, a migratory pathway, and a trap. An association of all of these components in an appropriate thermal environment is called a hydrocarbon system.

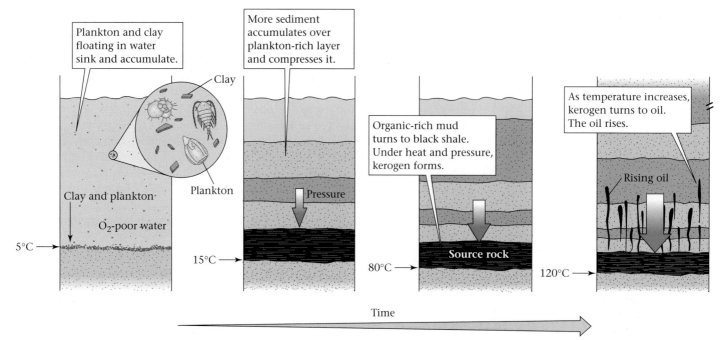

FIGURE 12.2 Plankton, algae, and clay settle out of water and become progressively buried and compacted, gradually being transformed into black organic shale. When heated for a long time, the organic matter in black shale is transformed into oil shale, which contains kerogen. Eventually, the kerogen transforms into oil and gas. The oil may then start to seep upward out of the shale. The red arrow indicates pressure, which increases as more sediment accumulates above.

FIGURE 12.3 The distribution of oil reserves around the world. The largest oil fields occur in the region surrounding the Persian Gulf.

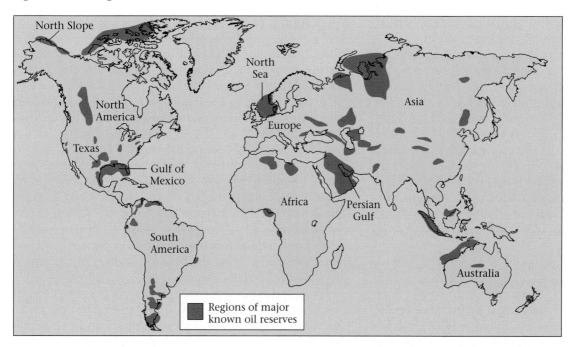

Source Rocks and Hydrocarbon Generation

Previously we saw that the chemicals that become oil and gas start out in algae and plankton cells. These accumulate along with clay to form an organic ooze, which when lithified becomes black organic shale. Geologists refer to organic-rich shale as a source rock because it is the source for the organic chemicals that ultimately become oil and gas. If black shale resides in the oil window, the organic material it contains transforms into kerogen, and then oil and gas. This process is called hydrocarbon generation.

Reservoir Rocks and Hydrocarbon Migration

Wells drilled into source rocks do not yield much oil because kerogen can't flow easily from the rock into the well; any organic matter in an oil shale remains trapped among the grains and can't move easily. So oil companies instead drill into **reservoir rocks,** rocks that contain (or could contain) an abundant amount of *easily accessible* oil and gas.

To be a reservoir rock, a rock must have high porosity and permeability. **Porosity** refers to the amount of open space (**pores**) in a rock. Pore space can hold oil or gas, much as the holes in a sponge can hold water. Shale has low porosity (< 10%), and thus is not a reservoir rock. Sandstone may have high porosity (> 30%) and thus could be a reservoir rock. **Permeability** refers to the degree to which pore spaces connect to each other. Even if a rock has high porosity, it is not necessarily permeable. In a permeable rock, the holes and cracks (pores) are linked, so a fluid can flow slowly through the rock, following a tortuous pathway. Keeping the concepts of porosity and permeability in mind, we can see that a poorly cemented sandstone makes a good reservoir rock, because it is both porous and permeable.

In an oil well, which is simply a hole drilled into the ground to a depth where it penetrates reservoir rock, oil flows from the reservoir rock into the well and then up to the ground surface. If the oil in the rock is under natural pressure, it may move by itself, but usually drillers must set up a pump to suck the oil up and out of the hole.

To fill the pores of a reservoir rock, oil and gas must first migrate (move) from the source rock into a reservoir rock, which they will do over millions of years of geologic time (▶Fig. 12.4). Why do hydrocarbons migrate? Oil and gas are less dense than

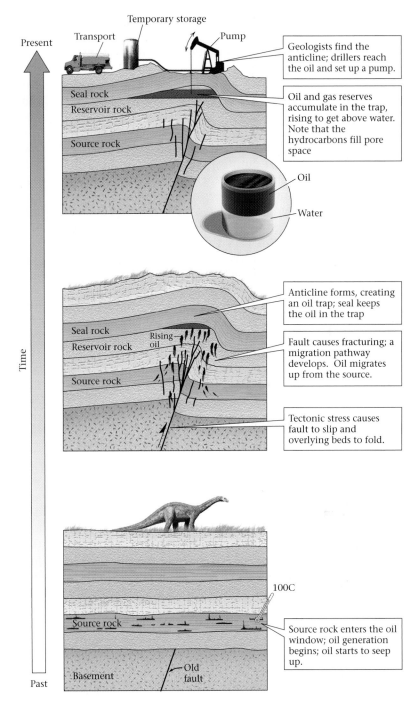

FIGURE 12.4 Initially, oil resides in the source rock. Oil gradually migrates out of the source rock and rises into the overlying water-saturated reservoir rock. Oil rises because it is buoyant relative to water—it tries to float on water. The oil is trapped beneath a seal rock. If gas exists, it floats to the top of the oil.

water, so they try to rise toward the Earth's surface to get above groundwater, just as salad oil rises above the vinegar in a bottle of salad dressing. Natural gas, being less dense, ends up floating above oil. In other words, buoyancy drives oil and gas upward. Typically, a hydrocarbon system must have a good migration pathway, such as a permeable set of fractures, in order for large volumes of hydrocarbons to move.

Traps and Seals

The existence of a reservoir rock alone does not create a reserve, because if hydrocarbons can flow easily into a reservoir rock, they can also flow out. If oil or gas escapes from the reservoir rock and ultimately reaches the Earth's surface, where it leaks away at an oil seep, none will be left underground to pump. Thus, for an oil reserve to exist, oil and gas must be *trapped* underground in the reservoir rock, by means of a geologic configuration called a **trap.**

An oil or gas trap has two components. First, a seal rock (a relatively impermeable rock such as shale, salt, or unfractured limestone) must lie above the reservoir rock and stop the hydrocarbons from rising farther. Second, the seal and reservoir rock bodies must be arranged in a geometry that collects the hydrocarbons in a restricted area. Geologists recognize several types of hydrocarbons trap geometries, four of which are described in ►Box 12.1.

12.5 OIL EXPLORATION AND PRODUCTION

Birth of the Oil Industry

In the United States during the first half of the nineteenth century, people collected "rock oil" (later called petroleum, from the Latin words *petra,* meaning "rock," and *oleum,* meaning "oil") at seeps and used it to grease wagon axles and to make patent medicines. But such oil was rare and expensive. In 1854, George Bissel, a New York lawyer, came to the realization that oil might have broader uses, particularly as fuel for lamps (to replace whale oil). Bissel and a group of investors asked Edwin Drake, a colorful character who had drifted among many professions, to find a way to drill for oil in rocks beneath a hill near Titusville, Pennsylvania, where an oily film floated on the water of springs. Using the phony title "Colonel" to add respectability, Drake hired drillers, built a wooden drilling rig, and obtained a steam-powered drill. Work was slow and the investors became discouraged, but the very day that a letter arrived ordering Drake to stop drilling, his drillers found that the hole, which had reached a depth of 21.2 m, had filled with oil. They set up a pump, and on August 27, 1859, for the first time in history, pumped oil out of the ground. No one had given much thought to the issue of storing the oil, so workers dumped it into empty whisky barrels. This first oil well yielded 10 to 35 barrels a day, which sold for about $20 a barrel.

Within a few years, thousands of oil wells had been drilled in many states, and by the turn of the twentieth century civilization had begun its addiction to oil. Initially, most oil went into the production of kerosene for lamps. Later, as electricity took over from kerosene as the primary source for illumination, gasoline derived from oil became the fuel of choice for the newly invented automobile, and was also used to fuel electric power plants. In its early years, the oil industry was in perpetual chaos. When wildcatters (people who search for new oil fields) discovered one, there would be a short-lived boom during which the price of oil could drop to pennies a barrel. Oil became a global industry governed by the complex interplay of politics, profits, supply, and demand.

The Modern Search for Oil

Wildcatters discovered the earliest oil fields either by blind luck or by searching for surface seeps. But when all known seeps had been drilled and blind luck became too risky, oil companies realized that finding new oil fields would require systematic exploration. The modern-day search for oil is a complex, sometimes dangerous, and often exciting procedure with many steps. Most modern-day exploration depends on the study of seismic-reflection images. These are produced by using small explosions, a special vibrating truck, or pulses of pressurized air to send seismic waves into the crust (see Interlude C). The waves reflect off the contacts between rock layers underground and return to the surface, where sensitive instruments called geophones record their arrival. A computer measures the time between the generation of the seismic waves and their return to the surface, and from this information defines the depth to the contacts that reflected the waves. The computer then constructs an image showing the arrangement of rock layers underground (►Fig. 12.6a).

If geological studies identify a trap, and if the geologic history of the region indicates the presence of good source rocks and reservoir rocks, geologists make a recommendation to drill. Once the decision has been made, drillers go to work. These days, they use rotary drills to grind a hole down through rock. A rotary drill consists of a rotating pipe tipped by a bit, a bulb of metal studded with industrial diamonds or hard metal prongs (►Fig. 12.6a, b). As the bit rotates, it scratches and gouges the rock, turning it into powder and chips. Drillers pump drilling mud, a slurry of water mixed with clay, down the center of the pipe. The mud squirts out of holes in the drill bit, cooling the bit and flushing rock cuttings up and out of the hole. The weight of the mud also keeps oil down in the hole and prevents gushers, fountains of oil formed when underground pressure causes the oil to rise out of the hole on its own. On completion of a hole, workers remove the drilling rig and set up a pump.

Types of Oil and Gas Traps

BOX 12.1
THE REST OF THE STORY

Geologists who work for oil companies spend much of their time trying to identify underground traps. No two traps are exactly alike, but we can classify most into the following four categories.

- *Anticline trap:* In some places, sedimentary beds are not horizontal, as they are when originally deposited, but have been bent by the forces involved in mountain building. These bends, as we have seen, are called folds. An anticline is a type of fold with an archlike shape (▶Fig. 12.5a; see Chapter 9). If the layers in the anticline include a source rock overlain by a reservoir rock that is overlain by a seal rock, then we have the recipe for an oil reserve. The oil and gas rise from the source rock, enter the reservoir rock, and rise to the crest of the anticline, where they are trapped by a seal.

- *Fault trap:* A fault is a fracture on which there has been sliding. If the slip on the fault crushes and grinds the adjacent rock to make an impermeable layer along the fault, then oil and gas may migrate upward along bedding in the reservoir rock until they stop at the fault surface (▶Fig. 12.5b). Alternatively, a fault trap develops if the slip on the fault juxtaposes an impermeable rock layer against a reservoir rock.

- *Salt-dome trap:* In some sedimentary basins, the sequence of strata contains a thick layer of salt, deposited when the basin was first formed and seawater covering the basin was shallow and very salty. Sandstone, shale, and limestone overlie the salt. The salt layer is not as dense as sandstone or shale, so it is buoyant and tends to rise up slowly through the overlying strata. Once the salt starts to rise, the weight of surrounding strata squeezes the salt out of the layer and up into a growing, bulbous **salt dome.** As the dome rises, it bends up the adjacent layers of sedimentary rock. Oil and gas in reservoir rock layers migrate upward until it is trapped against the boundary of the salt dome, for salt is not permeable (▶Fig 12.5c).

- *Stratigraphic trap:* In a stratigraphic trap, a tilted reservoir rock bed "pinches out" (thins and disappears along its length) between two impermeable layers. Oil and gas migrating upward along the bed accumulate at the pinch-out (▶Fig. 12.5d).

FIGURE 12.5 (a) Anticline trap. The oil and gas rise to the crest of the fold. (b) Fault trap. The oil and gas collect in tilted strata adjacent to the fault. (c) Salt-dome trap. The oil and gas collect in the tilted strata on the flanks of the dome. (d) Stratigraphic trap. The oil and gas collect where the reservoir layer pinches out.

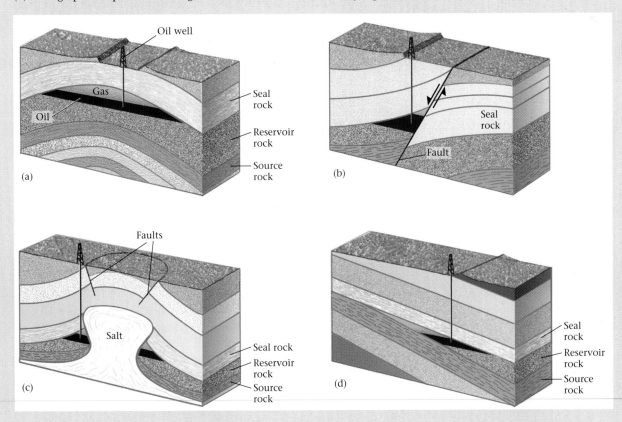

(a)

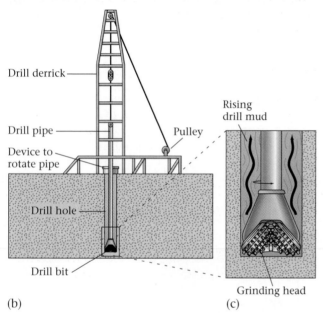

(b)   (c)

FIGURE 12.6 (a) A modern 3-D seismic-reflection image of the subsurface. Each vertical face of this block is a seismic-reflection profile. The colored bands represent beds of sedimentary rock. The horizontal slice shows a maplike image at a specified depth beneath the surface. (b) The derrick in this drilling platform is used to hoist the drill pipe, which comes in segments. (c) A drill bit, typically consisting of three diamond-studded parts, connects to the bottom of the pipe. The diameter of the bit is greater than the diameter of the pipe. Drilling mud is pumped down through the pipe, comes out through holes in the bit, and then rises between the pipe and the walls of the hole, thereby flushing cuttings out of the hole.

Once extracted directly from the ground, crude oil flows first into storage tanks and then into a pipeline or tanker, which transports it to a refinery. At a refinery, workers distill crude oil into several separate components

FIGURE 12.7 A distilling tower (or column) in an oil refinery.

by heating it gently in a vertical pipe called a distillation column (▶Fig. 12.7). Lighter molecules rise to the top of the column, while heavier molecules stay at the bottom. The heat may also "crack" larger molecules to make smaller ones.

12.6 ALTERNATIVE RESERVES OF HYDROCARBONS

So far, we've focused our discussion of hydrocarbons on reserves that can be pumped from the subsurface in the form of a liquid (oil). Hydrocarbons, however, also occur in other forms that are increasingly being used and that may provide energy sources when liquid oil starts to run out.

Tar Sands (Oil Sands) and Oil Shale

In several locations around the world, most notably Alberta (in western Canada) and Venezuela, vast reserves of very viscous, tarlike "heavy oil" exist. This heavy oil, also known as bitumen, has the consistency of molasses, and thus cannot be pumped directly from the ground; it can fill the pore spaces of sand or of poorly cemented sandstone, constituting up to 12% of the sediment or rock volume. Sand or sandstone containing high concentrations of bitumen is known as **tar sand** or oil sand.

The hydrocarbon system that leads to the generation of tar sands begins with the production and burial of a source rock in a large sedimentary basin. When subjected to temperatures of the oil window, the source rock yields oil and gas, which migrate into sandstone layers and then up the dip of tilted layers to the edge of the basin, where they become caught in stratigraphic traps (▶Box 12.1). Over time, microbes (bacteria) attack the oil reserve underground, digest lighter, smaller hydrocarbon molecules, and leave behind only the larger, tar molecules.

Production of usable oil from tar sand is difficult and expensive, but not impossible. It takes about 2 tons of tar sand to produce one barrel of oil. Oil companies mine near-surface deposits in open-pit mines, then heat the tar sand in a furnace to extract the oil. To extract oil from deeper deposits of tar sand, oil companies drill wells and pump steam or solvents down into the sand to liquefy the oil enough so that it can be pumped out.

Vast reserves of organic-rich shale that have not been subjected to temperatures of the oil window, or if they were, did not stay within the oil window long enough to complete the transformation to oil, still contain a high proportion of kerogen. Shale that contains at least 15% to 30% kerogen is called oil shale. During the 1850s, researchers developed techniques to produce liquid oil from oil shale. The process involves heating the oil shale to a temperature of 500°C; at this temperature, the shale decomposes, and the kerogen transforms into liquid hydrocarbon and gas. As is the case with tar sand, production of oil from oil shale is possible, but very expensive.

Natural Gas

Natural gas consists of volatile short-chain hydrocarbons, including methane, ethane, propane, and butane. It occurs in the pores of reservoir rock above oil, because it "floats" over the oil. Where temperatures in the subsurface are so high that oil molecules break apart to form gas, gas-only fields develop.

Gas burns more cleanly than oil. Burning gas produces primarily carbon dioxide and water, whereas burning oil also produces complex organic pollutants. Thus gas has become the preferred fuel for home cooking and heating. But transporting gas requires expensive high-pressure pipelines or container ships, so even though gas is much more abundant than oil, the world's population still consumes more oil than gas. However, in recent years many industries and electrical generation plants have switched to natural gas, so demand for gas has increased.

Gas Hydrate

Gas hydrate is a chemical compound consisting of a methane (CH_4) molecule surrounded by a cagelike arrangement of water molecules. An accumulation of gas hydrate occurs as a whitish solid that resembles regular water ice (▶Fig. 12.8a). Gas hydrate forms when anaerobic bacteria (bacteria that live in the absence of oxygen) eat organic matter, such as dead plankton, that has been incorporated into sea-floor sediment. When the bacteria digest organic matter, they produce methane as a byproduct, and the methane bubbles into the cold sea water that fills pore spaces in sediments. Under pressures found at water depths greater than 300 m, the methane dissolves in water and produces gas

FIGURE 12.8 (a) Gas hydrate samples (the white, icy material) brought up from the sea floor. (b) Chunks of coal. Since 1000 C.E., coal has been a major source of energy.

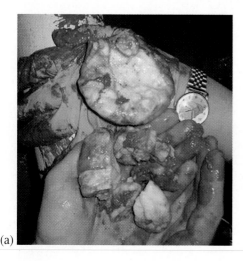
(a)

(b)

hydrate molecules. Gas hydrate occurs as layers interbedded with sediment, and/or as a cement holding together the sediment, at depths of between 90 and 900 m beneath the sea floor. Geologists estimate that an immense amount of methane lies trapped in gas hydrate layers. So far, however, techniques for safely recovering gas hydrate from the sea floor have not been devised.

12.7 COAL: ENERGY FROM THE SWAMPS OF THE PAST

Coal, a black, brittle, sedimentary rock that burns, consists of elemental carbon mixed with minor amounts of organic chemicals, quartz, and clay (▶Fig. 12.8b). Note that coal and oil do not have the same composition or origin. In contrast to oil, coal forms from plant material (wood, stems, leaves) that once grew in coal swamps, regions that resembled the wetlands of modern tropical to semitropical coastal areas (▶Fig. 12.9). Like oil and gas, coal is a fossil fuel because it stores solar energy that reached Earth long ago.

Significant coal deposits could not form until vascular land plants appeared in the late Silurian Period, about 420 million years ago. The most extensive deposits of coal in the world occur in Carboniferous-age strata (deposited between 286 and 354 million years ago). In fact, geolo-

gists coined the name *Carboniferous* because strata representing this interval of the geologic column contain so much coal. The abundance of Carboniferous coal reflects (1) the past position of the continents (during the Carboniferous Period, North America, Europe, and northern Asia straddled the equator, and thus had warm climates in which vegetation flourished) and (2) the height of sea level (shallow seas bordered by coal swamps covered vast parts of continental interiors). Extensive coal deposits can also be found in strata of Cretaceous age (about 65 to 144 million years ago).

The Formation of Coal

How do the remains of plants transform into coal? The vegetation of an ancient swamp must fall and be buried in an oxygen-poor environment, such as stagnant water, so that it can be incorporated in a sedimentary sequence without first reacting with oxygen or being eaten. Compaction and partial decay of the vegetation transforms it into **peat.** Peat, which contains about 50% carbon, itself serves as a fuel in many parts of the world, where deposits formed from moss and grasses in bogs during the last several thousand years; it can easily be cut out of the ground, and once dried, it will burn.

For peat to become coal, the peat must be buried deeply (4–10 km) by sediment. Such deep burial can happen where the surface of the continent gradually sinks, and a sedimentary basin forms that can collect sediment. At any given

FIGURE 12.9 This museum diorama depicts a Carboniferous coal swamp.

time, the type of sediment deposited at a location depends on sea level, which rises and falls over geologic time. As sea level rises, the shoreline migrates inland (transgression takes place), and the location of a coal swamp becomes submerged beneath deeper water. Marine silt and mud then bury the swamp (▶Fig. 12.10).

Eventually, many kilometers of sediment containing numerous peat layers accumulate. At depth in the pile, the weight of overlying sediment compacts the peat and squeezes out any remaining water. Then, because temperature increases with depth in the Earth, deeply buried peat gradually heats up. Heat accelerates chemical reactions that gradually destroy plant fiber and release elements such as hydrogen, nitrogen, and sulfur in the form of gas. These gases seep out of the reacting peat layer, leaving behind a residue concentrated with carbon. Once the proportion of carbon in the residue exceeds about 70%, we have coal. With further burial and higher temperatures, chemical reactions remove additional hydrogen, nitrogen, and sulfur, yielding progressively higher concentrations of carbon.

The Classification of Coal

Geologists classify coal according to the concentration of carbon. With increasing burial, peat transforms into a soft, dark-brown coal called lignite. At higher temperatures (about 100°–200°C), lignite, in turn, becomes dull, black bituminous coal, which contains 85% carbon, and at still higher temperatures (about 200°–300°C), bituminous coal is transformed into shiny, black anthracite coal, which contains 95% carbon. The progressive transformation of peat to anthracite coal, which occurs as the coal layer is buried more deeply and becomes warmer, reflects the completeness of chemical reactions that remove water, hydrogen, ni-

trogen, and sulfur from the organic chemicals of the peat and leave behind carbon (▶Fig. 12.11a–d). As the carbon content of coal increases, we say the **coal rank** increases.

Of note, the formation of anthracite coal requires high temperatures that develop only on the borders of mountain belts, where mountain-building processes can push thick sheets of rock along thrust faults and over the coal-bearing sediment, so the sediment ends up at depths of 8–10 km, at which temperatures reach 300°C. In the interiors of mountain belts, temperatures become even higher, and rock begins to undergo metamorphism, at which point almost all elements except carbon leave the coal. The remaining carbon atoms rearrange to form graphite.

FIGURE 12.11 The evolution of coal from peat. (a) A layer of peat accumulates beneath a coal swamp. (b) Later, after the peat has been substantially buried, it compacts, loses water, releases hydrogen, nitrogen, and sulfur, and becomes lignite. (c) Still later, after even more burial, the lignite compacts and alters to form bituminous coal. (d) In this mountain range, the coal bed has been folded, and faults have transported warmer rock from greater depth over the coal. As a result, the coal becomes even warmer and turns into anthracite coal.

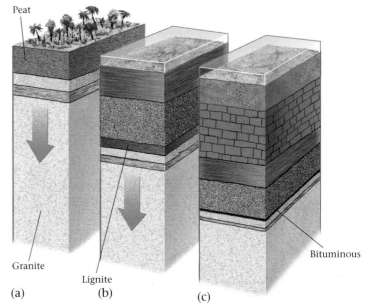

FIGURE 12.10 Sea level transgresses and regresses over time, with the result that a coal swamp along the coast migrates inland and the swamp's deposits eventually get buried by other strata. Notice that the floor of the basin gradually sinks, so there is room for all the strata. (See Fig. 5.33 for the consequences of progressive transgression and regression.)

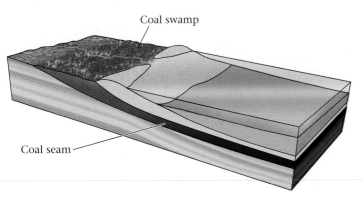

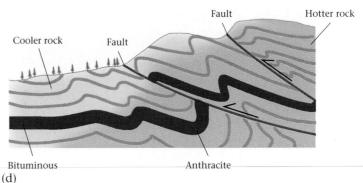

Finding and Mining Coal

Because the vegetation that eventually becomes coal was initially deposited in a sequence of sediment, coal occurs as sedimentary beds (seams, in mining parlance) interlayered with other sedimentary rocks (▶Fig. 12.12). To find coal, geologists search for sequences of strata that were deposited in tropical to semitropical, shallow-marine to terrestrial (fluvial or deltaic) environments—the environments in which a swamp could exist. The sedimentary strata of continents contain huge quantities of discovered coal, or coal reserves (▶Fig. 12.13).

The way in which companies mine coal depends on the depth of the coal seam. If the coal seam lies within 100 m of the ground surface, strip mining proves to be most economical. In strip mines, miners use a giant shovel called a drag line to scrape off soil and layers of sedimentary rock above the coal seam (▶Fig. 12.14a, b). Drag lines are so big that they could swallow a two-car garage. Once the drag line has exposed the seam, it then scrapes out the coal and dumps it into trucks or onto a conveyor belt. When the coal has been scraped out, the operator fills the hole with the rock that had been stripped to expose the coal and covers the rock back up with the saved topsoil on which grass or trees may eventually grow.

Deep coal can be obtained only by underground mining. Miners dig a shaft down to the depth of the coal seam and then create a maze of tunnels, using huge grinding machines that chew their way into the coal (▶Fig. 12.15).

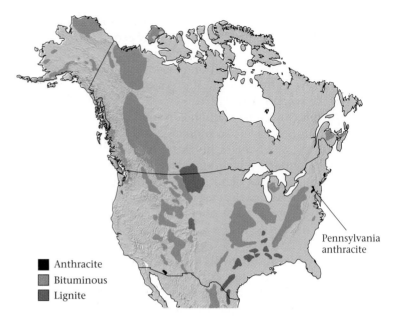

FIGURE 12.13 The distribution of coal reserves in North America. North America contains 27% of the world's reserves. Russia and Eastern Europe contain 28%. The rest of the world has less than half.

- ■ Anthracite
- ▨ Bituminous
- ▨ Lignite

Pennsylvania anthracite

FIGURE 12.12 Coal is found in sedimentary beds interlayered with other strata (sandstone, shale, and limestone).

Underground coal mining can be very dangerous, not only because the sedimentary rocks forming the roof of the mine are weak and can collapse, but also because methane gas released by chemical reactions in coal can accumulate in the mine, leading to the danger of a small spark triggering a deadly mine explosion. Unless they breathe through filters, underground miners also risk contracting black-lung disease from the inhalation of coal dust. The dust particles wedge into tiny cavities of the lungs and gradually cut off the oxygen supply or cause pneumonia.

Coalbed Methane and Coal Gasification

The natural process by which coal forms underground yields large quantities of methane, a type of natural gas. Over time, some of the gas escapes to the atmosphere, but vast amounts remain within the coal. Such **coalbed methane,** trapped in strata too deep to be reached by mining, is a valuable energy resource that has become a target for exploration in many regions of the world.

Obtaining coalbed methane from deep layers of strata involves drilling, rather than mining. Drillers penetrate a coal bed with a hole, and then start pumping out groundwater. As a result of pumping out water, the pressure in the vicinity of the drillhole decreases relative to the surrounding bed. Methane bubbles into the hole and then up to the ground surface, where condensers force it into tanks for storage.

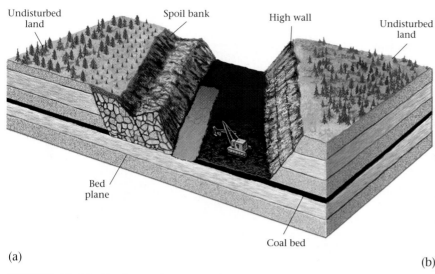

FIGURE 12.14 (a) The configuration of a coal strip mine. (b) A drag line.

Traditional burning coal produces clouds of smoke, containing fly ash (solid residue left after the carbon in coal has been burned) and noxious gases. This problem can be partially solved if coal is first transformed into gases, as well as solid byproducts, *before* burning. The gases burn quite cleanly. The process of producing clean-burning gases from solid coal is called **coal gasification.**

Coal gasification involves the following steps. First, pulverized coal is placed in a large container. Then, a mixture of steam and oxygen passes through the coal at high pressure. As a result, the coal heats up to a high temperature but does not ignite; under these conditions, chemical reactions break down the carbon molecules in coal and produce hydrogen gas and other gases such as carbon monoxide. Solid ash, as well as sulfur and mercury, concentrate at the bottom of the container and can be removed before the gases are burned.

Underground Coalbed Fires

Coal will burn not only in furnaces but also in surface and subsurface mines, as long as the fire has access to oxygen. Coal mining of the past two centuries has exposed coal to the air and has provided many more opportunities for fires to begin; once started, a coalbed fire that progresses underground (sucking in oxygen from joints in overlying rock) may be very difficult to extinguish.

Major coalbed fires can be truly disastrous. The most notorious of these fires in North America began as a result of trash burning in a mine near the town of Centralia, Pennsylvania. For the past forty years the fire has progressed underground, eventually burning coal beneath the town itself. The fire produces toxic fumes that rise through the ground and make the overlying landscape uninhabitable, and it also causes the land surface to collapse. Today, over 100 major coalbed fires are burning in northern China.

FIGURE 12.15 Machinery used in underground coal mining.

12.8 NUCLEAR POWER

How Does a Nuclear Power Plant Work?

So far, we have looked at fuels (oil, gas, and coal) that release energy when burned. During burning, a chemical reaction between the fuel and oxygen releases the potential energy stored in the chemical bonds of the materials.

Nuclear power, however, comes from a different process: the fission, or breaking, of the nuclear bonds that hold protons and neutrons together in the nucleus provides the energy. Fission splits an atom into smaller pieces.

Nuclear power plants were first built to produce electricity during the 1950s. A **nuclear reactor,** the heart of the plant, commonly lies within a dome-shaped shell (containment building) made of reinforced concrete (▶Fig. 12.16). The reactor contains nuclear fuel, pellets of concentrated uranium oxide or a comparable radioactive material, packed into metal tubes called fuel rods. Fission occurs when a neutron strikes a radioactive atom, causing it to split; for example, uranium-235 splits into barium-141 and krypton-92, plus three neutrons and energy. During fission, a tiny fraction of the matter composing the original atom transforms into thermal and electromagnetic energy. The neutrons released during the fission of one atom strike other atoms, thereby triggering more fission in a self-perpetuating process called a **chain reaction.** Pipes carry water close to the heat-generating fuel. The heat transforms the water into high-pressure steam. The pipes then carry this steam to a turbine, where it rotates fan blades; the rotation drives a dynamo that generates electricity. Eventually the steam goes into cooling towers, where it condenses back into water that can be reused in the plant or returned to the environment (▶Fig. 12.17).

The Geology of Uranium

^{235}U, an isotope of uranium containing 143 neutrons, is the most common fuel for conventional nuclear power plants. ^{235}U accounts for only about 0.7% of naturally occurring

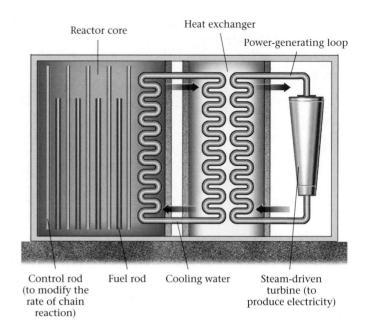

FIGURE 12.17 A nuclear reactor can be used to generate electricity.

uranium; most uranium consists of ^{238}U, an isotope with 145 neutrons. Thus, to make a fuel suitable for use in a power plant, the ^{235}U concentration in a mass of natural uranium must be increased by a factor of 2 or 3, an expensive process called enrichment.

Where does uranium come from? The Earth's radioactive elements, including uranium, probably developed during the explosion of a supernova before the existence of the solar system. Uranium atoms from this explosion became part of the nebula out of which the Earth formed and thus were incorporated into the planet. They gradually rose into the upper crust in granitic magma.

Even though granite contains uranium, it does not contain very much. But nature has a way of concentrating uranium. Hot water circulating through a pluton after intrusion dissolves the uranium and precipitates it, along with other elements, in veins. Uranium from veins typically occur in the mineral pitchblende (UO_2). Uranium may be further concentrated once plutons, and the associated uranium-rich veins, weather and erode at the ground surface. Sand derived from a weathered pluton washes down a stream, and as it does so, uranium-rich grains stay behind because they are so heavy, relative to quartz and feldspar grains. The world's richest uranium deposits, in fact, occur in ancient stream-bed gravels. Uranium deposits may also form when groundwater percolates through uranium-rich sedimentary rocks: the uranium dissolves in the water and moves with the water to another location,

FIGURE 12.16 This nuclear power plant has two reactors, each in its own containment building.

where it precipitates out of solution and fills the pores of the host sedimentary rock.

Problems with Nuclear Energy

Maintaining safety at nuclear power plants requires work. Operators must constantly cool the nuclear fuel with circulating water, and the rate of nuclear fission must be regulated by the insertion of boron-steel control rods, which absorb neutrons and thus decrease the number of collisions between neutrons and radioactive atoms. Without control rods, the number of neutrons dashing around in the nuclear fuel would progressively increase, causing the rate of fission and accompanying heat production to increase. Eventually, the fuel would become so hot that it would melt. Such a **meltdown** might cause a steam explosion that could breach the containment building and scatter radioactive debris into the air. (Note that a meltdown is *not* the same as an atomic bomb explosion.)

One near-meltdown has occurred since the beginning of the nuclear age—at the power plant in Chernobyl, Ukraine, in April 1986. The fuel pile became too hot, triggering a steam explosion that raised the roof of the containment building and spread fragments of the reactor and its fuel around the plant grounds. Within six weeks, twenty people had died from radiation sickness. Radioactive material entered the atmosphere and dispersed over Eastern Europe and Scandinavia; no one yet knows the full effects of this fallout on the health of exposed populations. These days, there is concern that even a safely operated plant could pose a hazard if it became the target of a terrorist attack.

Nuclear waste is the radioactive material produced in a nuclear plant. It includes spent fuel, which contains radioactive daughter products, as well as water and equipment that have come in contact with radioactive materials. Radioactive elements emit gamma rays and X-rays that can damage living organisms and cause cancer. Some radioactive material decays quickly (in decades to centuries), but some remains dangerous for thousands of years or more.

Nuclear waste cannot just be stashed in a warehouse or buried in a town landfill. If the waste were simply buried, groundwater passing through the dump site might transport radioactive elements into municipal water supplies or nearby lakes or streams. Ideally, waste should be sealed in containers that will last for thousands of years (the time needed for the short-lived radioactive atoms to undergo decay) and stored in a place where it will not come in contact with the environment. Finding an appropriate place is not easy.

So far, experts disagree about which is the best way to dispose of nuclear waste. The U.S. government favors storing waste at Yucca Mountain in the Nevada desert. The mountain consists of fairly dry rhyolite and the area is far from a population center.

12.9 OTHER ENERGY SOURCES

Geothermal Energy

As the name suggests, **geothermal energy** refers to the internal heat of the Earth. Geothermal energy exists because the Earth grows progressively hotter with depth. Where the geothermal gradient is steep (meaning temperature increases rapidly with depth), we find high temperatures at relatively shallow depths. Groundwater in such areas absorbs heat from the rock and becomes hot.

Power companies use geothermal energy to produce heat and electricity in two ways. In some places, they simply pump hot groundwater out of the ground and run it through pipes to heat houses and spas. Elsewhere, the groundwater is so hot that when it rises to the Earth's surface and decompresses, it turns to steam. This steam can drive turbines and generate electricity (►Fig. 12.18a, b).

In volcanic areas such as Iceland or New Zealand, geothermal energy provides a major portion of energy needs. But on a global basis, it doesn't, because few cities lie near geothermal resources. Furthermore, a geothermal energy supply can be destroyed when people pump groundwater out of the ground faster than it can be replenished.

Hydroelectric and Wind Power

As water flows downslope, its potential energy converts into kinetic energy. In a modern hydroelectric power plant, the water flow drives turbines, which in turn drive generators that produce electricity. In order to increase the rate and volume of water flow, engineers build dams to create a reservoir that retains water and raises it to a higher elevation—the water flows through pipes down to turbines at the foot of the dam (►Fig. 12.19).

At first glance, hydroelectric power seems ideal, because it produces no smoke or radioactive waste, and because reservoirs can also be used for flood control, irrigation, and recreation. But unfortunately, reservoirs may also bring unwanted changes to a region. Damming a fast-moving river may flood a spectacular canyon, eliminate exciting rapids, and destroy a river's ecosystem. Further, the reservoir traps sediment, so floodplains downstream lose their sediment and nutrient supply.

Not all hydroelectric power plants utilize river water. In a few places, engineers have employed the potential energy stored in ocean water at high tide. To do this, they build a floodgate dam across an inlet. Water flows into the inlet when the tide rises, only to be trapped when the gate is closed. After the tide has dropped outside the floodgate, the water retained by the floodgate flows back to sea via a pipe that carries it through a power-generating turbine.

Wind power has been used for millennia as a way to provide power for mills and water pumps. In recent decades,

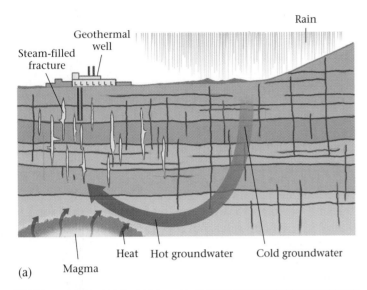

FIGURE 12.19 A hydroelectric dam. This example is the Grand Coulee Dam, located on the Columbia River in central Washington State; it is the third-largest producer of hydroelectric power in the world.

FIGURE 12.18 (a) This region can be used for geothermal energy production. Surface water sinks down into the ground as groundwater until it becomes heated by the hot ground (possibly by magma below). The hot water rises and, when it reaches shallow depths, turns to steam (as a result of decompression). The hot water may be pumped out. It turns to steam under ground-surface pressure and runs turbines. (b) A geothermal energy plant.

numerous wind farms have been established around the world to generate electricity (▶Fig. 12.20a). The electricity is clean, but wind production has a serious drawback: it requires construction of large, somewhat noisy towers. Not everyone wants a giant wind farm in their "backyard"— because of the visual effect, because of the noise, and because the towers pose a hazard for birds.

Solar Energy

The Sun drenches the Earth with energy in quantities that dwarf the amounts stored in fossil fuels. Were it possible to harness this energy directly, humanity would have a reliable and totally clean solution for powering modern technology. But using solar energy is not quite so simple. At present, energy consumers have two options for the direct use of solar energy: (1) A solar collector is a device that collects energy to produce heat. One class of solar collectors includes mirrors or lenses that focus light striking a broad area into a smaller area. On a small scale, such devices can be used for cooking. Another class of solar collectors consists of a black surface placed beneath a glass plate. The black surface absorbs light that has passed through the glass plate and heats up. The glass does not let the heat escape. When a consumer runs water between the glass and the black surface, the water heats up. (2) Photovoltaic cells (solar cells) convert light energy directly into electricity (▶Fig. 12.20b). Most photovoltaic cells consist of two wafers of silicon pressed together. One wafer contains atoms of arsenic, and the other wafer contains atoms of boron. When light strikes the cell, arsenic atoms release electrons that flow over to the boron atoms. If a wire loop connects the back side of one wafer to the back side of the other, this phenomenon produces an electrical current (▶Fig. 12.20c).

Ethanol

In recent years, farmers have begun to produce rapidly growing crops, such as corn and sugar cane, specifically for the purpose of producing biomass that, when fermented, yields ethanol (a type of alcohol). Ethanol can substitute for gasoline in car engines. The process of producing ethanol includes

FIGURE 12.20 (a) A wind farm in southwestern England. The towers are about 50 m high. (b) The roof of this house has been covered with photovoltaic cells. (c) Diagram illustrating the way in which a photovoltaic cell produces electricity. (d) Diagram illustrating the way in which a hydrogen fuel cell works.

the following steps. First, producers grind grain into a fine powder, mix it with water, and cook it to produce a mash of starch. Then, they add an enzyme to the mash, which converts the mash into sugar. The sugar, when mixed with yeast, ferments. Fermentation produces ethanol and CO_2. Finally, the fermented mash is then distilled to concentrate the ethanol.

Fuel Cells

In a fuel cell, chemical reactions produce electricity directly. Let's consider a hydrogen fuel cell. Hydrogen gas flows through a tube across an anode (a strip of platinum) that has been placed in a water solution containing an electrolyte (a substance that enables the solution to conduct electricity). At the same time, a stream of oxygen gas flows onto a separate platinum cathode that also has been placed into the solution. A wire connects the anode and the cathode to provide an electrical circuit (▶Fig. 12.20d). In this configuration, hydrogen reacts with oxygen to produce water and electricity. Fuel cells are efficient and clean. Their limitation lies in the need for a design that will protect the cells from damage by impact and that will enable them to store hydrogen, an explosive gas, in a safe way. Also, it takes a significant amount of energy to produce the hydrogen used in fuel cells.

12.10 ENERGY CHOICES, ENERGY PROBLEMS

The Oil Crunch

Energy usage in industrialized countries grew with dizzying speed through the mid-twentieth century, and during this time people came to rely increasingly on oil (▶Fig. 12.21a, b). Eventually, oil supplies within the borders of industrialized countries could no longer match the demand, and these countries began to import more oil than they produced themselves. Through the 1960s, oil prices remained low (about $1.80 a barrel), so this was not a problem. In 1973, however, a complex tangle of politics and war led the Organization of Petroleum-Exporting Countries (OPEC) to limit its oil exports. In the United States, fear of an oil shortage turned to panic, and motorists began lining up at gas stations, in many cases waiting for hours to fill their tanks. The price of oil rose to $18 a barrel, and newspaper headlines proclaimed an "Energy Crisis!" Governments in industrialized countries instituted new rules to encourage oil conservation. During the last two decades of the twentieth century, the oil market stabilized, though political events occasionally led to price jumps and short-term shortages. In 2006, the price of oil went over $75/bbl. Will a day come

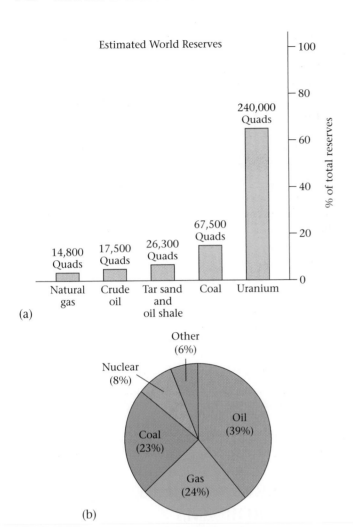

(a)

(b)

FIGURE 12.21 (a) Estimated world reserves of various energy resources. The numbers at the ends of each column are in "Quads," where 1 Quad = 1 quadrillion BTU. (1 BTU [British Thermal Unit] = 1,055 Joules = 252 calories; 1 barrel of oil produces 5.8 million BTU). (b) The proportion of world energy supply provided by different energy resources in 2003.

when shortages arise not because of an embargo or limitations on refining capacity but because there is no more oil to produce? To answer these questions, we need to examine the world oil supply. What we find there may be unnerving.

To understand the issues involved in predicting the future of energy supplies, we must first classify energy resources. We call a particular resource "renewable" if nature can replace it within a short time relative to a human life span (in months or, at most, decades). We call a resource "nonrenewable" if nature takes a very long time (hundreds to perhaps millions of years) to replenish it. Oil, gas, and coal are nonrenewable resources, in that the rate at which humans consume these materials far exceeds the rate at which nature replenishes them, so we will inevitably run out of oil. The question is, when?

Historians in the future may refer to our time as the **Oil Age,** because so much of our economy depends on oil. How long will this Oil Age last? Geologists estimate that as of 2000, there were about 850 to 1,000 billion barrels of proven reserves (oil that had been found). There may be an additional 2,000 billion barrels not yet found. Thus, the world probably holds between 1,000 and 3,000 billion barrels of obtainable oil (not counting oil in tar sands or oil shales). Presently, humanity guzzles oil at a rate of about 28 billion barrels per year. At this rate of consumption, oil supplies will last until sometime between 2050 and 2150. Already, geologists see the beginning of the end of the Oil Age, for the rate of consumption now exceeds the rate of discovery of new oil by a factor of 3. In the end, the Oil Age will probably have lasted less than three centuries. On a time line representing the four thousand years since the construction of the Egyptian pyramids, this looks like a very short blip (▶Fig. 12.22a, b). We may indeed be living during a unique interval of human history.

Can Other Energy Sources Meet the Need?

Are there alternatives to oil? Perhaps. Certainly the world contains vast fossil-fuel supplies in other forms, enough to last for centuries if they can be exploited in an economical and environmentally sound way. These supplies include tar sand, oil shale, natural gas, and coal. All told, perhaps 1.5 trillion barrels of hydrocarbons are trapped in tar sands and 3 trillion barrels trapped in oil shale. In addition, coal deposits could meet our energy needs for the next few centuries. But extracting tar sand, oil shale, and coal requires the construction of huge mines, which might scar the landscape. Natural gas is probably our best alternative in the near future.

Can nuclear power or hydroelectric power fill the need? Vast supplies of uranium, the fuel of traditional nuclear plants, remain untapped (Fig. 12.21a), and in addition nuclear engineers have designed alternative plants, powered by breeder reactors, that essentially produce their own fuel. But many people view nuclear plants with suspicion because of concerns about radiation, accidents, terrorism, and waste storage, and these concerns have stalled the industry. A substantial increase in hydroelectric power production is not likely, as most appropriate rivers have already been dammed, and industrialized countries have little appetite for taming any more. Similarly, the growth of geothermal-energy output seems limited.

Because of the problems that would result from relying more on coal, hydroelectric, and nuclear energy, researchers have been increasingly exploring "clean energy" options. One possibility is solar power. Unfortunately, affordable technologies for large-scale solar energy do not yet exist, and the idea of covering the landscape with solar

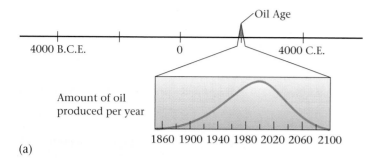

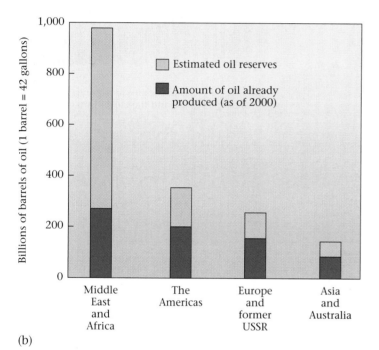

(a)

(b)

FIGURE 12.22 (a) The predicted history of the Oil Age, which may be a blip in human history like the Bronze Age. (b) The graph shows the amount of oil and gas reserves used and the estimated amounts still remaining.

pipelines or trucks sink into the subsurface and contaminate groundwater, and oil spills from ships create slicks that spread over the sea surface and foul the shoreline (▶Fig. 12.23). Coal and uranium mining also scar the land and can lead to the production of acid mine runoff, a dilute solution of sulfuric acid that forms when sulfur-bearing minerals such as pyrite (FeS_2) in mines react with rainwater. The runoff enters streams and kills fish and plants. Extensive coal mining may also cause the ground surface to sink. This happens when an underground mine collapses.

Numerous air-pollution issues also arise from the burning of fossil fuels, which sends soot, carbon monoxide, sulfur dioxide, nitrous oxide, and unburned hydrocarbons into the air. Coal, for example, commonly contains sulfur, primarily in the form of pyrite, which enters the air as sulfur dioxide (SO_2) when coal is burned. This gas combines with rainwater to form sulfuric acid (H_2SO_4), or acid rain. For this reason, many countries now regulate the amount of sulfur that coal can contain when it is burned.

But even if pollutants can be decreased, the burning of fossil fuels still releases carbon dioxide (CO_2) into the atmosphere. CO_2 is important because it traps heat in the Earth's atmosphere much like glass traps heat in a greenhouse—this is the **greenhouse effect.** Too much CO_2 may lead to a global increase in atmospheric temperature (global warming), which in turn may alter the distribution of climatic belts. We'll learn more about this issue in Chapter 19.

FIGURE 12.23 A marine oil spill. Volatile components evaporate, leaving a tar-like residue.

cells is not attractive. Similarly, we can turn to wind power for small-scale energy production in areas where strong gusts blow, but covering the landscape with windmills is no more appealing. Fusion power may be possible, but physicists and engineers have not yet figured out a way to harness it. Clearly society will be facing difficult choices in the not-so-distant future about where to obtain energy, and we will need to invest in the research needed to discover new alternatives.

Environmental Issues

Environmental concerns about energy resources begin right at the source. Oil drilling requires substantial equipment, the use of which can damage the land. Oil spills from

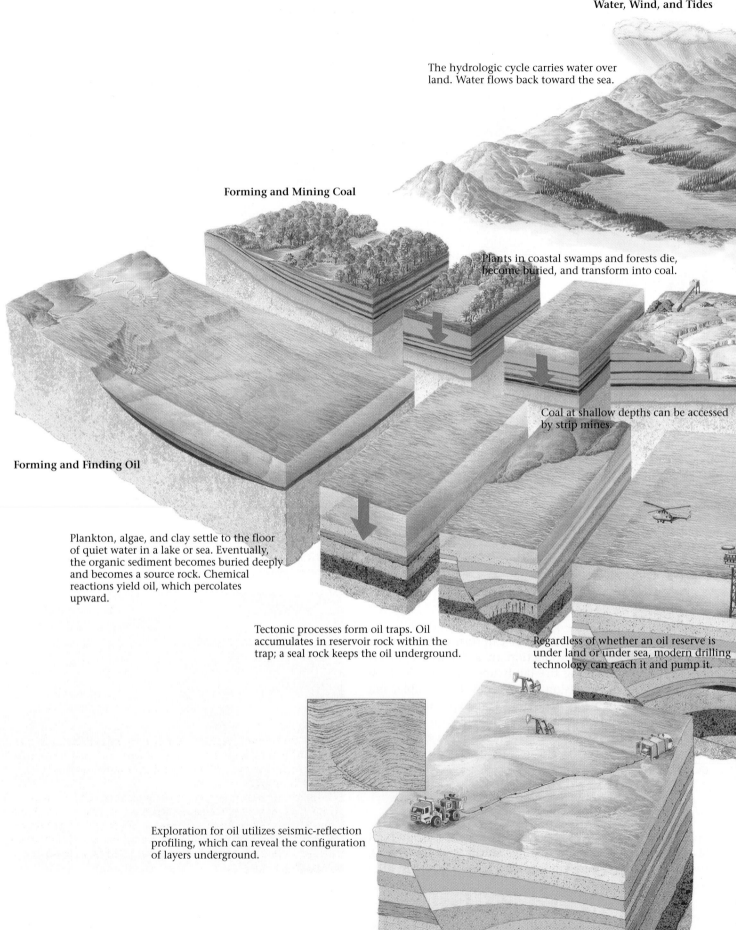

Water, Wind, and Tides

The hydrologic cycle carries water over land. Water flows back toward the sea.

Forming and Mining Coal

Plants in coastal swamps and forests die, become buried, and transform into coal.

Coal at shallow depths can be accessed by strip mines.

Forming and Finding Oil

Plankton, algae, and clay settle to the floor of quiet water in a lake or sea. Eventually, the organic sediment becomes buried deeply and becomes a source rock. Chemical reactions yield oil, which percolates upward.

Tectonic processes form oil traps. Oil accumulates in reservoir rock within the trap; a seal rock keeps the oil underground.

Regardless of whether an oil reserve is under land or under sea, modern drilling technology can reach it and pump it.

Exploration for oil utilizes seismic-reflection profiling, which can reveal the configuration of layers underground.

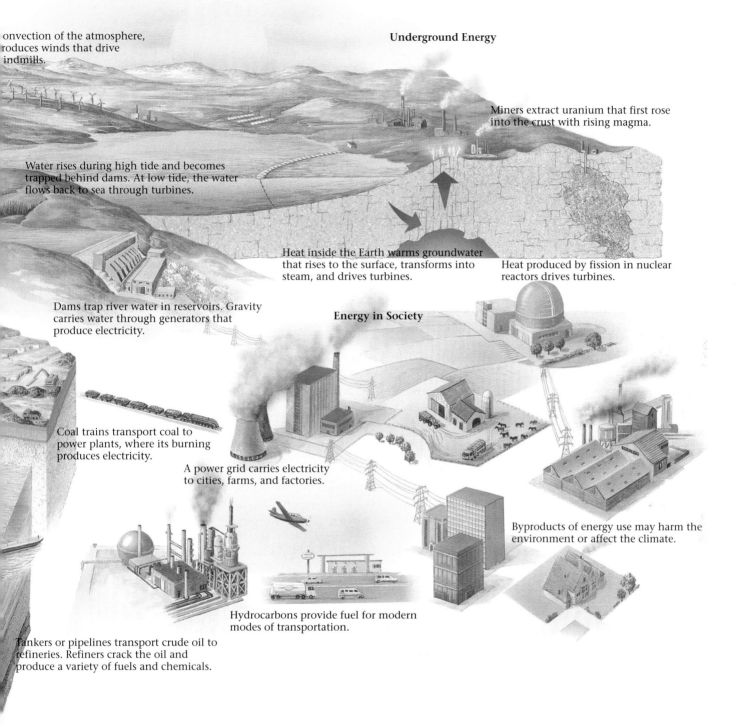

Underground Energy

Miners extract uranium that first rose into the crust with rising magma.

...onvection of the atmosphere, ...roduces winds that drive ...indmills.

Water rises during high tide and becomes trapped behind dams. At low tide, the water flows back to sea through turbines.

Heat inside the Earth warms groundwater that rises to the surface, transforms into steam, and drives turbines.

Heat produced by fission in nuclear reactors drives turbines.

Dams trap river water in reservoirs. Gravity carries water through generators that produce electricity.

Energy in Society

Coal trains transport coal to power plants, where its burning produces electricity.

A power grid carries electricity to cities, farms, and factories.

Byproducts of energy use may harm the environment or affect the climate.

Hydrocarbons provide fuel for modern modes of transportation.

Tankers or pipelines transport crude oil to refineries. Refiners crack the oil and produce a variety of fuels and chemicals.

Power from the Earth

Modern society, for better or worse, uses vast amounts of energy to produce heat, to drive modes of transportation, and to produce electricity (see Appendix). This energy comes either from geologic materials stored in the Earth, or from geologic processes happening at our planet's surface. For example, oil and gas fill the pores of reservoir rocks at depth below the surface, coal occurs in sedimentary beds, and uranium concentrates in ore deposits. A hydroelectric power plant taps into the hydrologic cycle, windmills operate because of atmospheric convection, and geothermal energy comes from hot groundwater. Ultimately, the

"energy" comes from the Sun, from gravity, from Earth's internal heat, and/or from nuclear reactions. Oil, gas, and coal are "fossil fuels" because the energy they store first came to Earth as sunlight, long ago.

As energy usage grows, easily obtainable energy resources dwindle, the environment can be degraded, and the composition of the atmosphere changed. The pattern of energy use that forms the backbone of society today may have to change radically in the not-so-distant future, if we wish to avoid a decline in living standards.

12.11 INTRODUCING MINERAL RESOURCES

In June 1845, over a year after leaving Missouri, James Marshall arrived by horse at Sutter's Fort, in central California, to make a new life. Marshall convinced Captain John Sutter, a former army officer, to finance the construction of a sawmill in the foothills of the Sierra Nevada Mountains. Marshall's scruffy crew finished the mill by the beginning of 1848. As Marshall stood admiring the new building, he noticed a glimmer of metal in the gravel that littered the bed of the adjacent stream. He picked it up, banged it between two rocks to test its hardness, and shouted, "Boys, by God, I believe I have found a gold mine!" For a short while, Marshall and Sutter managed to keep the discovery secret. But word of the gold soon spread, and within weeks the workers at Marshall's mill had disappeared into the mountains to seek their own fortunes.

All through that year, gold "fever" spread throughout California and eventually reached the East Coast. As a result, 1849 brought 40,000 prospectors to California. These "forty-niners," as they were called, had abandoned their friends and relatives on the gamble that they could strike it rich.

Gold-containing rocks are but one of many mineral resources in the Earth's upper crust that are of use to civilization. Without such resources, industrialized societies could not function. Geologists divide mineral resources into two categories: metallic mineral resources (rocks containing gold, copper, aluminum, iron, and so on) and nonmetallic mineral resources (building stone, gravel, sand, gypsum, phosphate, salt, and so on). Now we look at the nature of mineral resources, the geologic phenomena responsible for their formation, and the way people mine them. We conclude by considering limits to mineral reserves.

12.12 METALS, ORES, AND ORE DEPOSITS

Metal and Its Discovery

We use metals for many purposes—nails, wires, car bodies, girders—because of their properties. **Metals** are opaque, shiny, smooth solids that can conduct electricity and be bent, drawn into wire, or hammered into thin sheets. They look and behave quite differently from wood, plastic, meat, or rock. This is because, unlike other substances, the atoms that make up metals are held together by metallic bonds, which allow electrons to flow from atom to atom fairly easily. Despite the mobility of their electrons, metals are solids, so in metals, atoms lie fixed in a regular lattice defining a crystal structure.

The first metals that people used—copper, silver, and gold—can occur in rock as native metals. Native metals consist only of metal atoms, and thus look and behave like metal. Gold nuggets, for example, are chunks of native metal that have eroded free of bedrock (▶Fig. 12.24a, b). Fifteen thousand years ago, ice-age hunters collected nuggets of native metal from stream beds and pounded them together with stone hammers to make arrowheads, scrapers, and coins.

If we had to rely solely on native metals as our source of metal, we would have access to only a tiny fraction of our current metal supply. Most of the metal atoms we use today originated as ions bonded to nonmetallic elements in a great variety of minerals that themselves look nothing like metal. Only because of the chance discovery by some prehistoric genius that certain rocks, when heated to high temperatures in fire (a process called smelting), decompose to yield metal plus a nonmetallic residue called slag do we now have the ability to produce sufficient metal for the needs of industrialized society. The earliest smelters plied their trade in 4000 B.C.E. and produced copper. Iron smelting began about 1300 B.C.E.; aluminum smelting did not become widespread until about 1880 C.E.

What Is an Ore?

The minerals from which metals can be extracted are called ore minerals, or economic minerals. These minerals contain metal in high concentrations and in a form that can be

FIGURE 12.24 (a) Gold nuggets come from quartz veins. At an early stage, much of the quartz remains. (b) With time, most of the quartz breaks away, leaving only gold. The largest single nugget ever found weighed 71 kg (2,284 oz) and came from Australia.

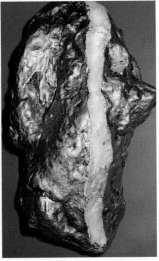

(a)   (b)

easily extracted. Galena (PbS), for example, is about 50% lead, so we consider it to be an ore mineral of lead (▶Fig. 12.25). We obtain most of our iron from the oxide minerals hematite and magnetite.

Geologists have identified different kinds of ore minerals. Many ore minerals are sulfides, in which the metal occurs in combination with sulfur (S), or oxides, in which the metal occurs in combination with oxygen (O). Quite a few ore minerals are colorful and come in interesting shapes, and some have a metallic luster (▶Fig. 12.26).

To obtain the metals needed for industrialized society, we mine **ore,** rocks containing native metals or a concentrated accumulation of ore minerals. To be an ore, a rock must not only contain ore minerals, it must also have a sufficient amount to make the rock worth mining. Iron constitutes about 6.2% of the continental crust's weight and makes up 30 to 60% of iron ore. The concentration of a useful metal in an ore determines the grade of the ore—the higher the concentration, the higher the grade. Whether or not an ore of a given grade is worth mining depends on the price of metal in the market.

How Do Ore Deposits Form?

Ore minerals do not occur uniformly through rocks of the crust—if they did, we would not be able to extract them economically. Fortunately for humanity, geologic processes concentrate these minerals in ore deposits. To put it simply, an **ore deposit** is an economically significant occurrence of ore. The various kinds of ore deposits differ from each other in terms of how they formed and which ore minerals they contain. Below, we introduce a few examples.

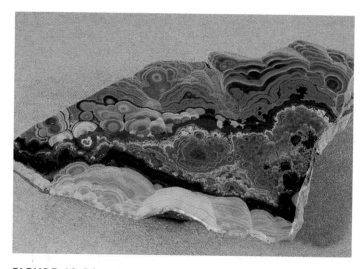

FIGURE 12.26 Colorful ore minerals containing copper. The blue mineral is azurite and the green is malachite.

Magmatic deposits. When a magma cools, sulfide ore minerals crystallize and then, because of their high density, sink to the bottom of the magma chamber, where they accumulate. An accumulation formed in this way is a magmatic deposit. When the magma freezes solid, the resulting igneous body may contain a solid mass of sulfide ore minerals at its base. Because of their composition, we consider these masses to be a type of massive-sulfide deposit (▶Fig. 12.27).

Hydrothermal deposits. Hydrothermal activity involves the circulation of hot-water solutions through a magma or through the rocks surrounding an igneous intrusion. These

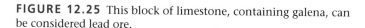

FIGURE 12.25 This block of limestone, containing galena, can be considered lead ore.

FIGURE 12.27 Heavy, metal crystals can sink to the bottom of a magma chamber to form a massive-sulfide deposit. The sulfide concentrate may become an ore body in the future.

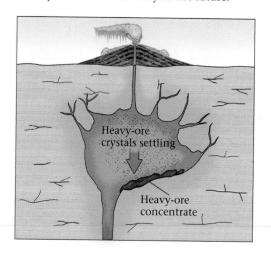

Heavy-ore crystals settling

Heavy-ore concentrate

fluids dissolve metal ions and carry them elsewhere. When the solution enters a region of lower pressure, lower temperature, different acidity, and/or different availability of oxygen, the metals come out of solution and form ore minerals that precipitate in fractures and pores, creating a hydrothermal deposit (▶Fig. 12.28). Such deposits may form within an igneous intrusion or in surrounding country rock. If the resulting ore minerals disperse through the intrusion, we call the deposit a disseminated deposit, but if they precipitate to fill cracks in preexisting rock, we call the deposit a vein deposit (veins are simply mineral-filled cracks).

In recent years, geologists have discovered that hydrothermal activity at the submarine volcanoes along mid-ocean ridges leads to the eruption of hot water, containing high concentrations of dissolved metal and sulfur, from a vent. When this hot water comes in contact with cold seawater, the dissolved components instantly precipitate as tiny crystals of metal sulfide minerals (▶Fig. 12.29) creating black clouds called black smokers. The minerals in the cloud eventually sink and form a pile of nearly pure ore minerals around the vent.

Secondary-enrichment deposits. Sometimes groundwater passes through ore-bearing rock long after the rock first formed. This groundwater dissolves some of the ore minerals and carries the dissolved ions away. When the water eventually flows into a different chemical environment (such as one with a different amount of oxygen or acid), it precipitates new ore minerals, commonly in concentrations that exceed that of the original deposit. A new ore deposit formed from metals that were dissolved and carried away from a preexisting ore deposit is called a secondary-enrichment deposit (▶Fig. 12.30a, b). Some of these deposits contain spectacularly beautiful copper-bearing carbonate minerals such as azurite and malachite (Fig. 12.26).

MVT ores. In recent years, geologists have recognized that some groundwater beneath mountain belts sinks several kilometers down into the crust and follows a curving flow path that eventually brings it back to the surface in the interior of a continent, perhaps hundreds of kilometers away from the mountain range. At the depths reached by the groundwater, temperatures become high enough that the water dissolves metals. When the water returns to the surface and enters cooler rock, these metals precipitate in ore minerals. Ore deposits formed in this way typically contain lead- and zinc-bearing minerals. They appear in dolomite beds of the Mississippi Valley region, and thus have come to be known as Mississippi Valley–type or MVT ore.

Sedimentary deposits of metals. Some ore minerals accumulate in sedimentary environments under special circumstances. For example, between 2.0 and 2.5 billion years ago, the atmosphere, which previously had contained little oxy-

FIGURE 12.29 Aprons of sulfide ore precipitate around "black smokers" along a mid-ocean ridge.

FIGURE 12.28 Water circulating through a granite pluton dissolves and redistributes metals, leading to formation of hydrothermal deposits.

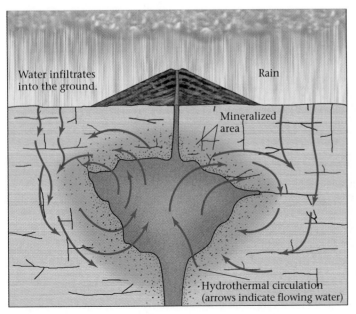

Water infiltrates into the ground.

Rain

Mineralized area

Hydrothermal circulation (arrows indicate flowing water)

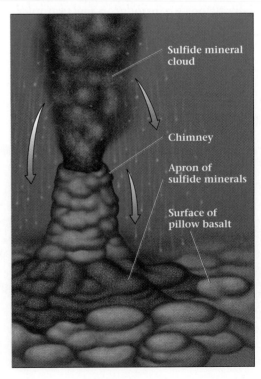

Sulfide mineral cloud

Chimney

Apron of sulfide minerals

Surface of pillow basalt

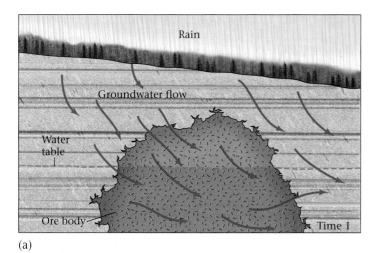

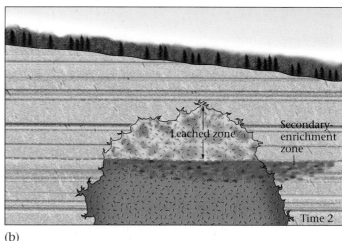

(a)

(b)

FIGURE 12.30 (a) In the process of secondary enrichment, water passing down through the ore body oxidizes the ore minerals, dissolves them, and carries the dissolved ions out of the rock. (b) The ore then reprecipitates just below the water table (the level at which pores in the rock are filled with water, where chemical conditions are less oxidizing).

gen, evolved into the oxygen-rich atmosphere we breathe today. This change affected the chemistry of sea water and led to the precipitation of iron-oxide minerals that settled as sediment on the sea floor. The resulting iron-rich sedimentary layers are known as banded-iron formation (BIF) (►Fig. 12.31a), because after lithification they consist of alternating beds of gray iron oxide (magnetite or hematite) and red beds of jasper (iron-rich chert).

The chemistry of seawater in some parts of the ocean today leads to the deposition of manganese-oxide minerals on the sea floor. These minerals grow into lumpy accumula-

tions known as manganese nodules (►Fig. 12.31b). Mining companies have begun to explore technologies for vacuuming up these nodules.

Residual mineral deposits. Recall from Chapter 5 that as rainwater sinks into the Earth, it leaches (dissolves) certain elements and leaves behind others, as part of the process of forming soil. In rainy tropical environments, the residuum left behind in soils after leaching includes concentrations of iron or aluminum. Locally, these metals become so concentrated that the soil itself becomes an ore deposit (►Fig. 12.32).

FIGURE 12.31 (a) Precambrian banded-iron formation near Ishpiming, Michigan. (b) Manganese nodules on the sea floor, viewed from above. Each nodule is a few centimeters in diameter.

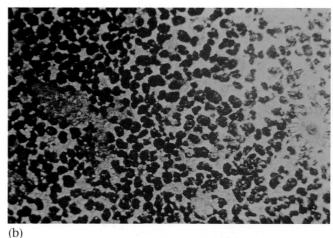

(a)

(b)

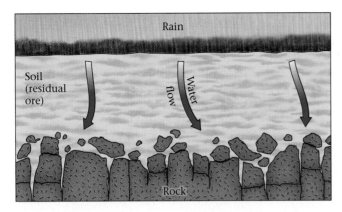

FIGURE 12.32 When rainwater sinks through the soil, it dissolves and removes many elements. A thick soil forms, containing a residuum of iron or aluminum.

We refer to such deposits as residual mineral deposits. Most of the aluminum ore mined today comes from bauxite, a residual mineral deposit created by the extreme leaching of granite.

Placer deposits. Ore deposits may develop when rocks containing native metals, such as gold, erode and create a mixture of sand grains and metal flakes or nuggets (pebble-sized fragments). The heavy metal grains accumulate in sand or gravel bars along the course of rivers, for the moving water carries away lighter mineral grains but can't move the metal grains as easily. Concentrations of metal grains in stream sediments are a type of placer deposit (▶Fig. 12.33). Panning further concentrates gold flakes or nuggets—

FIGURE 12.33 The process of forming a placer deposit. Ore-bearing rock is eroded, and clasts containing native metals fall into a stream. Sorting by the stream concentrates the metals.

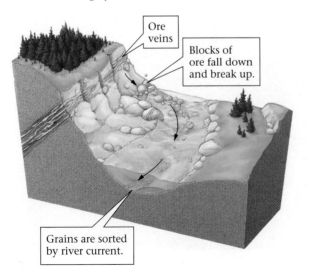

Ore veins

Blocks of ore fall down and break up.

Grains are sorted by river current.

swirling water in a pan causes the lighter sand grains to wash away, leaving the gold behind. Placer deposits may eventually be buried and lithify to become part of a new sedimentary rock.

Where Are Ore Deposits Found?

The Inca empire of fifteenth-century Peru boasted elaborate cities and temples, decorated with fantastic masks, jewelry, and sculpture made of gold. Then, around 1532, Spanish conquistadors arrived, led by commanders who quipped, "We Spaniards suffer from a disease that only gold can cure." The Incas, already weakened by civil war, were no match for the armor-clad Spaniards and their guns. Within six years the Inca empire had vanished and Spanish ships were transporting Inca treasure back to Spain. Why did the Incas possess so much gold? Or to ask the broader question, what geologic factors control the distribution of ore? We can find the answer once again by considering the consequences of plate tectonics.

Several of the ore-deposit types mentioned in this chapter occur in association with igneous rocks. As we learned in Chapter 4, igneous activity does not happen randomly around the Earth, but rather concentrates at convergent plate boundaries (in the overriding plate of a subduction zone), at divergent plate boundaries (along mid-ocean ridges), at continental rifts, or at hot spots. Thus, magmatic and hydrothermal deposits (and secondary enrichment deposits derived from these) occur along plate boundaries, along rifts, or at hot spots. Placer deposits are typically found in the sediments eroded from magmatic or hydrothermal deposits.

Consider the Inca gold. The Incas lived in the Andes Mountains, which had formed as a result of compression and volcanic activity where the Pacific Ocean floor subducts beneath the South American Plate. As the mountains rose, erosion stripped away surface rocks to expose the large granite plutons that had intruded into the continental crust beneath. The magma that froze to make the granite brought gold, copper, and silver atoms with it. Some of the gold precipitated along with quartz to form veins in the plutons. Inca miners quarried these veins and separated the gold, or panned for gold in the streams choked with sediment eroded from the plutons.

Some ore deposits are not a direct result of plate tectonics activity, and thus are not directly associated with plate boundaries. For example, banded iron formed along passive continental margins during the Precambrian—its occurrence today reflects the present distribution of Precambrian rocks, and thus most major exposures occur in shield areas of continents. Bauxite forms where that aluminum-rich bedrock occurs, thick soils form, and extreme leaching takes place. Thus, many bauxite deposits form on granite bedrock in stable continental areas that now lie in tropical regions.

12.13 ORE-MINERAL EXPLORATION AND PRODUCTION

Imagine an old prospector clanking through the desert with a broken-down donkey, eyeing the hillsides for "shows" of ore (exposures of ore minerals at the ground surface). If he finds a show of minerals, he pries out chunks of the rock with a pick, and the poor donkey hauls the rock back to town for an assay, a test to determine how much extractable metal the rock contains. Mining laws permitted a prospector to, literally, "stake a claim" by marking off an area of ground with stakes. The prospector would then have the exclusive right to dig up ore at that spot and sell it. What does a show look like? Typically, prospectors looked for milky-white quartz veins and/or exposures in which rocks were stained by the oxidation of metal-containing minerals (▶Fig. 12.34).

These days, large mining companies employ geologists to survey potential ore-bearing regions systematically. The geologists focus their studies on rocks that developed in settings appropriate for ore formation. Once such a region has been identified, they measure the local strength of Earth's magnetic field and the local pull of gravity. These measurements lead them to ore bodies, because ore minerals tend to be denser and more magnetic than average rocks. Geologists also sample rocks and soils to test for metal content, and may even analyze plants in the area to detect traces of metals, for plants absorb metals through their roots. Once geologists have identified a possible ore deposit, they drill holes to sample subsurface rock and to determine the ore deposit's shape and extent (▶Fig. 12.35). Ore-mineral exploration takes geologists into jungles, deserts, and tundras worldwide.

If calculations show that the mining of an ore deposit will yield a profit, and if environmental concerns can be accommodated, a company builds a mine. Mines can be below or above ground, depending on how close the ore deposit is to the surface. To make an open-pit mine (see chapter opening photo), workers first drill a series of holes into the solid bedrock and then fill the holes with high explosives. They must space the holes carefully and must set off the charges in a precise sequence, so that the bedrock shatters into appropriate-sized blocks for handling. When the dust settles, large front-end loaders dump the ore into giant trucks. The trucks transport waste rock (rock that

FIGURE 12.34 Stained rock is an indicator of ore. The stain develops when ore reacts with air and water. The photo encompasses about 1 m of section.

FIGURE 12.35 The three-dimensional shape of an ore body underground, and the workings that miners dig to access the ore body. Note that shafts are vertical and tunnels are horizontal.

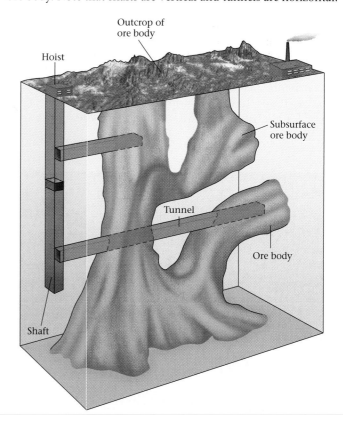

doesn't contain ore) to a tailings pile and the ore to a crusher, a giant set of moving steel jaws that smash rock into small fragments. Workers then separate ore minerals from other minerals and send the ore-mineral concentrate to a processing plant, where the ore undergoes smelting or treatment with acidic solutions to separate metal atoms from other atoms.

If the ore deposit lies more than about a hundred meters below the Earth's surface, miners must make an underground mine. To do so, miners sink a vertical shaft in which they install an elevator. At the level in the crust where the ore body appears, they build a maze of tunnels into the ore by drilling holes into the rock and then blasting. The rock removed must be carried back to the surface. Rock columns between the tunnels hold up the ceiling of the mine.

12.14 NONMETALLIC MINERAL RESOURCES

So far this chapter has focused on resources that contain metal. But society uses many other geological materials, sometimes called industrial minerals, as well. From the ground we get the stone used to make roadbeds and buildings, the chemicals for fertilizers, the gypsum in drywall, the salt filling salt shakers, and the sand used to make glass—the list is endless. This section looks at a few of these geological materials and explains where they come from.

Dimension Stone

The Parthenon, a colossal stone temple supported by forty-six carved columns, has stood atop a hill overlooking the city of Athens for almost 2,500 years. No wonder—stone, rock used for building and other practical purposes, outlasts nearly all other materials. We use stone to make facades, roofs, curbs and steps, and countertops and floors. The names that architects give to various types of stone may differ from the formal names used by geologists. For example, architects refer to any polished carbonate rock as marble, whether or not it has been metamorphosed. Likewise, they refer to any rock containing feldspar and quartz as granite, regardless of whether the rock has an igneous or a metamorphic texture.

To obtain intact slabs and blocks of rock (granite or marble)—known as dimension stone in the trade—for architectural purposes, workers carefully cut it out of the walls of quarries (▶Fig. 12.36a). Note that a quarry provides stone, whereas a mine supplies ore. To cut stone slabs, quarry operators split blocks from bedrock by hammering a series of wedges into the rock, or cut it off bedrock by using a wireline saw, a thermal lance, or a water jet. A wireline saw consists of a loop of braided wire moving between two pulleys. The movement of the wire drags the abrasive along the rock and grinds a slice into it. A thermal lance looks like a long blowtorch—a flame of burning diesel fuel stoked by high-pressure air pulverizes rock, and thereby cuts a slot. More recently, quarry operators have begun to use an abrasive water jet, which squirts out water and abrasives at very high pressure, to cut rock.

Crushed Stone and Concrete

Crushed stone forms the substrate of highways and railroads and is the raw material for manufacturing cement, concrete, and asphalt. In crushed-stone quarries (▶Fig. 12.36b), operators use high explosives to break up bedrock into rubble that they then transport by truck to a jaw crusher, which reduces the rubble into usable-size fractions.

Much modern construction utilizes mortar and concrete (▶Box 12.2), human-made rocklike materials formed

FIGURE 12.36 (a) An active quarrying operation, showing large blocks of cut stone; (b) a crushed-limestone quarry.

(a)

(b)

when a slurry composed of sand and/or gravel mixed with cement and water is allowed to harden. The hardening takes place when a complex assemblage of minerals grows by chemical reactions in the slurry; these minerals bind together the grains of sand or gravel in mortar or concrete. (Note that the word *mortar* refers to the substance that holds bricks or stone blocks together, whereas *concrete* refers to the substance that workers shape into roads or walls by spreading it out into a layer or by pouring it into a form.) The **cement** in mortar or concrete starts out as a powder composed of lime (CaO), quartz (SiO_2), aluminum oxide (Al_2O_3), and iron oxide (Fe_2O_3). Typically, lime accounts for 66% of cement, silica for 25%, and the remaining chemicals for about 9%.

It appears that the ancient Romans were the first to use cement—they made it from a mixture of volcanic glass and limestone. During recent centuries, most cement has been produced by heating specific types of limestone (which happened to contain calcite, clay, and quartz in the correct proportions) in a kiln up to a temperature of about 1,450°C; the heating releases CO_2 gas and produces "clinker," chunks consisting of lime and other oxide compounds. Manufacturers crush the clinker into cement powder and pack it in bags for transport. But natural limestone with the exact composition of cement is fairly rare, so most cement used today is Portland cement, made by intentionally mixing limestone, sandstone, and shale in just the right proportions to provide the proper chemical makeup. Isaac Johnson, an English engineer, came up with the recipe for Portland cement in 1844, and he named it after the town of Portland, England, because he thought it resembled rock exposed there.

Nonmetallic Minerals in Your Home

We use an astounding variety of nonmetallic geologic resources (▶Table 12.1) without ever realizing where they come from. Consider the materials in a house or apartment. The concrete foundation consists of cement, made from limestone mixed with sand or gravel. The bricks in the exterior walls originated as clay, formed from the chemical weathering of silicate rocks and perhaps dug from the floodplain of a stream. To make bricks, workers mold wet clay into blocks, which they then bake. Baking drives out water and causes metamorphic reactions that recrystallize the clay.

The glass used to glaze windows consists largely of silica, formed by first melting and then freezing pure quartz sand from a beach deposit or a sandstone formation. Gypsum board (drywall), used to construct interior walls, is a sandwich of gypsum powder between sheets of paper. Gypsum ($CaSO_4 \cdot 2H_2O$) occurs in evaporite strata precipitated from seawater or saline lake water. Evaporites provide other useful minerals as well, such as halite and borax. Truly, without the geologic resources of the Earth, modern society would grind to a halt.

12.15 GLOBAL MINERAL NEEDS

How Long Will Resources Last?

The average citizen of an industrialized country uses 25 kilograms (kg) of aluminum, 10 kg of copper, and 550 kg of iron and steel in a year's time (▶Fig. 12.37). If you combine

---

The Sidewalks of New York

BOX 12.2
THE HUMAN ANGLE

Untold tons of concrete have gone into the construction of New York City. In fact, with the exception of a few city parks, most of the walking space in the city consists of concrete. And concrete skyscrapers tower above this concrete plain. Where does all this concrete come from?

Much of the sand used in New York concrete was deposited during the last ice age. As vast glaciers moved southward over 14,000 years ago, they ground away the igneous and metamorphic rocks that constituted central and eastern Canada. These ancient rocks contained abundant quartz, and since quartz lasts a long time (it does not undergo chemical weathering easily), the sediment transported by the glaciers retained a large amount of quartz. Glaciers deposited this sediment in huge piles. As the glaciers melted, fast-moving rivers of meltwater washed the sediment, sorting sand from mud and pebbles. The sand was deposited by the meltwater rivers.

What about the cement? Cement contains a mixture of lime, derived from limestone, and other elements derived from shale and sandstone. The bedrock of New York, though, consists largely of schist and gneiss, not sedimentary rocks. Fortunately, a source of rocks appropriate for making cement lies up the Hudson River. A rock unit called the Rosendale Formation, which naturally contains exactly the right mixture of lime and silica needed to make a durable cement, crops out in low ridges just to the west of the river. Beginning in the late 1820s, workers began quarrying the Rosendale Formation for cement, creating a network of underground caverns. They then dumped the excavated rock into nearby kilns and roasted it to produce cement. The resulting powder was packed into barrels, loaded onto barges, and shipped downriver to New York. As demand for cement increased, operators eventually dug open-pit quarries from which they excavated other limestone and shale units, mixing them together in the correct proportion to make Portland cement.

The rocks making up limestone consist of cemented-together shell fragments and small, reeflike colonies of organisms. In other words, the lime in the concrete of New York sidewalks was originally extracted from seawater by living organisms over 350 million years ago.

TABLE 12.1 Common Nonmetallic Resources

Limestone	Sedimentary rock made of calcite; used for gravel or cement.
Crushed stone	Any variety of coherent rock (limestone, quartzite, granite, gneiss).
Siltstone	Beds of sedimentary rock, used to make flagstone.
Granite	Coarse igneous rock, used for dimension stone.
Marble	Metamorphosed limestone; used for dimension stone.
Slate	Metamorphosed shale; used for roofing shingles.
Gypsum	A sulfate salt precipitated from saltwater; used for wallboard.
Phosphate	From the mineral apatite; used for fertilizer.
Pumice	Frothy volcanic rock; used to decorate gardens and paths.
Clay	Very fine micalike mineral in sediment; used to make bricks or pottery.
Sand	From sandstone, beaches, or riverbeds; quartz sand is used for construction and for making glass.
Salt	From the mineral halite, formed by evaporating saltwater; used for food, melting ice on roads.
Sulfur	Occurs either as native sulfur, typically above salt domes, or in sulfide minerals; used for fertilizer and chemicals.

these figures with the quantities of energy resources and nonmetallic geologic resources a person uses, you get a total of about 15,000 kg (15 metric tons) of resources used per capita each year. Thus, the population of the United States consumes about 4 billion metric tons of geologic material per year. To supply this demand, workers must mine, quarry, or pump 18 billion metric tons of rock per year. By comparison, the Mississippi River transports 190 million metric tons of sediment per year into the Gulf of Mexico.

Mineral resources, like oil and coal, are nonrenewable resources. Once mined, an ore deposit or a limestone hill disappears forever. Natural geologic processes do not happen fast enough to replace the deposits as quickly as we use them.

Geologists have calculated reserves for various mineral deposits just as they have for oil. Judging from current definitions of reserves (which depend on today's prices) and rates of consumption, supplies of some metals may run out in only decades to centuries (▶Table 12.2). But these estimates may change as supplies become depleted and prices rise, making previously uneconomical deposits worth mining. And supplies could increase if geologists discover new reserves or if new mining techniques are developed. Further, increased efforts at conservation and recycling can cause a dramatic decrease in rates of consumption, and thereby stretch the lifetime of existing reserves.

Ore deposits do not occur everywhere, because their formation requires special geological conditions. As a result, some countries possess vast supplies, whereas others have none. In fact, no single country owns all the mineral resources it needs, so nations must trade with one another to maintain supplies, and global politics inevitably affects prices. Many wars have their roots in competition for mineral reserves, and it is no surprise that the outcomes of some wars have hinged on who controls these reserves.

Mining and the Environment

Mining leaves a big footprint in the Earth System. Some of the gaping holes that open-pit mining creates in the landscape have become so big that astronauts can see them from

FIGURE 12.37 Vast quantities of mineral resources are consumed in a year, as the diagram indicates. The numbers indicate the weight of the material used per person per year in an industrialized country.

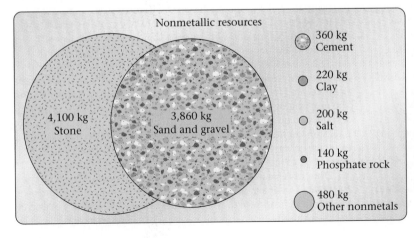

Nonmetallic resources

4,100 kg Stone

3,860 kg Sand and gravel

360 kg Cement

220 kg Clay

200 kg Salt

140 kg Phosphate rock

480 kg Other nonmetals

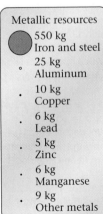

Metallic resources

550 kg Iron and steel

25 kg Aluminum

10 kg Copper

6 kg Lead

5 kg Zinc

6 kg Manganese

9 kg Other metals

TABLE 12.2 Lifetimes of Currently Known Ore Resources (in years)

Metal	World Resources	U.S. Resources
Iron	120	40
Aluminum	330	2
Copper	65	40
Lead	20	40
Zinc	30	25
Gold	30	20
Platinum	45	1
Nickel	75	less than 1
Cobalt	50	less than 1
Manganese	70	0
Chromium	75	0

space. Both open-pit and underground mining yield immense quantities of waste rock, which miners dump in tailings piles, some of which grow into artificial hills 200 meters high and many kilometers long. Lacking soil, tailings piles tend to remain unvegetated for a long time. Mining also exposes ore-bearing rock to the atmosphere, and since many ore minerals are sulfides, they react with rainwater to produce acid mine runoff, which can severely damage vegetation downstream (▶Fig. 12.38a). Ore processing tends to release noxious chemicals that can mix with rain and spread over the countryside, damaging ecosystems. Before the installation of modern environmental controls, smoke from ore-processing plants caused severe air pollution; plumes of smoke from the old smelters in Sudbury, Ontario, for example, created a wasteland for many kilometers downwind (▶Fig. 12.38b). In recent years, people have made efforts to reclaim mining spoils, and new technologies have been developed to extract metals in ways that are less deleterious to the environment and that treat waste more efficiently.

FIGURE 12.38 (a) The orange color in this acid mine runoff comes from dissolved iron in the water. (b) This unvegetated zone near Sudbury, Ontario, developed in response to acidic smelter smoke. A large tailings pile can be seen in the distance. Once the tall smoke stack came into use, the smoke blew further away, and the land in the foreground became revegetated.

(a)

(b)

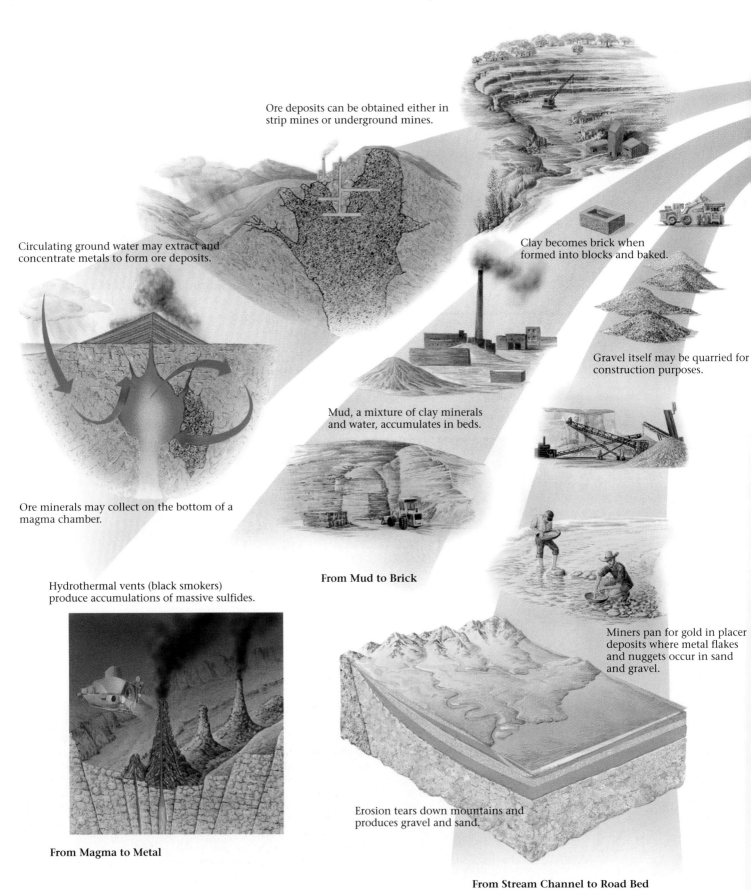

Ore deposits can be obtained either in strip mines or underground mines.

Circulating ground water may extract and concentrate metals to form ore deposits.

Clay becomes brick when formed into blocks and baked.

Gravel itself may be quarried for construction purposes.

Mud, a mixture of clay minerals and water, accumulates in beds.

Ore minerals may collect on the bottom of a magma chamber.

From Mud to Brick

Hydrothermal vents (black smokers) produce accumulations of massive sulfides.

Miners pan for gold in placer deposits where metal flakes and nuggets occur in sand and gravel.

Erosion tears down mountains and produces gravel and sand.

From Magma to Metal

From Stream Channel to Road Bed

Mining and processing ore has environmental consequences, including acid runoff, acid rain, and groundwater contamination.

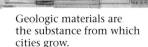

Geologic materials are the substance from which cities grow.

A mixture of lime, other elements, sand, and water, when allowed to harden, becomes concrete.

Mixed with water, spread into sheets, and wrapped in paper, gypsum makes wall board.

In quarries, operators dig up gypsum, crush it to powder, and ship it to factories.

Quarries extract limestone. Some becomes building stone, some crushed stone, and some is heated in a kiln to become lime.

Gypsum is a salt that precipitates when saline lakes evaporate. It grows as white or clear crystals.

Over millions of years, shells and shell fragments collect and eventually form beds of limestone.

From Lake Bed to Wall Board

Forming and Processing Earth's Mineral Resources

The raw material from which we manufacture the buildings, roads, wires, and coins of modern society were produced by geologic processes. For example, ore deposits, the concentrations of minerals that serve as a source for metal, form during a variety of magmatic or sedimentary processes. Limestone, a rock used for buildings and for making concrete, began as an accumulation of seashells. Brick began as clay, a byproduct of chemical weathering. And the gypsum of wall board began as an accumulation of salt along a desert lake. Metal, gravel, lime, and gypsum are all examples of Earth's mineral resources. We can use some mineral resources right from the Earth, simply by digging them out. But most only become usable after expensive processing.

Organisms extract ions from water and construct shells.

From Sea Floor to Sidewalk

CHAPTER SUMMARY

• Energy comes from a variety of sources: directly from the Sun; from tides, flowing water, or wind; from chemical reactions; from nuclear fission; and from Earth's internal heat. Plants use photosynthesis to store energy. Buried organic matter becomes fossil fuel.

• Oil and gas are hydrocarbons. The viscosity and volatility of a hydrocarbon depend on the length of its molecules.

• Oil and gas form from the bodies of plankton and algae, which settle out in a quiet-water, oxygen-poor depositional environment and form black organic shale. Later, chemical reactions at elevated temperatures convert the organic matter into kerogen, then oil.

• In order to create a usable oil reserve, oil must migrate from a source rock into a reservoir rock. Unless the reservoir rock is overlain by an impermeable seal rock, the oil will escape to the ground surface. The subsurface configuration of strata that leads to the entrapment of oil is called an oil trap.

• Substantial volumes of hydrocarbons exist in tar sand, oil shale, and gas hydrate.

• For coal to form, plant material must be deposited in an oxygen-poor environment so that it does not decompose. Compaction creates peat, which, when buried deeply and heated, transforms into coal.

• Coal is classified into ranks, based on the amount of carbon it contains. Coal occurs in beds, interlayered with other sedimentary rocks, and can be mined by either strip mining or underground mining.

• Coal-bed methane and coal gasification provide additional sources of energy.

• Nuclear power plants generate energy by using the heat released by the fission of uranium. The heat turns water into steam, and the steam drives turbines. Some economic uranium deposits occur as veins in igneous rock bodies, some are found in sedimentary beds.

• Nuclear reactors must be carefully controlled to avoid overheating or meltdown. The disposal of radioactive nuclear waste can create environmental problems.

• Geothermal energy uses Earth's internal heat to transform groundwater into steam that drives turbines; hydroelectric power plants use the potential energy of water; and solar cells convert sunlight to electricity.

• We now live in the Oil Age, but oil supplies may last only for another century.

• Use of energy resources has many negative environmental consequences.

• Industrial societies use many types of minerals, all of which must be extracted from the upper crust. We distinguish two general categories: metallic resources and nonmetallic resources.

• Metals come from ore. An ore is a rock containing native metals or ore minerals (minerals with a high proportion of metal) in sufficient quantities to be worth mining. An ore deposit is an accumulation of ore.

• Magmatic deposits form when sulfide ore minerals settle to the floor of a magma chamber. In hydrothermal deposits, ore minerals precipitate from hot-water solutions. Secondary-enrichment deposits form when groundwater carries metals away from a preexisting deposit. MVT deposits precipitate from groundwater that has passed long distances through the crust. Sedimentary deposits precipitate out of the ocean. Residual mineral deposits in soil are the result of severe leaching in tropical climates. Placer deposits develop when heavy metal grains accumulate in sediment along a stream.

• Many ore deposits are associated with igneous activity in subduction zones, along mid-ocean ridges, along continental rifts, or at hot spots.

• Nonmetallic resources include dimension stone for decorative purposes, crushed stone for cement and asphalt production, clay for brick making, sand for glass production, and many others. A large proportion of materials in your home have a geological ancestry.

• Mineral resources are nonrenewable. Many are now or may soon become in short supply.

KEY TERMS

cement (p. 357)
chain reaction (p. 342)
coal (p. 338)
coal gasification (p. 341)
coal rank (p. 339)
coalbed methane (p. 340)
energy (p. 329)
energy resource (p. 330)
fossil fuel (p. 330)
gas hydrate (p. 337)
geothermal energy (pp. 330, 343)
greenhouse effect (p. 347)
hydrocarbon (pp. 330–31)
hydrocarbon reserve (pp. 330–31)
kerogen (p. 331)
mineral resource (p. 330)
meltdown (p. 343)
metals (p. 350)

nuclear reactor (p. 342)
nuclear waste (p. 343)
Oil Age (p. 346)
oil shale (p. 331)
ore (p. 351)
ore deposit (p. 351)
peat (p. 338)
permeability (p. 333)
photosynthesis (p. 330)
pore (p. 333)
porosity (p. 333)
reservoir rock (p. 333)
resource (p. 329)
salt dome (p. 335)
source rock (p. 331)
tar sand (p. 337)
trap (p. 334)

REVIEW QUESTIONS

1. What are the fundamental sources of energy?

2. What is the source of the organic material in oil?

3. What is the oil window, and why does oil form only there?

4. How is organic matter trapped and transformed to create an oil reserve?

5. What are tar sand and oil shale, and how can oil be extracted from them?

6. What are gas hydrates, and where do they occur?

7. Where is most of the world's oil found? At present rates of consumption, how long will it last?

8. How is coal formed?

9. Explain how coal is transformed in rank from peat to anthracite.

10. Describe the operation of a nuclear reactor.

11. Where does uranium form in the Earth's crust? Where does it usually accumulate in minable quantities?

12. What are some of the drawbacks of nuclear energy?

13. What is geothermal energy? Why is it not more widely used?

14. What is the difference between a renewable and a nonrenewable resource?

15. What is the likely future of hydrocarbon production and use in the twenty-first century?

16. Why don't we use an average granite as a source for useful metals?

17. Describe various kinds of economic mineral deposits.

18. What procedures are used to locate and mine mineral resources today?

19. How is stone cut from a quarry?

20. What are the ingredients of cement? How is Portland cement made?

21. Will the supply of mineral resources run out? Can a country survive without importing minerals?

22. What are some environmental hazards of large-scale mining?

SUGGESTED READING

Brands, H. W. 2002. *The Age of Gold: The California Gold Rush and the New American Dream.* New York: Doubleday.

Cassedy, E. S., and P. Z. Grossman. 1998. *Introduction to Energy: Resources, Technology, and Society.* 2nd ed. Cambridge, UK: Cambridge University Press.

Conaway, C. F. 1999. *The Petroleum Industry: A Nontechnical Guide.* Tulsa, Okla.: PennWell.

Craddock, P., and J. Lang, eds. 2003. *Mining and Metal Production through the Ages.* London: British Museum Publications.

Craig, J. R., D. J. Vaughan, and B. J. Skinner. 1989. *Resources of the Earth.* Englewood Cliffs, N.J.: Prentice-Hall.

Deffeyes, K. S. 2003. *Hubbert's Peak: The Impending World Oil Shortage.* Princeton, N.J.: Princeton University Press.

Evans, A. M. 1992. *Ore Geology and Industrial Minerals: An Introduction.* 3rd ed. Malden, Mass.: Blackwell Science.

Evans, A. M. 1997. *An Introduction to Economic Geology and Its Environmental Impact.* Malden, Mass.: Blackwell Science.

Freese, B. 2004. *Coal: A Human History.* New York: Penguin USA.

Gluyas, J. G., and R. Swarbrick. 2003. *Petroleum Geoscience.* Malden, Mass.: Blackwell Science.

Guilbert, J. M., and C. F. Park, Jr. 1986. *The Geology of Ore Deposits.* New York: Freeman.

Hyne, N. J. 2001. *Nontechnical Guide to Petroleum Geology, Exploration, Drilling and Production.* 2nd ed. Tulsa, Okla.: PennWell.

Kesler, S. E. 1994. *Mineral Resources, Economics, and the Environment.* New York: Macmillan.

Manning, D. A. C. 1995. *Introduction to Industrial Minerals.* London: Chapman & Hall.

Schobert, H. H. 2001. *Energy & Society: An Introduction.* London, UK: Taylor & Francis.

Selley, R. C. 1997. *Elements of Petroleum Geology.* 2nd ed. New York: Academic Press.

Thomas, L. 2002. *Coal Geology.* New York: Wiley.

Yergin, D. 1993. *The Prize: The Epic Quest for Oil, Money and Power.* New York: Free Press.

THE VIEW FROM ABOVE A large limestone quarry cut into a forested landscape. The material gouged out of the ground at this location has been transferred elsewhere to make roadbeds and buildings. Use of mineral resources does make a mark on the landscape.

An Introduction to Landscapes, and the Hydrologic Cycle

This ***digital elevation map (DEM)*** *shows a portion of Oahu (Hawaii), oriented so that we are viewing the island obliquely from the south. The map was produced using data from satellites that bounce radar beams off the Earth's surface. Formation of this landscape first required the land surface to rise above sea level. Then, streams eroded numerous valleys. Erosion produced sediments that accumulated along the coast, building the flat land on which people settled.*

E.1 INTRODUCTION

It's no wonder that artists and writers across the ages have sought inspiration from the **landscape**—the character and shape of the land surface in a region—for landscapes display a diversity that rivals that of human emotion (►Fig. E.1a–d). Geologists, like artists and writers, savor the impression of a dramatic landscape, but on seeing one, they can't help but ask, "How did it come to be, and how will it change in the future?"

The subject of landscape development and evolution, and the **landforms** (individual shapes such as mesas, valleys, cliffs, and dunes) that constitute it, dominates many of the subsequent chapters of this book. This interlude, a general introduction to the topic, explains the driving forces behind landscape development, and identifies factors that control which landscape develops in a given locality. We also use this opportunity to describe the hydrologic cycle, the pathway water molecules follow as they move from ocean to air to land and back to ocean. We discuss the hydrologic cycle here because so many processes on or near Earth's surface involve water in its various forms.

E.2 SHAPING THE EARTH'S SURFACE

If the Earth's surface were totally flat, the great diversity of landscapes that embellish our vistas would not exist. But the surface isn't flat, because a variety of geologic processes cause portions of the surface to move up or down relative to adjacent regions. We refer to the upward movement of the land surface as **uplift,** and the sinking or downward movement of the land surface as **subsidence.**

When uplift or subsidence takes place, the elevation difference does not remain the same forever, because other components of the Earth System kick into action—material at higher elevations becomes unstable and susceptible to downslope movement (the tumbling or sliding of rock and sediment from higher elevations to lower ones); moving water, ice, and air cause erosion (the grinding away and removal of the Earth's surface); and where moving fluids slow down, deposition of sediment takes place. Downslope movement, erosion, and deposition redistribute rock and sediment, ultimately stripping it in from higher areas and collecting it in low areas. As a consequence, a great variety of both **erosional landforms** (those carved by erosion) and **depositional landforms** (those built from an accumulation of sediment) can form.

FIGURE E.1 Contrasting landscapes. (a) A tropical waterfall in Hawaii, where lush vegetation blunts the contours of the landscape, except where it's stripped clean by rushing water. (b) The desert ranges of the Mojave Desert, in southeastern California. Barren remnants of rocky cliffs protrude from fans of recently eroded and sparcely vegetated detritus. (c) The peaks of the Grand Tetons, in Wyoming. Fresh rock cliffs, carved by glaciers, penetrate the clouds. (d) A rock and sand seascape along the coast of Brazil, where the relentless pounding of Atlantic waves eats back into the land, grinds away at the massive "sugar loafs," and generates beach sand.

The energy that drives landscape evolution comes from three sources: **internal energy,** the heat within the Earth, which drives plate motions and mantle plumes that, in turn, cause displacement of the crust's surface; **external energy,** energy coming to the Earth from the Sun, which causes materials at the surface of the Earth to warm up; and **gravitational energy,** which pulls rock and water from higher to lower elevations. External energy and gravitation, working together, drive convective flow in the atmosphere and oceans and drive the hydrologic cycle. Landscape evolution, in fact, reflects a "battle" between (1) tectonic processes such as collision, convergence, and rifting, which create **relief** (an elevation difference between two locations) in an area, and (2) processes such as downslope movement, erosion, and deposition, which destroy relief by removing material from high areas and depositing it in low ones. If, in a particular region, the rate of uplift exceeds the rate of erosion, the land surface rises; if the rate of subsidence exceeds the rate of deposition, the land surface sinks. Without uplift and subsidence, Earth's surface would long ago have been beveled to a flat plain. Without erosion and deposition, high and low areas would have lasted for the entirety of Earth history.

How rapidly do uplift, subsidence, erosion, and deposition take place? The Earth's surface can rise or sink by as much as 3 m during a single major earthquake. But averaged over time, the rates of uplift and subsidence range between 0.01 and 10 mm per year (▶Fig. E.2a). Similarly, erosion can carve out several meters of **substrate,** the material just below the ground surface, during a single flood, storm, or landslide (▶Fig. E.2b). And deposition during a single event can create a layer of debris tens of meters thick in a matter of minutes to days. But averaged over time, erosional and depositional rates also vary between 0.10 and 10 mm per year. Although these rates seem small, a change in surface elevation of just 0.5 mm (the thickness of a fingernail) per year can yield a net change of 5 km in 10 million years. Uplift can build a mountain range, and erosion can whittle one down to near sea level—it just takes time!

E.3 FACTORS CONTROLLING LANDSCAPE DEVELOPMENT

Imagine traveling across a continent. On your journey, you pass plains, swamps, hills, valleys, mesas, and mountains. Some of these features are erosional landforms, in that they result from the breakdown and removal of rock or sediment and develop where **agents of erosion** such as water, wind, or ice carve into the substrate. (Of these three agents, water has the greatest effect on a global basis.) Other features are depositional landforms, in that they result from the deposition of sediment where the fluid carrying the sediment evaporates, slows down, or melts (see opening illustration). The specific landforms that develop at a given locality, which together make up the landscape, reflect several factors.

- *Eroding or transporting agent:* Water, ice, and wind all cause erosion and transport sediment. But the shapes of landforms formed by each are different, because of differences in the abilities of these agents to carve into the substrate and to carry debris.

- *Relief:* The elevation difference, or **relief,** between adjacent places in a landscape determines the height and steepness of slopes. Steepness, in turn, controls the velocity of ice or water flow and determines whether rock or soil stays in place or tumbles downslope.

- *Climate:* The average mean temperature and the volume of precipitation in a region determine whether running

(a)

(b)

FIGURE E.2 (a) Uplifted beach terraces along a coast appear where the coast is rising relative to sea level. Wave erosion eats into the land, creating a terrace. Eventually, the terrace rises out of the reach of the waves, which then cut a new one. (b) So much erosion can take place during a single hurricane that the foundations of houses built along the beach become undermined.

water, flowing ice, or wind serves as the main agent of erosion or deposition. The distribution of precipitation during the year also affects landscape development. For example, a short but torrential downpour may cause more erosion than a long-term light rain.

- *Substrate composition:* The material comprising the substrate determines how the substrate responds to erosion. For example, strong rocks can stand up to form steep cliffs, while soft sediment collapses to generate gentle slopes.

- *Life activity:* Some life activity weakens the substrate (by burrowing, wedging, or digesting), while some holds it together (by binding it with roots).

• *Time:* Landscapes evolve through time, in response to continued erosion and/or deposition. For instance, a gully that has just started to form in response to the flow of a stream does not look the same as a deep canyon that develops after the same stream has existed for a long time.

Although water, wind, and ice create most landscapes, human activities have had an increasingly important impact on the Earth's surface. We have dug pits (mines) where once there were mountains, built hills (tailings piles and landfills) where once there were valleys, and made steep slopes gentle and gentle slopes steep (▶Fig. E.3a–c). By constructing concrete walls, we modify the shapes of coastlines, change the courses of rivers, and fill new lakes (reservoirs). In cities, buildings and pavements completely seal the ground and cause water that might once have seeped down into the ground to spill into streams instead, increasing their flow. And in the country, agriculture, grazing, water usage, and deforestation substantially alter the rates at which natural erosion and deposition take place. For example, agriculture greatly increases the rate of erosion, because for much of the year farm fields have no vegetation cover.

E.4 THE HYDROLOGIC CYCLE

Nothing that is can pause or stay—
  The moon will wax, the moon will wane
  The mist and cloud will turn to rain,
  The rain to mist and cloud again,
  Tomorrow be today.
          —Henry Wadsworth Longfellow (1807–1882)

As is evident from the previous section, water in its various forms (liquid, gas, and solid) plays a major role in erosion and deposition on the Earth's surface. As we will see in succeeding chapters, it also modifies the upper crust underground, drives storms in the atmosphere, transports heat in the oceans, and hosts life everywhere. Our planet's water occupies several distinct reservoirs. Atmospheric water occurs as vapor, mist, or ice crystals. Surface water collects in oceans, lakes, streams, puddles, swamps, snowfields, and glaciers. Subsurface water dampens soil and rock near the surface, or sinks deeper to a realm where it fills pores and cracks. (The boundary defining the top of the subsurface realm in which water completely fills pores and cracks is called the **water table;** above the water table, pores and cracks contain some air. Geologists refer to subsurface water below the water table as **groundwater,** as discussed further in Chapter 16.) A significant amount of water resides in living organisms—about 50 to 65% of your body consists of water. The oceans contain 95% of Earth's free water (water not incorporated in mineral crystals), and thus make up the

(a)

(b)

(c)

FIGURE E.3 Human workings on a geological scale. (a) The pyramids of Egypt are essentially human-made hills that rise above the desert sands. They have lasted for thousands of years. (b) People routinely cut deep valleys through high ridges, such as this highway road cut in Colorado, to make a level grade for a road. (c) This stone dam holds back a reservoir in Colorado.

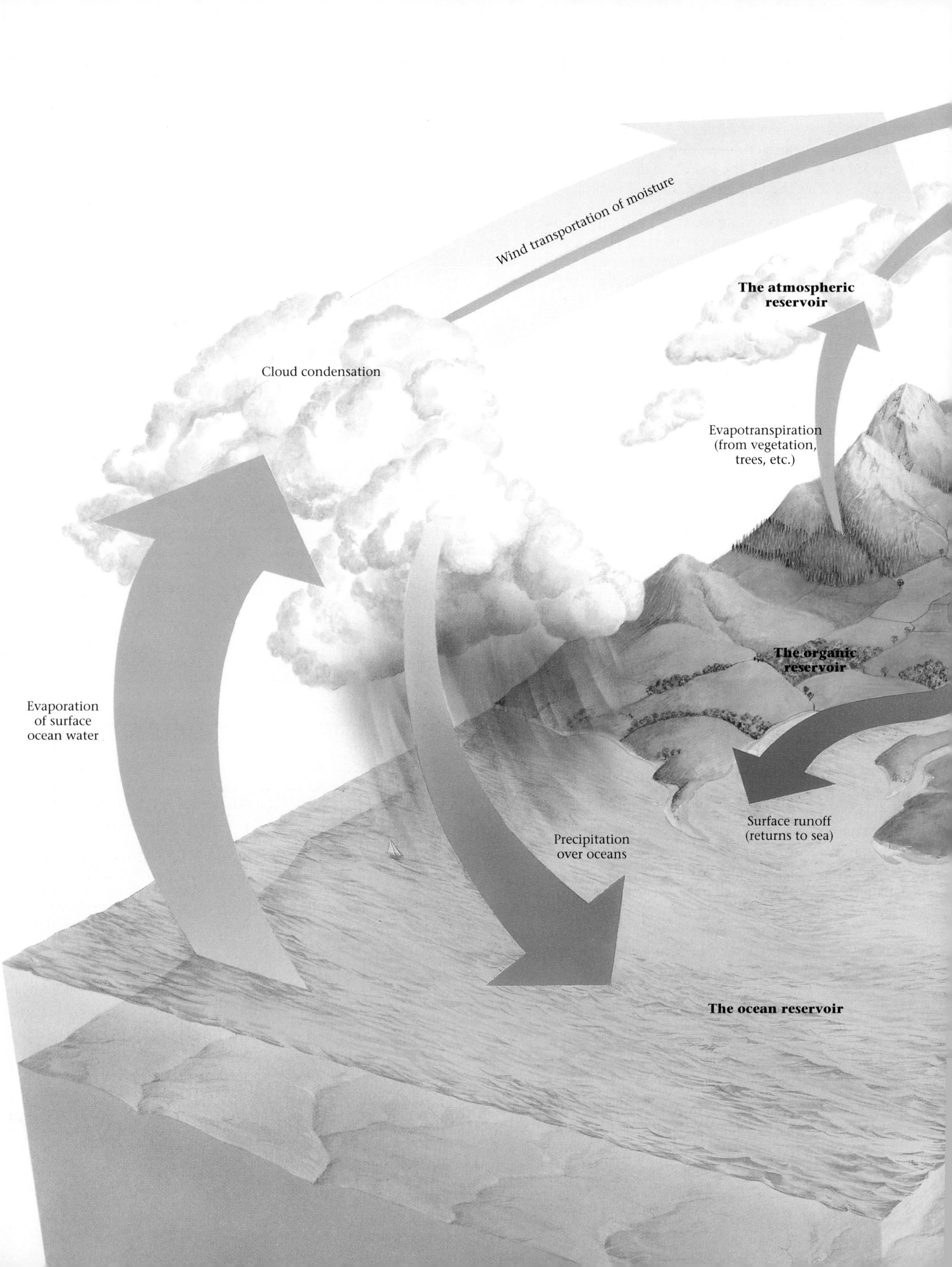

Wind transportation of moisture

Cloud condensation

The atmospheric reservoir

Evapotranspiration (from vegetation, trees, etc.)

The organic reservoir

Evaporation of surface ocean water

Surface runoff (returns to sea)

Precipitation over oceans

The ocean reservoir

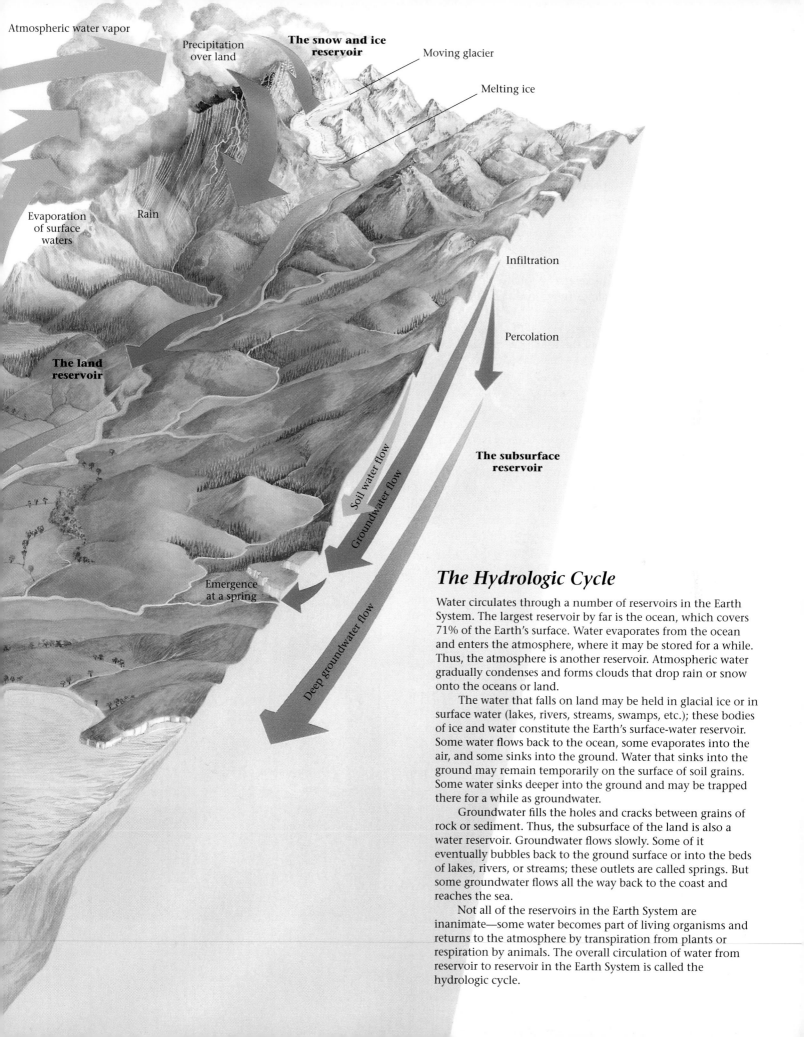

Atmospheric water vapor

Precipitation over land

The snow and ice reservoir

Moving glacier

Melting ice

Evaporation of surface waters

Rain

Infiltration

Percolation

The land reservoir

Soil water flow

Groundwater flow

The subsurface reservoir

Emergence at a spring

Deep groundwater flow

The Hydrologic Cycle

Water circulates through a number of reservoirs in the Earth System. The largest reservoir by far is the ocean, which covers 71% of the Earth's surface. Water evaporates from the ocean and enters the atmosphere, where it may be stored for a while. Thus, the atmosphere is another reservoir. Atmospheric water gradually condenses and forms clouds that drop rain or snow onto the oceans or land.

The water that falls on land may be held in glacial ice or in surface water (lakes, rivers, streams, swamps, etc.); these bodies of ice and water constitute the Earth's surface-water reservoir. Some water flows back to the ocean, some evaporates into the air, and some sinks into the ground. Water that sinks into the ground may remain temporarily on the surface of soil grains. Some water sinks deeper into the ground and may be trapped there for a while as groundwater.

Groundwater fills the holes and cracks between grains of rock or sediment. Thus, the subsurface of the land is also a water reservoir. Groundwater flows slowly. Some of it eventually bubbles back to the ground surface or into the beds of lakes, rivers, or streams; these outlets are called springs. But some groundwater flows all the way back to the coast and reaches the sea.

Not all of the reservoirs in the Earth System are inanimate—some water becomes part of living organisms and returns to the atmosphere by transpiration from plants or respiration by animals. The overall circulation of water from reservoir to reservoir in the Earth System is called the hydrologic cycle.

biggest reservoir. Water constantly flows from one reservoir to another, a never-ending process called the **hydrologic cycle** (see art on pp. 368–69). Perhaps without realizing it, Longfellow, an American poet fascinated with reincarnation, provided an accurate if somewhat romantic image of the hydrologic cycle.

What drives the hydrologic cycle? Simply put, the whole process takes place because energy from solar radiation bathes the Earth, and because gravity always pulls mass from higher elevations to lower elevations. To get a clearer sense of how the hydrologic cycle operates, let's follow the fate of seawater that has just reached the surface of the ocean. Solar radiation heats the water, and the increased thermal energy of the vibrating water molecules allows them to evaporate (break free from the liquid) and drift upward in a gaseous state to become part of the atmosphere. About 417,000 cubic km (102,000 cubic miles), or about 30% of the total ocean volume, evaporates every year. Atmospheric water vapor moves with the wind to higher elevations, where it cools, undergoes condensation (the molecules link together to form a liquid), and rains or snows. About 76% of this water precipitates (falls out of the air) directly back into the ocean. The remainder precipitates onto land. Most of this water becomes trapped temporarily in the soil, or in plants and animals, and soon returns directly to the atmosphere by evapotranspiration (the sum of evaporation from bodies of water, evaporation from the ground surface, and release of water from plants and animals). Rainwater that did not become trapped in the soil or in living organisms either enters lakes or rivers and ultimately flows back to the sea as surface water, becomes trapped in glaciers, or sinks deeper into the ground to become groundwater. Groundwater also flows and ultimately returns to the Earth's surface reservoirs. In sum, during the hydrologic cycle, water moves among the ocean, the atmosphere, reservoirs on or below the land, and living organisms.

The *average* length of time that water stays in a particular reservoir during the hydrologic cycle is called the residence time. Water in different reservoirs has different residence times. For example, a typical molecule of water remains in the oceans for 4,000 years or less, in lakes and ponds for 10 years or less, in rivers for 2 weeks or less, and in the atmosphere for 10 days or less. Groundwater residence times are highly variable and depend on how deep the groundwater flows. Water can stay underground for anywhere from 2 weeks to 10,000 years before it inevitably moves on to another reservoir.

E.5 LANDSCAPES OF OTHER PLANETS

The dynamic, ever-changing landscape on Earth contrasts markedly with those of other terrestrial planets. Each of the terrestrial planets and moons has its own unique surface landscape features, reflecting the interplay between the object's particular tectonic and erosional processes. Let's look at a few examples: the Moon, Mars, and Venus.

Our Moon has a static, pockmarked landscape generated exclusively by meteorite impacts and volcanic activity long ago. Because no plate tectonics occurs on the Moon, no new mountains form, and because no atmosphere or ocean exists, the Moon has no hydrologic cycle and undergoes no erosion from rivers, glaciers, or winds. Therefore, the lunar surface has remained largely unchanged for billions of years (▶Fig. E.4).

Landscapes on Mars differ from those of the Moon because Mars *does* have an atmosphere (though much less dense than that of Earth) whose winds generate huge dust storms, some of which obscure nearly the entire surface of the planet for months at a time. The landscapes of Mars also differ from the moon's because Mars probably once had surface water (▶Box E.1; ▶Fig. E.5). Thus, the Martian surface appears to have four kinds of materials: volcanic flows and deposits (primarily of basalt), debris from impacts, windblown sediment, and possibly water-laid sediment. Martian winds not only deposit sediment, they slowly erode impact craters and polish surface rocks.

Landscapes on Mars also differ from those on Earth, because Mars does not have plate tectonics. So, unlike Earth,

FIGURE E.4 This lunar landscape is virtually the same landscape that would have been visible 2 billion years ago, because the Moon's surface is static.

Water on Mars?

In 1877, an Italian astronomer named Giovanni Schiaparelli studied the surface of Mars with a telescope and announced that long, straight *canali* criss-crossed the planet's surface. *Canali* should have been translated into the English word "channel," but perhaps because of the recent construction of the Suez Canal, newspapers of the day translated the word into the English "canal," with the implication that the features had been constructed by intelligent beings. An eminent American astronomer began to study the "canals" and suggested that they had been built to carry water from polar ice caps to Martian deserts.

Late-twentieth-century satellite mapping of Mars showed that the "canals" do not exist—they were simply optical illusions. There are no lakes, oceans, rainstorms, or flowing rivers on the surface of Mars today. The atmosphere of Mars has such low density, and thus exerts so little pressure on the planet's surface, that any liquid water released at the surface in recent time would quickly evaporate. Thus, Mars has no hydrologic cycle the way the Earth does. But three crucial questions remain: Does liquid water ever form, even for short periods of time, on the Martian surface today? Did Mars ever have a significant amount of running water or standing water in the past? If the planet once had significant water, where is the water now? The question of the presence of water lies at the heart of the even more basic question: Given that the simplest life as we know it requires water, is there, or was there, life on Mars?

Many planetary geologists believe that the case for liquid water on Mars is quite strong. Much of the evidence comes from comparing landforms on the planet's surface with landforms of known origin on Earth. High-resolution images of Mars reveal a number of landforms that look as if they formed in response to flowing water. Examples include networks of channels resembling river networks on Earth (▶Fig. E.5), scour features, deep gullies, and streamlined deposits of sediment.

Studies by the *Odyssey* satellite in 2003, and by Mars rovers (*Spirit* and *Opportunity*) that landed on the planet in 2004, have added in-triguing new data to the debate. *Odyssey* detected hints that hydrogen, an element in water, exists beneath the surface of the planet over broad regions, and the Mars rovers have documented the existence of hematite and gypsum minerals that form in the presence of water. The rovers have also found sedimentary deposits that appear to have been deposited in water. Researchers speculate that Mars was much wetter in its past, perhaps billions of years ago. But since the atmosphere became less dense, the water evaporated and now lies hidden underground or trapped in polar ice caps.

FIGURE E.5 A riverlike network of channels on Mars, as photographed by a satellite. Small tributary channels join a large trunk channel.

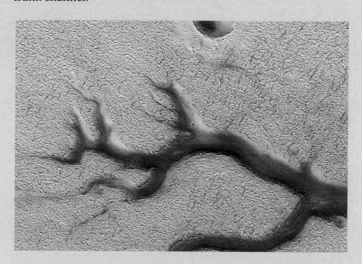

Mars has no mountain belts or volcanic arcs. In fact, most landscape features on Mars, with the exception of wind-related ones, are over three billion years old. Thermal activity also led to the eruption of gargantuan hot-spot volcanoes, such as the 22-km-high Olympus Mons (▶Fig. E.6). Mars also boasts the largest known canyon, the Valles Marineris, a gash over 3,000 km long and 8 km deep—no comparable feature exists on Earth. Because Mars has no vegetation and no longer has rain, its surface does not weather and erode like that of Earth, so it still bears the scars of impact by swarms of meteors earlier in the history of the solar system, scars that have long since disappeared on Earth.

Venus is closer in size to the Earth and may still have operating mantle plumes. Virtually the entire surface of Venus was covered by volcanic deposits about 300 to 1,600 Ma, making the planet's surface much younger than those of the Moon and Mars. Further, Venus has a dense atmosphere that protects it from impacts by smaller objects. Because there has been relatively little cratering since the resurfacing event, volcanic and tectonic features dominate the landscape of Venus (▶Fig. E.7). Satellites have used radar to reveal a variety of volcanic constructions (such as shield volcanoes, lava flows, calderas). Rifting on Venus produced faults. Liquid water cannot survive the scalding temperatures of Venus's surface, so no hydrologic cycle operates there and no life exists. Because of the density of the atmosphere, winds are too slow to cause much erosion or deposition.

After this brief side trip to other planets, let's now return to Earth. Our planet has the greatest diversity of landscapes in the solar system.

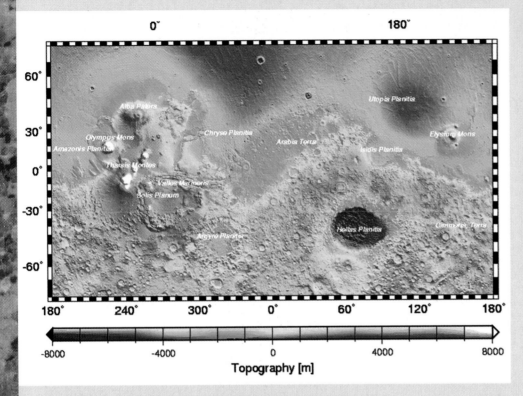

FIGURE E.6 A digital elevation model depicting the surface of Mars. Note the huge bulge of the Tharsis Ridge, the giant volcano of Olympic Mons, and the deep canyon of Valles Marineris.

FIGURE E.7 A radar map of the surface of Venus as it would appear if the atmosphere were removed. All these features are invisible to Earth-bound observers because of cloud cover. The colors represent elevation. The light tan is high, and the dark blue is low.

KEY TERMS

agents of erosion (p. 366)
depositional landforms (p. 364)
digital elevation map (DEM) (p. 364)
erosional landforms (p. 364)
external energy (p. 365)
gravitational energy (p. 365)
groundwater (p. 367)

hydrologic cycle (p. 370)
internal energy (p. 365)
landforms (p. 364)
landscape (p. 364)
relief (pp. 365, 366)
subsidence (p. 364)
substrate (p. 366)
uplift (p. 364)
water table (p. 367)

SUGGESTED READING

Ehlen, J. et al. (eds.) 2005. *Humans as Geologic Agents*. Boulder: Geological Society of America.

THE VIEW FROM SPACE: Satellites can provide images of the great variety of landscape features that decorate the Earth's surface. Textures, colors, and shapes all tell us about geologic processes happening at interfaces between the land, air, and sea. This image shows rugged hills, alluvial fans, sand dunes, and salt lakes in the Bogda Mountains and adjacent Turpan depression of China.

CHAPTER 13

Unsafe Ground: Landslides and Other Mass Movements

13.1 INTRODUCTION

It was Sunday, May 31, 1970, a market day, and thousands of people had crammed into the Andean town of Yungay, Peru, to shop. Suddenly they felt the jolt of an earthquake, strong enough to topple some masonry houses. But worse was to come. This earthquake also broke an 800-m-wide ice slab off the end of a glacier at the top of Nevado Huascarán, a nearby 6.6 km-high mountain peak. Gravity instantly pulled the ice slab 3.7 km down the mountain's steep slopes. As it tumbled, the ice disintegrated into a chaotic avalanche of chunks traveling at speeds of over 300 km per hour. Near the base of the mountain, most of the avalanche fed into a valley and thickened into a moving sheet as high as a ten-story building. Friction ripped up rocks and soil along the way and transformed the ice into water, which when mixed with rock and dust created 50 million cubic meters of mud, a slurry viscous enough to carry boulders larger than houses. This mass, sometimes floating on a compressed air cushion that allowed it to pass without disturbing the grass below, traveled over 14.5 km in less than four minutes.

At the mouth of the valley, most of the mass overran the village of Ranrahica before coming to rest, forming a dam that blocked the Santa

What goes up must come down. As a consequence, rock and regolith forming the substrate of hill slopes occasionally give way and slide downslope. The results can be disastrous, as happened when heavy rains caused mud to slip down a hill and flow over La Conchita, California, in 1995. Another mudslide occurred in 2005 (seen below), leading to several deaths.

373

(a)

(b)

FIGURE 13.1 (a) Before the May 1970 earthquake, the town of Yungay, Peru, perched on a hill within view of the ice-covered mountain Nevado Huascarán. (b) Three months after the earthquake, the town lay buried beneath debris. A landslide scar remains visible on the mountain.

River. But part of it shot up the sides of the valley and became airborne for several seconds, flying over the ridge bordering Yungay. As the town's inhabitants and visitors stumbled out of earthquake-damaged buildings, they heard a deafening roar and looked up to see the churning mud cloud bursting above the nearby ridge. The town was completely buried under several meters of mud and rock. When the dust had settled, only the top of the church and a few palm trees remained visible to show where Yungay once lay (►Fig. 13.1a, b). 18,000 people are forever entombed beneath the mass. Today, the site is a grassy meadow with a hummocky (irregular and lumpy) surface, spotted with crosses left by mourning relatives.

People often assume that the earth beneath them is *terra firma,* a solid foundation on which they can build

their lives. But the catastrophe at Yungay says otherwise—much of the Earth's surface is unsafe (unstable) ground, land capable of moving downslope in a matter of seconds to weeks. Geologists refer to the gravitationally caused downslope transport of rock, regolith (soil, sediment, and debris), snow, and ice as **mass movement,** or mass wasting. Like earthquakes, volcanic eruptions, storms, and floods, mass movements are a type of **natural hazard,** meaning a natural feature of the environment that can cause damage to living organisms and to buildings. Unfortunately, mass movement becomes more of a threat every year, because as the world's population grows, cities expand into areas of unsafe ground. But mass movement also plays a critical role in the rock cycle, for it's the first step in the transportation of sediment. And it plays a critical role in the evolution of landscapes, it's the most rapid means of modifying the shapes of slopes.

In this chapter, we look at the types, causes, and consequences of mass movement, and at the precautions society can take to protect people and property from its dangers. You might want to consider this information when selecting a site for your home or when voting on land-use propositions for your community.

13.2 TYPES OF MASS MOVEMENT

Though in everyday language people commonly refer to any mass-movement event as a "landslide," geologists and civil engineers tend to distinguish different types of mass movement based on four factors: the type of material involved (rock, regolith, or snow and ice), the velocity of the movement (fast, intermediate, or slow), the character of the moving mass (chaotic cloud, slurry, or coherent body), and the environment in which the movement takes place (subaerial or submarine). Below, we look at mass movements that occur on land, roughly in order from slow to very fast.

Creep and Solifluction

In temperate climates, the upper few centimeters of ground freeze during the winter, only to thaw again the following spring. Because water increases in volume by about 9.2% when it freezes, the water-saturated soil and underlying fractured rock expand outward, and particles in the regolith move out perpendicular to the slope during the winter. During the spring thaw, water becomes liquid again, and gravity makes the particles sink vertically and thus migrate downslope slightly. This gradual downslope movement of regolith is called **creep.** You can't see creep by staring at a hill slope because it occurs too slowly, but over a period of years, creep causes trees, fences, gravestones, walls, and foundations built on a hillside to tilt downslope. Notably,

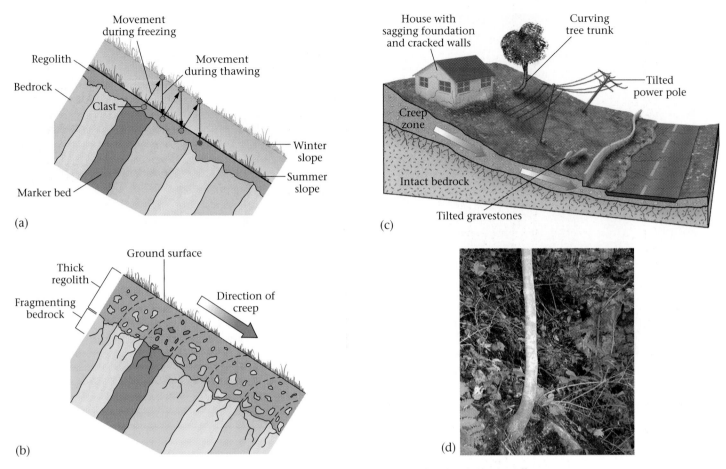

FIGURE 13.2 (a) Creep on hill slopes accompanies the annual freeze-thaw cycle. The clast originally came from the marker bed. It rose perpendicular to the slope when the slope froze and dropped down when the slope thawed. After three years, it has migrated downslope to the position shown. (b) As rock layers weather and break up, the resulting debris creeps downslope. (c) Soil creep causes walls to bend and crack, building foundations to sink, trees to bend, and power poles and gravestones to tilt. (d) Trees that grow in creeping soil gradually develop pronounced curves.

trees that continue to grow after they have been tilted display a pronounced curvature at their base (▶Fig. 13.2a–d).

In Arctic or high-elevation regions, regolith freezes solid to great depth during the winter. In the brief summer thaw, only the uppermost 1 to 3 m of the ground thaws. Since meltwater cannot sink into the permanently frozen ground, or permafrost, the melted layer becomes soggy and weak and flows slowly downslope in overlapping sheets. Geologists refer to this kind of creep, characteristic of tundra regions (cold, treeless regions), as **solifluction** (▶Fig. 13.3).

Slumping

Near Pacific Palisades, along the coast of southern California, Highway 1 runs between the beach and a 120-m-high cliff. Between March 31 and April 3, 1958, a 1-km-long section of the highway disappeared beneath a mass of regolith that had moved down the adjacent cliff. When the movement

stopped, the face of the cliff lay 200 m farther inland than it had before. It took weeks for bulldozers to uncover the road. During such slumping, a mass of regolith detaches from its substrate along a spoon-shaped sliding surface and slips semi-coherently downslope (▶Fig. 13.4a). We call the moving mass a slump, and the surface on which it slips a **failure surface** (▶Fig. 13.4b).

The distinct, curving step at the upslope edge of a slump, where the regolith detaches, is called a **head scarp.** Immediately below the head scarp, the land surface sinks below its previous elevation. Farther downslope, at the toe, or end, of the slump, the ground elevation rises as the slump rides up and over the preexisting land surface. Slumps come in all sizes, from only a few meters across to tens of kilometers across, and they move at speeds of millimeters per day to tens of meters per minute.

In recent years, geologists have increasingly recognized that huge slumps occur underwater, along the margins of

FIGURE 13.3 Solifluction on a hill slope in the tundra.

(a)

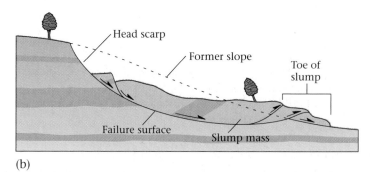

(b)

FIGURE 13.4 (a) A slump threatening a highway in California. (b) Note the curving failure surface in this slump. The dashed line indicates the slope's shape before slumping.

islands or of continents. These submarine slumps modify the shape of the sea floor significantly. For example, the broad apron of hummocky sea floor that surrounds Hawaii has been formed by countless slumps over millions of years—some of these slumps are more than 100 km across. Large submarine slumps may trigger immense tsunamis (▶Box 13.1).

Mudflows and Debris Flows

Rio de Janeiro, Brazil, originally occupied only the flatlands bordering beautiful crescent beaches between towering mountains. But in recent decades, its population has grown so much that the city has expanded up the steep sides of the mountains, and in many places densely populated communities of makeshift shacks cover the slopes. These communities, which have no storm drains, were built on the thick regolith that resulted from long-term weathering of bedrock in Brazil's tropical climate. In 1988, particularly heavy rains saturated the regolith, which turned into a viscous slurry of mud that flowed downslope. Whole communities disappeared overnight, replaced by a hummocky muddle of mud and debris. And at the base of the cliffs, the flowing mud caused high-rise buildings and homes to collapse (▶Fig. 13.6a).

In areas such as the hill-slope communities of Rio, where neither vegetation nor drainage systems protect the ground from rainfall, water mixes with regolith to create a slurry that moves downslope. If the slurry consists of just mud, it's a **mudflow,** but if the mud is mixed with larger rock fragments, it's a **debris flow.** The speed at which mud or debris flows can move depends on the slope angle and on the water content. Flows move faster if they are wetter (that is, less viscous) and if they move on steeper slopes. On a gentle slope, viscous mud flows like molasses, but on a steep slope, low-viscosity mud may move at tens of km per hour. Because mud and debris flows have greater viscosity than clear water,

they can carry large rock chunks, as well as houses and cars. They typically follow channels downslope, and at the base of the slope they spread out into a broad lobe.

Particularly devastating mudflows spill down the river valleys bordering volcanoes. These mudflows, known as **lahars,** consist of a mixture of volcanic ash and water from the snow and ice that melts in a volcano's heat or from heavy rains (▶Fig. 13.6b; see Chapter 7).

Landslides (Rock and Debris Slides)

In the early 1960s, engineers built a huge new dam across a river on the northern side of Monte Toc, in the Italian Alps, to create a reservoir for generating electricity. This dam, the Vaiont Dam, was an engineering marvel, a concrete wall rising 260 m (as high as an 85-story skyscraper) above the valley floor (▶Fig. 13.7a). Unfortunately, the dam's builders did not recognize the hazard posed by nearby Monte Toc. The side of Monte Toc facing the reservoir is underlain by dipping limestone beds interlayered with weak shale beds. These beds dipped parallel to the surface of the mountain and curved under the reservoir (▶Fig. 13.7b). As the reservoir filled, the flank of the mountain cracked, shook, and rumbled, and local residents began to call Monte Toc *la montagna che cammina* (the mountain that walks).

After several days of rain, Monte Toc began to rumble so much that on October 9, 1963, engineers lowered the water level in the reservoir. They thought the wet ground might slump a little into the reservoir, but no more than that, so no one ordered the evacuation of the town of Longarone, a few kilometers down the valley. Unfortunately, the engineers underestimated the problem. At 10:30 that evening, a huge chunk of Monte Toc—600 million tons of rock—detached from the mountain and slid downslope into the reservoir. Some debris rocketed up the opposite wall of the valley to a height of 260 m above the original reservoir level. The displaced water of the reservoir spilled over the top of the dam

The Storegga Slide and the North Sea Tsunamis

BOX 13.1
GEOLOGIC CASE STUDY

The Firth of Forth, a long inlet of the North Sea, forms the waterfront of Edinburgh, Scotland. At its western end, it merges with a broad plain in which mud and peat have been accumulating since the last ice age. Around 1865, geologists investigating this sediment discovered an unusual layer of sand containing bashed seashells, marine plankton, and torn-up fragments of substrate. This sand layer lies sandwiched between mud layers at an elevation of up to 4 m above the high-tide limit and 80 km inland from the shore. How could the sand layer have been deposited? The mystery simmered for many decades. During this time, layers of shelly sand, similar to the one found in Scotland, were discovered at many other localities along both sides of the North Sea (▶Fig. 13.5). In some cases, the layer occurred 20 m above the high-tide limit. Geologists studying the coast also found locations where coastal cliffs appeared to have been eroded by wave action at elevations well out of reach of normal storm waves.

While land-based geologists puzzled about these unusual sand beds and coastal erosional features, marine geologists investigating the continental shelf off the western coast of Norway discovered a region of very irregular sea floor underlaid with a jumble of chaotic blocks, some of which are 10 km by 30 km across and 200 m thick. When mapped out, the geologists concluded that a 290-km-long sector of the continental shelf had collapsed in a series of at least three submarine slides that together transported about 5,580 km³ of debris. The overall feature is called the Storegga Slide. Further studies show that the Storegga Slide formed during three movement events: one occurred 30,000 years ago, the second about 7,950 years ago, and the third about 6,000 years ago. Now the pieces of the puzzle were in place—slides the size of those constituting the Storegga Slide had displaced enough ocean water to create tsunamis.

Tsunamis produced by the second Storegga slide may coincide with the disappearance of Stone Age tribes along the North Sea coast, suggesting the tribes simply washed away. If such a calamity happened in the past, could it happen again, with submarine-slide-generated tsunamis inundating coastal cities with large populations? Geologists now realize that submarine landslides can trigger tsunamis every bit as devastating as earthquake- and volcano-generated tsunamis. A slide in 1929 along the coast of Newfoundland, for example, not only created a turbidity current that broke the trans-Atlantic telephone cable, but also generated tsunamis that washed away houses and boats along the coast of Newfoundland at elevations of up to 27 m above sea level. With a bit of looking, geologists have found huge boulders flung by tsunamis onto the land, layers of sand and gravel deposited well above the high-tide limit, and erosional features high up on shoreline cliffs along many coastal areas, even in areas (such as the Bahamas and southeastern Australia) far from seismic or volcanic regions. And submarine mapping shows that many large slumps occur all along continental shelves. It's no wonder that Edward Bryant, in his recent book on tsunamis, calls them "The Underrated Hazard."

FIGURE 13.5 Map of the North Sea region showing the location of sites (blue dots) where marine sand layers occur significantly above the high-tide limit. These were caused by tsunamis generated by movement of the Storegga Slide. The map shows the estimated position of the tsunamis at 2 hours, 4 hours, and 6 hours.

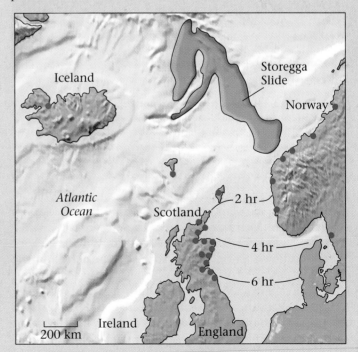

(a)

(b)

FIGURE 13.6 (a) The aftermath of a mudflow in Rio de Janeiro. (b) The aftermath of lahar that flowed down a stream valley after the 1980 Mt. St. Helens volcanic eruption.

(a)

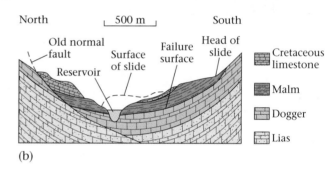

(b)

FIGURE 13.7 (a) The Vaiont Dam and the debris now behind it. The exposed failure surface is visible in the distance. (b) A cross section parallel to the face of Vaiont Dam before the landslide. Mount Toc is to the right. Note the failure surface at the base of the weak Malm Shale. The landslide completely filled the reservoir. Its surface after movement is indicated by the dashed line. The names Malm, Dogger, and Lias refer to epochs in the Jurassic Period.

and rushed down into the valley below. When the flood had passed, nothing remained of Longarone or its 1,500 inhabitants. Though the dam itself still stands, it holds back only debris and has never provided any electricity.

Geologists refer to such a sudden movement of rock and debris down a nonvertical slope as a **landslide.** If the mass consists only of rock, it may also be called a **rock slide** (the case in the Vaiont Dam disaster), and if it consists mostly of regolith, it may also be called a **debris slide.** Once a landslide has taken place, it leaves a landslide scar on the slope and forms a debris pile at the base of the slope.

Slides happen when bedrock and/or regolith detaches from a slope and shoots downhill on a failure surface (slide surface) roughly parallel to the slope surface. Thus, landslides generally occur where a weak layer of rock or sediment at depth below the ground parallels the land surface. (At the Vaiont Dam, the plane of weakness that would become the failure surface was a weak shale bed.) Slides may move at speeds of up to 300 km per hour; they are particularly fast when a cushion of air gets trapped beneath, so there is virtually no friction between the slide and its substrate, and the mass moves like a hovercraft. Rock and debris slides sometimes have enough momentum to climb the opposite side of the valley into which they fell.

Avalanches

In the winter of 1999, an unusual weather system passed over the Austrian Alps. First it snowed; then the temperature warmed and the snow began to melt. But then the weather turned cold again, and the melted snow froze into a hard, icy crust. This cold snap ushered in a blizzard that blanketed the ice crust with tens of centimeters (1–2 ft.) of snow. On the mountaintops the wind built the snow into huge over-

FIGURE 13.8 (a) Aftermath of the 1999 avalanches in the Austrian Alps. (b) Trees that were flattened by an avalanche, now exposed after the snow melted.

(a)

(b)

FIGURE 13.9 Successive rock falls have littered the base of this sandstone cliff with boulders. Note the talus at the base of the cliff.

hanging drifts called cornices. Skiers delighted in the bounty of white, but not for long. No one knows how it started—perhaps a gust of wind or a sudden noise was enough—but the world witnessed the aftermath. With the frozen snow layer underneath acting as a failure surface, the heavy layer of new snow began to slide down the mountain, accelerating as it moved and then disintegrating and mixing with air. It became a roaring cloud—an avalanche—traveling at hurricane speeds and flattening everything in its path. Trees and ski lodges toppled like toothpicks before the mass finally reached the valley floor and came to a halt (▶Fig. 13.8a, b).

Avalanches are turbulent clouds of debris mixed with air that rush down steep hill slopes at high velocity. If the debris consists of snow, as in the Austrian avalanche, it's a snow avalanche. If it consists of fragments of rock and dust, it's a debris avalanche. The moving air-debris mass is denser than clear air and thus hugs the ground and acts like an extremely strong and viscous wind that can knock down and blow away anything in its path. As illustrated by the Austrian example, snow avalanches pose a particular threat when frozen snow layers get buried and thus can become a failure surface for the overlying snow. Typically, avalanches happen again and again in the same area, creating pathways, called avalanche chutes, in which no mature trees grow. Avalanches of sediment, mixed with water, can occur underwater. These are more commonly referred to as turbidity currents (see Chapter 5).

Rock Falls and Debris Falls

Rock falls and **debris falls,** as their names suggest, occur when a mass free-falls from a steep (vertical) cliff (▶Fig. 13.9). Friction and collision with other rocks may bring some blocks to a halt before they reach the bottom of the slope; these blocks pile up to form a **talus,** a sloping apron of rocks along the base of the cliff (Fig. 13.9). Rock or debris that has fallen a long way can reach speeds of 300 km per hour, and may have so much momentum that it keeps going when it touches bottom and triggers a debris avalanche that can cross a valley floor and rise up the other side. Like avalanches, large, fast rock falls push the air in front of them, creating a short blast of hurricanelike wind. The wind alone from a 1996 rock fall in Yosemite National Park flattened over 2,000 trees. Commonly, rock falls happen when a rock separates from a cliff face along a joint.

Most rock falls involve only a few blocks detaching from a cliff face and dropping into the talus. But some falls dislodge immense quantities of rock. In September 1881, a 600-m-high crag of slate, undermined by quarrying, suddenly collapsed onto the town of Elm in a valley of the Swiss Alps. Over 10 million cubic meters of rock fell to the valley floor, burying Elm and its 115 inhabitants to a depth of 10 to 20 m.

FIGURE 13.10 Perfectly intact rock is rare at the surface of the Earth. Most outcrops, such as this one on the western coast of Ireland, are highly jointed.

13.3 WHY DO MASS MOVEMENTS OCCUR?

In order for the mass movements described above to take place, the following conditions must develop. First, materials at the Earth's surface must be weakened by fracturing and weathering. Second, relief (the difference in elevation between two adjacent locations) must develop, for where relief exists, gravity can pull material from a higher elevation down to a lower elevation. Let's look at these conditions in more detail.

Weakening the Surface: Fragmentation and Weathering

If the Earth's surface were covered with intact (unbroken) rock, mass movements would be of little concern, for intact rock has great strength and could form stalwart mountain faces that would never tumble, even if they were vertical. But the rock of Earth's upper crust has been broken by jointing and faulting, and in many locations the surface has a cover of regolith created by the weathering of rock in Earth's corrosive atmosphere. Fragmented rock and regolith are much weaker than intact rock and can indeed collapse fairly easily in response to Earth's gravitational pull (▶Fig. 13.10). Thus, jointing, faulting, and weathering make mass movements possible.

Why are regolith and fractured rock so much weaker than intact rock? Intact rock is held together by the strong chemical bonds within mineral crystals, by mineral cement, or by the interlocking of grains. In contrast, a joint-

bounded or fault-bounded block is held in place only by friction between the block and its surroundings. Regolith is unconsolidated; that is, it consists of unattached grains. Dry regolith holds together because of friction between adjacent grains and/or because weak electrical charges cause grains to attract each other. Slightly wet regolith holds together because of water's surface tension, caused by the attraction of water molecules to each other.

Slope Stability: The Battle between Downslope Force and Resistance Force

Mass movements do not take place on all slopes, and even on slopes where such movements are possible, they occur only occasionally. Geologists distinguish between stable slopes, on which sliding is unlikely, and unstable slopes, on which sliding will likely happen. When material starts moving on an unstable slope, we say that slope failure has occurred. Whether a slope fails or not depends on the balance between two forces—the downslope force, caused by gravity, and the resistance force, which inhibits sliding. If the downslope force exceeds the resistance force, the slope fails and mass movement results.

Imagine a block sitting on a slope. We can represent the gravitational attraction between this block and the Earth by an arrow (a vector) that points straight down, toward the Earth's center of gravity. This arrow can be resolved into two components, the downslope force parallel to the slope and the normal force perpendicular to the slope. The resistance force can be represented by an arrow pointing uphill. If the downslope force exceeds the resistance force, then the block moves; otherwise, it stays in place (▶Fig. 13.11a, b). Note that for a given mass, the magnitude of the downslope force

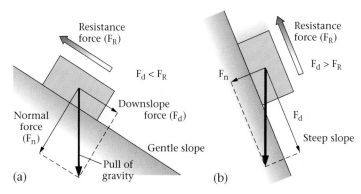

FIGURE 13.11 (a) Gravity, represented by the black arrow, pulls a block toward the center of the Earth. The gravitational force has two components, the downslope force parallel to the slope and the normal force perpendicular to the slope. On gentle slopes, the normal force is larger than the downslope force. The resistance force, caused by friction and represented by an arrow pointing upslope, is larger in this example than the downslope force. (b) If the slope angle increases, the normal force becomes smaller than the downslope force. If the downslope force then becomes greater than the resistance force, then the block starts to move.

increases as the slope angle increases, so downslope forces are greater on steeper slopes.

What causes the resistance force? As we saw above, chemical bonds in mineral crystals, cement, and the jigsaw-puzzle-like interlocking of crystals hold intact rock in place, friction holds an unattached block in place, electrical charges and friction hold dry regolith in place, and surface tension holds slightly wet regolith in place.

Because of resistance force, granular debris tends to pile up and create the steepest slope it can without collapsing. The angle of this slope is called the **angle of repose,** and for most dry unconsolidated materials (such as dry sand) it typically has a value of between 30° and 37°. The angle depends partly on the shape and size of grains, which determine the amount of friction across boundaries. For example, larger angles of repose (up to 45°) tend to form on slopes composed of large, irregularly shaped grains, for these grains interlock with each other (▶Fig. 13.12a–c). Wet sand has a much steeper angle of repose than does dry sand, as you know if you've had the opportunity to build a sand castle.

In many locations, the resistance force is less than might be expected because a weak surface exists at some depth below ground level. This weak surface separates unstable rock and debris above from the stronger substrate below. Geologists recognize several different kinds of weak surfaces that are likely to become failure surfaces (▶Fig. 13.13a–c). These include: wet clay layers; wet, unconsolidated sand layers; surface-parallel joints (also known as exfoliation joints); weak bedding planes; and metamorphic foliation planes.

Failure surfaces that dip parallel to the slope are particularly likely to fail because the downslope force is parallel to

the surface. As an example, consider the 1959 landslide that occurred in Madison Canyon, in southwestern Montana. On August 17 of that year, shock waves from a strong earthquake jarred the region. The southern wall of the canyon is underlain by metamorphic rock with a strong foliation that provided a plane of weakness. When the ground vibrated, rock detached along a foliation plane and tumbled downslope. Unfortunately, twenty-eight campers lay sleeping on the valley floor. They were awakened by the hurricanelike winds blasting in front of the moving mass, but seconds later were buried under 45 m of rubble.

Fingers on the Trigger: Factors Causing Slope Failure

What triggers an individual mass-wasting event? In other words, what causes the balance of forces to change so that the downslope force exceeds the resistance force, and a slope suddenly fails? Here, we look at various phenomena—natural and human-made—that trigger slope failure.

Shocks, Vibrations, and Liquefaction: Earthquake tremors, the passing of large trucks, or blasting in construction sites may cause a mass that was on the verge of moving to actually start moving. For example, an earthquake-triggered slide dumped debris into Lituya Bay, in southeastern Alaska, in 1958. The debris displaced the water in the bay, creating a 300 meter-high (1200 foot-high) splash that washed the slopes on the opposite side of the bay clean of their forests and carried fishing boats many kilometers out to sea. The vibrations of an earthquake break bonds that hold a mass in place and/or cause the mass and the slope to separate slightly, thereby decreasing friction. As a consequence,

FIGURE 13.12 The angle of repose is the steepest slope that a pile of unconsolidated sediment can have and remain stable. Angles of repose depend on the size and shape of grains. (a) Fine, well-rounded sand has a small angle of repose. (b) Coarse, angular sand has a larger angle. (c) Large, irregularly shaped pebbles have a large angle of repose.

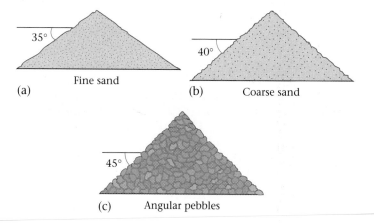

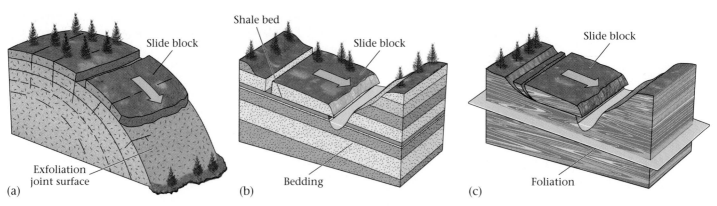

FIGURE 13.13 Different kinds of surfaces become failure surfaces in different geologic settings. (a) In exfoliated massive granite, exfoliation joints become failure surfaces. (b) In sedimentary rock, bedding planes become failure surfaces. (c) In metamorphic rock, foliation planes, especially schistosity (the parallel alignment of mica flakes), become failure surfaces.

the resistance force decreases, and the downslope force sets the mass in motion. Shaking can also cause **liquefaction** of wet sediment by destroying the cohesion between the grains.

Changing Slope Angles, Slope Loads, and Slope Support: Factors that make a slope steeper or heavier may cause the slope to fail (▶Fig. 13.14a–c). For example, when a river eats away at the base of the slope, or a contractor excavates at the base of a slope, the slope becomes steeper and the downslope force increases. But while this happens, the resistance force stays the same. Therefore, if the excavation continues, the downslope force will eventually exceed the resistance force and the slope will fail.

Addition of water to regolith not only makes the regolith heavier, thereby increasing the downslope force, but at the same time weakens failure surfaces, thereby de-

creasing resistance force. Therefore, heavy rain can trigger mass movement. The largest observed landslide in U.S. history—the Gros Ventre slide, which took place in 1925 on the flank of Sheep Mountain, near Jackson Hole, Wyoming—illustrates this phenomenon (▶Fig. 13.15a–c). Almost 40 million cubic meters of rock, soil, and forest detached from the side of the mountain and slid 600 m down a slope, filling the valley and creating a 75-m-high natural dam across the Gros Ventre River, after river erosion had removed support at the base of the hill.

In some cases, excavation results in the formation of an overhang. When such **undercutting** has occurred, rock making up the overhang eventually breaks away from the slope and falls. Overhangs commonly develop along seacoasts and rivers, where the water cuts into a slope (▶Fig. 13.16a, b).

FIGURE 13.14 Slope angles may become steeper, making the slopes unstable. (a) A river can cut into the base of a slope, steepening the sides of the valley. (b) Cutting terraces in a hill slope creates a steeper slope. (c) Adding debris to the top of an unconsolidated sediment pile may cause the angle of repose to be exceeded.

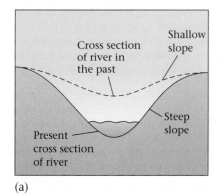

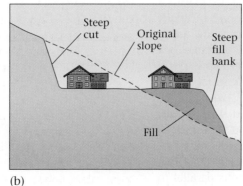

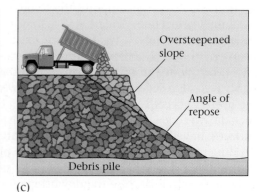

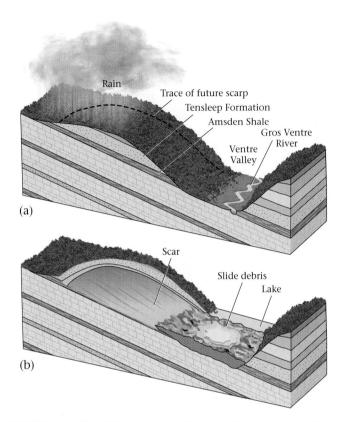

FIGURE 13.15 (a) The huge Gros Ventre slide took place after heavy rains had seeped into the ground, weakening the Amsden Shale and making the overlying Tensleep Formation heavier. The slope was already unstable because the Gros Ventre River had cut down to the shale, and the bedding planes dipped parallel to the slope. (b) After the slide moved, it filled the river valley and dammed the river, creating a lake. A huge landslide scar formed on the hill slope. (c) The Gros Ventre slide.

Changing the Slope Strength: The stability of a slope depends on the strength of the material constituting it. If the material weakens with time, the slope becomes weaker and eventually collapses. Three factors influence the strength of slopes: weathering, vegetation cover, and water.

With time, chemical weathering produces weaker minerals, and physical weathering breaks rocks apart. Thus, a formerly intact rock composed of strong minerals is transformed into a weaker rock or into regolith.

We've seen that thin films of water cause cohesion between grains. Water in larger quantities, though, decreases cohesion, because it fills pore spaces entirely and keeps grains apart. Saturation of regolith with water during a rainstorm weakens the regolith so much that it may begin to move downslope as a slurry. If the water weakens a specific subsurface layer, then the layer becomes a failure surface. Similarly, if groundwater rises above a weak failure surface, overlying rock or regolith may start to slide.

In the case of slopes underlain with regolith, vegetation tends to strengthen the slope, because the roots hold otherwise unconsolidated grains together. Also, plants absorb water from the ground, thus keeping it from turning into slippery mud. The removal of vegetation therefore has the net result of making slopes more susceptible to downslope mass movement. In 2003, terrifying wildfires, stoked by strong winds, destroyed the ground-covering vegetation in many areas of California. When heavy rains followed, the barren ground of this hilly region became saturated with water and turned into mud, which then flowed downslope, damaging and destroying many homes and roads. Deforestation in tropical rain forests, similarly, leads to catastrophic mass wasting of the forest's substrate (▶Fig. 13.17).

Tectonic Setting: Most unstable ground on Earth ultimately owes its existence to the activity of plate tectonics. As we've seen, plate tectonics causes uplift, generating relief, and

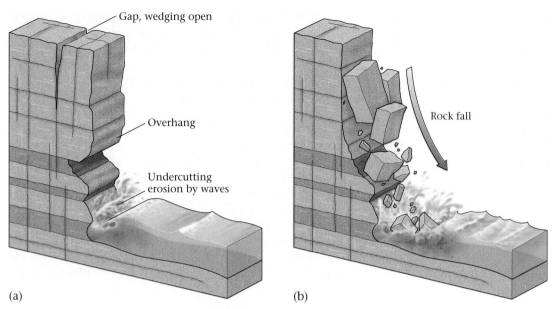

FIGURE 13.16 (a) Undercutting by waves removes the support beneath an overhang. (b) Eventually, the overhang breaks off along joints, and a rock fall takes place.

FIGURE 13.17 Deforestation makes slopes more susceptible to mass movement, as shown in this example from Puebla, Mexico. The slide destroyed the small village of Acalama, killing all but 30 of its 150 to 200 residents.

causes faulting, which fragments the crust. And, of course, earthquakes on plate boundaries trigger devastating landslides. Spend a day along the steep slopes of the Alpine Fault, a plate boundary that transects New Zealand, and you can hear mass movement in progress: during heavy rains, rockfalls and landslides clatter with astounding frequency, as if the mountains were falling down around you.

Coastal California is particularly prone to mass wasting. Terrible slumps have sent multi-million dollar homes tumbling and have closed highways and railroads for weeks. A number of phenomena combine to cause California's landslides and mudflows: (1) California sits astride an active plate boundary, so faulting not only breaks and weakens crustal materials, but also jars them with seismic energy. (2) Wildfires and human-caused deforestation have removed vegetation that could stabilize slopes. (3) Torrential winter rains saturate the regolith with water. (4) Development of cities and suburbs has oversteepened slopes and has affected water content of the substrate.

13.4 HOW CAN WE PROTECT AGAINST MASS-MOVEMENT DISASTERS?

Identifying Regions at Risk

Clearly, landslides, mudflows, and slumps are natural hazards we cannot ignore. Too many of us live in areas where mass wasting has the potential to kill people and destroy

property. In many cases, the best solution is avoidance: don't build, live, or work in an area where mass movement will take place. But avoidance is possible only if we know where the hazards are.

To pinpoint dangerous regions, geologists look for landforms known to result from mass movements, for where these movements have happened in the past, they might happen again in the future. Features such as slump head scarps, swaths of forest in which trees have been flattened and point downslope, piles of loose debris at the base of hills, and hummocky land surfaces all indicate recent mass wasting.

Geologists may also be able to detect regions that are beginning to move downslope (▶Fig. 13.18). For example, roads, buildings, and pipes on top of unstable ground begin to crack. Power lines may be too tight or too loose because the poles to which they are attached move together or apart. Visible cracks form on the ground at the potential

head of a slump, while the ground may bulge up at the toe of the slump. Subsurface cracks may drain the water from an area and kill off vegetation, while another area may sink and form a swamp. Slow movements cause trees to develop pronounced curves at their bases. In some cases, the activity of land masses moving too slowly for people to perceive can be documented with sensitive surveying techniques that can detect a subtle tilt of the ground or changes in distance between nearby points.

Even without evidence of recent movement, a danger may still exist, for just because a steep slope hasn't collapsed in the recent past doesn't mean it won't in the future. In recent years, geologists have begun to identify such potential hazards by using computer programs that evaluate factors that trigger mass wasting. They then produce maps that portray the degree of risk for a certain location. These factors include the following: slope steepness; strength of substrate;

FIGURE 13.18 The features shown here indicate that a large slump is beginning to develop. Note the cracks at the site of the growing head scarp, which drain water and kill trees. Power-line poles crossing the unstable ground bend, and the lines become overtight. Fences and roads that straddle the scarp begin to break up. Houses that straddle the scarp begin to crack, and their foundations sink.

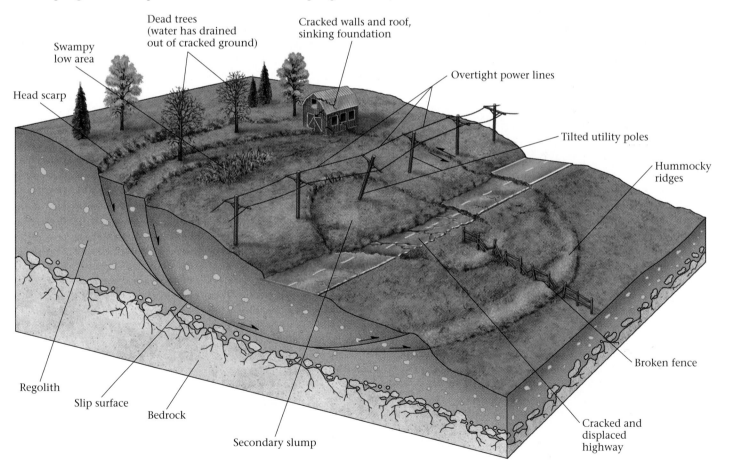

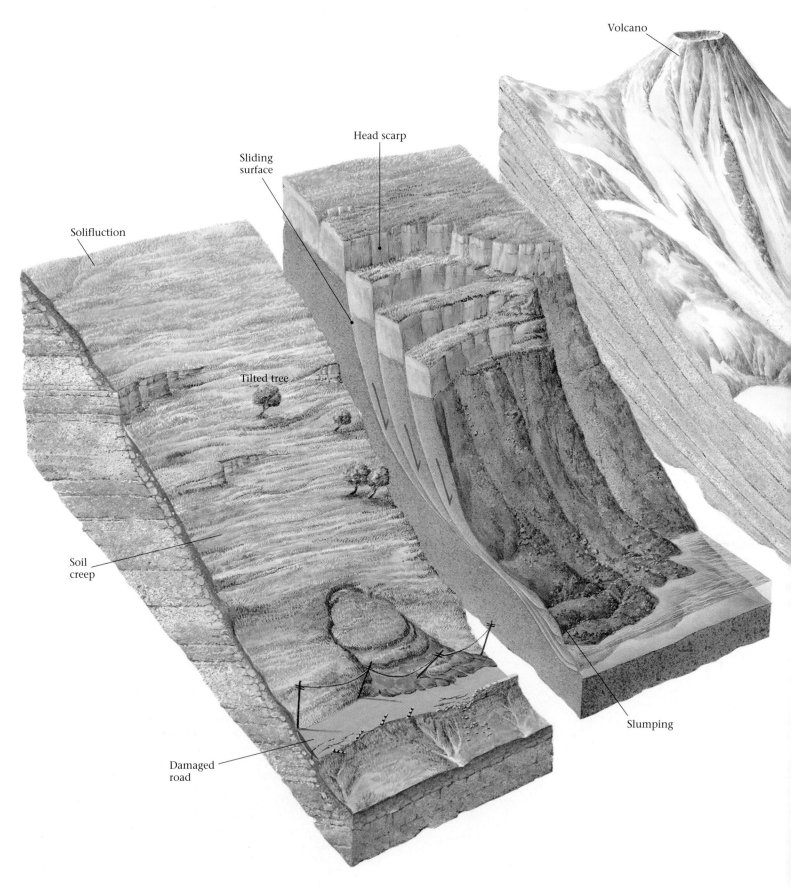

Solifluction

Sliding surface

Head scarp

Volcano

Tilted tree

Soil creep

Damaged road

Slumping

Mass Movement

In Earth's gravity field, what goes up must come down—sometimes with disastrous consequences. Rock and regolith are not infinitely strong, so every now and then slopes or cliffs give way in response to gravity, and materials slide, tumble, or career

downslope. This downslope movement, called mass movement, or mass wasting, is the first step in the process of erosion and sediment formation. The resulting debris may eventually be carried away by water, ice, or wind.

The kind of mass wasting that takes place at a given location reflects the composition of the slope (is it composed of weak soil,

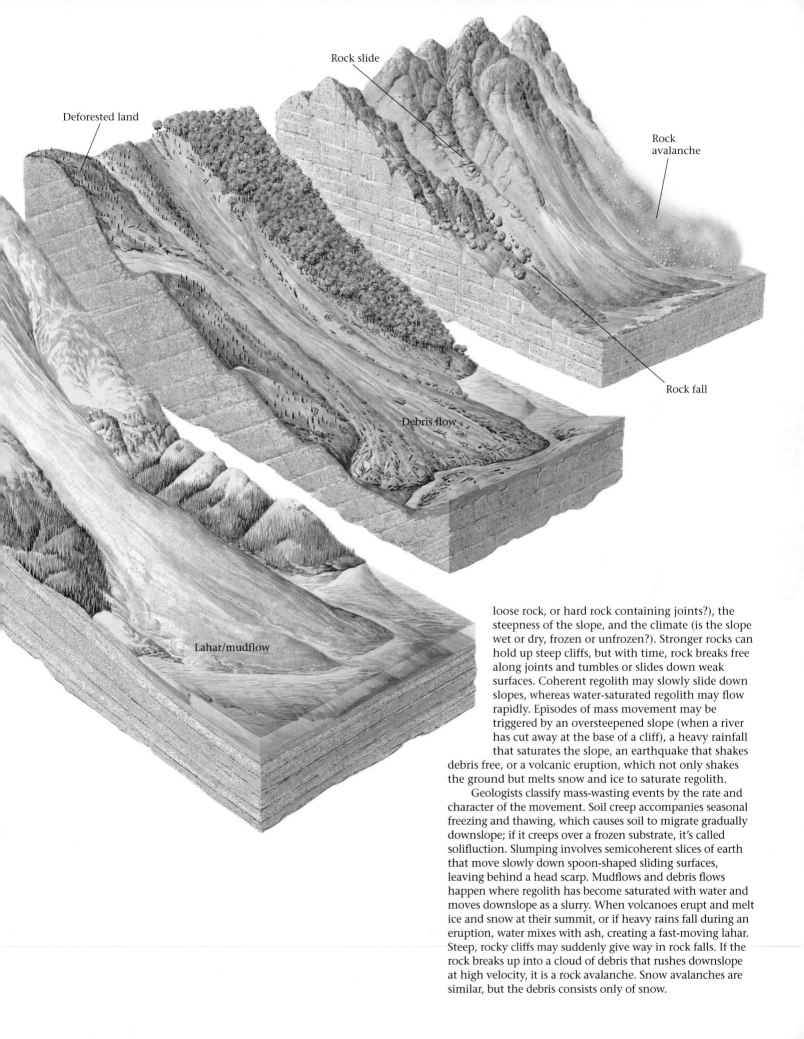

Deforested land

Rock slide

Rock
avalanche

Rock fall

Debris flow

Lahar/mudflow

loose rock, or hard rock containing joints?), the
steepness of the slope, and the climate (is the slope
wet or dry, frozen or unfrozen?). Stronger rocks can
hold up steep cliffs, but with time, rock breaks free
along joints and tumbles or slides down weak
surfaces. Coherent regolith may slowly slide down
slopes, whereas water-saturated regolith may flow
rapidly. Episodes of mass movement may be
triggered by an oversteepened slope (when a river
has cut away at the base of a cliff), a heavy rainfall
that saturates the slope, an earthquake that shakes
debris free, or a volcanic eruption, which not only shakes
the ground but melts snow and ice to saturate regolith.

Geologists classify mass-wasting events by the rate and
character of the movement. Soil creep accompanies seasonal
freezing and thawing, which causes soil to migrate gradually
downslope; if it creeps over a frozen substrate, it's called
solifluction. Slumping involves semicoherent slices of earth
that move slowly down spoon-shaped sliding surfaces,
leaving behind a head scarp. Mudflows and debris flows
happen where regolith has become saturated with water and
moves downslope as a slurry. When volcanoes erupt and melt
ice and snow at their summit, or if heavy rains fall during an
eruption, water mixes with ash, creating a fast-moving lahar.
Steep, rocky cliffs may suddenly give way in rock falls. If the
rock breaks up into a cloud of debris that rushes downslope
at high velocity, it is a rock avalanche. Snow avalanches are
similar, but the debris consists only of snow.

degree of water saturation; orientation of bedding, joints, or foliation relative to the slope; nature of vegetation cover; potential for heavy rains; potential for undercutting to occur; and likelihood of earthquakes. From such hazard-assessment studies, geologists compile landslide-potential maps, which rank regions according to the likelihood that a mass movement will occur. In any case, common sense suggests that you should avoid building on or below particularly dangerous slide-prone slopes.

Preventing Mass Movements

In areas where a hazard exists, people can take certain steps to remediate the problem and stabilize the slope (▶Fig. 13.19a–i).

• *Revegetation:* Since bare ground is much more susceptible to downslope movement than vegetated ground, stability in deforested areas will be greatly enhanced if owners replant the region with vegetation that sends down deep roots.

• *Regrading:* An oversteepened slope can be regraded so that it does not exceed the angle of repose.

• *Reducing subsurface water:* Because water weakens material beneath a slope and adds weight to the slope, an unstable situation may be remedied either by improving drainage so that water does not enter the subsurface in the first place, or by removing water from the ground.

• *Preventing undercutting:* In places where a river undercuts a cliff face, engineers can divert the river. Similarly, along coastal regions they may build an offshore breakwater or pile riprap (loose boulders or concrete) along the beach to absorb wave energy before it strikes the cliff face.

• *Constructing safety structures:* In some cases, the best way to prevent mass wasting is to build a structure that stabilizes a potentially unstable slope or protects a region downslope from debris if a mass movement does occur. For example, civil engineers can build retaining walls or bolt loose slabs of rock to more coherent masses in the substrate in order to stabilize highway embankments. The danger from rock falls can be decreased by covering a road cut with chainlink fencing or by spraying road cuts with shotcrete, a cement that coats the wall and prevents water infiltration and consequent freezing and thawing. Highways at the base of an avalanche chute can be covered by an avalanche shed, whose roof keeps debris off the road.

• *Controlled blasting of unstable slopes:* When it is clear that unstable ground threatens a particular region, the best solution may be to blast the unstable ground or snow loose at a time when its movement can do no harm.

CHAPTER SUMMARY

• Rock or regolith on unstable slopes has the potential to move downslope under the influence of gravity. This process, called mass movement or mass wasting, plays an important role in the erosion of hills and mountains.

• Slow mass movement, caused by the freezing and thawing of regolith, is called creep. In places where slopes are underlain by permafrost, solifluction causes a melted layer of regolith to flow down slopes. During slumping, a semicoherent mass of material moves down a spoon-shaped failure surface. Mudflows and debris flows occur where regolith has become saturated with water and moves downslope as a slurry.

• Landslides (rock and debris slides) move very rapidly down a slope; the rock or debris breaks apart and tumbles. During avalanches, debris mixes with air and moves downslope as a turbulent cloud. And in a debris fall or rock fall, the material free-falls down a vertical cliff.

• Intact, fresh rock is too strong to undergo mass movement. Thus, for mass movement to be possible, rock must be weakened by fracturing (joint formation) or weathering.

• Unstable slopes start to move when the downslope force exceeds the resistance force that holds material in place. The steepest angle at which a slope of unconsolidated material can remain without collapsing is the angle of repose.

• Downslope movement can be triggered by shocks and vibrations, a change in the steepness of a slope, a change in the strength of a slope, deforestation, weathering, or heavy rain.

• Geologists produce landslide-potential maps to identify areas susceptible to mass movement. Engineers can help prevent mass movements using a variety of techniques.

KEY TERMS

angle of repose (p. 381)
avalanche (p. 379)
creep (p. 374)
debris fall (p. 379)
debris flow (p. 376)
debris slide (p. 378)
failure surface (p. 375)
head scarp (p. 375)
lahar (p. 376)
landslide (p. 378)

liquefaction (p. 382)
mass movement (mass wasting) (p. 374)
mudflow (p. 376)
natural hazard (p. 374)
rock fall (p. 379)
rock slide (p. 378)
solifluction (p. 375)
talus (p. 380)
undercutting (p. 382)

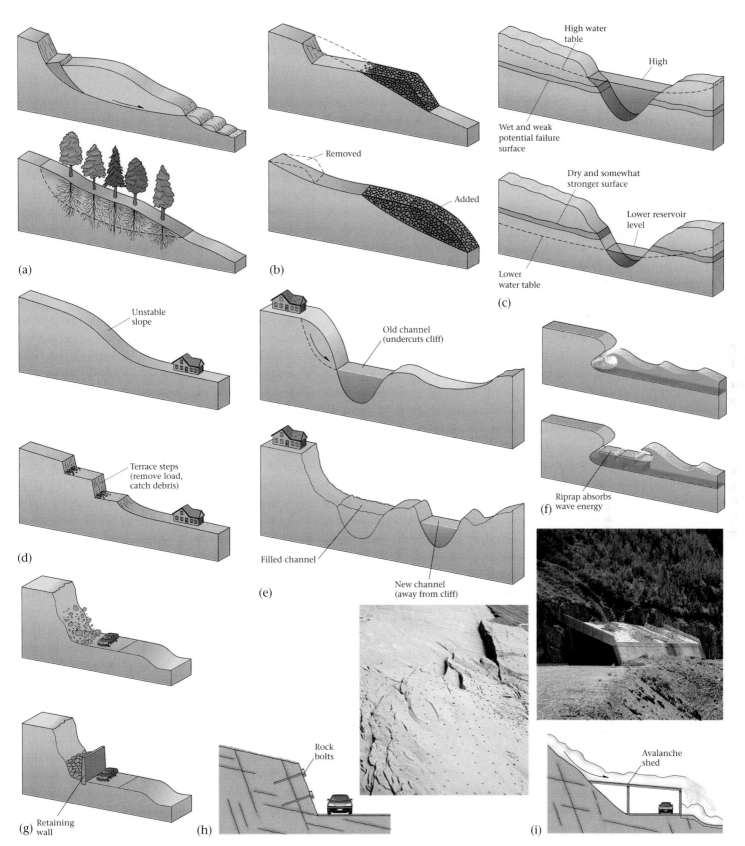

FIGURE 13.19 A variety of remedial steps can stabilize unstable ground. (a) Revegetation removes water, and tree roots bind regolith. (b) Redistributing the mass on a slope eases the load where necessary, adds support where necessary, and decreases slope angles. (c) Lowering the level of groundwater (the water table) may allow a failure surface to dry out. (d) Terracing a steep slope may decrease the load and provide benches to catch debris. (e) Relocating a river channel stops undercutting, and filling the old channel adds support. (f) Riprap absorbs wave energy along the coast. (g) A retaining wall traps falling rock. (h) Bolting rock to a steep cliff face holds loose blocks in place. (i) An avalanche shed diverts avalanche debris over a roadway.

REVIEW QUESTIONS

1. What factors distinguish the various types of mass movement?

2. How does a slump differ from creep? How does it differ from a mudflow or debris flow?

3. How does a rock or debris slide differ from a slump? What conditions trigger a snow avalanche?

4. How does a small amount of water between grains hold material together? How does this change when the sediment becomes oversaturated?

5. What force is responsible for downslope movement? What force helps resist that movement?

6. How does the angle of repose change with grain size? How does it change with water content?

7. What factors trigger downslope movement?

8. How do geologists predict whether an area is susceptible to mass wasting?

9. What steps can people take to reduce the risk of mass wasting?

SUGGESTED READING

Brabb, E. E., and B. L. Harrod, eds. 1989. *Landslides: Extent and Economic Significance.* Brookfield, Va.: Balkema.

Cornforth, D. 2005. *Landslides in Practice: Investigation, Analysis, and Remedial/Preventative Options in Soils.* New York: Wiley.

Dikau, R., D. Brunsden, and L. Schrott, eds. 1996. *Landslide Recognition: Identification, Movement, and Causes.* Chichester, UK: Wiley.

Evans, S. G., and J. V. Degraff, eds. 2003. *Catastrophic Landslides: Effects, Occurrence, and Mechanisms.* Boulder, Colo.: Geological Society of America.

Slosson, J. E., A. G. Keene, and J. A. Johnson, eds. 1993. *Reviews in Engineering Geology.* Vol. 9, *Landslides/Landslide Mitigation.* Boulder, Colo.: Geological Society of America.

Streams and Associated Flooding: The Geology of Running Water

Water that falls on land drains back to the sea via streams. The character of an individual stream depends on a variety of factors, such as slope, water volume, and sediment load. Here, we see part of the Genesee River, which drains the hills of Central New York State and carries the water over waterfalls, on its way to Lake Ontario. The energy of flowing water has cut a canyon into beds of shale and siltstone.

14.1 INTRODUCTION

By the 1880s, Johnstown, built along the Conemaugh River in scenic western Pennsylvania, had become a significant industrial town with numerous steel-making factories. Recognizing the attraction of the region as a summer retreat from the heat and pollution of nearby Pittsburgh, speculators built a mud and gravel dam across the river, upstream of Johnstown, to trap a pleasant reservoir of cool water. A group of industrialists and bankers bought the reservoir and established the exclusive South Fork Hunting and Fishing Club, a cluster of lavish fifteen-room "cottages" on the shore. Unfortunately, the dam had been poorly designed, and debris blocked its spillway (a passageway for surplus water), setting the stage for a monumental tragedy. On May 31, 1889, torrential rain drenched Pennsylvania, and the reservoir surface rose until water flowed over the dam and down its face. Despite frantic attempts to strengthen the dam, the soggy structure abruptly collapsed, and the reservoir emptied into the Conemaugh River Valley. A 20-m-high wall of water roared downstream and slammed into Johnstown, transforming bridges and buildings into twisted wreckage (▶Fig. 14.1). When the water subsided, 2,300 people lay dead, and Johnstown became the focus of national sympathy. It took years for the town to recover,

FIGURE 14.1 During the disastrous 1889 flood in Johnstown, Pennsylvania, the force of flowing water was able to move and tumble sturdy buildings.

and many residents simply picked up and left. Despite several lawsuits, no one payed a penny of restitution, but the South Fork Hunting and Fishing Club abandoned its property.

The unlucky inhabitants of Johnstown experienced one of the more destructive consequences of running water, water that flows on the land surface. The Conemaugh River overflowed when the dam broke, causing a **flood,** an event during which the volume of water in a stream becomes so great that water covers areas outside the stream's normal limits. But even when not in flood, running water relentlessly modifies the surface of our planet as it digs into the land, transports debris, and deposits sediment.

Much of the work of running water takes place in **streams,** ribbons of water that flow down **channels,** or troughs, cut into the land. (Note that geologists may call any channelized body of flowing water a stream, regardless of its size. In common usage, though, a large stream is a river.) Streams remove, or drain, excess water (runoff) from the landscape and carry it eventually to the sea just as culverts drain water from parking lots. In the process, streams transport sediment and nutrients, provide a home for living organisms, and modify the landscape. For human society, streams supply avenues for commerce, water for agriculture, and sources for power.

Earth is the only planet in the solar system that currently has streams of running water. (Mars may have had them in the past; see Interlude E.) In this chapter, we see how streams operate in the Earth System. We first learn about the origin of running water and the architecture of streams. Then we examine how running water ultimately deposits sediment. Finally, we look at the landforms that erosion and/or deposition generate in response to stream

flow, and we conclude with a consideration of flooding and its effects on human society.

14.2 DRAINING THE LAND

Forming Streams and Drainage Networks

Where does the water in a stream come from? To answer this question, recall our discussion of the hydrologic cycle in Interlude E. During the cycle, atmospheric water eventually condenses and falls back to the Earth's surface as rain or snow that accumulates in various reservoirs. Some rain or snow remains on the land as surface water (in puddles, swamps, lakes, snowfields, and glaciers), some flows downslope as a thin film called **sheetwash,** and some sinks into the ground where it either becomes trapped in soil (as soil moisture) or descends below the water table to become groundwater. Streams can receive input of water from all of these reservoirs (▶Fig. 14.2). Specifically, gravity pulls surface water (including meltwater) downhill into stream channels, and the pressure exerted by the weight of new rainfall squeezes existing soil moisture or groundwater back out of the ground and into stream channels.

Running water collects in stream channels, because a channel is lower than the surrounding area and gravity always moves material from higher to lower elevation. But how does a stream channel form in the first place? The process of channel formation begins when sheetwash starts flowing downslope. Like any flowing fluid, sheetwash erodes its substrate (the material it flows over). The efficiency of such erosion depends on the velocity of the flow—faster flows erode more effectively. In nature, the ground is not perfectly planar, not all substrate has the same resistance to erosion, and the amount of vegetation that covers and protects the ground varies with location. Thus, the velocity of sheetwash also varies with location. Where the flow happens to be a bit faster, or the substrate a little weaker, erosion scours (digs) a channel (▶Fig. 14.3a, b). Since this channel is lower than the surrounding ground, sheetwash in adjacent areas starts to head toward it. With time, the extra flow deepens the channel relative to its surroundings, a process called **downcutting,** and creates a stream.

As its flow increases, a stream channel begins to lengthen up its slope, a process called **headward erosion** (▶Fig. 14.3c, d). Headward erosion occurs because the flow is more intense at the entry to the channel (upslope) than in the surrounding sheetwashed areas. As the main channel lengthens, new side channels form nearby. These side channels, or **tributaries,** merge with the main channel, because once a channel forms, the surrounding land slopes into it. An array of linked streams evolves, with the smaller tributaries flowing into a single or trunk stream. The array of

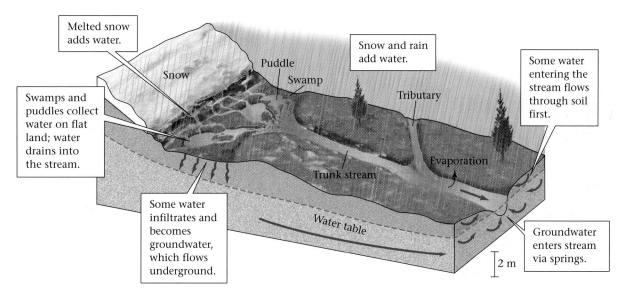

FIGURE 14.2 Excess surface water comes from rain, melting ice or snow, and groundwater springs. Where the ground is flat, the water accumulates in puddles or swamps, but on sloping ground, it flows downslope, collecting in natural troughs called streams.

interconnecting streams together constitute the **drainage network.**

Like transportation networks, drainage networks reach into all corners of a region to provide conduits for the removal of water. The configuration of tributaries and trunk streams defines the map pattern of a drainage network. This pattern depends on the shape of the landscape and the composition of the substrate. Geologists recognize several types of networks on the basis of their map pattern:

- *Dendritic:* When rivers flow over a fairly uniform substrate with a fairly uniform initial slope, they develop a dendritic network, which looks like the pattern of branches connecting to the trunk of a deciduous tree (▶Fig. 14.4a).

- *Radial:* Drainage networks forming on the surface of a cone-shaped mountain flow outward from the mountain peak, like spokes on a wheel. Such a pattern defines a radial network (▶Fig. 14.4b).

- *Rectangular:* In places where a rectangular grid of fractures (vertical joints) breaks up the ground, channels form along the preexisting fractures, and streams join each other at right angles, creating a rectangular network (▶Fig. 14.4c).

- *Trellis:* In places where a drainage network develops across a landscape of parallel valleys and ridges, major tributaries flow down a valley and join a trunk stream that cuts across the ridges. The resulting map pattern resembles a garden trellis, so the arrangement of streams constitutes a trellis network (▶Fig. 14.4d).

Drainage Basins and Divides

A drainage network collects water from a broad region, variously called a drainage basin, catchment, or **watershed,** and feeds it into the trunk stream, which carries the water away. The highland, or ridge, that separates one watershed from another is a **drainage divide** (▶Fig. 14.5). A continental divide separates drainage that flows into one ocean from drainage that flows into another. For example, if you straddle the North American continental divide and pour a cup of water out of each hand, the water in one cup flows to the Atlantic, and the water in the other flows to the Pacific. Three divides bound the Mississippi drainage basin, which drains the interior of the United States (▶Fig. 14.6).

Streams That Last, Streams That Don't: Permanent and Ephemeral Streams

Some streams flow all year long, whereas others flow for only part of the year; in fact, some flow only for a brief time after a heavy rain. The character of a stream depends on the depth of the water table. If the bed, or floor, of a stream lies below the water table, then the stream flows year-round (▶Fig. 14.7a). In such **permanent streams,** found in humid or temperate climates, water comes not only from upstream or from surface runoff, but also from springs through which groundwater seeps. But if the bed of a stream lies above the water table, then water flows only when the rate at which water enters the stream channel exceeds the rate at which water infiltrates the ground below (▶Fig. 14.7b). Such streams can be permanent only if supplied by

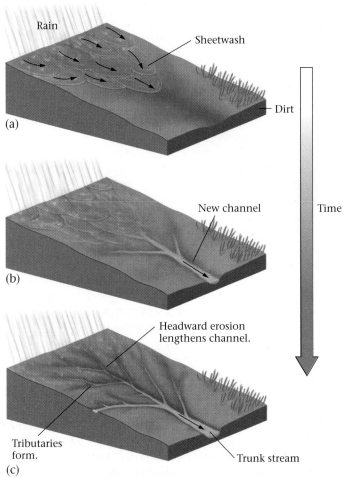

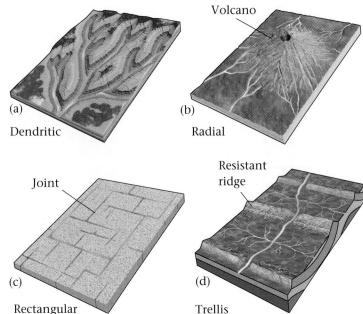

FIGURE 14.4 (a) Dendritic patterns resemble the branches of a tree and form on land with a uniform substrate. (b) A radial network drains a conical mountain, in this case, a volcano. (c) Rectangular patterns develop where a gridlike array of vertical joints controls drainage. (d) Trellis patterns (resembling a garden trellis) form where drainage networks cross a landscape in which ridges of hard rock separate valleys of soft rock. In this example, the alternation is due to folding of the rock layers.

FIGURE 14.3 (a) Drainage on a slope first occurs when sheetwash—overlapping films or sheets of water—moves downslope. (b) Where the sheetwash happens to move a little faster, it scours a channel. (c) The channel grows upslope, a process called headward erosion, and new tributary channels form. The interconnecting streams compose a drainage network. (d) Headward erosion in Canyonlands National Park, Utah, where the main canyon and its tributaries are slowly cutting upstream.

FIGURE 14.5 A drainage divide is a relatively high ridge that separates one drainage basin from another.

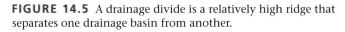

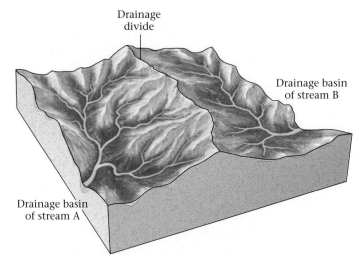

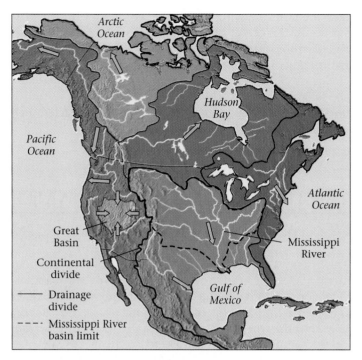

FIGURE 14.6 The Mississippi drainage basin (surrounded by dashed line) is one of several drainage basins in North America. The continental divide separates basins that drain into the Atlantic (and waters connected to the Atlantic) from basins that drain into the Pacific.

14.3 DISCHARGE AND TURBULENCE

Imagine two streams—a larger one in which water flows slowly, and a smaller one in which water flows rapidly. Which stream carries more water? To answer this question, geologists calculate the stream's discharge. Technically speaking, we define **discharge** as the volume of water passing through an imaginary cross section of the stream in a unit of time. We can calculate stream discharge by using a simple formula: $D = A_c \times v_a$. In this formula, A_c is the area of the stream, as measured specifically in an imaginary plane perpendicular to the stream flow, and v_a is the *average* velocity with which water moves in the downstream direction. Stream discharge can be determined at a stream-gauging station, where instruments measure the velocity and depth of the water (▶Fig. 14.8).

Different streams have different average discharges, depending on the size of the drainage basin and the climate. The Amazon River drains a huge rainforest and has the largest average discharge in the world—about 200,000 m³/s, or 15% of the total amount of runoff on Earth. The next-largest stream, the Congo River, has an average discharge of 40,000 m³/s, whereas the "mighty" Mississippi's is only 17,000 m³/s. The discharge of a given stream varies along its length. For example, the discharge in a temperate region *increases* in the downstream direction, because each tributary that enters the stream adds more water, whereas the discharge in an arid region may *decrease* downstream, as progressively more water seeps into the ground or evaporates. Human activity can affect discharge. If people divert the river's water for irrigation, the river's discharge decreases downstream. Finally, the discharge at a given location can vary with time: in a temperate climate, a stream's discharge during the spring may be double or triple the amount during a hot summer, and a flood may increase the discharge to more than a hundred times normal.

The average velocity of stream water (v_a) can be difficult to calculate, because the water doesn't all travel at the same

abundant water from upstream. In dry climates with intermittent rainfall and high evaporation rates, water in the channel entirely sinks into the ground, and the stream dries up when the supply of water stops. Streams that do not flow all year are called **ephemeral streams.** Ephemeral streams only flow during rainstorms or after spring thaws. A dry ephemeral stream bed (channel floor) is called a dry wash, wadi, or arroyo.

FIGURE 14.7 (a) If the bed of a stream channel lies below the water table, then springs add water to the stream, and the stream contains water even during periods when there is no rainfall. Such streams are permanent. (b) If the stream bed lies above the water table, then the stream flows only during rainfall or spring thaws, when water enters the stream faster than it can infiltrate into the ground. Such streams are ephemeral. In desert regions, a dry stream bed is called a dry wash.

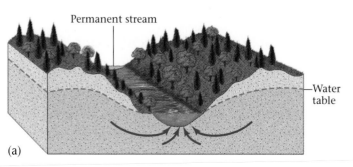

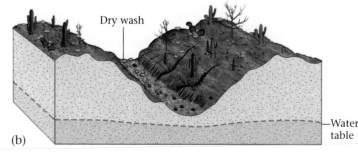

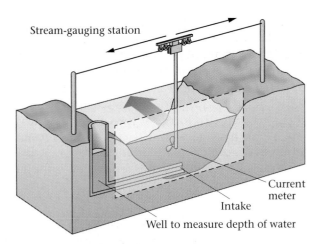

FIGURE 14.8 Geologists obtain information needed to calculate discharge at a stream-gauging station. First, they make a survey of the channel so they know its shape. Then they measure its depth, using a well, and its velocity, using a current meter (either a propeller that spins in response to moving water as shown or, more recently, Doppler radar). The current meter must take measurements at various points in the stream, because velocity changes with location.

velocity for two reasons. First, friction along the sides and floor of the stream slows the flow. Thus, water near the channel walls or the stream bed (the floor of the stream) moves more slowly than water in the middle of the flow. In

FIGURE 14.9 In a turbulent flow, because of shearing between volumes of water and because of movement over a rough surface, the water swirls in curving paths and becomes caught in eddies.

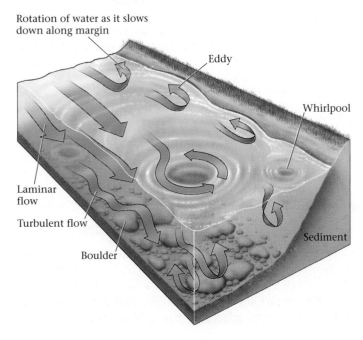

fact, the fastest-moving part of the stream flow lies near the surface in the center of the channel. Second, turbulence—the twisting, swirling motion of a fluid—can create eddies (whirlpools) in which water curves and flows upstream or circles in place (▶Fig. 14.9). Turbulence develops because the shearing motion of one water volume against its neighbor causes the neighbor to spin, and because obstacles, such as boulders, deflect water volumes.

14.4 THE WORK OF RUNNING WATER

Slicing through the Earth: Erosional Processes

The energy that makes running water move comes from gravity. As water heads downslope from a higher to a lower elevation, the gravitational potential energy stored in water transforms into kinetic energy. About 3% of this energy goes into the work of eroding the walls and beds of stream channels. Running water causes erosion by means of:

- *Scouring:* Running water removes loose fragments of sediment, a process called scouring.
- *Breaking and lifting:* The push of flowing water can break chunks of solid rock off the channel floor or walls. In addition, the flow of a current over a clast can cause the clast to rise, or lift off the substrate.
- *Abrasion:* Clean water has little erosive effect, but sand-laden water acts like sandpaper and grinds or rasps away at the channel floor and walls, a process called abrasion. In places where turbulence creates long-lived whirlpools, abrasion by sand or gravel carves a bowl-shaped depression, called a pothole, into the floor of the stream (▶Fig. 14.10a).
- *Dissolution:* Running water dissolves soluble minerals as it passes, and carries the minerals away in solution.

The efficiency of erosion depends on the velocity and volume of water and on its sediment content. A large volume of fast-moving, turbulent sandy water causes more erosion than a trickle of quiet, clear water. Thus, most erosion takes place during floods, which supply streams with large volumes of fast-moving, sediment-laden water.

Transport in Streams: Sediment Loads

The Mississippi River received the nickname "Big Muddy" for a reason—its water can become chocolate brown, because of all the clay and silt it carries. All streams carry sediment, though not in the same quantities. Geologists refer to the total volume of sediment carried by a stream as its sediment load. The sediment load consists of:

(a)

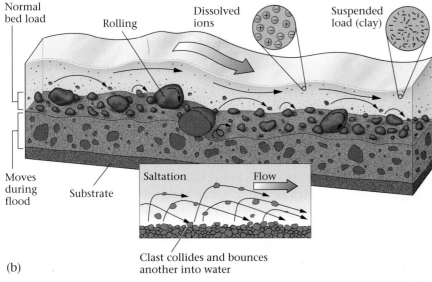

(b)

FIGURE 14.10 (a) A polished red sandstone canyon wall in northern Arizona. (b) Streams transport sediment in many forms. Dissolved load consists of ions in solution. Suspended load consists of tiny grains distributed through the water. Bed load consists of grains that undergo saltation (they bounce along the bed) or grains that roll along the bed. Part of the bed load stays in place during times of normal flow, but begins to move during a flood.

- *Dissolved load:* Running water dissolves soluble minerals in the sediment or rock of its substrate, and groundwater seeping into a stream through the channel walls brings dissolved minerals with it. These ions constitute a stream's dissolved load.

- *Suspended load:* The suspended load of a stream consists of tiny solid grains (silt or clay size) that swirl along with the water without settling to the floor of the channel (▶Fig. 14.10b).

- *Bed load:* The bed load of a stream consists of large particles (such as sand, pebbles, or cobbles) that bounce or roll along the stream floor. Typically, bed-load movement involves saltation, a process during which grains on the channel floor get knocked into the water column momentarily, follow a curved trajectory downstream, and gradually sink to the bed again, where they strike other grains and knock them into the water column.

When describing a stream's ability to carry sediment, geologists specify its competence and capacity. The **competence** of a stream refers to the maximum particle size it carries; a stream with high competence can carry large particles, whereas one with low competence can carry only small particles. A fast-moving, turbulent stream has greater competence (it can carry bigger particles) than a slow-moving stream, and a stream in flood has greater competence than a stream with normal flow. In fact, the huge boulders that litter the bed of a mountain creek move only during floods. The **capacity** of a stream refers to the total quantity of sediment it can carry. A stream's capacity depends on both its competence and its discharge.

Depositional Processes: When Streams Lose Their Loads

A raging torrent of water can carry coarse and fine sediment—the finer clasts rush along with the water as suspended load, while the coarser clasts bounce and tumble as bed load. If the flow velocity decreases, either because the gradient (downstream slope) of the stream bed becomes shallower or because the channel broadens out and friction between the bed and the water increases, then the competence of the stream decreases and sediment settles out. The size of the clasts that settle at a particular locality depends on the decrease in flow velocity. For example, if the stream slows by a small amount, only large clasts settle; if the stream slows by a greater amount, medium-sized clasts settle; and if the stream slows to almost a standstill, the fine grains settle. Thus, coarser sediment tends to settle out further upstream, where the gradient of the stream is steeper and water flows faster, whereas finer grains settle out farther downstream, where the water flows more slowly. Because of this process of sediment sorting, stream deposits tend to be segregated by size—gravel accumulates in one location, and mud in another. Geologists refer to sediments transported by a stream as fluvial deposits (from the Latin *fluvius*, meaning "river") or **alluvium.** Fluvial deposits may accumulate along the stream bed in elongate mounds called **bars** (▶Fig. 14.11a, b). Some stream channels make broad curves known as meanders—water slows along the inner edge of a meander, so crescent-shaped point bars bordering the shoreline develop. During floods, a stream may overtop the banks of its channel and spread out over its **floodplain,** a broad flat area bordering the stream. Friction slows the water on the floodplain, so a sheet of silt and mud settles out. Where a stream empties, at its mouth, into a standing body of water, the water slows and a wedge of sediment called a **delta,** accumulates.

14.5 HOW STREAMS CHANGE ALONG THEIR LENGTH

Longitudinal Profiles

In 1803, the United States under President Thomas Jefferson's leadership bought the Louisiana Territory, a vast tract of land encompassing the western half of the Mississippi drainage basin. At the time, the geography of the territory was a mystery. To fill the blank on the map, Jefferson asked Meriwether Lewis and William Clark to lead a voyage of exploration across the Louisiana Territory to the Pacific.

Lewis and Clark, along with about forty men, began their expedition at the mouth of the Missouri River where it joins the Mississippi. At this juncture, the Missouri is a wide, languid stream of muddy water. The group found the lower reaches (segments) of the Missouri, where the channel is deep and the water smooth, to be easy going. But the farther upstream they went, the more difficult their voyage became, for the stream gradient became progressively steeper, and its discharge became less. When Lewis and Clark reached the site of what is now Bismarck, North Dakota, they had to abandon their original boats and haul smaller vessels up rapids (turbulent water that plunges over a steep, bouldery bed) and around waterfalls, where water drops over an escarpment. When they reached what is now southwestern Montana, they deserted these boats as well and followed stream valleys on foot or by horseback, struggling up steep gradients until they reached the continental divide.

If Lewis and Clark had been able to plot a graph showing their elevation above sea level relative to their distance along the Missouri, they would have found that the **longitudinal profile** of the Missouri, a cross-sectional image showing the variation in the river's elevation *along its length,* was roughly a concave-up curve (▶Fig. 14.12). This curve illustrates that a stream's gradient is steeper near its headwaters (source) than near its mouth. Real longitudinal profiles are not perfectly smooth curves, but rather display little plateaus and steps, representing interruptions by lakes or waterfalls. Near its headwaters, an idealized stream flows down deep valleys or canyons; near its mouth, it flows over nearly horizontal plains.

The Base Level

Streams progressively deepen their channels by downcutting, but there is a depth below which a stream cannot downcut.

FIGURE 14.11 (a) Recently deposited gravel in the bed of a steep mountain stream, at Denali National Park, Alaska. The large cobbles and boulders were deposited during ferocious floods. (b) Point bars of mud deposited along a gentle, slowly moving stream in Brazil.

(a)

(b)

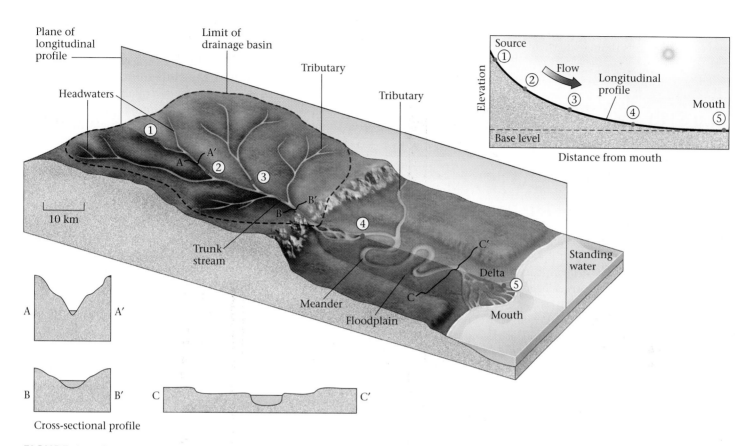

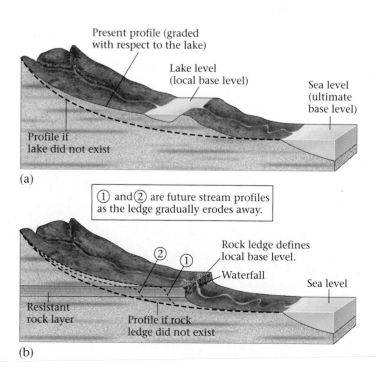

FIGURE 14.12 A drainage network collects water from a broad drainage basin, or watershed, via numerous tributaries. These carry water to a trunk stream and eventually to a standing body of water. Points 1–5 refer to locations along the longitudinal profile (inset). The cross-sectional profiles show how river valley shapes change along the length of a river.

The lowest elevation a stream channel's floor can reach at a given locality is the **base level** of the stream. A local base level occurs upstream of a drainage network's mouth, whereas the ultimate base level (that is, the lowest elevation along the drainage network's longitudinal profile) is determined by sea level. The trunk stream cannot downcut deeper than sea level; if it did, it would have to flow upslope to enter the sea.

Lakes or reservoirs can act as local base levels along a stream, for where the stream enters such standing bodies of water, it almost slows to a halt and cannot downcut further (▶Fig. 14.13a). A ledge of resistant rock can also act as a local base level, for the stream level cannot drop below the ledge until the ledge erodes away (▶Fig. 14.13b). Finally, where a tributary joins a larger stream, the channel of the larger stream acts as the base level for the tributary. Local base levels

FIGURE 14.13 (a) A lake acts as a local base level. The longitudinal profile of the stream upstream of the lake lies above the profile for a "graded stream" (one that can deposit as much sediment as it removes). Eventually, headward erosion of the stream below the lake will cause the lake to drain. (b) A resistant rock ledge also acts as a local base level. With time, the ledge will erode and the waterfall will migrate upstream. Sea level is the ultimate base level for a drainage network.

do not last forever, because running water eventually removes obstructions that create them.

14.6 STREAMS AND THEIR DEPOSITS IN THE LANDSCAPE

Valleys and Canyons

About 10 million years ago, a large block of crust, the region now known as the Colorado Plateau (located in Arizona, Utah, Colorado, and New Mexico), began to rise. Before the rise, the Colorado River had been flowing over a plain not far above sea level, causing little erosion. But as the land uplifted, the river began to downcut steadily. Soon its channel lay as much as 1.6 km below the surface of the plateau, and the Grand Canyon came into existence (▶Fig. 14.14). The formation of the Grand Canyon illustrates a general phenomenon. In regions where the land surface lies well above the base level, a stream can create a deep trough, much deeper than the channel itself, in the land; if the walls of the trough slope gently, the landform is a valley; if they slope steeply, the landform is a canyon.

Whether stream erosion produces a valley or a canyon depends on the rate at which downcutting occurs relative to the rate at which mass wasting causes the walls on either side of the stream to collapse. In places where a stream erodes down through its substrate faster than the walls of the stream collapse, a steep-walled canyon ("slot canyon") develops. Such canyons typically form in hard rock, which can hold up steep cliffs for a long time (▶Fig. 14.15a). In places where the walls collapse as fast as the stream downcuts, landslides and slumps gradually cause the slope of the walls to approach the angle of repose. When this happens, the stream channel lies at the floor of a valley whose cross-sectional shape resembles the letter V (▶Fig. 14.15b); this landform is called a V-shaped valley. Where the walls of the stream consist of alternating layers of hard and soft rock, the walls develop a stair-step shape such as that of the Grand Canyon (▶Fig. 14.15c).

In places where active downcutting occurs, the valley floor remains relatively clear of sediment, for the stream—especially when it floods—carries away sediment that has fallen or slumped into the channel from the stream walls. But if the stream's base level rises or its discharge decreases, the valley floor fills with alluvium

FIGURE 14.14 Erosion by the Colorado River and its tributaries has cut a deep gash, the Grand Canyon, into the rock of the Colorado Plateau.

FIGURE 14.15 (a) If downcutting by a stream happens faster than mass wasting alongside the stream, as is typical when streams erode through hard rock, a slot canyon forms. The canyon widens with time as the stream undercuts the walls. (b) If mass wasting takes place as fast as downcutting occurs, a V-shaped valley develops. (c) In regions where the stream downcuts through alternating hard and soft layers, a stair-step canyon forms.

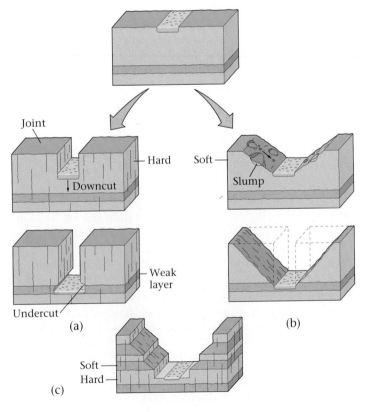

(stream sediment), producing an alluvium-filled valley (▶Fig. 14.16a, b). The surface of the alluvium deposit becomes a broad floodplain. If the stream's base level then drops again and/or the discharge increases, the stream will start to cut down into its own alluvium, a process that generates **stream terraces** bordering the present floodplain (▶Fig. 14.16c).

Rapids and Waterfalls

Recall that when Lewis and Clark struggled up the Missouri River, they came to reaches that could not be navigated by boat. Their path was blocked by rapids, particularly turbulent water with a rough surface (▶Fig. 14.17). Rapids form where water flows over steps or large clasts in the channel floor. Rapids also form where the channel abruptly narrows or its gradient changes, suddenly changing the velocity of the water. The turbulence in rapids can create eddies, waves,

FIGURE 14.17 Rapids in the Grand Canyon (raft for scale). These form where a flood from a side canyon has dumped debris into the channel of the Colorado River.

and whirlpools that roil and churn the water into whitewater, a mixture of bubbles and water. Modern-day whitewater boaters and rafters thrill to the unpredictable movement of a boat caught in rapids.

A waterfall forms where the gradient of a stream becomes so steep that the water literally free-falls down the stream bed (▶Fig. 14.18). The energy of falling water may scour a depression, called a plunge pool, at the base of the waterfall. Though a waterfall may appear to be a permanent

FIGURE 14.16 The evolution of alluvium-filled valleys. (a) Stream erosion creates a valley. (b) Later, a rise in the base level or a decrease in discharge allows the valley to fill with alluvium. (c) If the base level later falls or discharge increases, the stream downcuts through the alluvium and a new, lower floodplain develops. The remnants of the original alluvial plain remain as a pair of terraces, one on each side of the new floodplain.

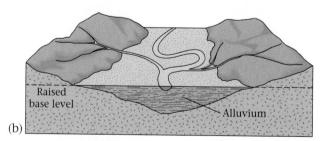

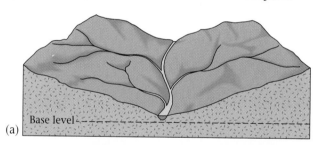

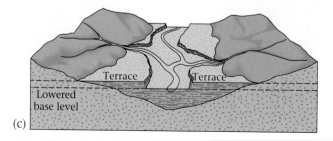

(a)

(b)

(c)

FIGURE 14.18 Iguaçu Falls, along the Brazil/Argentina border.

entity of the landscape, all waterfalls eventually disappear as headward erosion slowly eats back the resistant ledge until the stream reaches grade. We can see a classic example of headward erosion at Niagara Falls. As water flows from Lake Erie to Lake Ontario, it drops over a 55-m-high ledge of hard Silurian dolostone, which overlies a weak shale. Erosion of the shale undercuts the dolostone. Gradually, the overhang of dolostone becomes unstable and collapses, with the result that the waterfall migrates upstream. Before the industrial age, the edge of Niagara Falls cut upstream at an average rate of 1 m per year, but since then, the diversion of water from the Niagara River into a hydroelectric power station has decreased the rate of headward erosion to half that (▶Fig. 14.19a, b).

Alluvial Fans and Braided Streams

Where a fast-moving stream abruptly emerges from a mountain canyon into an open plain at the range front, the water that was once confined to a narrow channel can spread out over a broad surface. As a consequence, the water slows and abruptly drops its sedimentary load, and a gently sloping apron of sediment (sand, gravel, and cobbles) called an **alluvial fan**, accumulates (▶Fig. 14.20a). The stream then divides into an array of small channels that spread out over the fan.

In some localities, streams carry abundant coarse sedi-ment during floods but cannot carry this sediment during normal flow. Thus, during normal flow, the sediment settles out and the channel becomes choked with sediment. As a consequence, the stream divides into numerous strands weaving back and forth between elongate mounds or bars of gravel and sand. The result is a **braided stream**—the name emphasizes that the streams intertwine like strands of hair in a braid (▶Fig. 14.20b).

Meandering Streams and Their Floodplains

A riverboat cruising along the lower reach of the Mississippi River cannot sail in a straight line, for the river channel winds back and forth in a series of snakelike curves called **meanders** (▶Fig. 14.21a). In fact, the boat has to go 500 km along the river channel to travel 100 km as the crow flies. Meandering streams such as the lower Mississippi form where running water travels over a broad floodplain underlain with a soft substrate, in a region where the river has a very gentle gradient. The development of meanders increases the volume of the stream by increasing its length.

How do meanders form and evolve? Even if a stream starts out with a straight channel, natural variations in the water depth cause the fastest-moving current to swing back and forth. The water erodes the side of the stream more effectively where it flows faster, so it begins to cut away faster

FIGURE 14.19 (a) Niagara Falls exists because Lake Erie lies at a higher elevation than Lake Ontario. The Lockport Dolostone, a resistant layer, serves as a local base level for Lake Erie. The Niagara Escarpment is located along the outcrop belt of the dolostone, and Niagara Falls first formed where the outlet of Lake Erie flowed over the escarpment. With time, the falls have cut upstream, at about 1 m per year or less, creating Niagara Gorge. (b) The American Falls of Niagara Falls.

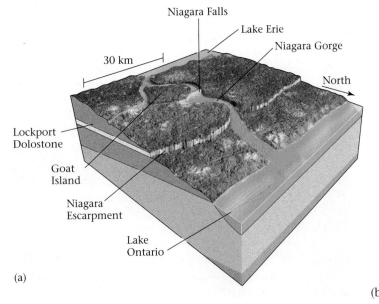

(a)

(b)

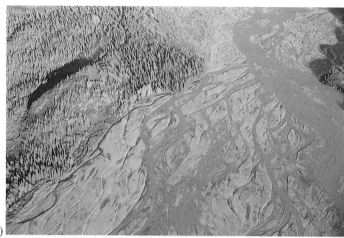

(a)   (b)

FIGURE 14.20 (a) An alluvial fan in Death Valley, California. (b) A braided stream, carrying meltwater from a glacier, near Denali, Alaska.

on the outer arc of the curve. Thus, each curve begins to migrate sideways and grow more pronounced until it becomes a meander. On the outside edge of a meander, erosion continues to eat away at the channel wall, creating a cut bank. On the inside edge, water slows down so that its competence decreases and sediment accumulates in a wedge-shaped deposit called a point bar. With continued erosion, a meander may curve through more than 180°, so that the cut bank at the meander's entrance approaches the cut bank at its end, leaving a meander neck, a narrow isthmus of land separating the portions of the meander.

People building communities along a riverbank may assume that the shape of a meander remains fixed for a long time—it doesn't. In a natural meandering river system, the river channel migrates back and forth across the floodplain. When erosion eats through a meander neck, a straight reach called a cutoff develops. The meander that has been cut off is called an oxbow lake if it remains filled with water, or an abandoned meander if it dries out (►Fig. 14.21b).

Most meandering stream channels cover only a relatively small portion of a broad **floodplain** (►Fig. 14.21c). In many cases, a floodplain terminates at its sides along a bluff, or escarpment. During a flood, water spills out from the stream channel onto the floodplain, and large floods may cover the entire region from bluff to bluff. As the water leaves the channel, friction between the ground and the thin sheet of water moving over the floodplain slows down the flow. This slowdown decreases the competence of the running water, so sediment settles out along the edge of the channel. Over time, the accumulation of this sediment creates a pair of low ridges, called **natural levees,** on either side of the stream. Natural levees may grow so large that the floor of the channel may become higher than the surface of the floodplain.

Deltas: Deposition at the Mouth of a Stream

At the mouth of the Nile in North Africa, the river divides into a fan of small streams called **distributaries,** and the stripe of fertile alluvium broadens into a triangular patch. The Greek historian Herodotus noted that this triangular patch resembles the shape of the Greek letter delta (Δ), and so the region became known as the Nile Delta.

Deltas develop where the running water of a stream enters standing water, the current slows, the stream loses competence, and sediment settles out. Geologists refer to any wedge of sediment formed at a river mouth as a delta, even though relatively few have the triangular shape of the Nile Delta (►Fig. 14.22a–d). Some deltas define arclike lobes, whereas others consist of many elongate lobes that protrude into the sea; the latter are called bird's-foot deltas, because they resemble the scrawny toes of a bird. The existence of several toes indicates that the main course of the river in the delta has shifted on several occasions. These shifts occur when a toe builds so far out into the sea that the slope of the stream becomes too gentle to allow the river to flow. At this point, the river overflows a natural levee upstream and begins to flow in a new direction, an event called an avulsion. The distinct lobes of the Mississippi Delta, a bird's-foot delta, suggest that avulsion has happened several times during the past 9,000 years (►Fig. 14.23). New Orleans, built along one of the Mississippi's distributaries, may eventually lose its riverfront, for a break in a levee upstream of the city could divert the Mississippi into the Atchafalaya River channel.

The shape of a delta depends on many factors. Deltas that form where the strength of the river current exceeds that of ocean currents have a bird's-foot shape, since the sediment can be carried far offshore. In contrast, deltas that

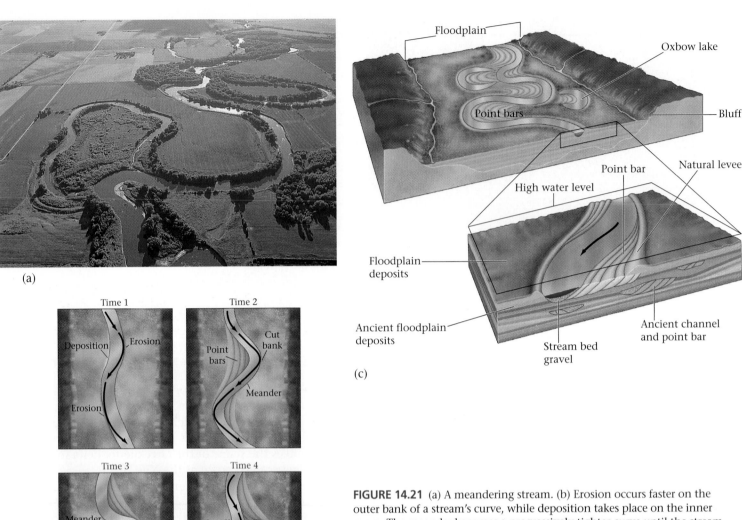

FIGURE 14.21 (a) A meandering stream. (b) Erosion occurs faster on the outer bank of a stream's curve, while deposition takes place on the inner curve. The meander becomes a progressively tighter curve until the stream cuts through the meander neck and forms a cutoff, thereby isolating an oxbow lake. (c) The landforms of a meandering stream. The detail of a stream channel shows a natural levee, the structure of a point bar, and floodplain deposits. Note that alluvium below the present-day channel includes ancient channels and point bars, surrounded by fine-grained floodplain deposits.

form where the ocean currents are strong have a Δ shape, for the ocean currents redistribute sediment in bars running parallel to the shore. And in places where waves and currents are strong enough to remove sediment as fast as it arrives, a river has no delta at all.

Through time, the sediment of a delta compacts, and the lithosphere beneath the delta slowly subsides. As a consequence, the surface of a delta slowly sinks. In a natural delta, distributaries provide sediment that fills the resulting space so that the delta's surface remains at or just above sea level. But if people build artificial levees to constrain the river to its channel, sediment gets carried directly to the seaward edge of the delta and the delta's interior "starves" (does not receive sediment). When this happens, the land surface drops below sea level. Because of this process, much of New Orleans lies below sea level, and when Hurricane Katrina caused artificial levees to fail in 2005, the city flooded extensively (see Chapter 15).

14.7 THE EVOLUTION OF DRAINAGE

Beveling Topography

Landscapes undergoing erosion by streams change over time. To understand how, imagine a region where uplift of the land surface has just formed a mountain range.

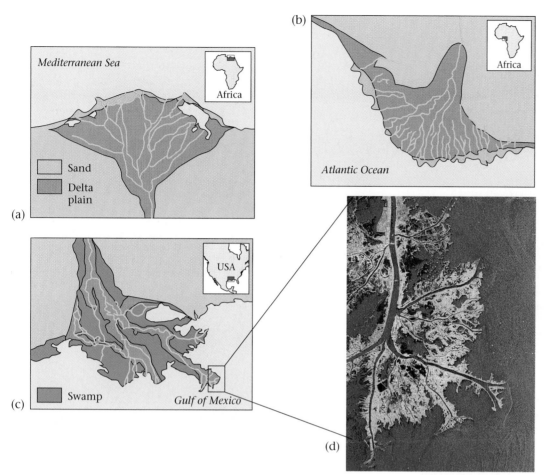

FIGURE 14.22 (a) The Nile is a Δ-shaped delta. (b) The Niger is an arclike delta. (c) The Mississippi is a bird's-foot delta. (d) In this satellite photo of the end of the active tip of the Mississippi Delta, you can see the sediment beneath the surface of the water.

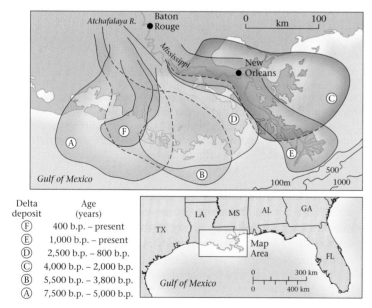

Delta deposit	Age (years)
Ⓕ	400 b.p. – present
Ⓔ	1,000 b.p. – present
Ⓓ	2,500 b.p. – 800 b.p.
Ⓒ	4,000 b.p. – 2,000 b.p.
Ⓑ	5,500 b.p. – 3,800 b.p.
Ⓐ	7,500 b.p. – 5,000 b.p.

FIGURE 14.23 The map shows the different, dated lobes of the Mississippi Delta and the different channels that served as their source. Inset shows a present-day view of the delta, relative to Louisiana. A major flood could divert the main flow of the Mississippi into the channel of the Atchafalaya River, in which case a new delta would form to the west of the Mississippi's present mouth.

Running water soon cuts stream channels, and a drainage network that transports water from the mountain range to the sea develops. At first, the streams in the mountain range have steep gradients, drop over numerous rapids and water-falls, and flow in deep valleys. As time passes, however, slumps and landslides move debris down mountainsides and into stream channels, and floods carry the debris away. In addition, downcutting by sediment-laden water in the channels themselves gradually lowers channel beds. Overall, erosion and transport transforms the rugged mountains into low, rounded hills. As this happens, once-narrow valleys broaden into wide floodplains with gentle gradients. As more time passes, even the hills erode away, and the landscape evolves into a plain at an elevation close to the streams' base level (▶Fig. 14.24a–d). Geologists sometimes refer to such plains as peneplains.

At any stage during the evolution of an uplifted landscape into a low-lying plain, the base level of the streams may become relatively lower. This can happen if renewed mountain building uplifts the land over which the streams flow, or if global sea level drops. When the base level drops, the streams start to downcut again to form new canyons or valleys. Geologists refer to the process of renewed downcutting

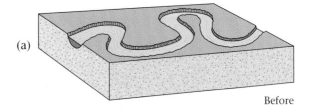

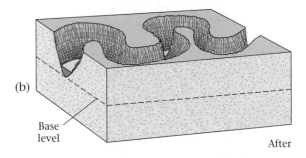

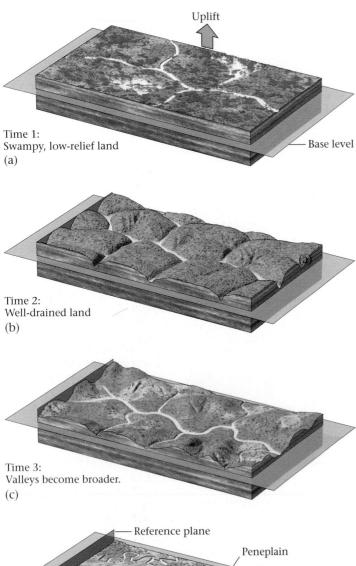

Time 1:
Swampy, low-relief land
(a)

Time 2:
Well-drained land
(b)

Time 3:
Valleys become broader.
(c)

Time 4:
A new, low-relief landscape
(d)

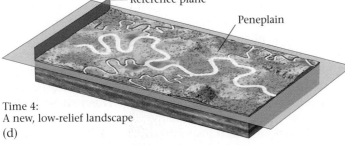

FIGURE 14.24 (a) A fluvial landscape is first uplifted, so that the base level lies at a lower elevation than does the stream channel. (b) Then, the stream cuts down into the plain, leaving remnants of the plain as flat-topped hills between valleys. (c) Later, the landscape consists of rounded hills dissected by tributaries that feed a trunk stream flowing on a floodplain near the base level. Valleys are V-shaped. (d) Still later, only a few remnant hills are left, for most of the landscape has been beveled to form a new peneplain, nearly at sea level. The decreased height of the land surface above the reference plane indicates the thickness of land that has eroded away.

FIGURE 14.25 (a) A stream forms meanders while it flows across a plain. (b) Uplift of the land over which the stream flows causes the meanders to cut down and carve out canyons that meander like the stream. (c) The goosenecks of the San Juan River, in Utah, illustrate incised meanders.

by a stream as **stream rejuvenation.** In cases where a stream had a meandering course before rejuvenation, renewed downcutting may preserve the shapes of meanders, so that the new canyon formed by rejuvenation follows the preexisting meanders (▶Fig. 14.25 a, b). The goosenecks of the San Juan River in southern Utah illustrate this geometry (▶Fig. 14.25c).

Stream Piracy

"Stream piracy" sounds like pretty violent stuff. In reality, **stream piracy,** or stream capture, simply refers to a situation in which headward erosion causes one stream to

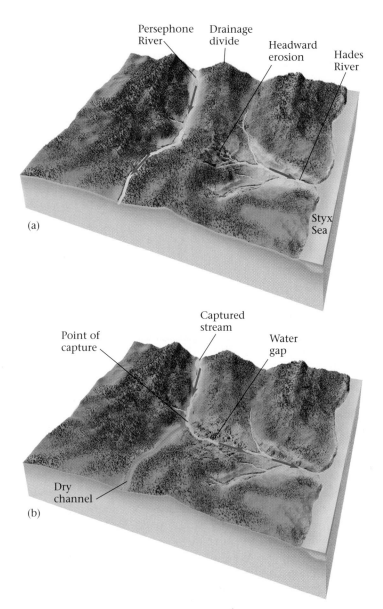

(a)

Persephone River · Drainage divide · Headward erosion · Hades River · Styx Sea

(b)

Point of capture · Captured stream · Water gap · Dry channel

FIGURE 14.26 (a) A drainage divide separates the Hades River drainage from the Persephone River drainage. Headward erosion by the Hades River gradually breaches the drainage divide, creating a water gap. (b) When the source of the Hades River reaches the channel of the Persephone River, Hades (the pirate stream) captures Persephone and carries off its water to the Styx Sea. As a result, the former channel of the Persephone River becomes a dry canyon or channel.

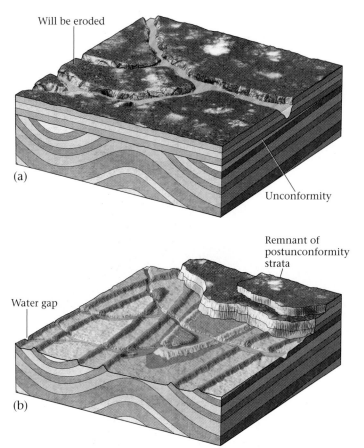

(a)

Will be eroded · Unconformity

(b)

Water gap · Remnant of postunconformity strata

FIGURE 14.27 (a) A superposed stream first establishes its geometry while flowing over uniform, flat layers above an unconformity. (b) The stream gradually erodes away the layers and exposes underlying rock with a different structure (in this example, the older strata are folded). The drainage is "superposed" (laid down) on the folded rocks, and appears to ignore structural control.

intersect the course of another stream. When this happens, the pirate stream "captures" the water in the stream it intersects, and the water of the captured stream starts flowing into the pirate stream (▶Fig. 14.26a, b).

Superposed and Antecedent Streams

In some locations, the structure and topography of the landscape does *not* appear to control the path or course of a

stream. As an example, imagine a stream that carves a deep canyon straight across a strong mountain ridge—why didn't the stream find a way around the ridge? We distinguish two types of streams that cut across resistant topographic highs:

• *Superposed streams:* Imagine a region in which drainage initially forms on a layer of soft, flat strata that unconformably overlies folded strata. Streams carve channels into the flat strata; when they eventually erode down through the unconformity and start to downcut into the folded strata, they maintain their earlier course, ignoring the structure of the folded strata. Geologists call such streams **superposed streams,** because their preexisting geometry has been laid down on the rock structure (▶Fig. 14.27a, b).

• *Antecedent streams:* In some cases, tectonic activity (such as subduction or collision) causes a mountain range to rise up beneath an already established stream. If the

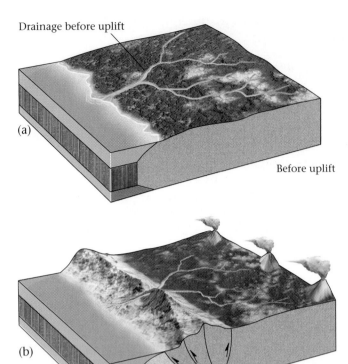

Drainage before uplift

(a)

Before uplift

(b)

Antecedent drainage
cuts through uplift.

New course

(c)

Diverted drainage;
older drainage is
diverted by uplift.

FIGURE 14.28 (a) An antecedent stream flows across the land to the sea. (b) A mountain range develops across the path of the stream. If the stream erosion keeps pace with the rate of uplift, the stream cuts across the mountain range and is an antecedent stream. (c) If uplift happens faster than erosion, the stream is diverted and flows along the front of the range. This stream is not antecedent.

stream downcuts as fast as the range rises, it can maintain its course and will cut right across the range. Geologists call such streams **antecedent streams** (from the Greek *ante,* meaning "before"), to emphasize that they existed before the range uplifted. Note that if the range rises faster than the stream downcuts, the new highlands

divert (change) the stream's course so that it flows along the range face (▶Fig. 14.28a–d).

14.8 RAGING WATERS

[And Enhil, the ruler of the gods, said,] "The earth bellows like a herd of wild oxen. The clamor of human beings disturbs my sleep. Therefore, I want Adad [god of the skies] to cause heavy rains to pour down upon the Earth, both day and night. I want a great flood to come like a thief upon the Earth, steal the food of these people and destroy their lives."

—from the EPIC OF GILGAMESH (written in Sumeria, c. 2100 B.C.E.)

The Inevitable Catastrophe

Up to now, this chapter has focused on the variety of features and processes of a river system (see art on pp. 410–11). Now we turn our attention to the havoc that a stream can cause when it floods. Floods can be catastrophic—they can strip land of forests and buildings, they can bury land in mud and silt, and they can submerge cities. A flood occurs when the volume of water flowing down a stream exceeds the volume of the stream channel, so water rises out of the normal channel and spreads out over the floodplain or delta plain (by breaking through levees), or it fills a canyon to a greater depth than normal. Because of its increased discharge, a stream in flood flows faster, is more turbulent, has greater competence, and exerts more pressure on structures in its path than does a stream when flows are normal.

Floods happen (1) during abrupt, heavy rains, when water falls on the ground faster than it can infiltrate and thus immediately becomes surface runoff; (2) after a long period of continuous rain, when the ground has become saturated with water and can hold no more; (3) when heavy snows from the previous winter melt rapidly in response to a sudden hot spell; or (4) when an artificial or natural dam holding back a lake suddenly collapses and the contents of the lake suddenly flow downstream. In all these cases, the discharge increases dramatically.

Geologists find it convenient to divide floods into two general categories. A **seasonal flood** gradually submerges a floodplain and covers a broad region (▶Fig. 14.29a–d). Such floods typically take hours to days to develop, allowing people living in the floodplain to evacuate before the water gets too deep. These floods, which sometimes are known as floodplain floods, generally occur during the rainy season or when spring melting takes place. Because seasonal floods affect a broad area and may last for days or weeks, they can cause staggering losses of life and property—a 1931 flood of the Yangtze River in China, for example, destroyed fields and disrupted transportation, leading to a famine that killed 3.7 million people. During a **flash flood,** water in a stream rises so fast that it may be impossible for people to escape from the path of the flood. Flash floods typically happen as a result of unusually intense local rainfall, or as a result of a dam collapse. They tend to be more of a problem in narrow valleys or canyons (and thus may also be known as canyon floods), where water cannot spread out over a broad area and thus rapidly becomes very deep. In fact, a flash flood may arrive as a wall of water, slamming downstream with great force. Because the flash floods contain water from only a single storm or single reservoir, the supply of water that feeds them is limited, so they tend to last only from a few minutes to a few hours.

Case Study: A Seasonal Flood (Midwestern United States)

In the spring of 1993, the jet stream, the high-altitude (10–15 km high) wind current that controls weather systems, drifted southward. For weeks, the jet stream's cool, dry air formed an invisible wall that trapped warm, moist air from the Gulf of Mexico over the central United States. When this moist air rose to higher elevations, it cooled, and the water it held condensed and fell as rain, rain, and more rain. In fact, almost a whole year's supply of rain fell in just that spring—some regions received 400% more than usual. Eventually, the ground became saturated and could no longer absorb additional water, so the excess

(a)

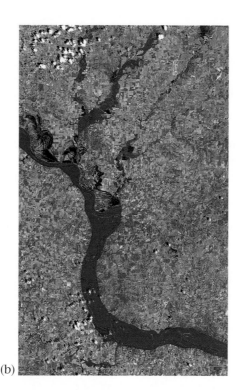

(b)

FIGURE 14.29 (a) Satellite photo showing the Mississippi and Missouri Rivers during a time of drought as seen from high altitude. (b) The same area during the 1993 flood. Notice how the floodplains of the rivers are totally submerged. (c) Great Falls, Montana, submerged by floodwaters in 1975. (d) After floodwaters recede (from an orchard in Arizona), they leave a layer of mud and silt.

(c)

(d)

River Systems

Rivers, or streams, drain the landscape of surface runoff. Typically, an array of connected streams called a drainage network develops, consisting of a trunk stream into which numerous tributaries flow. The land drained is the network's watershed. A stream starts from a source, or headwaters, in the mountains, perhaps collecting water from rainfall or from melting ice and snow. In the mountains, streams carve deep, V-shaped valleys and tend to have steep gradients. For part of its course, a river may flow over a steep, bouldery bed, forming rapids. It may drop off an escarpment, creating a waterfall. Rivers gradually erode landscapes and carry away debris, so after a while, if there is no renewed uplift, mountains evolve into gentle hills. Through time, rivers can bevel once-rugged mountain ranges into nearly flat plains.

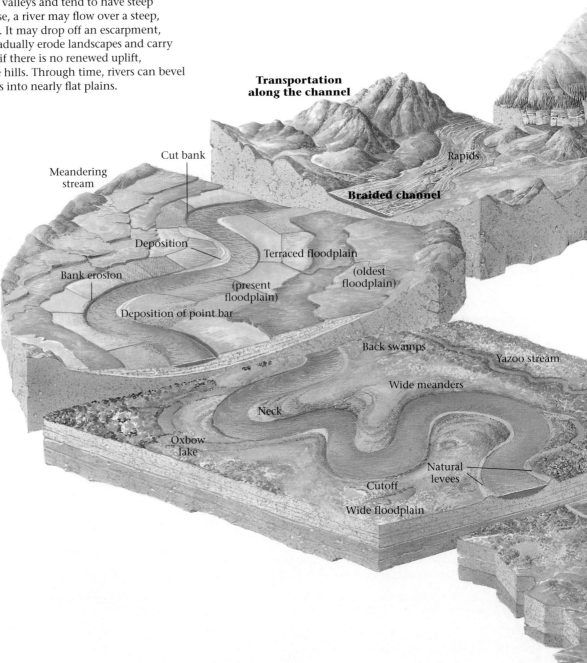

Transportation along the channel

Rapids

Cut bank

Meandering stream

Braided channel

Deposition

Terraced floodplain

Bank erosion

(present floodplain)

(oldest floodplain)

Deposition of point bar

Back swamps

Yazoo stream

Wide meanders

Neck

Oxbow lake

Natural levees

Cutoff

Wide floodplain

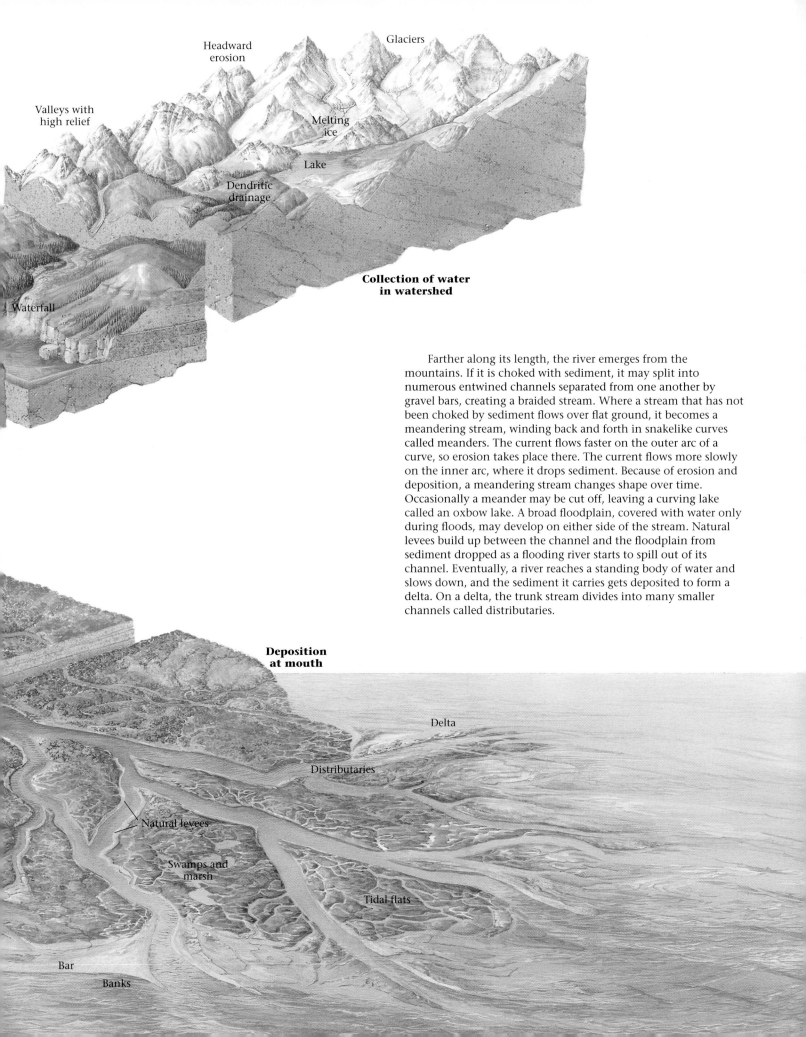

**Collection of water
in watershed**

Glaciers

Headward
erosion

Valleys with
high relief

Melting
ice

Lake

Dendritic
drainage

Waterfall

Farther along its length, the river emerges from the mountains. If it is choked with sediment, it may split into numerous entwined channels separated from one another by gravel bars, creating a braided stream. Where a stream that has not been choked by sediment flows over flat ground, it becomes a meandering stream, winding back and forth in snakelike curves called meanders. The current flows faster on the outer arc of a curve, so erosion takes place there. The current flows more slowly on the inner arc, where it drops sediment. Because of erosion and deposition, a meandering stream changes shape over time. Occasionally a meander may be cut off, leaving a curving lake called an oxbow lake. A broad floodplain, covered with water only during floods, may develop on either side of the stream. Natural levees build up between the channel and the floodplain from sediment dropped as a flooding river starts to spill out of its channel. Eventually, a river reaches a standing body of water and slows down, and the sediment it carries gets deposited to form a delta. On a delta, the trunk stream divides into many smaller channels called distributaries.

**Deposition
at mouth**

Delta

Distributaries

Natural levees

Swamps and
marsh

Tidal flats

Bar

Banks

entered the region's streams, which carried it into the Missouri and Mississippi rivers. Eventually, the water in these rivers rose above the height of levees and spread out over the floodplain. By July, parts of nine states were under water (Fig. 14.29a, b).

The roiling, muddy flood uprooted trees, cars, and even coffins (which floated up from inundated graveyards). All barge traffic along the Mississippi came to a halt, bridges and roads were undermined and washed away, and towns along the river were submerged in muddy water. For example, in Davenport, Iowa, the riverfront district and the baseball stadium were covered with 4 m (14 feet) of water. In Des Moines, Iowa, 250,000 residents lost their supply of drinking water when floodwaters contaminated the municipal water supply with raw sewage and chemical fertilizers. Rowboats replaced cars as the favored mode of transportation in towns where only the rooftops remained visible.

For seventy-nine days, the flooding continued. When the water finally subsided, it left behind a thick layer of silt and mud, filling living rooms and kitchens in floodplain towns and burying crops in floodplain fields. In the end, more than 40,000 square km of the floodplain had been submerged, fifty people died, at least 55,000 homes were destroyed, and countless acres of crops were buried. Officials estimated that the flood caused over $12 billion in damage.

Case Study: A Flash Flood (Big Thompson Canyon)

On a typical sunny day in the Front Ranges of the Rocky Mountains, north of Denver, the Big Thompson River seems quite harmless. Clear water, dripping from melting ice and snow higher in the mountains, flows down its course through a narrow canyon, frothing around boulders. In places, vacation cabins, campgrounds, and motels line the river, for the pleasure of tourists. The landscape seems immutable, but, as is the case with so many geologic features, permanence is an illusion.

On July 31, 1976, easterly winds blew warm, moist air from the Great Plains toward the Rocky Mountain front. As this air rose over the mountains, towering thunderheads built up, and at 7:00 P.M. rain began to fall. It poured, in quantities that even old-timers couldn't recall. In a little over an hour, 7.5 inches (19 cm) drenched the watershed of the Big Thompson River. The river's discharge grew to more than four times the maximum recorded at any time during the previous century. The river rose quickly, in places reaching depths several meters above normal. Turbulent water swirled down the canyon at up to 8 m per second and churned up so much sand and mud that it became a viscous slurry. Slides of rock and soil tumbled down the steep slopes bordering the river and fed the torrent with even more sediment. The water undercut house foundations and washed the houses away, along with their inhabitants. Roads and bridges disappeared

(►Fig. 14.30). Boulders that had stood like landmarks for generations bounced along in the torrent like beachballs, striking and shattering other rocks along the way; the largest rock known to be moved by the flood weighed 275 tons. Cars drifted downstream until they finally wrapped like foil around obstacles. When the flood subsided, the canyon had changed forever, and 144 people had lost their lives.

Ice-Age Mega-Floods

Perhaps the greatest floods recorded in the geologic record happen when natural ice dams burst. The Great Missoula Floods of about 11,000 years ago illustrate this phenomenon. These floods occurred at the end of the last ice age, when a glacier acted like a dam, holding back a large lake called Glacial Lake Missoula. When the glacier melted and the dam suddenly broke, the lake abruptly drained, and water roared over what is now eastern Washington, eventually entering the Columbia River Valley and flowing on out to the Pacific Ocean. When the glacier then grew again, the dam reformed, trapping a new lake, which drained during a subsequent failure. These floods created the channeled scablands of eastern Washington, a region where the soil and regolith have been stripped off the land surface, leaving barren, craggy rock (►Fig. 14.31).

Living with Floods

Mark Twain once wrote of the Mississippi that we "cannot tame that lawless stream, cannot curb it or confine it, cannot say to it, 'go here or go there,' and make it obey." Was Twain right? Since ancient times, people have attempted

FIGURE 14.30 During the 1976 Big Thompson River flood, this house was carried off its foundation and dropped on a bridge.

to confine rivers to a set course so as to prevent undesired flooding. In the twentieth century, flood-control efforts intensified as the population living along rivers increased. For example, since the passage of the 1927 Mississippi River Flood Control Act (drafted after a disastrous flood took place that year), the U.S. Army Corps of Engineers has labored to control the Mississippi. First, engineers built about 300 dams along the river's tributaries so that excess runoff could be stored in reservoirs and later be released slowly. Second, they built artificial levees of sand and mud, and built concrete floodwalls. Artificial levees and floodwalls increase the channel's volume and protect portions of the floodplain (▶Fig. 14.32a, b).

But although the Corps' strategy worked for floods up to a certain size, it was insufficient to handle the 1993 flood. Because of the volume of water drenching the Midwest during the spring and early summer of that year, the reservoirs were filled to capacity, and additional runoff headed downstream. The river rose until it spilled over the top of some artificial levees and undermined others. Undermining occurs when rising floodwaters increase the water pressure on the river side of the levee, forcing water through sand under the levee. In susceptible areas, water begins to spurt out of the ground on the dry side of the levee, thereby washing away the levee's support. The levee finally becomes so weak that it collapses, and water fills in the area behind it.

Sooner or later a flood comes along that can breach a river's levees, allowing water to spread out over the floodplain. Defensive efforts merely delay the inevitable, for it is unfeasible and too expensive to build levees high enough to handle all conceivable floods. And in some cases, building artificial levees may be counterproductive, since they constrain water to a smaller area and thus make floodwaters rise

to a higher level than they would if they were free to spread over a wide floodplain.

There are other ways to prevent floods besides building levees and reservoirs. For example, transforming portions of floodplains back into natural wetlands helps prevent floods, for wetlands absorb water like a sponge. A solution to some flooding in some cases may lie in the removal, rather than the construction, of levees. Property may also be kept safe by defining floodways, regions likely to be flooded, and then by moving or abandoning buildings located there. Even the simple act of moving levees farther away from the river and creating natural habitats in the resulting floodways would decrease flooding damage immensely.

When making decisions about investing in flood-control measures, mortgages, or insurance, planners need a basis for defining the hazard or risk posed by flooding. If floodwaters submerge a locality every year, a bank officer would be ill advised to approve a loan that would promote building there. But if floodwaters submerge the locality very rarely, then the loan may be worth the risk. Geologists characterize the risk of flooding in two ways. The **annual probability** of flooding indicates the likelihood that a flood of a given size or larger will happen at a specified locality during any given year. For example, if we say that a flood of a given size has an annual probability of 1%, then we mean that there is a 1 in 100 chance that a flood of at least this size will happen in any given year. The **recurrence interval** of a flood of a given size is defined as the *average* number of years between successive floods of at least this size. For example, if a flood of a given size happens once in 100 years, *on average,* then it is assigned a recurrence interval of 100 years and is called a 100-year-flood.

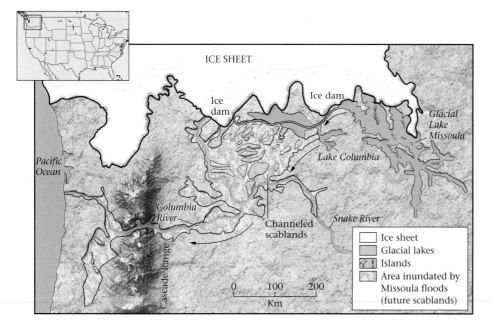

FIGURE 14.31 The map illustrates Glacial Lake Missoula, blocked by an ice dam. When the dam melted away and finally broke, the lake drained westward, eroding away all sediment cover and leaving behind scoured basalt—a region now known as the channeled scablands.

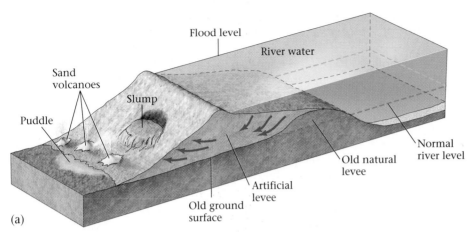

(a)

(b)

FIGURE 14.32 (a) When the water level on the river side of the levee is much higher than on the dry floodplain, pressure causes water to infiltrate the ground and flow through this artificial levee. The water spurts out of the ground on the dry side of the levee, generating sand volcanoes. Water saturates the levee, so the face of the levee slumps. The levee eventually collapses. (b) A concrete floodwall at Cape Girardeau, Missouri. When floods threaten, a crane drops a gate into the slot to hold out the Mississippi River. High-water marks are indicated by black lines.

Note that annual probability and recurrence interval are related:

$$\text{annual probability} = \frac{1}{\text{recurrence interval}}.$$

For example, the annual probability of a 50-year-flood is 1/50, which can also be written as 0.02 or 2%.

Note again that the recurrence interval is simply an average determined by studying the past history of flooding in a stream. Some floodplain residents have the impression that since a 100-year flood just happened, another won't happen for 100 years. This is a false impression. When we

say that a given flood is a "100-year flood," we simply mean that a flood of such a size has a 1% chance of happening in a given year. We do *not* mean that such floods happen regularly at 100-year intervals. Two 100-year floods can happen in the same year, can happen 80 years apart, or can happen 210 years apart.

The recurrence interval for a flood along a particular river reflects the size of a flood. For example, a 100-year flood is larger than a 2-year flood, and happens less frequently (▶Fig. 14.33a). To define this relationship, geologists construct graphs that plot flood discharge on the vertical axis against either recurrence interval or the year on the horizontal axis (▶Fig. 14.33b, c).

Knowing the discharge during a flood of a specified annual probability, and knowing the shape of the river channel and the elevation of the land bordering the river, hydrologists can predict the extent of land that will be submerged by such a flood. Such data, in turn, permit hydrologists to produce flood-hazard maps. In the United States, the Federal Emergency Management Agency (FEMA) produces Flood Insurance Rate Maps that show the 1% annual probability (100-year) flood area and the 0.2% annual probability (500-year) flood risk zones (▶Fig. 14.34).

14.9 RIVERS: A VANISHING RESOURCE?

As *Homo sapiens* evolved from hunter-gatherers into farmers, areas along rivers became attractive places to settle. Considering the multitudinous resources that rivers provide, it's no coincidence that early civilizations gathered in river valleys and on floodplains. Unfortunately, over time, humans have increasingly tended to abuse or overuse the Earth's rivers. Here we note four pressing environmental issues.

Pollution. The capacity of some rivers to carry pollutants has long been exceeded, transforming them into deadly cesspools. Pollutants include raw sewage and storm drainage from urban areas, spilled oil, toxic chemicals from industrial sites, and excess fertilizer and animal waste from agricultural fields.

Dam construction. In 1950, there were about 5,000 large (over 15 m high) dams worldwide, but today there are over 38,000. Damming rivers has both positive and negative results. Reservoirs provide irrigation water and hydroelectric power, and they trap some floodwaters and create popular recreation

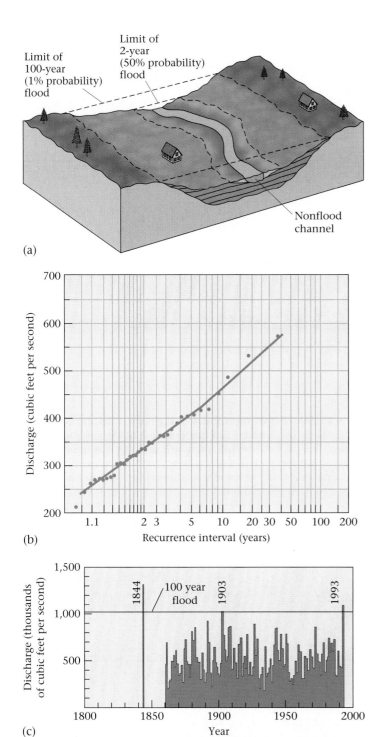

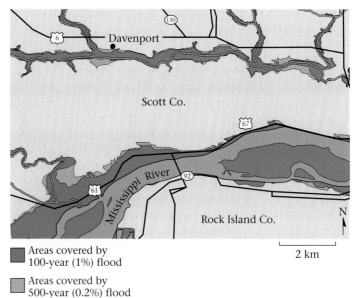

FIGURE 14.34 A flood hazard map for a region near Davenport, Iowa, as prepared by FEMA. It shows areas likely to be flooded.

FIGURE 14.33 (a) A 100-year flood covers a larger area than a 2-year flood. (b) A flood-frequency graph shows the relationship between the recurrence interval and discharge for an idealized river. (c) The peak discharge of the Mississippi River as measured at St. Louis, Missouri. Each bar represents the largest discharge of a given year. The horizontal line represents the discharge of a 100-year flood.

areas. But sometimes their construction destroys wild rivers (the whitewater streams of hilly and mountainous areas) and alters the ecosystem of a drainage network by providing barriers to migrating fish, by decreasing the nutrient supply to organisms downstream, by removing the source of sediment for the delta, and by eliminating seasonal floods that replenish nutrients in the landscape.

Overuse of water. Because of growing populations, our thirst for river water continues to increase, but the supply of water does not. The use of water has grown especially in response to the "Green Revolution" of the 1960s, during which huge new tracts of land came under irrigation. Today, 65% of the water taken out of rivers is used for agriculture, 25% for industry, and 9% for drinking and sewage transport. Civilization needed three times as much river water in 1995 as it did in 1950.

As a result, in some places human activity consumes the entire volume of a river's water, so that the channel contains little more than a saline trickle, if that, at its mouth. For example, except during unusually wet years, the Colorado River contains almost no water where it crosses the Mexican border, for huge pipes and canals carry the water instead to Phoenix and Los Angeles (▶Fig. 14.35). In the case of the Colorado River, states along its banks have established legal agreements that divide up the river's water. Unfortunately, the agreements were written during wet years when the river had unusually large discharge. Thus, the amount of water specified in the agreements actually exceeds the amount of water the river carries in most years.

The Effects of Urbanization and Agriculture on Flooding

Human society has had a major impact on the world's rivers. People have introduced pollution into streams, have restricted flow by building dams, and have diverted flow into canals and ditches that carry it to farms and cities. Notably, overuse of water by people has resulted in the long-term drying up of many rivers around the world. But although the consumption of water for agricultural and industrial purposes decreases the overall supply of river water, urbanization may actually increase the short-term supply. Cities cover the ground with impermeable concrete or blacktop, so rainfall does not soak into the ground but rather runs into storm sewers and then into streams, causing local flooding. Stream discharge during a rainfall thus increases much more rapidly than it would without urbanization. Similarly, although damming rivers decreases the amount of silt a river carries downstream, agriculture may increase the sediment supply: agriculture decreases the vegetative cover on the land, so that when it rains, soil washes into streams.

FIGURE 14.35 The Central Arizona Project canal, shunting water from the Colorado River to Phoenix.

CHAPTER SUMMARY

• Streams are bodies of water that flow down channels and drain the land surface. Channels develop when sheet-wash cuts into the substrate and concentrates the water flow; they grow by headward erosion.

• Drainage networks consist of many tributaries that flow into a trunk stream.

• Permanent streams exist where the water table lies above the bed of the channel. Where the water table lies below the channel bed, streams are ephemeral.

• The discharge of a stream is the volume of water passing a point in a second.

• Streams erode the landscape by scouring, lifting, abrading, and dissolving. Sediment in a stream consists of dissolved, suspended, and bed loads. The total quantity of sediment carried by a stream is its capacity. Competence is the maximum particle size a stream can carry. When stream water slows, it deposits alluvium.

• Longitudinal profiles of streams are concave up, for a stream has steeper gradients at its headwaters than near its mouth. Streams cannot cut below the base level.

• Streams cut valleys or canyons. Where a stream has a bed littered with large rocks, rapids develop. Where a stream plunges off a vertical face, a waterfall forms.

• Meandering streams wander back and forth across a floodplain. Such a stream erodes its outer bank and builds out sediment into a point bar on the inner bank. Eventually,

THE VIEW FROM SPACE The Ganges River drains the base of the Himalaya Mountains and carries sediment into the Indian Ocean. The river divides into many meandering channels and has built out a huge delta. These low-lying lands are home to millions of people, but many of these areas flood during the monsoon season, or during typhoons.

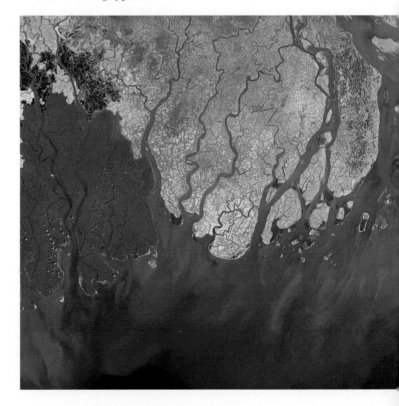

a meander may be cut off and turn into an oxbow lake. Natural levees form on either side of the river channel.

• Where streams or rivers flow into standing water, they deposit deltas.

• With time, fluvial erosion can bevel landscapes to a nearly flat plain. If the base level drops or the land surface rises, stream rejuvenation takes place.

• If an increase in rainfall or spring melting causes more water to enter a stream than the channel can hold, a flood results. Some floods submerge broad floodplains. Flash floods happen very rapidly. Officials try to prevent floods by building reservoirs and levees.

• We can specify the likelihood of flooding by giving the annual probability, or the recurrence interval.

KEY TERMS

alluvial fan (p. 402)	flood (p. 392)
alluvium (p. 398)	floodplain (pp. 398, 403)
annual probability (p. 413)	headward erosion (p. 392)
antecedent stream (p. 408)	longitudinal profile (p. 398)
bar (p. 398)	meander (p. 402)
base level (p. 399)	natural levee (p. 403)
braided stream (p. 402)	permanent stream (p. 393)
capacity (p. 397)	recurrence interval (p. 413)
channel (p. 392)	seasonal flood (p.408)
competence (p. 397)	sheetwash (p. 392)
delta (pp. 398, 403)	stream piracy (p. 406)
discharge (p. 395)	stream rejuvenation (p. 406)
distributary (p. 403)	stream terrace (p. 401)
downcutting (p. 392)	stream (p. 392)
drainage divide (p. 393)	superposed stream (p. 407)
drainage network (p. 393)	tributary (p. 392)
ephemeral stream (p. 395)	watershed (p. 393)
flash flood (p. 409)	

REVIEW QUESTIONS

1. What role do streams have during the hydrologic cycle? Indicate various sources of water in streams.

2. Describe the four different types of drainage networks. What factors contribute to the formation of each?

3. What factors determine whether a stream is permanent or ephemeral?

4. How does discharge vary according to the stream's length, climate, and position along the stream course?

5. Why is average downstream velocity always less than maximum downstream velocity?

6. Describe how streams and running water erode the Earth.

7. What are three components of sediment load in a stream?

8. Distinguish between a stream's competence and its capacity.

9. What factors determine the position of the base level?

10. Why do canyons form in some places and valleys in others?

11. How does a braided stream differ from a meandering stream?

12. Describe how meanders form and are cut off.

13. Describe how deltas grow and develop. How do they differ from alluvial fans?

14. How does a stream-eroded landscape evolve as time passes?

15. How are superposed and antecedent drainages similar? How are they different?

16. What human activities tend to increase flood risk and damage?

17. What is the recurrence interval of a flood? Why can't someone say that "the hundred-year flood happened last year, so I'm safe for another hundred years"?

SUGGESTED READING

Leopold, L. B. 1994. *A View of the River.* Cambridge, Mass.: Harvard University Press.

———. *Water, Rivers, and Creeks.* Sausalito, Calif.: University Science Books.

Leopold, L. B., M. G. Wolman, and J. P. Miller. 1995. *Fluvial Processes in Geomorphology.* Garden City, N.Y.: Dover.

Schumm, S. A. 2003. *The Fluvial System.* Caldwell, N.J.: The Blackburn Press.

Smith, K., and R. Ward. 1998. *Floods: Physical Processes and Human Impacts.* New York: Wiley.

Restless Realm: Oceans and Coasts

15.1 INTRODUCTION

A thousand kilometers from the nearest shore, two scientists and a pilot wriggle through the entry hatch of the research submersible *Alvin*, ready for a cruise to the floor of the ocean and, they hope, back. *Alvin* consists of a superstrong metal sphere connected to a cigar-shaped tube (▶Fig. 15.1). The sphere protects its crew from the immense water pressures in the deep ocean, and the tube holds motors and oxygen tanks. When the hatch seals, *Alvin* starts to sinks like a stone, reaching the sea floor two hours later. Then the cramped explorers turn on outside lights to reveal a stark vista of loose sediment, black rock, and the occasional sea creature. For the next few hours the scientists take photographs and use a robotic arm to collect samples. When finished, they release ballast and rise like a bubble.

Alvin dives began in the 1970s, but humans have explored the ocean for tens of centuries. In fact, mapmakers have known that the ocean spanned the entire globe since Ferdinand Magellan's round-the-world voyage of 1519–1522. But systematic study of the ocean only began in the late nineteenth century, with the voyage of a converted British navy ship, H.M.S. *Challenger,* whose on-board scientists spent four years, beginning in 1872, dredging

Land, sea, and sky all come together at the shore. A seal takes advantage of this special place, basking in the sun on the sandy beach of Kangaroo Island, South Austraila.

FIGURE 15.1 The *Alvin*, a submersible used to explore the ocean floor.

rocks from the sea floor, analyzing water composition, collecting specimens of marine organisms, and measuring water depths and currents.

The fields of oceanography (the study of ocean water and its movements), marine geology (the study of the ocean floor), and marine biology (the study of life forms in the sea) expanded rapidly during the latter half of the twentieth century as new technology became available. A fleet of modern research ships invaded the seas. These ships tow instruments that measure characteristics of water, produce seismic-reflection profiles, and provide platforms from which researchers drill into the sea floor. In recent years, satellites circling the globe have collected even more data, permitting researchers to produce global maps of the ocean floor and of variations in seawater characteristics in startling detail. And while ships and satellites investigate the open oceans, land-based geologists continued to explore the processes that modify its margins.

When seen from space, Earth glows blue, for the oceans cover 70.8% of its surface. The sea provides the basis for life, tempers Earth's climate, and spawns its storms. It is a vast reservoir for water and chemicals that cycle into the atmosphere and crust, and for sediment washed off the continents. In this chapter, we first learn the fundamental characteristics of ocean basins and seawater. Then we focus on the landforms that develop along the **coast,** the region where the land meets the sea (and where over 60% of the global population lives today). Finally, we consider how to cope with the hazards of living on the coast.

15.2 LANDSCAPES BENEATH THE SEA

The oceans exist because oceanic lithosphere and continental lithosphere differ markedly from each other in terms of composition and thickness (▶Fig. 15.2; see Chapter 1). The

surface of the denser and thinner oceanic lithosphere lies deeper than the surface of the relatively buoyant, thicker continental lithosphere, creating oceanic basins (low areas) that collect water. On the present-day map of the world, the ocean encircles the globe. For purposes of reference, however, cartographers divide the ocean into several major parts, with somewhat arbitrary boundaries and significantly different volumes (▶Fig. 15.3a, b).

Have you ever wondered what the ocean floor would look like if all the water evaporated? Marine geologists can now provide a clear image of the ocean's **bathymetry,** or variation in depth, based originally on sonar measurements and more recently on measurements made by satellites. Such studies indicate that the ocean contains broad bathymetric provinces, distinguished from one another by their water depth.

Continental Shelves, Slopes, and Rises

Imagine you're in a submersible cruising just above the floor of the western half of the North Atlantic. If you start at the shoreline of North America and head east, you will find that a **continental shelf,** a relatively shallow portion of the ocean in which water depth does not exceed 500 m, fringes the continent. Across the width of the shelf, about 100 to 500 km, the ocean floor slopes seaward at only about 0.3°, an almost imperceptible amount. At its eastern edge, the continental shelf merges with the **continental slope,** which descends to depths of nearly 4 km at an angle of about 2°. From about 4 km down to about 4.5 km, a province called the **continental rise,** the angle decreases until at 4.5 km deep, you find yourself above a vast, nearly horizontal plain called the **abyssal plain.**

Broad continental shelves, like that of eastern North America, form along **passive continental margins,** margins that are not plate boundaries and thus host few earthquakes (▶Fig. 15.4a; see Chapter 2). Passive margins originate after rifting breaks a continent in two; when rifting stops and sea-floor spreading begins, the stretched lithosphere at the boundary between the ocean and continent gradually cools and sinks. Sediment washed off the continent, as well as the shells of marine creatures, buries the sinking crust, slowly producing a pile of sediment up to 20 km thick. The flat surface of this sedimentary pile constitutes the continental shelf.

If you were to take your submersible to the western coast of South America and cruise out into the Pacific, you would find a very different continental margin. After crossing a narrow continental shelf, the sea floor falls off at the relatively steep angle of 3.5° down to a depth of over 8 km. South America does not have a broad continental shelf because it is an **active continental margin,** a margin that coincides with a plate boundary and thus

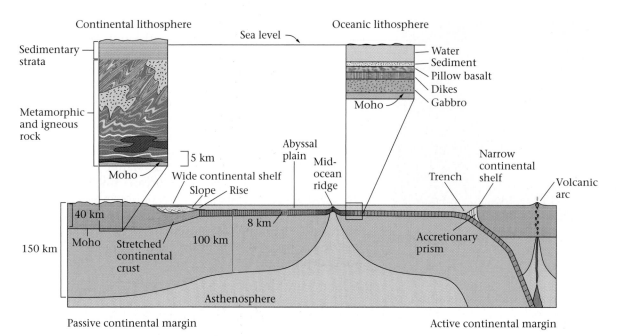

FIGURE 15.2 The bathymetric provinces of the sea floor. At a passive continental margin, a thick wedge of sediment accumulates in an ocean basin over continental lithosphere that had been stretched and thinned during the rifting that formed the basin. The flat surface of this sedimentary wedge creates a wide continental shelf. At an active continental margin, here a convergent plate boundary, a narrow continental shelf forms over an accretionary prism. Insets: These vertical slices through continental and oceanic lithosphere illustrate that oceanic lithosphere is thinner, and that the two kinds differ in composition.

hosts many earthquakes (▶Fig. 15.4b). In the case of South America, the edge of the Pacific Ocean is a convergent plate boundary. The narrow shelf along a convergent plate boundary forms where an apron of sediment spreads out over the top of an accretionary prism, the pile of material scraped off the downgoing subducting plate. Here, the continental slope corresponds to the face of the accretionary prism.

At many locations, relatively narrow and deep valleys called submarine canyons dissect continental shelves and slopes. The largest submarine canyons start offshore of major rivers, and for good reason: rivers cut into the continental shelf at times when sea level was low and the shelf was exposed. But river erosion cannot explain the total depth of these canyons—some slice almost 1,000 m down into the continental margin, far greater than the maximum sea-level change. Submarine exploration demonstrates that much of the erosion of submarine canyons results from the flow of turbidity currents, avalanches of sediment mixed with water (see Chapter 5). When turbidity currents finally reach the base of the continental slope, turbidites (composed of graded beds) accumulate and build up into a submarine fan (▶Fig. 15.4c).

The Bathymetry of Oceanic Plate Boundaries

You can see all three types of plate boundaries by studying the bathymetry of the ocean floor. Sea-floor spreading at a divergent boundary yields a mid-ocean ridge, a 2-km-high submarine mountain belt (▶Fig. 15.5a, b). Because crust stretches and breaks as sea-floor spreading continues, the axis of a ridge includes escarpments, created by normal faulting. Oceanic transform faults, strike-slip faults along which one plate shears sideways past another, typically link segments of mid-ocean ridges (Fig. 15.5a; see Chapter 2). Transforms consist of narrow belts of steep escarpments that connect, along their length, with fracture zones that can be traced into the oceanic plate away from the ridge axis. Subduction at convergent boundaries yields a trench, a deep, elongate trough bordering a volcanic arc. Many trenches are over 8 km deep—the deepest point in the ocean, −11,035 m, lies in the Mariana Trench of the western Pacific.

Abyssal Plains and Seamounts

As oceanic crust ages and moves away from the axis of the mid-ocean ridge, two changes take place. First, the lithosphere cools, and as it does so, its surface sinks. Second, a blanket of

(a)

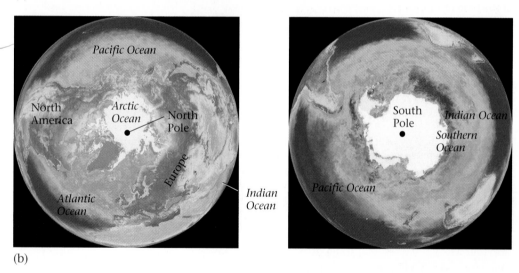

(b)

FIGURE 15.3 (a) The oceans of the world. A SeaWiFS satellite image showing the biologic productivity of the land and sea. In the sea, productivity is indicated by chlorophyll concentration; on land, it is indicated by vegetation cover. (b) The Earth's oceans as viewed looking down on the poles. Note that the Southern Hemisphere (map on the right) is mostly ocean.

sediment gradually accumulates and covers the basalt of the oceanic crust. This blanket consists mostly of microscopic plankton shells and fine flakes of clay, which slowly fall like snow from the ocean water and settle on the sea floor. Because the ocean crust gets progressively older away from the ridge axis, sediment thickness increases away from the ridge axis (see Fig. 15.4c). Eventually, over old sea floor, the sediment buries the escarpments that had formed at the mid-ocean ridge, yielding a flat, featureless surface of the abyssal plain.

Numerous distinct, localized high areas rise above surrounding ocean depths (see Chapter 2 and Figure 15.4a). These high areas result from hot-spot volcanic activity. If the rocks formed by this activity protrude above sea level, they build an oceanic island. Oceanic islands that lie over hot spots host active volcanoes, whereas volcanoes on those islands that have moved off the hot spot are extinct. In warm climates, coral reefs grow and surround oceanic islands. With time, oceanic islands erode, and the seafloor beneath them

(a)

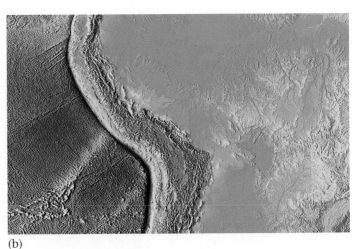

(b)

FIGURE 15.4 (a) Digital bathymetric map portraying the surface of the western Atlantic sea floor. Maps like this one are constructed by computer analysis of satellite data. This map shows a passive continental margin, here a portion of the eastern coast of North America, with a broad continental shelf. Several seamounts protrude from the sea floor east of the continental shelf. (b) A digital bathymetric map of an active continental margin, here the subduction zone on the western coast of South America. (c) Submarine fans accumulate at the base of submarine canyons.

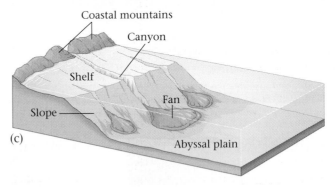

(c)

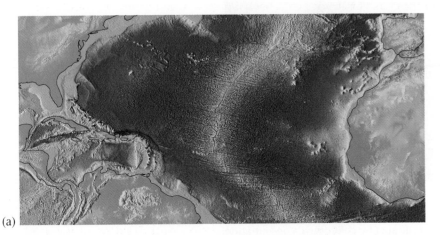

(a)

FIGURE 15.5 (a) A segment of a mid-ocean ridge, showing transform faults that link segments of the ridge. (b) The sea floor slopes away from a mid-ocean ridge and gradually flattens out to become an abyssal plain. Sediment increases in thickness away from the ridge axis, because the sea floor gets older as it moves away from the ridge axis.

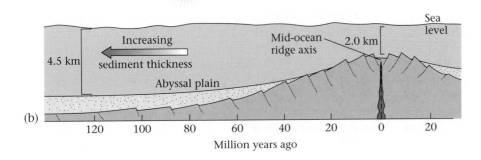

(b)

ages and sinks. As a result, each island's peak eventually submerges. Submarine volcanoes that never emerged from the sea and islands that sank below sea level are called **seamounts.** A seamount that submerges after being overgrown by a reef will have a flat top, and can also be called a guyot. In places where hot-spot igneous activity was particularly voluminous, a broad oceanic plateau, underlain by flood basalt, forms.

15.3 OCEAN WATER AND CURRENTS

Composition

If you've ever had a chance to swim in the ocean, you may have noticed that you float much more easily in ocean water than you do in freshwater. That's because ocean water contains an average of 3.5% dissolved salt (▶Fig. 15.6); in contrast, typical freshwater contains only 0.02% salt. The dissolved ions fit between water molecules without changing the volume of the water, so adding salt to water increases the water's density, and you float higher in a denser liquid.

Most cations in sea salt—sodium (Na^+), potassium (K^+), calcium (Ca^{2+}), and magnesium (Mg^{2+})—come from the chemical weathering of rocks, and the anions, chloride (Cl^-) and sulfate (SO_4^{-2}), from volcanic gases. There's so much salt in the ocean that if all the water suddenly evaporated, a 60-m-thick layer of salt would coat the ocean floor. This layer would consist of about 75% halite (NaCl), with lesser amounts of gypsum ($CaSO_4 \cdot H_2O$), anhydrite ($CaSO_4$), and other salts. Oceanographers refer to the concentration of salt in water as salinity. Although ocean salinity averages 3.5%,

measurements from around the world demonstrate that salinity varies with location, ranging from about 1.0% to about 4.1%. Salinity reflects the balance between the addition of freshwater by rivers or rain and the removal of freshwater by evaporation, for when seawater evaporates, salt stays behind. Salinity may also depend on water temperature, for warmer water can hold more salt in solution than can cold water.

Temperature

When the *Titanic* sank after striking an iceberg in the North Atlantic, most of the unlucky passengers and crew who jumped or fell into the sea died within minutes, because the seawater temperature at the site of the tragedy approached freezing, and cold water removes heat from a body very rapidly. Yet swimmers can play for hours in the Caribbean, where sea-surface temperatures reach 28°C (83°F). Though the *average* global sea-surface temperature hovers around 17°C, it ranges between freezing near the poles to almost 35°C in restricted tropical seas. The correlation of average temperature with latitude exists because the intensity of solar radiation varies with latitude.

Water temperature in the ocean also varies markedly with depth. Waters warmed by the Sun are less dense and tend to remain at the surface. An abrupt boundary, below which water temperatures decrease sharply, reaching near freezing at the sea floor, appears at a depth of about 300 m in the tropics. There is no pronounced boundary in polar seas, since surface waters there are already so cold.

Currents: Rivers in the Sea

Since first setting sail on the open ocean, people have known that the water of the ocean does not stand still, but rather flows or circulates at velocities of up to several kilometers per hour in fairly well-defined streams called **currents.** When the skippers of sailing ships planned their routes from Europe to North America, they paid close attention to the directions of currents, for sailing against a current slowed down the voyage substantially. If they headed due west at a high latitude, they would find themselves battling an eastward-flowing current, the Gulf Stream. Oceanographic studies made since the *Challenger* expedition demonstrate that circulation in the sea occurs at two levels: surface currents affect the upper hundred meters of water, and deep currents keep even water at the bottom of the sea in motion.

Surface currents occcur in all the world's oceans (▶Fig. 15.7). They result from interaction between the sea surface and the wind—as moving air molecules shear across the surface of the water, the friction between air and water drags the water along. If we look at a map that shows global wind patterns along with oceanic currents, we can

FIGURE 15.6 The composition of average seawater. The expanded part of the graph shows the proportions of ions in the salt of seawater.

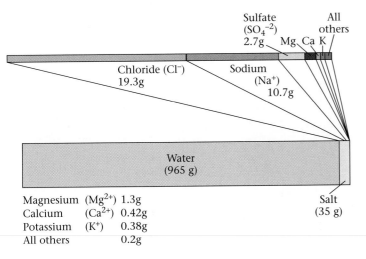

Magnesium	(Mg^{2+})	1.3g
Calcium	(Ca^{2+})	0.42g
Potassium	(K^+)	0.38g
All others		0.2g

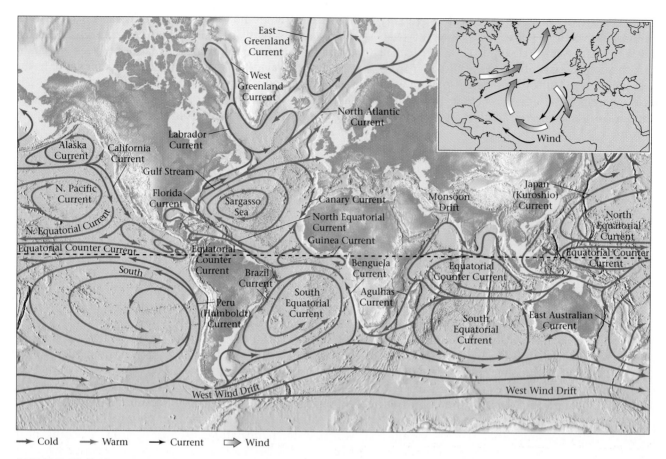

→ Cold → Warm → Current ⇨ Wind

FIGURE 15.7 The major surface currents of the world's oceans. Inset: The relationship between the prevailing wind direction and the North Atlantic Current.

see this relationship (▶Fig. 15.7 inset). But the movement of water resulting from wind shear does not exactly parallel the movement of the wind. This is a consequence of Earth's rotation, which generates the **Coriolis effect** (▶ Box 15.1). This phenomenon causes surface currents in the Northern Hemisphere to veer toward the right and surface currents in the Southern Hemisphere to veer toward the left of the average wind direction.

Deep currents exist because at various locations, water must sink or rise in the vertical direction. Oceanographers have now identified downwelling zones, places where near-surface water sinks, and upwelling zones, places where subsurface water rises.

What causes upwelling and downwelling? First, along coastal regions, these two phenomena exist because as the wind blows, it drags surface water along. If surface water moves toward the coast, then an oversupply of water develops along the shore and excess water must sink—that is, downwelling takes place (▶Fig. 15.8a). Alternatively, if surface water moves away from the coast, then a deficit of water develops near the coast and water rises to fill in the gap—upwelling takes place (▶Fig. 15.8b).

Upwelling of subsurface water also occurs along the equator because the winds blow steadily from east to west. The Coriolis effect causes water to deflect to the right in the Northern Hemisphere and to the left in the Southern Hemisphere. Upwelling replaces this deficit and causes the surface water at the equator to be cooler and rich in nutrients. The nutrients foster an abundance of life in equatorial water.

FIGURE 15.8 (a) Where surface water moves toward shore, it downwells to make room for more water. (b) Where surface water moves offshore, deep water upwells to replace the water that flowed away.

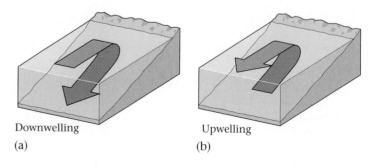

Downwelling
(a)

Upwelling
(b)

Upwelling and downwelling can also be driven by contrasts in water density, caused by differences in temperature and salinity; we refer to the rising and sinking of water driven by density contrasts as **thermohaline circulation.** During thermohaline circulation, denser water (colder and/or saltier) sinks, whereas water that is less dense (warmer and/or less salty) rises. As a result, the water in polar regions sinks and flows back along the bottom of the ocean toward the equator. This process divides the ocean vertically into a number of distinct water masses, which mix only very slowly with one another. In the Atlantic Ocean, for example, the Antarctic Bottom Water sinks along the coast of Antarctica, and the North Atlantic Deep Water sinks in the north polar region (▶Fig. 15.10).

The combination of surface currents and thermohaline circulation, like a conveyor belt, moves water and heat among the various ocean basins (▶Fig. 15.11). This transport of heat moderates global climate.

BOX 15.1
SCIENCE TOOLBOX

The Coriolis Effect

Imagine you are spinning a playground merry-go-round counterclockwise around a vertical axis at a rate of 10 revolutions per minute. The circumference of the outer edge of the merry-go-round is 5 m. Thus, Emma, a child sitting at the outer edge, moves at a velocity of 50 m per minute, whereas David, a child sitting at the center, spins around an axis but moves at zero velocity. If Emma were to try throwing a ball to David by aiming directly along a radius, the ball would seem to veer to the right of the radius and miss David, because the ball is not only moving in the direction parallel to a radius line, but is also moving in the direction parallel to the edge of the circle. If David were to throw a ball along a radius to Emma, this ball would miss Emma because the revolution of the merry-go-round moves her relative to the ball's trajectory (▶Fig. 15.9a, b).

The rotation of the Earth creates the same phenomenon. Earth spins counterclockwise around its axis, so a cannon shell fired along a line of longitude from the North Pole toward the equator veers to the right (west) because the Earth is moving faster to the east at the equator (▶Fig. 15.9c). A cannon shell fired parallel to a line of longitude from the equator to the North Pole veers to the right (east), because as it moves north, it is traveling east faster than the land beneath it (▶Fig. 15.9d). Similarly, a cannon shell fired from the equator to the South Pole veers to the left (east).

In 1835, a French engineer named Gaspard Gustave de Coriolis (1792–1843) proposed that a similar effect would cause the deflection of winds and currents on the surface of the Earth. Because of this Coriolis effect, north-flowing currents in the Northern Hemisphere deflect to the east, whereas south-flowing currents deflect to the west. The opposite is true in the Southern Hemisphere.

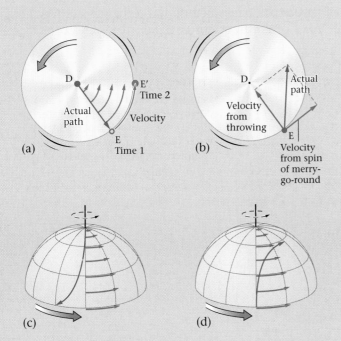

FIGURE 15.9 The Coriolis effect. (a, b) The velocity of a point on the rim of this spinning merry-go-round is greater than the velocity at the center. A ball thrown from point D to point E would follow a straight line, but while the ball is in the air, point E moves to point E'. Relative to the surface of the merry-go-round, the ball looks like it follows a curved path—but remember, the ball goes straight; it's the surface that moves underneath the ball. A ball aimed from the rim to the center won't go straight to the center. Again, since the merry-go-round is moving under the ball, the ball appears to follow a curved path with respect to the surface of the merry-go-round. (c, d) The same phenomenon happens on Earth. A projectile shot from the pole to the equator in the Northern Hemisphere deflects to the west, whereas a projectile shot from the equator to the pole deflects to the east relative to the moving Earth.

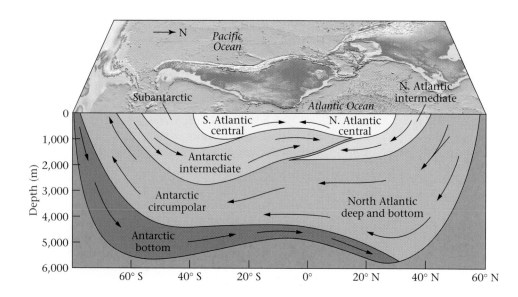

FIGURE 15.10 Because of variations in density, primarily caused by variations in temperature, the oceans are vertically stratified into moving water masses. Each mass has a name. Note that Antarctic bottom water, which sinks down from the surface along the chilly shores of Antarctica, flows northward along the floor of the Atlantic at least as far as the equator.

15.4 THE TIDES GO OUT . . . THE TIDES COME IN . . .

A ship captain seeking to float a ship over reefs, a fisherman hoping to set sail from a shallow port, a tourist eager to harvest shellfish from nearshore mud—all must pay attention to the rise and fall of sea level, a vertical movement called a **tide,** if they are to be successful. If the tide is too low, ships run aground on reefs or stay trapped in harbors. If the tide is too high, a beach may become too narrow to permit access. The tidal reach, meaning the difference between sea level at high tide and sea level at low tide, depends on location. The largest tidal reach on Earth, 16.8 m (54.6 feet), occurs in the Bay of Fundy, on the Atlantic coast of Canada. The intertidal

zone, the region submerged at high tide and exposed at low tide, is a fascinating ecological niche.

During a rising tide, or flood tide, the shoreline (the boundary between water and land) moves inland, whereas during the falling tide, or ebb tide, the shoreline moves seaward. The horizontal distance over which the shoreline migrates between high and low tides depends on the slope of the shore surface (▶Fig. 15.12a-c). Where the slope is gentle and the tidal reach is large, the position of the shoreline can move a long way during a tidal cycle—at low tide a broad tidal flat lies exposed to the air in the intertidal zone (Fig. 15.12b). Such settings can be hazards. For example, in February 2004 fifteen shellfish hunters lost their lives along the coast of northwestern England; they were far offshore,

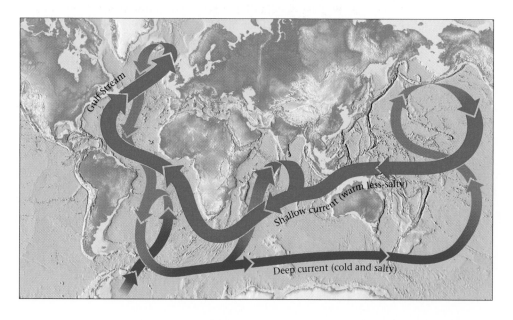

FIGURE 15.11 The exchange between upwelling deep water and downwelling surface water creates a global conveyor belt that circulates water throughout the entire ocean. A complete cycle takes hundreds of years to millennia.

searching for cockles in the mud, when the flood tide came in. Arrival of a flood tide can create a visible wall of water, or tidal bore, that ranges from a few centimeters high to a couple of meters high, and moves at up to 35 km/h—faster than a person can run.

Tides are caused by a **tide-generating force** that is due in part to the gravitational attraction of the Sun and Moon, and in part to centrifugal force caused by the revolution of the **Earth-Moon system** around its center of mass. (Please see an oceanography book, or *Earth: Portrait of a Planet,* for an explanation of this complex concept.) Gravitational pull by the Moon contributes most of the gravitational part of the tide-generating force. The Sun, even though it is larger, is so far away that its contribution is only 46% that of the Moon. Tide-generating forces create two bulges in the global ocean, making this envelope of water more oval shaped than the solid Earth (▶Fig. 15.12d). One bulge, the sublunar bulge, lies on the side of the Earth closer to the Moon. The other, the secondary bulge, lies on the opposite (far) side of the Earth, 12,000 km—the diameter of the Earth—farther from the Moon. A depression in the global ocean surface separates the two bulges. When a shore location lies under a tidal bulge, it experiences a high tide; when it passes under a depression, it experiences low tide.

If the Earth's solid surface were smooth and completely submerged beneath the ocean, so that there were no continents or islands, the timing of tides would be fairly simple to understand. Because the Earth spins on its axis once a day, we would predict two high tides and two low tides at a given point per day. But the story isn't quite that simple—many other factors (such as shapes of basins, air pressure, the position of the Sun) affect the timing and magnitude of tides, so tidal patterns vary with location and time.

15.5 WAVE ACTION

Waves make the ocean surface a restless, ever-changing vista. They develop because of the shear between the molecules of air in the wind and the molecules of water at the

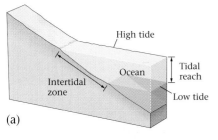

(a)

High tide
Ocean
Intertidal zone
Tidal reach
Low tide

(b)

(c)

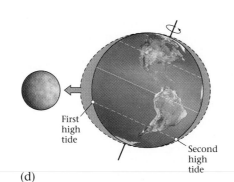

First high tide

Second high tide

(d)

FIGURE 15.12 (a) The tidal reach is the difference between the high and low tide. (b) Mont-Saint-Michel, along the western coast of France, is an island during high tide. (c) At low tide Mont-Saint-Michel is surrounded by tidal flats. (d) A larger tidal bulge appears on the side of the Earth closer to the Moon, and a smaller tidal bulge on the opposite side. Because the Earth spins beneath the bulges, each point on a completely water-covered planet would experience two high tides per day. (But because of land masses and other factors, tides at a given location are more complex.) Since the Earth's axis is tilted with respect to the Moon's orbit, the two high tides in a given day are not the same magnitude.

surface of the sea. When you watch a wave travel across the open ocean, you may get the impression that the whole mass of water constituting the wave moves with the wave. But drop a cork overboard and watch it bob up and down and back and forth; it does not move along with a wave. Within a wave, away from shore, a particle of water moves in a circular motion, as viewed in cross section. The diameter of the circle is greatest at the ocean's surface, where it equals the amplitude of the wave. With increasing depth, though, the diameter of the circle decreases until, at a depth equal to about half the wavelength (the horizontal distance between two wave troughs), there is no wave movement at all (▶Fig. 15.13). Submarines traveling below this **wave base** cruise through smooth water, while ships toss about on the surface above.

The character of waves in the open ocean depends on the strength of the wind (how fast the air moves) and on the fetch of the wind (over how long a distance it blows). When the wind first begins to blow, it creates ripples in the water surface, pointed waves whose amplitude (the height from rest to crest or rest to trough) and wavelength are small. With continued blowing over a long fetch, swells—larger waves with amplitudes of 2–10 m and wavelengths of 40–500 m—begin to build. Hurricane wave amplitudes may grow to over 25 m. Overlapping of waves and interaction between waves and currents may build localized rogue waves over 35 m (100 ft.) high that can submerge the deck of an ocean liner.

Waves have no effect on the ocean floor, as long as the floor lies below the wave base. However, near the shore, where the wave base just touches the floor, it causes a slight

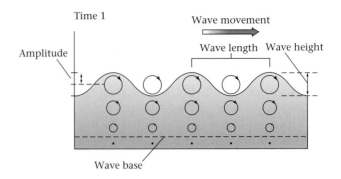

FIGURE 15.13 Within a deep-ocean wave, water molecules follow a circular path. The diameter of the circle decreases with depth to the wave base, below which the wave has no effect. When a wave passes, the shape of the water surface changes, but water does not move as a mass. Note that the amplitude is one-half the wave height.

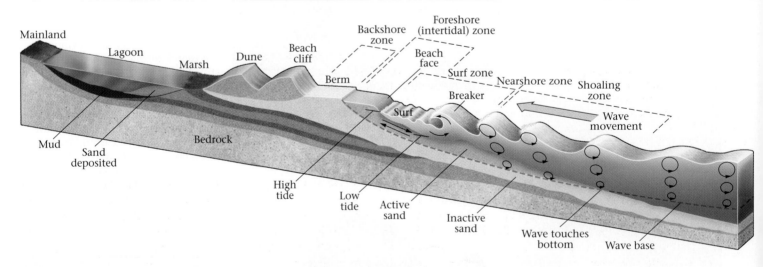

FIGURE 15.14 This profile shows the various landforms of a beach, as well as a cross section of a barrier island. As a wave approaches the shore, it touches the bottom of the sea, at a depth of about half the wavelength. Because of friction, the wave slows down and the wavelength decreases, so the wave height must increase. Because the bottom of the wave moves more slowly than the top, the wave builds up into a breaker that carries water up onto the beach, with the top of the wave falling over the bottom. The water washing up on the beach is swash, and the water rushing back is backwash (indicated by arrows).

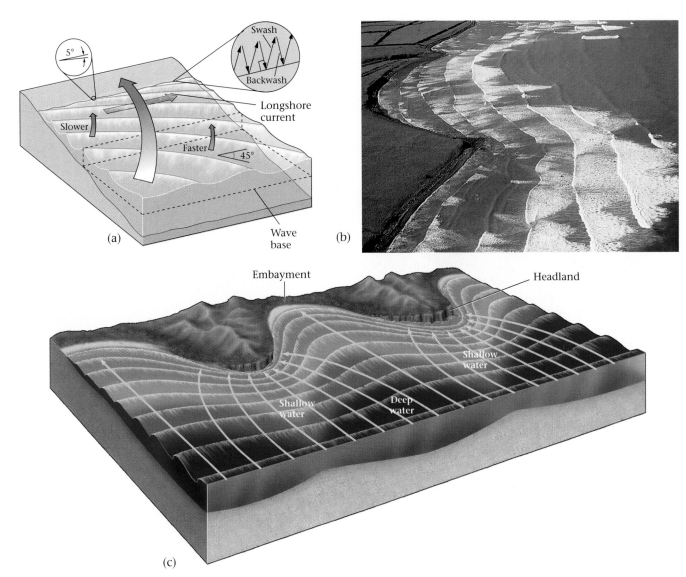

FIGURE 15.15 (a) Wave refraction occurs when waves approach the shore at an angle. The part of the wave that touches bottom first slows down, then the rest of the wave catches up. As a result, the wave bends so that it's nearly parallel with the shore. However, because the wave hits the shore at an angle, water moving parallel to the shore creates a longshore current. (b) Wave refraction on a beach. (c) Like a lens, wave refraction focuses wave energy on a headland, so erosion occurs; and it disperses wave energy in embayments, so deposition occurs.

back-and-forth motion of sediment. Closer to shore, as the water gets shallower, friction between the wave and the sea floor slows the deeper part of the wave, and the motion of water in the wave becomes more elliptical (▶Fig. 15.14). Eventually, water at the top of the wave curves over the base, and the wave becomes a breaker, ready for surfers to ride. Breakers crash onto the shore in the surf zone, sending a surge of water up the beach. This upward surge, or swash, continues until friction brings motion to a halt. Then gravity draws the water back down the beach as backwash.

Waves may make a large angle with the shoreline as they're coming in, but they bend as they approach the shore, a phenomenon called **wave refraction;** right at the shore, their crests make no more than about a 5° angle with the shoreline (▶Fig. 15.15a). To understand why this happens, imagine a wave approaching the shore so that its crest initially makes an angle of 45° with the shoreline. The end of the wave closest to the shore touches bottom first and slows down because of friction, while the end farther offshore continues to move at its original velocity, swinging the whole wave around so that it's more parallel with the shoreline.

Though refraction decreases the angle at which a wave rolls onto shore, the wave may still arrive at an

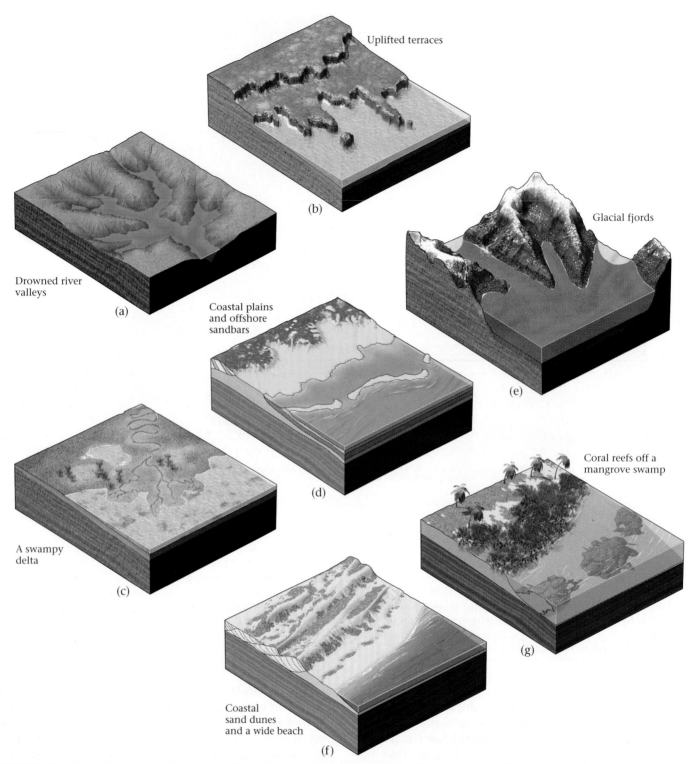

FIGURE 15.16 A wide variety of coastal landforms have developed on Earth. (a) Drowned river valleys, formed where sea level rises and floods valleys, create complex, irregular coastlines. (b) Uplifted terraces develop where the coastline rises relative to sea level and creates escarpments. (c) Swampy deltas form where a sediment-laden stream deposits sediment along the coast. (d) Along sandy coastal plains, large beaches and offshore bars appear. (e) Glacial fjords develop where sea level rises and floods a glacially carved valley. (f) Coastal dunes form where there is a large sand supply and strong wind. (g) In tropical environments, mangrove swamps grow along the shore, protected from wave action by offshore coral reefs.

(a)

(b)

FIGURE 15.17 (a) A pebble and cobble beach, Olympic Peninsula, Washington. The clasts were derived from nearby cliffs. (b) A sand beach, western coast of Puerto Rico.

angle, so there is a component of motion parallel to the shore. This **longshore current** can carry swimmers away from where they entered the water. Also because of wave refraction, wave energy is focused on headlands (places where higher land protrudes into the sea), and is weaker in embayments (places set back from the sea). Thus, erosion happens at headlands, forming a cliff, whereas deposition takes place in embayments, forming a beach (▶Fig. 15.15b, c).

Waves pile water up on the shore incessantly. As the excess water moves back to the sea, it creates a strong, localized seaward flow perpendicular to the beach called a rip current. Rip currents are the cause of many drownings every year along beaches, because they suddenly carry unsuspecting swimmers away from the beach.

15.6 WHERE LAND MEETS SEA: COASTAL LANDFORMS

Tourists along the Amalfi coast of Italy thrill to the sound of waves crashing on rocky shores. But on the coast of Cape Cod, sunbathers lie on seemingly endless sand beaches. Large, domelike mountains rise directly from the sea in Rio de Janeiro, Brazil, but a 100-m-high vertical cliff marks the boundary between the Nullarbor Plain of southern Australia and the Great Southern Ocean. As these examples illustrate, coasts, the belt of land bordering the sea, vary dramatically in terms of topography and associated landforms (▶Fig. 15.16a–g).

Beaches and Tidal Flats

For millions of vacationers, the ideal holiday includes a trip to a **beach,** a gently sloping fringe of sediment along the shore. Some beaches consist of pebbles or boulders, whereas others consist of sand grains (▶Fig. 15.17a, b). Cobble beaches exist only where nearby cliffs or streams supply large rock fragments, for storm waves smash cobbles against one another with enough force to shatter them—with time, cobbles break into smaller fragments. Sediment on a beach tends to be no smaller than sand sized, however, because the swash and backwash of waves winnows out finer sediment and carries it into deeper, quieter water, where it settles.

The composition of sand itself varies from beach to beach, because different sands come from different sources. Sands derived from the weathering and erosion of silicic-to-intermediate rocks consist mainly of quartz; other minerals in these rocks chemically weather to form clay, which washes away in waves. Beaches made from the erosion of limestone or of recent corals and shell beds consist of carbonate sand, including lots of sand-sized chips of shells. And beaches derived by the recent erosion of basalt may have black sand, made of tiny basalt grains.

A beach profile, a cross section drawn perpendicular to the shore, illustrates the shape of a beach (see Fig. 15.14). Starting from the sea and moving landward, a beach consists of a foreshore zone, or intertidal zone, across which the tide rises and falls. The beach face, a steeper, concave part of the foreshore zone, forms where the swash of the waves actively scours the sand. The backshore zone extends from a small step, or escarpment, cut by high-tide swash to the front of the dunes or cliffs that lie farther in-

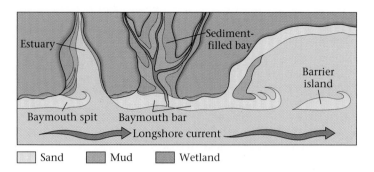

☐ Sand	▨ Mud	▨ Wetland

FIGURE 15.18 Beach drift can generate sand spits and baymouth bars. Sedimentation fills in the region behind a baymouth bar. As a result, the shoreline gets smoother with time.

FIGURE 15.19 Tidal flats are broad muddy areas submerged only at high tide. At low tide, boats at anchor rest in the mud of this tidal flat along the coast of Wales.

shore. The backshore zone includes one or more berms, horizontal to landward-sloping terraces that received sediment during a storm.

Geologists commonly refer to beaches as "rivers of sand," to emphasize that beach sand moves along the coast over time—it is not a permanent substrate. Wave action at the shore moves an active sand layer on the sea floor on a daily basis. Inactive sand, buried below this layer, moves only during severe storms or not at all. Where waves hit the beach at an angle, the swash of each successive wave moves active sand up the beach at an angle to the shoreline, but the backwash moves this sand straight down the slope of the beach, back towards the sea, because of gravity. This sawtooth mo-

tion causes sand to migrate gradually along the beaches, a process called **beach drift** (see Fig. 15.15a). Beach drift, which happens in association with the longshore drift of water, can transport sand hundreds of kilometers along a coast in a matter of centuries. Where the coastline indents landward, beach drift stretches beaches out into open water to create a **sand spit.** Some sand spits grow across the opening of a bay to form a baymouth bar (▶Fig. 15.18).

The scouring action of waves piles sand up in a narrow ridge away from the shore called an **offshore bar,** which

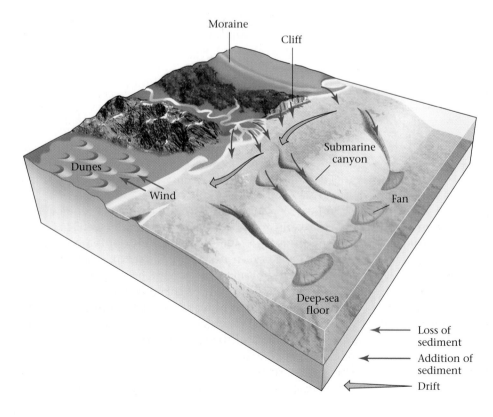

FIGURE 15.20 The sediment budget along a coast. Sediment is brought into the system by rivers, by the erosion of cliffs and moraines, and by wind. Sediment moves along the coast as a result of beach drift. Sediment leaves the system by being blown off the beach, by sinking into deeper water, or by being carried out by the longshore current.

FIGURE 15.21 (a) The major landforms of a rocky shore include cliffs, sea caves, wave-cut notches, sea stacks, sea arches, wave-cut benches, and tombolos. Beaches tend to collect in embayments, whereas erosion happens at headlands. (b) Erosion by waves creates a wave-cut notch. Eventually, the overhanging rock collapses into the sea to form gravel on the wave-cut bench. (c) A wave-cut notch exposed along a rocky shore. (d) A wave-cut bench at the foot of the cliffs at Etrétat, France.

parallels the shoreline. In regions with an abundant sand supply, offshore bars rise above the mean high-water level and become **barrier islands.** The water between a barrier island and the mainland becomes a quiet-water **lagoon,** a body of shallow seawater separated from the open ocean.

Though developers have covered some barrier islands with expensive resorts, barrier islands are temporary features in the time frame of centuries to millennia. Wind and waves pick up sand from the ocean side of the barrier island and drop it on the lagoon side, causing the island to migrate landward. Storms may breach barrier islands and create an inlet (a narrow passage of water). Finally, beach drift gradually transports the sand of barrier islands and modifies their shape.

Tidal flats, regions of mud and silt exposed or nearly exposed at low tide but totally submerged at high tide, develop in regions protected from strong wave action (see Fig. 15.12b; ▶Fig. 15.19). They are typically found along the margins of lagoons or on shores protected by barrier islands. Here, mud and silt accumulate to form thick, sticky layers.

Because of the movement of sediment, the sediment budget (the difference between sand supplied and sand removed) plays an important role in determining the long-term evolution of a beach. Let's look at how the budget works for a small segment of beach (▶Fig. 15.20). Sand may be supplied to the segment from local rivers or by wind from nearby dune fields; it may also be brought from just offshore by waves or from far away by beach drift. (In

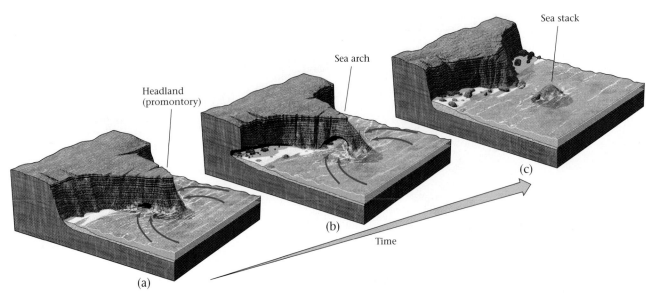

FIGURE 15.22 The erosion of a headland. (a) At first, wave refraction causes wave energy to attack the sides of a promontory, making a sea cave on either side. (b) Gradually erosion breaks through the promontory to create a sea arch. (c) The arch finally collapses, leaving a sea stack.

fact, the large quantity of sand along beaches of the southeastern United States may have originated in Pleistocene glacial outwash far to the north.) Some of the sand from a stretch of beach may be removed by beach drift, while some gets carried offshore by waves, where it either settles locally or tumbles down a submarine canyon into the deep sea. If the lost sand cannot be replaced, the beach segment grows narrower, whereas if the supply of sand exceeds the amount that washes away, the beach becomes wider.

Rocky Coasts

More than one ship has met its end smashed and splintered in the spray and thunderous surf of a rocky coast, where bedrock cliffs rise directly from the sea (see Fig. 15.17a; ▶Fig. 15.21a). Lacking the protection of a beach, rocky coasts feel the full impact of ocean breakers. The water pressure generated during the impact of a breaker can pick up boulders and smash them together until they shatter, and it can squeeze air into cracks, creating enough pressure to wedge rocks free. Further, because of its turbulence, the water hitting a cliff face carries suspended sand, and thus can abrade the cliff. The combined effects of shattering, wedging, and abrading, together called wave erosion, gradually undercut a cliff face and make a **wave-cut notch** (▶Fig. 15.21b, c). Undercutting continues until the overhang becomes unstable and breaks away at a joint, building a pile of rubble at the base of the cliff that waves immediately attack and break up. By this process, wave erosion cuts away at a rocky coast, so that the cliff gradually migrates in-

land. Such cliff retreat leaves behind a **wave-cut bench,** or platform, which becomes visible at low tide (▶Fig. 15.21d).

Many rocky coasts have an irregular shoreline, with headlands protruding into the sea and embayments set back from the sea. Such irregular coastlines tend to change over time: as wave energy focuses on headlands and disperses in embayments (a result of wave refraction). Eventually, erosion removes debris at headlands and sediment accumulates in embayments (see Fig. 15.15c); thus, over time the shoreline may become less irregular.

A headland erodes in stages (▶Fig. 15.22a–c). Because of refraction, waves curve and attack the sides of a headland, slowly eating through it to create a sea arch connected to the mainland by a narrow bridge (▶Fig. 15.23a). Eventually the arch collapses, leaving isolated sea stacks just offshore (▶Fig. 15.23b). Once formed, a sea stack protects the adjacent shore from waves. Therefore, sand collects in the lee of the stack, slowly building a tombolo, a narrow ridge of sand that links the sea stack to the mainland.

Coastal Wetlands

Let's move now from the crashing waves of rocky coasts to the gentlest type of shore, the **coastal wetland,** a vegetated flat-lying stretch of coast that floods with shallow water but does not feel the impact of strong waves. In temperate climates, coastal wetlands include swamps (wetlands dominated by trees), marshes (wetlands dominated by grasses; ▶Fig. 15.24a), and bogs (wetlands dominated by moss and shrubs).

In tropical or semitropical climates (between 30° north and 30° south of the equator), mangrove swamps thrive in wetlands (▶Fig. 15.24b). Mangrove tree roots can filter salt

(a)

(b)

FIGURE 15.23 (a) A sea arch exposed along a rocky coast of southern Australia. Another arch once bridged the gap to the cliff on the left, but it collapsed, stranding several tourists. (b) These sea stacks along the southern coast of Australia, together with others, comprise a tourist attraction called the Twelve Apostles. Unfortunately in July 2005, one of the sea stacks suddenly collapsed.

out of water, so the trees have evolved the ability to survive in freshwater or saltwater. Some mangrove species form a broad network of roots above the water surface, making the plant look like an octopus standing on its tentacles, and some send up small protrusions from roots that rise above the water and allow the plant to breathe. Dense stands of mangroves counter the effects of stormy weather and thus prevent coastal erosion.

Estuaries

Along some coastlines, a relative rise in sea level causes the sea to flood river valleys that merge with the coast, result-ing in **estuaries,** where seawater and river water mix. You can recognize an estuary on a map by the dendritic pattern of its river-carved coastline (▶Fig. 15.25). Oceanic and flu-vial water interact in two ways within an estuary. In quiet estuaries, protected from wave action or river turbulence, the water becomes stratified, with denser oceanic saltwater flowing upstream as a wedge beneath less dense fluvial freshwater. In turbulent estuaries, such as Chesapeake Bay, oceanic and fluvial water combine to create nutrient-rich brackish water with a salinity between that of oceans and rivers. Estuaries are complex ecosystems inhabited by unique species of shrimp, clams, oysters, worms, and fish that can tolerate large changes in salinity.

FIGURE 15.24 Examples of coastal wetlands: (a) a salt marsh; (b) a mangrove swamp.

(a)

(b)

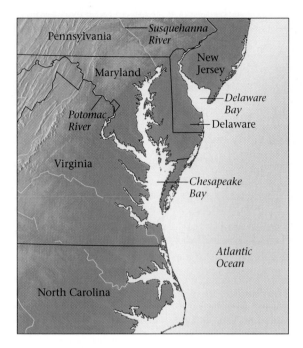

FIGURE 15.25 Chesapeake Bay, a large estuary along the East Coast of the United States, formed when sea level rose and flooded the Potomac and Susquehanna river valleys and the mouths of their tributaries.

Fjords

During the last ice age, glaciers carved deep valleys in coastal mountain ranges. When the ice age came to a close, the glaciers melted away, leaving deep, U-shaped valleys (see Chapter 18). The water stored in the glaciers, along with the water within the vast ice sheets that covered continents during the ice age, flowed back into the sea and caused sea level to rise.

The rising sea filled the deep valleys, creating **fjords,** or flooded glacial valleys. Coastal fjords are fingers of the sea surrounded by mountains; because of their deep water and steep walls of polished rock, they are distinctively beautiful (▶Fig. 15.26a, b). Some of the world's most spectacular fjords decorate the western coasts of Norway, British Columbia, and New Zealand. Smaller examples appear along the coast of Maine and southeastern Canada.

Coral Reefs

In the Undersea National Park of the Virgin Islands, visitors swim through colorful growths of living coral (▶Fig. 15.27a). Some corals look like brains, others like elk antlers, still others like delicate fans. Sea anemones, sponges, and clams grow on and around the coral. Though at first glance coral looks like a plant, it is actually a colony of tiny invertebrates related to jellyfish. An individual coral animal, or polyp, has a tubelike body with a head of tentacles. Corals obtain part of their livelihood by filtering nutrients out of seawater; the remainder comes from algae that live on the corals' tissue. Algae and coral have a symbiotic (mutually beneficial) relationship, in that the algae photosynthesize and provide nutrients and oxygen to the corals, while the corals provide carbon dioxide and nutrients for the algae.

Coral polyps secrete calcite shells, which gradually build into a mound of solid limestone, whose top surface lies from just below the low-tide level down to a depth of about 60 m. At any given time, only the surface of the mound lives—the mound's interior consists of shells from previous generations of coral. The realm of shallow water underlain by coral mounds, associated organisms, and debris makes up a **coral reef** (▶Fig. 15.27b). Reefs absorb wave energy and thus are

FIGURE 15.26 (a) The subsurface shape of a fjord, a drowned U-shaped glacial valley. (b) Fjords in Norway provide spectacular scenery.

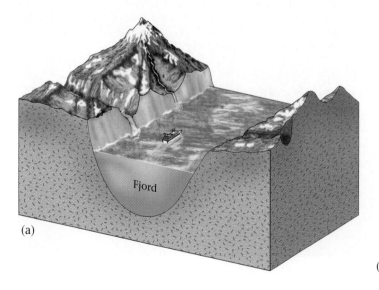

(a)

(b)

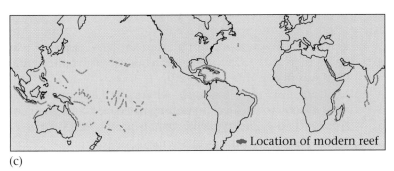

(c)

FIGURE 15.27 (a) Corals and other organisms make up a reef. (b) A coral reef bordering an island in the Caribbean Sea. (c) The distribution of coral reefs on Earth today.

living buffer zones that protect coasts from erosion. Corals need clear, well-lit, warm (18°–30°C) water with normal oceanic salinity, so coral reefs only grow along clean coasts at latitudes of less than about 30° (▶Fig. 15.27c).

Marine geologists distinguish three different kinds of coral reef, based on their geometry (▶Fig. 15.28a–c). A fring-ing reef forms directly along the coast, a barrier reef develops offshore (separated from the coast by a lagoon), and an atoll makes a circular ring surrounding a lagoon. As Charles Darwin first recognized back in 1859, coral reefs associated with islands in the Pacific start out as fringing reefs and then later become barrier reefs and finally atolls. Darwin suggested,

FIGURE 15.28 The progressive change from a fringing reef around a young volcanic island to a ring-shaped atoll. (a) The reef begins to grow around the volcano. (b) The volcano subsides as the sea floor under it ages, so the reef is now a ring, separated from a small island (the peak of the volcano) by a lagoon. (c) The volcano has subsided completely, so that only an atoll surrounding a lagoon remains. When the lagoon fills with debris, and together with the atoll finally sinks below sea level, the result is a guyot.

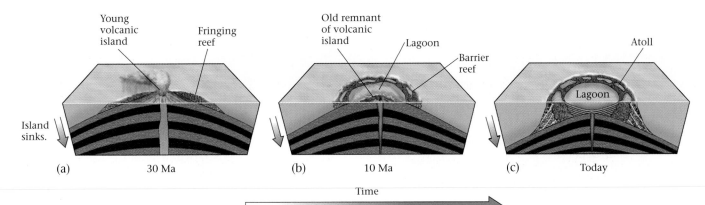

correctly, that this progression reflects the continued growth of the reef as the island around which it formed gradually sank. Eventually, the reef itself sinks too far below sea level to remain alive and becomes the cap of a guyot.

15.7 CAUSES OF COASTAL VARIABILITY

Plate Tectonic Setting

The tectonic setting of a coast plays a role in determining whether the coastline is marked by steep cliffs or broad plains (see art on pp. 440–41). Along an active margin, compression squeezes the crust and pushes it up, producing mountains such as the Andes along the western coast of South America. Along a passive margin, the cooling and sinking of the lithosphere may create a broad **coastal plain,** a flat land that merges with the continental shelf, as exists along the Gulf Coast and southeastern Atlantic coast of the United States. But not all passive margins have coastal plains. At some, the margin of the rift that gave birth to the passive margin remains at a high elevation, even tens of millions of years after rifting ceased. For example, highlands formed during recent rifting border the Red Sea, whereas highlands initiated during Cretaceous rifting persist along portions of the Brazilian coast.

Relative Sea-Level Changes (Emergent and Submergent Coasts)

Sea level, relative to the land surface, changes over geologic time. For example, during the last ice age, because so

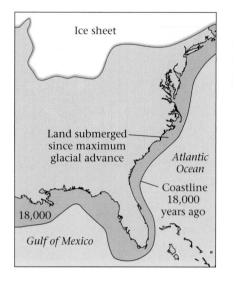

FIGURE 15.29 During the last ice age, sea level fell, and much of North America's continental shelf lay exposed to the air.

much water was trapped in glaciers on land, sea level was lower than it is today. As a consequence, the continental shelf was dry land (▶Fig. 15.29).

Geologists refer to coasts where the land is rising or rose, relative to sea level, as **emergent coasts.** At emergent coasts, steep slopes typically border the shore. A series of steplike terraces form along some emergent coasts (▶Fig. 15.30a, b). These terraces reflect episodic changes in relative sea level. Those coasts at which the land sinks relative to sea level become **submergent coasts** (▶Fig. 15.31a, b). At submergent coasts, landforms include estuaries and fjords that developed when the sea flooded coastal valleys. Chesapeake Bay in eastern North America demonstrates the consequences of submergence.

FIGURE 15.30 (a) Wave erosion creates a wave-cut bench along an emergent coast. (b) After uplift, the land rises, and the bench becomes a terrace.

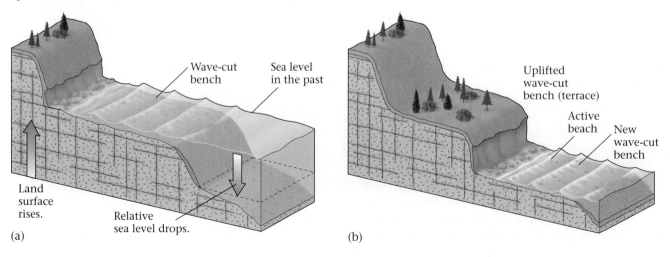

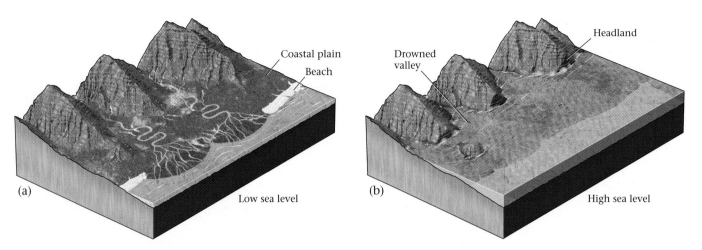

FIGURE 15.31 (a) A coast before sea level rises. Rivers drain valleys onto a coastal plain. (b) As a submergent coast forms, sea level rises and floods the valleys, and waves erode the headlands.

Sediment Supply and Climate

The quantity and character of sediment supplied to a shore affects its character. That is, erosional coasts where the sea washes sediment away faster than it can be supplied recede landward and may become rocky, whereas accretionary coasts that receive more sediment than erodes away grow seaward and develop broad beaches.

Climate also affects the character of a coast. Shores that enjoy generally calm weather erode less rapidly than those constantly savaged by storms. A sediment supply large enough to generate an accretionary coast in a calm environment may be insufficient to prevent the development of an erosional coast in a stormy environment. The climate also affects biological activity along coasts. For example, in the warm water of tropical climates, mangrove swamps flourish along the shore, and coral reefs form offshore. The reefs may build into a broad carbonate platform such as appears in the Bahamas today. In temperate climates, salt marshes develop, and in arctic regions, the coast may be a stark environment of lichen-covered rock and barren sediment.

15.8 COASTAL PROBLEMS AND SOLUTIONS

Contemporary Sea-Level Changes

People tend to view a shoreline as a permanent entity. But in fact, shorelines are ephemeral geologic features. On a time scale of hundreds to thousands of years, a shoreline moves inland or seaward depending on whether relative sea level rises or falls. In places where sea level is rising today, shoreline towns will eventually be submerged. If present rates of sea-level rise along the East Coast of the United States continue, major coastal cities such as Washington, New York, Miami, and Philadelphia may be inundated within the next millennium (▶Fig. 15.32).

FIGURE 15.32 A possible sea-level rise in the future may flood major cities of the northeastern United States. The Washington–New York corridor would lie underwater.

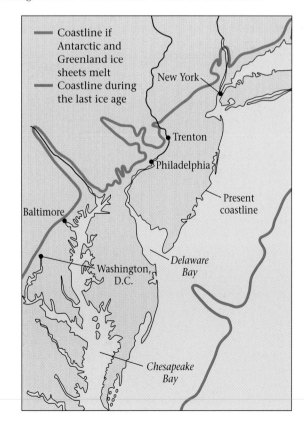

Oceans and Coasts

The oceans of the world provide a diverse array of environments illustrating the full complexity of the Earth System. Tectonic processes and surface processes constantly battle with each other to produce submarine and subaerial landscapes.

Water in the ocean circulates in currents that transport heat from equator to pole. Interactions between the atmosphere and the ocean build waves that ripple the surface. Waves erode shorelines and transport sediment. Sea-floor features define the location of plate boundaries and hot spots. Coastal landforms depend on tectonic setting, climate, and sediment supply. Specifically, passive margins differ markedly from active convergent margins; equatorial coasts differ from sandy coasts. A great variety of organisms inhabit all these realms.

Wave erosion cuts notches at the base of cliffs and bevels wave-cut benches.

Along sandy shores, sand build beaches, sand spits, and bars.

Turbidities flowing down submarine canyons produce submarine fans.

In tropical environments, mangroves live along the shore and coral reefs grow offshore.

Along rocky coasts, sea cliffs, sea arches, and sea stacks evolve.

At a passive margin, a broad continental shelf develops. Submarine slumping may occur along the shelf.

The ocean teems with life

At divergent plate boundaries, a mid-ocean ridge rises. Transform faults, marked by fracture zones, link segments of the ridge.

The Global Conveyor

Surface winds drive surface currents.

Cold water sinks at polar regions.

0
1
2
3 Km
4
5
6

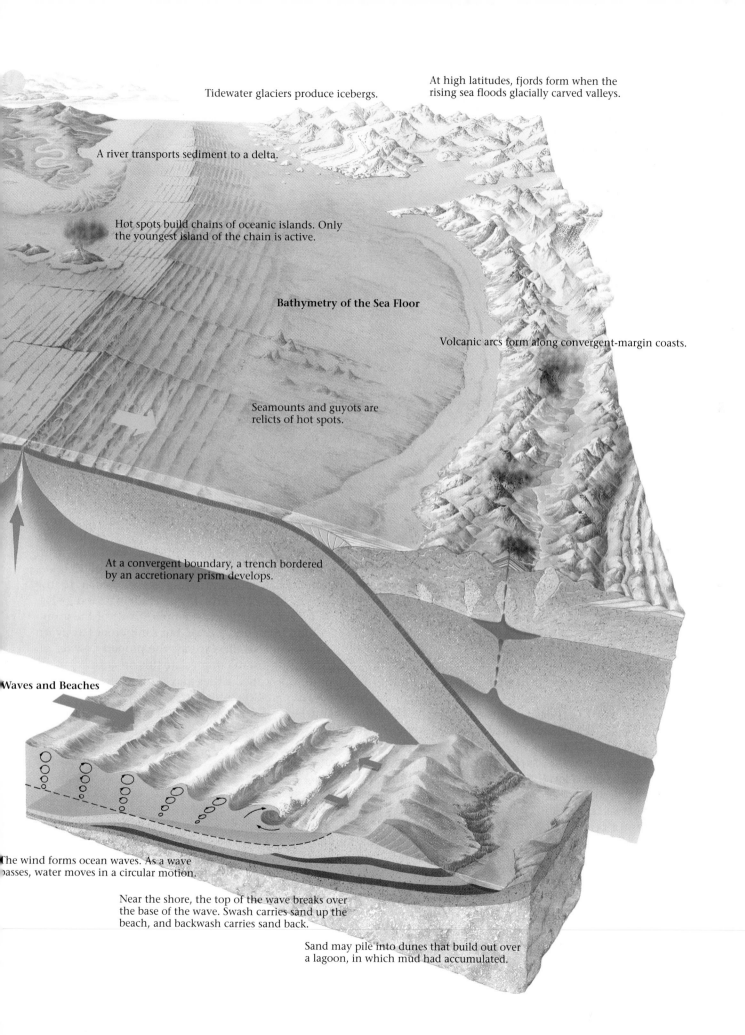

Tidewater glaciers produce icebergs.

At high latitudes, fjords form when the rising sea floods glacially carved valleys.

A river transports sediment to a delta.

Hot spots build chains of oceanic islands. Only the youngest island of the chain is active.

Bathymetry of the Sea Floor

Volcanic arcs form along convergent-margin coasts.

Seamounts and guyots are relics of hot spots.

At a convergent boundary, a trench bordered by an accretionary prism develops.

Waves and Beaches

The wind forms ocean waves. As a wave passes, water moves in a circular motion.

Near the shore, the top of the wave breaks over the base of the wave. Swash carries sand up the beach, and backwash carries sand back.

Sand may pile into dunes that build out over a lagoon, in which mud had accumulated.

(a)

(b)

FIGURE 15.33 (a) Damaged beachfront homes after a hurricane in Florida. (b) Hurricane Isabel (September 2003) eroded away the foundation of this house at Kitty Hawk, North Carolina, causing it to collapse.

Beach Destruction—Beach Protection?

In a matter of hours, a storm can radically alter a landscape that took centuries or millennia to form. The backwash of storm waves sweeps vast quantities of sand seaward, leaving the beach a skeleton of its former self. The surf submerges barrier islands and shifts them toward the lagoon. Waves and wind together rip out mangrove swamps and salt marshes and break up coral reefs, thereby destroying the organic buffer that normally protects the coast and leaving it vulnerable to erosion for years to come. Of course, major storms also destroy human constructions: erosion undermines shoreside buildings, causing them to collapse into the sea; wave impacts smash buildings to bits; and the storm surge—very high water levels created when storm winds push water toward the shore—floats buildings off their foundations (▶Fig. 15.33a, b).

But even less dramatic events, such as the loss of river sediment, a gradual rise in sea level, a change in the shape of a shoreline, or the destruction of coastal vegetation, can alter the balance between sediment accumulation and sediment removal from a beach, leading to beach erosion. In some places, beaches retreat landward at rates of 1 to 2 m per year, forcing homeowners literally to pick up and move their houses.

In many parts of the world, beachfront property has great value. But if a hotel loses its beach sand, it probably won't stay in business. Thus, property owners often construct artificial barriers to protect their stretch of coastline or to shelter the mouth of a harbor from waves. These barriers alter the natural movement of sand in the beach system and thus change the shape of the beach, sometimes with undesirable results. For example, people may build groins, concrete or stone walls protruding perpendicular to the shore, to prevent beach drift from removing sand (▶Fig. 15.34a). Sand accumulates on the updrift side of the groin, forming a long triangular wedge, but sand erodes away on the downdrift side. Needless to say, the property owner on the downdrift side doesn't appreciate this process.

A pair of walls called jetties may protect the entrance to a harbor (▶Fig. 15.34b). But jetties erected at the mouth of a river channel effectively extend the river into deeper water, and thus may lead to the deposition of an offshore sandbar. Engineers may also build an offshore wall called a breakwater, parallel or at an angle to the beach, to prevent the full force of waves from reaching a harbor. With time, however, sand builds up in the lee of the breakwater and the beach grows seaward, clogging the harbor (▶Fig. 15.34c). And to protect expensive shoreside homes, people build seawalls out of riprap (large stone or concrete blocks) or reinforced concrete on the landward side of the backshore zone. Seawalls reflect wave energy that crosses the beach back to sea. Unfortunately, this process increases the rate of erosion at the foot of the seawall, and thus during a large storm the seawall may be undermined so that it collapses (▶Fig. 15.35).

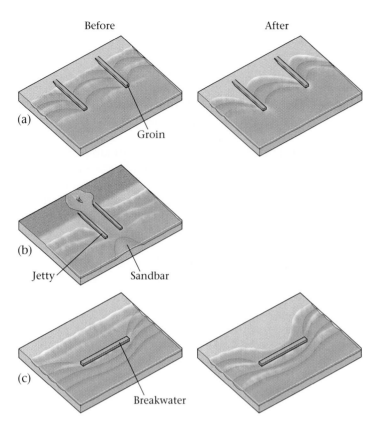

FIGURE 15.34 (a) The construction of groins creates a sawtooth beach. (b) Jetties extend a river farther into the sea, but may result in the deposition of a sandbar at the jetties' ends. (c) A breakwater causes the beach to build out in the lee.

In some places, people have given up trying to decrease the rate of beach erosion, and instead have worked to increase the rate of sediment supply. To do this, they truck or ship in vast quantities of sand to replenish a beach. This procedure, called beach nourishment, can be hugely expensive and at best provides only a temporary fix, for the backwash and beach drift that removed the sand in the first place continue unabated as long as the wind blows and the waves break. Clearly, beach management remains a controversial issue, for beachfront properties are expensive, but the shore is, geologically speaking, a temporary feature whose shape can change radically with the next storm.

Pollution and the Destruction of Organic Coasts

Bad cases of beach pollution create headlines. Because of longshore current, garbage dumped in the sea in an urban area may drift along the shore and be deposited on a tourist beach far from its point of introduction. For example, hospital waste from New York City has washed up on beaches tens of kilometers to the south. Oil spills—most commonly from

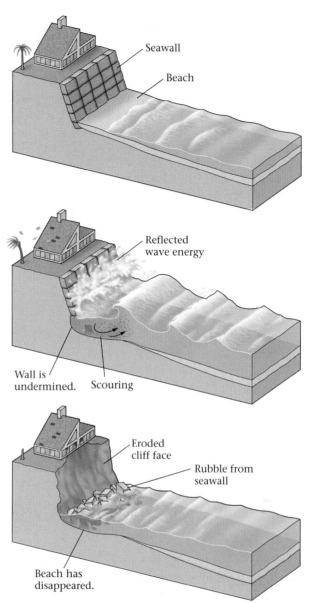

FIGURE 15.35 A seawall protects the sea cliff under most conditions, but during a severe storm the wave energy reflected by the seawall helps scour the beach. As a result, the wall may be undermined and collapse.

ships that flush their bilges but also from tankers that have run aground or foundered in stormy seas—have contaminated shorelines at several places around the world.

Coasts in which living organisms control landforms along the shore are called **organic coasts.** These coasts are particularly susceptible to changes in the environment. The loss of such landforms can increase a coast's vulnerability to erosion and, considering that they provide spawning grounds for marine organisms, can upset the food chain of the global ocean.

In wetlands and estuaries, sewage, chemical pollutants, and agricultural runoff cause havoc. Toxins settle along with clay and concentrate in the sediments, where they contaminate burrowing marine life and then move up the food chain. Fertilizers and sewage that enter the sea with runoff increase the nutrient content of water, triggering algae blooms that absorb oxygen and therefore kill animal and plant life. Coastal wetlands face destruction by development—they have been filled or drained to be converted into farmland or suburbia and have been used as garbage dumps. In most parts of the world, between 20% and 70% of coastal wetlands were destroyed in the last century.

Reefs, which depend on the health of delicate coral polyps, can be devastated by even slight changes in the environment. Pollutants and hydrocarbons, for example, will poison them. Organic sewage fosters algae blooms that rob water of dissolved oxygen and suffocate the coral. And agricultural runoff or suspended sediment introduced to coastal water during beach-nourishment projects reduces the light, killing the algae that live in the coral, and clogs the pores that coral polyps use to filter water. Changes in water temperature or salinity caused by dumping waste water from power plants into the sea or by global warming of the atmosphere also destroy reefs, for reef-building organisms are very sensitive to temperature changes. People can destroy reefs directly by dragging anchors across reef surfaces, by touching reef organisms, or by quarrying reefs to obtain construction materials. In the last decade, marine biologists have noticed that reefs around the world have lost their color and died. This process, called reef bleaching, may be due to the removal or death of symbiotic algae in response to the warming of seawater, or may be a result of the dust carried by winds from desert or agricultural areas.

Hurricanes—A Coastal Calamity

Global-scale convection of the atmosphere, influenced by the Coriolis effect, causes currents of warm air to flow steadily from east to west in tropical latitudes. As the air flows over the ocean, it absorbs moisture. Warm air has low density, so the damp air over tropical ocean regions eventually begins to rise like a balloon. As the air rises, it cools, and the water vapor it contains condenses to form clouds (mists of very tiny water droplets). If the air contains sufficient moisture, the clouds grow into a cluster of large thunderstorms, which consolidate to form a single, very large storm. Because of the Coriolis effect, this large storm evolves into a rotating swirl called a tropical disturbance. If the disturbance remains over warm ocean water, as can happen in late summer and early fall, warm moist air continues to feed the storm, fostering more growth. Eventually a closed loop of rapidly circulating clouds forms, and the tropical disturbance becomes a tropical depression. Additional nourishment causes the tropical depression to spiral even faster and grow broader, until it becomes a tropical storm and receives a name. If a tropical storm becomes large enough, with sustained wind speeds exceeding 119 km per hour (74 mph), it becomes a hurricane.

Formally defined, a **hurricane** (named for the Carib god of evil), is a huge rotating storm formed in the tropical latitudes (south of 20°N) of the Atlantic Ocean, where strong prevailing winds blow over warm water. It resembles a giant counterclockwise spiral in map view (▶Fig. 15.36a) and hosts sustained winds that exceed 119 km per hour (74 mph). Hurricanes, which range from several hundred km up to 1500 km (930 miles) wide, first drift westward toward the Caribbean or North America. As a whole, a hurricane moves at speeds of up to 60 km per hour (37 mph), but in some cases, the storm may stall and hover over a given location for a few days. Hurricanes may eventually turn north and head into the North Atlantic or into the interior of North America, where they die when they run out of a supply of warm water (▶Fig. 15.36b). Occasionally, hurricanes cross Central America and hook north over the Pacific. Similar storms form in tropical latitudes of other oceans—those of the western Pacific are called typhoons, and those formed in the Indian Ocean are called cyclones.

A typical hurricane (or typhoon or cyclone) consists of several spiral arms extending inward to a central zone of relative calm known as the hurricane's **eye** (▶Fig. 15.36c). A rotating vertical cylinder of clouds, the eye wall, surrounds the eye. Winds spiral toward the eye, so like an ice skater who spins faster when she brings her arms inward, the winds accelerate toward the interior of the storm and are fastest along the eye wall. Thus, hurricane-force winds affect a belt that is only 15% to 35% as wide as the whole storm (▶Fig. 15.36d). On the side of the eye where winds blow in the same direction as the whole storm is moving, the ground speed of winds is greatest, because the storm's overall speed adds to the rotational motion.

Meteorologists classify hurricanes using the Saffir-Simpson scale. Category 1 storms are the weakest, with winds between 119 and 153 km per hour (74 to 95 mph), and Category 5 storms are the strongest, with winds of over 250 km per hour (155 mph). The fastest winds yet to be measured in an Atlantic hurricane topped 300 km per hour (190 mph). In the last decade, there have been an average of 8 or 9 Atlantic hurricanes per year. New evidence suggests that in the past 35 years, the proportion of Category 4 and 5 hurricanes, typhoons, and cyclones has increased and now averages 18 per year, globally. This increase may be due to the warming of ocean waters.

Hurricanes pose extreme danger in the open ocean, because their winds cause huge waves to build, and thus have led to the foundering of countless ships. They also wreak havoc in coastal regions, and even inland, though they die out rapidly after moving onshore. Hurricanes cause coastal damage in several ways:

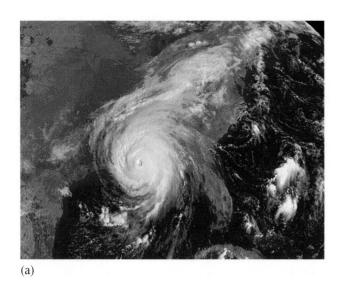

(a)

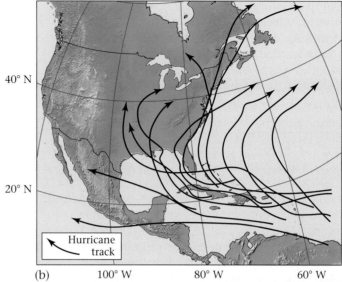

(b)

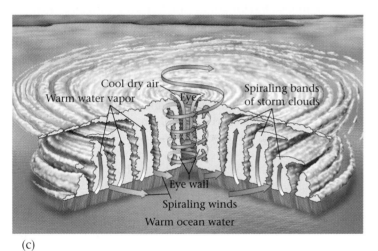

(c)

(d)

FIGURE 15.36 (a) Satellite image showing the immensity of Hurricane Katrina as it crossed over the Gulf coast of the United States. (b) The tracks of typical hurricanes during the past decade. (c) A three-dimensional sketch illustrating the architecture of a hurricane. (d) A wind-swath map of Hurricane Katrina. The red area was affected by hurricane-force winds, and the gold area was affected by tropical-storm-force winds. Note how the hurricane grew after entering the Gulf of Mexico.

- *Wind:* Winds of weaker hurricanes tear off branches and smash windows. Stronger hurricanes uproot trees, rip off roofs, and collapse walls.

- *Waves:* Winds shearing across the sea surface during a hurricane generate huge waves. In the open ocean, these waves, which may be up to 30 m high, can capsize ships. Near shore, waves batter and erode beaches, rip boats from moorings, and destroy coastal property.

- *Storm Surge:* Rising air in a hurricane causes a region of extremely low air pressure beneath. This decrease in

pressure causes the surface of the sea to bulge upward over an area with a diameter of 60 to 80 km. Sustained winds blowing in an onshore direction build this bulge even higher. When the hurricane reaches the coast, this bulge of water, called **storm surge,** swamps the land. If the bulge hits the land at high tide, the sea surface will be especially high and will affect a broader area. Wind-driven waves on top of the storm surge can batter buildings hundreds of meters or farther inland.

- *Rain, Stream Flooding, and Landslides:* Rain drenches the Earth's surface beneath a hurricane. In places, half a meter or more of rain falls in a single day. Rain causes streams to flood, even far inland, and by saturating the ground, can trigger landslides, especially in coastal areas.

- *Disruption of Social Structure:* When the storm passes, the hazard is not over. By disrupting transportation and communication networks, breaking water mains, and washing away sewage-treatment plants, hurricane damage creates severe obstacles to search and rescue, and can lead to the spread of disease, fire, and looting.

Nearly all hurricanes that reach the coast cause death and destruction, but some are truly catastrophic. Storm surge from a 1970 cyclone making landfall on the low-lying delta lands of Bangladesh led to an estimated 500,000 deaths. Storm surge from a hurricane that struck Galveston, Texas, in 1900 caused up to 12,000 deaths, and Hurricane Andrew in 1992 leveled extensive areas of southern Florida. It caused over $30 billion in damage, and left 250,000 people homeless. Hurricane Katrina, in 2005, stands as the most expensive hurricane to strike the United States in historic time. Because of where Katrina passed, worst-case damage-prediction scenarios came to be reality, particularly in the historic city of New Orleans. Let's look at this storm's history.

Tropical Storm Katrina came into existence over the Bahamas and headed west. Just before landfall in southeastern Florida, winds strengthened and the storm became Hurricane Katrina. This hurricane sliced across the southern tip of Florida, causing several deaths and millions of dollars in damage. It then entered the Gulf of Mexico and passed directly over the Loop Current, an eddy (circular flow) of summer-heated water from the Caribbean that had entered the Gulf of Mexico (see Fig. 15.36d). Water in the Loop Current reaches temperatures of 32°C (90°F), and thus stoked the storm, injecting it with a burst of energy sufficient for the storm to morph into a Category 5 monster whose swath of hurricane-force winds reached a width of 325 km (200 miles). When it entered the central Gulf of Mexico, Katrina turned north and began to bear down on the Louisiana-Mississippi coast. The eye of the storm passed just east of New Orleans, and then across the coast of Mississippi. Storm surges broke records, in places rising 7.5 m (25 ft) above sea level, and they washed coastal communities off the map along a broad swath of the Gulf Coast (▶Fig. 15.37a, b). In addition to the devastating wind and surge damage, Katrina led to the drowning of New Orleans.

To understand what happened to New Orleans, we must consider the city's environmental history. New Orleans grew on the Mississippi Delta, between the banks of the Mississippi River on the south and Lake Pontchartrain (actually a bay of the Gulf of Mexico) on the north. The oldest part of the town, a neighborhood called the French Quarter, was built on the relatively high land of the Mississippi's natural levee. Younger parts of the city, however, spread out over the lower delta plain. As the decades passed, people modified the surrounding delta landscape by draining wetlands, by constructing artificial levees that confined the Mississippi River, and by extracting groundwater. Sediment beneath the delta compacted,

FIGURE 15.37 (a) An air photo, looking straight down at the coast of Mississippi, following Hurricane Katrina. The distribution of debris and the near-total destruction of buildings shows the extent to which storm surge washed inland. (b) Close-up of Hurricane Katrina storm damage.

(a)

(b)

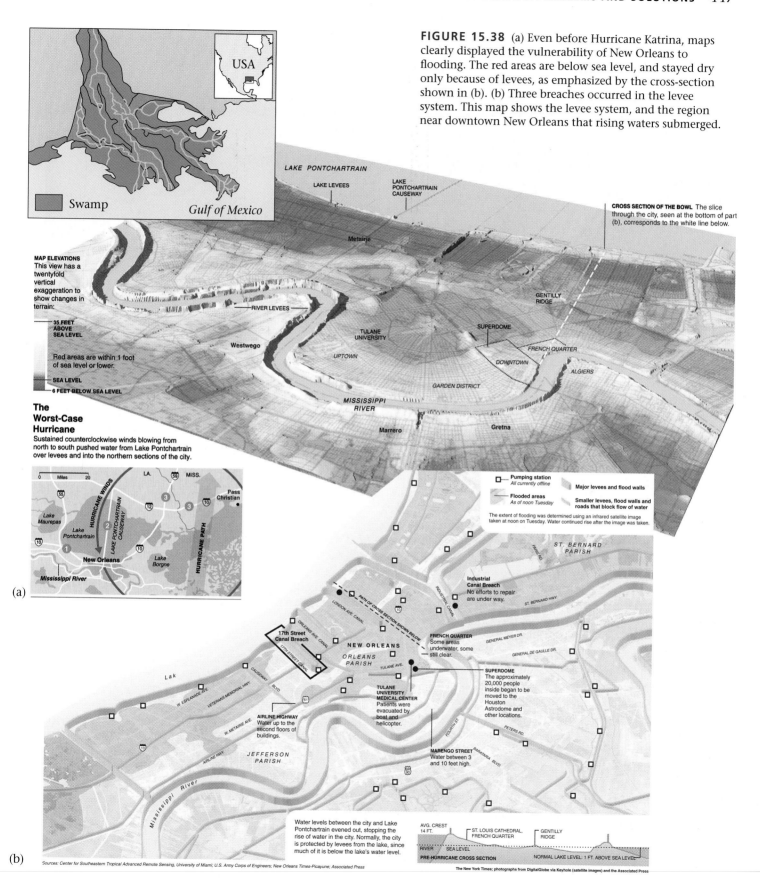

FIGURE 15.38 (a) Even before Hurricane Katrina, maps clearly displayed the vulnerability of New Orleans to flooding. The red areas are below sea level, and stayed dry only because of levees, as emphasized by the cross-section shown in (b). (b) Three breaches occurred in the levee system. This map shows the levee system, and the region near downtown New Orleans that rising waters submerged.

USA

Swamp

Gulf of Mexico

LAKE PONTCHARTRAIN

LAKE LEVEES

LAKE PONTCHARTRAIN CAUSEWAY

CROSS SECTION OF THE BOWL The slice through the city, seen at the bottom of part (b), corresponds to the white line below.

Metairie

MAP ELEVATIONS
This view has a twentyfold vertical exaggeration to show changes in terrain:

35 FEET ABOVE SEA LEVEL

Red areas are within 1 foot of sea level or lower.

SEA LEVEL

6 FEET BELOW SEA LEVEL

RIVER LEVEES

GENTILLY RIDGE

SUPERDOME

TULANE UNIVERSITY

FRENCH QUARTER

Westwego

UPTOWN

DOWNTOWN

ALGIERS

GARDEN DISTRICT

MISSISSIPPI RIVER

The Worst-Case Hurricane
Sustained counterclockwise winds blowing from north to south pushed water from Lake Pontchartrain over levees and into the northern sections of the city.

Marrero

Gretna

0 Miles 20

LA. MISS.

HURRICANE WINDS

Pass Christian

Lake Maurepas

Lake Pontchartrain

LAKE PONTCHARTRAIN CAUSEWAY

New Orleans

Lake Borgne

HURRICANE PATH

Mississippi River

(a)

Pumping station
All currently offline

Flooded areas
As of noon Tuesday

Major levees and flood walls

Smaller levees, flood walls and roads that block flow of water

The extent of flooding was determined using an infrared satellite image taken at noon on Tuesday. Water continued rise after the image was taken.

ST. BERNARD PARISH

Industrial Canal Breach
No efforts to repair are under way.

ST. BERNARD HWY

PATH OF CROSS SECTION SHOWN BELOW

LONDON AVE. CANAL

ORLEANS AVE. CANAL

17th Street Canal Breach

17TH STREET CANAL

NEW ORLEANS

ORLEANS PARISH

FRENCH QUARTER
Some areas underwater, some still clear.

GENERAL MEYER DR.

GENERAL DE GAULLE DR.

TULANE AVE.

Lak

CAUSEWAY BLVD

W. ESPLANADE AVE.

VETERANS MEMORIAL HWY

TULANE UNIVERSITY MEDICAL CENTER
Patients were evacuated by boat and helicopter.

SUPERDOME
The approximately 20,000 people inside began to be moved to the Houston Astrodome and other locations.

PETERS RD.

AIRLINE HIGHWAY
Water up to the second floors of buildings.

W. METAIRIE AVE.

AIRLINE HWY

JEFFERSON PARISH

FOURTH ST

BARATARIA BLVD

MARENGO STREET
Water between 3 and 10 feet high.

Mississippi River

Water levels between the city and Lake Pontchartrain evened out, stopping the rise of water in the city. Normally, the city is protected by levees from the lake, since much of it is below the lake's water level.

AVG. CREST 14 FT.

ST. LOUIS CATHEDRAL, FRENCH QUARTER

GENTILLY RIDGE

RIVER SEA LEVEL

PRE-HURRICANE CROSS SECTION

NORMAL LAKE LEVEL: 1 FT. ABOVE SEA LEVEL

(b)

Sources: Center for Southeastern Tropical Advanced Remote Sensing, University of Miami; U.S. Army Corps of Engineers; New Orleans Times-Picayune; Associated Press

The New York Times; photographs from DigitalGlobe via Keyhole (satellite images) and the Associated Press

(a)   (b)

FIGURE 15.39 (a) Water flowing across the levees bordering the 17th Street Canal. (b) An aerial photo showing a portion of the flooded city. Large sections of the highway had been completely submerged.

and the delta's surface has been starved of new sediment, so large areas of the delta sank below sea level. Today, most of New Orleans lies in a bowl-shaped depression as much as 2 m (7 ft) below sea level—the hazard implicit in this situation had been recognized for years (▶Fig. 15.38a).

The winds of Hurricane Katrina ripped off roofs, toppled trees, smashed windows, and triggered the collapse of weaker buildings, but their direct consequences were not catastrophic. However, when the winds blew storm surge into Lake Pontchartrain, its water level rose beyond most expectations and pressed against the system of artificial levees and flood walls that had been built to protect New Orleans. This system had been built to withstand a Category 3 storm, but when Katrina hit, it still had Category 4 strength (▶Fig. 15.38b), and upkeep of the levee system had been neglected for many years. Thus, hours after the hurricane eye had passed, the high water of Lake Pontchartrain finally found a weakness along the levee bordering the 17th Street Canal and pushed out a 100-m-long section. Preliminary studies suggest that the foundation of the levee along the canal had not been driven deeply enough, so that the base of the levee was anchored in a very weak bed of plant debris, remnants of an ancient swamp, rather than in the firmer sand below—under pressure, the base of the wall just gave way. Breaks eventually formed in a few other locations as well. So, a day after the hurricane was over, New Orleans began to flood. As the water line climbed the walls of houses, brick by brick, residents fled first upstairs, then to their attics, and to their roofs. Water spread across the city until finally the bowl of New Orleans filled to the same level as Lake Pontchartrain and 80% of the city was submerged (▶Fig. 15.39a, b). The disaster grew to be of national significance,

as the trapped population sweltered without food, drinking water, or adequate shelter. With no communications, no hospitals, and few police, the city almost descended into anarchy. It took days for outside relief to reach the city, and by then, many had died and much of New Orleans, a cultural landmark and major port, had become uninhabitable.

CHAPTER SUMMARY

• The landscape of the sea floor depends on the character of the underlying crust. For example, wide continental shelves form over passive-margin basins. Abyssal plains develop on old, cool oceanic lithosphere. Seamounts and guyots formed above hot spots and then drifted off.

• The salinity, temperature, and density of seawater vary with location and depth.

• Water in the oceans circulates in currents. Surface currents are driven by the wind and are deflected in their path by the Coriolis effect. The vertical upwelling and downwelling of water create deep currents. Some of this movement is thermohaline circulation, a consequence of variations in temperature and salinity.

• Tides—the daily rise and fall of sea level—are caused by a tide-generating force, mostly driven by the gravitational pull of the Moon.

• Waves are caused by friction where the wind shears across the surface of the ocean. Water particles follow a circular motion, in a vertical plane, as a wave passes. Waves refract (bend) when they approach the shore because of frictional drag with the sea floor.

- Sand on beaches moves with the swash and backwash of waves. If there is a longshore current, the sand gradually moves along the beach and may extend outward from headlands to form sand spits.

- At rocky coasts, waves grind away at rocks, yielding such features as wave-cut beaches and sea stacks. Some shores are wetlands where marshes or mangrove swamps grow. Coral reefs grow along coasts in warm, clear water.

- The differences in coasts reflect their tectonic setting, whether sea level is rising or falling, sediment supply, and climate.

- To protect beach property, people build groins, jetties, breakwaters, and seawalls.

- Human activities have led to the pollution of coasts. Reef bleaching has become dangerously widespread.

KEY TERMS

abyssal plain (p. 419)
active continental margin (p. 419)
barrier island (p. 433)
bathymetry (p. 419)
beach (p. 431)
beach drift (p. 432)
coast (p. 419)
coastal plain (p. 438)
coastal wetland (p. 434)
continental rise (p. 419)
continental shelf (p. 419)
continental slope (p. 419)
coral reef (p. 436)
Coriolis effect (p. 424)
current (p. 423)
Earth-Moon system (p. 427)
emergent coast (p. 438)
estuary (p. 435)
eye (p. 444)
fjord (p. 436)

hurricane (p. 444)
lagoon (p. 433)
longshore current (p. 431)
offshore bar (p. 432)
organic coast (p. 443)
passive continental margin (p. 419)
sand spit (p. 432)
seamount (p. 423)
submergent coast (p. 438)
storm surge (p. 445)
thermohaline circulation (p. 425)
tide (p. 426)
tide-generating force (p. 427)
wave base (p. 428)
wave refraction (p. 429)
wave-cut bench (platform) (p. 434)
wave-cut notch (p. 434)

REVIEW QUESTIONS

1. How much of the Earth's surface is covered by oceans?

2. Describe the typical topography of a passive continental margin, from the shoreline to the abyssal plain.

3. Where does the salt in the ocean come from? How does the salinity in the ocean vary?

4. What factors control the direction of surface currents in the ocean? What is the Coriolis effect, and how does it affect oceanic circulation? Explain thermohaline circulation.

5. What causes the tides?

6. Describe the motion of water molecules in a wave. How does wave refraction cause longshore currents?

7. Describe the components of a beach profile.

8. How does beach sand migrate as a result of longshore currents? Explain the sediment budget of the coast.

9. Describe how waves affect a rocky coast, and how such coasts evolve.

10. What is an estuary? Why is it such a delicate ecosytem? What is the difference between an estuary and a fjord?

11. Describe the different kinds of reefs, and how a reef surrounding an oceanic island changes with time.

12. Explain the difference between emergent and submergent coasts.

13. In what ways do people try to modify or "stabilize" coasts? How do the actions of people threaten the natural systems of coastal areas?

SUGGESTED READING

Ballard, R. D., and W. Hively. 2002. *The Eternal Darkness: A Personal History of Deep-Sea Exploration.* Princeton, N.J.: Princeton University Press.

Bird, E. C. 2001. *Coastal Geomorphology: An Introduction.* New York: Wiley.

Davis, R. A. 1997. *The Evolving Coast.* New York: Holt.

Emanuel, K. 2005. *Divine Wind: The History and Science of Hurricanes.* Oxford, UK: Oxford University Press.

Erickson, J. 2003. *Marine Geology.* London, UK: Facts on File.

Garrison, T. 2002. *Oceanography: An Invitation to Marine Science,* 4th ed. Pacific Grove, Calif.: Wadsworth/Thomson.

Kunzig, R. 1999. *The Restless Sea: Exploring the World beneath the Waves.* New York: Norton.

Seibold, E., and W. H. Berger. 1995. *The Sea Floor: An Introduction to Marine Geology.* 3rd ed. New York: Springer-Verlag.

Sverdrup, K. A., A. B. Duxbury, and A. C. Duxbury. 2002. *An Introduction to the World's Oceans.* 7th ed. New York: McGraw-Hill.

Viles, H., and T. Spencer. 1995. *Coastal Problems: Geomorphology, Ecology and Society at the Coast.* New York: Wiley.

Woodroffe, C. D. 2002. *Coasts: Form, Process and Evolution.* Cambridge, UK: Cambridge University Press.

A Hidden Reserve: Groundwater

Groundwater can turn a desert green, as shown by these circular irrigated fields sprouting in the sands of Jordan. A water well lies at the center of each circle.

16.1 INTRODUCTION

Imagine Rosa May Owen's surprise when, on May 8, 1981, she looked out her window and discovered that a large sycamore tree in the backyard of her Winter Park, Florida, home had suddenly disappeared. It wasn't a particularly windy day, so the tree hadn't blown over—it had just vanished! When Owen went outside to investigate, she found that more than the tree had disappeared. Her whole backyard had become a deep, gaping hole. The hole continued to grow for a few days until finally it swallowed Owen's house and six other buildings, as well as the deep end of the municipal swimming pool, part of a road, and the stock of expensive Porsches in a car dealer's lot (▶Fig. 16.1a).

What had happened in Winter Park? The bedrock beneath the town consists of limestone, a fairly soluble rock. **Groundwater,** the water that resides under the surface of the Earth, had gradually dissolved the limestone, carving open rooms, or caverns, underground. On May 8, the roof of a cavern underneath Owen's backyard began to collapse, forming a circular depression called a **sinkhole.** The sycamore tree and the rest of the neighborhood simply dropped down into the sinkhole. It would have taken too much effort to fill in the hole with soil, so the community allowed it to fill with water, and

(a)

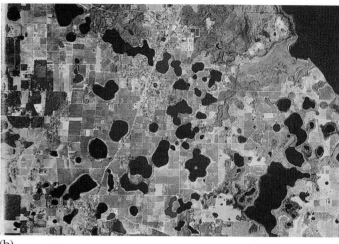

(b)

FIGURE 16.1 (a) The Winter Park, Florida, sinkhole as seen from a helicopter. (b) Numerous sinkhole lakes dot central Florida, as seen from high altitude.

now it's a circular lake, the centerpiece of a pleasant municipal park. Similar lakes appear throughout central Florida (▶Fig. 16.1b).

The Winter Park sinkhole is one of the more dramatic reminders that significant quantities of water reside underground. Whereas we can easily see Earth's surface water (in lakes, rivers, streams, marshes, and oceans) and atmospheric water (in clouds and rain), groundwater lies hidden beneath the surface in the pores and cracks found within sediment or rock. Nevertheless, groundwater has increasingly become a major supply of water for homes, agriculture, and industry. Groundwater accounts for about two-thirds of the world's freshwater supply. This chapter provides a basic picture of where groundwater comes from, how it flows, and how it interacts with the rock and sediment through which it flows.

16.2 WHERE DOES GROUNDWATER RESIDE?

The Underground Reservoir of the Hydrologic Cycle

As we saw in Interlude E, water moves among various reservoirs (the ocean, the atmosphere, rivers and lakes, groundwater, living organisms, soil, and glaciers) during the hydrologic cycle. Of the water that falls on land, some evaporates directly back into the atmosphere, some gets trapped in glaciers, and some becomes runoff that enters a network of streams and lakes that drains to the sea. The remainder sinks, via the process of infiltration, into the ground; in effect, the upper part of the crust behaves like a giant sponge that can soak up water.

Of the water that does infiltrate, some descends only into the soil and wets the surfaces of grains and organic material making up the soil. This water, called **soil moisture,** later evaporates back into the atmosphere or gets sucked up by the roots of plants and then transpires back into the atmosphere. But some water sinks deeper and fills available spaces in cracks and between grains of sediment or rock; this water, as well as water trapped in rock at the time the rock formed, makes up groundwater. Groundwater slowly flows underground for anywhere from a few months to tens of thousands of years before returning to the surface to pass once again into other reservoirs of the hydrologic cycle.

Porosity: The Home of Groundwater

Contrary to popular belief, only a small proportion of groundwater flows freely in the underground lakes and streams of cavern networks. Most groundwater resides within the pore space of what might at first glance look like solid rock or sediment. Generally speaking, a **pore** is any open space (as opposed to solid material) within a body of sediment or rock, and **porosity** refers to the total volume of empty space in a material, usually expressed as a percentage. For example, if we say that a piece of sandstone has 30% porosity, then 30% of a block of what looks like solid sandstone actually consists of open space.

Geologists further distinguish between primary and secondary porosity. Primary porosity consists of space that remains between solid grains or crystals after sediment accumulates or rocks form (▶Fig. 16.2a). For example, in sand or gravel, primary porosity exists for the simple reason that rounded clasts can't fit together tightly. Secondary porosity refers to new pore space in rocks, produced some time after the rock first formed. For example, when rocks fracture, the opposing walls of the fracture do not fit together tightly, so narrow spaces remain in between (▶Fig. 16.2b). Secondary porosity also forms when groundwater flowing through rock dissolves and removes minerals, thereby producing openings called solution cavities.

16.3 PERMEABILITY: THE EASE OF FLOW

If solid rock completely surrounds a pore, the water in the pore cannot flow to another location. For groundwater to flow, therefore, pores must be linked by conduits (openings). **Permeability,** a measure of the ease with which fluids can flow through a porous material, depends on the degree to which pores are interconnected (▶Fig. 16.3). Water flows easily through a permeable material, whereas water flows slowly or not at all through an impermeable material. Porosity and permeability are not the same. A material whose pores are isolated from one another can have high porosity but low permeability.

FIGURE 16.2 (a) Isolated, nonconnected pores in an impermeable material. (b) Fractures in a rock provide secondary porosity.

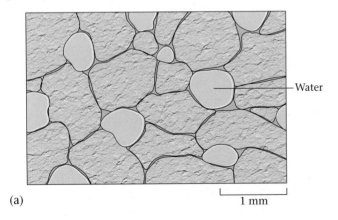

(a)

— Water

1 mm

(b)

20 cm

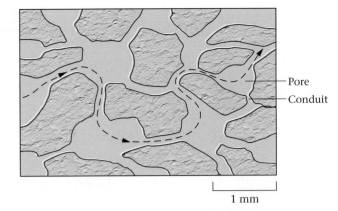

Pore
Conduit

1 mm

FIGURE 16.3 Pores connected to one another by a network of conduits in permeable material. The dashed line indicates the flow path.

The permeability of a material depends on several factors.

- *Number of available conduits:* As the number of conduits increases, permeability increases.

- *Size of the conduits:* More fluids can travel through wider conduits than through narrower ones.

- *Straightness of the conduits:* Water flows more rapidly through straight conduits than it does through crooked ones.

Note that the factors that control permeability in rock or sediment resemble those that control the ease with which traffic moves through a city. Traffic can flow quickly through cities with many straight, multilane boulevards, whereas it flows slowly through cities with only a few narrow, crooked streets.

Aquifers and Aquitards

With the concept of permeability in mind, hydrogeologists (geologists who study groundwater) distinguish between **aquifers,** sediment or rocks that transmit water easily, and **aquitards,** sediment or rocks that do not transmit water easily and thereafter retard the motion of water. Aquifers that intersect the surface of the Earth are called unconfined aquifers, because water can percolate directly from the surface down into the aquifer, and water in the aquifer can rise to the surface. Aquifers that are separated from the surface by an aquitard are called confined aquifers, as the water they contain is isolated from the surface (▶Fig. 16.4).

16.4 THE WATER TABLE

Are porous rocks and sediments filled with water right up to the ground surface everywhere? The answer is no. Geologists define a boundary, called the **water table,** above which pore spaces contain mostly air and below which they contain only water

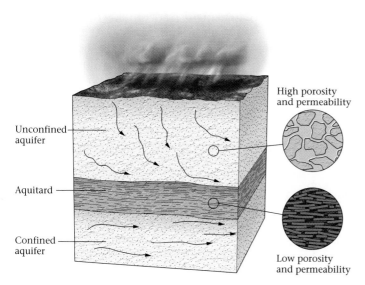

FIGURE 16.4 An aquifer is a high-porosity, high-permeability rock. If it has access to the ground surface, it's an unconfined aquifer. If it's trapped below an aquitard, a rock with low permeability, it's a confined aquifer.

(▶Fig. 16.5). In technical jargon, the region of the subsurface above the water table is called the unsaturated zone; the region below the water table is the saturated zone. Material in the unsaturated zone can be damp, but pores are not full. Typically, surface tension, the electrostatic attraction of water molecules to mineral surfaces, causes water to seep up from the water table (just as water rises in a thin straw), filling pores in the capillary fringe, a thin layer between the saturated and unsaturated zones.

The depth of the water table in the subsurface varies greatly with location. The surface of a permanent stream, lake, or marsh, for example, defines the water table at that location, for water saturates the soil or rock below (▶Fig. 16.5b, c). Elsewhere, the water table lies hidden below the ground surface: in humid regions, it typically lies within a few meters of the ground surface, whereas in arid regions, it may lie tens of meters to over 200 m below the surface.

Rainfall affects the water-table depth—the level sinks during a dry season. If the water table drops below the floor of a river or lake, the river or lake dries up, because the water

FIGURE 16.5 (a) The geometry of the water table, illustrating the saturated zone, the unsaturated zone, and the capillary fringe. (b) The surface of a permanent pond is the water table. (c) During the dry season, the water table can drop substantially, causing the pond to dry up.

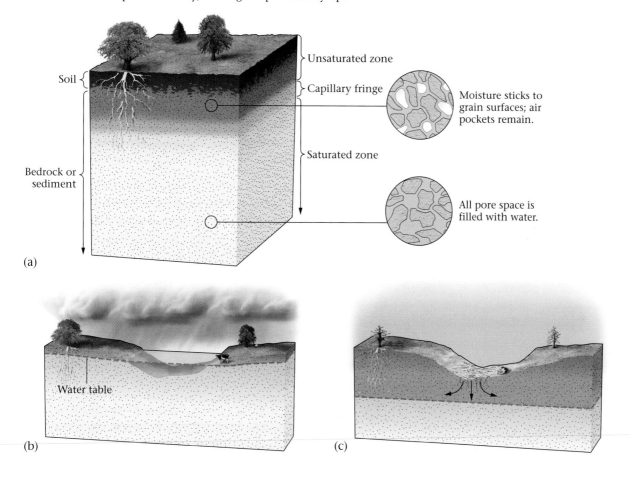

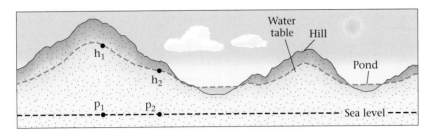

FIGURE 16.6 The shape of a water table beneath hilly topography. Note that the water table can be above sea level. Point h_1 on the water table is higher than point h_2, relative to a reference elevation (sea level).

it contains infiltrates into the ground. Arid regions have no permanent streams, for the water table lies below the stream bed. Streams in such regions flow only during storms, when rain falls at a faster rate than the water can infiltrate.

Topography of the Water Table

In hilly regions, the water table is not a planar surface. Rather its shape mimics, in a subdued way, the shape of the overlying topography (▶Fig. 16.6). This means that the water table lies at a higher elevation beneath hills than it does beneath valleys. But the relief (the vertical distance between the highest and lowest elevations) of the water table is not so great as that of the overlying land, so the surface of the water table tends to be smoother than that of the landscape.

At first thought, it may seem surprising that the elevation of the water table varies as a consequence of ground-surface topography. After all, when you pour a bucket of water into a pond, the surface of the pond immediately

FIGURE 16.7 The configuration of a perched water table. A lens of groundwater lies above, and the regional water table lies at greater depth.

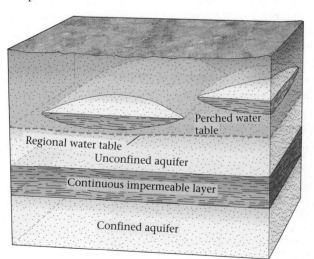

adjusts to remain horizontal. The elevation of the water table varies because groundwater moves so slowly through rock and sediment that it cannot *quickly* assume a horizontal surface. When it rains on a hill and water infiltrates down to the water table, the water table rises a little. When it doesn't rain, the water table sinks slowly, but so slowly that rain will fall again and make the water table rise before the water table has had time to sink very far.

Perched Water Tables

In some locations, an isolated lens or layer of an impermeable material lies within the unsaturated zone between the ground surface and the regional water table. Where this happens, downward infiltrating water gets trapped to form an isolated mound or layer that lies above the regional water table. Such groundwater is, in effect, "perched" on an aquitard, like a bird that is perched on a branch above the ground. We refer to the top surface of groundwater that has been trapped above the regional water table as a **perched water table** (▶Fig. 16.7).

16.5 GROUNDWATER FLOW

Groundwater Flow Paths

What happens to water that has infiltrated down into the ground and now lies below the water table? Does it just sit, unmoving, like the water in a stagnant puddle, or does it flow and eventually find its way back to the surface? Countless measurements confirm that groundwater enjoys the latter fate: groundwater indeed flows, and in some cases it moves great distances underground.

In the unsaturated zone, the region between the ground surface and the water table, water percolates straight down, like the water passing through a drip coffee maker, for this water moves only in response to the downward pull of gravity. But in the saturated zone, the region below the water table, water flow is more complex, for in addition to the downward pull of gravity, water responds to differences in pressure. Pressure may cause groundwater to flow sideways, or even upward. (If you've ever watched water spray up from a fountain, you've seen how pressure can push water upward.) Thus, to understand the nature of groundwater flow, we must first understand the origin of pressure in groundwater. For simplicity, we'll consider only the case of groundwater in an unconfined aquifer.

Pressure in groundwater at a specific point underground is caused by the weight of all the overlying water from that point up to the water table. (The weight of overlying rock does not contribute to the pressure in groundwater, for the contact points between mineral grains bear the rock's weight.) Thus, a point at a greater depth below the water table feels more pressure than does a point at lesser depth. If the water table is horizontal, the pressure acting on an imaginary horizontal reference plane at a specified depth below the water table is the same everywhere. But if the water table is not horizontal, as shown in Figure 16.6, the pressure at points on a horizontal reference plane at depth changes with location. For example, the pressure acting at point p_1, which lies below the hill in Figure 16.6, is greater than the pressure acting at point p_2, which lies below the valley, even though both p_1 and p_2 are at the same elevation (sea level, in this case).

Both the elevation of a volume of groundwater and the pressure within the water provide energy that, if given the chance, will cause the water to flow. Physicists refer to such stored energy as "potential energy" (see Appendix). The potential energy available to drive the flow of a volume of groundwater at a given location is called the **hydraulic head.** To measure the hydraulic head at a point in an aquifer, hydrogeologists drill a vertical hole down to the point and then insert a pipe in the hole. The height above a reference elevation (for example, sea level) to which water rises in the pipe represents the hydraulic head—water rises higher in the pipe where the head is higher. As a rule, *groundwater flows from regions where it has higher hydraulic head to regions where it has lower hydraulic head*. In simple terms, this statement generally implies that groundwater flows from locations where the water table is higher to locations where the water table is lower.

To calculate how hydraulic head changes with location underground, hydrogeologists take into account both the effect of gravity and the effect of pressure. They conclude that groundwater flows along concave-up curved paths, as illustrated in cross section (▶Fig. 16.8). These curved paths eventually take groundwater from regions where the water table is high (under a hill) to regions where the water table is low (below a valley), but because of flow path shape, some groundwater may flow deep down into the crust—in some cases, to depths of 10 km—along the first part of its path, and then may flow back up, toward the ground surface, along the final part of its path. The location where water enters the ground and flows down is called the **recharge area;** the location where groundwater flows back up to the surface is called the **discharge area** (Fig. 16.8).

Rates of Groundwater Flow: Darcy's Law

Flowing water in an ocean current moves at up to 3 km per hour (over 26,000,000 m per year); water in a steep river channel can reach speeds of up to 30 km per hour (over 260,000,000 m per year). In contrast, groundwater moves at a snail's pace—typical rates range between 0.01 and 1.4 m per day (about 4 to 500 m per year). Groundwater moves much more slowly than surface water for two reasons. First, ground-

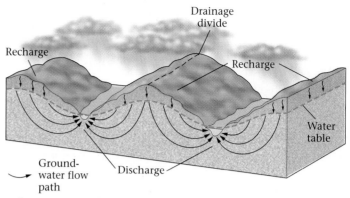

FIGURE 16.8 The flow lines from the recharge area to the discharge area curve through the substrate. In fact, some groundwater descends to great depth and then rises back to the surface.

water moves by percolating through a complex, crooked network of tiny conduits, so it must travel a much greater distance than it would if it could follow a straight path. Second, friction and/or electrostatic attraction between groundwater and conduit walls slows down the water flow.

Why is the rate of groundwater flow so variable? In the mid-nineteenth century, a French engineer named Henry Darcy studied this problem. He discovered that the rate at which groundwater flows between two points in the subsurface depends on two factors: the permeability of the material between the two points and a quantity called the hydraulic gradient. **Hydraulic gradient** is defined as the difference in hydraulic head between two points, divided by the distance between the two points as measured along the flow path. This relation, we depict in ▶Fig. 16.9, can be written as a formula: hydraulic gradient = $\Delta h/j$, where Δh is the difference in hydraulic head and the j is the distance along the flow path. Note that, for the region close to the water table, we can simplistically picture hydraulic gradient as the slope of the water table (meaning the difference in elevation between two points on the water table, divided by the horizontal distance between the two points). Where the water table has a steep slope, the hydraulic gradient is greater than where the water table has a gentle slope. Hydrogeologists now refer to Darcy's discovery as **Darcy's law.**[1] According to Darcy's law, groundwater flows faster in a more permeable material than in a less permeable material, and groundwater flows faster where the water table has a steep slope than where the water table has a gentle slope.

---

[1]Darcy's law can be written formally as: $Q = K(\Delta h/j)A$. In this equation, Q is the flow rate (or discharge), K is the hydraulic conductivity (a number that depends on characteristics of both the groundwater and of the material through which the groundwater flows), $\Delta h/j$ is the hydraulic gradient, and A is the cross-section area through which the groundwater flows.

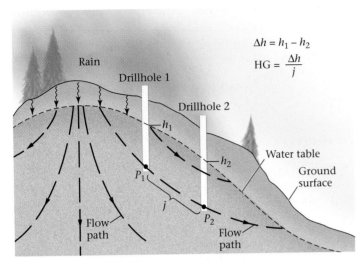

$$\Delta h = h_1 - h_2$$

$$HG = \frac{\Delta h}{j}$$

FIGURE 16.9 A hydraulic gradient (HG) is the change in hydraulic head per unit of distance between two points along the flow path.

FIGURE 16.10 In this ordinary well, the water has to be brought up by bucket.

16.6 TAPPING THE GROUNDWATER SUPPLY

Groundwater can be brought to the ground surface at wells or springs. **Wells** are holes that people dig or drill to obtain water. **Springs** are natural outlets from which groundwater flows. Wells and springs provide a welcome source of water but must be treated with care if they are to last.

Wells

In an **ordinary well,** the base of the well penetrates an aquifer below the water table. Water from the pore space in the aquifer seeps into the well and fills it; the water surface in the well is the water table. Drilling into an aquitard, or into rock that lies above the water table, will not supply water, and thus yields a dry well. Some ordinary wells are seasonal, in that they only function during the rainy season, when the water table rises. During the dry season, the water table lies below the base of the well, so the well is dry.

To obtain water from an ordinary well, you either pull water up in a bucket or pump the water out (▶Fig. 16.10). As long as the rate at which groundwater fills the well exceeds the rate at which water is removed, the level of the water table near the well remains about the same. However, if users pump water out of the well too fast, then the water table sinks down around the well, a process called drawdown, so that the water table becomes a downward-pointing, cone-shaped surface called a **cone of depression** (▶Fig. 16.11). Drawdown may cause shallower wells that have been drilled nearby to run dry.

An **artesian well,** named for the province of Artois in France, penetrates confined aquifers, in which water is under enough pressure to cause the water to rise on its own to a level above the surface of the aquifer. If this level lies below the ground surface, the well is a nonflowing

FIGURE 16.11 The base of an ordinary well penetrates below the water table. If groundwater is extracted faster than it can be replaced, a cone of depression forms around the well. Pumping by the big well, in this example, may lower the water table sufficiently to cause the small well to become dry.

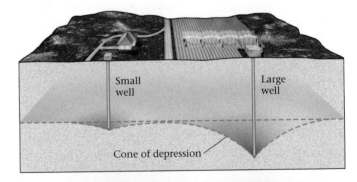

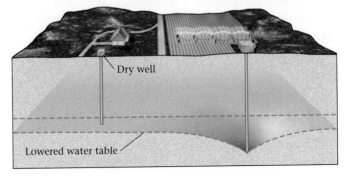

artesian well. But if the level lies above the ground surface, the well is a flowing artesian well, and water actively fountains out of the ground. Artesian wells occur in special situations where a confined aquifer lies beneath a sloping aquitard.

We can understand why artesian wells exist if we look first at the configuration of a city water supply (►Fig. 16.12a). Water companies pump water into a high tank that has a significant hydraulic head relative to the surrounding areas. If the water were connected by a water main to a series of vertical pipes, pressure caused by the elevation of the water in the high tank would make the water rise in the pipes until it reached an imaginary surface, called a potentiometric surface, that lies above the ground. This pressure drives water through water mains to household water systems without requiring pumps. In an artesian system, water enters a tilted, confined aquifer that intersects the ground in the hills of a high-elevation recharge area (►Fig. 16.12b). The confined groundwater flows down to the adjacent plains, which lie at a lower elevation. The potentiometric

surface to which the water would rise were it not confined lies above this aquifer; in fact, it may lie above the ground surface of the plains. Pressure in the confined aquifer pushes water up a well. Where the potentiometric surface lies below ground, the well will be a nonflowing artesian well, but where the surface lies above the ground, the well will be a flowing artesian well.

Springs

More than one town has grown up around a **spring,** a place where groundwater naturally flows or seeps onto the Earth's surface, for springs provide fresh, clear groundwater for drinking or irrigation without the expense of drilling or digging. Springs form under a variety of conditions:

- Where the ground surface intersects the water table in a discharge area (►Fig. 16.13a): such springs typically occur in valley floors where they may add water to lakes or streams.
- Where downward-percolating water runs into an impermeable layer and migrates along the top surface of the layer to a hillslope (►Fig. 16.13b).
- Where a particularly permeable layer or zone intersects the surface of a hill (►Fig. 16.13c): water percolates down through the hill and then migrates along the permeable layer to the hill face.
- Where a network of interconnected fractures channels groundwater to the surface of a hill (►Fig. 16.13d).
- Where flowing groundwater collides with a steep, impermeable barrier, and pressure pushes it up to the ground along the barrier (►Fig. 16.13e): faulting can create such barriers by juxtaposing impermeable rock against permeable rock.
- **Artesian springs** form if the ground surface intersects a natural fracture (joint) that taps a confined aquifer in which the pressure is sufficient to drive the water to the surface (►Fig. 16.13f and Box 16.1).
- Where a perched water table intersects the surface of a hill (►Fig. 16.13g, h).

Springs can provide water in regions that would otherwise be uninhabitable. For example, oases in deserts may develop around a spring (►Box 16.1).

16.7 HOT SPRINGS AND GEYSERS

Hot springs, springs that emit water ranging in temperature from about 30° to 104°C, are found in two geologic settings. First, they occur where very deep groundwater, heated in warm bedrock at depth, flows up to the ground surface. This water brings heat with it as it rises. Such hot

FIGURE 16.12 (a) The configuration of a city water supply. Water will rise in vertical pipes up to the level of the potentiometric surface. (b) The configuration of an artesian system. Artesian wells flow if the potentiometric surface lies above the ground surface. Nonflowing artesian wells form where the potentiometric surface lies below the ground.

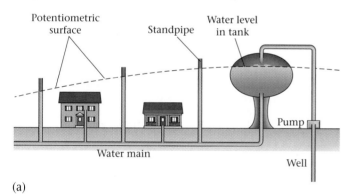

(a)

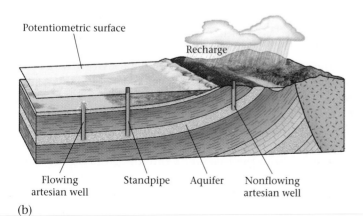

(b)

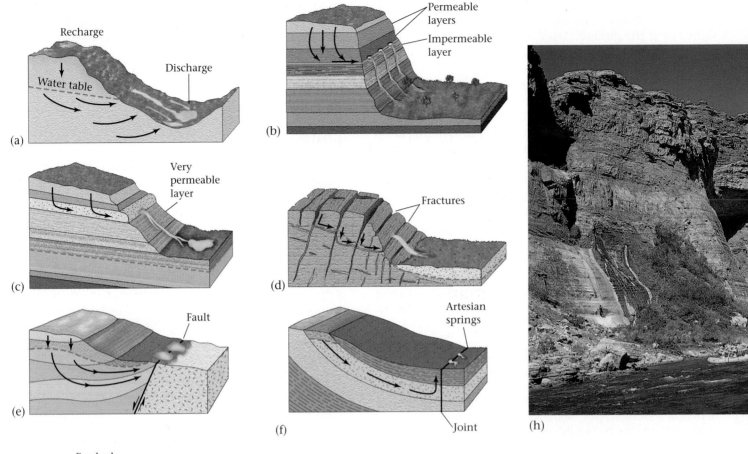

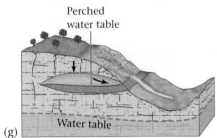

FIGURE 16.13 Springs form (a) where groundwater rises in a discharge area; (b) where groundwater has been forced to migrate along an impermeable barrier; (c) where a particular permeable layer transmits water to the surface of a hill; (d) where a network of interconnected fractures channels water to the hill face; (e) where groundwater collides with a steep impermeable barrier, and pressure pushes it up to the ground along the barrier. (f) An artesian spring forms where water from a confined aquifer migrates up a joint; (g) springs also form where a perched water table intersects the surface of a hill; (h) photo of a spring arising from a perched water table intersecting a wall of the Grand Canyon.

springs form in places where faults or fractures provide a high-permeability conduit for deep water, or where the water emitted in a discharge region followed a trajectory that first carried it deep into the crust (▶Fig. 16.15a). Second, hot springs develop in **geothermal regions,** places where magma and/or very hot rock resides close to the Earth's surface (▶Fig. 16.15b). In hot springs, groundwater is a steaming tea of water and dissolved minerals. Hot groundwater contains more dissolved minerals because water becomes a more effective solvent when hot. People use the water emitted at hot springs to fill relaxing mineral baths (Fig. 16.15a).

Numerous distinctive geologic features form in geothermal regions as a result of the eruption of hot water (▶Fig. 16.16a–d). In places where the hot water rises into soils rich in volcanic ash and clay, a viscous slurry forms and fills bubbling mud pots. Bubbles of steam rising through the slurry cause it to splatter about in goopy drops. Where geothermal waters spill out of natural springs and then cool, dissolved minerals in the water precipitate, forming colorful mounds or terraces of travertine and other minerals. Geothermal waters may accumulate in brightly colored pools—the gaudy greens, blues, and oranges of these pools come from thermophilic (heat-loving) bacteria and archaea that eat the sulfur-containing minerals dissolved in the groundwater.

The most spectacular consequence of geothermal waters is a **geyser** (from the Icelandic word for "gush"), a fountain of steam and hot water that erupts episodically from a vent in the ground (Fig. 16.16d). To understand why a geyser erupts, we first need a picture of its underground plumbing.

Oases

BOX 16.1
THE HUMAN ANGLE

The Sahara Desert of northern Africa is now one of the most barren and desolate places on Earth, for it lies in a climatic belt where rain seldom falls. But it wasn't always that way. During the last ice age, when glaciers covered parts of northern Europe on the other side of the Mediterranean, the Sahara enjoyed a more temperate climate, and the water table was high enough that permanent streams dissected the landscape. In fact, using ground-penetrating radar, geologists can detect these streams today—they look like ghostly valleys beneath the sand (▶Fig. 16.14a). When the climate warmed after the ice age, rainfall diminished, the water table sank below the floor of most stream beds, and the streams dried up.

Today, the water of the Sahara region lies locked in a vast underground aquifer composed of porous sandstone. Recharge into this aquifer comes from highlands bordering the desert, from occasional downpours, and from the Nile River (particularly Lake Nasser, behind the Aswan High Dam). In general, the water of the aquifer can only be obtained by drilling deep wells, but locally, water spills out at the surface—either because folding brings the aquifer particularly close to the ground so that valley floors intersect the water table, or be-

cause artesian pressure pushes groundwater up along joints or faults (▶Fig. 16.14b). In either case, the aquifer feeds springs that quench the thirst of desert and tropical plants and create an oasis, an island of green in the sand sea (▶Fig. 16.14c). Oases became important stopping points along caravan routes, allowing both people and camels to replenish water supplies.

In some oases, people settled and used the groundwater to irrigate date palms and other crops. For example, the Bahariya Oasis, about 400 km southwest of Cairo, Egypt, hosted a town of perhaps 30,000 between 300 B.C.E. and 300 C.E. During that time, the water table lay only 5 m below the ground and could be easily reached by shallow wells. Today, as a result of changing climates and centuries of use, the water table lies 1,500 m below the ground, almost out of reach. Bahariya's glorious past came to light in 1996, quite by accident. A man was riding his donkey in the desert near the oasis when the ground beneath the donkey suddenly caved in. The guard had inadvertently opened the roof into a tomb filled with over 150 mummies, along with thousands of well-preserved artifacts. The site has since come to be known as the Valley of the Mummies.

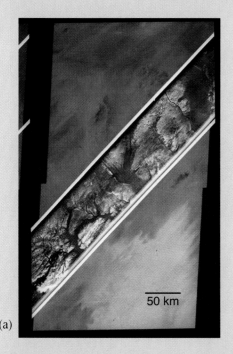

50 km

(a)

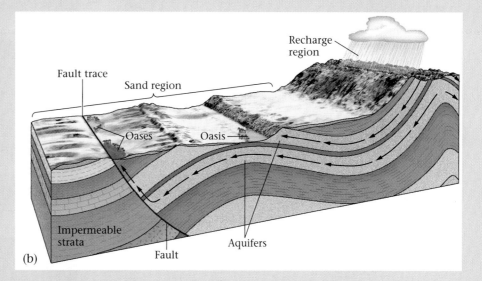

(b)

FIGURE 16.14 (a) A satellite image, taken using ground-penetrating radar, showing a long-abandoned drainage network now buried by Saharan sand. The gray strip shows darker channels and lighter high areas under the sand. The orange area is the sand surface that you see without radar. (b) This subsurface configuration of aquifers leads to the formation of an oasis, where groundwater reaches the surface. (c) An oasis in the Sahara.

(c)

(a)

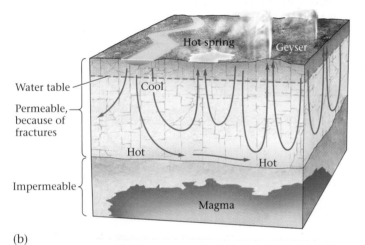

(b)

FIGURE 16.15 (a) In the city of Bath, England, the Romans built an elaborate spa around an artesian hot spring. The spring formed along a fault that tapped a supply of deep groundwater that still flows today. The baths of Bath have remained popular through the ages, and are currently undergoing renovation. (b) Geysers and hot springs form where groundwater, heated at depth, rises to the surface.

Beneath a geyser lies a network of irregular fractures in very hot rock; groundwater sinks and fills these fractures. Adjacent hot rock then "superheats" the water: it raises the temperature above the temperature at which water at a pressure of 1 atmosphere will boil. Eventually, this superhot water rises through a conduit to the surface. When some of this water transforms into steam, the resulting expansion causes water higher up to spill out of the conduit at the ground surface. When this spill happens, pressure in the conduit, from the weight of overlying water, suddenly decreases. A sudden drop in pressure causes the superhot water at depth instantly to turn to steam, and this steam quickly rises,

(a)

(b)

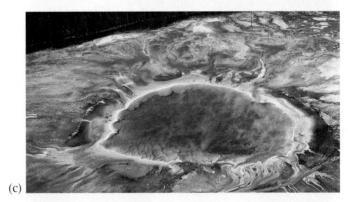

(c)

(d)

FIGURE 16.16 Features of geothermal regions. (a) Mud pots in Yellowstone National Park; (b) mounds of siliceous minerals precipitated around the Grotto Geyser, in Yellowstone; (c) colorful bacteria- and archaea-laden pools, Yellowstone; (d) Old Faithful geyser at Yellowstone. The geyser erupts somewhat predictably.

ejecting all the water and steam above it out of the conduit in a geyser eruption. Once the conduit empties, the eruption ceases, and the conduit fills once again with water that gradually heats up, starting the eruptive cycle all over again.

Hot springs are found in many localities around the world: at Hot Springs, Arkansas, where deep groundwater rises to the surface; in Yellowstone National Park, above the magma chamber of a continental hot spot; around the Salton Sea in southern California, where the mid-ocean ridge of the Gulf of California merges with the San Andreas Fault; in the Geysers Geothermal Field of California, an area of significant geothermal power generation formed above a felsic magma intrusion; in Iceland, which has grown on top of an oceanic hot spot along the Mid-Atlantic Ridge; and in Rotorua, New Zealand, which lies in an active volcanic field above a subduction zone.

16.8 GROUNDWATER USAGE PROBLEMS

Since prehistoric times, groundwater has been an important resource that people have relied on for drinking, irrigation, and industry. Groundwater feeds the lushness of desert oases in the Sahara, the amber grain in the North American high plains, and the growing cities of arid regions. Though groundwater accounts for about 95% of the liquid freshwater on the planet, accessible groundwater cannot be replenished quickly in important locations, and this leads to shortages. In modern times, the problem has been exacerbated by the contamination of existing groundwater. Such pollution, caused when toxic wastes and other impurities infiltrate down to the water table, may be invisible to us but may ruin a water supply for generations to come. In this section, we'll take a look at problems associated with the use of groundwater supplies.

Depletion of Groundwater Supplies

Is groundwater a renewable resource? In a time frame of 10,000 years, the answer is yes, for the hydrologic cycle will eventually resupply depleted reserves. But in a time frame of 100 to 1,000 years—the span of a human lifetime or a civilization—groundwater in many regions may be viewed as a nonrenewable resource. By pumping water out of the ground at a rate faster than nature replaces it, people effectively mine the groundwater supply. In fact, in portions of the desert Sunbelt region of the United States, supplies of young groundwater have already been exhausted, and deep wells now extract 10,000-year-old groundwater. But such ancient water has been in rock so long that some of it has become an unusable soup of dissolved minerals. A number of other problems accompany the depletion of groundwater.

- *Lowering the water table:* When we extract groundwater from wells at a rate faster than it can be resupplied, the water table drops. First, a cone of depression forms locally around the well; then the water table gradually becomes lower in a broad region. As a consequence, existing wells, springs, and rivers dry up (►Fig. 16.17a, b). To continue tapping into the water supply, we must drill progressively deeper.

The water table can also drop when people divert surface water from the recharge area. Such a problem has developed in the Everglades of southern Florida, a huge swamp where, before the expansion of Miami and the development of agriculture, the water table lay at the ground surface. Diversion of water from the Everglades' recharge area into canals has significantly lowered the water table, causing parts of the Everglades to dry up.

- *Reversing the flow direction of groundwater:* The cone of depression that develops around a well creates a local slope to the water table. The resulting hydraulic gradient may be large enough to reverse the flow direction of nearby groundwater (►Fig. 16.18a, b). Such reversals can allow pollutants, seeping out of a septic tank, to contaminate a well.

- *Saline intrusion:* In coastal areas, fresh groundwater lies in a layer above saltwater that entered the aquifer from the adjacent ocean (►Fig. 16.18c, d). (Saltwater is denser than freshwater, so the fresh groundwater floats above it.) If people pump water out of a well too quickly, the boundary between the saline water and the fresh groundwater rises. And if this boundary rises above the base of the well, then the well will start to yield useless saline water.

- *Pore collapse and land subsidence:* When groundwater fills the pore space of a rock, it holds the grains of the rock or regolith apart, for water cannot be compressed. The

FIGURE 16.17 (a) Before a water table is lowered, a large swamp exists. (b) Pumping by a nearby city causes the water table to sink, so the swamp dries up.

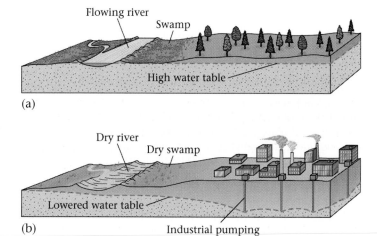

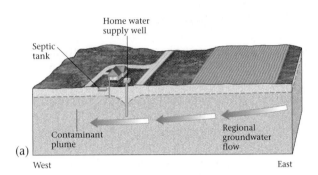

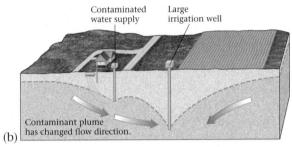

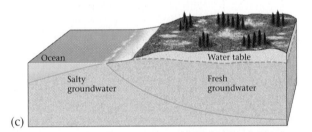

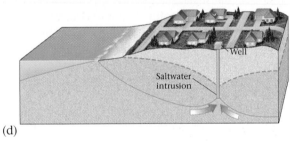

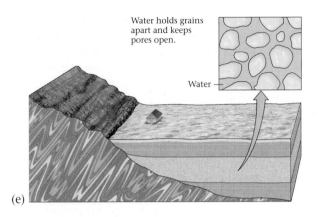

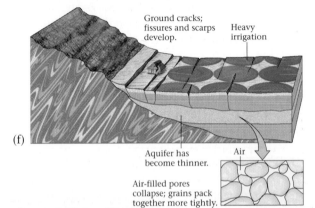

FIGURE 16.18 (a) Before pumping, effluent from a septic tank drifts west with the regional groundwater flow. (b) After pumping, it drifts east into the well, in response to the local slope of the water table. (c) Before pumping, fresh groundwater forms a large lens over salty groundwater. (d) Pumping too fast sucks saltwater from below into the well. (e, f) Pore space collapses when water is removed. The pore collapse makes the land subside, as indicated by fissures and cracked houses. (g) The Leaning Tower of Pisa's foundation was destabilized by groundwater removal. (h) Evidence of land subsidence in the San Joaquin Valley, California. Former ground elevations are marked on the pole. (i) The canals of Venice provide transportation routes in a city slowly subsiding below sea level.

extraction of water from a pore eliminates the support holding the grains apart, because the air that replaces the water *can* be compressed. As a result, the grains pack more closely together. Such pore collapse permanently decreases the porosity and permeability of a rock, and thus lessens its value as an aquifer (▶Fig. 16.18e, f).

Pore collapse also decreases the volume of the aquifer, with the result that the ground above the aquifer sinks. Such land subsidence may cause fissures at the surface to develop and the ground to tilt. Buildings constructed over regions undergoing land subsidence may themselves tilt, or their foundations may crack. The Leaning Tower of Pisa, in Italy, tilts because the removal of groundwater caused its foundation to subside (▶Fig. 16.18g). In the San Joaquin Valley of California, the land surface subsided by 9 m between 1925 and 1975, because water was removed to irrigate farm fields (▶Fig. 16.18h). In coastal areas, land subsidence may even make the land surface sink below sea level. The flooding of Venice, Italy, is due in part to land subsidence accompanying the withdrawal of groundwater (▶Fig. 16.18i).

To avoid such problems, communities have sought to prevent groundwater depletion, either by directing surface water into recharge areas or by pumping surface water back into the ground. For example, some communities have excavated to lower the land surface of a park, so that the park acts as a catchment for storm water. This water can then infiltrate down into the soil.

Groundwater Quality and Contamination

Much of the world's groundwater is crystal clear, pure enough to drink right out of the ground, for rocks and sediment are natural filters capable of efficiently removing suspended solids (mud and solid waste) from groundwater. But in some places, groundwater either is unusable or must be treated before being used, even though it has not been contaminated by human-created materials. Specifically, groundwater quality can be affected by the presence of natural dissolved ions that enter solution as groundwater flows through rock or sediment. For example, groundwater that has passed through salt-containing strata may become salty and unsuitable for irrigation or drinking. Groundwater that has passed through limestone or dolomite contains dissolved calcium (Ca^{+2}) and magnesium (Mg^{+2}) ions; this water, called hard water, can be a problem because carbonate minerals precipitate from it to form "scale" that clogs pipes. Groundwater that has passed through iron-bearing rocks may contain dissolved iron oxide that precipitates to form rusty stains. Some groundwater contains dissolved hydrogen sulfide, which comes out of solution when the groundwater rises to the surface; hydrogen sulfide is a poisonous gas that has a rotten-egg smell. In recent years, concern has grown about arsenic in groundwater. Arsenic,

a highly toxic chemical, enters groundwater when arsenic-bearing minerals dissolve.

Groundwater Contamination

In addition to natural chemicals added to groundwater by interaction with the rock through which it flows, human activities have increasingly added a great variety of unhealthy materials (pesticides, gasoline and oil, acids, radioactive wastes, exotic organic chemicals, bacteria, and viruses) to groundwater. Such **groundwater contamination** may be due to leaking septic systems and landfills, leaking fuel and chemical storage tanks, the intentional pumping of waste down wells, the leaching of chemicals from minerals exposed by mining, and the infiltration of chemicals and wastes washed from city streets, agricultural fields, and livestock facilities (▶Fig. 16.19a–c). Typically, contamination disperses from its source to form a diffuse cloud of contaminants mixed with, or dissolved in, groundwater. Geologists refer to such a cloud as a contaminant plume. The concentration of contamination tends to decrease toward the margins of the plume.

The best way to avoid contamination is to prevent contaminants from entering groundwater in the first place. This can be done by locating potential sources of contamination on impermeable bedrock so that they are isolated from the aquifer. If such a site is not available, the storage area should be lined with a thick layer of clay, for the clay not only acts as an aquitard, but it can hold on to contaminants. For this reason, environmental engineers commonly place landfills on top of a liner of packed clay, or place waste in durable, sealed containers. Government agencies have studied various options to store the containers safely. One option involves stockpiling them in tunnels cut into salt domes, for salt is impermeable. Another option involves hiding waste containers in tunnels above the water table. For example, American nuclear waste may be stored in a network of tunnels 300 m beneath Yucca Mountain, Nevada. This mountain consists of dry, fairly impermeable tuff, below which the water table lies at great depth.

Fortunately, in some cases, natural processes can clean up groundwater contamination. Chemicals may be absorbed by clay, oxygen in the water may oxidize the chemicals, and bacteria in the water may metabolize the chemicals, thereby turning them into harmless substances. Where contaminants do make it into an aquifer, environmental engineers drill test wells to determine which way and how fast the contaminant plume is flowing; once they know the flow path, they can close wells in the path to prevent consumption of contaminated water. Alternatively, engineers attempt to clean the groundwater by drilling a series of extraction wells to pump it out of the ground. If the contaminated water does not rise fast enough, engineers drill injection wells to force clean water or steam into the ground beneath the contaminant plume

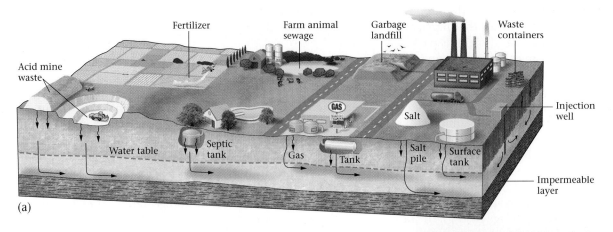

(a)

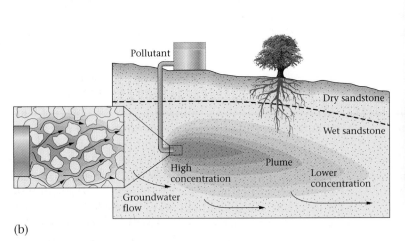

(b)

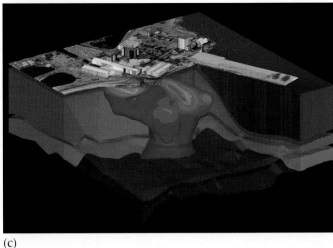

(c)

FIGURE 16.19 (a) The various sources of groundwater contamination. (b) A contaminant plume as seen in cross section. (c) Modern computer analysis can produce three-dimensional images of subsurface plumes. In this image, showing a coastal industrial facility, the ground has been removed to reveal the plume beneath. Red color indicates the greatest concentration of the contaminant.

(▶Fig. 16.20a, b). The injected fluids then push the contaminated water up into the extraction wells. More recently, environmental engineers have begun exploring techniques of bioremediation: injecting oxygen and nutrients into a contaminated aquifer to foster growth of bacteria that can react with and break down molecules of contaminants. Needless to say, cleaning techniques are expensive and generally only partially effective.

Unwanted Effects of Rising Water Tables

We've seen the negative consequences of sinking water tables, but what happens when the water table rises? Is that necessarily good? Sometimes, but not always. If the water table rises above the floor of a house's basement, water seeps through the foundation and floods the basement floor. Catastrophic damage occurs when a rising water table

weakens the base of a hill slope. The addition of water into pore space buoys up the overlying rock or regolith, and therefore makes the slope unstable and likely to slip. Thus, rising water tables can trigger landslides and slumps (▶Fig. 16.21).

16.9 CAVES AND KARST: A SPELUNKER'S PARADISE

Dissolution and the Development of Caves

In 1799, as legend has it, a hunter by the name of Hutchins was tracking a bear through the woods of Kentucky when the bear suddenly disappeared on a hill slope. Baffled,

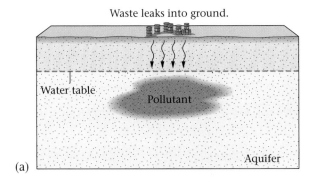

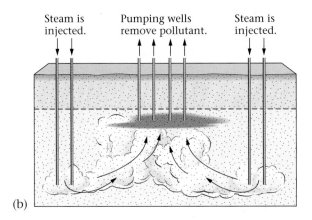

FIGURE 16.20 (a) Leaky drums of chemicals introduce pollutants into the groundwater. (b) Steam injected beneath the contamination drives the contaminated water upward in the aquifer, where pumping wells remove it.

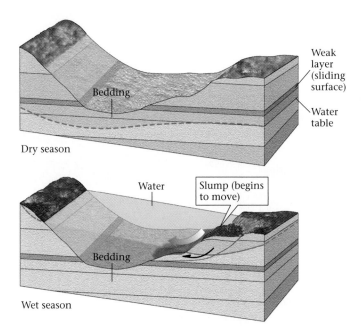

FIGURE 16.21 When the water table rises, material above a weak sliding surface begins to slump, and a landslide may result.

Hutchins plunged through the brambles trying to sight his prey. Suddenly he felt a draft of surprisingly cool air flowing down the slope from uphill. Now curious, Hutchins climbed up the hill and found a dark portal into the hill slope beneath a ledge of rocks. Bear tracks were all around—was the creature inside? Hutchins returned later with a lantern and stepped into the passageway. It led into a large, open room. Hutchins had discovered Mammoth Cave, an immense network—over 540 km in total length—of tunnel-like passages connecting underground rooms.

Large networks such as Mammoth Cave develop primarily in limestone bedrock, as a consequence of the dissolution of limestone by groundwater. Slightly acidic groundwater reacts with calcite to produce HCO_3^{-1} and Ca^{+2} ions, which readily dissolve. Of note, groundwater develops acidity because the water that ultimately becomes groundwater absorbed carbon dioxide (CO_2) partly from the atmosphere when it fell as rain, and more from organic-rich soil as it percolated down to the water table. When it absorbs CO_2, water transforms into carbonic acid.

Geologists have debated about the depth at which limestone caves form in the subsurface. Acidic rainwater clearly dissolves limestone near the ground surface, as indicated by the presence of extensive pitting on limestone bedrock surfaces and along joints in limestone bedrock. And some caves may also form at depth below the water table. It appears, however, that most dissolution takes place in limestone that lies just below the water table, for in this interval the acidity of the groundwater remains high, the mixture of groundwater and newly introduced rainwater is undersaturated (meaning it can dissolve more ions), and groundwater flow is fastest. The association between cave formation and the water table helps explain why openings in a cave system align along the same horizontal plane.

Cave Networks

Caves in limestone usually occur as part of a network. Networks include rooms, or chambers, which are large, open spaces sometimes with cathedral-like ceilings, and tunnel-shaped or slot-shaped passages. (See art on pp. 466–67.) The shape of the cave network reflects variations in permeability and in the composition of the rock from which the caves formed. Larger open spaces develop where the limestone was most soluble and where groundwater flow was fastest. Thus, in a sequence of strata, caves develop preferentially in the more soluble limestone beds. Passages in cave networks typically follow preexisting joints, for the joints provide secondary porosity along which groundwater can flow faster

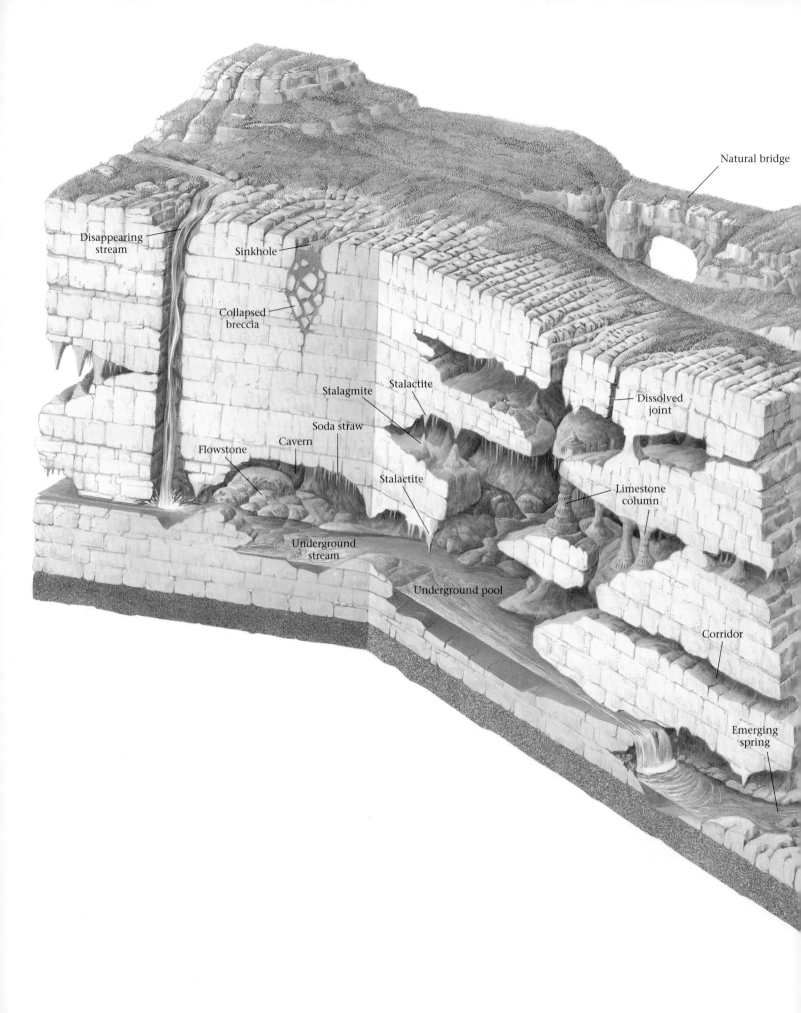

Natural bridge

Disappearing
stream

Sinkhole

Collapsed
breccia

Stalagmite

Stalactite

Soda straw

Dissolved
joint

Flowstone

Cavern

Stalactite

Limestone
column

Underground
stream

Underground pool

Corridor

Emerging
spring

Caves and Karst Landscapes

Limestone, a sedimentary rock made of the mineral calcite, is soluble in acidic water. Much of the water that falls to the ground as rain, or seeps through the ground as groundwater, tends to be acidic, so in regions of the Earth where bedrock consists of limestone, we find signs of dissolution. Underground openings that develop by dissolution are called caves or caverns. Some of these may be large, open rooms, whereas others are long, narrow passages. Underground lakes and streams may form on the floor. A cave's location depends on the orientation of bedding and joints, for these features localize the flow of groundwater.

Caves originally form at or near the water table (the subsurface boundary between rock or sediment in which pores contain air, and rock or sediment in which pores contain water). As the water table drops, caves empty of most water and become filled with air. In many locations, groundwater drips from the ceiling of a cave or flows along its walls. As the water evaporates and thus loses its acidity (because of the evaporation of dissolved carbon dioxide), new calcite precipitates. Over time, this calcite builds into cave formations, or speleothems, such as stalactites, stalagmites, columns, and flowstone.

Distinctive landscapes, called karst landscapes, develop at the Earth's surface over limestone bedrock. In such regions, the ground may be rough where rock has dissolved along joints; where the roofs of caves collapse, sinkholes develop. If a surface stream flows through an open joint into a cave network, we say that the stream is "disappearing." The water from such streams may flow underground for a ways and then reemerge elsewhere as a spring. In some places, the collapse of subsurface openings leaves behind natural bridges.

(►Fig. 16.22). Because joints commonly occur in orthogonal systems (consisting of two sets of joints oriented at right angles to each other; see Chapter 9), passages form a grid.

Precipitation and the Formation of Speleothems

When the water table drops below the level of a cave, the cave becomes an open space filled with air. Acidic, calcite-containing water then emerges from the rock above the cave and drips from the ceiling or trickles down the walls. As soon as this water reenters the air, it evaporates a little and releases some of its dissolved carbon dioxide. As a result, calcium carbonate (limestone) precipitates out of the water. Rock formed by such precipitation is a type of travertine called dripstone, and the various intricately shaped formations that grow in caves by the accumulation of dripstone are called **speleothems.**

Cave explorers, or spelunkers, and geologists have developed a detailed nomenclature for different kinds of speleothems (►Fig. 16.23). Where water drips from the ceiling of the cave, the precipitated limestone adds to the tip of an iciclelike cone called a stalactite. Initially, calcite precipitates around the outside of the drip, forming a delicate, hollow stalactite called a soda straw. But eventually, the soda straw fills up, and water migrates down the margin of the cone to form a more massive, solid stalactite. Where the drips hit the floor, the resulting precipitate builds an upward-pointing cone called a stalagmite. If the process of dripstone formation in a cave continues long enough, stalagmites merge with overlying stalactites to create lime-

FIGURE 16.23 Large speleothems in a cave.

stone columns (►Fig. 16.24). If the groundwater flows along the surface of a wall, it drapes the wall with clothlike sheets of limestone called flowstone. Thin sheets of dripstone and flowstone tend to be translucent and, when lit from behind, glow with an eerie amber light.

The Formation of Karst Landscapes

Limestone bedrock underlies most of the Kras Plateau in Slovenia, along the eastern coast of the Adriatic Sea. The name "Kras," which means "bare, rocky ground," is apt because of the rough, unvegetated land surface in the region. Throughout the area, the bedrock contains abundant caves; the landscape is pockmarked by deep circular-to-elongate depressions, or sinkholes, which, as we have seen, can form where the roof of a cave collapses. Such sinkholes are called collapse sinkholes. (Sinkholes can also form where acidic water accumulates in a pool at the surface, seeps down, and

FIGURE 16.22 Joints act as conduits for water in cave networks. Thus, caves and passageways lie along joints.

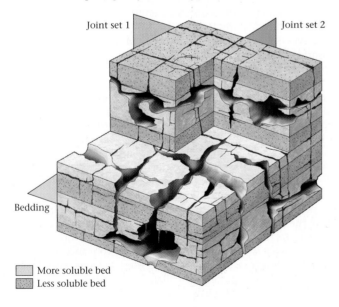

Joint set 1

Joint set 2

Bedding

☐ More soluble bed
▨ Less soluble bed

FIGURE 16.24 The evolution of a soda straw stalactite into a limestone column.

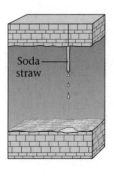
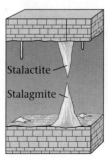
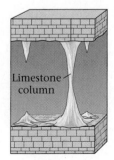

Soda straw

Stalactite

Stalagmite

Limestone column

Time ⟶

dissolves bedrock below. Eventually, a solution pit develops in the bedrock, and the overlying soil sinks downward, forming a depression called a solution sinkhole.) The sinkholes of the Kras Plateau are separated from one another by hills or walls of bedrock (▶Fig. 16.25a–d). Locally, where most of a cave collapsed, a natural bridge spans the cave remnant. Where the water table rises above the floor of a sinkhole, the sinkhole fills to become a lake. And where surface streams intersect cracks or holes that link to the caves below, the water disappears into the subsurface and becomes an underground stream. Such disappearing streams reemerge from a cave entrance downstream.

Geologists refer to landscapes such as the Kras Plateau in which surface features reflect the dissolution of bedrock below as **karst landscapes,** from the Germanized version of "kras." Karst landscapes form in a series of stages (▶Fig. 16.26a–c).

- *The establishment of a water table in limestone:* The story of a karst landscape begins after the formation of a thick interval of limestone. If relative sea level drops, a water table can develop in the limestone below the ground surface. If, however, the limestone gets buried deeply, it must first be uplifted and exposed by erosion before it can contain the water table.

- *The formation of a cave network:* Once the water table has been established, dissolution begins and a cave network develops.

- *A drop in the water table:* If the water table later becomes lower, either because of a decrease in rainfall or because nearby rivers cut down through the landscape and drain the region, newly formed caves dry out. Downward-percolating groundwater emerges from the roofs of the caves; dripstone and flowstone precipitate.

- *Roof collapse:* If rocks fall off the roof of a cave for a long time, the roof eventually collapses. Such collapse creates sinkholes and troughs, leaving behind hills, ridges, and natural bridges of limestone.

Some karst landscapes contain many round sinkholes separated by hills. The giant Arecibo radio telescope in Puerto Rico, for example, consists of a dish formed by smoothing the surface of a 300-m-wide round sinkhole (Fig. 16.25b). In

FIGURE 16.25 (a) Numerous sinkholes of a karst landscape. (b) The Arecibo radio telescope in Puerto Rico was built in a sinkhole. (c) Natural Bridge, Virginia. (d) A disappearing stream.

(a)

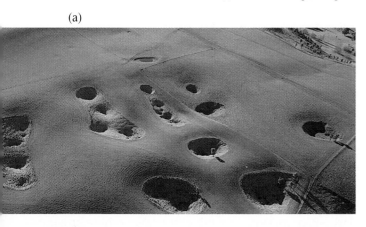

(b)

(c)

(d)

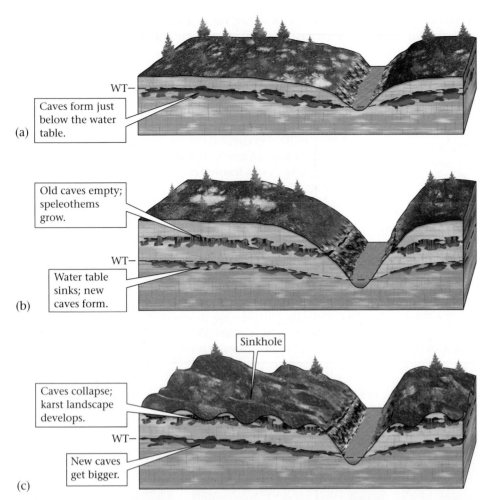

FIGURE 16.26 The formation of caves and a karst landscape. (a) Dissolution takes place near the water table (WT) in an uplifted sequence of limestone. (b) Downcutting by an adjacent river lowers the water table, and the caves empty. Speleothems grow on the cave walls. (c) After roof collapse, the landscape becomes pockmarked by sinkholes.

regions where vertical joints control roof collapse, steep-sided residual bedrock towers remain between sinkholes. A karst landscape with such spires is called tower karst. The surreal collection of pinnacles constituting the tower karst landscape in the Guilin region of China has inspired generations of artists to portray them on scroll paintings (▶Fig. 16.27a, b).

Life in Caves

Despite their lack of light, caves are not sterile, lifeless environments. Caves that are open to the air provide a refuge for bats as well as for various insects and spiders.

Similarly, fish and crustaceans enter caves where streams flow in or out. Species living in caves have evolved some unusual characteristics. For example, cave fish lose their pigment and in some cases their eyes (▶Fig. 16.28a). Recently, explorers discovered caves in Mexico in which warm, mineral-rich groundwater currently flows. Colonies of bacteria metabolize sulfur-containing minerals in this water, and create thick mats of living ooze in the complete darkness of the cave. Long gobs of this bacteria slowly drip from the ceiling. Because of the mucuslike texture of these drips, they have come to be known as snotites (▶Fig. 16.28b).

(a)   (b)

FIGURE 16.27 (a) Tower karst in China. (b) Painting of tower karst by an unknown Chinese artist.

FIGURE 16.28 Life in caves: (a) blind fish and (b) snotites (gobs of bacteria).

(a)

(b)

CHAPTER SUMMARY

• During the hydrologic cycle, water infiltrates the ground and fills the pores and cracks in rock and sediment. This sub-surface water is called groundwater. The amount of open space in rock or sediment is its porosity; the degree to which pores are interconnected, so that water can flow through, defines its permeability.

• Geologists classify rock and sediment according to their permeability. Aquifers are relatively permeable, and aquitards are relatively impermeable.

• The water table is the surface in the ground above which pores contain mostly air, and below which pores are filled with water. The shape of a water table is a subdued imitation of the shape of the overlying land surface.

• Groundwater flows wherever the water table has a hydraulic gradient, and moves slowly from recharge areas to discharge areas. Darcy's law shows that the rate of movement depends on permeability and on the hydraulic gradient.

• Groundwater can be extracted in wells. An ordinary well penetrates below the water table, but in an artesian well, water rises on its own. Pumping water out of a well too fast causes drawdown, yielding a cone of depression. At a spring, groundwater exits the ground on its own.

• Hot springs and geysers release hot water to the Earth's surface. This water may have been heated by residing very deep in the crust, or by the proximity of a magma chamber.

• Groundwater is a precious resource, used for municipal water supplies, industry, and agriculture. In recent years, some regions have lost their groundwater supply because of overuse or contamination.

• When limestone dissolves just below the water table, underground caves are the result. Soluble beds and joints determine the location and orientation of caves. If the water table drops, caves empty out. Limestone precipitates out of water dripping from cave roofs, and creates speleothems (such as stalagmites and stalactites). Regions that contain abundant caves, some of which have collapsed to form sinkholes, are called karst landscapes.

KEY TERMS

aquifer (p. 452)
aquitard (p. 452)
artesian springs (p. 457)
artesian well (p. 456)

cone of depression (p. 456)
Darcy's law (p. 455)
discharge area (p. 455)
geothermal region (p. 458)

geyser (p. 458)
groundwater (p. 450)
groundwater contamination (p. 463)
hot spring (p. 457)
hydraulic gradient (p. 455)
hydraulic head (p. 455)
karst landscape (p. 469)
ordinary well (p. 456)
perched water table (p. 454)

permeability (p. 452)
pore (p. 451)
porosity (p. 451)
recharge area (p. 455)
sinkhole (p. 450)
soil moisture (p. 451)
speleothem (p. 468)
spring (pp. 456, 457)
water table (p. 452)
well (p. 456)

REVIEW QUESTIONS

1. How do porosity and permeability differ?

2. What factors affect the level of the water table? What factors affect the flow direction of the water below the water table?

3. How does the rate of groundwater flow compare with that of moving ocean water or river currents?

4. What does Darcy's law tell us about how the hydraulic gradient and permeability affect groundwater flow?

5. How does excessive pumping affect the local water table?

6. How is an artesian well different from an ordinary well?

7. Why do hot springs form and geysers erupt?

8. Is groundwater a renewable or nonrenewable resource?

9. Describe some of the ways in which human activities can adversely affect the water table.

10. What are some sources of groundwater contamination? How can it be prevented?

11. Describe the process leading to the formation of caves and the speleotherms within caves.

SUGGESTED READING

Glennon, R. J. 2002. *Water Follies: Groundwater Pumping and the Fate of America's Fresh Waters:* Washington, D.C.: Island Press.

Gunn, J. 2004. *Encyclopedia of Caves and Karst Science.* London: Fitzroy Dearborn.

Schwartz, F. W., and H. Zhang. 2002. *Introduction to Groundwater Hydrology.* New York: Wiley.

White, W. B. 1997. *Geomorphology and Hydrology of Karst Terrains.* Oxford, UK: Oxford University Press.

CHAPTER 17

Dry Regions: The Geology of Deserts

The bare hills are cut out with sharp gorges, and over their stone skeletons scanty earth clings . . . A white light beat down, dispelling the last tract of shadow, and above hung the burnished shield of hard, pitiless sky.
—CLARENCE KING (1842–1901, 1ST DIRECTOR OF THE U.S. GEOLOGICAL SOCIETY, DESCRIBING A DESERT)

In a desert, there is very little vegetation because it rarely rains. But not all deserts are seas of sand. Here, in the Sonoran Desert of northwestern Mexico, saguaro cacti stand guard over a landscape of stony plains and barren rock cliffs.

17.1 INTRODUCTION

For generations, nomadic traders have used camels to traverse the Sahara Desert in northern Africa (▶Fig. 17.1a). The Sahara, the world's largest desert, receives so little rainfall that it has hardly any surface water or vegetation. So camels must be able to walk for up to three weeks without drinking or eating. They can survive these journeys because they sweat relatively little, thereby conserving their internal water supply; they have the ability to metabolize their own body fat (up to 20 kg of fat make up the animal's hump alone) to produce new water; and they can withstand severe dehydration (loss of water). Most mammals die after losing only 10–15% of their body fluid, for their blood plasma dries up, but camels can survive 30% dehydration with no ill effects. Camels do get thirsty, though. After a marathon trek across the desert, a camel may guzzle up to 100 liters of water in less than ten minutes.

The survival challenges faced by a camel emphasize that deserts are lands of extremes—extreme dryness, heat, cold, and, in some places, beauty. Desert vistas include everything from sand seas to sagebrush plains, cactus-covered hills to endless stony pavements. Although less populated than other regions on Earth, deserts cover a significant percentage

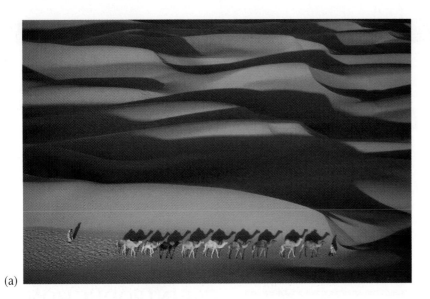

(a)

FIGURE 17.1 (a) Camels, as in this Sahara caravan in Mauritania, survive harsh desert conditions by storing water in fat reserves and by sweating little, if at all. (b) The global distribution of deserts. Note that the largest lie in the subtropical belts. Arrows indicate prevailing wind directions.

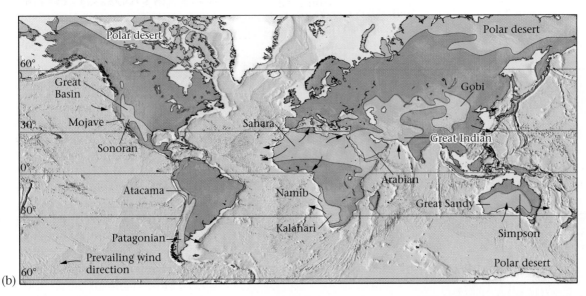

(b)

(about 25%) of the land surface, and thus constitute an important component of the Earth System (►Fig. 17.1b). In this chapter, we take a look at the desert landscape. We learn why deserts occur where they do, and how erosion and deposition shape their surface. We conclude by exploring life in the desert and by examining the problem of desertification, the gradual transformation of temperate lands into desert.

17.2 WHAT IS A DESERT?

Formally defined, a **desert** is a region that is so arid (dry) that it contains no permanent streams (except for rivers that bring water in from temperate regions elsewhere) and sup-

ports vegetation on no more than 15% of its surface. In general, desert conditions exist where less than 25 cm of rain falls per year. But rainfall alone does not determine the aridity of a region. Aridity also depends on whether rainfall occurs sporadically or continuously during the year, and on rates of evaporation. If all the rain in a region falls during isolated downpours once every few years, the region becomes a desert because the intervals of drought last so long that plants and permanent streams cannot survive. Similarly, if high temperatures and dry air cause evaporation rates from the ground to exceed the rate at which rainfall wets the ground, the region becomes a desert even if it gets more than 25 cm per year of rain.

Note that the definition of a desert depends on a region's aridity, not on its temperature. Geologists distinguish between cold deserts, where temperatures generally stay below

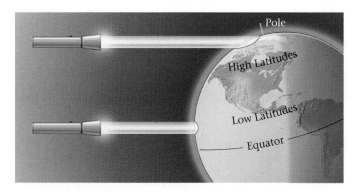

FIGURE 17.2 At low latitudes, rays of sunlight (here represented by the flashlight beam) strike the Earth straight on and spread over a small area. At high latitudes, rays of sunlight strike the Earth obliquely and spread over a broad area. Thus, a square meter of the Earth's surface at a high latitude receives less light energy then does a square meter at a low latitude.

about 20°C, and hot deserts, where summer daytime temperatures exceed 35°C. Cold deserts exist at high latitudes where the Sun's rays strike the Earth obliquely and thus don't provide much energy (▶Fig. 17.2), at high elevations where the air is too thin to hold much heat, or in lands adjacent to cold oceans, where the cold water absorbs heat from the air above. Hot deserts are found at low latitudes where the Sun's rays strike the desert at a high angle, at low elevations where dense air can hold a lot of heat, and in regions distant from the cooling effect of cold ocean currents. The hottest recorded temperatures on Earth occur in low-latitude, low-elevation deserts—58°C (136°F) in Libya and 56°C (133°F) in Death Valley, California.

17.3 TYPES OF DESERTS

Each desert on Earth has unique characteristics of landscape and vegetation that distinguish it from others. Geologists group deserts into five different classes, based on the environment in which the deserts form.

• *Subtropical deserts:* Subtropical deserts (e.g., the Sahara, Arabian, Kalahari, and Australian) form because of the pattern of convection cells in the atmosphere. At the equator, the air becomes warm and humid, for sunlight is intense and water rapidly evaporates from the ocean. The hot, moisture-laden air rises to great heights above the equator. As this air rises, it expands and cools, and can no longer hold as much moisture. Water condenses and falls in downpours that feed the lushness of the equatorial rain forest. The now-dry air, now at high altitude, spreads laterally north or south. When this air reaches latitudes of 20° to 30°, a region called the subtropics, it has become cold and dense enough to sink (▶Fig. 17.3; ▶Box 17.1 provides further explanation). Because the air is dry, no clouds form, and intense solar radiation strikes the Earth's surface. The sinking dry air condenses and heats up, soaking up any moisture present. In the regions swept by this hot air on its journey back to the equator, evaporation rates greatly exceed rainfall rates, and thus the regions become deserts.

• *Deserts formed in rain shadows:* As winds blow at low elevation over the ocean toward the shore, the air in the wind absorbs moisture and becomes quite humid. Upon reaching a coastal mountain range, this humid air must rise (▶Fig. 17.5). As the air rises, it expands and cools. The water it contains condenses and falls as rain on the seaward flank of the mountains, nourishing a coastal rain forest. When the air finally reaches the inland side of the mountains, it has lost all its moisture

FIGURE 17.3 Rising air at the equator loses its moisture by raining over rain forests. When the air sinks over the subtropics, it warms and absorbs water. Thus, rainfall rarely occurs in the subtropics. Note that, for simplicity, the tilt of the Earth is not shown.

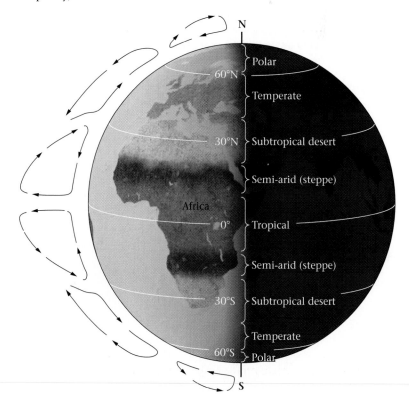

Convection in the Atmosphere and Generation of Prevailing Wind

BOX 17.1
THE REST OF THE STORY

Light energy from the Sun passes straight down through the atmosphere until it reaches the ground, for air is transparent. The ground, however, consists of opaque material which absorbs some light energy and heats up. Warm ground radiates heat back up into the air, in the form of infrared waves. This infrared radiation heats air at the base of the atmosphere, causing the air to expand and become less dense. Warm air, because of its lower density, is buoyant and rises. As this happens, denser, cold air sinks to take its place. A pattern of circular flow, known as convection, develops (see Appendix). The same process happens in a pot of water when you place it on a stove.

Because the Earth is a sphere not all areas receive the same amount of incoming solar energy, or insolation: portions of the Earth's surface hit by direct rays of the Sun receive more energy per square meter than portions hit by oblique rays (see Fig. 17.2). Higher latitudes thus receive less energy than lower latitudes. (Because of the tilt of Earth's axis, the amount of solar radiation that any point on the surface receives changes during the year, and for that reason, we have seasons).

The contrast in the amount of solar radiation received by different latitudes means that polar regions are cooler at the surface than are equatorial regions. In 1735, George Hadley, a British mathematician, realized that this contrast could cause air to circulate on a global scale by convection. Specifically, he suggested that warm air at the equator would rise and flow toward the poles, to be replaced by cool polar air, which would flow to the equator at lower elevations (▶Fig. 17.4a). Hadley's proposal, however, did not take into account an important factor, namely the Earth's rotation and the resulting Coriolis effect. The Coriolis effect, as we learned in Chapter 15, refers to the deflection that an object is subject to as it moves from the circumference to the center of a rotating disk or from the equator to the pole of a rotating sphere (and vice versa) (▶Fig. 17.4b).

The moving atmosphere is subject to the Coriolis effect. Northward-moving high-altitude air in the Northern Hemisphere deflects to the east, so by the latitude of 30°N, it is basically moving due east and cannot make it the rest of the way to the pole. By the time it reaches this latitude, the air has also cooled significantly and thus starts to sink. This means that a global convection cell (a current that looks like a loop in cross section) develops that conveys warm air north from the equator to the subtropics (latitude 30°), where it cools and sinks. When the sinking air reaches low elevations, it divides, some moving back toward the equator near the surface and some moving north near the surface. A place where sinking air separates into two flows that move in opposite directions is called a divergence zone.

Meanwhile, cool air from the polar region moves south near the surface and deflects to the west. At about a latitude of 60°, the near-surface polar air collides, or converges, with the northward-moving, near-surface mid-latitude air. This air must rise, because there is nowhere for the extra air to go but up. A place where two surface air flows meet so that air has to rise is called a convergence zone. The convergence zone at latitude 60° is called the polar front. The air that rises along the polar front divides at the top of the troposphere, the atmosphere's lowest layer, with some air heading toward the equator at high altitude to complete a mid-latitude convection cell, and some heading toward the pole to create a polar convection cell. Because of seasonal changes on Earth, the exact position of the polar front changes during the year.

In sum, because of the Coriolis effect, circulating air in the troposphere splits into three globe-encircling convection cells in each hemisphere. The low-latitude cells, extending from the equator to a latitude of about 30°, are called Hadley cells, in honor of George Hadley. The mid-latitude cells are called Ferrel cells, in honor of the American meteorologist William Ferrel, who proposed them. The high-latitude cells are simply called polar cells (▶Fig. 17.4c). The global-scale flow in the six major convective cells on Earth creates belts in which surface air generally flows in a consistent direction. Such air flows are called prevailing surface winds—these are labeled on Figure 17.4c.

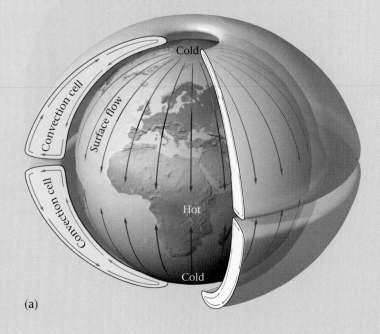

(a)

FIGURE 17.4 (a) If the Earth did not rotate, two simple convection cells would be established in the atmosphere, one stretching from the equator to the North Pole and one stretching from the equator to the South Pole. (b) Because of the Coriolis effect, a rocket sent north from the equator curves to the right (east). This is because the rocket not only has a northward velocity driven by its engines but also has an eastward velocity caused by the Earth's rotation at the launch site. (c) Because of the Coriolis effect, atmospheric circulation breaks into three convection cells within each hemisphere, named, from equator to pole, Hadley, Ferrel, and polar.

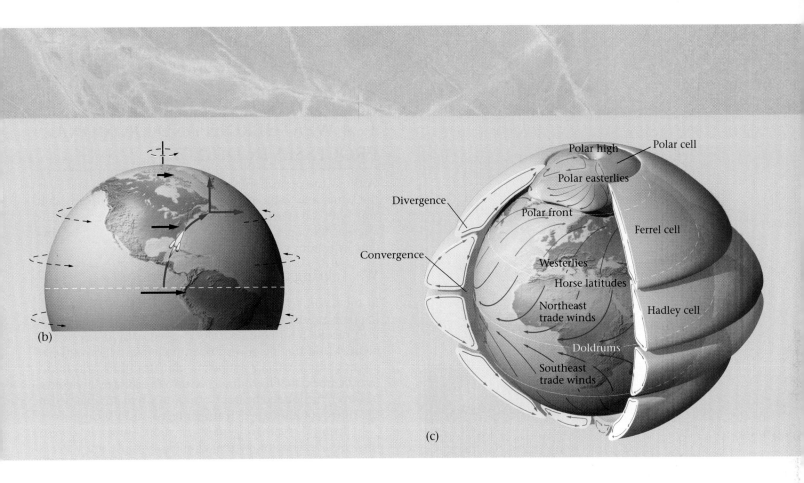

(b)

(c)

and can no longer provide rain. As a consequence, a **rain shadow** forms, and the land beneath the rain shadow becomes a desert. A rain-shadow desert can be found east of the Cascade Mountains in Washington State.

- *Coastal deserts formed along cold ocean currents:* Cool ocean water cools the overlying air by absorbing heat, and decreases the capacity of the air to hold moisture. For example, the cold Humboldt Current, which carries water northward from Antarctica to the western coast of South America, absorbs water from the breezes that blow east over the coast. Thus, rain rarely falls on the coastal areas of Chile and Peru. As a result, this region hosts a desert landscape, including one of the driest deserts in the world, the Atacama (▶Fig. 17.6a–c). Portions of this narrow (less than 200 km wide) desert, which lies between the Pacific coast on the west and the Andes on the east, received no rain at all between 1570 and 1971.

- *Deserts formed in the interiors of continents:* As air masses move across a continent, they lose moisture by dropping rain, even in the absence of a coastal mountain range. Thus, when an air mass reaches the interior of a large continent such as Asia, it has grown quite dry, so the land beneath becomes arid. The largest example of such a continental-interior desert, the Gobi, lies in central Asia, over 2,000 km away from the nearest ocean.

FIGURE 17.5 When forced to rise by mountains, moist air cools. As this happens, the moisture condenses and rain falls, nourishing coastal rain forests, so by the time the air reaches the inland side of the mountains, it no longer holds enough moisture to rain. Deserts form in the rain shadow of mountains.

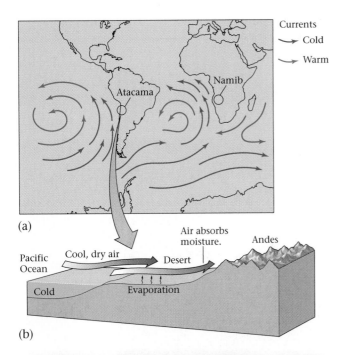

FIGURE 17.6 (a) Currents bringing cold water up from the Antarctic cool the air along the southwestern coasts of South America and Africa. (b) The cool, dry air absorbs moisture from the adjacent coastal land, keeping it dry, so coastal deserts form. (c) The Atacama Desert of South America is the driest place in the world.

• *Deserts of the polar regions:* So little precipitation falls in Earth's polar regions (north of the Arctic Circle, at 66°30′N, and south of the Antarctic Circle, at 66°30′S) that these areas are, in fact, arid. Polar regions are dry, in part, for the same reason that the subtropics are dry (the global pattern of air circulation means that the air flowing over these regions is dry) and, in part, for the same reason that coastal areas along cold currents are dry (cold air holds little moisture).

The distribution of deserts around the world changes through geologic time because of plate tectonics, for plate movements determine the latitude of land masses, the position of land masses relative to the coast, and the proximity of land masses to mountain ranges. Because of continental drift, some regions that were deserts in the past are temperate or tropical regions now, and vice versa.

17.4 WEATHERING AND EROSIONAL PROCESSES IN DESERTS

Without the protection of foliage to catch rainfall and slow the wind, and without roots to hold regolith in place, rain and wind can attack and erode the land surface of deserts. As a result, hillslopes are typically bare, and plains can be covered with stony debris or drifting sand.

Weathering and Soil Formation

In the desert, as in temperate climates, physical weathering happens primarily when joints (natural fractures) split rock into pieces. Joint-bounded blocks eventually break free of bedrock and tumble down slopes, fragmenting into smaller pieces as they fall. In temperate climates, thick soil forms over bedrock, so it lies buried beneath the surface. In deserts, however, the jointed bedrock commonly remains exposed at the surface of hillslopes, creating rugged, rocky escarpments.

Chemical weathering happens more slowly in deserts than in temperate or tropical climates, because less water is available to react with rock. Still, rain or dew provides enough moisture for *some* weathering to occur. This water seeps into rock and leaches (dissolves and carries away) calcite, quartz, and various salts. Leaching effectively rots the rock by transforming it into a poorly cemented aggregate. Over time, the rock will crumble and form a pile of unconsolidated sediment, susceptible to transport by water or wind.

Although enough rain falls in deserts to leach chemicals out of rock, there is insufficient water to flush the dissolved minerals entirely away. Thus, when the water percolates down and dries up, minerals precipitate in regolith below the surface. If calcite precipitates, it cements loose grains together, forming solid calcrete. The growth of calcrete can occur rapidly enough to incorporate tools that prospectors abandoned.

Shiny **desert varnish,** a dark, rusty brown coating of iron oxide, manganese oxide, and clay, covers the surface of rock in deserts. Desert varnish was once thought to form when water from rain or dew seeped into a rock, dissolved iron and magnesium ions, and carried the ions back to the surface of the rock by capillary action. More recent studies, however, have shown that desert varnish forms when windborne dust settles on the surface of the rock; in the presence of moisture, microorganisms (bacteria) extract elements from the dust and transform it into iron or manganese oxide. The oxides bind together clay flakes. Such varnish won't form in humid climates, because rain washes the ions away too fast.

(a)   (b)

FIGURE 17.7 (a) Desert varnish, made from iron oxide, manganese oxide, and clay, is a dark coating on desert rock surfaces. Native American artists created petroglyphs, images on the rock surface, by chipping through the varnish to reveal the lighter rock beneath. (b) In the Painted Desert of Arizona, the different colors of the rock layers are due, in part, to the oxidation state of iron.

Desert varnish takes a long time to form. In fact, the thickness of a desert varnish layer provides an approximate estimate of how long a rock has been exposed at the ground surface. In past centuries, Native Americans used desert-varnished rock as a medium for art: by chipping away the varnish to reveal the underlying lighter-colored rock, they were able to create figures or symbols on a dark background. The resulting drawings are called petroglyphs (▶Fig. 17.7a).

Because of the lack of plant cover in deserts, variations in bedrock color stand out. Locally derived soils typically retain the color of the bedrock from which they were derived. Slight variations in the concentration of iron, or in the degree of iron oxidation, in adjacent beds result in spectacular color bands in rock layers and the thin soils derived from them. The Painted Desert of northern Arizona earned its name from the brilliant and varied hues of its shale (▶Fig. 17.7b).

Water Erosion

Although rain rarely falls in deserts, when it does come, it can radically alter a landscape in a matter of minutes. Since deserts lack plant cover, rainfall, sheetwash, and stream flow all are extremely effective agents of erosion. It may seem surprising, but water generally causes more erosion than does wind in most deserts.

Water erosion begins with the impact of raindrops, which eject sediment into the air. On a hill, the ejected sediment lands downslope. The ground quickly becomes saturated with water during a heavy rain, so water starts flowing across the surface, carrying the loose sediment with it. Within minutes after a heavy downpour begins, dry stream

channels fill with a turbulent mixture of water and sediment, which rushes downstream as a flash flood. When the rain stops, the water sinks into the stream bed's gravel and disappears—such streams are called ephemeral streams (see Chapter 14). Because of the relatively high viscosity of the water, owing to its load of suspended sediment, and the velocity and turbulence of the flow, flash floods in deserts cause intense erosion. They undercut cliffs and transport boulders and gravel downstream.

Flash floods carve steep-sided channels into the ground. Scouring of bedrock walls by sand-laden water may polish the walls and create grooves. Dry stream channels in desert regions of the western United States are called **dry washes,** or arroyos. In the Middle East and North Africa they are called wadis (▶Fig. 17.8).

Wind Erosion

In temperate and humid regions, plant cover protects the ground surface from the wind, but in deserts, the wind has direct access to the ground. In hot deserts, gusts of hot air feel like blasts from a furnace. Wind, just like flowing water, can carry sediment either as suspended load or as surface load. The wind's **suspended load** consists of fine-grained sediment, such as dust and silt, that floats in the air and moves with it (▶Fig. 17.9). The suspended sediment can be carried so high into the atmosphere and so far downwind that it may move completely out of its source region.

Moderate to strong winds can roll and bounce sand grains along the ground, a process called **saltation** (▶Fig. 17.10). Saltation begins when turbulence caused by wind

FIGURE 17.8 Gravel and sand are left behind on the floor of a dry wash after a flash flood. The wash has steep walls because downcutting happens so fast and the walls consist of hard rock.

FIGURE 17.9 Dust clouds form in deserts when turbulent air carries very fine sediment in suspension.

shearing along the ground surface lifts sand grains. The grains move downwind, following an asymmetric, archlike trajectory, but eventually return to the ground, where they strike other sand grains, causing the new grains to bounce up and drift or roll downwind. The collisions between sand grains make the grains rounded and frosted. Saltating grains generally rise no more than 0.5 m. But where sand bounces on bedrock during a desert sandstorm, they may rise 2 m and can strip the paint off a car. Saltating grains comprise the wind's **surface load.**

The size of clasts that wind can carry depends on the wind velocity. Wind, therefore, does an effective job of sorting sediment, sending dust-sized particles skyward and sand-sized particles bouncing along the ground,

while pebbles and larger grains remain behind. In some cases, wind carries away so much fine sediment that pebbles and cobbles become concentrated at the ground surface (▶Fig. 17.11). An accumulation of coarser sediment left behind when fine-grained sediment blows away is called a **lag deposit.**

In many locations, the desert surface resembles a tile mosaic in that it consists of separate stones that fit together tightly, forming a fairly smooth surface layer above a soil composed of silt and clay. Such natural mosaics constitute **desert pavement** (▶Fig. 17.12a, b). Typically, desert varnish coats the top surfaces of the stones forming desert pavement. Geologists have come up with several explanations for the origin of desert pavements: (1) They are

FIGURE 17.11 The progressive development of a lag deposit. Pebbles in a desert are distributed through a matrix of finer sediment. With time, wind blows the finer sediment away, and the pebbles concentrate on the ground surface and may settle together to create a lag deposit. The deposit acts like armor, protecting the substrate from further erosion.

FIGURE 17.10 During saltation, sand grains roll and bounce along the ground surface. As they bounce, they follow parabolic paths.

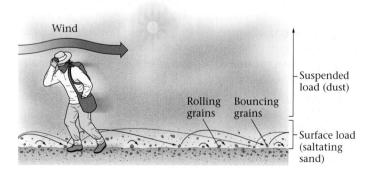

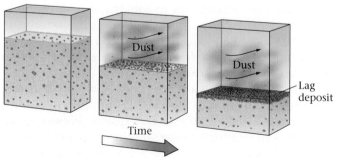

(a)

Desert
pavement

Soil

Old
alluvium

(b)

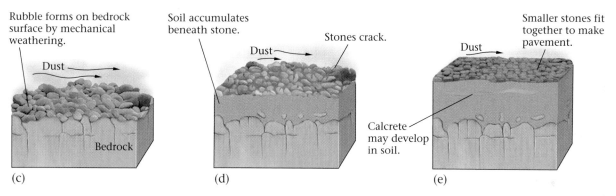

Rubble forms on bedrock
surface by mechanical
weathering.

Dust

Bedrock

(c)

Soil accumulates
beneath stone.

Dust

Stones crack.

(d)

Smaller stones fit
together to make
pavement.

Dust

Calcrete
may develop
in soil.

(e)

FIGURE 17.12 (a) A well-developed desert pavement in Arizona. (b) Photo of a trench dug in Arizona, showing how the desert pavement lies on top of a fine-grained sediment. This pavement was lifted up from the surface of a now-buried alluvial fan. (c) Desert pavement forms in stages. First, loose pebbles and cobbles collect at the surface. These can be formed by mechanical weathering, as shown, or by deposition of alluvium. (d) Dust settles among stones and builds up a layer of soil beneath the stone layer. Stones crack into smaller pieces. (e) A durable, mosaiclike pavement forms.

lag deposits, formed when wind blows away the fine sediment between clasts, so that the clasts can settle down and fit together. (2) They form when sheetwash during heavy rainfalls washes away the fine sediment between larger clasts. (3) They form in response to wetting and drying of desert soils. According to this hypothesis, clays in the soil absorb water and expand during heavy rains. The expansion pushes larger stones upward. When the soils dry and shrink, the stones settle down and fit together while the finer-grained sediment sinks between the stones and disappears. (4) They form when wind-blown dust slowly sifts down onto the stones, and then washes down between the stones. In this model, the pavement is "born at the surface," meaning that the stones forming the pavement were never buried, but have been progressively lifted up as sedi-ment collects and builds up beneath (▶Fig. 17.12c–e). Recent studies lend most support to this last model.

Just as sand blasting cleans the grime off the surface of a building, windblown sand and dust grind away at surfaces in the desert. Over long periods of time, such wind abrasion creates smooth faces, or facets, on pebbles, cobbles, and boulders. If a rock rolls or tips relative to the prevailing wind direction after it has been faceted on one side, or if the wind shifts direction, a new facet with a different orientation forms, and the two facets join at a sharp edge. Rocks whose surface has been faceted by the wind are faceted rocks, or **ventifacts** (▶Fig. 17.13a–d). Recent exploration has revealed landscapes of ventifacts on Mars, where immense dust storms occasionally envelop the planet's entire surface.

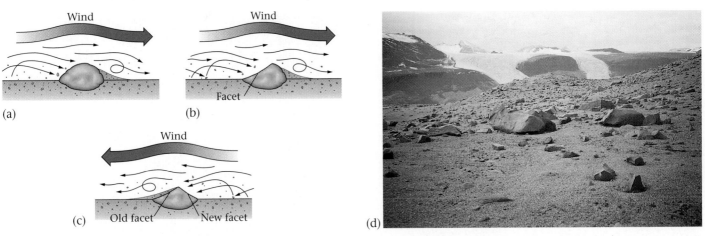

FIGURE 17.13 The progressive development of a ventifact. (a) Wind-blown sand and dust abrade the face of a rock. (b) Slowly, erosion carves a smooth surface called a facet. (c) Later, the wind shifts direction, and a new facet forms. The two facets join at a sharp edge. (d) Ventifacts formed by strong winds in Antarctica.

Over time, in regions where the substrate consists of soft sediment, wind picks up and removes so much sediment that the land surface sinks. The process of lowering the land surface by wind erosion is called **deflation.** Shrubs can stabilize a small patch of sediment with their roots, so after deflation a forlorn shrub with its residual pedestal of soil stands isolated above a lowered ground surface. In some places, deflation scours a deep, bowl-like depression called a blowout (▶Fig. 17.14).

FIGURE 17.14 A blowout, made where wind erosion has hollowed out a topographic depression.

17.5 DEPOSITIONAL ENVIRONMENTS IN DESERTS

We've seen that erosion relentlessly eats away at bedrock and sediment in deserts. Where does the debris go? Below, we examine the various desert settings in which sediment accumulates.

Talus Aprons and Alluvial Fans

With time, joint-bounded blocks of rock break off rock ledges and cliffs on the sides of hills. Under the influence of gravity, the resulting debris tumbles downslope and accumulates as a **talus apron** at the base of a hill. Talus aprons can survive for a long time in desert climates, so we typically see them fringing the base of cliffs in deserts (▶Fig. 17.15). The angular clasts constituting talus aprons gradually become coated in desert varnish.

Flash floods can carry sediment downstream and away from eroding cliffs in an ephemeral stream channel. When the turbulent water flows out into a plain at the mouth of a canyon, it spreads out over a broader surface and slows. As a consequence, sediment in the water settles out. The resulting lenses of sediment cause the channel that has emerged from the mountains to subdivide into a number of subchannels (distributaries) that diverge outward in a broad fan. The fan of distributaries spreads the sediment, or alluvium, out into a broad **alluvial fan,** a wedge- or apron-shaped pile of sediment (▶Fig. 17.16). Alluvial fans emerging from adjacent valleys may merge and overlap along the front of a mountain range, creating an elongate wedge of sediment called a bajada. Over long periods, the sediment of bajadas fills in adjacent valleys to depths of several kilometers.

FIGURE 17.15 This talus apron along the base of a desert cliff formed from rocks that broke off and tumbled down the cliff.

FIGURE 17.16 The sediment constituting this alluvial fan in Death Valley was carried to the mountain front during flash floods. The water drops the sediment at the mountain front, because the water slows down.

Playas and Salt Lakes

Water from a flash flood may make it out to the center of an alluvium-filled basin, but if the supply of water is relatively small, it quickly sinks into the permeable alluvium without accumulating as a standing body of water. During a particularly large storm or an unusually wet spring, however, a temporary lake may develop over the low part of a basin. During drier times, such desert lakes evaporate entirely, leaving behind a dry flat lake bed known as a **playa** (▶Fig. 17.17). Over time, a smooth crust consisting of clay and various salts (halite, gypsum, borax, and other minerals) accumulates on the surface of playas. Some of these minerals have industrial uses and thus have been mined.

Where sufficient water flows into a desert basin, it creates a permanent lake. If the basin is an interior basin, with no outlet to the sea, the lake becomes very salty, because although its water escapes by evaporation in the desert sun, its salt cannot. The Great Salt Lake, in Utah, exemplifies the consequences of this process. Even though the streams feeding the lake are fresh enough to drink, their water contains trace amounts of dissolved salt ions. Because the lake has no outlet, these ions have become concentrated in the lake over time, making it even saltier than the ocean.

Deposition from the Wind

As mentioned earlier, wind carries two kinds of sediment loads—a suspended load of dust-sized particles and a surface load of sand. Much of the dust is carried out of the desert and accumulates elsewhere, forming layers of fine-grained sedi-

ment called **loess.** Sand, however, cannot travel far, and accumulates within the desert in piles called **dunes,** ranging in size from less than a meter to over 300 m high. In favorable locations, dunes accumulate to form vast sand seas hundreds of meters thick. We'll look at dunes in more detail later in this chapter.

FIGURE 17.17 A playa, a basin covered with clay and salt, develops where a shallow desert lake dries up.

FIGURE 17.18 (a) The process of cliff retreat in a desert. Rocks break off the cliff along joints that run parallel to the cliff face. The cliff face therefore maintains the same shape and orientation, even though erosion causes the cliff to retreat from point A to point B. Stair-step cliffs appear where beds of strong rock are interlayered with thin beds of rock. (b) Because of cliff retreat, a once-continuous layer of rock evolves into a series of isolated remnants. (c) These buttes were carved from massive red sandstone layers in Monument Valley, Arizona. (d) These hoodoos, in Bryce Canyon, Utah, were made from the erosion of multicolored layers of sandstone, siltstone, and shale.

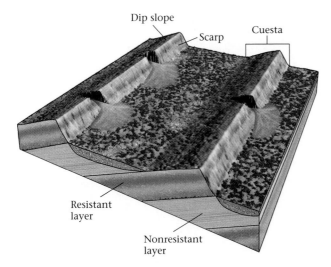

FIGURE 17.19 Asymmetric ridges called cuestas appear where the strata in a region are not horizontal. A joint-controlled cliff forms the steep side, while a dip slope makes up the gentle side. The surface of a dip slope, by definition, is parallel to the bedding of strata beneath.

17.6 DESERT LANDSCAPES

The popular media commonly portray deserts as endless seas of sand, piled into dunes that hide the occasional palm-studded oasis. In reality, vast sand seas are merely one type of desert landscape. Some deserts are broad rocky plains, others sport a stubble of cacti and other hardy desert plants, and still others contain intricate rock formations that look like medieval castles. In this section, we'll see how the erosional and depositional processes described above lead to the formation of such contrasting landscapes.

Rocky Cliffs and Mesas

In hilly desert regions, the lack of soil exposes rocky ridges and cliffs. As noted earlier, cliffs erode when rocks split away along vertical joints. When this happens, the cliff face retreats but retains roughly the same form. The process, commonly referred to as cliff retreat, or scarp retreat, occurs in fits and

starts—a cliff may remain unchanged for decades or centuries, and then suddenly a block of rock falls off and crumbles into rubble at the foot of the cliff (▶Fig. 17.18a). Cliffs formed from stratified rocks commonly develop a steplike shape; strong layers (sandstone or limestone) become vertical cliffs, and weak layers (shale) become rubble-covered slopes.

With continued erosion and cliff retreat, a plateau of rock slowly evolves into a cluster of isolated hills, ridges, or columns (▶Fig. 17.18b). Flat-lying strata or flat-lying layers of volcanic rocks erode to make flat-topped hills. These go by different names, depending on their size. Large examples (with a top surface area of several square km) are **mesas,** from the Spanish word for "table." Medium-sized examples are **buttes** (▶Fig. 17.18c). Small examples, whose height greatly exceeds their top surface area, are **chimneys.** Erosion of strata has resulted in the skyscraperlike buttes of Monument Valley, Arizona, and the stark cliffs of Canyonlands National Park. Bryce Canyon National Park in Utah contains countless chimneys of brightly colored shale and sandstone—locally, these chimneys are called hoodoos (▶Fig. 17.18d). Natural arches, such as those at Arches National Monument, form when erosion along joints leaves narrow walls of rock. When the lower part of the wall erodes while the upper part remains, an arch results (see art on pp. 486–87).

In places where bedding dips at an angle to horizontal, flat-topped mesas and buttes don't form; rather, asymmetric ridges called cuestas develop. A joint-controlled cliff forms the steep front side of a cuesta, while the tilted top surface of a resistant bed forms the gradual slope on the backside (▶Fig. 17.19). If the bedding dip is steep to near vertical, a narrow symmetrical ridge, called a hogback, forms.

With progressive cliff retreat on all sides of a hill, finally all that remains of the hill is a relatively small island of rock, surrounded by alluvium-filled basins. Geologists refer to such islands of rock by the German word **inselberg** ("island mountain"; ▶Fig. 17.20). Depending on the rock type or the orientation of stratification in the rock, and on rates of erosion, inselbergs may be sharp-crested, plateaulike, or loaf-shaped (steep sides and a rounded crest) (▶Fig. 17.21).

FIGURE 17.20 Inselbergs are small islands of rock surrounded by pediments. Alluvium gradually covers the pediments. Alternating mountain ridges and sediment-filled valleys comprise a basin-and-range landscape.

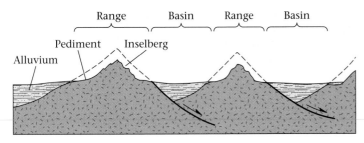

FIGURE 17.21 Uluru (Ayers Rock) in Central Australia is an inselberg composed of vertical beds of red sandstone. It is 3.6 km long and 360 m high. Such loaf-shaped inselbergs are also known as bornhardts.

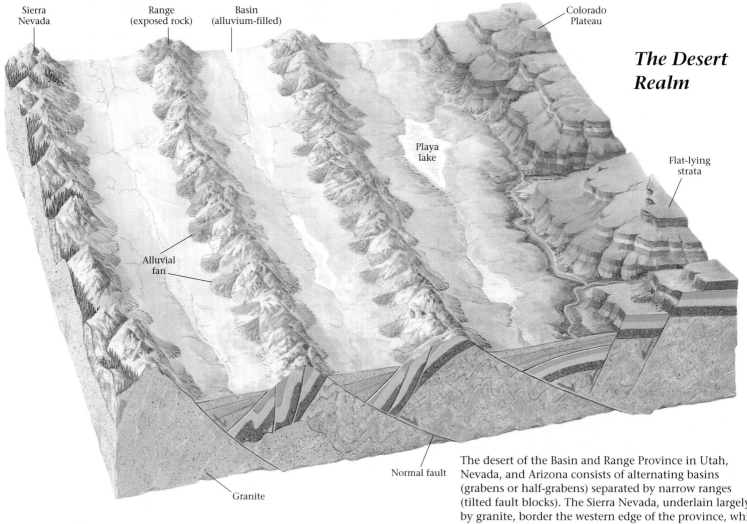

Sierra Nevada

Range (exposed rock)

Basin (alluvium-filled)

Colorado Plateau

The Desert Realm

Playa lake

Flat-lying strata

Alluvial fan

Normal fault

Granite

The desert of the Basin and Range Province in Utah, Nevada, and Arizona consists of alternating basins (grabens or half-grabens) separated by narrow ranges (tilted fault blocks). The Sierra Nevada, underlain largely by granite, border the western edge of the province, while the Colorado Plateau, underlain by flat-lying sedimentary strata, borders the eastern edge. The overall climate of the region is dry. Because of the great variety of elevations and rock types, the region hosts different desert landscapes.

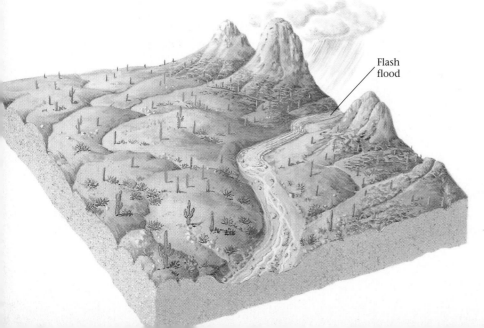

Except in the case of large rivers, such as the Nile or the Colorado, which bring water into a desert region from a more temperate region, streams in deserts fill with water only after heavy rains. At other times, the stream channels are dry. These channels are called dry washes, arroyos, or wadis. When there is a heavy rain, water cannot be absorbed into the ground fast enough, so runoff enters dry washes and fills them very quickly, creating a flash flood. The turbulent, muddy water of a flash flood can transport even large boulders. This flash flood is rushing down a stream in the Sonoran Desert of Arizona.

Flash flood

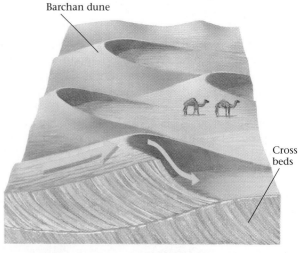

Barchan dune

Cross beds

Where there is a large supply of sand, a variety of sand dunes develop. The geometry of a particular sand dune (such as barchan, longitudinal, or star) depends on the sand supply and the wind. Inside sand dunes, we find cross beds.

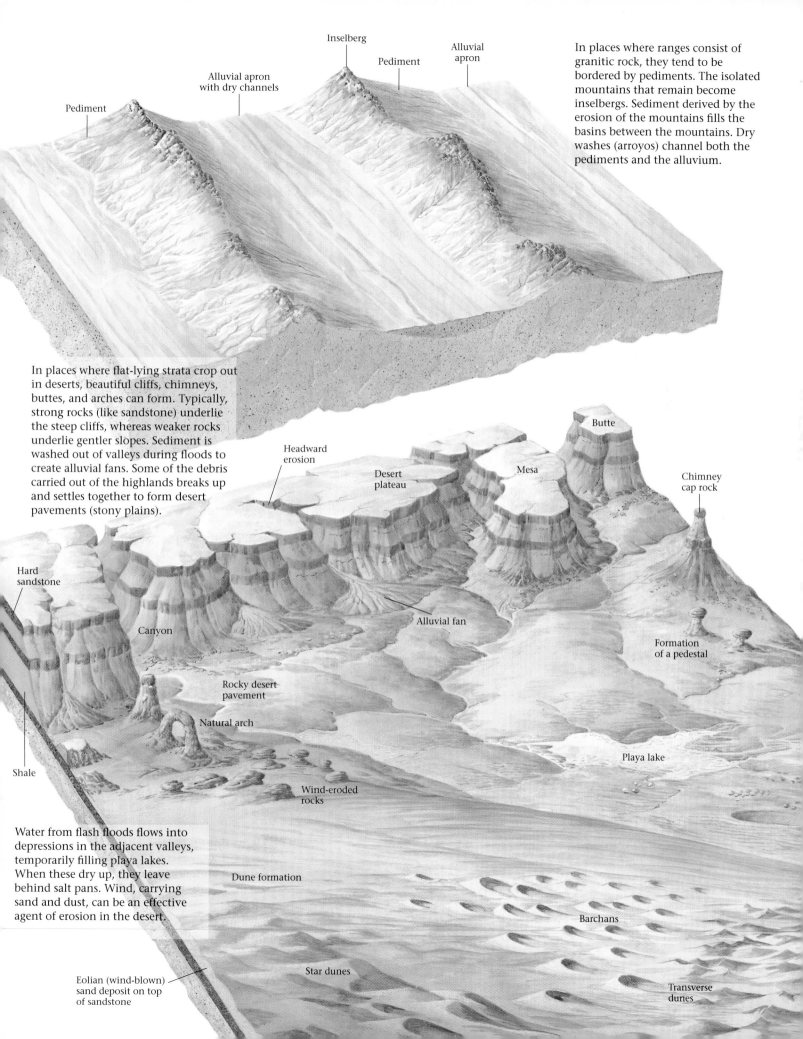

Inselberg

Pediment

Alluvial apron

Alluvial apron with dry channels

Pediment

Pediment

In places where ranges consist of granitic rock, they tend to be bordered by pediments. The isolated mountains that remain become inselbergs. Sediment derived by the erosion of the mountains fills the basins between the mountains. Dry washes (arroyos) channel both the pediments and the alluvium.

In places where flat-lying strata crop out in deserts, beautiful cliffs, chimneys, buttes, and arches can form. Typically, strong rocks (like sandstone) underlie the steep cliffs, whereas weaker rocks underlie gentler slopes. Sediment is washed out of valleys during floods to create alluvial fans. Some of the debris carried out of the highlands breaks up and settles together to form desert pavements (stony plains).

Headward erosion

Desert plateau

Mesa

Butte

Chimney cap rock

Hard sandstone

Canyon

Alluvial fan

Formation of a pedestal

Rocky desert pavement

Natural arch

Shale

Playa lake

Wind-eroded rocks

Water from flash floods flows into depressions in the adjacent valleys, temporarily filling playa lakes. When these dry up, they leave behind salt pans. Wind, carrying sand and dust, can be an effective agent of erosion in the desert.

Dune formation

Barchans

Eolian (wind-blown) sand deposit on top of sandstone

Star dunes

Transverse dunes

Blowing sand

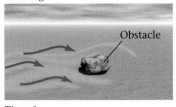  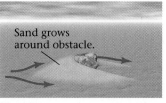 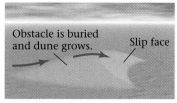

Time 1   Time 2   Time 3   Time 4

FIGURE 17.22 Progressive stages in the growth of a small sand dune.

Stony Plains

The coarse sediment eroded from desert mountains and ridges washes into the lowlands and builds out to form gently sloping alluvial fans. The surfaces of these gravelly piles are strewn with pebbles, cobbles, and boulders, and are dissected by dry washes (wadis or arroyos). Portions of these stony plains evolve into desert pavements.

Pediments

When travelers began trudging through the desert Southwest of the United States during the nineteenth century, they

FIGURE 17.23 The various kinds of sand dunes.

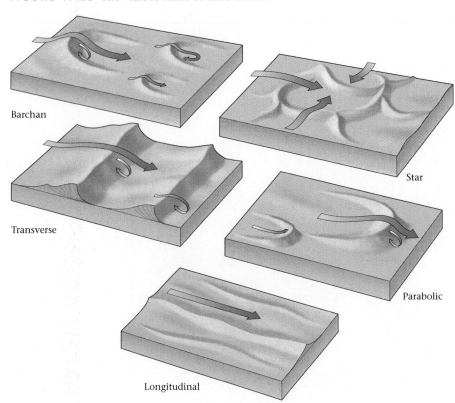

Barchan

Transverse

Longitudinal

Star

Parabolic

found that in many locations the wheels of their wagons were rolling over flat or gently sloping bedrock surfaces. These bedrock surfaces extended outward like ramps from the steep cliffs of a mountain range on one side to alluvium-filled valleys on the other (see Fig. 17.20). Geologists now refer to such surfaces as **pediments.** A pediment is a consequence of erosion, left behind as a mountain front gradually retreats. Pediments develop when sheetwash during floods carries sediment away from the mountain front. As it moves, the sediment grinds away the bedrock that it tumbles over.

Seas of Sand: The Geometry of Dunes

A sand dune is a pile of sand deposited by a moving current. Dunes in deserts form because of the wind. Some start to form where sand becomes trapped on the windward side of an obstacle, such as a rock or a shrub. Gradually the sand builds downwind into the lee of the obstacle (▶Fig. 17.22). Once initiated, the dune itself affects the wind flow, and sand accumulates on the lee (downwind) side of the dune. Here, sand slides down the lee surface of the dune, so this surface is aptly named the slip face.

In places where abundant sand accumulates, sand seas bury the landscape. The wind builds the sand into dunes that display a variety of shapes and sizes, depending on the character of the wind and the sand supply (▶Fig. 17.23). Where the sand is relatively scarce and the wind blows steadily in one direction, beautiful crescents called barchan dunes develop, with the tips of the crescents pointing downwind. If the wind shifts direction frequently, a group of crescents pointing in different directions overlap one another, creating a constantly changing star dune. Where enough sand accumulates to bury the ground surface completely and only moderate winds blow, sand piles into simple, wavelike shapes called transverse dunes. The crests of transverse dunes lie perpendicular to the wind direction. Strong winds may break through transverse dunes and change them into parabolic dunes, whose ends point in the upwind direction. Finally, if there is abundant sand and a strong, steady wind, the sand streams into longitudinal dunes, whose axis lies parallel to the wind direction.

In a sand dune, sand saltates up the windward side of the dune, blows over the crest of the dune, and then settles on the steeper, lee face of the dune. The slope of this face attains the angle of repose, the slope angle of a freestanding pile of sand. As sand collects on this surface, it may become unstable and slide down the slope—as we've noted, geologists refer to the lee side of a dune as the slip face. As progressively more sand accumulates on the slip face, the crest of the dune migrates downwind, and former slip faces become preserved inside the dune. In cross section, these slip faces appear as cross beds (►Fig. 17.24a–d). The surfaces of dunes are not, in general, smooth surfaces, but rather are covered with delicate ripples.

17.7 DESERT PROBLEMS

In the modern era, natural droughts (periods of unusually low rainfall), aggravated by overpopulation, overgrazing, careless agriculture, and diversion of water supplies, have transformed semi-arid grasslands into true deserts, leading to tragic famines that have killed millions of people. **Desertification,** the process of transforming nondesert areas to desert, has accelerated.

The consequences of desertification have devastated the Sahel, the belt of semi-arid land that fringes the southern margin of the Sahara Desert. In the past, the Sahel provided sufficient vegetation to support a small population of nomadic people and animals. But during the second half of the twentieth century, large numbers of people migrated into the Sahel to escape overcrowding in central Africa. The immigrants began to farm and to maintain large herds of cattle and goats. Plowing and overgrazing removed soil-preserving grass and caused the soil to dry out. In addition, the trampling of animal hooves compacted the ground so it could no longer soak up water. In the 1960s and again in the 1980s, a series of natural droughts hit the region, bringing catastrophe (►Fig. 17.25a, b). Wind erosion stripped off the remaining

(a)

(c)

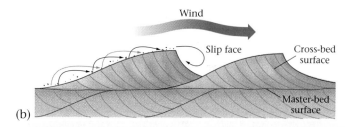

(b)

(d)

FIGURE 17.24 (a) A sand dune with surface ripples; (b) cross bedding inside a dune; (c) cross beds exposed in the Mesozoic sandstone of Zion National Park, in Utah; (d) closeup of the contact between the top of one set of cross beds and the bottom of another.

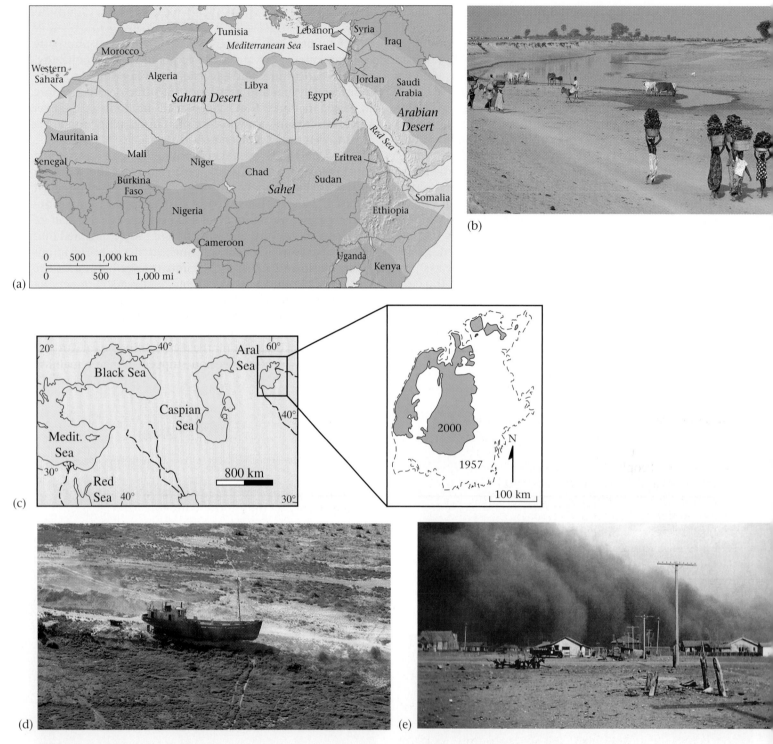

FIGURE 17.25 (a) The Sahel is the semi-arid land along the southern edge of the Sahara. As a result of grazing and agriculture, the vegetation in this region has vanished, and large parts have undergone desertification. (b) Drought in the Sahel has brought deadly consequences. Here, residents seek water from a dwindling pond. (c) The Aral Sea in central Asia has started to dry up because of diversion of rivers that once flowed into it. What was once the floor of the sea is now a parched land. The shorelines in 1957 and 2000 are shown. (d) Fishing vessels now lie stranded in the Aral Sea area. (e) During the Dust Bowl days in central Oklahoma (the 1930s), dust storms stripped valuable topsoil off the land.

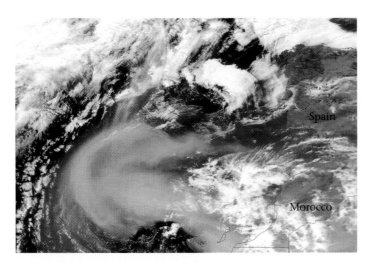

FIGURE 17.26 In this satellite image, we see a huge dust cloud blowing across the Atlantic. The dust originated in the Sahara.

topsoil. Without vegetation, the air grew drier, and the semi-arid grassland of the Sahel became desert, with mass starvation as the result.

Other regions on Earth are developing the same problem. The Aral Sea in Kazakhstan, for example, has started to dry up. Diversion of the rivers that once fed the sea have so diminished its water supply that the area of the sea has shrunk dramatically. Boats that once plied its waters now lie as rusting hulks in a sea of dust (▶Fig. 17.25c, d).

Desertification does not only happen in less industrialized nations. People in the western Great Plains of the United States and Canada suffered from the problem beginning in 1933, the fourth year of the Great Depression. Banks had failed, workers had lost their jobs, the stock market had crashed, and hardship burdened all. No one needed yet another disaster—but that year, even nature turned hostile. All through the fall, so little rain fell in the plains of Texas and Oklahoma that the region's grasslands and croplands browned and withered, and the topsoil turned to powdery dust. Then, on November 12 and 13, strong storms blew eastward across the plains. Without vegetation to protect it, the wind tore at the ground, stripped off the topsoil, and sent it skyward to form rolling black clouds that literally blotted out the sun (▶Fig. 17.25e). People caught in the resulting dust storm found themselves choking and gasping for breath. When the dust finally settled, it had buried houses and roads under huge drifts, and dirtied every nook and cranny. The dust blew east as far as New England, where it turned the snow brown. What had once been rich farmland in the southwestern plains turned into a wasteland that soon acquired a nickname, the Dust Bowl.

Why did the fertile soils of the southern Great Plains suddenly dry up? The causes were complex; some were natural and some were human-induced. Drought episodically visits certain regions, but drought alone doesn't inevitably lead to dust-bowl conditions. They may result from human activity.

Typically, the Great Plains region has a semi-arid climate in which only thin soil develops. But the plains were settled in the 1880s and 1890s, unusually wet years. Not realizing its true character, far more people moved into the region than it could sustain, and the land was farmed too intensively. When farmers used steel plows, they destroyed the fragile grassland root systems that held the thin soil in place. And when the drought of the 1930s came, it was catastrophic.

Desertification has a potentially dangerous side effect—blowing dust. In recent years, satellite images have revealed that wind-blown dust from deserts can travel across oceans and affect regions on the other side. For example, dust blown off the Sahara can traverse the Atlantic and settle over the Caribbean (▶Fig. 17.26).

The Dust Bowl of the 1930s and the fate of the Sahel in Africa remind us of how fragile the Earth's green blanket of vegetation really is. Global climate changes can shift climatic belts sufficiently to transform agricultural regions into deserts. Unfortunately, if the current trend in climate change continues, our present agricultural belts could someday become new Saharas.

CHAPTER SUMMARY

- Deserts generally receive less than 25 cm of rain per year. Vegetation covers no more than 15% of their surface.

- Subtropical deserts form between 20° and 30° latitude, rain-shadow deserts are found on the inland side of mountain ranges, coastal deserts are located on the land adjacent to cold ocean currents, continental-interior deserts exist in land-locked regions far from the ocean, and polar deserts form at high latitude.

- In deserts, chemical weathering happens slowly, so rock bodies tend to erode primarily by physical weathering. Desert varnish forms on rock surfaces, and soils tend to accumulate soluble minerals.

- Water causes significant erosion in deserts, mostly during heavy downpours. Flash floods carry large quantities of sediments down ephemeral streams.

- Wind causes significant erosion in deserts, for it picks up dust and silt and carries them as suspended load, and causes sand to saltate. Where wind blows away finer sediment, a lag deposit remains. Wind-blown sediment abrades the ground.

- Desert pavements are mosaics of varnished stones armoring the surface of the ground.

- Talus aprons form when rock fragments accumulate at the base of a slope. Alluvial fans form at a mountain front where water in ephemeral streams deposits sediment. When temporary desert lakes dry up, they leave playas.

- In some desert landscapes, erosion causes cliff retreat, eventually resulting in the formation of mesas and inselbergs.

Pediments of nearly flat or gently sloping bedrock surround some inselbergs.

- Where there is abundant sand, the wind builds it into dunes. Common types include barchan, star, transverse, parabolic, and longitudinal dunes.
- Changing climates and land abuse may cause desertification, the transformation of semi-arid land into deserts. Wind-blown dust from deserts may waft across oceans.

KEY TERMS

alluvial fan (p. 482)
butte (p. 485)
chimney (p. 485)
deflation (p. 482)
desert (p. 474)
desert pavement (p. 480)
desert varnish (p. 478)
desertification (p. 489)
dry wash (p. 479)
dunes (p. 483)
inselberg (p. 485)

lag deposit (p. 480)
loess (p. 483)
mesa (p. 485)
pediment (p. 488)
playa (p. 483)
rain shadow (p. 477)
saltation (p. 479)
surface load (p. 480)
suspended load (p. 479)
talus apron (p. 482)
ventifact (p. 481)

REVIEW QUESTIONS

1. What factors determine whether a region can be classified as a desert?

2. Explain why deserts form.

3. Have today's deserts always been deserts? Keep in mind the consequences of plate tectonics.

4. How do weathering processes in deserts differ from those in temperate or humid climates?

5. Describe how water modifies the landscape of a desert. Be sure to discuss both erosional and depositional landforms.

6. Explain the ways in which desert winds transport sediment.

7. Explain how the following features form: (a) desert varnish; (b) desert pavement; (c) ventifacts.

8. Describe the process of formation of the different types of depositional landforms that develop in deserts.

9. Describe the process of cliff (scarp) retreat and the landforms that result from it.

10. What are the various types of sand dunes? What factors determine which type of dune develops?

11. What is the process of desertification, and what causes it? How can desertification in Africa affect the Caribbean?

SUGGESTED READING

Abrahams, A. D., and A. J. Parsons, eds. 1994. *Geomorphology of Desert Environments*. London, UK: Chapman and Hall.

Bagnold, R. A. 2005. *The Physics of Blown Sand and Desert Dunes*. Mineola, N.Y.: Dover Publications.

Cooke, R. U., A. Warren, A. Goudie, 1992. *Desert Geomorphology*. Boca Raton, Fla:. CRC Press.

Dregne, H. E. 1983. *Desertification of Arid Lands*. Chur, Switzerland: Harwood Academic.

Livingston, I., S. Stokes, and A. S. Goudie, eds. 2000. *Aeolian Environments, Sediment and Landforms*. New York: Wiley.

Tchakerian, V. P., ed. 1995. *Desert Aeolian Processes*. London, UK: Chapman and Hall.

THE VIEW FROM SPACE Erosion of rock produces sand, which, under appropriate conditions in arid climates, can build into large dunes. These dunes, some of which are 300 m high, developed in the Namib Desert of Namibia, southwestern Africa.

Amazing Ice: Glaciers and Ice Ages

18.1 INTRODUCTION

There's nothing like a good mystery, and one of the most puzzling in the annals of geology came to light in northern Europe early in the nineteenth century. When farmers of the region prepared their land for spring planting, they occasionally broke their plows by running them into large boulders buried randomly through otherwise fine-grained sediment. Many of these boulders did not consist of local bedrock, but rather came from outcrops hundreds of kilometers away. Because the boulders had apparently traveled so far, they came to be known as **erratics** (from the Latin *errare*, "to wander").

The mystery of the wandering boulders became a subject of great interest to early nineteenth-century geologists, who realized that deposits of *un*sorted sediment (containing a variety of different clast sizes) could not be examples of typical stream alluvium, for running water sorts sediment by size. Most attributed the deposits to a vast flood that they imagined had somehow spread a slurry of boulders, sand, and mud across the continent. In 1837, however, a young Swiss geologist named Louis Agassiz proposed a radically different interpretation. Agassiz often hiked among **glaciers** (slowly flowing masses of ice that survive the summer melt) in the Alps near

Glaciers are rivers or sheets of ice. They carve beautiful landscapes and deposit hills of sediment. During an ice age, glaciers covered vast areas of continents.

FIGURE 18.1 During the ice age, glaciers covered much of North America and Europe, so that much of these continents resembled present-day Antarctica, where glacial ice extends from the foot of the Transantarctic Mountains (seen here on the horizon) to the far side of Antarctica, over 3,000 km away.

his home. He observed that glacial ice could carry enormous boulders as well as sand and mud, because ice is solid and has the strength to support the weight of rock. Agassiz realized that because flowing solid ice does not sort sediment, glaciers leave behind unsorted sediment when they melt. Working from these observations, he proposed that the mysterious erratic-bearing sediments of Europe were deposits left by ice sheets, vast glaciers that had once covered much of the continent. In Agassiz's view, Europe had at one time been in the grip of an **ice age,** a time when the climate was significantly colder and glaciers grew (▶Fig. 18.1).

Agassiz's radical proposal faced intense criticism for the next two decades. But he didn't back down, and instead challenged his opponents to visit the Alps and examine the sedimentary deposits that glaciers had left behind. By the late 1850s, most doubters had changed their minds, and the geological community concluded that the notion that Europe had once endured Arctic-like climates was correct. Later in life, Agassiz traveled to the United States and documented many glacier-related features in North America's landscape, proving that an ice age had affected vast areas of the planet.

Glaciers, which have many forms, cover only about 10% of the land on Earth today (▶Fig. 18.2a, b), but during the most recent ice age, which ended only about 11,000 years ago, as much as 30% of the continents had a coating of ice. New York City, Montréal, and many of the great cities of Europe occupy land that once lay beneath hundreds of meters to a few kilometers of ice.

The work of Louis Agassiz brought the subject of glaciers and ice ages into the realm of geologic study and led people to recognize that major climate changes happen in Earth history. In this chapter, after looking at the nature of ice, we see how glaciers form and why they move. Next, we consider how glaciers modify landscapes by erosion and deposition. A substantial portion of the chapter concerns the Pleistocene ice ages, for their impact on the landscape can still be seen today, but we briefly introduce ice ages that happened earlier in Earth history too. We conclude by considering hypotheses to explain why ice ages happen.

18.2 THE NATURE OF GLACIERS

Ice: A Rock Made from Water

Ice consists of solid water, formed when liquid water cools below its freezing point. We can consider a single ice crystal to be a mineral specimen: it is a naturally occurring, inorganic

FIGURE 18.2 (a) A piedmont glacier near the coast of Greenland. (b) A large valley glacier in Pakistan.

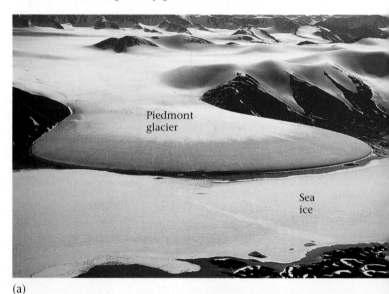

Piedmont glacier

Sea ice

(a)

Valley glacier

(b)

solid, with a definite chemical composition (H₂O) and a regular crystal structure. Ice crystals have a hexagonal form, so snowflakes grow into six-pointed stars (▶Fig. 18.3a). We can think of a layer of fresh snow as a layer of sediment, and a layer of snow that has been compacted so that the grains stick together as a layer of sedimentary rock (▶Fig. 18.3b). We can also think of the ice that appears on the surface of a pond as an igneous rock, for it forms when "molten ice" (liquid water) solidifies. Glacier ice, in effect, is a metamorphic rock. It develops when preexisting ice recrystallizes in the solid state, meaning that the molecules in solid water rearrange to form new crystals. Typically, glacial ice contains air bubbles (▶Fig. 18.3c).

Like glass, ice has a high albedo, meaning that it reflects light well—so well, in fact, that if you walk on ice without eye protection, you risk blindness from the glare. Ice differs from most other familiar materials in that its solid form is less dense than its liquid form—the architecture of an ice crystal holds water molecules apart. Ice, therefore, floats on water. This unusual characteristic has major implications for the Earth System. If ice didn't float, ice in oceans would sink, leaving room for new ice to form until the oceans froze solid.

Categories of Glaciers

Glaciers, formally defined, are streams or sheets of recrystallized ice that last all year long and flow under the influence of gravity. Today, they highlight coastal and mountain scenery in Alaska, western North America, the Alps of Europe, the Southern Alps of New Zealand, the Himalayas of Asia, and the Andes of South America, and they cover most of Greenland and Antarctica. Geologists distinguish between two main categories, mountain glaciers and continental glaciers.

Mountain glaciers (also called alpine glaciers) exist in mountainous regions (▶Fig. 18.4a). Topographical features of the mountains control their shape; overall, mountain glaciers flow from higher elevations to lower elevations. Mountain glaciers include cirque glaciers, which fill bowl-shaped depressions, or **cirques,** on the flank of a mountain; valley glaciers, rivers of ice that flow down valleys (Fig. 18.2b); mountain ice caps, mounds of ice that submerge peaks and ridges at the crest of a mountain range (▶Fig. 18.4b); and piedmont glaciers (Fig. 18.2a), fans or lobes of ice that form where a valley glacier emerges from a valley and spreads out into the adjacent plain. Mountain glaciers range in size from a few hundred meters to a few hundred kilometers long.

Continental glaciers are vast ice sheets that spread over thousands of square kilometers of continental crust. Continental glaciers now exist only on Antarctica and Greenland (▶Fig. 18.5). (Remember that Antarctica is a continent, so the ice beneath the South Pole rests on solid ground; in contrast, the ice at the North Pole is part of a thin sheet of sea ice floating on the Arctic Ocean.) Continental glaciers flow outward from their thickest point (where ice reaches a thickness of 3.5 km) and become thin toward their margins, where they may be only a few hundred meters thick. The front edge of the glacier may divide into several tongue-shaped lobes, because not all of the glacier flows at the same speed.

Geologists also find it valuable to distinguish between types of glaciers on the basis of the thermal conditions in which the glaciers exist. Temperate glaciers occur in regions where atmospheric temperatures become high enough for the glacial ice to be at or near its melting temperature throughout much of the year. Polar glaciers occur in regions where atmospheric temperatures stay so low that the glacial

FIGURE 18.3 (a) The hexagonal shape of snowflakes. (b) Snowdrifts carved by the wind in Antarctica. Note the person for scale. (c) Thin section of glacial ice, showing the texture of ice crystals and air bubbles.

2 mm

(a)

(b)

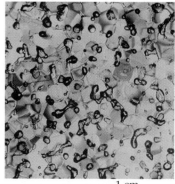

1 cm

(c)

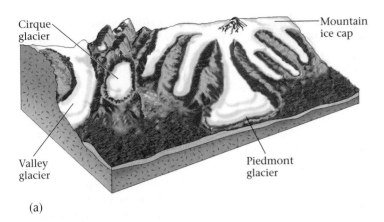

FIGURE 18.4 (a) The various kinds of mountain glaciers. (b) A mountain ice cap in Greenland. Note the ice flowing out of the ice cap in a big valley glacier.

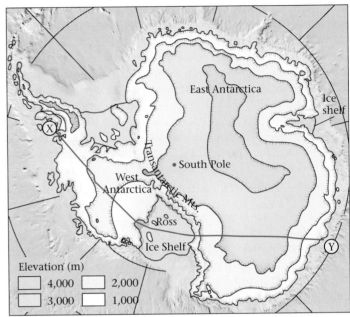

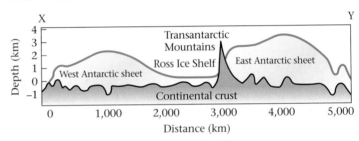

FIGURE 18.5 A map and cross section of the Antarctic ice sheet. Note that the Transantarctic Mountains divide the ice sheet in two. The Ross Ice Shelf lies between the East Antarctic and West Antarctic sheets. Valley glaciers carry ice from the ice sheets down to the shelf.

ice remains well below melting temperature throughout the entire year.

Forming a Glacier

In order for a glacier to form, three conditions must be met. First, the local climate must be sufficiently cold that winter snow does not melt entirely away during the summer. Second, there must be, or must have been sufficient snowfall for a large amount to accumulate. And third, the slope of the surface on which the snow accumulates must be gentle enough that the snow does not slide away in avalanches, and must be protected enough that the snow doesn't blow away.

Glaciers develop in polar regions because even though relatively little snow falls today, temperatures remain so low that most ice and snow survive all year. Glaciers develop in mountains, even at low latitudes, because temperature decreases with elevation, so that at high eleva-

tions, the mean temperature stays low enough for ice and snow to survive all year. Since the temperature of a region depends on latitude, the specific elevation at which mountain glaciers form also depends on latitude. In Earth's present-day climate, glaciers form only at elevations above 5 km between 0° and 30° latitude and down to sea level at 60° to 90° latitude (▶Fig. 18.6). Thus, you can see high-latitude glaciers from a ship, but you have to climb the highest mountains of the Andes to find glaciers at the equator.

The transformation of snow to glacier ice takes place slowly, as younger snow progressively buries older snow. Freshly fallen snow consists of delicate hexagonal crystals with sharp points. The crystals do not fit together tightly, so this snow contains about 90% air. With time, the points of the snowflakes become blunt, because they either sublimate (evaporate directly into vapor) or melt, and the snow packs more tightly. As snow becomes buried, the weight of the overlying snow increases pressure, which

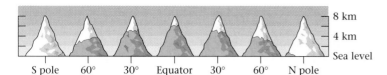

FIGURE 18.6 The snow line depends on latitude.

causes remaining points of contact between snowflakes to melt. Gradually, the snow transforms into a packed granular material called firn, which contains only about 25% air (▶Fig. 18.7a, b). Melting of firn grains at contact points produces water that crystallizes in the spaces between grains until eventually the firn transforms into a solid mass of glacial ice composed of interlocking ice crystals. Such glacial ice, which may still contain up to 20% air trapped in bubbles, has a bluish color. The transformation of fresh snow to glacier ice can take as little as tens of years in regions with abundant snowfall, or as long as thousands of years in regions with little snowfall.

The Movement of Glacial Ice

How do glaciers move? Some move when meltwater accumulates at their bases, so that the mass of the glacier slides partially on a layer of water or on a slurry of water and sediment. During this kind of movement, known as basal sliding, the water or wet slurry holds the glacier above the underlying rock and reduces friction between the glacier and its substrate (▶Fig. 18.8a).

Glaciers also move by means of internal flow, during which the mass of ice slowly changes shape internally, without breaking apart or completely melting. By studying ice deformation with a microscope, geologists have determined that internal flow involves two processes. First, ice crystals become plastically deformed by the rearrangement of water molecules within the crystal lattice (during this process, some ice crystals change shape, whereas others disappear to be replaced by new ones); second, if conditions allow very thin films of water to form on the surfaces of ice crystals, the crystals can slide past one another (▶Fig. 18.8b).

Note that the plastic deformation of ice in glaciers resembles the plastic deformation of metamorphic rock in mountain belts. However, metamorphic rock deforms plastically only at depths greater than 10 to 15 km in the crust, whereas ice deforms plastically at depths below about 60 m (▶Fig. 18.8c). In other words, the brittle-plastic transition in ice lies at a depth of about 60 m. Above this depth, large cracks can form in ice, but below this depth, cracks cannot form because ice is too plastic. A large crack that develops by brittle deformation in a glacier is called a **crevasse.** In large glaciers, crevasses can be hundreds of meters long and tens of meters deep (▶Fig. 18.9).

Why do glaciers move? Ultimately, because the pull of gravity can cause weak ice to flow (▶Fig. 18.10a). Glaciers flow in the direction in which their surfaces slope. Thus, valley glaciers flow down their valleys, and continental ice sheets spread outward from their thickest point. To picture the movement of a continental ice sheet, imagine that a thick pile of ice builds up, and gravity causes the top of the pile to push down on the ice at the base. Eventually, the basal ice can no longer support the weight of the overlying ice and begins to deform plastically. When this happens, the basal ice starts squeezing out to the side, carrying the overlying ice with it. The greater the volume of ice that builds up, the wider the sheet of ice can become. You've seen a similar process of gravitational spreading if you've ever poured honey onto a plate. The honey can't

FIGURE 18.7 (a) Snow compacts and melts to form firn, which recrystallizes to yield ice. (b) The size of ice crystals increases with depth in a glacier, where some crystals grow at the expense of others.

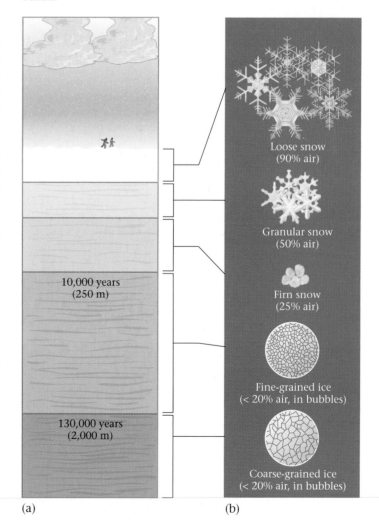

Loose snow
(90% air)

Granular snow
(50% air)

Firn snow
(25% air)

Fine-grained ice
(< 20% air, in bubbles)

Coarse-grained ice
(< 20% air, in bubbles)

10,000 years
(250 m)

130,000 years
(2,000 m)

(a)   (b)

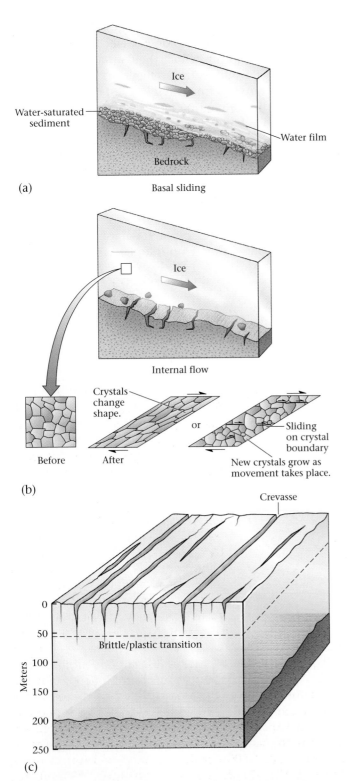

(a) Basal sliding

(b)

Crystals change shape.

Before After or Sliding on crystal boundary

New crystals grow as movement takes place.

Crevasse

Brittle/plastic transition

(c)

FIGURE 18.8 (a) Warmer glaciers are wet at the base and move by means of basal sliding. (b) Colder glaciers flow in the solid state. In some cases, the ice crystals stretch and rotate; in other cases, the ice recrystallizes and some crystals slide past each other, especially if there are thin films of water between grains. (c) Crevasses form in the brittle ice above the brittle-plastic transition in glaciers.

FIGURE 18.9 Crevasses in an Antarctic glacier. The crevasses are up to 10 m wide.

build up into a narrow column because it's too weak. Rather, it flows laterally away from the point where it lands to form a wide, thin layer (►Fig. 18.10b).

Glaciers generally flow at rates of between 10 and 300 m per year. Not all parts of a glacier move at the same rate. For example, friction between rock and ice slows a glacier, so the center of a valley glacier moves faster than its margins, and the top of a glacier moves faster than its base (►Fig. 18.11a, b). If water builds up beneath a valley glacier to the point where it lifts the glacier off its substrate, the glacier undergoes a surge and flows much faster for a limited time (rarely more than a few months) until the water escapes. During surges, glaciers have been clocked at speeds of 20 to 110 m per day.

Glacial Advance and Retreat

Glaciers resemble bank accounts: snowfall adds to the account, whereas ablation—the removal of ice by sublimation (the evaporation of ice into water vapor), melting (the transformation of ice into liquid water), and calving (the breaking off of chunks of ice at the edge of the glacier)—subtracts from the account (►Fig. 18.12). Snowfall adds to the glacier in the **zone of accumulation,** whereas ablation subtracts in the **zone of ablation;** the boundary between these two zones is the **equilibrium line.**

The leading edge or margin of a glacier is called its toe, or terminus (►Fig. 18.13a). If the rate at which ice accumulates in the zone of accumulation exceeds the rate at which ablation occurs below the equilibrium line, then the toe moves forward into previously unglaciated regions. Such a change is called a **glacial advance** (►Fig. 18.13b). In the case of mountain glaciers, the position of a toe moves downslope during an advance, whereas in the case of continental gla-

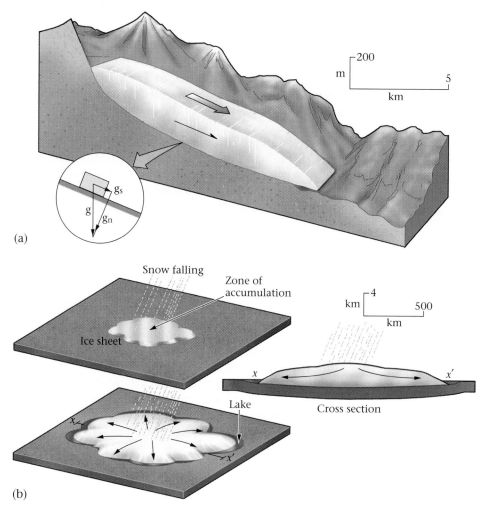

(a)

(b)

FIGURE 18.10 (a) Gravity drives the downslope motion of glaciers. Gravity can be portrayed by an arrow (vector) that points straight down. One component (g_s) is parallel to the slope, and drives the flow, whereas the other (g_n) is perpendicular to the slope. (b) The gravitational spreading of an ice sheet resembles honey spreading across a table.

ice crystals follow curved trajectories (Fig. 18.13a–c). For this reason, rocks picked up by ice at the base of the glacier slowly move to the surface.

Ice in the Sea

On the moonless night of April 14, 1912, the great ocean liner *Titanic* plowed through the calm but frigid waters of the North Atlantic on her maiden voyage from Southampton, England, to New York. Although radio broadcasts from other ships warned that icebergs, large blocks of ice floating in the water, had been sighted in the area and might pose a hazard, the ship sailed on, its crew convinced that they could see and avoid the biggest bergs, and that smaller ones would not be a problem for the steel hull of this "unsinkable" vessel. But, in a story now retold countless times, their confidence was fatally wrong. At 11:40 P.M., while the first-class passengers danced, the *Titanic* struck a large iceberg. Lookouts had seen the ghostly mass of frozen water only minutes earlier and had alerted the ship's pilot, but the ship had been unable to turn fast enough to avoid disaster. The force of the blow split the steel hull that spanned five of the ship's sixteen water-tight compartments: the ship could stay afloat if four compartments flooded, but the flooding of five meant it would sink.

ciers, the toe moves outward, away from the glacier's origin. If the rate of ablation equals the rate of accumulation, then the position of the toe remains fixed. But if the rate of ablation exceeds the rate of accumulation, then the position of the toe moves back toward the origin of the glacier. Such a change is called a **glacial retreat** (▶Fig. 18.13c). During a mountain glacier's retreat, the position of the toe moves upslope. It's important to realize that when a glacier retreats, it's only the *position* of the toe that moves back toward the origin, for ice continues to flow toward the toe. Glacial ice cannot flow back toward the glacier's origin.

One final point before we leave the subject of glacial flow. Beneath the zone of accumulation, a crystal of ice gradually moves down toward the base of the glacier as new ice accumulates above it. However, beneath the zone of ablation, a crystal of ice gradually moves up toward the surface of the glacier, as overlying ice ablates. Thus, as a glacier flows,

At about 2:15 A.M., the bow disappeared below the water, and the stern rose until the ship protruded nearly vertically from the water. Without water to support its weight, the hull buckled and split in two—the stern section fell back down onto the water and momentarily bobbed horizontally before following the bow, settling downward through over 3.5 km of water to the silent sea floor below. Because of an inadequate number of lifeboats and the inability of the crew to load passengers efficiently, only 705 passengers survived; 1,500 perished in the frigid waters of the Atlantic. The *Titanic* remained lost until 1985, when a team of oceanographers located the sunken hull and photographed its eerie form where it lies, upright in the mud.

Where do icebergs, such as the one responsible for the *Titanic*'s demise, originate? In high latitudes, mountain glaciers and continental ice sheets flow down to the sea. Valley glaciers whose terminus lies in the water are called tidewater

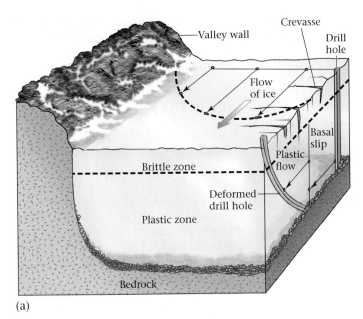

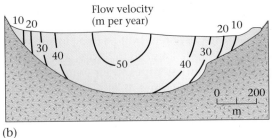

FIGURE 18.11 (a) Different parts of a glacier may flow at different velocities. The vector lengths indicate the velocity of flow. (b) This cross section of a glacier shows measured velocities of flow.

glaciers. Some of these may protrude far enough into the ocean to become ice tongues (▶Fig. 18.14). Continental glaciers entering the sea become broad, flat sheets called ice shelves. Four-fifths of the ice lies below the sea surface, so in shallow water, an ice shelf remains grounded (▶Fig. 18.15). But where the water becomes deep enough, the ice floats. At the boundary between glacier and ocean, blocks of ice from tidewater glaciers or ice shelves calve off and tumble into the water with an impressive splash. If a free-floating chunk rises 6 m above the water and is at least 15 m long, it is formally called an **iceberg.** Since four-fifths of the ice lies below the surface of the sea, the base of a large iceberg may actually be a few hundred meters below the surface (▶Fig. 18.16).

Not all ice floating in the sea originates as glaciers on land. In polar climates, the surface of the sea itself freezes, forming **sea ice** (▶Fig. 18.17). Some sea ice, such as that covering the interior of the Arctic Ocean, floats freely, but some protrudes outward from the shore. Icebreakers can crunch through sea ice that is up to 2.5 m thick. To do this, the icebreaker rides up on the ice and crushes it. Vast areas of ice shelves and of sea ice have been disintegrating in recent years, probably because of global warming. For example, ice-free regions have been forming in the Arctic Ocean during the summers. Some geologists, in fact, predict that the Arctic Ocean may be entirely ice-free during the summer within the next one hundred years. Similarly, the ice shelf in Antarctica has been decreasing in area rapidly.

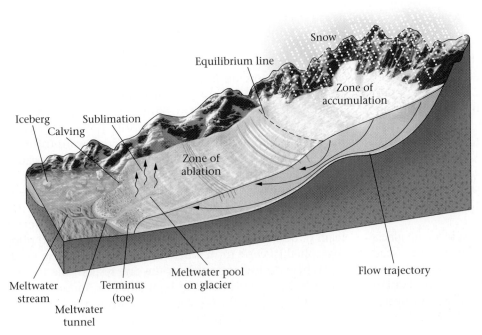

FIGURE 18.12 The zone of ablation, zone of accumulation, and equilibrium line. The arrows illustrate how a crystal of ice flows down in the zone of accumulation and up in the zone of ablation, so that it follows a curved trajectory. Note that ice at the base of the glacier can flow up and over obstacles as long as the surface of the glacier has a downhill slope.

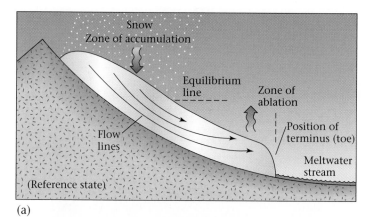

(a)

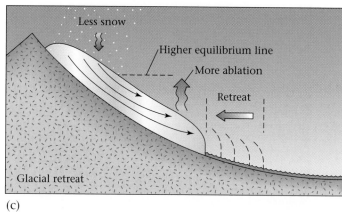

(c)

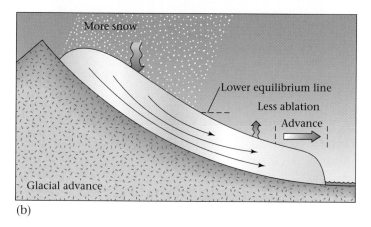

(b)

FIGURE 18.13 (a) The position of a glacier's toe represents the balance between the amount of ice that forms beneath the zone of accumulation and the amount of ice lost in the zone of ablation. (b) Glacial advance and (c) glacial retreat. Notice that ice always flows toward the toe of the glacier regardless of whether the toe advances or retreats, as indicated by the curving flow line.

FIGURE 18.15 Where a glacier reaches the sea, the ice stays grounded in shallow water and floats in deep water. Icebergs calve off the front of such tidewater glaciers. Sediment known as drop stones falls off the base of icebergs and collects on the sea floor.

FIGURE 18.14 An ice tongue protruding into the Ross Sea along the coast of Antarctica.

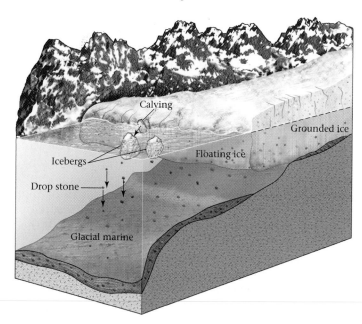

FIGURE 18.16 An artist's rendition of an iceberg. This image is a composite of photographs spliced together. It conveys a sense of the relative proportions of ice above and below the surfce of the sea.

FIGURE 18.17 Broken-up sea ice off the coast of Antarctica.

18.3 CARVING AND CARRYING BY ICE

Glacial Erosion and Its Products

The Sierra Nevada mountains of California consist largely of granite that formed during the Mesozoic Era in the crust beneath a volcanic arc. During the past 10–20 million years, the land surface slowly rose and erosion stripped away overlying rock to expose the granite. The style of erosion formed rounded domes. Then, during the last ice age, valley glaciers cut deep, steep-sided valleys into the range. In the process, some of the domes were cut in half, leaving a rounded surface on one side and a steep cliff on the other. Half Dome, in Yosemite National Park, formed in this way (▶Fig. 18.18a); its steep cliff has challenged many rock climbers. Such glacial erosion also produces the knife-edge ridges and pointed spires of high mountains and the broad expanses where rock outcrops have been stripped of overlying sediment and polished smooth (▶Fig. 18.18b).

Glaciers erode their substrate in several ways (▶Fig. 18.19a, b). During glacial incorporation, ice surrounds and incorporates debris. During glacial plucking (or glacial quarrying), a glacier breaks off and then carries away fragments

of bedrock. Plucking occurs when ice freezes onto rock that has already started to crack and separate from its substrate, so that movement of the ice lifts off pieces of the rock. At the toe of a glacier, ice may actually bulldoze sediment and trees slightly before flowing over them.

As glaciers flow, clasts embedded in ice act like the teeth of a giant rasp and grind away the substrate. This process, called glacial abrasion, produces very fine sediment called rock flour, just as sanding wood produces sawdust. Rasping by embedded sand yields shiny, glacially polished surfaces. Rasping by large clasts produces long gouges, grooves, or scratches (1 cm to 1 m across) called **glacial striations** (▶Fig. 18.19c). As you might expect, striations run parallel to the flow direction of the ice.

Let's now look more closely at the erosional features associated with a mountain glacier (▶Fig. 18.20a–f). Freezing and thawing during the fall and spring help fracture the rock bordering the head of the glacier (the ice edge high in the mountains). This rock falls on the ice or gets picked up at the base of the ice, and moves downslope with the glaciers. As a consequence, a bowl-shaped depression, or cirque, develops on the side of the mountain. If the ice later melts, a lake called a tarn may form at the base of the cirque. An arête (French for "ridge"), a residual knife-edge ridge of rock, separates two adjacent cirques. A pointed mountain peak surrounded by at least three cirques is called a **horn.** The

(a)

(b)

FIGURE 18.18 Products of glacial erosion. (a) Half Dome, in Yosemite National Park, California. Before glaciation, the mountain was a complete dome; then glacial erosion truncated one side. (b) A glacially polished surface on bedrock in New York City.

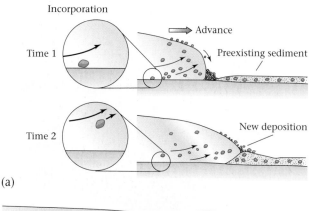

(a)

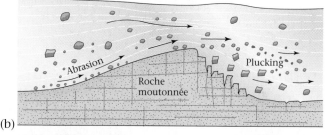

(b)

(c)

FIGURE 18.19 The various kinds of glacial erosion. (a) Plowing and incorporation; (b) plucking and abrasion; (c) glacial striations in downtown Victoria, British Columbia (Canada).

Matterhorn, a peak in Switzerland, is a particularly beautiful example of a horn (Fig. 18.20e).

Glacial erosion severely modifies the shape of a valley. To see how, compare a river-eroded valley with a glacially eroded valley. If you look along the length of a river in unglaciated mountains, you'll see that it flows down a V-shaped valley, with the river channel forming the point of the V. The V develops because river erosion occurs only in the channel, and mass wasting causes the valley slopes to approach the angle of repose. But if you look down the length of a glacially eroded valley, you'll see that it resembles a U, with steep walls. A **U-shaped valley** forms because the combined processes of glacial abrasion and plucking not only lower the floor of the valley, they also bevel its sides.

Glacial erosion in mountains also modifies the intersections between tributaries and the trunk valley. In a river system, tributaries cut side valleys that merge with the trunk valley, such that the mouths of the tributary valleys lie at the same elevation as the trunk valley. The ridges (spurs) be-

tween valleys taper to a point where they join the trunk valley floor. During glaciation, tributary glaciers flow down side valleys into a trunk glacier. But the trunk glacier cuts the floor of its valley down to a depth that far exceeds the depth cut by the tributary glaciers. Thus, when the glaciers melt away, the mouths of the tributary valleys perch at a higher elevation than the floor of the trunk valley. Such side valleys are called **hanging valleys.** The water in post-glacial streams that flow down a hanging valley must cascade over a spectacular waterfall to reach the post-glacial trunk stream (Fig. 18.20f). Trunk glaciers, as they erode, also chop off the ends of spurs, creating truncated spurs.

FIGURE 18.20 (a) A mountain landscape before glaciation. The V-shaped valleys are the result of river erosion; the floors of the tributary valleys are at the same elevation as the trunk valley where they intersect. (b) During glaciation, the valleys fill with ice. (c) After glaciation, the region contains U-shaped valleys, hanging valleys, truncated spurs, and horns. (d) A U-shaped glacial valley. (e) The Matterhorn in Switzerland. (f) A waterfall spilling out of a hanging valley. Truncated spurs occur on both sides of the waterfall.

Now let's look at the erosional features produced by continental ice sheets. To a large extent, these depend on the nature of the preglacial landscape. Where an ice sheet spreads over a region of low relief such as the Canadian Shield, glacial erosion creates a vast region of polished, flat striated surfaces. Where an ice sheet spreads over a hilly area, it deepens valleys and smooths hills. Notably, glacially eroded hills end up being elongate in the direction of flow and are asymmetric; glacial rasping smooths and bevels the upstream part of the hill, creating a gentle slope, whereas glacial plucking eats away at the downstream part, making a steep slope. Ultimately, the hill's profile resembles that of a sheep lying in a meadow—such a hill is called a **roche moutonnée,** from the French for "sheep rock" (▶Fig. 18.21).

Fjords: Submerged Glacial Valleys

As noted earlier, where a valley glacier meets the sea, the glacier's base remains in contact with the ground until the water depth exceeds about four-fifths of the glacier's thickness, at which point the glacier floats. Thus, glaciers can continue carving a U-shaped valley even below sea level. In addition, during an ice age, water extracted from the sea becomes locked in the ice sheets on land, so sea level drops significantly. Therefore, the floors of valleys cut by coastal glaciers during the last ice age were cut much deeper than present sea level. Today, the sea has flooded these deep valleys, producing **fjords** (see Chapter 15). In the spectacular fjord-land regions along the coasts of Norway, New Zealand, Chile, and Alaska, the walls of submerged U-shaped valleys rise straight from the sea as vertical cliffs up to 1,000 m high (▶Fig. 18.22). A lake filling a glacially carved valley is also called a fjord.

FIGURE 18.22 One of the many spectacular fjords of Norway.

18.4 DEPOSITION ASSOCIATED WITH GLACIATION

Types of Glacial Sedimentary Deposits

If you drill through the soil throughout much of the upper midwestern and northeastern United States and adjacent parts of Canada, the drill penetrates a layer of sediment deposited during the last ice age. A similar story holds for much of northern Europe. Thus, many of the world's richest agricultural regions rely on soil derived from sediment deposited by glaciers during the ice age. Notably, this sediment buries a pre–ice age landscape, as frosting fills the irregularities on a cake. Preglacial valleys may be completely filled with sediment.

Several different types of sediment can be deposited in glacial environments. All of these types together constitute glacial drift. (The term dates from pre-Agassiz studies of glacial deposits, when geologists thought that the sediment had "drifted" into place during an immense flood.) Specifically, glacial drift includes the following:

- *Till:* Glaciers consist of solid ice, so, like a conveyor belt, they can carry sediment of any size in the direction of flow. The sediment load of a glacier consists of sediment dropped on the ice at the origin of a glacier, sediment dropped along the margins of a glacier from adjacent slopes, and sediment plucked off the substrate on which the glacier flows (▶Fig. 18.23a). Sediment transported by ice and deposited on, beneath, or at the toe of a glacier is **glacial till.** Glacial till is unsorted (▶Fig. 18.23b).

- *Erratics:* Glacial erratics are cobbles and boulders that have been dropped by a glacier. Some protrude from till piles, and others rest on glacially polished surfaces (▶Fig. 18.24a).

FIGURE 18.21 A roche moutonnée. The glacier flowed from right to left.

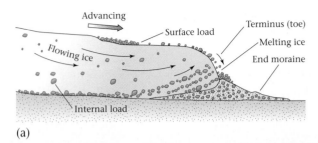

(a)

(b)

FIGURE 18.23 (a) The surface load plus the internal load accumulates at the toe of the glacier to constitute an end moraine; in effect, glaciers act like conveyor belts, constantly transporting more sediment toward the toe, regardless of whether the glacier advances or retreats. (b) The unsorted sediment constituting glacial till in Ireland.

• *Glacial marine:* Where a sediment-laden glacier flows into the sea, icebergs calve off the toe and raft clasts out to sea. As the icebergs melt, they drop the clasts, which settle into the muddy sediment on the sea floor. Pebble- and larger-size clasts deposited in this way are called drop stones. Sediment consisting of ice-rafted clasts mixed with marine sediment makes up glacial marine.

• *Glacial outwash:* Till deposited by a glacier at its toe may be picked up and transported by meltwater streams that sort the sediment. The clasts are deposited by a braided stream network in a broad area of gravel and sand bars called an outwash plain. This sediment is known as glacial outwash (▶Fig. 18.24b).

• *Glacial lake-bed sediment:* Streams transport fine clasts, including rock flour, away from the glacial front. This sediment eventually settles in meltwater lakes, forming a thick layer of glacial lake-bed sediment. This sediment commonly contains varves. A **varve** is a pair of thin sediment layers deposited during a single year. One layer consists of silt brought in during spring floods, and the

other of clay deposited in winter when the lake's surface freezes over and the water is standing still (▶Fig. 18.24c).

• *Loess:* When the warmer air above ice-free land beyond the toe of a glacier rises, the cold, denser air from above the glacier rushes in to take its place. A strong wind therefore blows at the margin of a glacier. This wind picks up fine clay and silt and transports it away from the glacier's toe. Where the winds die down, the sediment settles and forms a thick layer. This sediment, called **loess,** sticks together because of the electrical charges on clay flakes. Thus, steep escarpments develop by erosion of loess deposits (▶Fig. 18.24d).

Sediment Accumulations on Ice

Because ice is a solid, sediment can collect on the surface of a glacier. Specifically, where sediment is plentiful, the surface sedimentary load of a glacier may build into a ridge called a **moraine.** (We use this term both for ridges of sediment on a glacier and, as we discuss in the next section, for accumulations of ice-transported sediment left behind when ice melts.) Geologists specifically refer to a stripe of sediment that accumulates along the side of a glacier as a **lateral moraine** (▶Fig. 18.25a). If flowing meltwater runs along the edge of a glacier and redistributes this sediment, a stratified sequence of sediment called a **kame** forms. Where two valley glaciers merge, the debris constituting two lateral moraines merges to become a **medial moraine,** a stripe running down the interior of a glacier (▶Fig. 18.25b). Trunk glaciers into which numerous tributary glaciers may flow can contain several parallel medial moraines.

Depositional Landforms of Glacial Environments

Picture a group of hunters, dressed in deerskin, standing at the toe of a continental glacier in what is now southern Canada, waiting for an unwary woolly mammoth to wander by. It's a sunny summer day 12,000 years ago, and milky, sediment-laden streams gush from tunnels and channels at the base of the glacier, and pour off the top of the glacier, as the ice melts. No mammoths venture by today, so the bored hunters climb to the top of the glacier for a view. The climb isn't easy, partly because of the incessant wind, and partly because deep crevasses interrupt their path. Reaching the top of the ice sheet, the hunters look northward, and the glare almost blinds them. Squinting, they see the white of snow, and where the snow has blown away, they see the rippled, glassy surface of bluish ice (see Fig. 18.1). Here and there, a rock protrudes from the ice. Now looking southward, they survey a stark landscape of low, sinuous ridges separated by hummocky (bumpy) plains. Braided streams, which carry meltwater out across this landscape, flow through the hummocky plains and supply a number of lakes. Dust fills the air because of the wind.

(a)

(b)

FIGURE 18.24 (a) A glacial erratic in Central Park, New York City. (b) Braided streams choked with glacial outwash near the toe of a glacier in Alaska. (c) Note the alternating light and dark layers in this 20 cm-thick cross section of varved glacial lake bed sediment. (d) A small escarpment cut into glacial loess deposits in Illinois.

(c)

(d)

All the landscape features that the hunters observe as they look southward form by deposition in glacial environments, both mountain and continental (▶Fig. 18.26 a, b). The low, sinuous ridges, called **end moraines,** develop when the toe of a glacier stalls in one position for a while; the ice keeps flowing to the toe, and, like a giant conveyor belt, transports sediment with it; the sediment accumulates in a pile at the toe (see Fig. 18.23a). The end moraine at the farthest limit of glaciation is called the **terminal moraine.**

(The ridge of sediment that makes up Long Island, New York, and continues east-northeast into Cape Cod, Massachusetts, is part of the terminal moraine of the ice sheet that covered New England and eastern Canada during the last ice age; ▶Fig. 18.27.) End moraines that form when a glacier stalls for a while as it recedes are recessional moraines.

Till that has been released at the base of a flowing glacier and remains after the glacier has melted away is called lodgment till. Clasts in lodgment till may be aligned and

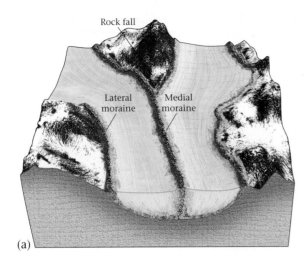

(a)

(b)

FIGURE 18.25 (a) Lateral and medial moraines on a glacier. (b) Medial moraines on a glacier near Alyeska, Alaska. The source of the moraine on the right is off the image.

scratched during their movement or as ice flows over them. The flow of the glacier may mold till and other subglacial sediment into streamlined, elongate hills called **drumlins** (from the Gaelic word for "hills"). Drumlins tend to be asymmetric along their length, with a gentle downstream slope, tapered in the direction of flow, and a steeper up-

stream slope (▶Fig. 18.28a, b). The till left behind during rapid recession forms a thin, hummocky layer on the land surface. This till, together with lodgment till, forms a landscape feature known as ground moraine.

The hummocky surface of moraines reflects partly the variations in the amount of sediment supplied by the glacier,

FIGURE 18.26 (a) The depositional landforms resulting from glaciation. (b) The setting in which various types of moraines form.

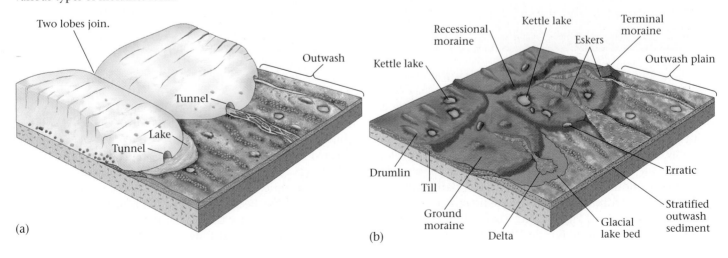

(a)

(b)

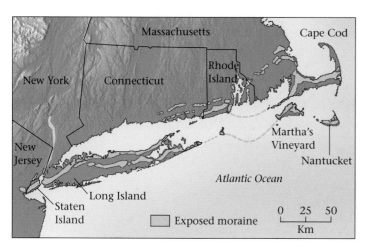

FIGURE 18.27 The moraines that constitute Long Island and Cape Cod.

and partly the occurrence of **kettle holes,** circular depressions made when blocks of ice calve off the toe of the glacier, become buried by sediment, and then melt (▶Fig. 18.29a). A land surface with many kettle holes separated by round hills of till displays knob-and-kettle topography (▶Fig. 18.29b, c).

Sediment-choked water pours out of tunnels in the ice. This water feeds large braided streams that sort till and redeposit it as glacial outwash, stratified layers and lenses of gravel

FIGURE 18.29 (a) When this ice block melts away, a kettle hole will form. (b, c) Knob-and-kettle topography on a moraine in Yellowstone Park.

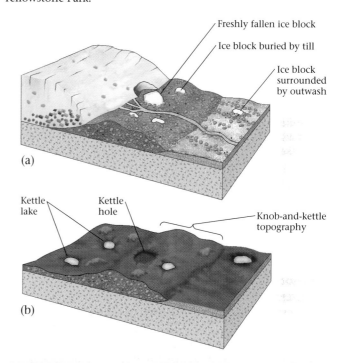

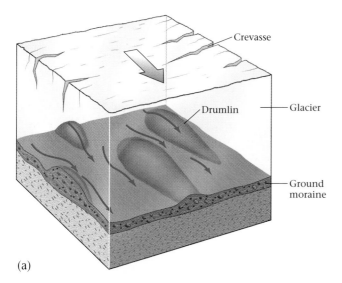

FIGURE 18.28 (a) The formation of a drumlin beneath a glacier. (b) Drumlins near Rochester, New York.

Glaciers and Glacial Landforms

Continental ice sheet

Crevasses

Ice shelf

Lower sea level

Drop stones

Iceberg

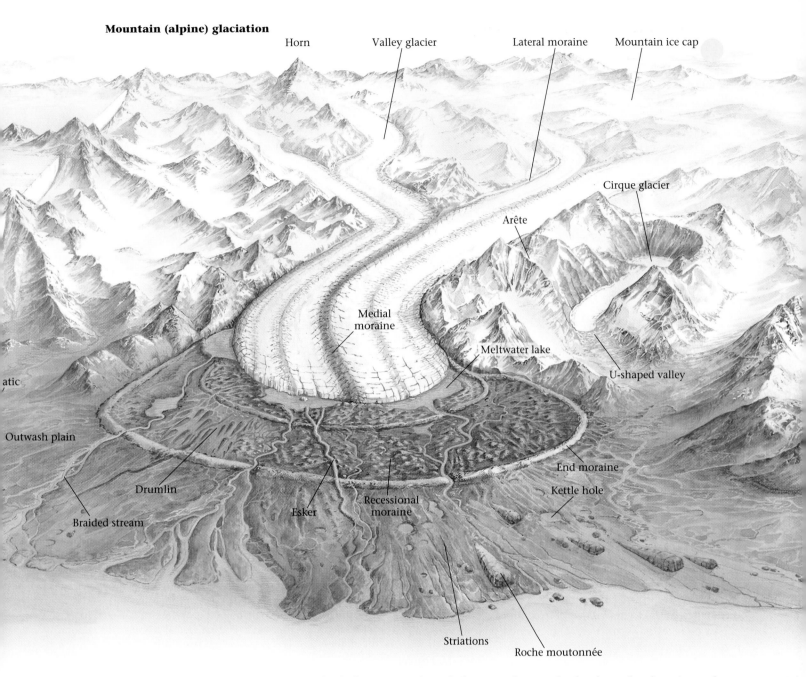

Mountain (alpine) glaciation

Horn

Valley glacier

Lateral moraine

Mountain ice cap

Cirque glacier

Arête

Medial moraine

Meltwater lake

U-shaped valley

...atic

Outwash plain

Drumlin

End moraine

Kettle hole

Esker

Recessional moraine

Braided stream

Striations

Roche moutonnée

Glaciers are rivers or sheets of ice that last all year and slowly flow. Continental glaciers, vast sheets of ice up to a few kilometers thick, covered extensive areas of land during times when Earth had a colder climate. Continental glaciers form when snow accumulates at high latitudes, then, when buried deeply enough, packs together and recrystallizes to make glacial ice. Ice, though solid, is weak, and thus ice sheets spread over the landscape, like syrup over a pancake. At the peak of the last ice age, ice sheets covered almost all of Canada, much of the United States, northern Europe, and parts of Russia.

The upper part of a sheet is brittle and may crack to form crevasses. Because ice sheets store so much of the Earth's water, sea level becomes lower during an ice age. When a glacier reaches the sea, it becomes an ice shelf. Rock that the glacier has plucked up along the way is carried out to sea with the ice; when the ice melts, the rocks fall to the sea floor as drop stones. At the edge of the shelf, icebergs calve off and float away.

A second class of glaciers, called mountain, or alpine, glaciers, exist in mountainous areas because snow can last all year at high elevations. During the ice age, mountain glaciers grew and flowed out onto the land surface beyond the mountain front. The glacier at the right has started to recede after formerly advancing and covering more of the land. Glacial recession may happen when the climate warms, so ice melts away faster at the toe (terminus) of the glacier than can be added at the source. In front of the glacier, you can find consequences of glacial erosion such as striations on bedrock and roches moutonnées.

When the glacier pauses, till (unsorted glacial sediment) accumulates to form an end moraine. Meltwater lakes gather at the toe, and streams carry sediment and deposit it as glacial outwash. Sediment that accumulates in ice tunnels, exposed when the glacier melts, make up sinuous ridges called eskers. Even when the toe remains fixed in position for a while, the ice continues to flow, and thus molds underlying sediment into drumlins. Ice blocks buried in till melt to form kettle holes. (Though the examples shown here were left after the melting away of a piedmont glacier, most actually form during continental glaciation.)

In the mountains, the glacier is confined to a valley. Sediment falling on it from the mountains creates lateral and medial moraines. Glaciers carve distinct landforms in the mountains, such as cirques, arêtes, horns, and U-shaped valleys.

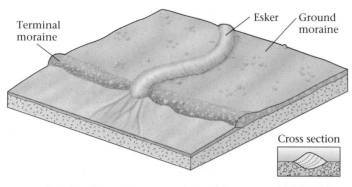

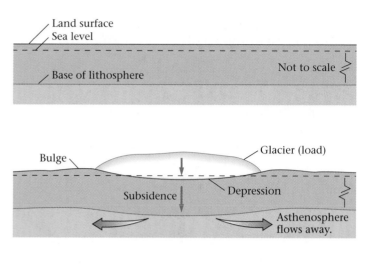

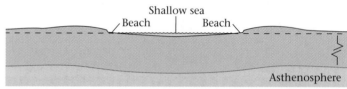

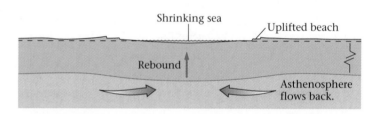

FIGURE 18.30 Eskers are snakelike ridges of sand and gravel that form when sediment fills meltwater tunnels at the base of a glacier.

FIGURE 18.31 Cross sections illustrating the concept of glacial rebound. Before glaciation, the surface of the lithosphere is flat. The weight of the glacier pushes the lithosphere down below sea level. The asthenosphere flows out of the way below, and lithosphere on either side bulges up. When the glacier melts, the depression fills with water. During glacial rebound, the floor of the shallow sea rises, and beaches rise along its shore.

and sand. Thus, outwash plains form between recessional moraines and beyond the terminal moraine. Some of the melt water pools to form lakes in lowlands between recessional moraines, or in ice-margin lakes between the glacier's toe and the nearest recessional moraine. Even long after the glacier has melted away, the lowlands between recessional moraines persist as lakes or swamps. When drained, glacial lake beds provide particularly fertile soil for agriculture. When a glacier eventually melts away, ridges of sorted sand and gravel, that were deposited in subglacial meltwater tunnels, snake across the ground moraine. These ridges are called **eskers** (▶Fig. 18.30). Sediments in glacial outwash plains and in eskers serve as important sources of sand and gravel for road building and construction.

18.5 OTHER CONSEQUENCES OF CONTINENTAL GLACIATION

Ice Loading and Glacial Rebound

When a large ice sheet (more than 50 km in diameter) grows on a continent, its weight causes the surface of the lithosphere to sink. In other words, ice loading causes glacial sub-

sidence. Lithosphere, the relatively rigid outer shell of the Earth, is able to sink because the underlying asthenosphere is soft enough to flow slowly out of the way.

What happens when continental ice sheets melt away? Gradually, the surface of the underlying continent rises back up, a process called glacial rebound, and the asthenosphere flows back underneath to fill the space (▶Fig. 18.31). This process doesn't take place instantly, because the asthenosphere flows so slowly (at rates of a few millimeters per year). It takes thousands of years for ice-depressed continents to rebound. Thus, glacial rebound is still taking place in some regions that were burdened by ice during the last ice age. Where rebound affects coastal areas, beaches along the shoreline rise several meters above sea level and become terraces.

Sea-Level Changes: The Glacial Reservoir in the Hydrologic Cycle

More of the Earth's surface and near-surface freshwater is stored in glacial ice than in any other reservoir. During the last ice age, when glaciers covered almost three times as much land area as they do today, they held significantly more water. In effect, water from the ocean reservoir transferred to the glacial reservoir and remained trapped on land. As a consequence, sea level dropped by as much as 100 m, and extensive areas of continental shelves became exposed as the coastline migrated seaward, in places by more than 100 km (▶Fig. 18.32a–c). People and animals migrated into the newly exposed coastal plains. The drop in sea level also created land bridges across the Bering Strait between North America and northeastern Asia and between Australia and Indonesia, providing convenient migration routes for animals and people.

Ice Dams, Drainage Reversals, and Lakes

When ice freezes over a sewer opening in a street, neither meltwater nor rain can enter the drain, and the street floods. Ice sheets play a similar role in glaciated environments. The ice may block the course of a river, leading to the formation of a lake. In addition, the weight of a glacier changes the tilt of the land surface and therefore the gradient of streams, and glacial sediment may fill preexisting valleys. In sum, continental glaciation destroys preexisting drainage networks. While the glacier exists, streams find different routes and carve out new valleys. By the time the glacier melts away, these new streams have become so well established that old river courses may remain abandoned. Glaciation during the last ice age profoundly modified North America's drainage. Before the ice age, several major rivers drained much of the interior of the continent to the north, into the Arctic Ocean (▶Fig. 18.33a). The ice sheet

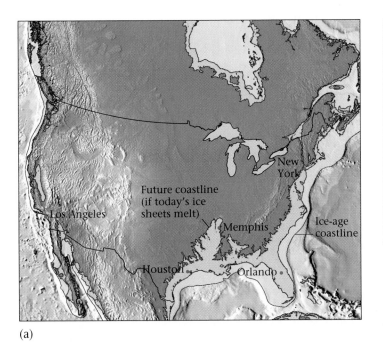

(a)

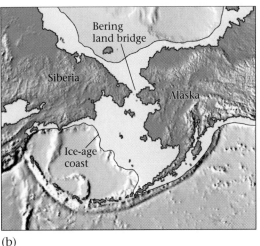

(b)

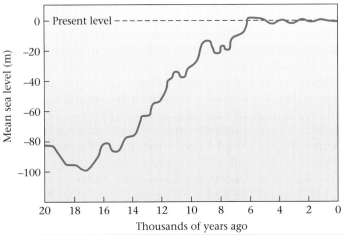

(c)

FIGURE 18.32 (a) This map shows the coastline of North America during the last ice age, and the coastline should present-day ice sheets melt. Note that much of the continental shelf was exposed during the last ice age. (b) A land bridge existed across the Bering Strait between Asia and North America during the last ice age. (c) The graph shows the change in sea level during the past 20,000 years.

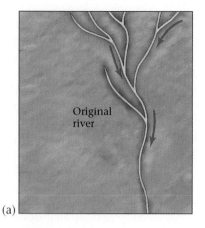
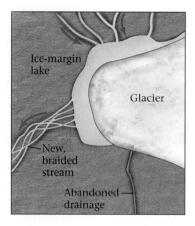
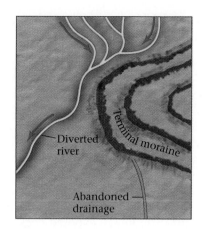

(a)

(b)

FIGURE 18.33 (a) When a glacier advances on the course of a river, the glacier blocks the drainage and causes a new stream to form. After the glacier melts away, the river remains diverted. (b) A map showing the extent of Glacial Lake Agassiz, at a time when it was an ice-margin lake. Lake Winnipeg is a remnant of Lake Agassiz.

buried this drainage network and diverted the flow into the Mississippi-Missouri network, which became larger.

Ice-margin lakes, as we've mentioned, develop where glaciers block drainage and meltwater fills the depressed land surface at the front of the glacier. Between about 11,700 and 9,000 years ago, as the most recent phase of the ice age came to a close and North America's vast continental glacier retreated back toward Hudson Bay, a huge ice-margin lake formed to the south of the receding glacier (▶Fig. 18.33b). This lake, now known as Glacial Lake Agassiz, was up to 1,100 km (680 mi) wide, and covered over 250,000 sq km (~100,000 sq mi)—it was larger than all the present Great Lakes combined. The main body of the lake covered extensive areas of Manitoba and Ontario; elongate bays extended

northwest into Saskatchewan and south into the Red River Valley of Minnesota and North Dakota. Eventually, the lake may have been left behind as the glacier continued to recede, so that near the end of its life, ice-free land existed on the northeast side of the lake.

Geologists reconstructed the history of Glacial Lake Agassiz by studying the present-day landscape. Broad plains define the region that was once the lake floor. In places, the linear groove-like troughs, thought to be the tracks of icebergs which dragged on the lake floor, indent the surface of these plains. Beach terraces, scarps, and small deltas, now high and dry, define what was once the shore of the lake.

When ice dams that held back large ice-margin lakes melted and broke, the contents of the lakes drained in a matter of hours to days, yielding immense floodwaters that stripped the land of soil and left behind huge ripple marks. As we discussed in Chapter 14, Glacial Lake Missoula, in Montana, filled when glaciers advanced and blocked the outlet of a large valley. When the glaciers retreated, the ice dam broke, releasing immense torrents that scoured eastern Washington, creating a barren, soil-free landscape called the channeled scablands.

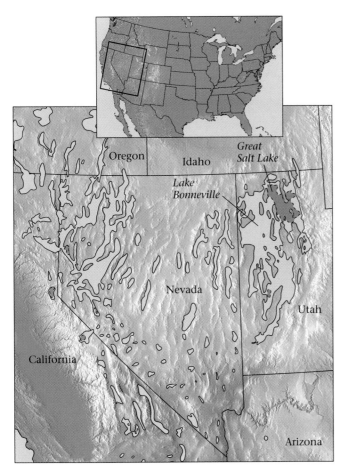

FIGURE 18.34 The distribution of pluvial lakes in the Basin and Range Province during the last ice age.

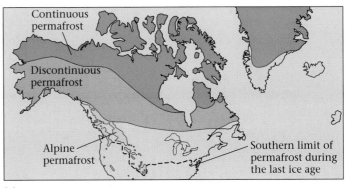

(a)

(b)

FIGURE 18.35 (a) The present-day distribution of periglacial environments in the Northern Hemisphere. (b) Patterned ground in the Northwest Territories, Canada, as viewed from a low-flying plane. The polygons are about 10 m across. The straight lines on the left side of the photo are caribou tracks.

Pluvial Features

During ice ages, regions to the south of continental glaciers were wetter than they are today. Fed by enhanced rainfall, lakes accumulated in low-lying land even at a great distance from the ice front. Such **pluvial lakes** (from the Latin *pluvia,* for "rain") flooded interior basins of the Basin and Range Province in Utah and Nevada (▶Fig. 18.34). The largest pluvial lake, Lake Bonneville, covered almost a third of western Utah. When this lake suddenly drained after a natural dam holding it back broke, it left a bathtub ring of shorelines rimming the Wasatch Mountains near Salt Lake City. Today's Great Salt Lake itself is but a small remnant of Lake Bonneville.

18.6 PERIGLACIAL ENVIRONMENTS

In polar latitudes today, and in regions adjacent to the fronts of continental glaciers during the last ice age, the mean annual temperature stays low enough (below –5°C) that soil moisture and groundwater freeze and, except in the upper few meters, stay solid all year. Such permanently frozen ground, or **permafrost,** may extend to depths of 1,500 m below the ground surface. Regions with widespread permafrost but without a blanket of snow or ice are called periglacial environments (the Greek *peri* means "around," or "encircling"; periglacial environments appear around the edges of glacial environments) (▶Fig. 18.35a).

The upper few meters of permafrost may melt during the summer months, only to refreeze again when winter comes. As a consequence of the freeze-thaw process, the ground splits into pentagonal or hexagonal shapes, creating a landscape of **patterned ground** (▶Fig. 18.35b). Water fills the gaps between the cracks and freezes to create wedge-shaped walls of ice.

Permafrost presents a unique challenge to people who live in polar regions or who work to extract resources from

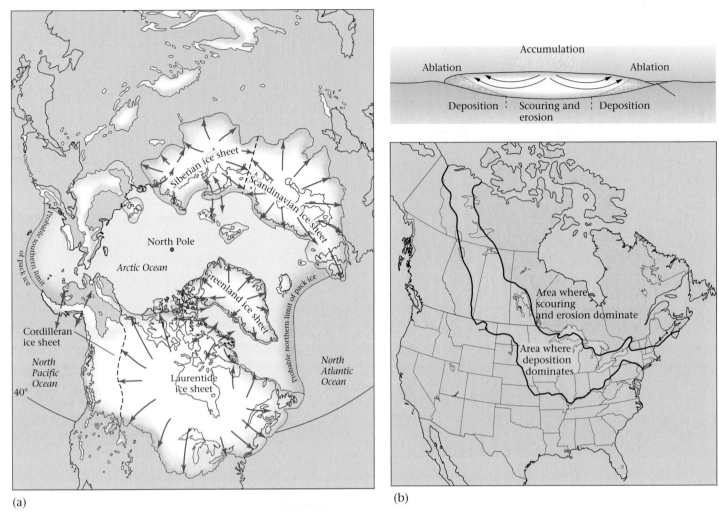

(a)

(b)

FIGURE 18.36 (a) A simplified map of the distribution of major ice sheets during the Pleistocene Epoch. The arrows indicate the flow trajectories of the ice sheets. (b) Top: A continental glacier scours and erodes the land surface beneath its center, while at the margins it deposits sediment. Map: The Laurentide ice sheet scoured and eroded the land in northern and eastern Canada, and deposited sediment in western Canada and the midwestern United States.

these regions. For example, heat from a building may warm and melt underlying permafrost, creating a mire into which the building settles. For this reason, buildings in permafrost regions must be placed on stilts, so that cold air can circulate beneath them to keep the ground frozen. When geologists discovered oil on the northern coast of Alaska, oil companies faced the challenge of shipping the oil to markets outside of Alaska. After much debate over the environmental impact, they built the trans-Alaska pipeline, which now carries oil for 1,000 km to a seaport in southern Alaska (see Chapter 12). The oil must be warm during transport, or it would be too viscous to flow. Thus, to prevent the warm pipeline from melting underlying permafrost, it had to be built on a frame that holds it above the ground for its entire length.

A broad threat to permafrost may be developing because of global climate change. Recent measurements suggest that extensive regions of permafrost in Alaska are starting to melt.

18.7 THE PLEISTOCENE ICE AGES

The Pleistocene Glaciers

Today, most of the land surface in New York City lies hidden beneath concrete and steel, but in the city's Central Park it's still possible to see land in a seminatural state. If you stroll through the park and study the rock outcrops, you'll find that their top surfaces are smooth and polished, and in places have been grooved and scratched (Fig. 18.18b). You are seeing evidence that an ice sheet once scraped along this ground, rasping and gouging the bedrock as it moved. Geologists estimate that the ice sheet that overrode the New York City area may have been 250 m thick, enough to bury the Empire State Building up to the 75th floor.

Glacial features such as those on display in Central Park led Louis Agassiz to propose the idea that vast continental

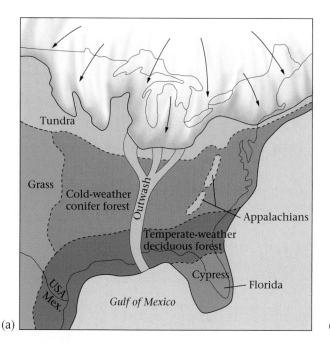

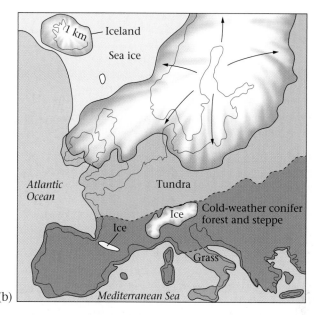

FIGURE 18.37 (a) Pleistocene climatic belts in North America; (b) Pleistocene climatic belts in Europe.

glaciers advanced over substantial portions of Europe, North America, and Asia during a great ice age. Since Agassiz's day, thousands of geologists, by mapping out the distribution of glacial deposits and landforms, have gradually defined the extent of ice-age glaciers and a history of their movement.

The fact that these glacial features decorate the surface of the Earth today means that the most recent ice age occurred fairly recently during Earth history (▶Box 18.1). This ice age, responsible for the glacial landforms of North America and Eurasia, happened mostly during the Pleistocene Epoch, which began 1.8 million years ago (see Chapter 11), so it is commonly known as the Pleistocene ice age. (More recent evidence suggests that glaciations began as early as 3 million years ago.) Geologists use the name Holocene to refer to the time since the last glaciation (that is, the time period that began 11,000 years ago).

Based on their mapping of glacial striations and deposits, geologists have determined where the great Pleistocene ice sheets originated and flowed. In North America, the Laurentide ice sheet started to grow over northeastern Canada, then merged with the Keewatin ice sheet, which originated in northwestern Canada. Together, these ice sheets eventually covered all of Canada east of the Rocky Mountains and extended southward across the border as far as southern Illinois (▶Fig. 18.36a). At their maximum, they attained a thickness of 2 to 3 km; each thinned toward the toe. In northeastern Canada, the ice sheet eroded the land surface. Further south and west, it deposited sediment

(▶Fig. 18.36b). These ice sheets also eventually merged with the Greenland ice sheet to the northeast and the Cordilleran ice sheet to the west. The Cordilleran covered the mountains of western Canada as well as the southern third of Alaska.

During the Pleistocene ice age, mountain ice caps and valley glaciers also grew in the Rocky Mountains, the Sierra Nevada, and the Cascade Mountains. In Eurasia, a large ice sheet formed in northernmost Europe and adjacent Asia, and gradually covered all of Scandinavia and northern Russia. This ice sheet flowed southward across France until it reached the Alps and merged with Alpine mountain glaciers; it also covered all of Ireland and almost all of the United Kingdom. A smaller ice sheet grew in eastern Siberia and expanded in the mountains of central Asia. In the Southern Hemisphere, Antarctica remained ice covered, and mountain ice caps expanded in the Andes, but there were no continental glaciers in South America, Africa, or Australia.

In addition to continental ice sheets, sea ice in the Northern Hemisphere expanded to cover all of the Arctic Ocean and parts of the North Atlantic. Sea ice surrounded Iceland and approached Scotland, and also fringed most of western Canada and southeastern Alaska.

Life and Climate in the Pleistocene World

During the Pleistocene ice age, all climatic belts shifted southward (▶Fig. 18.37a, b). Geologists can document this

BOX 18.1
THE HUMAN ANGLE

So You Want to See Glaciation?

Though the last of the Pleistocene continental ice sheet that once covered much of North America vanished about 6,000 years ago, you can find evidence of its power quite easily. The Great Lakes, along the U.S.-Canadian border, the Finger Lakes and drumlins of New York, the low-lying moraines and outwash plains of Illinois, and the polished outcrops of southern Canada all formed in response to the movement of this glacier. But if you want to see continental glaciers in action today, you must trek to Greenland or Antarctica.

Mountain glaciers are easier to reach. A trip to the mountains of western North America (including Alaska), the Alps of France or Switzerland, the Andes of South America, or the mountains of southern New Zealand will bring you in contact with live glaciers. You can even spot glaciers from the comfort of a cruise ship.

Some of the most spectacular glacial landscapes in North America formed during the Pleistocene Epoch, when mountain glaciers were more widespread. These are now on display in national parks.

- *Glacier National Park (Montana):* This park, which borders Waterton Lakes National Park in Canada, displays giant cirques, U-shaped valleys, hanging valleys, terminal moraines, and fifty small relicts of formerly larger glaciers, all in a mountainous terrain that reaches elevations of over 3 km. Unfortunately, these glaciers are melting away quickly!

- *Yosemite National Park (California):* A huge U-shaped valley, carved into the Sierra Nevada granite batholith, makes up the centerpiece of this park. Waterfalls spill out of hanging valleys bordering the valley.

- *Voyageurs National Park (Minnesota):* This park lacks the high peaks of mountainous parks, but shows the dramatic consequences of glacial scouring and deposition on the Canadian Shield. The low-lying landscape, dotted with lakes, contains abundant polished surfaces, glacial striations, and erratics, along with moraines, glacial lake beds, and outwash plains.

- *Acadia National Park (Maine):* During the last ice age, the continental ice sheet overrode low bedrock hills and flowed into the sea along the coast of Maine. This park provides excellent examples of the consequences. Its hills were scoured and shaped into large roches moutonnées by glacial flow. Some of the deeper valleys have now become small fjords.

- *Glacier Bay National Park (Alaska):* In Glacier Bay, huge tidewater glaciers fringe the sea, creating immense ice cliffs from which icebergs calve off. Cruise ships bring tourists up to the toes of these glaciers. More adventurous visitors can climb the coastal peaks and observe lateral and medial moraines, crevasses, and the erosional and depositional consequences of glaciers that have already retreated up the valley.

shift by examining fossil pollen, which can survive for thousands of years, preserved in the sediment of bogs.

Fossils also tell us that numerous species of now-extinct large mammals inhabited the Pleistocene world (▶Fig. 18.38). Giant mammoths and mastodons, relatives of the elephant, along with woolly rhinos, musk oxen, reindeer, giant ground sloths, bison, lions, saber-toothed cats, giant cave bears, and hyenas wandered forests and tundra in North America. Early humanlike species were already foraging in the woods by the beginning of the Pleistocene Epoch, and by the end modern *Homo sapiens* lived on every continent except Antarctica, and had discovered fire and invented tools.

Chronology of the Pleistocene Ice Ages

Louis Agassiz assumed that only one ice age had affected the planet. But close examination of the stratigraphy of glacial deposits on land revealed that paleosol (ancient soil preserved in the stratigraphic record), as well as beds containing fossils of warmer-weather animals and plants, separated distinct layers of glacial sediment. This observation suggested that between episodes of glacial deposition, glaciers receded and temperate climates prevailed. During the second half of the twentieth century, when modern methods for dating geological materials became available, the difference in ages between the different layers of glacial sediment could be confirmed. Clearly, ice-age glaciers had advanced and then retreated more than once. Times during which the glaciers grew and covered substantial areas of the continents are called glacial periods, or glaciations, and times between glacial periods are called interglacial periods, or interglacials.

Using the on-land sedimentary record, geologists recognized five Pleistocene glaciations in Europe (named, in order of increasing age: Würm, Riss, Mindel, Gunz, and Donau) and traditionally four in the midwestern United States (Wisconsinan, Illinoian, Kansan, and Nebraskan, named after the southernmost states in which their till was deposited; ▶Fig. 18.39). Since the mid-1980s, geologists no longer recognize Nebraskan and Kansan; they are lumped together as "pre-Illinoian." The four-stage chronology of North American glaciation, however, was turned on its head in the 1960s, when geologists began to study submarine sediment containing the fossilized shells of microscopic marine plankton. Because the assemblage of plankton species living in warm water is not the same as the assemblage living in cold water, geologists can track changes in the temperature of the ocean by studying plankton fossils. Researchers found that the succession of plankton in sediment of the last 2 million

FIGURE 18.38 Examples of now-extinct large mammals that roamed the countryside during the Pleistocene Epoch.

years, assuming that cold water indicates a glacial period and warm water an interglacial period, records twenty to thirty different glacial advances. The four traditionally recognized glaciations probably represent only the largest of these—sediments deposited on land by other glaciations were eroded and redistributed during subsequent glaciations, or were eroded by streams and wind during interglacials.

Geologists refined their conclusions about the frequency of Pleistocene glaciations by examining the composition of fossil shells. Shells of many plankton species consist of calcite ($CaCO_3$). The oxygen in the shells includes two isotopes, a heavier one (^{18}O) and a lighter one (^{16}O). The ratio of these isotopes tells us about the water temperature in which the plankton grew. This is because as water gets colder plankton incorporate a higher proportion of ^{18}O into their shells.

FIGURE 18.39 Pleistocene deposits in the United States.

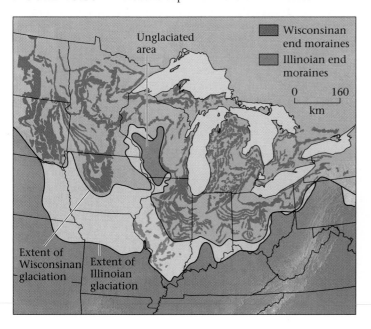

Thus, intervals in the stratigraphic record during which plankton shells have a large ratio of $^{18}O/^{16}O$ define times when Earth had a colder, glacial climate. The record also indicates that twenty to thirty of these glaciations occurred during the last 3 million years (▶Fig. 18.40).

Older Ice Ages During Earth History

So far, we've focused on the Pleistocene ice age, because of its importance in developing Earth's present landscape. Was this the only ice age during Earth history, or do ice ages happen frequently? To answer such questions, geologists study the stratigraphic record and search for ancient glacial deposits that have hardened into rock. These deposits, called tillites, consist of larger clasts distributed throughout a

FIGURE 18.40 This time column shows the variations in oxygen-isotope ratios from marine sediment that define twenty to thirty glaciations and interglacials during the Pleistocene Epoch. The green bands represent the approximate boundaries of the traditional glacial stages recognized on land in the midwestern United States. Note that the traditional names "Kansan" and "Nebraskan" are no longer used, and have been replaced by "Pre-Illinoian."

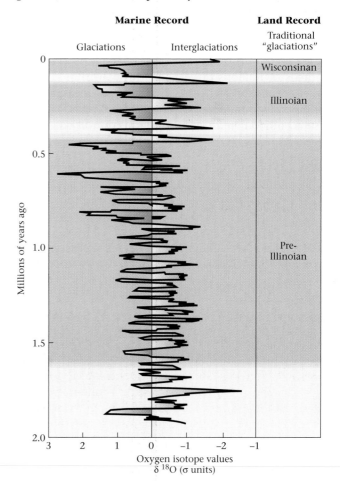

matrix of sandstone and mudstone (▶Fig. 18.41a). In many cases, tillites are deposited on glacially polished surfaces.

By using the stratigraphic principles described in Chapter 10, geologists have determined that tillites were deposited about 280 million years ago in Permian time. (These are the deposits Alfred Wegener studied when he argued in favor of continental drift; ▶Fig. 18.41b.) Tillites were also deposited about 600 to 700 million years ago (at the end of the Proterozoic Eon), about 2.2 billion years ago (near the beginning of the Proterozoic), and perhaps about 2.7 billion years ago (in the Archean Eon). Strata deposited at other times in Earth history do not contain tillites. Thus, it appears that glacial advances and retreats have not occurred steadily throughout Earth history, but rather are restricted to specific time intervals, or ice ages, of which there are four or five: Pleistocene, Permian, late Proterozoic, early Proterozoic, and perhaps Archean.

Of particular note, some tillites of the late Proterozoic event were deposited at equatorial latitudes, suggesting that on three occasions, for at least a short time, the continents worldwide were largely glaciated, and the sea may have been covered worldwide by ice. Geologists refer to the ice-encrusted planet as **snowball Earth.**

18.8 THE CAUSES OF ICE AGES

Ice ages occur only during restricted intervals of Earth history, hundreds of millions of years apart. But within an ice age, glaciers advance and retreat with a frequency measured in tens of thousands to hundreds of thousands of years. Thus, there must be both long-term and short-term controls on glaciation.

Long-Term Causes

Plate tectonics probably provides some long-term controls on the timing of ice ages. For example, the drift of conti-

nents that occurs because of plate motions may determine whether or not continents lie in appropriate latitudes and configurations to be glaciated. The arrangement of continents and volcanic arcs may also affect the global pattern of ocean currents. This pattern in turn, affects climate. Finally, rates of sea-floor spreading affect global sea level, and thus determine whether or not continents are above sea level and susceptible to ice formation.

The concentration of carbon dioxide in the atmosphere probably also plays a major role in determining whether an ice age can occur. Carbon dioxide is a greenhouse gas—it traps infrared radiation rising from the Earth—so if the concentration of CO_2 increases, the atmosphere becomes warmer and if the concentration decreases, the atmosphere becomes cooler. Ice sheets cannot form during periods when the atmosphere has a relatively high concentration of CO_2, even if other factors favor glaciation. What might cause long-term changes in CO_2 concentration? Possibilities include: changes in the population of living organisms, for many organisms extract CO_2; changes in the amount of chemical weathering on land, for weathering absorbs CO_2; and changes in the amount of volcanic activity, for volcanoes release gas. Thus, ice ages may have been triggered by mountain building events, which exposed more rock to weathering, and by major stages in evolution, which produced new species that extracted CO_2 from the atmosphere. For example, the appearance of coal swamps at the end of Paleozoic may have removed CO_2, for plants incorporate CO_2. The low CO_2 levels might have triggered glaciations of Pangaea.

Short-Term Causes

Now we've seen how the stage can be set for an ice age to occur, but why do glaciers advance and retreat periodically *during* an ice age? In 1920, Milutin Milankovitch, a Serbian astronomer and geophysicist, came up with an explanation. Milankovitch studied how the Earth's orbit changes shape

FIGURE 18.41 (a) This time column shows pre-Pleistocene glaciations during Earth history. (b) The distribution of Permian glacial features on a reconstruction of Gondwana, the southern part of the Pangaea supercontinent that existed at the time.

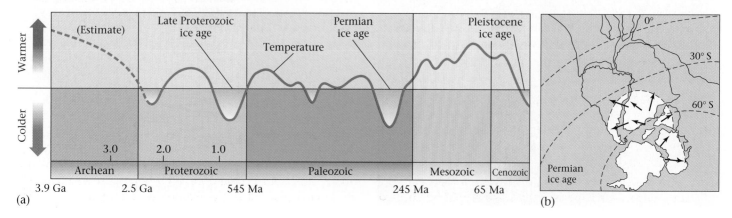

and how its axis changes orientation through time, and he calculated the frequency of these changes. In particular, he evaluated three aspects of Earth's movement around the Sun.

- *Orbital eccentricity:* Milankovitch showed that the Earth's orbit gradually changes from a more circular shape to a more elliptical shape. This eccentricity cycle takes around 100,000 years (►Fig. 18.42a).
- *Tilt of Earth's axis:* We have seasons because the Earth's axis is not perpendicular to the plane of its orbit. Milankovitch calculated that over time, the tilt angle varies between 22.5° and 24.5°. This cycle takes 41,000 years (►Fig. 18.42b).
- *Precession of Earth's axis:* If you've ever set a top spinning, you've probably noticed that its axis gradually traces a conical path. This motion, or wobble, is called precession (►Fig. 18.42c). Milankovitch determined that the Earth's axis wobbles over the course of about 23,000 years. Precession determines the relationship between the timing of the seasons and the position of Earth along its orbit around the Sun.

Milankovitch showed that precession, along with variations in orbital eccentricity and in tilt combine to affect the total annual amount of insolation (exposure to the Sun's rays) and the seasonal distribution of insolation that the Earth receives at the high latitudes. For example, high-latitude regions receive more insolation when the Earth's axis is almost perpendicular to its orbital plane than when its axis is greatly tilted. According to Milankovitch, glaciers tend to advance during times of cool summers at 65°N, which occur roughly 100,000, 40,000, and 20,000 years apart. When geologists began to study the climate record, they found climate cycles with the frequency that Milankovitch predicted. These climate cycles are now called **Milankovitch cycles** (►Fig. 18.42d).

Orbit and tilt changes are probably not the whole story, however. Geologists suggest that several other factors may also come into play in order to trigger a glacial advance.

- *A changing albedo:* When snow remains on land throughout the year, or clouds form in the sky, the albedo (reflectivity) of the Earth increases, so Earth's surface reflects incoming sunlight and thus becomes even cooler.
- *Interrupting the global heat conveyor:* As the climate cools, evaporation rates from the sea decrease, so seawater does not become so salty. And decreasing salinity might stop the system of thermohaline currents that brings warm water to high latitudes (see Chapter 15). Such changes can be very abrupt.
- *Biological processes that change CO_2 concentration:* A bloom of algae in the ocean may have amplified climate

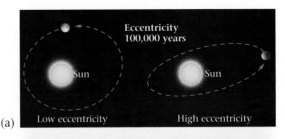

(a)

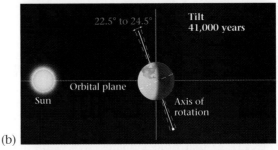

(b)

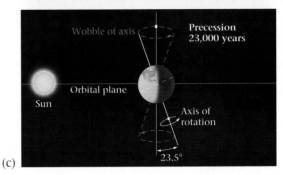

(c)

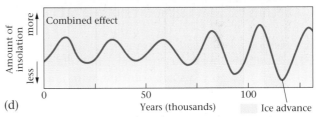

(d)

FIGURE 18.42 The Milankovitch cycles affect the amount of insolation (exposure to the Sun's rays) at high latitudes. (a) Variations in insolation caused by changes in orbital shape; (b) variations caused by changes in the tilt angle of Earth's axis; (c) variations caused by the precession of Earth's axis. (d) When the effects of eccentricity, tilt, and precession are combined, we see that there are distinct warm and cold periods; cold periods occur when there is less insolation.

changes by decreasing the concentration of carbon dioxide in the atmosphere.

A Model for Pleistocene-Ice-Age History

Long-term cooling in the Cenozoic Era. Taking all of the above causes into account, we can now propose a scenario for the events that led to the Pleistocene glacial advances. Our

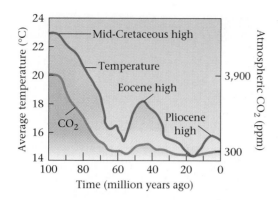

FIGURE 18.43 The graph shows the gradual cooling of Earth's atmosphere since the Cretaceous Period.

story begins in the Eocene Epoch, about 55 million years ago (►Fig. 18.43). At that time, climates were warm and balmy, not only in the tropics but even above the Arctic Circle. At the end of the Middle Eocene (37 million years ago), the climate began to cool. These long-term climate changes may have been caused, in part, by changes in the pattern of oceanic currents that happened because of plate tectonics. For example, in the Eocene, the collision of India with Asia cut off warm equatorial currents that had been flowing in the Tethys Sea. Changes to atmospheric circulation and temperature may also have happened at this time—the uplift of the Himalayas and Tibet diverted winds in a way that, models suggest, cooled climate. This uplift also exposed more rock to chemical weathering, perhaps leading to extraction of CO_2 from the atmosphere. A decrease in the concentration of this greenhouse gas would contribute to atmospheric cooling.

The sudden appearance of the Laurentide ice sheet about 2 to 3 million years ago coincides with another well-known plate-tectonic event, the closing of the gap between North and South America by the growth of the Isthmus of Panama. When this land bridge formed, it separated the waters of the Caribbean from those of the tropical Pacific for the first time. And when this happened, warm currents that previously flowed out of the Caribbean into the Pacific were blocked and diverted northward to merge with the Gulf Stream. This current transfers warm water from the Caribbean up the Atlantic Coast of North America and ultimately to the British Isles. As the warm water moves up the Atlantic Coast, it generates warm, moisture-laden air that provides a source for the snow that falls over New England, eastern Canada, and Greenland. In other words, the Arctic has long been cold enough for ice caps, but until the Gulf Stream was diverted northward by the growth of Panama, there was no source of moisture to make abundant snow and ice.

Short-term advances and retreats in the Pleistocene Epoch. Once the Earth's climate had cooled overall, short-term processes such as the Milankovitch cycles led to periodic advances and retreats of the glaciers. To understand how, let's look at a possible case history of a single advance and retreat of the Laurentide ice sheet. (Note that such models remain the subject of vigorous debate.)

- *Stage 1:* During the overall cooler climates of the late Cenozoic Era, the Earth reaches a point in the Milankovitch cycle when the average mean temperature in temperate latitudes drops. With lower temperatures, not all of winter's snow melts away during the summer. Eventually, snow covers the entire region of northern Canada even during the summer. Because of the snow's high albedo, it reflects sunlight, so the region grows still colder and even more snow accumulates. Precipitation rates are high, because evaporation off the Gulf Stream provides moisture. Finally, the snow at the base of the pile turns to ice, and the ice begins to spread outward under its own weight. A new continental glacier has been born.

- *Stage 2:* The ice sheet continues to grow as more snow piles up in the zone of accumulation. And as the ice sheet grows, the atmosphere continues to cool because of the albedo effect. But now, the weight of the ice loads the continent and makes it sink, so the elevation of the glacier decreases, and its surface approaches the equilibrium line. Also, the temperature becomes cold enough that the Atlantic Ocean in high latitudes begins to freeze. As the sea ice covers the ocean, the amount of evaporation decreases, so the source of snow is cut off and the amount of snowfall diminishes. The glacial advance pretty much chokes on its own success. The decrease in the glacier's elevation (leading to warmer summer temperatures) on the ice surface, as well as the decrease in snowfall, causes ablation to occur faster than accumulation, and the glacier begins to retreat.

- *Stage 3:* As the glacier retreats, temperatures gradually increase, and the sea ice begins to melt. The supply of water to the atmosphere from evaporation increases once again, but with the warmer temperatures and lower elevations, this water precipitates as rain during the summer. The rain drastically accelerates the rate of ice melting, and the retreat progresses quite rapidly.

18.9 WILL THERE BE ANOTHER GLACIAL ADVANCE?

What does the future hold? Considering the periodicity of glacial advances and retreats during the Pleistocene Epoch, we may be living in an interglacial period. Pleistocene interglacials lasted about 10,000 years, and since the present interglacial began about 11,000 years ago, the time seems ripe for a new glaciation. The Earth actually had a brush with ice-age conditions between the 1300s and the mid-1800s, when average annual temperatures in the Northern Hemisphere fell suf-

FIGURE 18.44 Skaters (c. 1600) on the frozen canals of the Netherlands during the little ice age.

FIGURE 18.45 A couple of hundred years ago, glacial ice filled the cirque in the background and all of the valley up to the height of the lateral moraine in the foreground. In this 2003 photo, most of the ice of this Alaskan glacier has vanished.

ficiently for mountain glaciers to advance significantly. During this period, now known as the little ice age, sea ice surrounded Iceland, and canals froze in the Netherlands, leading to that country's tradition of skating (▶Fig. 18.44). The little ice age did not become a full-fledged ice age because during the past 150 years, temperatures have warmed. You can see evidence for this warming by visiting a mountain glacier—most mountain glaciers have retreated substantially (▶Fig. 18.45). Some researchers suggest that this global-warming trend is due to the addition of carbon dioxide to the atmosphere from the burning of forests and the use of fossil fuels (see Chapter 19).

CHAPTER SUMMARY

- Glaciers are streams or sheets of recrystallized ice that survive for the entire year and flow in response to gravity. Mountain glaciers exist in high regions and fill cirques and valleys. Continental glaciers (ice sheets) spread over substantial areas of the continents.

- Glaciers form when snow accumulates over a long period of time. With progressive burial, the snow first turns to firn, and then to ice.

- Glacial ice moves by basal sliding over water or wet sediment, and/or by internal flow. In general, glacial ice moves tens of meters per year.

- Whether the toe of a glacier stays fixed in position, advances farther from the glacier's origin, or retreats back toward the origin depends on the balance between the rate at which snow builds up in the zone of accumulation and the rate at which glaciers melt or sublimate in the zone of ablation.

- Glacial ice can flow over sediment or incorporate sediment. The clasts embedded in glacial ice act like a rasp that abrades the substrate.

- Mountain glaciers carve numerous landforms, including cirques, arêtes, horns, U-shaped valleys, hanging valleys, and truncated spurs. Glacially carved valleys that fill with water when sea level rises after an ice age are fjords.

- Moraines are piles or ridges of glacial till. Till is unsorted sediment, which accumulates because glaciers can transport sediment of all sizes.

- Glacial depositional landforms include moraines, knob-and-kettle topography, drumlins, kames, eskers, meltwater lakes, and outwash plains.

- Continental crust subsides as a result of ice loading. When the glacier melts away, the crust rebounds.

- When water is stored in continental glaciers, sea level drops. When glaciers melt, sea level rises.

- During past ice ages, the climate in regions south of the continental glaciers was wetter, and pluvial lakes formed. Permafrost exists in periglacial environments.

- During the Pleistocene ice age, large continental glaciers covered much of North America, Europe, and Asia.

- The stratigraphy of glacial deposits indicates that glaciers advanced and retreated many times during the Pleistocene.

- Long-term causes of ice ages include plate tectonics and changes in the concentration of CO_2 in the atmosphere. Short-term causes include the Milankovitch cycles (caused by periodic changes in Earth's orbit and tilt).

KEY TERMS

cirque (p. 495)
continental glacier (ice sheet)
    (p. 495)
crevasse (p. 497)
drumlin (p. 508)
end moraine (p. 507)
equilibrium line (p. 498)
erratic (p. 493)
esker (p. 512)
fjord (p. 505)
glacial advance (p. 498)
glacial retreat (p. 499)
glacial striations (p. 502)
glacial till (p. 505)
glacier (p. 493)
hanging valley (p. 503)
horn (p. 502)
ice age (p. 494)
iceberg (p. 500)
kame (p. 506)

kettle hole (p. 509)
lateral moraine (p. 506)
loess (p. 506)
medial moraine (p. 506)
Milankovitch cycles (p. 521)
moraine (p. 506)
mountain (alpine) glacier
    (p. 495)
patterned ground (p. 515)
permafrost (p. 515)
pluvial lake (p. 515)
roche moutonnée (p. 505)
sea ice (p. 500)
snowball Earth (p. 520)
terminal moraine (p. 507)
U-shaped valley (p. 503)
varve (p. 506)
zone of ablation (p. 498)
zone of accumulation (p. 498)

THE VIEW FROM SPACE Glaciers and glacially carved features dominate the landscape of southeastern Alaska, as seen in this infrared image. Here we see Hubbard glacier as it enters Yakutat Bay. Where the glacier meets the sea, large blocks of ice calve off and float away. Many tributary glaciers can still be seen, but some glaciers have melted away, leaving behind long fjords.

REVIEW QUESTIONS

1. What evidence did Louis Agassiz offer to support the idea of an ice age?

2. How do mountain glaciers and continental glaciers differ in terms of dimensions, thickness, and patterns of movement?

3. Describe the transformation from snow to ice.

4. Explain how arêtes, cirques, and horns form.

5. Describe the mechanisms that enable glaciers to move, and explain why they move.

6. Explain how the balance between ablation and accumulation determines whether a glacier advances or retreats.

7. How does a glacier transform a V-shaped river valley into a U-shaped valley? Discuss how hanging valleys develop.

8. Describe the various kinds of glacial deposits. Be sure to note the materials from which the deposits are made and the landforms that result from deposition.

9. How does the lithosphere respond to the weight of glacial ice?

10. How was the world different during the glacial advances of the Pleistocene ice age? Be sure to mention the relation between glaciations and sea level.

11. Were there ice ages before the Pleistocene? If so, when?

12. What are some of the long-term causes that lead to ice ages? What are the short-term causes that trigger glaciations and interglacials?

SUGGESTED READING

Alley, R. B. 2002. *The Two-Mile Time Machine: Ice Cores, Abrupt Climate Change, and Our Future*. Princeton, N.J.: Princeton University Press.

Bennett, M. R., and N. F. Glasser. 1996. *Glacial Geology: Ice Sheets and Landforms*. New York: Wiley.

Dawson, A. G. 1992. *Ice Age Earth: Late Quaternary Geology and Climate*. New York: Routledge, Chapman, and Hall.

Erickson, J. 1996. *Glacial Geology: How Ice Shapes the Land*. New York: Facts on File.

Fagan, B. 2002. *The Little Ice Age: How Climate Made History, 1350–1850*. New York: Basic Books.

Hambrey, M. J., and J. Alean. 1992. *Glaciers*. Cambridge: Cambridge University Press.

MacDougall, J. D. 2004. *Frozen Earth: The Once and Future Story of Ice Ages*. Berkeley: University of California Press.

Menzies, J. 2002. *Modern and Past Glacial Environments*. Woburn, MA.: Butterworth-Heinemann.

Post, A., and E. R. Lachapelle. 2000. *Glacier Ice*. Seattle: University of Washington Press.

Global Change in the Earth System

All we in one long caravan
are journeying since the world began,
we know not whither, we know . . . all must go.
—BHARTRIHARI (INDIAN POET, C. 500 C.E.)

This view, taken from an airplane landing at Chicago's O'Hare Airport, emphasizes the extent to which the Earth's surface has changed. Fifteen thousand years ago, the view would have been the surface of a glacier. Five hundred years ago, it would have been a vast tall-grass prairie. One hundred years ago, it would have been a checkerboard of farm fields. Today, most of the land has been covered by a layer of concrete.

19.1 INTRODUCTION

Would the Earth's surface have looked the same in the Jurassic Period as it does today? Definitely not! In the Jurassic, the North Atlantic Ocean was a narrow sea and the South Atlantic Ocean didn't exist at all, so most dry land connected to form a single vast continent (▶Fig. 19.1a). Today, both parts of the Atlantic are wide oceans, and the Earth has seven separate continents (▶Fig. 19.1b). Moreover, in the Jurassic, the call of the wild rumbled from the throats of dinosaurs; today, the largest land animals are mammals. In essence, what we see of the Earth today is just a snapshot, an instant in the life story of a constantly changing planet. This idea arguably stands as geology's greatest philosophical contribution to humanity's understanding of its surroundings.

Why does the Earth change so much through time? Ultimately, it's because the Earth's asthenosphere is warm and soft enough to flow, because the Sun is close enough to heat the Earth's surface, and because gravity causes heavy objects to fall and buoyant ones to rise. If the asthenosphere could not flow and gravitational force did not exist, internal processes such as sea-floor spreading, subduction, and continental drift would not occur. Without the warmth of the Sun, there could be no liquid water, advanced

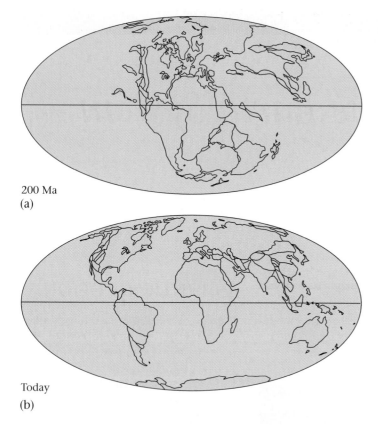

200 Ma
(a)

Today
(b)

FIGURE 19.1 A comparison of (a) a map of the Earth in the Jurassic with (b) a map of today's Earth, emphasizing the change that has resulted from continental drift.

life, wind, rivers, or glaciers. Without gravity, the wind, rivers, and glaciers would not move. And without wind, rivers, or glaciers, processes such as weathering and erosion would not occur and thus landscapes would not evolve and the rock cycle would not take place.

Changes that take place on Earth also reflect complex interactions among geologic and biological phenomena. For example, photosynthetic organisms affect the composition of the atmosphere by providing oxygen, and atmospheric composition in turn determines the nature of chemical weathering in rocks. For purposes of discussion, we refer to the global interconnecting web of physical and biological phenomena on Earth as the **Earth System,** and we define **global change** as the transformations or modifications of both physical and biological components of the Earth System through time.

Geologists distinguish among different types of global change, based on the rate or way in which change progresses with time. Gradual change takes place over long periods of geologic time (millions to billions of years); catastrophic change takes place relatively rapidly in the context of geologic time (seconds to millennia). Unidirectional change involves transformations that never repeat;

cyclic change repeats the same steps over and over, though not necessarily with the same results. Some types of cyclic change are periodic in that the cycles happen with a definable frequency.

In this chapter, we begin by reviewing examples of global change involving phenomena discussed earlier in the book. Then we look at an example of a **biogeochemical cycle,** the cyclic exchange of chemicals among living and nonliving reservoirs, for some kinds of global change are due to changes in the proportions of chemicals held in different reservoirs through time. Finally, we focus on **global climate change,** the transformations or modifications in Earth's climate through time. We conclude this chapter, and the book, by considering hypotheses that describe the ultimate global change—the end of the Earth.

19.2 UNIDIRECTIONAL CHANGES

The Evolution of the Solid Earth

Recall from Chapter 1 that Earth began as a fairly homogenous mass, formed by the coalescence of planetesimals. But the homogeneous proto-Earth did not last long—within about one hundred million years of its birth the planet began to melt, yielding liquid iron alloy that sank rapidly to the center to form the core (▶Fig. 19.2a, b). This process of differentiation represents major unidirectional change: it produced a layered, onionlike planet, with an iron-alloy core surrounded by a rocky mantle.

Soon after differentiation, a Mars-sized proto-planet appears to have collided with the newborn Earth. This collision caused a catastrophic change—a significant portion of the Earth and the colliding object fragmented and vaporized, creating a ring of debris that coalesced to form the Moon (▶Fig. 19.2c–e). After the collision, the Earth's mantle was probably partially molten, and its surface became a sea of magma. The Earth continued to endure intense bombardment by asteroids and comets until about 3.9 Ga, so any crust that had formed prior to 3.9 Ga was largely pulverized or melted. Eventually, however, bombardment ceased and our planet gradually cooled, permitting a crust to form at its surface and plate tectonics to begin operating. Over time, igneous activity produced lasting continental crust.

The Evolution of the Atmosphere and Oceans

Like its surface, the Earth's atmosphere has also changed over time. Volcanic activity released large quantities of gases. More gases may have arrived when comets collided with our new planet. Eventually, Earth accumulated an early atmosphere composed dominantly of carbon dioxide (CO_2) and

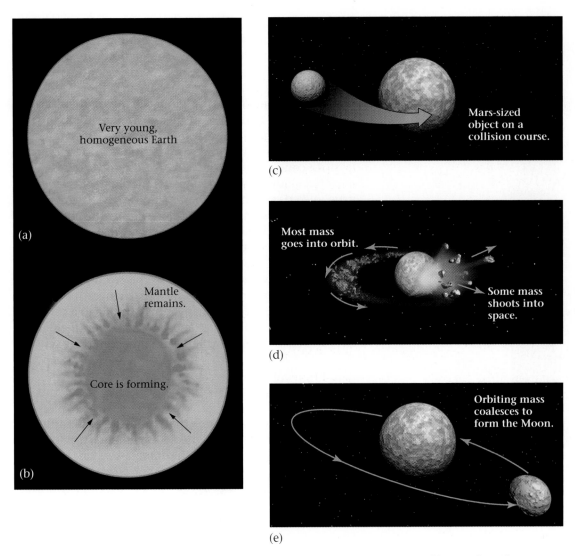

FIGURE 19.2 (a) When it first formed, the Earth was probably homogeneous. (b) Soon thereafter, the iron in the Earth melted and sank to the center. When this differention, a unidirectional change, was complete, the Earth had a distinct core and mantle. (c–e) Then a Mars-sized body collided with Earth, sending off fragments that coalesced to form the Moon. All these phenomena radically changed the Earth.

water (H_2O). Other gases, such as nitrogen (N_2), composed only a minor proportion of the early atmosphere. When the Earth's surface cooled, however, water condensed and fell as rain, collecting in low areas to form oceans. This may have first happened before 4.0 Ga, but had certainly happened by 3.8 Ga. Gradually, CO_2 dissolved in the oceans and was absorbed by chemical-weathering reactions on land, so its concentration in the atmosphere decreased. Nitrogen, which doesn't react with other chemicals, was left behind. Thus, the atmosphere's composition changed to become dominated by nitrogen. Photosynthetic organisms appeared by 3.2 Ga. But it probably wasn't until after 2.5 Ga that oxygen (O_2) became a significant proportion of the atmosphere.

The Evolution of Life

During most of the Hadean Eon, Earth's surface was probably lifeless, for carbon-based organisms could not survive the high temperatures of the time. The fossil record indicates that life had appeared at least by 3.8 billion years ago and has undergone unidirectional change (evolution) in fits and starts ever since (see Interlude D). Though simple organisms such as archaea and bacteria still exist, life evolution during the late Proterozoic and early Phanerozoic yielded multicellular plants and animals (▶Fig. 19.3). Life now inhabits regions from a few kilometers below the surface to a few kilometers above, yielding a diverse biosphere.

FIGURE 19.3 New species of life have evolved through geologic time. Though some of the simplest still exist, more complex organisms have appeared more recently.

19.3 PHYSICAL CYCLES

The Supercontinent Cycle

During Earth history, the map of the planet's surface has constantly changed. At times, almost all continental crust merged to form a single supercontinent, but usually the crust is distributed among several smaller continents. The process of change during which supercontinents form and later break apart is the **supercontinent cycle.** Geologists have found evidence that at least three or four times during the past three billion years of Earth history, supercontinents existed. The most recent one, Pangaea, formed at the end of the Paleozoic Era.

The Sea-Level Cycle

Global sea level rose and fell by as much as 300 m during the Phanerozoic, and likely did the same in the Precambrian. When sea level rises, the shoreline migrates inland, and low-lying plains in the continents become submerged. During periods of particularly high sea level, more than half of Earth's continental area can be covered by shallow seas; at such times, sediment buries continental regions, thereby changing their surface. When sea level falls, the continents become dry again, and regional unconformities develop. For example, the sedimentary strata of the midwestern United States record at least six continent-wide advances and retreats of the sea.

After studying stratigraphy around the world, geologists have pieced together a chart defining the succession of global transgressions and regressions during the Phanerozoic Eon. This global sedimentary cycle chart may largely reflect the cycles of **eustatic** (worldwide) **sea-level change** (▶Fig. 19.4). However, the chart probably does not give us an exact image of sea-level change, because the sedimentary record reflects other factors as well, such as changes in sediment supply. Eustatic sea-level changes may be due to a variety of factors, including advances and retreats of continental glaciers and changes in the volume of mid-ocean ridge systems.

The Rock Cycle

We learned early in this book that the crust of the Earth consists of three rock types: igneous, sedimentary, and metamorphic. Atoms making up the minerals of one rock type may later become part of another rock type. This process is the rock cycle. Each stage in the rock cycle changes the Earth by redistributing and modifying material.

FIGURE 19.4 To some extent, the transgressions and regressions indicated by this sedimentary sequence chart reflect global (eustatic) sea-level change. Here, two versions of the sequence chart are shown, each produced by a different author—the shape of the curve is still a subject for research.

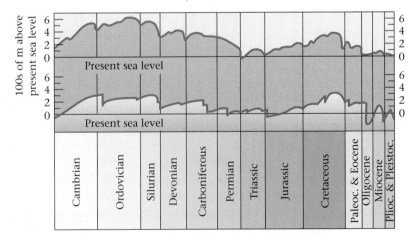

19.4 BIOGEOCHEMICAL CYCLES

A **biogeochemical cycle** involves the passage of a chemical among nonliving and living reservoirs in the Earth System, mostly on or near the surface. Nonliving reservoirs include the atmosphere, the crust, and the ocean; living reservoirs include plants, animals, and microbes. Although a great variety of chemicals participate in biogeochemical cycles, here we look at only two: water (H_2O) and carbon (C).

Some stages in a biogeochemical cycle may take only hours, some may take thousands of years, and others may take millions of years. Because chemicals can cycle rapidly, the transfer of a chemical from reservoir to reservoir during these cycles doesn't really seem like a change in the Earth in the way that the movement of continents or the metamorphism of rock seems like a change. In fact, for intervals of time, biogeochemical cycles attain a steady-state condition, meaning that the proportions of a chemical in different reservoirs remain fairly constant even though there is a constant flux (flow) of the chemical among reservoirs. When we speak of global change in a biogeochemical cycle, we mean a change in the relative proportions of a chemical held in different reservoirs at a given time.

The Hydrologic Cycle

As we learned in Interlude E, the hydrologic cycle involves the movement of water from reservoir to reservoir on or near the surface of the Earth. The hydrologic cycle is an example of a biogeochemical cycle, in that a chemical (H_2O) passes through both nonliving and living entities—the oceans, the atmosphere, surface water, groundwater, glaciers, soil, and living organisms. Global change in the hydrologic cycle occurs when a change in global climate alters the ratio between the amount of water held in the ocean and the amount held in continental ice sheets. For example, during an ice age, water that had been stored in oceans moves into glacial reservoirs.

The Carbon Cycle (the Movement of a Greenhouse Gas)

Most carbon in the near-surface realm of Earth originally bubbled out in the form of CO_2 gas released at volcanoes (▶Fig. 19.5). Once it enters the atmosphere, it can be removed in various ways. Some dissolves in seawater to form bicarbonate (HCO_3^-) ions, whereas some is absorbed by photosynthetic organisms that convert it into sugar and other organic chemicals. This carbon enters the food chain and ultimately makes up the flesh, fat, and sinew of animals. Some of the reactions that take place when rock undergoes chemical weathering incorporate atmospheric CO_2, and thus also remove carbon from the atmosphere.

Some carbon returns directly to the atmosphere, in the form of CO_2 or methane (CH_4), by the respiration of animals, by the flatulence of animals, or by the decay of dead organisms. But some can be stored for long periods of time in fossil fuels (oil and coal), in organic shale, in methane hydrates, or in limestone. But this carbon either returns to the atmosphere in the form of CO_2, as a result of the burning of fossil fuels

FIGURE 19.5 In the carbon cycle, carbon transfers among various reservoirs at or near the Earth's surface. Red arrows indicate release to the air, and green arrows indicate absorbtion from air.

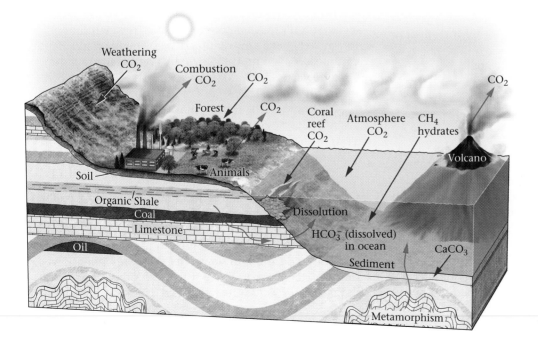

and the metamorphism of rocks containing carbonate, or returns to the sea after undergoing dissolution in river water or groundwater.

The concentrations of carbon dioxide and methane in the atmosphere play an essential role in controlling Earth's climate because these gases, along with several other trace gases (such as water), are **greenhouse gases,** meaning that they prevent heat from escaping into space from the atmosphere, much as glass traps heat in a greenhouse. An increase in their concentration warms the atmosphere, whereas a decrease cools it down.

19.5 GLOBAL CLIMATE CHANGE

How often have you seen a newspaper proclaim "Record High Temperatures!" Does this mean that the climate, the average range of weather conditions for a given region, is changing? Is it something we should be worried about? Perhaps. Global climate change, the transformations or modifications in Earth's climate through time, is indeed important because it affects sea level and the distribution and character of climatic belts and, therefore, the distribution of habitats, agricultural lands, and landscapes. A rise in the average global near-surface atmospheric temperature of only a few degrees might melt enough polar ice to cause a sea-level rise that would in turn flood coastal population centers and also cause devastating droughts.

For purposes of discussion, we distinguish between long-term climate change, which takes place over millions of years, and short-term climate change, which takes place over tens of thousands of years. If the average atmospheric and sea-surface temperature rises, we have **global warming;** if it falls, we have **global cooling.** Some changes are great enough to cause oceanic islands and large regions of continents to be submerged by shallow seas, or to be covered by ice. Others are subtle, creating only a slight latitudinal shift in vegetation belts and a sea-level change measured in meters or less.

Methods of Study

Geologists and climatologists are working hard to define the kinds of climate changes that can occur, the rates at which these changes take place, and the effects they may have on society. There are two basic approaches to studying global climate change: (1) researchers measure past climate change, as indicated by the stratigraphic record, to document the magnitude of change that is possible and the rate at which such change occurred; (2) researchers make computer programs to calculate how factors such as atmospheric composition, topography, ocean currents, and Earth's orbit affect the climate. The resulting global climate models provide insight

into when and why changes took place in the past and whether they will happen in the future.

Let's look first at how geologists study the **paleoclimate** (past climate), so as to document climate changes throughout Earth history. Any feature whose character depends on the climate and whose age can be determined serves as a clue to defining paleoclimate. Here are some examples:

- *The stratigraphic record:* The nature of sedimentary strata deposited at a certain location reflects the climate at that location. For example, a bed of coal represents deposits of a warm climate, whereas a bed of till represents deposits of a glacial climate.

- *Paleontological evidence:* Different assemblages of species survive in different climatic belts. Thus, the succession of species in a sedimentary sequence provides clues to the changes in climate at that site. For example, a record of short-term climate change can be obtained by studying the succession of plankton fossils in sea-floor sediments, for cold-water species of plankton are different from warm-water species. Fossil plant pollen preserved in the mud of bogs also provides information about paleoclimate, because different plant species live in different climates (▶Fig. 19.6a, b). For example, studies show that spruce forests, indicative of cool climates, have slowly migrated north since the ice age (▶Fig. 19.6d, e).

- *Oxygen-isotope ratios:* Two isotopes of an element have the same atomic number but different atomic weights (see Appendix). Geologists have found that the ratio of ^{18}O to ^{16}O in glacial ice indicates the atmospheric temperature in which the snow that made up the ice formed: the ratio is larger in snow that forms in warmer air, but smaller in snow that forms in colder air. Because of this relationship, the isotope ratio measured in a succession of ice layers in a glacier indicates temperature change through time. Researchers have now obtained ice cores down to a depth of almost 3.3 km in Antarctica; this record spans about 740,000 years. Similar data has been obtained from cores in Greenland (▶Fig. 19.7a). For a number of reasons, the $^{18}O/^{16}O$ ratio in the $CaCO_3$ making up plankton shells also gives geologists an indication of past temperatures. Thus, measurement of oxygen-isotope ratios in drill cores of marine sediment extends the record of temperature change back through millions of years (▶Fig. 19.7b, c).

- *Growth rings:* If you've ever looked at a tree stump, you'll have noticed the concentric rings visible in the wood. Each ring represents one year of growth, and the thickness of the ring indicates the rate of growth in a given year. Trees grow faster during warmer, wetter years and more slowly during cold, dry years. Thus, the succession of ring widths provides an easily calibrated record of climate during the lifetime of the tree. Growth rings in corals and shells can provide similar information.

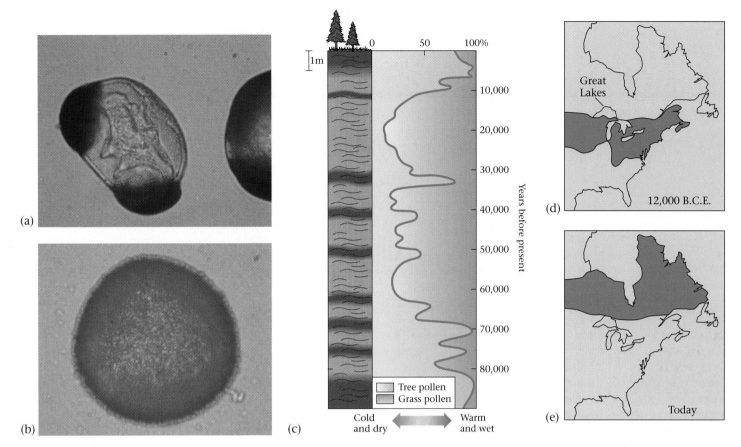

FIGURE 19.6 Changes in the assemblage of pollen in sediment indicate a shift in climate belts. (a) Spruce pollen from a cold-climate coniferous forest. (b) Hemlock pollen from a warm-climate deciduous forest. (c) This model shows how the proportion of tree pollen relative to grass pollen can change in a sedimentary sequence through time. Tree pollen indicates cooler and drier conditions, whereas grass pollen indicates warmer and wetter conditions. (d) Pollen data suggest that about 12,000 years ago, spruce forests (green areas) lay south of the Great Lakes. (e) Today, they are found north of the Great Lakes.

Long-Term Climate Change

Using the variety of techniques described above, as well as others, geologists have reconstructed an approximate record of global climate, represented by mean temperature and rainfall, for geologic time. The record shows that at some times in the past, the Earth's atmosphere was significantly warmer than it is today, whereas at other times it was significantly cooler. The warmer periods have come to be known as **greenhouse** (or hothouse) **periods** and the colder as **icehouse periods.** (The more familiar term, *ice age,* refers to the times during an icehouse period when the Earth was cold enough for ice sheets to advance and cover substantial areas of the continents.) As the chart in ▶Figure 19.8 shows, there have been at least five major icehouse periods during geologic history.

What caused long-term global climate change? The answer may lie in the complex relationships among the vari-ous geologic and biogeochemical cycles of the Earth system, as described earlier. Following are some likely influences on long-term global change:

- *Positions of continents:* Continental drift influences the climate by controlling the pattern of oceanic currents, which redistribute heat around the planet's surface. Drift also determines whether the land is at high or low latitudes (and thus how much solar radiation strikes it), and whether or not there are large continental interior regions where extremely cold winter temperatures can develop.

- *Volcanic activity:* A long-term global increase in volcanic activity may contribute to long-term global warming, because it increases the concentration of greenhouse gases in the atmosphere.

- *The uplift of land surfaces:* Tectonic events that lead to the long-term uplift of the land affect atmospheric CO_2

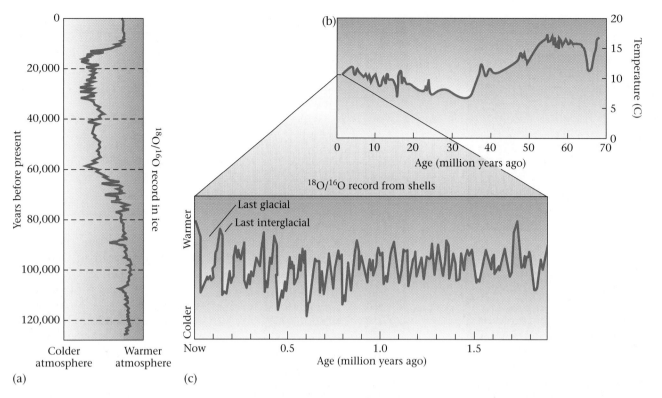

FIGURE 19.7 (a) The $^{18}O/^{16}O$ ratio in a 2-km-long ice core drilled in the ice cap of Greenland varies with depth in the core, indicating that atmospheric temperature varies over time. Smaller ratios mean a colder atmosphere. (b) The ratio of ^{18}O to ^{16}O in the calcite of fossil plankton shells of deep-marine sediment shows variations in temperature for the past 70 million years. (c) The detailed plankton record of temperature for the past 2 million years. The decreases in the $^{18}O/^{16}O$ ratio correspond with glacial advances.

concentration, because such events expose land to weathering, and chemical-weathering reactions absorb CO_2. Such uplift will also affect atmospheric circulation and rainfall rates.

- *The formation of coal, oil, or organic shale:* At various times during Earth history, environments suitable for coal or oil formation have been particularly widespread. Such formation removes CO_2 from the atmosphere and stores it underground.

- *Life evolution:* The appearance of, or extinction of, certain life forms may have impacted climate significantly by removing CO_2 from the atmosphere.

Short-Term Climate Change

The record of the past million years gives a sense of the magnitude and duration of short-term climate change. During this period, there have been about five major and twenty to thirty minor episodes of glaciation, separated by interglacial periods. If we focus on the last 15,000 years, we see that overall the temperature has increased, but there are still notable ups and downs (▶Fig. 19.9a).

As a result of the warming that began 15,000 years ago, the glaciers retreated for about 4,500 years. Then there was a return to colder conditions for a few more thousand years. This interval of cooler temperature is named the Younger Dryas, after an Arctic flower that became widespread during the time. The climate then warmed, reaching a peak at 5,000 to 6,000 years ago, a period called the Holocene maximum, when average temperatures were about 2°C above temperatures of today (▶Box 19.1). This warming peak led to increased evaporation and therefore precipitation, making the Middle East unusually wet and fertile—conditions that may partially account for the rise of civilization in this part of the world.

The temperature dipped to a low about 3,000 years ago, before returning to a high during the Middle Ages, a time called the Medieval warm period. During this time, when temperatures were 0.5 to 0.8° above those of today, Vikings established self-supporting agricultural settlements along the coast of Greenland (▶Fig. 19.9b). The temperature dropped

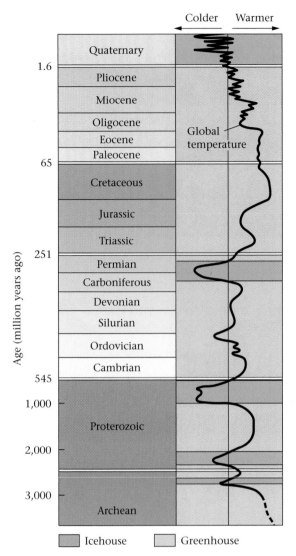

Colder — Warmer

Global temperature

Icehouse   Greenhouse

FIGURE 19.8 The chart shows the timing of icehouse and greenhouse (or hothouse) periods during Earth history.

again from 1500 to about 1800, a period known as the little ice age, when Alpine glaciers advanced and the canals of the Netherlands froze over in winter (▶Fig. 19.9c, d). Overall, the climate has warmed since the end of the little ice age, and today it is as warm as it was during the medieval warm period.

Geologists have suggested that several factors explain short-term climate change. These include:

- *Fluctuations in solar radiation and cosmic rays:* The amount of energy produced by the Sun varies with the sunspot cycle. This cycle involves the appearance of large numbers of sunspots (black spots thought to be magnetic storms on the Sun's surface) about every 9 to 11.5 years. There may be longer-term cycles that have not yet been identified. This variation in energy may affect the climate. Recently, some researchers have speculated that changes in the rate of influx of cosmic rays may affect climate, perhaps by generating clouds. Specifically, recent research suggests that cosmic rays striking the atmosphere produce clusters of ions that serve as condensation nuclei around which water molecules congregate, thus forming the mist droplets making up clouds.

- *Changes in Earth's orbit and tilt:* As Milankovitch first recognized in 1920, the change in the tilt of Earth's axis, the Earth's precession cycle, and changes in the eccentricity of its orbit together cause the amount of summer heat in high latitudes to vary, and cause the overall amount of heat reaching Earth to vary (see Chapter 18). These changes correlate with observed ups and downs in atmospheric and oceanic temperature.

- *Changes in volcanic emissions:* Not all of the sunlight that reaches the Earth penetrates its atmosphere and warms the ground. Some is reflected by the atmosphere. The degree of reflectivity, or albedo, of the atmosphere increases not only if cloud cover increases, as we have

Global Climate Change and the Birth of Legends

BOX 19.1
THE HUMAN ANGLE

Some geologists argue that myths passed down from the early days of civilization may have their roots in global climate change. For example, recent evidence suggests that earlier than 7,600 years ago, the region that is now the Black Sea contained a much smaller freshwater lake surrounded by settlements. At the end of the most recent ice-age glacial advance, the ice sheets melted and sea level rose, and the Mediterranean eventually broke through a dam at the site of the present Bosporus Strait. Researchers suggest that seawater from

the Mediterranean spilled into the Black Sea basin via a waterfall two hundred times larger than Niagara Falls. This influx of water caused the lake level to rise by perhaps 10 cm per day. Within a year, 155,000 square km (60,000 square miles) of populated land had become submerged beneath hundreds of meters of water. This flooding presumably led to a huge human migration. Its timing has suggested to some researchers that it may have inspired the Babylonian *Epic of Gilgamesh* (c. 2000 B.C.E.) and, later, the biblical epic of Noah's Ark.

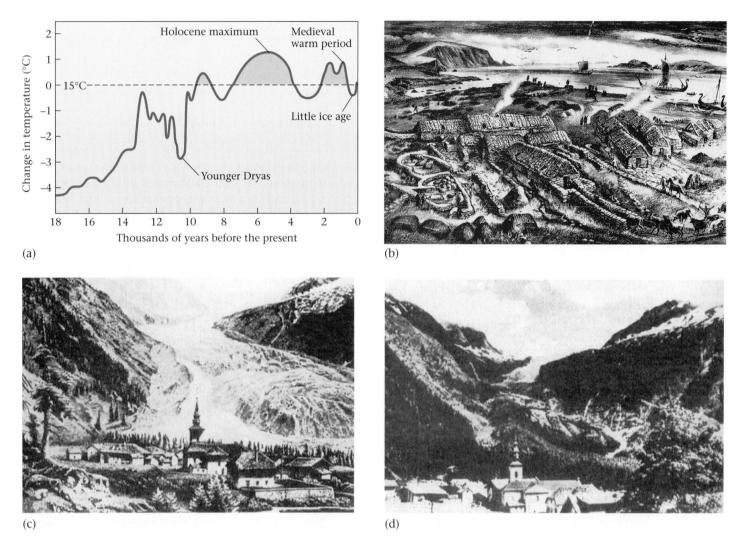

FIGURE 19.9 (a) The past 15,000 years (the Holocene Epoch) experienced several periods of warming and cooling. (b) Coastal Greenland was settled by the Vikings during one of the warmer periods, when the region could support agriculture. (c) Glaciers that formed during the little ice age persisted into the nineteenth century. Here we see a glacier in the French Alps as it appeared in 1850. (d) By the second half of the twentieth century (1966), the glacier had almost disappeared.

seen, but also if the concentration of volcanic aerosols (microscopic liquid droplets) in the atmosphere increases.

- *Changes in ocean currents:* Recent studies suggest that the configuration of currents can change quite quickly, and that this configuration affects the climate.

- *Changes in surface albedo:* Regional-scale changes in the nature of continental vegetation cover, and/or in the proportion of snow and ice on our planet's surface, would affect our planet's albedo. Increasing albedo causes cooling, whereas decreasing albedo causes warming.

- *Abrupt changes in concentrations of greenhouse gases:* A sudden change in greenhouse gas concentration in the atmosphere could affect climate. One such change might happen if some of the methane hydrate that crystallized

in sediment of the sea floor suddenly melted, releasing CH_4 to the atmosphere. Algal blooms and reforestation (or deforestation) conceivably could also change CO_2 concentrations significantly.

Catastrophic Climate Change and Mass-Extinction Events

The changes that happen on Earth almost instantaneously are called catastrophic changes. For example, a volcanic eruption, an earthquake, a tsunami, or a landslide can change a local landscape in seconds or minutes. But such events affect only relatively small areas. Can such catastrophes happen on a global scale? In the past decades, geoscientists have come to the conclusion that the answer is yes. The stratigraphic record

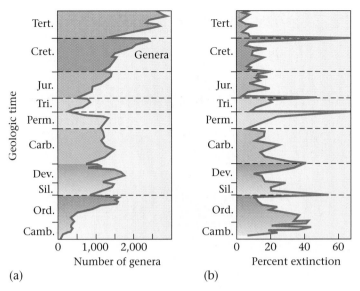

FIGURE 19.10 (a) During a mass-extinction event, biodiversity on Earth, as indicated by the number of fossil species (grouped together in genera), suddenly decreases. (b) Paleontologists can calculate the percentage of species that became extinct during a given event.

shows that Earth history includes several **mass-extinction events,** when large numbers of species abruptly vanished (▶Fig. 19.10a). Some of these define boundaries between geologic periods. A mass-extinction event decreases the biodiversity (the number of different species that exist at a given time) of life on Earth (▶Fig. 19.10b). The most notable examples occurred at the end of the Permain and at the end of the Cretaceous.

Geologists speculate that some mass-extinction events reflect a catastrophic change in the planet's climate, brought about by unusually voluminous volcanic eruptions or by the impact of a comet or asteroid with the Earth (▶Fig. 19.11). Either of these events could eject enough debris into the atmosphere to block sunlight (see Chapter 11). Without the warmth of the Sun, winterlike or nightlike conditions would last for weeks to years, long enough to disrupt the food chain. Either event, in addition, could eject aerosols that would turn into global acid rain, scatter hot debris that would ignite forest fires, or give off chemicals that, when dissolved in the ocean, would make the ocean either toxic or so nutritious that oxygen-consuming algae could thrive.

FIGURE 19.11 The K-T (Cretaceous-Tertiary) boundary event may have been caused by the impact of a large comet or asteroid. Here, we see an artist's rendering of the impact and its aftermath.

19.6 ANTHROPOGENIC CHANGES IN THE EARTH SYSTEM

During the Stone Age, the human population worldwide was less than 10 million. By the dawn of civilization, 4000 B.C.E., it was still, at most, a few tens of millions. But by the beginning of the nineteenth century, revolutions in industrial methods, agriculture, medicine, and hygiene had substantially lowered death rates and raised living standards, so that the population began to grow at accelerating rates—it took tens of thousands of years to grow from Stone Age populations to 1 billion people worldwide in 1850, but it took only eighty years to double again, reaching 2 billion in 1930. Now, the doubling time is only forty-four years, and the human population passed the 6 billion mark just before the year 2000 (▶Fig. 19.12).

As the population grows and the standard of living improves, *per capita* usage of resources increases; we use land for agriculture and grazing, forests for wood, rock and dirt for construction, oil and coal for energy or plastics, and ores for metals. Without a doubt, our usage of resources has affected the Earth System profoundly, and thus humanity has become a major agent of global change. Here, we examine some of these anthropogenic impacts.

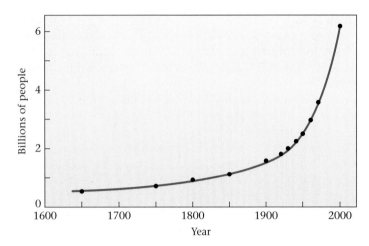

FIGURE 19.12 The human population has increased dramatically during the past two centuries. Currently, it doubles every forty-four years.

(a)

(b)

FIGURE 19.13 (a) A giant, power-driven shovel can move more dirt and rock in a day than a stream can move in a decade. (b) Agriculture and development radically change the landscape of a region. Prairies of the midwestern United States have been replaced by giant farm fields.

The Modification of Landscapes

Every time we pick up a shovel and move a pile of dirt, we redistribute a portion of the Earth's crust. In the last century, the pace of Earth movement has accelerated, for now we have shovels in coal mines that can move 300 cubic meters of coal in a single scoop, trucks that can carry 200 tons of ore in a single load, and tankers that can transport 50 million liters of oil during a single journey. In North America, human activity now moves more sediment each year than rivers do. The extraction of rock during mining, the building of levees and dams along rivers or of sea walls along the coast, and the construction of highways and cities all involve the redistribution of Earth materials (▶Fig. 19.13a). In addition, people clear and plow fields, drain and fill wetlands, and pave over the land surface (▶Fig. 19.13b; see also chapter-opening photo). All these activities change the landscape.

The Modification of Ecosystems

In undisturbed areas, the **ecosystem** of a region (an interconnected network of organisms and the physical environment in which they live) is the product of evolution for an extended period of time. The ecosystem's flora (plant life) include species that have adapted to living together in that particular climate and on the substrate available, while its fauna (animal life) can survive local climate conditions and utilize local food supplies. Human-caused deforestation, overgrazing, agriculture, and urbanization disrupt ecosystems and lead to a decrease in biodiversity.

Archaeological studies have found that the earliest example of human modification of an ecosystem occurred in the Stone Age, when hunters played a major role in causing the mass extinction of many species of large mammals (mammoths, giant sloths, giant bears). Today, less than 5% of Europe retains its original habitats. The same number can be applied to the eastern United States, which lost its original forest and prairie. Tropical rain forests cover less than about half the area worldwide that they covered before the dawn of civilization, and they are disappearing at a rate of about 1.8% per year (▶Fig. 19.14a, b). Much of this loss comes from slash-and-burn agriculture during which farmers and ranchers destroy forest to make open land for farming and grazing (▶Fig. 19.14c). Unfortunately, the heavy rainfall of tropical regions removes nutrients from the soil, making the soil useless in just a few years. Overgrazing by domesticated animals can remove vegetation so completely that some grasslands have undergone desertification. And urbanization replaces the natural land surface with concrete or asphalt, a process that completely destroys an ecosystem and radically changes the amount of rain that infiltrates the land surface to become groundwater.

Pollution

The environment has always contained contaminants such as soot, dust, and the byproducts of organisms. But when human populations grew, urbanization, industrial and agricultural activity, the production of electricity, and modern modes of transportation greatly increased both the quantity and diversity of contaminants that entered the air, surface water, and groundwater. These contaminants, or **pollution,** include both natural and synthetic materials (in liquid, solid, or gaseous form). They have become a problem because they are produced at a higher rate than can be naturally absorbed or modified by the Earth System. For example, although small quantities of sewage can be absorbed by

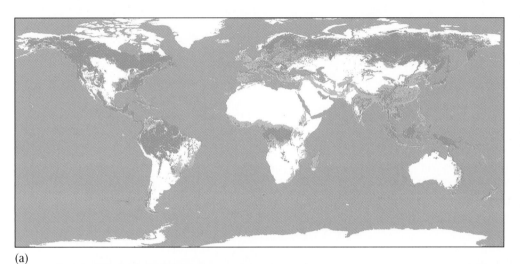

(a)

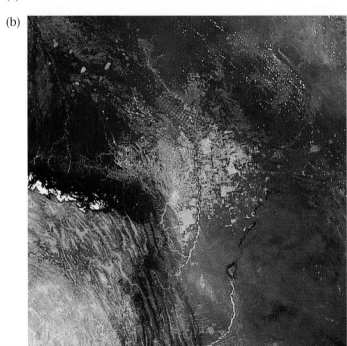

(b)

(c)

FIGURE 19.14 (a) The area of the Earth covered by rain forest shrank steadily during the past century. On this map, the dark green areas indicate existing forest (including high-latitude scrub forest), while the light green areas indicate regions that were forested 8,000 years ago. Note how tropical rain forests are shrinking. (b) A satellite image of Bolivia showing forest (dark green) being replaced by fields (light yellow). (c) Much of the loss is due to slash-and-burn agriculture.

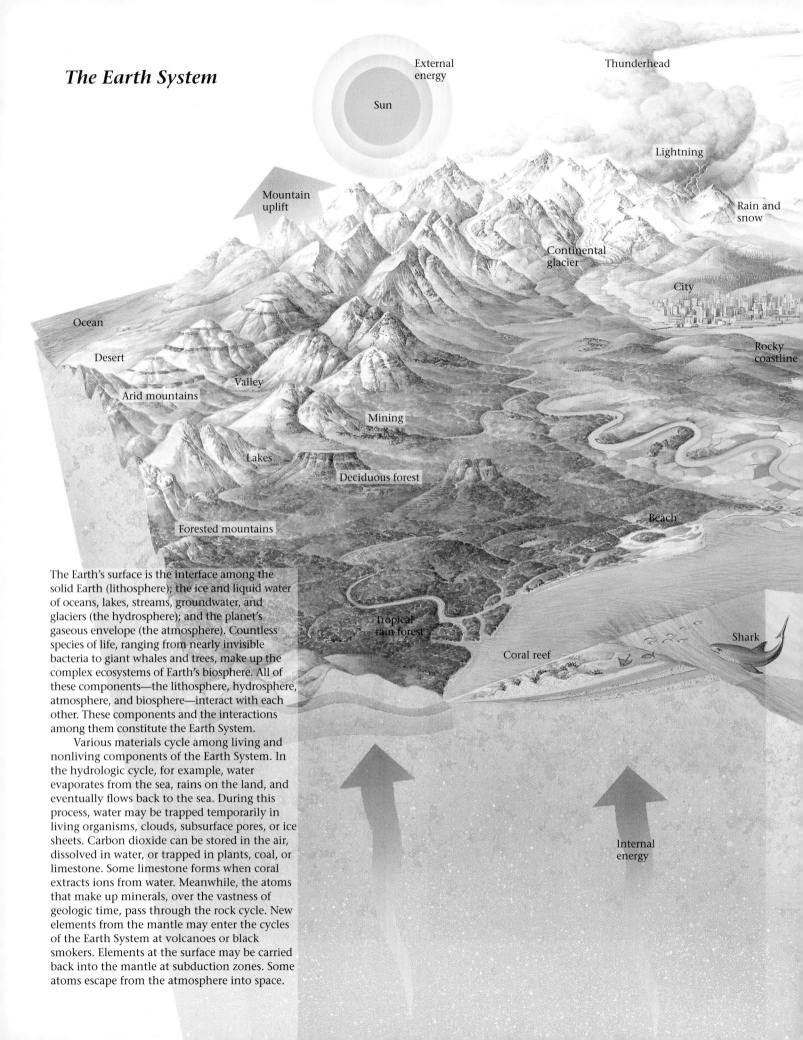

The Earth System

External energy

Sun

Thunderhead

Lightning

Mountain uplift

Rain and snow

Continental glacier

City

Ocean

Rocky coastline

Desert

Valley

Arid mountains

Mining

Lakes

Deciduous forest

Beach

Forested mountains

The Earth's surface is the interface among the solid Earth (lithosphere); the ice and liquid water of oceans, lakes, streams, groundwater, and glaciers (the hydrosphere); and the planet's gaseous envelope (the atmosphere). Countless species of life, ranging from nearly invisible bacteria to giant whales and trees, make up the complex ecosystems of Earth's biosphere. All of these components—the lithosphere, hydrosphere, atmosphere, and biosphere—interact with each other. These components and the interactions among them constitute the Earth System.

Various materials cycle among living and nonliving components of the Earth System. In the hydrologic cycle, for example, water evaporates from the sea, rains on the land, and eventually flows back to the sea. During this process, water may be trapped temporarily in living organisms, clouds, subsurface pores, or ice sheets. Carbon dioxide can be stored in the air, dissolved in water, or trapped in plants, coal, or limestone. Some limestone forms when coral extracts ions from water. Meanwhile, the atoms that make up minerals, over the vastness of geologic time, pass through the rock cycle. New elements from the mantle may enter the cycles of the Earth System at volcanoes or black smokers. Elements at the surface may be carried back into the mantle at subduction zones. Some atoms escape from the atmosphere into space.

Tropical rain forest

Coral reef

Shark

Internal energy

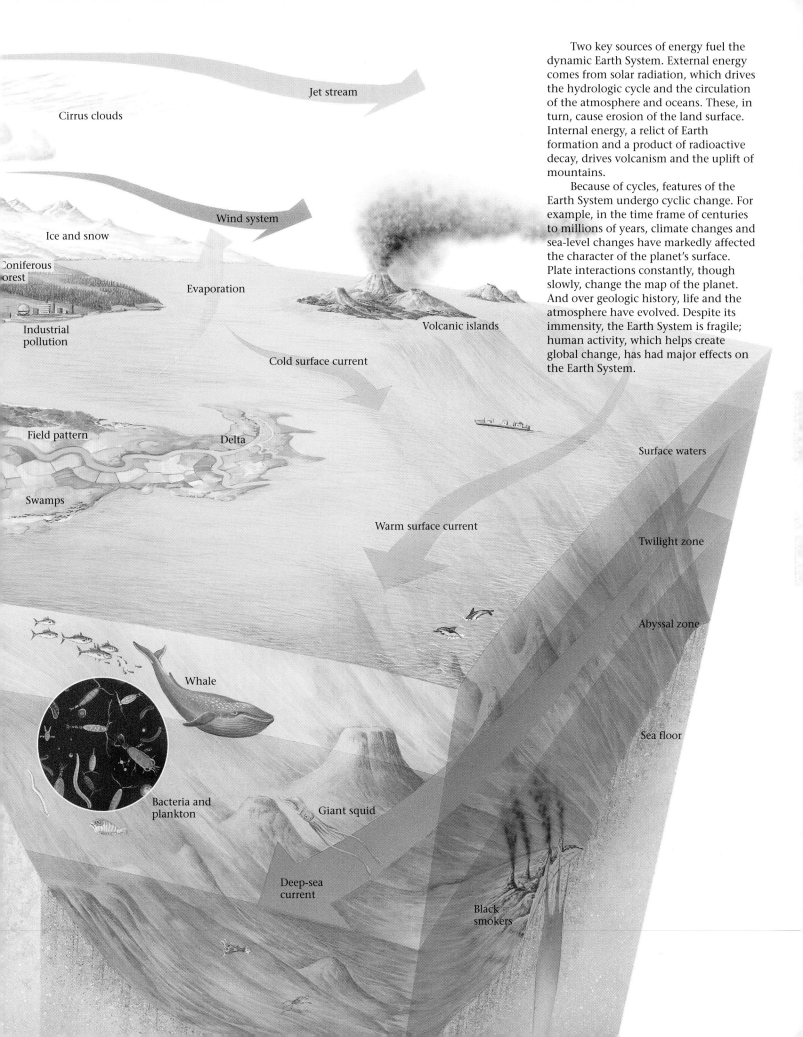

Jet stream

Cirrus clouds

Ice and snow

Coniferous forest

Wind system

Industrial pollution

Evaporation

Volcanic islands

Cold surface current

Field pattern

Delta

Surface waters

Swamps

Warm surface current

Twilight zone

Abyssal zone

Whale

Sea floor

Bacteria and plankton

Giant squid

Deep-sea current

Black smokers

Two key sources of energy fuel the dynamic Earth System. External energy comes from solar radiation, which drives the hydrologic cycle and the circulation of the atmosphere and oceans. These, in turn, cause erosion of the land surface. Internal energy, a relict of Earth formation and a product of radioactive decay, drives volcanism and the uplift of mountains.

Because of cycles, features of the Earth System undergo cyclic change. For example, in the time frame of centuries to millions of years, climate changes and sea-level changes have markedly affected the character of the planet's surface. Plate interactions constantly, though slowly, change the map of the planet. And over geologic history, life and the atmosphere have evolved. Despite its immensity, the Earth System is fragile; human activity, which helps create global change, has had major effects on the Earth System.

clay minerals in the soil or destroyed by bacterial metabolism, large quantities overwhelm natural controls and can accumulate into destructive concentrations. Further, because many contaminants are not produced in nature, they are not easily removed by natural processes. Pollution of the Earth System is a type of global change, because it redistributes and reformulates materials. Some key problems associated with this change include the following.

- *Smog:* The term was originally coined to refer to the dank, dark air that resulted when smoke from the burning of coal mixed with fog in London and other industrial cities. Another kind of smog, called photochemical smog, is the ozone-rich brown haze that blankets cities when exhaust from cars and trucks reacts with air in the presence of sunlight.

- *Water contamination:* We dump a great variety of chemicals into surface water and groundwater. Examples include gasoline, other organic chemicals, radioactive waste, acids, fertilizers—the list could go on for pages. These chemicals affect biodiversity.

- *Acid runoff:* Dissolution of sulfide-containing minerals in ores or coal by groundwater or stream water makes the water acidic and toxic to life forms.

- *Acid rain:* When rain passes through air that contains sulfur-containing aerosols (emitted from power plants), the water dissolves the sulfur, creating sulfuric acid, or **acid rain.** Wind can carry aerosols far from a power plant, so acid rain can damage a broad region (▶Fig. 19.15).

- *Radioactive materials:* Nuclear weapons, nuclear energy, and medical waste transfer radioactive materials from rock to Earth's surface environment.

- *Ozone depletion:* When emitted into the atmosphere, human-produced chemicals, most notably chlorofluorocarbons (CFCs), react with ozone in the stratosphere. This reaction, which happens most rapidly on the surfaces of tiny ice crystals in polar stratospheric clouds, destroys ozone molecules, thus creating an **ozone hole** over high-latitude regions, particularly during the spring (▶Fig. 19.16). Note that the "hole" is not really an area where no ozone is present, but rather is a region where atmospheric ozone has been reduced substantially. Ozone holes have dangerous consequences, for they affect the ability of the atmosphere to shield the Earth's surface from harmful ultraviolet radiation.

Effects on the Climate: The Global-Warming Issue

During the past two centuries, industry, energy production, and agriculture have significantly altered the rate at which greenhouse gases, such as carbon dioxide (CO_2) and meth-

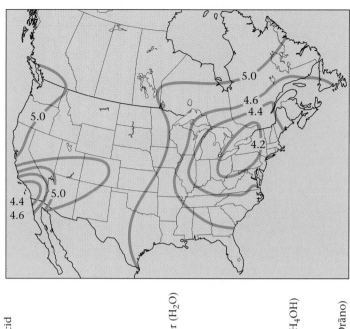

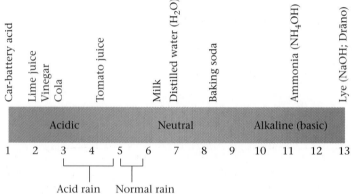

FIGURE 19.15 Acid rain has affected large portions of the United States. The map contours pH numbers, indicating the concentration of hydrogen ions in a solution (according to the formula pH = $-\log[H^+]$; the square brackets mean "concentration"). Note that very acidic rain falls in the Northeast and Southwest. The scale gives a sense of what the numbers mean. A solution with a pH of 7 is neutral; acidic solutions have a pH less than 7; alkaline solutions have a pH greater than 7.

ane (CH_4), are added to the atmosphere. In fact, the rate of addition has exceeded the rate at which these gases can be absorbed by dissolution in the ocean, by incorporation in plants, or by chemical-weathering reactions. Effectively, by burning fossil fuels at the rate that we do, we transfer CO_2 from underground reservoirs (oil and coal deposits) back into the atmosphere. In 1800, the mean concentration of CO_2 in the atmosphere was 280 parts per million (ppm), in 1900 it was 295 ppm, and by 2000 it had reached 370 ppm (▶Fig. 19.17). At the same time, the decay of organic material in rice paddies and the flatulence of cows have released enough methane to change the concentration of this organic chemical in the atmosphere.

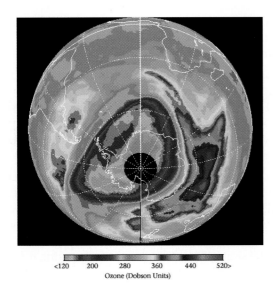

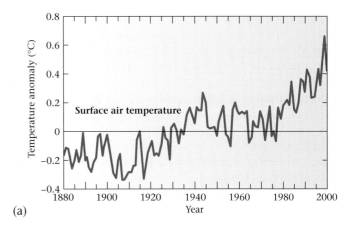

(a)

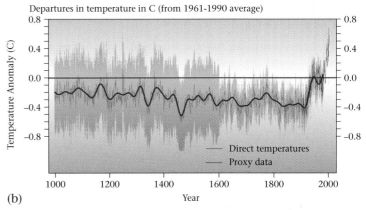

(b)

FIGURE 19.16 The colors on this map show the quantity of ozone in the atmosphere when ozone levels are at a minimum. A significant ozone hole has formed over the Antarctic region, where the amount of ozone has decreased by as much as 50%. The shape and size of the hole vary over time. One Dobson unit is the amount of ozone needed to make a 0.01-mm-thick layer at the surface of the Earth.

We might expect this increase in greenhouse gases to cause global warming, and indeed global mean temperature appears to have risen by about 0.9° between 1880 and 2000 (▶Fig. 19.18a, b). This may not seem like much, but by comparison, the magnitude of change between the last ice age and now was only a few degrees. Similarly, mean ocean temperature has been rising: in 2000, temperatures at the North Pole were the warmest in four centuries, and a 15-km-by-5-km patch of open water appeared at the pole. Summer ice in the Arctic Ocean may totally disappear in the next one hundred years.

FIGURE 19.18 (a) The average global surface atmospheric temperature varies year by year, but overall, there has been a noticeable increase since about 1920. Here, the difference in global temperature, relative to an arbitrary reference value, is plotted as a function of time. (b) A reconstruction of Northern Hemisphere temperature for the past 1,000 years, based on measurements of tree rings and ice cores. The 0° reference line is the 1920–1980 mean. Note that climate seemed to be cooling slowly between 1000 and 1900, but since 1900 it has increased substantially.

FIGURE 19.17 The concentration of CO_2 in air has been measured directly since 1958. Concentrations at earlier times can be determined by analyzing air bubbles trapped in glacier ice. A graph of variations in concentration through time shows that the concentration has increased dramatically since the Industrial Revolution.

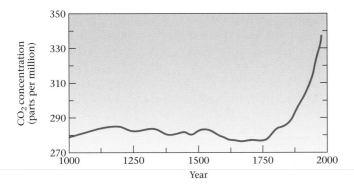

Though the large majority of researchers have concluded that global warming is really happening, not all do. Some argue that it would have happened anyway, in the context of natural cycles of short-term climate change. But of note, a recent study of ice-core data characterizing Earth's climate during the past few hundred thousand years suggests that the Earth's climate started on a cooling trend about 10,000 years ago and that, if only solar and volcanic phenomena controlled climate, we should now be heading into another ice age. Instead, according to this study, temperatures began to deviate from the cooling trend about 8,000 years ago, and we have followed an overall warming trend ever since. The timing of this warming trend suggests it results from human activity that produce greenhouse gases.

There is even more disagreement about whether global warming will continue in the future and what its consequences might be, because predictions depend on computer models, and not all researchers agree on how to represent the factors that affect climate in these models. In the worst-case scenario, global warming will continue into the future at the present rate, so that by 2050—within the lifetime of many readers of this book—the average annual temperature will have increased in some parts of the world by 1.5° to 2.0°C. At these rates, by 2150 global temperatures may be 5° to 11° warmer than present—the warmest since the Eocene Epoch, 40 million years ago. The effects of such a change are controversial, but according to some climate models, the following events might happen.

- *A shift in climate belts:* Temperate climates would move to higher latitudes, and vegetation belts would follow this trend (▶Fig. 19.19a, b).

- *Stronger storms:* An increase in average ocean temperatures would mean that more of the ocean could evaporate when a tropical depression passed over. This evaporation

would nourish stronger hurricanes. In nondesert areas, there might be more precipitation and, therefore, flooding.

- *A rise in sea level:* The melting of ice sheets in polar regions and the expansion of water in the sea as the water warms would make sea level rise enough to flood coastal wetlands and communities and damage deltas. Measurements suggest that there has been a rise of almost 12 cm in the past century (▶Fig. 19.20a, b).

- *An interruption of the oceanic heat conveyor:* Oceanic currents play a major role in transferring heat across latitudes. If global warming melts polar ice, the resulting freshwater would dilute surface ocean water at high latitudes. This water could not sink, and thus thermohaline circulation would be shut off (see Chapter 15), preventing it from conveying heat.

FIGURE 19.19 The distribution of climate belts, indicated by vegetation type, will change if the global temperature rises. (a) Distribution of vegetation types today. (b) Distribution of vegetation types if the climate warms by just a couple of degrees.

FIGURE 19.20 (a) Estimates of the average global sea level between 1880 and 1990 suggest that there has been a rise of about 12 cm. (b) All models of future sea-level change predict that there will be a rise. In the worst-case model, sea level could rise by almost 1 m in the next century.

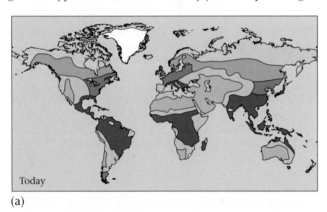

Today

(a)

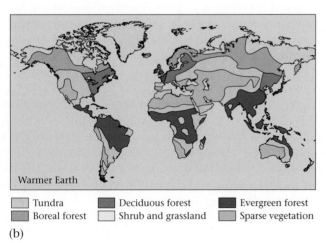

Warmer Earth

| ▢ Tundra | ▢ Deciduous forest | ▢ Evergreen forest |
| ▢ Boreal forest | ▢ Shrub and grassland | ▢ Sparse vegetation |

(b)

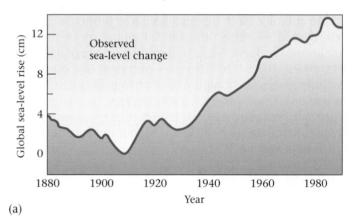

(a)

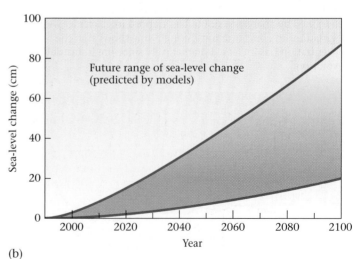

(b)

19.7 THE FUTURE OF THE EARTH: A SCENARIO

Most of the discussion in this book has focused on the past, for the geologic record preserved in rocks tells us of earlier times. Let's now bring this book to a close by facing the opposite direction and speculating what the world might look like in geologic time to come.

In the geologic near term, the future of the world depends largely on human activities. Whether the Earth System undergoes a major disruption and shifts to a new equilibrium, whether a catastrophic mass-extinction event takes place, or whether society achieves **sustainable growth** (meaning an ability to prosper within the constraints of the Earth System) will depend on our own foresight and ingenuity. Projecting thousands of years into the future, we might well wonder if the Earth will return to ice-age conditions, with glaciers growing over major cities and the continental shelf becoming dry land, or if the ice age is over for good because of global warming. No one really knows for sure.

If we project millions of years into the future, it is clear the map of the planet will change significantly because of the continuing activity of plate tectonics. For example, during the next 50 million years or so, the Atlantic Ocean will probably become bigger, the Pacific Ocean will shrink, and the western part of California will migrate northward. Eventually, Australia will crush against the southern margin of Asia, and the islands of Indonesia will be flattened in between.

Predicting the map of the Earth beyond that is hard, because we don't know for sure where new subduction zones will develop. Most likely, subduction of the Pacific Ocean will lead to the collision of the Americas with Asia, to produce a supercontinent ("Amasia"). A subduction zone eventually will form on one side of the Atlantic Ocean, and the ocean will be consumed. As a consequence, the eastern margin of the Americas will collide with the western margin of Europe and Africa. The sites of major cities—New York, Miami, Rio de Janeiro, Buenos Aires, London—will be incorporated in a collisional mountain belt, and likely will be subjected to metamorphism and igneous intrusion before being uplifted and eroded. Shallow seas may once again cover the interiors of continents and then later retreat, and glaciers may once again cover the continents—it happened in the past, so it could happen again. And if the past is the key to the future, biological evolution may have introduced new species to the biosphere, and there is no way to predict what these species will be like.

And what of the end of the Earth? Geologic catastrophes resulting from asteroid and comet collisions will undoubtedly occur in the future as they have in the past. We can't predict when the next strike will come, but unless the object can be diverted, Earth is in for another radical readjustment of surface conditions. But it's not likely that such collisions will destroy our planet. Rather, astronomers predict that the end of the Earth will occur some 5 billion years from now, when the Sun begins to run out of nuclear fuel. When this happens, thermal pressure caused by fusion reactions will no longer be able to prevent the Sun from collapsing inward, because of the immense gravitational pull of its mass. Were the Sun a few times larger than it is, the collapse would trigger a supernova explosion that would blast matter of the Universe out into space to form a new nebula. But since the Sun is not that large, the thermal energy generated when its interior collapses inward will heat the gases of its outer layers sufficiently to cause them to expand. As a result, the Sun will become a red giant, a huge star whose radius would grow beyond the orbit of Earth (▶Fig. 19.21). Our planet will then vaporize, and its atoms will join an expanding ring of gas—the ultimate global change. If this happens, the atoms that once made Earth and all its inhabitants through geologic time may eventually be incorporated in a future solar system, where the cycle of planetary formation and evolution will begin anew.

FIGURE 19.21 In about 5 billion years, the Sun will grow to become a red giant. When that happens, the Earth will first dry out, then vaporize. Initially, it may look like a giant comet. Perhaps some of its atoms will become part of a nebula that someday condenses to form a new sun and planetary system.

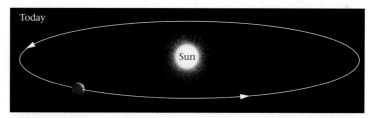

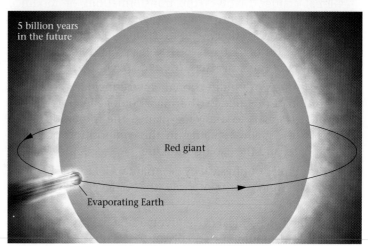

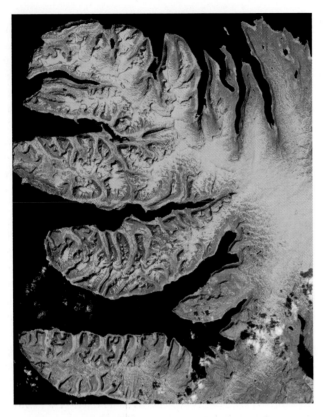

THE VIEW FROM SPACE Numerous fjords make western Iceland's coast resemble the shape of a leaf. The glaciers that carved these fjords vanished over 10,000 years ago. Their growth and demise is one manifestation of climate change on Earth.

CHAPTER SUMMARY

• We refer to the global interconnecting web of physical and biological phenomena on Earth as the Earth System. Global change involves the transformations or modifications of physical and biological components of the Earth System through time. Unidirectional change results in transformations that never repeat; cyclic change involves repetition of the same steps over and over.

• Examples of unidirectional change include the gradual evolution of the solid Earth from a homogeneous collection of planetesimals to a layered planet, the formation of the oceans, the gradual change in the composition of the atmosphere, and the evolution of life.

• Examples of physical cycles that take place on Earth include the supercontinent cycle, the sea-level cycle, and the rock cycle.

• A biogeochemical cycle involves the passage of a chemical among nonliving and living reservoirs. Examples in-

clude the hydrologic cycle and the carbon cycle. Global change occurs when factors change the relative proportions of the chemical in different reservoirs.

• Tools for documenting global climate change include the stratigraphic record, paleontology, oxygen-isotope ratios, and growth rings.

• Studies of long-term climate change show that at times in the past the Earth experienced greenhouse (warmer) periods; at other times there were icehouse (cooler) periods. Factors leading to long-term climate change include the positions of continents, volcanic activity, the uplift of land, and the formation of materials that remove CO_2, an important greenhouse gas.

• Short-term climate change can be seen in the record of the last million years. In fact, during only the past 15,000 years, we see that the climate has warmed and cooled a few times. Causes of short-term climate change include fluctuations in solar radiation and cosmic rays, changes in Earth's orbit and tilt, changes in reflectivity, and changes in ocean currents.

• Mass extinction, a catastrophic change in biodiversity, may be caused by the impact of a comet or asteroid or by intense volcanic activity.

• During the last two centuries, humans have changed landscapes, modified ecosystems, and added pollutants to the land, air, and water at rates faster than the Earth System can process.

• The addition of CO_2 and CH_4 to the atmosphere may be causing global warming, which could shift climate belts and lead to a rise in sea level.

• In the future, in addition to climate change, the Earth will witness a continued rearrangement of continents resulting from plate tectonics, and will likely suffer the impact of asteroids and comets. The end of the Earth may come when the Sun runs out of fuel in about 5 billion years and becomes a red giant.

KEY TERMS

acid rain (p. 540)
biogeochemical cycle (pp. 526, 529)
Earth System (p. 526)
ecosystem (p. 536)
eustatic sea-level change (p. 528)
global change (p. 526)
global climate change (p. 526)
global cooling (p. 530)
global warming (p. 530)

greenhouse gases (p. 530)
greenhouse (hot-house) period (p. 531)
icehouse period (p. 531)
mass-extinction event (p. 535)
ozone hole (p. 540)
paleoclimate (p. 530)
pollution (p. 537)
supercontinent cycle (p. 528)
sustainable growth (p. 543)

REVIEW QUESTIONS

1. How have the Earth's crust and atmosphere changed since they first formed?
2. What processes control the rise and fall of sea level on Earth?
3. How does carbon cycle through the various Earth systems?
4. How do paleoclimatologists study ancient climate change?
5. Contrast icehouse and greenhouse conditions.
6. What are the possible causes of long-term climatic change?
7. What factors explain short-term climatic change?
8. Give some examples of events that cause catastrophic change.
9. Give some examples of how humans have changed the Earth.
10. What is the ozone hole, and how does it affect us?
11. Describe how carbon dioxide–induced global warming takes place, and how humans may be responsible. What effects might global warming have on the Earth System?
12. What are some of the likely scenarios for the long-term future of the Earth?

SUGGESTED READING

Alvarez, W. 1997. *T. Rex and the Crater of Doom*. Princeton, N.J.: Princeton University Press.

Burroughs, W. J. 2001. *Climate Change: A Multidisciplinary Approach*. Cambridge, UK: Cambridge University Press.

Collier, M., and R. H. Webb. 2002. *Floods, Droughts, and Climate Change*. Tucson: University of Arizona Press.

Ehlen, J., et al. 2005. *Humans as Geologic Agents*. Boulder, CO: Geological Society of America.

Flannery, T. 2002. The Future Eaters: *An Ecological History of the Australian Lands and People*. New York: Grove Press.

Flannery, T. 2006. *The Weather Makers: How Man is Changing the Climate and What it Means for Life on Earth*. Washington, D. C.: Atlantic Monthly Press.

Harvey, D. 1999. *Global Warming: The Hard Science*. Upper Saddle River, N.J.: Prentice-Hall.

Holland, H. D., and U. Petersen. 1995. *Living Dangerously: The Earth, Its Resources, and the Environment*. Princeton, N.J.: Princeton University Press.

Houghton, J. T., et al., eds. 2001. *Climate Change 2001: The Scientific Basis*. A report of the Intergovernmental Panel on Climate Change. Cambridge, UK: Cambridge University Press.

Kump, L. R., J. F. Kasting, and R. G. Crane. 1999. *The Earth System*. Upper Saddle River, N.J.: Prentice-Hall.

Leggett, J. K. 2001. *The Carbon War: Global Warming and the End of the Oil Era*. New York: Routledge.

MacKenzie, F. T. 1998. *Our Changing Planet: An Introduction to Earth System Science and Global Environmental Change*. Upper Saddle River, N.J.: Prentice-Hall.

Officer, C., and J. Page. 1993. *Tales of the Earth: Paroxysms and Perturbations of the Blue Planet*. New York: Oxford University Press.

Ruddiman, W. F. 2005. *Plows, Plagues, and Petroleum: How Humans Took Control of the Climate*. Princeton, N. J.: Princeton University Press.

Speth, J. G. 2004. *Red Sky at Morning: America and the Crisis of the Global Environment*. New Haven, Conn.: Yale University Press.

Turco, R. P. 1997. *Earth under Siege: From Air Pollution to Global Change*. Oxford, UK: Oxford University Press.

Van Andel, T. H. 1994. *New Views on an Old Planet: A History of Global Change*. Cambridge, UK: Cambridge University Press.

Weart, S. R. 2003. *The Discovery of Global Warming*. Cambridge, Mass.: Harvard University Press.

Scientific Background: Matter and Energy

a.I INTRODUCTION

In order to understand the formation and evolution of the Universe, as well as descriptions of materials that constitute the Earth and processes that shape the Earth, we must first understand some basic facts about matter, energy, and heat (►Fig. a.1). The following synopsis highlights key topics from physics and chemistry that are an essential background to the rest of the book.

a.1 DISCOVERING THE NATURE OF MATTER

Matter takes up space—you can feel it and see it. We use the word **matter** to refer to any material making up the Universe. The amount of matter in an object is its **mass.** An object with a greater mass contains more matter—for example, a large tree contains more matter than a blade of grass. There's a subtle but important distinction between mass and weight. **Weight** depends on the amount of an object's mass but also on the strength of gravity. An astronaut has the same mass on both the Earth and the Moon, but weighs a different amount in each place. Since the amount of gravitational pull exerted by an object depends on its mass, and the Moon has about one-sixth the mass of the Earth, an astronaut weighing 68 kilograms (150 pounds) on Earth would weigh only about 11 kilograms (25 pounds) on the Moon. Thus, lunar explorers have no trouble jumping great distances even when burdened by space suits and oxygen tanks.

What does matter consist of? Early philosophers deduced that the Earth and the plants and animals on it must be formed from simpler components, just as bread is made from a measured mixture of primary ingredients. They initially thought that the "primary ingredients" making up matter included only earth, air, fire, and water. Then a philosopher named Democritus (c. 460–370 B.C.E.) argued that if you were able to keep dividing matter into progressively smaller pieces, you would eventually end up with nothing, and since it doesn't seem possible to make something out of nothing, there must be a smallest piece of matter that can't be subdivided further. He proposed the name **atom** for these

FIGURE a.1 A lightning storm over mountains. The solid rock, the air, and the clouds consist of matter. The lightning, a rapidly moving current of electrons, is a manifestation of energy.

smallest pieces, based on the Greek word *atomos*, which means "indivisible."

Our modern understanding of matter didn't become established until the seventeenth century, when chemists such as Robert Boyle (1627–1691) recognized that certain substances, such as hydrogen, oxygen, carbon, and sulfur, cannot break down into other substances, whereas others, like water and salt, *can* break down into other substances

FIGURE a.2 A compound, salt, can be subdivided to form two elements, sodium metal and chlorine gas. Neither sodium nor chlorine can be divided further.

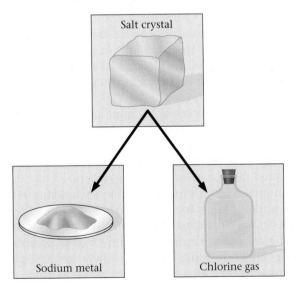

(▶Fig. a.2). For example, water breaks down into oxygen and hydrogen. Substances that can't be broken down came to be known as **elements,** whereas those that can be broken down came to be known as **compounds.** An English schoolteacher, John Dalton (1766–1844), adopted the word atom for the smallest piece of an element that maintains the property of the element, and suggested that compounds consisted of combinations of different atoms. Then in 1869, a Russian chemist named Dmitri Mendeléev (1834–1907) realized that groups of elements share similar characteristics. Mendeléev organized the elements into a chart we now call the **periodic table of the elements** (▶Fig. a.3). In the figure, elements within each column of the table behave similarly. For example, all elements in the right-hand column are inert gases, meaning that they can't combine with other elements to form compounds. But Mendeléev and his contemporaries didn't know what caused the similarities and differences. An understanding of the cause would have to wait until twentieth-century physicists discovered the internal structure of the atom.

a.2 A MODERN VIEW OF MATTER

Atoms

In modern terminology, an element is a substance composed only of atoms of the same kind. Ninety-two different elements occur naturally on Earth, but physicists have created more than a dozen new elements in the laboratory. Each element has a name (such as nitrogen, hydrogen, sulfur) and a **symbol,** an abbreviation of its English or Latin name (N = nitrogen, H = hydrogen, Fe = iron, Ag = silver).

Work in the late 1800s and early 1900s demonstrated that, contrary to the view of Democritus, atoms actually can be divided. Ernest Rutherford, a British physicist, made this key discovery in 1910 when he shot a beam of atoms at a gold foil and found, to his amazement, that only a tiny fraction of the atoms bounced back; most of the mass in the beam passed through the foil as if it were invisible. This result could mean only one thing. Most of the mass in an atom clusters in a dense ball at the atom's center, and this ball is surrounded by a cloud that contains very little mass, so an atom as a whole consists mostly of empty space.

Physicists now refer to the dense ball at the center of the atom as the **nucleus,** and the low-density cloud surrounding the nucleus as the **electron cloud.** Further study in the first half of the twentieth century led to the conclusion that the nucleus contains two types of subatomic particles: **neutrons,** which have a neutral electrical charge, and **protons,** which have a positive electrical charge (▶Fig. a.4a–d). The electron cloud consists of negatively charged

Alkali metals			Symbol →	He 2	Atomic number													Inert gases

Legend box:

He	2
Helium	
4.002	

Symbol / Atomic number / Name / Atomic weight

H 1 Hydrogen 1.007																	He 2 Helium 4.002

Nonmetals

H 1																	He 2
Hydrogen 1.007																	Helium 4.002
Li 3 Lithium 6.941	Be 4 Beryllium 9.0121											B 5 Boron 10.811	C 6 Carbon 12.011	N 7 Nitrogen 14.006	O 8 Oxygen 15.999	F 9 Fluorine 18.998	Ne 10 Neon 20.179
Na 11 Sodium 22.989	Mg 12 Magnesium 24.305	Transition elements (metals)										Al 13 Aluminum 26.981	Si 14 Silicon 28.085	P 15 Phosphorus 30.973	S 16 Sulfur 32.066	Cl 17 Chlorine 35.452	Ar 18 Argon 39.948
K 19 Potassium 39.098	Ca 20 Calcium 40.078	Sc 21 Scandium 44.955	Ti 22 Titanium 47.88	V 23 Vanadium 50.941	Cr 24 Chromium 51.996	Mn 25 Manganese 54.938	Fe 26 Iron 55.847	Co 27 Cobalt 58.933	Ni 28 Nickel 58.693	Cu 29 Copper 63.546	Zn 30 Zinc 65.39	Ga 31 Gallium 69.723	Ge 32 Germanium 72.61	As 33 Arsenic 74.921	Se 34 Selenium 78.96	Br 35 Bromine 79.904	Kr 36 Krypton 83.80
Rb 37 Rubidium 85.467	Sr 38 Strontium 87.62	Y 39 Yttrium 88.905	Zr 40 Zirconium 91.224	Nb 41 Niobium 92.906	Mo 42 Molybdenum 95.94	Tc 43 Technetium 98.907	Ru 44 Ruthenium 101.07	Rh 45 Rhodium 102.905	Pd 46 Palladium 106.42	Ag 47 Silver 107.868	Cd 48 Cadmium 112.411	In 49 Indium 114.82	Sn 50 Tin 118.710	Sb 51 Antimony 121.757	Te 52 Tellurium 127.60	I 53 Iodine 126.904	Xe 54 Xenon 131.29
Cs 55 Cesium 132.905	Ba 56 Barium 137.327	La 57 Lanthanum 138.905	Hf 72 Hafnium 178.49	Ta 73 Tantalum 180.947	W 74 Tungsten 183.85	Re 75 Rhenium 186.207	Os 76 Osmium 190.2	Ir 77 Iridium 192.22	Pt 78 Platinum 195.08	Au 79 Gold 196.966	Hg 80 Mercury 200.59	Tl 81 Thallium 204.383	Pb 82 Lead 207.2	Bi 83 Bismuth 208.980	Po 84 Polonium 208.982	At 85 Astatine 209.987	Rn 86 Radon 222.017
Fr 87 Francium 223.019	Ra 88 Radium 226.025	Ac 89 Actinium 227.027															

Ce 58 Cerium 140.115	Pr 59 Praseodymium 140.907	Nd 60 Neodymium 144.24	Pm 61 Promethium 144.912	Sm 62 Samarium 150.36	Eu 63 Europium 151.965	Gd 64 Gadolinium 157.25	Tb 65 Terbium 158.925	Dy 66 Dysprosium 162.50	Ho 67 Holmium 164.930	Er 68 Erbium 167.26	Tm 69 Thulium 168.934	Yb 70 Ytterbium 173.04	Lu 71 Lutetium 174.967
Th 90 Thorium 232.038	Pa 91 Protactinium 231.035	U 92 Uranium 238.028	Np 93 Neptunium 237.048	Pu 94 Plutonium 244.064	Am 95 Americium 243.061	Cm 96 Curium 247.070	Bk 97 Berkelium 247.070	Cf 98 Californium 251.079	Es 99 Einsteinium 252.083	Fm 100 Fermium 257.095	Md 101 Mendelevium 258.10	No 102 Nobelium 259.100	Lr 103 Lawrencium 262.11

FIGURE a.3 The modern periodic table of the elements. The columns group elements with related properties. For example, inert gases are listed in the column on the right. Metals are found in the central and left parts of the chart.

particles, **electrons,** which have only $\frac{1}{1,836}$ times the mass of protons. Remember that opposite charges attract; the positive charge on the nucleus attracts the negative charge of the electrons, so the nucleus holds on to the electron cloud. (For simplicity, think of a positive charge as the "+" end of a battery and a negative charge as the "−" end.) The mass of a neutron approximately equals the sum of the mass of a proton and the mass of an electron. In the past few decades, physicists have found that protons and neutrons are made up of myriad even smaller particles, the smallest of which is called a **quark.**

Electron clouds have a complex internal structure. Electrons are grouped in intervals called **orbitals,** energy levels, or shells. Some shells have a spherical shape, whereas others resemble dumbbells, rings, or groups of balls—for simplicity, we portray shells as circles in cross section (▶Fig. a.5). Electrons in inner shells are concentrated near the nucleus, whereas those of outer shells predominate farther from the nucleus. Successively higher shells lie at progressively greater distances from the nucleus. Each shell can only contain a specific number of electrons: the lowest shell can contain only two electrons, whereas the next several shells each can hold 8. Electrons fill the lowest shells first, so that atoms with a small number of electrons have only the innermost shells; the outer shells do not exist unless there are electrons to fill them.

The outermost shell of electrons, in effect, defines the outer edge of an atom, so an atom with many occupied electron shells is larger than one with few occupied shells. (Argon, for example, with 18 electrons, contains more occupied electron shells than helium, with 2 electrons, so an argon atom is bigger than a helium atom.) If we picture the nucleus of a carbon atom as being the size of an orange, the electrons of the outermost shell would lie at a distance of over 1 km from the orange.

Atoms are so small that they can't be seen with even the strongest light microscopes. (However, by using some clever techniques, scientists have been able, in recent years, to create images of very large atoms; ▶Fig. a.6.) Atoms are so small, in fact, that 1 gram (0.036 ounces) of helium contains 6.02×10^{23} atoms; in other words, a quantity of helium weighing little more than a postage stamp contains 602,000,000,000,000,000,000,000 atoms! The number of atoms of helium in a small balloon is approximately the same as the number of balloons it would take to replace the entirety of Earth's atmosphere.

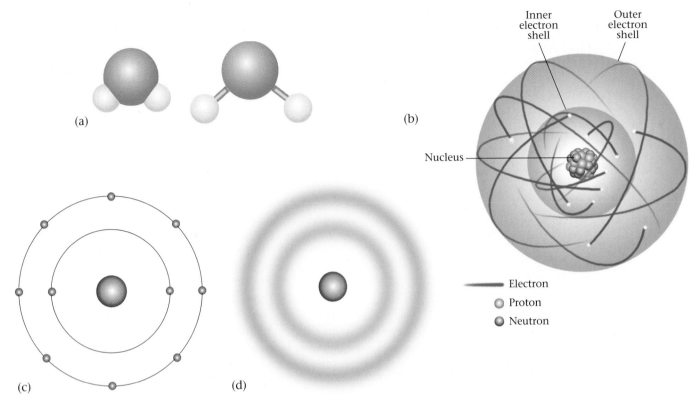

FIGURE a.4 Different ways of portraying atoms. (a) Two ways of portraying a water molecule. The large ball is oxygen, and the small ones are hydrogen. The "sticks" represent chemical bonds. (b) An image of an atom with a nucleus surrounded by electrons. (c) This diagram shows the number of electrons in the inner shells. (d) An alternative depiction of electron shells, implying that the electrons constitute a cloud. In reality, electrons do not follow simple circular orbits.

FIGURE a.5 This schematic drawing of a neon atom shows the two complete electron shells. The inner shell contains two electrons, the outer one eight. The "shells" merely represent the most likely location for an electron to be.

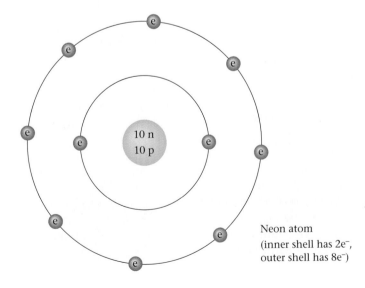

Atomic Number, Atomic Mass, and Isotopes

We distinguish atoms of different elements from each other by their **atomic number,** the number of protons in their nucleus. For example, hydrogen's atomic number is 1, oxygen's is 8, lead's is 82, and uranium's is 92. We write the atomic number as a subscript to the left of the element's symbol (e.g., $_1H$, $_8O$, $_{82}Pb$, $_{92}U$). With the exception of the most common hydrogen nuclei, all atomic nuclei also contain neutrons. In smaller atoms, the number of neutrons equals the number of protons, but in larger atoms, the number of neutrons exceeds the number of protons. Subatomic particles are held together in a nucleus by **nuclear bonds.**

Atomic weight (or **atomic mass**) defines the amount of matter in a single atom. For a given element, the atomic weight *approximately* equals the number of protons plus the number of neutrons. (Precise atomic weights are actually slightly greater than this sum, because neutrons have slightly more mass than protons, and because the electrons have mass.) For example, helium contains 2 protons and

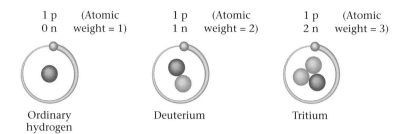

FIGURE a.7 The three isotopes of hydrogen: ordinary hydrogen, deuterium, and tritium.

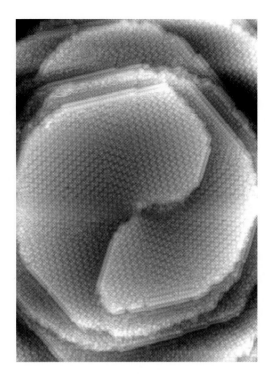

FIGURE a.6 An image of atoms taken with a special microscope that uses electrons to illuminate the surface. Each tiny spot is an atom.

2 neutrons, and thus has an atomic weight of about 4, whereas oxygen contains 8 protons and 8 neutrons and thus has an atomic weight of about 16. We indicate the weight of an atom by a superscript to the left of the element's symbol (^{4}He, ^{16}O).

Some elements occur in more than one form, and these differ in atomic weight. For example, uranium 235 ($^{235}_{92}$U) contains 92 protons and 143 neutrons, whereas uranium 238 ($^{238}_{92}$U) contains 92 protons and 146 neutrons. Note that both forms of uranium have the *same* atomic number—they must, if they are to be considered the same element. But they have different quantities of neutrons and thus differ in atomic weight. Multiple versions of the same element, which differ from one another in atomic weight, are called **isotopes** of the element (▶Fig. a.7).

Ions: Atoms with a Charge

If the number of electrons (negatively charged particles) exactly equals the number of protons (positively charged particles) in an atom, then the atom is electrically neutral. But if the numbers aren't equal, then the atom has a net electrical charge and is called an **ion.** For example, if an atom has two fewer electrons than protons, then we say it has a charge of +2, and if it has two more electrons than protons, it has a charge of −2. We write the charge as a superscript to the right of the symbol (for example, Na$^+$). Ions with a negative charge

are **anions** (pronounced ANN-eye-ons), and ions with a positive charge are **cations** (pronounced CAT-eye-ons). Ions form because atoms "prefer" to have a complete outer electron shell. Thus, an atom may give up or take on electrons in order for its outer shell to contain the proper number of electrons. As is the case with neutral atoms, the size of an ion depends on the number of shells containing electrons. Note that oxygen ions are larger than silicon ions, even though silicon has a larger atomic number, because oxygen has gained electrons and thus uses more electron shells than does silicon, for silicon has lost electrons (▶Fig. a.8).

FIGURE a.8 Relative sizes of common ions making up materials in rocks at the Earth's surface.

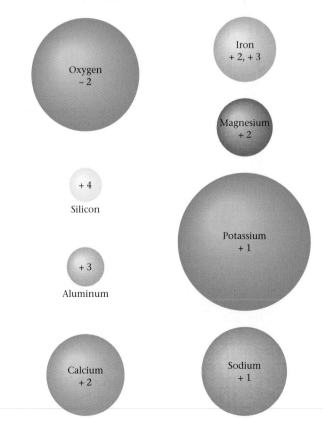

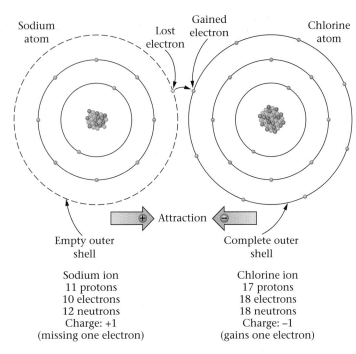

FIGURE a.9 The sodium atom has an unfilled outer shell—the shell has room for eight electrons but only has one. The chlorine atom also has an unfilled outer shell—it's missing one electron. The sodium atom gives up its outer electron, and thus has one more proton than electron (i.e., a net positive charge), whereas the chlorine gains an electron and thus has one more electron than proton (a net negative charge). The two ions attract each other.

Molecules and Chemical Bonds

Most of the materials we deal with in everyday life—oxygen, water, plastic—are not composed of isolated atoms. Rather, most atoms tend to stick, or bond, to other atoms; two or more atoms stuck together constitute a **molecule.** Hydrogen gas, for example, consists of H_2 molecules; note that a hydrogen molecule consists of two of the same kind of atom. But many materials contain different kinds of atoms bonded together. As we noted earlier, such materials are called compounds. For example, common table salt is a compound containing sodium and chlorine atoms bonded together. Compounds generally differ markedly from the elements that make them up—salt bears no resemblance at all to pure sodium (a shiny metal) or pure chlorine (a noxious gas) (see Fig. a.2). A molecule is the smallest identifiable piece of a compound, containing the correct proportion of the compound's elements. For example, a molecule of water consists of two atoms of hydrogen and one atom of oxygen. We represent water by the **chemical formula** H_2O, a concise recipe that indicates the relative proportions of different elements in the molecule (see Fig. a.4a).

Chemical bonds act as the glue that holds atoms together to form molecules and holds molecules together to form larger pieces of a material. Chemical bonding results from the interaction among the electrons of nearby atoms, and can take place in four different ways. (Note that chemical bonds are not the same as the nuclear bonds that hold together protons and neutrons in a nucleus.)

- *Ionic bonds:* As an inviolate rule of nature, "like" electrical charges repel (two positive charges push each other away), whereas "unlike" electrical charges attract (a negative charge sticks to a positive charge). Bonds that form in this way are called **ionic bonds** (▶Fig. a.9). For example, in a molecule of salt, positively charged sodium ions (Na^+) attract negatively charged chloride (Cl^-) ions. ("Chloride" is the name given to ions of chlorine.)

- *Covalent bonds:* The atoms of carbon making up a diamond do not transfer electrons to one another, but rather share electrons. Bonding that involves the sharing of electrons is called **covalent bonding** (▶Fig. a.10). Because of the sharing, the electron shells of all the carbon atoms in a diamond are complete, and all the carbon atoms have a neutral charge. Water molecules also exist because of covalent bonding: in a water molecule, two hydrogen atoms are covalently bonded to one oxygen atom.

- *Metallic bonds:* In metals, electrons of the outer shells move easily from atom to atom and bind the atoms to each other. We call this type of bonding **metallic bonding** (▶Fig. a.11). Because outer-shell electrons move so freely, metals conduct electricity easily—when you connect a metal wire to an electrical circuit, a current of electrons flows through the metal.

- *Bonds resulting from the polarity of atoms or molecules:* Chemists long recognized that some materials break or split so easily that they must be held together by particularly weak chemical bonds. Eventually, they

FIGURE a.10 Covalent bonding. Carbon only has four electrons in its outer shell, which has a capacity of eight. Thus, the carbon atom shares electrons with four other carbon atoms—it is covalently bonded to the other atoms.

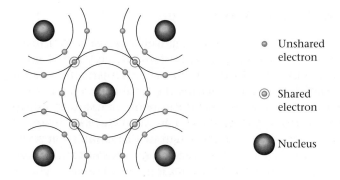

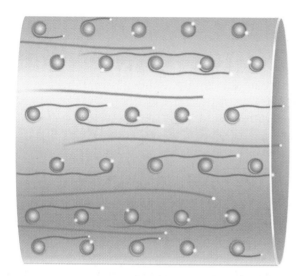

FIGURE a.11 In a metallically bonded material, nuclei and their inner shells of electrons float in a "sea" of free electrons. Sometimes the electrons orbit the nuclei, but at other times they stream through the metal.

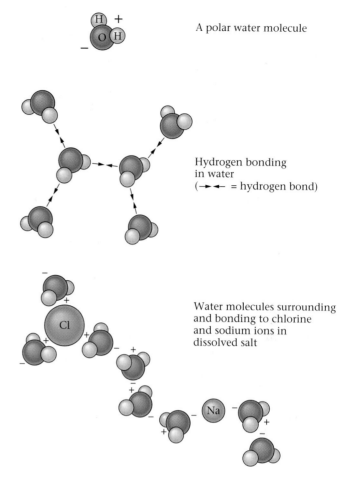

A polar water molecule

Hydrogen bonding in water (→◄ = hydrogen bond)

Water molecules surrounding and bonding to chlorine and sodium ions in dissolved salt

FIGURE a.12 Water is a polar molecule, because the two hydrogen atoms lie on the same side of the oxygen atom. Thus, water molecules tend to attract each other (this attraction creates surface tension, and causes water to form drops). Similarly, the polar molecules surround chlorine and sodium ions when salt dissolves in water.

realized that these bonds are due to the permanent or temporary polarity of molecules. **Polarity** means that the molecule has a positive charge on one side and a negative charge on the other. Polarity-related bonds form because the negative side of one molecule attracts the positive side of another.

Water molecules are polar, and they attract each other. This attraction is called a **hydrogen bond** (▶Fig. a.12). The polarity of water molecules makes water a good solvent (meaning that it can dissolve substances), because the polar water molecules attract and surround ions of soluble materials and pull them apart.

Johannes van der Waals (1837–1923), a Dutch physicist, discovered another type of weak chemical bonding that depends on polarity. This type, now known as **van der Waals bonding,** links one covalently bonded molecule to another. The bonds exist because electrons temporarily cluster on one side of each molecule, giving it a polarity.

The Forms of Matter

Matter exists in one of four states: **solids,** which can maintain their shape for a long time without the restraint of a container; **liquids,** which flow fairly quickly and assume the shape of the container they fill (it's possible for a liquid to fill only part of its container); **gases,** which expand to fill the entire container they have been placed in and will disperse in all directions when not restrained; and **plasma,** an unfamiliar, gaslike mixture of positive ions and free electrons that exists only at very high temperatures (▶Fig. a.13a–d).

a.3 THE FORCES OF NATURE

In our everyday experience, we constantly see or feel the effect of force. Forces can squash objects; speed them up and slow them down; tear, stretch, spin, and twist them; and make them float or sink. Isaac Newton, the great British scientist who effectively founded the field of physics, was the first to describe the way forces work. He defined a **force** as simply the push or pull that causes the velocity (speed) of a mass to change in magnitude and/or direction.

We can distinguish between two categories of force. The first, which includes force applied by the movement of a mass (a hand, a hammer, the wind, waves), is called **mechanical force,** or contact force. When you push a block across the floor, you are applying a mechanical force. The second category, which includes force resulting from the action of an invisible agent, is called a **field force,** or noncontact

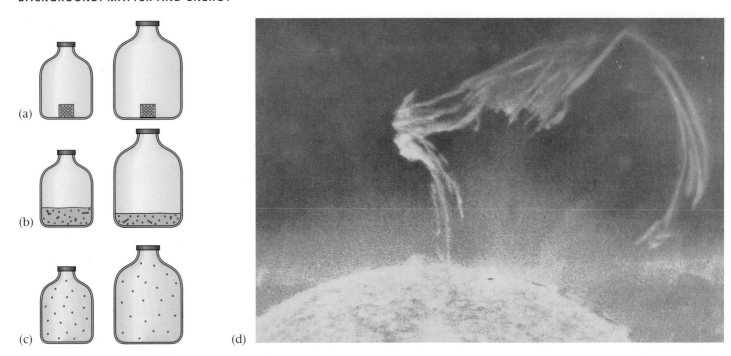

FIGURE a.13 The states of matter. (a) A solid block sitting in a bottle retains its shape regardless of the size of the container. (b) A liquid conforms to the shape of the container. If the container changes shape, the liquid also changes shape, so that it stays constant in density (mass per unit volume). (c) A gas will expand to fill whatever volume it occupies, and thus can change density if the volume changes. (d) The Sun contains plasma, a gaslike material that is so hot that electrons have been stripped from the nuclei.

force. If you drop a book, the invisible force of gravity pulls the book toward the floor.

Physicists further recognize four types of field forces: gravity, electromagnetic, strong nuclear, and weak nuclear. **Gravity** is a force of attraction between any two masses, and can act over large distances. The magnitude of gravitational attraction depends on the size of the masses and the distance between them. We feel a much stronger gravitational pull to the planet we walk on than we do to a baseball. **Electromagnetic force,** the force associated with electricity and magnetism, is stronger than gravity, but operates only between materials that have electrical charges or are magnetic, and operates only over short distances. Like gravity, electromagnetic force depends on the distance between objects, but unlike gravity it can be either attractive or repulsive—as mentioned previously, like charges repel and unlike charges attract. Nuclear forces operate within the nucleus of an atom.

a.4 ENERGY

To a physicist, **energy** is the ability to do work, or, in other words, the ability to apply a force that moves a mass by some distance (**work** = force × distance). According to this defini-

tion, gasoline serves as an energy source because we can burn gasoline to move heavy cars and trucks (a type of work).

Kinetic and Potential Energy

We classify energy into two basic types: **kinetic energy,** the energy of motion; and **potential energy,** the energy stored in a material. A boulder sitting at the top of a hill has potential energy, because the force of gravity will do work on it as the boulder topples down. If the boulder falls down, part of its potential energy converts into kinetic energy (▶Fig. a.14). Some of this kinetic energy can be transferred to a second boulder if the original boulder strikes another boulder, causing a contact force that starts the second boulder moving. Similarly, the gas in a car's tank has potential energy, which converts to kinetic energy when the gas burns and molecules start moving quickly. This kinetic energy starts the cylinders in the engine moving up and down. Now we briefly look at where energy comes from.

Energy from Chemical Reactions

A **chemical reaction** is a process whereby one or more compounds (the reactants) come together or break apart to form new compounds and/or ions (the products). Many chemical reactions produce energy. In the case of burning

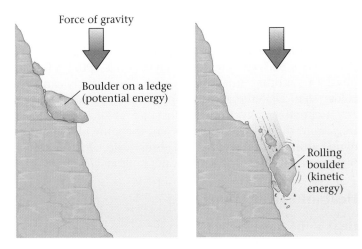

FIGURE a.14 A boulder sitting on a ledge in a gravitational field has potential energy. When the boulder starts to roll, this potential energy converts to kinetic energy.

gasoline, gasoline molecules react with oxygen molecules in the air to form carbon dioxide molecules and water molecules, plus energy. We can describe this reaction in a chemical formula, that uses symbols for the molecules:

$$2C_8H_{18} + 25O_2 \rightarrow 16CO_2 + 18H_2O + \text{energy}$$

Note that reactants appear on the left side of the formula (C_8H_{18} is a typical compound in gasoline), and products on the right side. The arrow indicates the direction in which the reaction proceeds. The energy comes from the breaking of chemical bonds—more energy is stored in the chemical bonds of gasoline and oxygen than in the chemical bonds of carbon dioxide and water.

Energy from Field Forces

When you place an object in a field force, the force acts on the object, and the object, unless it is restrained, moves. In a gravity field, as described earlier, a boulder resting on top of a hill has potential energy (it is restrained by the ground), but when the boulder starts rolling, it has kinetic energy. Similarly, an iron bar taped to a table near a magnet has potential energy, but an unrestrained bar moving toward the magnet has kinetic energy.

Fission and Fusion: Energy from Breaking or Making Atoms

During **nuclear fission,** a large atomic nucleus splits into two or more smaller nuclei (not all of the same element) and other byproducts, including free neutrons (▶Fig. a.15). Such fission is one form of nuclear decay. Other forms include: (1) the ejection of an electron from one of the neutrons in the nucleus, transforming the neutron

into a proton; (2) the ejection of a couple of protons or neutrons from the nucleus. Nuclear decay produces new atoms with different atomic numbers from that of the original atom. Put another way, one element can transform into another by means of fission reactions. In effect, the ultimate goal of medieval alchemists, to turn lead into gold, can now be achieved in modern atom smashers, though at a significant cost.

When fission or other nuclear decay reactions occur, some of the matter constituting the original atom is transformed into a huge amount of energy (heat and electromagnetic radiation), as defined by Einstein's famous equation: $E = mc^2$, where E is energy, m is mass, and c is the speed of light. The realization that fission releases vast quantities of energy led to the rush to build atomic weapons during and after World War II. Nuclear fission provides the energy in atomic bombs, nuclear power plants, and nuclear submarines.

Nuclear fusion results when two nuclei slam together at such high velocity that they get close enough for nuclear forces to bind them together, thereby creating a new and larger atom of a different element (▶Fig. a.16). In a manner of speaking, fusion is the opposite of fission. Fusion reactions also produce huge amounts of energy, because during fusion some of the matter converts into energy, according to Einstein's equation. Fusion reactions generate the heat in stars; in the Sun, for example, hydrogen atoms fuse together to form helium. Fusion reactions also generate the explosive energy of a hydrogen bomb (a thermonuclear device). Such reactions can take place only at extremely high temperatures (over 1,000,000°C). In fact, in order to get a hydrogen bomb to blow up, you need to use an atomic bomb as the trigger, because only an atomic explosion can generate high enough temperatures to start fusion. So far, no one has figured out how to sustain a controlled fusion reaction on Earth economically for the purpose of generating electricity.

FIGURE a.15 A uranium atom splits during nuclear fission. Note that the two atoms produced by the reaction are not of the same element.

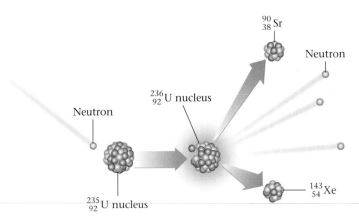

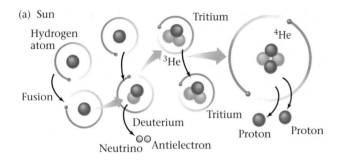

(a) Sun

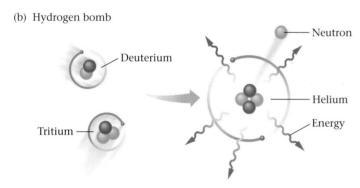

(b) Hydrogen bomb

FIGURE a.16 (a) In the Sun, four hydrogen atoms fuse in sequence to form a helium atom. (b) In a hydrogen bomb, a deuterium and a tritium atom fuse to form a helium atom plus a free neutron.

a.5 HOT AND COLD

Thermal Energy and Temperature

The atoms and molecules that make up an object do not stay rigidly fixed in place, but rather jiggle and jostle with respect to each other. This vibration creates **thermal energy**—the faster the atoms move, the greater the thermal energy and the hotter the object. In simple terms, the thermal energy in a substance represents the sum of the kinetic energy of all the substance's atoms. (This includes the back-and-forth displacements that an atom makes as it vibrates as well as the movement of an atom from one place to another.)

When we say that one object is hotter or colder than another, we are describing its temperature. **Temperature** represents the average kinetic energy of atoms in the material and is a measure of warmth relative to some standard. In everyday life, we generally use the freezing or boiling point of water at sea level as the standard. When using the **Celsius (centigrade) scale,** we arbitrarily set the freezing point of water (at sea level) as 0°C and the boiling point as 100°C; whereas in the **Fahrenheit scale,** we set the freezing point as 32°F and the boiling point as 212°F.

The coldest a substance can be is the temperature at which its atoms or molecules stand still. We call this temperature **absolute zero,** or 0K (pronounced "zero kay"), where **K** stands for **Kelvin** (after Lord Kelvin [1824–1907], a British physicist), another unit of temperature. You simply can't get colder than absolute zero, meaning that you can't extract any thermal energy from a substance at 0K (–273.15°C). Degrees in the Kelvin scale represent the same increment of temperature as degrees in the Celsius scale. On the Kelvin scale, ice melts at 273.15K and water boils at 373.15K.

Heat and Heat Transfer

Heat is the thermal energy transferred from one object to another. Heat can be measured in **calories.** One thousand calories can heat 1 kilogram of water by 1°C. There are four ways in which heat transfer takes place in the Earth System: radiation, conduction, convection, and advection.

Radiation is the process by which electromagnetic waves transmit heat into a body or out of a body (▶Fig. a.17a). For example, when the Sun heats the ground during the day, radiative heating takes place. Similarly, when heat rises from the ground at night, radiative heating is occurring—in the opposite direction.

Conduction takes place when you stick the end of an iron bar in a fire (▶Fig. a.17b). The iron atoms at the fire-licked end of the bar start to vibrate more energetically; they gradually incite atoms farther from the flame to start jiggling, and these atoms in turn set atoms even farther along in motion. In this way, heat slowly flows down the bar until you feel it with your hand. Note that even though the iron atoms vibrate more as the bar heats up, the atoms remain locked in their position within the solid. Thus, conduction does not involve actual movement of atoms from one place to another. If you place two bars of different temperatures in contact with each other, heat conducts from the hot bar into the cold one until both end up at the same temperature.

Convection takes place when you set a pot of water on a stove (▶Fig. a.17c). The heat from the stove warms the water at the base of the pot (it makes the molecules of water vibrate faster and move around more). As a consequence, the density of the water at the base of the pot decreases, for as you heat a liquid, the atoms move away from each other and the liquid expands. For a time, cold water remains at the top of the pot, but eventually the warm, less dense water becomes buoyant relative to the cold, dense water. In a gravitational field, a buoyant material rises (like a styrofoam ball in a pool of water) if the material above is weak enough to flow out of the way. Since liquid water can flow easily, hot water rises. When this happens, cold water sinks to take its place. The new volume of cold water then heats up and rises. Thus, during convection, the flow of the material itself carries heat.

The trajectory of flow defines **convective cells;** in a convective cell, water follows a loop, with warm water rising and cold water sinking. Convection occurs in the atmosphere, the ocean, and the interior of the Earth.

Advection, a less familiar process, happens when heat moves with a fluid, flowing through cracks and pores within a solid material (▶Fig. a.17d). The heat brought by the fluid conductively heats up the solid that the fluid passes through. Advection takes place, for example, if you pass hot water through a sponge and the sponge itself gets hot. In the Earth, advection occurs where molten rock rises through the crust beneath a volcano and heats up the crust in the process or where hot water circulates through cracks in rocks beneath geysers and hot springs.

FIGURE a.17 The four processes of heat transfer. (a) Radiation occurs when sunlight strikes the ground. (b) Conduction occurs when you heat the end of an iron bar in a flame. Heat flows from the hot region toward the cold region as vibrating atoms cause their neighbors to vibrate. (c) Convection takes place when moving fluid carries heat with it. Hot water at the bottom of the pot rises while cool water sinks, setting up a convective cell. (d) During advection, a hot liquid (such as molten rock) rises into cooler material. Heat conducts from the hot liquid into the cooler material.

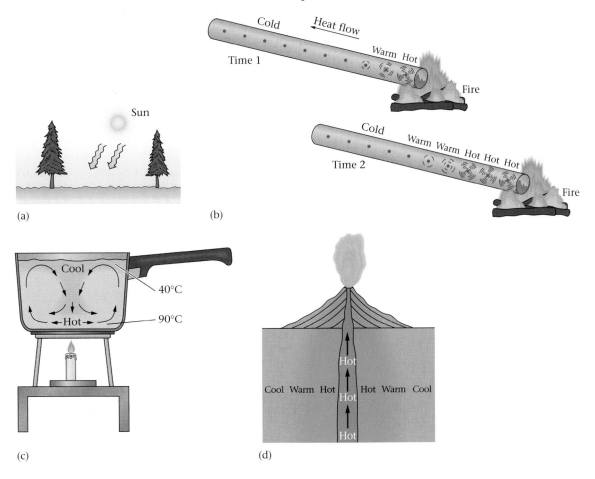

Metric Conversion Chart

Length

1 kilometer (km) = 0.6214 mile (mi)
1 meter (m) = 1.094 yards = 3.281 feet
1 centimeter (cm) = 0.3937 inch
1 millimeter (mm) = 0.0394 inch
1 mile (mi) = 1.609 kilometers (km)
1 yard = 0.9144 meter (m)
1 foot = 0.3048 meter (m)
1 inch = 2.54 centimeters (cm)

Area

1 square kilometer (km^2) = 0.386 square mile (mi^2)
1 square meter (m^2) = 1.196 square yards (yd^2)
= 10.764 square feet (ft^2)
1 square centimeter (cm^2) = 0.155 square inch (in^2)
1 square mile (mi^2) = 2.59 square kilometers (km^2)
1 square yard (yd^2) = 0.836 square meter (m^2)
1 square foot (ft^2) = 0.0929 square meter (m^2)
1 square inch (in^2) = 6.4516 square centimeters (cm^2)

Volume

1 cubic kilometer (km^3) = 0.24 cubic mile (mi^3)
1 cubic meter (m^3) = 264.2 gallons
= 35.314 cubic feet (ft^3)
1 liter (1) = 1.057 quarts
= 33.815 fluid ounces
1 cubic centimeter (cm^3) = 0.0610 cubic inch (in^3)
1 cubic mile (mi^3) = 4.168 cubic kilometers (km^3)
1 cubic yard (yd^3) = 0.7646 cubic meter (m^3)
1 cubic foot (ft^3) = 0.0283 cubic meter (m^3)
1 cubic inch (in^3) = 16.39 cubic centimeters (cm^3)

Mass

1 metric ton = 2,205 pounds
1 kilogram (kg) = 2.205 pounds
1 gram (g) = 0.03527 ounce
1 pound (lb) = 0.4536 kilogram (kg)
1 ounce (oz) = 28.35 grams (g)

Pressure

1 kilogram per square
centimeter (kg/cm^2)* = 0.96784 atmosphere (atm)
= 0.98066 bar
= 9.8067×10^4 pascals (Pa)
1 bar = 10 megapascals (Mpa)
= 1.0×10^5 pascals (Pa)
= 29.53 inches of mercury (in a barometer)
= 0.98692 atmosphere (atm)
= 1.02 kilograms per square centimeter (kg/cm^2)
1 pascal (Pa) = 1 $kg/m/s^2$
1 pound per square inch = 0.06895 bars
= 6.895×10^3 pascals (Pa)
= 0.0703 kilogram per square centimeter

Temperature

To change from Fahrenheit (F) to Celsius (C):
$$°C = \frac{(°F - 32°)}{1.8}$$

To change from Celsius (C) to Fahrenheit (F):
°F = (°C × 1.8) + 32°

To change from Celsius (C) to Kelvin (K):
K = °C + 273.15

To change from Fahrenheit (F) to Kelvin (K):
$$K = \frac{(°F - 32°)}{1.8} + 273.15$$

---

*Note: Because kilograms are a measure of mass whereas pounds are a unit of weight, pressure units incorporating kilograms assume a given gravitational constant (g) for Earth. In reality, the gravitational constant for Earth varies slightly with location.

Glossary

aa A lava flow with a rubbly surface.

abandoned meander A meander that dries out after it was cut off.

ablation The removal of ice at the toe of a glacier by melting, sublimation (the evaporation of ice into water vapor), and/or calving.

abrasion The process in which one material (such as sand-laden water) grinds away at another (such as a stream channel's floor and walls).

absolute age Numerical age (the age specified in years).

absolute plate velocity The movement of a plate relative to a fixed point in the mantle.

abyssal plain A broad, relatively flat region of the ocean that lies at least 4.5 km below sea level.

Acadian orogeny A convergent mountain-building event, that occurred around 400 million years ago, during which continental slivers accreted to the eastern edge of the North American continent.

accreted terrane A block of crust that collided with a continent at a convergent margin and stayed attached to the continent.

accretionary coast A coastline that receives more sediment than erodes away.

accretionary orogen An orogen formed by the attachment of numerous buoyant slivers of crust to an older, larger continental block.

accretionary prism A wedge-shaped mass of sediment and rock scraped off the top of a downgoing plate and accreted onto the overriding plate at a convergent plate margin.

acid mine runoff A dilute solution of sulfuric acid, produced when sulfur-bearing minerals in mines react with rainwater, that flows out of a mine.

acid rain Precipitation in which air pollutants react with water to make a weak acid that then falls from the sky.

active continental margin A continental margin that coincides with a plate boundary.

active fault A fault that has moved recently or is likely to move in the future.

active sand The top layer of beach sand, which moves daily because of wave action.

active volcano A volcano that has erupted within the past few centuries and will likely erupt again.

adiabatic cooling The cooling of a body of air or matter without the addition or subtraction of thermal energy (heat).

adiabatic heating The warming of a body of air or matter without the addition or subtraction of heat.

aerosols Tiny solid particles or liquid droplets that remain suspended in the atmosphere for a long time.

aftershocks The series of smaller earthquakes that follow a major earthquake.

air The mixture of gases that make up the Earth's atmosphere.

air-fall tuff Tuff formed when ash settles gently from the air.

air mass A body of air, about 1,500 km across, that has recognizable physical characteristics.

air pressure The push that air exerts on its surroundings.

albedo The reflectivity of a surface.

Alleghenian orogeny The orogenic event that occurred about 270 million years ago when Africa collided with North America.

alloy A metal containing more than one type of metal atom.

alluvial fan A gently sloping apron of sediment dropped by an ephemeral stream at the base of a mountain in arid or semi-arid regions.

alluvium Sorted sediment deposited by a stream.

alluvium-filled valley A valley whose floor fills with sediment.

amber Hardened (fossilized) ancient sap or resin.

amphibolite facies A set of metamorphic mineral assemblages formed under intermediate pressures and temperatures.

amplitude The height of a wave from crest to trough.

Ancestral Rockies The late Paleozoic uplifts of the Rocky Mountain region; they eroded away long before the present Rocky Mountains formed.

angiosperm A flowering plant.

angle of repose The angle of the steepest slope that a pile of uncemented material can attain without collapsing from the pull of gravity.

angularity The degree to which grains have sharp or rounded edges or corners.

angular unconformity An unconformity in which the strata below were tilted or folded before the unconformity developed; strata below the unconformity therefore have a different tilt than strata above.

anhedral grains Crystalline mineral grains without well-formed crystal faces.

Antarctic bottom water mass The mass of cold, dense water that sinks along the coast of Antarctica.

antecedent stream A stream that cuts across an uplifted mountain range; the stream must have existed before the range uplifted and must then have been able to downcut as fast as the land was rising.

anthracite coal Shiny black coal formed at temperatures between 200° and 300°C. A high-rank coal.

anticline A fold with an arch-like shape in which the limbs dip away from the hinge.

anticyclone The clockwise flow of air around a high-pressure mass.

Antler orogeny The Late Devonian mountain-building event in which slices of deep-marine strata were pushed eastward, up and over the shallow-water strata on the western coast of North America.

anvil cloud A large cumulonimbus cloud that spreads laterally at the tropopause to form a broad, flat top.

aphanitic A textural term for fine-grained igneous rock.

apparent polar-wander path A path on the globe along which a magnetic pole appears to have wandered over time; in fact, the continents drift, while the magnetic pole stays fairly fixed.

aquiclude Sediment or rock that transmits no water.

aquifer Sediment or rock that transmits water easily.

aquitard Sediment or rock that does not transmit water easily and therefore retards the motion of the water.

archaea A kingdom of "old bacteria," now commonly found in extreme environments like hot springs. (Also called "archaeobacteria.")

Archean The middle Precambrian Eon.

Archimedes' principle The mass of the water displaced by a block of material equals the mass of the whole block of material.

arête A residual knife-edge ridge of rock that separates two adjacent cirques.

argillaceous sedimentary rock Sedimentary rock that contains abundant clay.

arroyo The channel of an ephemeral stream; dry wash; wadi.

artesian well A well in which water rises on its own.

ash fall Ash that falls to the ground out of an ash cloud.

ash flow An avalanche of ash that tumbles down the side of an explosively erupting volcano.

assimilation The process of magma contamination in which blocks of wall rock fall into a magma chamber and dissolve.

asthenosphere The layer of the mantle that lies between 100–150 km and 350 km deep; the asthenosphere is relatively soft and can flow when acted on by force.

atm A unit of air pressure that approximates the pressure exerted by the atmosphere at sea level.

atmosphere A layer of gases that surrounds a planet.

atoll A coral reef that develops around a circular reef surrounding a lagoon.

atomic number The number of protons in the nucleus of a given element.

atomic weight The number of protons plus the number of neutrons in the nucleus of a given element. (Also known as atomic mass.)

aurora australis The same phenomenon as the aurora borealis, but in the Southern Hemisphere.

aurora borealis A ghostly curtain of varicolored light that appears across the night sky in the Northern Hemisphere when charged particles from the Sun interact with the ions in the ionosphere.

avalanche A turbulent cloud of debris mixed with air that rushes down a steep hill slope at high velocity; the debris can be rock and/or snow.

avalanche chute A downslope hillside pathway along which avalanches repeatedly fall, consequently clearing the pathway of mature trees.

avulsion The process in which a river overflows a natural levee and begins to flow in a new direction.

axial plane The imaginary surface that encompasses the hinges of successive layers of a fold.

axial trough A narrow depression that runs along a mid-ocean ridge axis.

backscattered light Atmospheric scattered sunlight that returns back to space.

backshore zone The zone of beach that extends from a small step cut by high-tide swash to the front of the dunes or cliffs that lie farther inshore.

backswamp The low marshy region between the bluffs and the natural levees of a floodplain.

backwash The gravity-driven flow of water back down the slope of a beach.

bajada An elongate wedge of sediment formed by the overlap of several alluvial fans emerging from adjacent valleys.

Baltica A Paleozoic continent that included crust that is now part of today's Europe.

banded-iron formation (BIF) Iron-rich sedimentary layers consisting of alternating gray beds of iron oxide and red beds of iron-rich chert.

bar (1) A sheet or elongate lens or mound of alluvium; (2) a unit of air-pressure measurement approximately equal to 1 atm.

barchan dune A crescent-shaped dune whose tips point downwind.

barrier island An offshore sand bar that rises above the mean high-water level, forming an island.

barrier reef A coral reef that develops offshore, separated from the coast by a lagoon.

basal sliding The phenomenon in which meltwater accumulates at the base of a glacier, so that the mass of the glacier slides on a layer of water or on a slurry of water and sediment.

basalt A fine-grained mafic igneous rock.

base level The lowest elevation a stream channel's floor can reach at a given locality.

basement Older igneous and metamorphic rocks making up the Earth's crust beneath sedimentary cover.

basement uplift Uplift of basement rock by faults that penetrate deep into the continental crust.

base metals Metals that are mined but not considered precious. Examples include copper, lead, zinc, and tin.

basin A fold or depression shaped like a right-side-up bowl.

Basin and Range Province A broad, Cenozoic continental rift that has affected a portion of the western United States in Nevada, Utah, and Arizona; in this province, tilted fault blocks form ranges, and alluvium-filled valleys are basins.

batholith A vast composite, intrusive, igneous rock body up to several hundred km long and 100 km wide, formed by the intrusion of numerous plutons in the same region.

bathymetric map A map illustrating the shape of the ocean floor.

bathymetric profile A cross section showing ocean depth plotted against location.

bathymetry Variation in depth.

bauxite A residual mineral deposit rich in aluminum.

baymouth bar A sandspit that grows across the opening of a bay.

beach drift The gradual migration of sand along a beach.

beach erosion The removal of beach sand caused by wave action and longshore currents.

beach face A steeply concave part of the foreshore zone formed where the swash of the waves actively scours the sand.

bedding Layering or stratification in sedimentary rocks.

bed load Large particles, such as sand, pebbles, or cobbles, that bounce or roll along a stream bed.

bedrock Rock still attached to the Earth's crust.

Bergeron process Precipitation involving the growth of ice crystals in a cloud at the expense of water droplets.

berm A horizontal or landward-sloping terrace in the backshore zone of a beach that receives sediment during a storm.

big bang A cataclysmic explosion that scientists suggest represents the formation of the Universe; before this event, all matter and all energy were packed into one volumeless point.

biochemical sedimentary rock Sedimentary rock formed from material (such as shells) produced by living organisms.

biodiversity The number of different species that exist at a given time.

biogeochemical cycle The exchange of chemicals between living and nonliving reservoirs in the Earth System.

bioremediation The injection of oxygen and nutrients into a contaminated aquifer to foster the growth of bacteria that will ingest or break down contaminants.

biosphere The region of the Earth and atmosphere inhabited by life; this region stretches from a few km below the Earth's surface to a few km above.

bioturbation   The mixing of sediment by burrowing animals such as clams and worms.

bituminous coal   Dull, black intermediate-rank coal formed at temperatures between 100° and 200°C.

black-lung disease   Lung disease contracted by miners from the inhalation of too much coal dust.

black smoker   The cloud of suspended minerals formed where hot water spews out of a vent along a mid-ocean ridge; the dissolved sulfide components of the hot water instantly precipitate when the water mixes with seawater and cools.

blind fault   A fault that does not intersect the ground surface.

blocking temperature   The temperature below which isotopes in a mineral are no longer free to move, so the radiometric clock starts.

blowout   A deep, bowl-like depression scoured out of desert terrain by a turbulent vortex of wind.

blue shift   The phenomenon in which a source of light moving toward you appears to have a higher frequency.

body waves   Seismic waves that pass through the interior of the Earth.

bog   A wetland dominated by moss and shrubs.

bolide   A solid extraterrestrial object such as a meteorite, comet, or asteroid that explodes in the atmosphere.

bornhardt   An inselberg with a loaf geometry, like that of Uluru (Ayers Rock) in central Australia.

Bowen's reaction series   The sequence in which different silicate minerals crystallize during the progressive cooling of a melt.

braided stream   A sediment-choked stream consisting of entwined subchannels.

breaker   A water wave in which water at the top of the wave curves over the base of the wave.

breakwater   An offshore wall, built parallel or at an angle to the beach, that prevents the full force of waves from reaching a harbor.

breccia   Coarse sedimentary rock consisting of angular fragments; or rock broken into angular fragments by faulting.

breeder reactor   A nuclear reactor that produces its own fuel.

brine   Water that is not fresh but is less salty than seawater; brine may be found in estuaries.

brittle deformation   The cracking and fracturing of a material subjected to stress.

brittle-ductile transition (brittle-plastic transition)   The depth above which materials behave brittlely and below which materials behave ductilely (plastically); this transition typically lies between a depth of 10 and 15 km in continental crustal rock, and 60 m deep in glacial ice.

buoyancy   The upward force acting on a less dense object immersed or floating in denser material.

butte   A medium-size, flat-topped hill in an arid region.

caldera   A large circular depression with steep walls and a fairly flat floor, formed after an eruption as the center of the volcano collapses into the drained magma chamber below.

caliche   A solid mass created where calcite cements the soil together (also called calcrete).

calving   The breaking off of chunks of ice at the edge of a glacier.

Cambrian explosion of life   The remarkable diversification of life, indicated by the fossil record, that occurred at the beginning of the Cambrian Period.

Canadian Shield   A broad, low-lying region of exposed Precambrian rock in the Canadian interior.

canyon   A trough or valley with steeply sloping walls, cut into the land by a stream.

capillary fringe   The thin subsurface layer in which water molecules seep up from the water table by capillary action to fill pores.

carbonate rocks   Rocks containing calcite and/or dolomite.

carbon-14 dating   A radiometric dating process that can tell us the age of organic material containing carbon originally extracted from the atmosphere.

cast   Sediment that preserves the shape of a shell it once filled before the shell dissolved or mechanically weathered away.

catabatic winds   Strong winds that form at the margin of a glacier where the warmer air above ice-free land rises and the cold, denser air from above the glaciers rushes in to take its place.

catastrophic change   Change that takes place either instantaneously or rapidly in geologic time.

catchment   *Drainage network.*

cement   Mineral material that precipitates from water and fills the spaces between grains, holding the grains together.

cementation   The phase of lithification in which cement, consisting of minerals that precipitate from groundwater, partially or completely fills the spaces between clasts and attaches each grain to its neighbor.

Cenozoic   The most recent era of the Phanerozoic Eon, lasting from 65 Ma up until the present.

chalk   Very fine-grained limestone consisting of weakly cemented plankton shells.

change of state   The process in which a material changes from one phase (liquid, gas, or solid) to another.

channel   A trough dug into the ground surface by flowing water.

channeled scablands   A barren, soil-free landscape in eastern Washington, scoured clean by a flood unleashed when a large glacial lake drained.

chatter marks   Wedge-shaped indentations left on rock surfaces by glacial plucking.

chemical sedimentary rocks   Sedimentary rocks made up of minerals that precipitate directly from water solution.

chemical weathering   The process in which chemical reactions alter or destroy minerals when rock comes in contact with water solutions and/or air.

chert   A sedimentary rock composed of very fine-grained silica (cryptocrystalline quartz).

Chicxulub crater   A circular excavation buried beneath younger sediment on the Yucután peninsula; geologists suggest that a meteorite landed there 65 Ma.

chimney   (1) A conduit in a magma chamber in the shape of a long vertical pipe through which magma rises and erupts at the surface; (2) an isolated column of strata in an arid region.

cinder cone   A subaerial volcano consisting of a cone-shaped pile of tephra whose slope approaches the angle of repose for tephra.

cinders   Fragments of glassy rock ejected from a volcano.

cirque   A bowl-shaped depression carved by a glacier on the side of a mountain.

cirrus cloud   A wispy cloud that tapers into delicate, feather-like curls.

clastic (detrital) sedimentary rock   Sedimentary rock consisting of cemented-together detritus derived from the weathering of preexisting rock.

cleavage   (1) The tendency of a mineral to break along preferred planes; (2) a type of foliation in low-grade metamorphic rock.

cleavage planes   A series of surfaces on a crystal that form parallel to the weakest bonds holding the atoms of the crystal together.

cliff (or scarp) retreat   The change in the position of a cliff face caused by erosion.

climate   The average weather conditions, along with the range of conditions, of a region over a year.

cloud   A mist of tiny water droplets in the sky.

coal rank A measurement of the carbon content of coal; higher-rank coal forms at higher temperatures.

coal reserve The quantities of discovered, but not yet mined, coal in sedimentary rock of the continents.

coal swamp A swamp whose oxygen-poor water allows thick piles of woody debris to accumulate; this debris transforms into coal upon deep burial.

coastal plain Low-relief regions of land adjacent to the coast.

cold front The boundary at which a cold air mass pushes underneath a warm air mass.

collision The process of two buoyant pieces of lithosphere converging and squashing together.

columnar jointing A type of fracturing that yields roughly hexagonal columns of basalt; columnar joints form when a dike, sill, or lava flow cools.

comet A ball of ice and dust, probably remaining from the formation of the solar system, that orbits the Sun.

compaction The phase of lithification in which the pressure of the overburden on the buried rock squeezes out water and air that was trapped between clasts, and the clasts press tightly together.

composite volcano *Stratovolcano.*

compositional banding A type of metamorphic foliation, found in gneiss, defined by alternating bands of light and dark minerals.

compressibility The degree to which a material's volume changes in response to squashing.

compression A push or squeezing felt by a body.

compressional waves Waves in which particles of material move back and forth parallel to the direction in which the wave itself moves.

conchoidal fractures Smoothly curving, clamshell-shaped surfaces along which materials with no cleavage planes tend to break.

condensation The process of gas molecules linking together to form a liquid.

condensation nuclei Preexisting solid or liquid particles, such as aerosols, onto which water condenses during cloud formation.

cone of depression The downward-pointing, cone-shaped surface of the water table in a location where the water table is experiencing drawdown because of pumping at a well.

confined aquifer An aquifer that is separated from the Earth's surface by an overlying aquitard.

conglomerate Very coarse-grained sedimentary rock consisting of rounded clasts.

consuming boundary *Convergent plate boundary.*

contact The boundary surface between two rock bodies (as between two stratigraphic formations, between an igneous intrusion and adjacent rock, between two igneous rock bodies, or between rocks juxtaposed by a fault).

contact metamorphism *Thermal metamorphism.*

contaminant plume A cloud of contaminated groundwater that moves away from the source of the contamination.

continental crust The crust beneath the continents.

continental divide A highland separating drainage that flows into one ocean from drainage that flows into another.

continental-drift hypothesis The idea that continents have moved and are still moving slowly across the Earth's surface.

continental glacier A vast sheet of ice that spreads over thousands of square km of continental crust.

continental-interior desert An inland desert that develops because by the time air masses reach the continental interior, they have lost all of their moisture.

continental lithosphere Lithosphere topped by continental crust; this lithosphere reaches a thickness of 150 km.

continental margin A continent's coastline.

continental rift A linear belt along which continental lithosphere stretches and pulls apart.

continental rifting The process by which a continent stretches and splits along a belt; if it is successful, rifting separates a larger continent into two smaller continents separated by a divergent boundary.

continental rise The sloping sea floor that extends from the lower part of the continental slope to the abyssal plain.

continental shelf A broad, shallowly submerged region of a continent along a passive margin.

continental slope The slope at the edge of a continental shelf, leading down to the deep sea floor.

continental volcanic arc A long curving chain of subaerial volcanoes on the margin of a continent adjacent to a convergent plate boundary.

contour lines Lines on a map along which a parameter has a constant value; for example, all points along a contour line on a topographic map are at the same elevation.

control rod Rods that absorb neutrons in a nuclear reactor and thus decrease the number of collisions between neutrons and radioactive atoms.

convection Heat transfer that results when warmer, less dense material rises while cooler, denser material sinks.

convergence zone A place where two surface air flows meet so that air has to rise.

convergent margin *Convergent plate boundary.*

convergent plate boundary A boundary at which two plates move toward each other so that one plate sinks (subducts) beneath the other; only oceanic lithosphere can subduct.

coral reef A mound of coral and coral debris forming a region of shallow water.

core The dense, iron-rich center of the Earth.

core-mantle boundary An interface 2,900 km below the Earth's surface separating the mantle and core.

Coriolis effect The deflection of objects, winds, and currents on the surface of the Earth owing to the planet's rotation.

cornice A huge, overhanging drift of snow built up by strong winds at the crest of a mountain ridge.

correlation The process of defining the age relations between the strata at one locality and the strata at another.

cosmic rays Nuclei of hydrogen and other elements that bombard the Earth from deep space.

cosmology The study of the overall structure of the Universe.

country rock (wall rock) The preexisting rock into which magma intrudes.

crater (1) A circular depression at the top of a volcanic mound; (2) a depression formed by the impact of a meteorite.

craton A long-lived block of durable continental crust commonly found in the stable interior of a continent.

cratonic platform A province in the interior of a continent in which Phanerozoic strata bury most of the underlying Precambrian rock.

creep The gradual downslope movement of regolith.

crevasse A large crack that develops by brittle deformation in the top 60 m of a glacier.

critical mass A sufficiently dense and large mass of radioactive atoms in which a chain reaction happens so quickly that the mass explodes.

cross section A diagram depicting the geometry of materials underground as they would appear on an imaginary vertical slice through the Earth.

crude oil Oil extracted directly from the ground.

crust The rock that makes up the outermost layer of the Earth.

crustal root   Low-density crustal rock that protrudes downward beneath a mountain range.

crystal   A single, continuous piece of a mineral bounded by flat surfaces that formed naturally as the mineral grew.

crystal form   The geometric shape of a crystal, defined by the arrangement of crystal faces.

crystal habit   The general shape of a crystal or cluster of crystals that grew unimpeded.

crystal lattice   The orderly framework within which the atoms or ions of a mineral are fixed.

crystalline   Containing a crystal lattice.

cuesta   An asymmetric ridge formed by tilted layers of rock, with a steep cliff on one side cutting across the layers and a gentle slope on the other side; the gentle slope is parallel to the layering.

cumulonimbus cloud   A rain-producing puffy cloud.

cumulus cloud   A puffy, cotton-ball-shaped cloud.

current   (1) A well-defined stream of ocean water; (2) the moving flow of water in a stream.

cut bank   The outside bank of the channel wall of a meander, which is continually undergoing erosion.

cutoff   A straight reach in a stream that develops when erosion eats through a meander neck.

cyanobacteria   Blue-green algae; a type of archaea.

cycle   A series of interrelated events or steps that occur in succession and can be repeated, perhaps indefinitely.

cyclone   (1) The counterclockwise flow of air around a low-pressure mass; (2) the equivalent of a hurricane in the Indian Ocean.

cyclothem   A repeated interval within a sedimentary sequence that contains a specific succession of sedimentary beds.

Darcy's law   A mathematical equation stating that a volume of water, passing through a specified area of material at a given time, depends on the material's permeability and hydraulic gradient.

daughter isotope   The decay product of radioactive decay.

day   The time it takes for the Earth to spin once on its axis.

debris avalanche   An avalanche in which the falling debris consists of rock fragments and dust.

debris flow   A downslope movement of mud mixed with larger rock fragments.

debris slide   A sudden downslope movement of material consisting only of regolith.

decompression melting   The kind of melting that occurs when hot mantle rock rises to shallower depths in the Earth so that pressure decreases while the temperature remains unchanged.

deep current   An ocean current at a depth greater than 100 m.

deep-focus earthquake   An earthquake that occurs at a depth between 300 and 670 km; below 670 km, earthquakes do not happen.

deflation   The process of lowering the land surface by wind abrasion.

deformation   A change in the shape, position, or orientation of a material, by bending, breaking, or flowing.

dehydration   Loss of water.

delta   A wedge of sediment formed at a river mouth when the running water of the stream enters standing water, the current slows, the stream loses competence, and sediment settles out.

delta plain   The low, swampy land on the surface of a delta.

delta-plain flood   A flood in which water submerges a delta plain.

dendritic network   A drainage network whose interconnecting streams resemble the pattern of branches connecting to a deciduous tree.

dendrochronologist   A scientist who analyzes tree rings to determine the geologic age of features.

density   Mass per unit volume.

denudation   The removal of rock and regolith from the Earth's surface.

deposition   The process by which sediment settles out of a transporting medium.

depositional landform   A landform resulting from the deposition of sediment where the medium carrying the sediment evaporates, slows down, or melts.

desert   A region so arid that it contains no permanent streams except for those that bring water in from elsewhere, and has very sparse vegetation cover.

desertification   The process of transforming nondesert areas into desert.

desert pavement   A mosaic-like stone surface forming the ground in a desert.

desert varnish   A dark, rusty-brown coating of iron oxide and magnesium oxide that accumulates on the surface of the rock.

detachment fault   A nearly horizontal fault at the base of a fault system.

detritus   The chunks and smaller grains of rock broken off outcrops by physical weathering.

dewpoint temperature   The temperature at which air becomes saturated so that dew can form.

differential stress   A condition causing a material to experience a push or pull in one direction of a greater magnitude than the push or pull in another direction; in some cases, differential stress can result in shearing.

differential weathering   What happens when different rocks in an outcrop undergo weathering at different rates.

diffraction   The splitting of light into many tiny beams that interfere with one another.

dike   A tabular (wall-shaped) intrusion of rock that cuts across the layering of country rock.

dimension stone   An intact block of granite or marble to be used for architectural purposes.

dipole   A magnetic field with a north and south pole, like that of a bar magnet.

dipole field (for Earth)   The part of the Earth's magnetic field, caused by the flow of liquid iron alloy in the outer core, that can be represented by an imaginary bar magnet with a north and south pole.

dip-slip fault   A fault in which sliding occurs up or down the slope (dip) of the fault.

dip slope   A hill slope underlain by bedding parallel to the slope.

disappearing stream   A stream that intersects a crack or sinkhole leading to an underground cavern, so that the water disappears into the subsurface and becomes an underground stream.

discharge   The volume of water in a conduit or channel passing a point in one second.

discharge area   A location where groundwater flows back up to the surface, and may emerge at springs.

disconformity   An unconformity parallel to the two sedimentary sequences it separates.

displacement (or offset)   The amount of movement or slip across a fault plane.

disseminated deposit   A hydrothermal ore deposit in which ore minerals are dispersed throughout a body of rock.

dissolution   A process during which materials dissolve in water.

dissolved load   Ions dissolved in a stream's water.

distillation column   A vertical pipe in which crude oil is separated into several components.

distributaries   The fan of small streams formed where a river spreads out over its delta.

divergence zone A place where sinking air separates into two flows that move in opposite directions.

divergent plate boundary A boundary at which two lithosphere plates move apart from each other; they are marked by mid-ocean ridges.

diversification The development of many different species.

DNA (deoxyribonucleic acid) The complex molecule, shaped like a double helix, containing the code that guides the growth and development of an organism.

doldrums A belt with very slow winds along the equator.

dome Folded or arched layers with the shape of an overturned bowl.

Doppler effect The phenomenon in which the frequency of wave energy appears to change when a moving source of wave energy passes an observer.

dormant volcano A volcano that has not erupted for hundreds to thousands of years but does have the potential to erupt again in the future.

downcutting The process in which water flowing through a channel cuts into the substrate and deepens the channel relative to its surroundings.

downdraft Downward-moving air.

downgoing plate (or slab) A lithosphere plate that has been subducted at a convergent margin.

downslope force The component of the force of gravity acting in the downslope direction.

downslope movement The tumbling or sliding of rock and sediment from higher elevations to lower ones.

downwelling zone A place where near-surface water sinks.

drag fold A fold that develops in layers of rock adjacent to a fault during or just before slip.

drainage divide A highland or ridge that separates one watershed from another.

drainage network (or basin) An array of interconnecting streams that together drain an area.

drawdown The phenomenon in which the water table around a well drops because the users are pumping water out of the well faster than it flows in from the surrounding aquifer.

drilling mud A slurry of water mixed with clay that oil drillers use to cool a drill bit and flush rock cuttings up and out of the hole.

dripstone Limestone (travertine in a cave) formed by the precipitation of calcium carbonate out of groundwater.

drop stone A rock that drops to the sea floor once the iceberg that was carrying the rock melts.

drumlin A streamlined, elongate hill formed when a glacier overrides glacial till.

dry-bottom (polar) glacier A glacier so cold that its base remains frozen to the substrate.

dry wash The channel of an ephemeral stream when empty of water.

dry well (1) A well that does not supply water because the well has been drilled into an aquitard or into rock that lies above the water table; (2) a well that does not yield oil, even though it has been drilled into an anticipated reservoir.

ductile (plastic) deformation The bending and flowing of a material (without cracking and breaking) subjected to stress.

dune A pile of sand generally formed by deposition from the wind.

dust storm An event in which strong winds hit unvegetated land, strip off the topsoil, and send it skyward to form rolling dark clouds that block out the Sun.

dynamic metamorphism Metamorphism that occurs as a consequence of shearing alone, with no change in temperature or pressure.

dynamo A power plant generator in which water or wind power spins an electrical conductor around a permanent magnet.

dynamothermal metamorphism Metamorphism that involves heat, pressure, and shearing.

earthquake A vibration caused by the sudden breaking or frictional sliding of rock in the Earth.

earthquake belt A relatively narrow, distinct belt of earthquakes that defines the position of a plate boundary.

earthquake engineering The design of buildings that can withstand shaking.

earthquake zoning The determination of where land is relatively stable and where it might collapse because of seismicity.

Earth System The global interconnecting web of physical and biological phenomena involving the solid Earth, the hydrosphere, and the atmosphere.

ebb tide The falling tide.

eccentricity cycle The cycle of the gradual change of the Earth's orbit from a more circular to a more elliptical shape; the cycle takes around 100,000 years.

ecliptic The plane defined by a planet's orbit.

ecosystem An environment and its inhabitants.

eddy An isolated, ring-shaped current of water.

effusive eruption An eruption that yields mostly lava, not ash.

Ekman spiral The change in flow direction of water with depth, caused by the Coriolis effect.

Ekman transport The overall movement of a mass of water, resulting from the Eckman spiral, in a direction 90° to the wind direction.

elastic strain A change in shape of a material that disappears instantly when stress is removed.

electromagnet An electrical device that produces a magnetic field.

electron microprobe A laboratory instrument that can focus a beam of electrons on a small part of a mineral grain in order to create a signal that defines its chemical composition.

El Niño The flow of warm water eastward from the Pacific Ocean that reverses the upwelling of cold water along the western coast of South America and causes significant global changes in weather patterns.

embayment A low area of coastal land.

emergent coast A coast where the land is rising relative to sea level or sea level is falling relative to the land.

end moraine (terminal moraine) A low, sinuous ridge of till that develops when the terminus (toe) of a glacier stalls in one position for a while.

energy The capacity to do work.

energy resource Something that can be used to produce work; in a geologic context, a material (such as oil, coal, wind, flowing water) that can be used to produce energy.

eon The largest subdivision of geologic time.

epeirogenic movement The gradual uplift or subsidence of a broad region of the Earth's surface.

epeirogeny An event of epeirogenic movement; the term is usually used in reference to the formation of broad mid-continent domes and basins.

ephemeral (intermittent) stream A stream whose bed lies above the water table, so that the stream flows only when the rate at which water enters the stream from rainfall or meltwater exceeds the rate at which water infiltrates the ground below.

epicenter The point on the surface of the Earth directly above the focus of an earthquake.

epicontinental sea A shallow sea overlying a continent.

epoch An interval of geologic time representing the largest subdivision of a period.

equant A term for a grain that has the same dimensions in all directions.

equatorial low The area of low pressure that develops over the equator because of the intertropical convergence zone.

equinox One of two days out of the year (September 22 and March 21) in which the Sun is directly overhead at noon at the equator.

era An interval of geologic time representing the largest subdivision of the Phanerozoic Eon.

erg Sand seas formed by the accumulation of dunes in a desert.

erosion The grinding away and removal of Earth's surface materials by moving water, air, or ice.

erosional coast A coastline where sediment is not accumulating and wave action grinds away at the shore.

erosional landform A landform that results from the breakdown and removal of rock or sediment.

erratic A boulder or cobble that was picked up by a glacier and deposited hundreds of kilometers away from the outcrop from which it detached.

esker A ridge of sorted sand and gravel that snakes across a ground moraine; the sediment of an esker was deposited in subglacial meltwater tunnels.

estuary An inlet in which seawater and river water mix, created when a coastal valley is flooded because of either rising sea level or land subsidence.

Eubacteria The kingdom of "true bacteria."

euhedral crystal A crystal whose faces are well formed and whose shape reflects crystal form.

eukaryotic cell A cell with a complex internal structure, capable of building multicellular organisms.

eustatic sea-level change A global rising or falling of the ocean surface.

evaporate To change from liquid to vapor.

evapotranspiration The sum of evaporation from bodies of water and the ground surface and transpiration from plants and animals.

exfoliation The process by which an outcrop of rock splits apart into onion-like sheets along joints that lie parallel to the ground surface.

exhumation The process (involving uplift and erosion) that returns deeply buried rocks to the surface.

exotic terrane A block of land that collided with a continent along a convergent margin and attached to the continent; the term "exotic" implies that the land was not originally part of the continent to which it is now attached.

expanding Universe theory The theory that the whole Universe must be expanding because galaxies in every direction seem to be moving away from us.

explosive eruptions Violent volcanic eruptions that produce clouds and avalanches of pyroclastic debris.

external process A geomorphologic process—such as downslope movement, erosion, or deposition—that is the consequence of gravity or of the interaction between the solid Earth and its fluid envelope (air and water). Energy for these processes comes from gravity and sunlight.

extinction The death of the last members of a species so that there are no parents to pass on their genetic traits to offspring.

extinct volcano A volcano that was active in the past but has now shut off entirely and will not erupt in the future.

extraordinary fossil A rare fossilized relict, or trace, of the soft part of an organism.

extrusive igneous rock Rock that forms by the freezing of lava above ground, after it flows or explodes out (extrudes) onto the surface and comes into contact with the atmosphere or ocean.

eye The relative calm in the center of a hurricane.

eye wall A rotating vertical cylinder of clouds surrounding the eye of a hurricane.

facies (1) Sedimentary: a group of rocks and primary structures indicative of a given depositional environment; (2) Metamorphic: a set of metamorphic mineral assemblages formed under a given range of pressures and temperatures.

fault A fracture on which one body of rock slides past another.

fault-block mountains An outdated term for a narrow, elongate range of mountains that develops in a continental-rift setting as normal faulting drops down blocks of crust, or tilts blocks.

fault breccia Fragmented rock in which angular fragments were formed by brittle fault movement; fault breccia occurs along a fault.

fault creep Gradual movement along a fault that occurs in the absence of an earthquake.

fault gouge Pulverized rock consisting of fine powder that lies along fault surfaces; gouge forms by crushing and grinding.

faulting Slip events along a fault.

fault scarp A small step on the ground surface where one side of a fault has moved vertically with respect to the other.

fault system A grouping of numerous related faults.

fault trace (or line) The intersection between a fault and the ground surface.

felsic An adjective used in reference to igneous rocks that are rich in elements forming feldspar and quartz.

Ferrel cells The name given to the middle-latitude convection cells in the atmosphere.

fetch The distance across a body of water along which a wind blows to build waves.

fine-grained A textural term for rock consisting of many fine grains or clasts.

firn Compacted granular ice (derived from snow) that forms where snow is deeply buried; if buried more deeply, firn turns into glacial ice.

fission track A line of damage formed in the crystal lattice of a mineral by the impact of an atomic particle ejected during the decay of a radioactive isotope.

fissure A conduit in a magma chamber in the shape of a long crack through which magma rises and erupts at the surface.

fjord A deep, glacially carved, U-shaped valley flooded by rising sea level.

flank eruption An eruption that occurs when a secondary chimney, or fissure, breaks through the flank of a volcano.

flash flood A flood that occurs during unusually intense rainfall or as the result of a dam collapse, during which the floodwaters rise very fast.

flexing The process of folding in which a succession of rock layers bends and slip occurs between the layers.

flocculation The clumping together of clay suspended in river water into bunches that are large enough to settle out.

flood An event during which the volume of water in a stream becomes so great that it covers areas outside the stream's normal channel.

flood basalt Vast sheets of basalt that spread from a volcanic vent over an extensive surface of land; they may form where a rift develops above a continental hot spot, and where lava is particularly hot and has low viscosity.

floodplain The flat land on either side of a stream that becomes covered with water during a flood.

floodplain flood A flood during which a floodplain is submerged.

flood stage The stage when water reaches the top of a stream channel.

flood tide The rising tide.

floodway A mapped region likely to be flooded, in which people avoid constructing buildings.

flow fold A fold that forms when the rock is so soft that it behaves like weak plastic.

flowstone A sheet of limestone that forms along the wall of a cave when groundwater flows along the surface of the wall.

fluvial deposit Sediment deposited in a stream channel, along a stream bank, or on a floodplain.

flux Flow.

focus The location where a fault slips during an earthquake (hypocenter).

fog A cloud that forms at ground level.

fold A bend or wrinkle of rock layers or foliation; folds form as a consequence of ductile deformation.

fold axis An imaginary line that, when moved parallel to itself, can trace out the shape of a folded surface.

fold-thrust belt An assemblage of folds and related thrust faults that develop above a detachment fault.

foliation Layering formed as a consequence of the alignment of mineral grains, or of compositional banding in a metamorphic rock.

foraminifera Microscopic plankton with calcitic shells, components of some limestones.

foreland sedimentary basin A basin located under the plains adjacent to a mountain front, which develops as the weight of the mountains pushes the crust down, creating a depression that traps sediment.

foreshocks The series of smaller earthquakes that precede a major earthquake.

foreshore zone The zone of beach regularly covered and uncovered by rising and falling tides.

formation *Stratigraphic formation.*

fossil The remnant, or trace, of an ancient living organism that has been preserved in rock or sediment.

fossil assemblage A group of fossil species found in a specific sequence of sedimentary rock.

fossil correlation A determination of the stratigraphic relation between two sedimentary rock units, reached by studying fossils.

fossil fuel An energy resource such as oil or coal that comes from organisms that lived long ago, and thus stores solar energy that reached the Earth then.

fossiliferous limestone Limestone consisting of abundant fossil shells and shell fragments.

fossilization The process of forming a fossil.

fractional crystallization The process by which a magma becomes progressively more silicic as it cools, because early-formed crystals settle out.

fracture zone A narrow band of vertical fractures in the ocean floor; fracture zones lie roughly at right angles to a mid-ocean ridge, and the actively slipping part of a fracture zone is a transform fault.

fresh rock Rock whose mineral grains have their original composition and shape.

friction Resistance to sliding on a surface.

fringing reef A coral reef that forms directly along the coast.

front The boundary between two air masses.

frost wedging The process in which water trapped in a joint freezes, forces the joint open, and may cause the joint to grow.

fuel rod A metal tube that holds the nuclear fuel in a nuclear reactor.

Fujita scale A scale that distinguishes among tornadoes on the basis of wind speed, path dimensions, and possible damage.

Ga Billions of years ago (abbreviation).

gabbro A coarse-grained, intrusive mafic igneous rock.

Gaia The term used for the Earth System, with the implication that it resembles a complex living entity.

galaxy An immense system of hundreds of billions of stars.

gene An individual component of the DNA code that guides the growth and development of an organism.

genetics The study of genes and how they transmit information.

geocentric Universe concept An ancient Greek idea suggesting that the Earth sat motionless in the center of the Universe while stars and other planets and the Sun orbited around it.

geochronology The science of dating geologic events in years.

geode A cavity in which euhedral crystals precipitate out of water solutions passing through a rock.

geographical pole The locations (north and south) where the Earth's rotational axis intersects the planet's surface.

geologic column A composite stratigraphic chart that represents the entirety of the Earth's history.

geologic history The sequence of geologic events that has taken place in a region.

geologic map A map showing the distribution of rock units and structures across a region.

geologic time The span of time since the formation of the Earth.

geologic time scale A scale that describes the intervals of geologic time.

geology The study of the Earth, including our planet's composition, behavior, and history.

geotherm The change in temperature with depth in the Earth.

geothermal energy Heat and electricity produced by using the internal heat of the Earth.

geothermal gradient The rate of change in temperature with depth.

geothermal region A region of current or recent volcanism in which magma or very hot rock heats up groundwater, which may discharge at the surface in the form of hot springs and/or geysers.

geyser A fountain of steam and hot water that erupts periodically from a vent in the ground in a geothermal region.

glacial abrasion The process by which clasts embedded in the base of a glacier grind away at the substrate as the glacier flows.

glacial advance The forward movement of a glacier's toe when the supply of snow exceeds the rate of ablation.

glacial drift Sediment deposited in glacial environments.

glacial incorporation The process by which flowing ice surrounds and incorporates debris.

glacial marine Sediment consisting of ice-rafted clasts mixed with marine sediment.

glacial outwash Coarse sediment deposited on a glacial outwash plain by meltwater streams.

glacially polished surface A polished rock surface created by the glacial abrasion of the underlying substrate.

glacial plucking (or quarrying) The process by which a glacier breaks off and carries away fragments of bedrock.

glacial rebound The process by which the surface of a continent rises back up after an overlying continental ice sheet melts away and the weight of the ice is removed.

glacial retreat The movement of a glacier's toe back toward the glacier's origin; glacial retreat occurs if the rate of ablation exceeds the rate of supply.

glacial subsidence   The sinking of the surface of a continent caused by the weight of an overlying glacial ice sheet.

glacial till   Sediment transported by flowing ice and deposited beneath a glacier or at its toe.

glaciation   A period of time during which glaciers grew and covered substantial areas of the continents.

glacier   A river or sheet of ice that slowly flows across the land surface and lasts all year long.

glass   A solid in which atoms are not arranged in an orderly pattern.

glassy igneous rock   Igneous rock consisting entirely of glass, or of tiny crystals surrounded by a glass matrix.

glide horizon   The surface along which a slump slips.

global change   The transformations or modifications of both physical and biological components of the Earth System through time.

global circulation   The movement of volumes of air in paths that ultimately take it around the planet.

global climate change   Transformations or modifications in Earth's climate over time.

global cooling   A fall in the average atmospheric temperature.

global positioning system (GPS)   A satellite system people can use to measure rates of movement of the Earth's crust relative to one another, or simply to locate their position on the Earth's surface.

global warming   A rise in the average atmospheric temperature.

gneiss   A compositionally banded metamorphic rock typically composed of alternating dark- and light-colored layers.

Gondwana   A supercontinent that consisted of today's South America, Africa, Antarctica, India, and Australia. Also called Gondwanaland.

graben   A down-dropped crustal block bounded on either side by a normal fault dipping toward the basin.

gradualism   The theory that evolution happens at a constant, slow rate.

grain   A fragment of a mineral crystal or of a rock.

grain rotation   The process by which rigid, inequant mineral grains distributed through a soft matrix may rotate into parallelism as the rock changes shape owing to differential stress.

granite   A coarse-grained intrusive silicic igneous rock.

granulite facies   A set of metamorphic mineral assemblages formed at very high pressures and temperatures.

gravitational spreading   A process of lateral spreading that occurs in a material because of the weakness of the material; gravitational spreading causes continental glaciers to grow and mountain belts to undergo orogenic collapse.

graywacke   An informal term used for sedimentary rock consisting of sand-size up to small-pebble-size grains of quartz and rock fragments all mixed together in a muddy matrix; typically, graywacke occurs at the base of a graded bed.

greenhouse conditions (greenhouse period)   Relatively warm global climate leading to the rising of sea level for an interval of geologic time.

greenhouse effect   The trapping of heat in the Earth's atmosphere by carbon dioxide and other greenhouse gases, which absorb infrared radiation; somewhat analogous to the effect of glass in a greenhouse.

greenhouse gases   Atmospheric gases, such as carbon dioxide and methane, that regulate the Earth's atmospheric temperature by absorbing infrared radiation.

greenschist facies   A set of metamorphic mineral assemblages formed under relatively low pressures and temperatures.

greenstone   A low-grade metamorphic rock formed from basalt; if foliated, the rock is called greenschist.

Greenwich mean time (GMT)   The time at the astronomical observatory in Greenwich, England; time in all other time zones is set in relation to GMT.

Grenville orogeny   The orogeny that occurred about 1 billion years ago and yielded the belt of deformed and metamorphosed rocks that underlie the eastern fifth of the North American continent.

groin   A concrete or stone wall built perpendicular to a shoreline in order to prevent beach drift from removing sand.

ground moraine   A thin, hummocky layer of till left behind on the land surface during a rapid glacial recession.

groundwater   Water that resides under the surface of the Earth, mostly in pores or cracks of rock or sediment.

group   A succession of stratigraphic formations that have been lumped together, making a single, thicker stratigraphic entity.

growth ring   A rhythmic layering that develops in trees, travertine deposits, and shelly organisms as a consequence of seasonal changes.

gusher   A fountain of oil formed when underground pressure causes the oil to rise on its own out of a drilled hole.

guyot   A seamount that had a coral reef growing on top of it, so that it is now flat-crested.

gymnosperm   A plant whose seeds are "naked," not surrounded by a fruit.

gyre   A large circular flow pattern of ocean surface currents.

Hadean   The oldest of the Precambrian eons; the time between Earth's origin and the formation of the first rocks that have been preserved.

Hadley cells   The name given to the low-latitude convection cells in the atmosphere.

hail   Falling ice balls from the sky, formed when ice crystallizes in turbulent storm clouds.

hail streak   An approximately 2-by-10-km stretch of ground, elongate in the direction of a storm, onto which hail has fallen.

half-graben   A wedge-shaped basin in cross section that develops as the hanging-wall block above a normal fault slides down and rotates; the basin develops between the fault surface and the top surface of the rotated block.

half-life   The time it takes for half of a group of a radioactive element's isotopes to decay.

halocline   The boundary in the ocean between surface-water and deep-water salinities.

hamada   Barren rocky highlands in a desert.

hanging valley   A glacially carved tributary valley whose floor lies at a higher elevation than the floor of the trunk valley.

hanging wall   The rock or sediment above an inclined fault plane.

hard water   Groundwater that contains dissolved calcium and magnesium, usually after passing through limestone or dolomite.

head   (1) The elevation of the water table above a reference horizon; (2) the edge of ice at the origin of a glacier.

headland   A place where a hill or cliff protrudes into the sea.

head scarp   The distinct step along the upslope edge of a slump where the regolith detached.

headward erosion   The process by which a stream channel lengthens up its slope as the flow of water increases.

headwaters   The beginning point of a stream.

heat   Thermal energy resulting from the movement of molecules.

heat capacity   A measure of the amount of heat that must be added to a material to change its temperature.

heat flow   The rate at which heat rises from the Earth's interior up to the surface.

heat-transfer melting Melting that results from the transfer of heat from a hotter magma to a cooler rock.

heliocentric Universe concept An idea proposed by Greek philosophers around 250 B.C.E. suggesting that all heavenly objects including the Earth orbited the Sun.

Hercynian orogen The late Paleozoic orogen that affected parts of Europe; a continuation of the Alleghenian orogen.

heterosphere A term for the upper portion of the atmosphere in which gases separate into distinct layers on the basis of composition.

hiatus The interval of time between deposition of the youngest rock below an unconformity and deposition of the oldest rock above the unconformity.

high-altitude westerlies Westerly winds at the top of the troposphere.

high-grade metamorphic rocks Rocks that metamorphose under relatively high temperatures.

high-level waste Nuclear waste containing greater than 1 million times the safe level of radioactivity.

hinge The portion of a fold where curvature is greatest.

hogback A steep-sided ridge of steeply dipping strata.

Holocene The period of geologic time since the last glaciation.

Holocene climatic maximum The period from 5,000 to 6,000 years ago, when Holocene temperatures reached a peak.

homosphere The lower part of the atmosphere, in which the gases have stirred into a homogenous mixture.

hoodoo The local name for the brightly colored shale and sandstone chimneys found in Bryce Canyon National Park in Utah.

horn A pointed mountain peak surrounded by at least three cirques.

hornfels Rock that undergoes metamorphism simply because of a change in temperature, without being subjected to differential stress.

horse latitudes The region of the subtropical high in which winds are weak.

horst The high block between two grabens.

hot spot A location at the base of the lithosphere, at the top of a mantle plume, where temperatures can cause melting.

hot-spot track A chain of now-dead volcanoes transported off the hot spot by the movement of a lithosphere plate.

hot-spot volcano An isolated volcano not caused by movement at a plate boundary, but rather by the melting of a mantle plume.

hot spring A spring that emits water ranging in temperature from about 30° to 104°C.

hummocky surface An irregular and lumpy ground surface.

hurricane A huge rotating storm, resembling a giant spiral in map view, in which sustained winds blow over 119 km per hour.

hurricane track The path a hurricane follows.

hyaloclastite A rubbly extrusive rock consisting of glassy debris formed in a submarine or sub-ice eruption.

hydration The absorption of water into the crystal structure of minerals; a type of chemical weathering.

hydraulic conductivity The coefficient K in Darcy's law; hydraulic conductivity takes into account the permeability of the sediment or rock as well as the fluid's viscosity.

hydraulic gradient The slope of the water table.

hydrocarbon A chain-like or ring-like molecule made of hydrogen and carbon atoms; petroleum and natural gas are hydrocarbons.

hydrocarbon system The association of source rock, migration pathway, reservoir rock, seal, and trap geometry that leads to the occurrence of a hydrocarbon reserve.

hydrologic cycle The continual passage of water from reservoir to reservoir in the Earth System.

hydrolysis The process in which water chemically reacts with minerals and breaks them down.

hydrosphere The Earth's water, including surface water (lakes, rivers, and oceans), groundwater, and liquid water in the atmosphere.

hydrothermal deposit An accumulation of ore minerals precipitated from hot-water solutions circulating through a magma or through the rocks surrounding an igneous intrusion.

hypsometric curve A graph that plots surface elevation on the vertical axis and the percentage of the Earth's surface on the horizontal axis.

ice age An interval of time in which the climate was colder than it is today, glaciers occasionally advanced to cover large areas of the continents, and mountain glaciers grew; an ice age can include many glacials and interglacials.

iceberg A large block of ice that calves off the front of a glacier and drops into the sea.

icehouse period A period of time when the Earth's temperature was cooler than it is today and ice ages could occur.

ice-margin lake A meltwater lake formed along the edge of a glacier.

ice-rafted sediment Sediment carried out to sea by icebergs.

ice sheet A vast glacier that covers the landscape.

ice shelf A broad, flat region of ice along the edge of a continent formed where a continental glacier flowed into the sea.

ice stream A portion of a glacier that travels much more quickly than adjacent portions of the glacier.

ice tongue The portion of a valley glacier that has flowed out into the sea.

igneous rock Rock that forms when hot molten rock (magma or lava) cools and freezes solid.

ignimbrite Rock formed when deposits of pyroclactic flows solidify.

inactive fault A fault that last moved in the distant past and probably won't move again in the near future, yet is still recognizable because of displacement across the fault plane.

inactive sand The sand along a coast that is buried beneath a layer of active sand and moves only during severe storms or not at all.

incised meander A meander that lies at the bottom of a steep-walled canyon.

index minerals Minerals that serve as good indicators of metamorphic grade.

induced seismicity Seismic events caused by the actions of people (e.g. filling a reservoir, that lies over a fault, with water).

industrial minerals Minerals that serve as the raw materials for manufacturing chemicals, concrete, and wallboard, among other products.

inequant A term for a mineral grain whose length and width are not the same.

inertia The tendency of an object at rest to remain at rest.

infiltrate Seep down into.

injection well A well in which a liquid is pumped down into the ground under pressure so that it passes from the well back into the pore space of the rock or regolith.

inner core The inner section of the core 5,155 km deep to the Earth's center at 6,371 km, and consisting of solid iron alloy.

inselberg An isolated mountain or hill in a desert landscape created by progressive cliff retreat, so that the hill is surrounded by a pediment or an alluvial fan.

insolation Exposure to the Sun's rays.

interglacial A period of time between two glaciations.

interior basin A basin with no outlet to the sea.

interlocking texture The texture of crystalline rocks in which mineral grains fit together like pieces of a jigsaw puzzle.

internal process A process in the Earth System, such as plate motion, mountain building, or volcanism, ultimately caused by Earth's internal heat.

intertidal zone The area of coastal land across which the tide rises and falls.

intertropical convergence zone The equatorial convergence zone in the atmosphere.

intraplate earthquakes Earthquakes that occur away from plate boundaries.

intrusive contact The boundary between country rock and an intrusive igneous rock.

intrusive igneous rock Rock formed by the freezing of magma underground.

ionosphere The interval of Earth's atmosphere, at an elevation between 50 and 400 km, containing abundant positive ions.

iron catastrophe The proposed event very early in Earth history when the Earth partly melted and molten iron sank to the center to form the core.

isobar A line on a map along which the air has a specified pressure.

isograd (1) A line on a pressure-temperature graph along which all points are taken to be at the same metamorphic grade; (2) A line on a map making the first appearance of a metamorphic index mineral.

isostasy (or **isostatic equilibrium**) The condition that exists when the buoyancy force pushing lithosphere up equals the gravitational force pulling lithosphere down.

isostatic compensation The process in which the surface of the crust slowly rises or falls to reestablish isostatic equilibrium after a geologic event changes the density or thickness of the lithosphere.

isotherm Lines on a map or cross section along which the temperature is constant.

isotopes Different versions of a given element that have the same atomic number but different atomic weights.

jet stream A fast-moving current of air that flows at high elevations.

jetty A manmade wall that protects the entrance to a harbor.

joints Naturally formed cracks in rocks.

joint set A group of systematic joints.

Jovian A term used to describe the outer gassy, Jupiter-like planets (gas-giant planets).

kame A stratified sequence of lateral-moraine sediment that's sorted by water flowing along the edge of a glacier.

karst landscape A region underlain by caves in limestone bedrock; the collapse of the caves creates a landscape of sinkholes separated by higher topography, or of limestone spires separated by low areas.

kerogen The waxy molecules into which the organic material in shale transforms on reaching about 100°C. At higher temperatures, kerogen transforms into oil.

kettle hole A circular depression in the ground made when a block of ice calves off the toe of a glacier, becomes buried by till, and later melts.

knob-and-kettle topography A land surface with many kettle holes separated by round hills of glacial till.

K-T boundary event The mass extinction that happened at the end of the Cretaceous Period, 65 million years ago, possibly because of the collision of an asteroid with the Earth.

lag deposit The coarse sediment left behind in a desert after wind erosion removes the finer sediment.

lagoon A body of shallow seawater separated from the open ocean by a barrier island.

lahar A thick slurry formed when volcanic ash and debris mix with water, either in rivers or from rain or melting snow and ice on the flank of a volcano.

landslide A sudden movement of rock and debris down a non-vertical slope.

landslide-potential map A map on which regions are ranked according to the likelihood that a mass movement will occur.

land subsidence Sinking elevation of the ground surface; the process may occur over an aquifer that is slowly draining and decreasing in volume because of pore collapse.

La Niña Years in which the El Niño event is not strong.

lapilli Marble-to-plum-sized fragments of pyroclastic debris.

Laramide orogeny The mountain-building event that lasted from about 80 Ma to 40 Ma, in western North America; in the United States, it formed the Rocky Mountains as a result of basement uplift and the warping of the younger overlying strata into large monoclines.

latent heat of condensation The heat released during condensation, which comes only from a change in state.

lateral moraine A strip of debris along the side margins of a glacier.

laterite soil Soil formed over iron-rich rock in a tropical environment, consisting primarily of a dark-red mass of insoluble iron and/or aluminum oxide.

Laurentia A continent in the early Paleozoic Era composed of today's North America and Greenland.

Laurentide ice sheet An ice sheet that spread over northeastern Canada during the Pleistocene ice age(s).

lava Molten rock that has flowed out onto the Earth's surface.

lava dome A dome-like mass of rhyolitic lava that accumulates above the eruption vent.

lava flows Sheets or mounds of lava that flow onto the ground surface or sea floor in molten form and then solidify.

lava lake A large pool of lava produced around a vent when lava fountains spew forth large amounts of lava in a short period of time.

lava tube The empty space left when a lava tunnel drains; this happens when the surface of a lava flow solidifies while the inner part of the flow continues to stream downslope.

leach To dissolve and carry away.

leader A conductive path stretching from a cloud toward the ground, along which electrons leak from the base of the cloud, and which provides the start for a lightning flash to the ground.

lightning flash A giant spark or pulse of current that jumps across a gap of charge separation.

light year The distance that light travels in one Earth year (about 6 trillion miles or 9.5 trillion km).

lignite Low-rank coal that consists of 50% carbon.

limb The side of a fold, showing less curvature than at the hinge.

limestone Sedimentary rock composed of calcite.

liquification The process by which wet sediment becomes a slurry; liquification may be triggered by earthquake vibrations.

lithification The transformation of loose sediment into solid rock through compaction and cementation.

lithologic correlation A correlation based on similarities in rock type.

lithosphere The relatively rigid, nonflowable, outer 100- to 150-km-thick layer of the Earth; constituting the crust and the top part of the mantle.

little ice age A period of cooler temperatures, between 1500 and 1800 C.E., during which many glaciers advanced.

local base level A base level upstream from a drainage network's mouth.

lodgment till A flat layer of till smeared out over the ground when a glacier overrides an end moraine as it advances.

loess Layers of fine-grained sediments deposited from the wind; large deposits of loess formed from fine-grained glacial sediment blown off outwash plains.

longitudinal (seif) dune A dune formed when there is abundant sand and a strong, steady wind, and whose axis lies parallel to the wind direction.

longitudinal profile A cross-sectional image showing the variation in elevation along the length of a river.

longshore current A current that flows parallel to a beach.

lower mantle The deepest section of the mantle, stretching from 670 km down to the core-mantle boundary.

low-grade metamorphic rocks Rocks that underwent metamorphism at relatively low temperatures.

low-velocity zone The asthenosphere underlying oceanic lithosphere in which seismic waves travel more slowly, probably because rock has partially melted.

luster The way a mineral surface scatters light.

L-waves (*love waves*) Surface seismic waves that cause the ground to ripple back and forth, creating a snake-like movement.

Ma Millions of years ago (abbreviation).

macrofossil A fossil large enough to be seen with the naked eye.

mafic A term used in reference to magmas or igneous rocks that are relatively poor in silica and rich in iron and magnesium.

magma Molten rock beneath the Earth's surface.

magma chamber A space below ground filled with magma.

magma contamination The process in which flowing magma incorporates components of the country rock through which it passes.

magmatic deposit An ore deposit formed when sulfide ore minerals accumulate at the bottom of a magma chamber.

magnetic anomaly The difference between the expected strength of the Earth's magnetic field at a certain location and the actual measured strength of the field at that location.

magnetic declination The angle between the direction a compass needle points at a given location and the direction of true north.

magnetic field The region affected by the force emanating from a magnet.

magnetic field lines The trajectories along which magnetic particles would align, or charged particles would flow, if placed in a magnetic field.

magnetic force The push or pull exerted by a magnet.

magnetic inclination The angle between a magnetic needle free to pivot on a horizontal axis and a horizontal plane parallel to the Earth's surface.

magnetic reversal The change of the Earth's magnetic polarity; when a reversal occurs, the field flips from normal to reversed polarity, or vice versa.

magnetic-reversal chronology The history of magnetic reversals through geologic time.

magnetization The degree to which a material can exert a magnetic force.

magnetometer An instrument that measures the strength of the Earth's magnetic field.

magnetosphere The region protected from the electrically charged particles of the solar winds by Earth's magnetic field.

magnetostratigraphy The comparison of the pattern of magnetic reversals in a sequence of strata, with a reference column showing the succession of reversals through time.

manganese nodules Lumpy accumulations of manganese-oxide minerals precipitated onto the sea floor.

mantle The thick layer of rock below the Earth's crust and above the core.

mantle plume A column of very hot rock rising up through the mantle.

marble A metamorphic rock composed of calcite and transformed from a protolith of limestone.

mare The broad darker areas on the Moon's surface, which consist of flood basalts that erupted over 3 billion years ago and spread out across the Moon's lowlands.

marginal sea A small ocean basin created when sea-floor spreading occurs behind an island arc.

maritime tropical air mass A mass of air that originates over tropical or subtropical oceanic regions.

marsh A wetland dominated by grasses.

mass-extinction event A time when vast numbers of species abruptly vanish.

mass movement (or **mass wasting**) The gravitationally caused downslope transport of rock, regolith, snow, or ice.

matrix Finer-grained material surrounding larger grains in a rock.

meander A snake-like curve along a stream's course.

meandering stream A reach of stream containing many meanders (snake-like curves).

meander neck A narrow isthmus of land separating two adjacent meanders.

mean sea level The average level between the high and low tide over a year at a given point.

mechanical weathering *Physical weathering.*

medial moraine A strip of sediment in the interior of a glacier, parallel to the flow direction of the glacier, formed by the lateral moraines of two merging glaciers.

Medieval warm period A period of high temperatures in the Middle Ages.

melt Molten (liquid) rock.

meltdown The melting of the fuel rods in a nuclear reactor that occurs if the rate of fission becomes too fast and the fuel rods become too hot.

melting curve The line defining the range of temperatures and pressures at which a rock melts.

melting temperature The temperature at which the thermal vibration of the atoms or ions in the lattice of a mineral is sufficient to break the chemical bonds holding them to the lattice, so a material transforms into a liquid.

meltwater lake A lake fed by glacial meltwater.

Mercalli intensity scale An earthquake characterization scale based on the amount of damage that the earthquake causes.

mesa A large, flat-topped hill (with a surface area of several square km) in an arid region.

mesopause The boundary that marks the top of the mesosphere of Earth's atmosphere.

mesosphere The cooler layer of atmosphere overlying the stratosphere.

Mesozoic The middle of the three Phanerozoic eras; it lasted from 245 Ma to 65 Ma.

metal A solid composed almost entirely of atoms of metallic elements; it is generally opaque, shiny, smooth, and malleable, and can conduct electricity.

metallic bond A chemical bond in which the outer atoms are attached to each other in such a way that electrons flow easily from atom to atom.

metamorphic aureole The region around a pluton, stretching tens to hundreds of meters out, in which heat transferred into the country rock and metamorphosed the country rock.

metamorphic facies A set of metamorphic mineral assemblages indicative of metamorphism under a specific range of pressures and temperatures.

metamorphic foliation A fabric defined by parallel surfaces or layers that develop in a rock as a result of metamorphism; schistocity and gneissic layering are examples.

metamorphic mineral assemblage A group of minerals that form in a rock as a result of metamorphism.

metamorphic rock Rock that forms when preexisting rock changes into new rock as a result of an increase in pressure and temperature and/or shearing under elevated temperatures; metamorphism occurs without the rock first becoming a melt or a sediment.

metamorphic zone The region between two metamorphic isograds, typically named after an index mineral found within the region.

metamorphism The process by which one kind of rock transforms into a different kind of rock.

metasomatism The process by which a rock's overall chemical composition changes during metamorphism because of reactions with hot water that bring in or remove elements.

meteoric water Water that falls to Earth from the atmosphere as either rain or snow.

meteorite A piece of rock or metal alloy that fell from space and landed on Earth.

micrite Limestone consisting of lime mud (i.e., very fine-grained limestone).

microfossil A fossil that can be seen only with a microscope or an electron microscope.

mid-latitude (wave) cyclone The circulation of air around large, low-pressure masses.

mid-ocean ridge A 2-km-high submarine mountain belt that forms along a divergent oceanic plate boundary.

migmatite A rock formed when gneiss is heated high enough so that it begins to partially melt, creating layers, or lenses, of new igneous rock that mix with layers of the relict gneiss.

Milankovitch cycles Climate cycles that occur over tens to hundreds of thousands of years, because of changes in Earth's orbit and tilt.

mine A site at which ore is extracted from the ground.

mineral A homogenous, naturally occurring, solid inorganic substance with a definable chemical composition and an internal structure characterized by an orderly arrangement of atoms, ions, or molecules in a lattice. Most minerals are inorganic.

mineral classes Groups of minerals distinguished from each other on the basis of chemical composition.

mineral resources The minerals extracted from the Earth's upper crust for practical purposes.

Mississippi Valley–type (MVT) ore An ore deposit, typically in dolostone, containing lead- and zinc-bearing minerals that precipitated from groundwater that had moved up from several km depth in the upper crust; such deposits occur in the upper Mississippi Valley.

Moho The seismic-velocity discontinuity that defines the boundary between the Earth's crust and mantle.

Mohs hardness scale A list of ten minerals in a sequence of relative hardness, with which other minerals can be compared.

mold A cavity in sedimentary rock left behind when a shell that once filled the space weathers out.

monocline A fold in the land surface whose shape resembles that of a carpet draped over a stair step.

monsoon A seasonal reversal in wind direction that causes a shift from a very dry season to a very rainy season in some regions of the world.

moraine A sediment pile composed of till deposited by a glacier.

mountain front The boundary between a mountain range and adjacent plains.

mountain (alpine) glacier A glacier that exists in or adjacent to a mountainous region.

mountain ice cap A mound of ice that submerges peaks and ridges at the crest of a mountain range.

mouth The outlet of a stream where it discharges into another stream, a lake, or a sea.

mudflow A downslope movement of mud at slow to moderate speed.

mud pot A viscous slurry that forms in a geothermal region when hot water or steam rises into soils rich in volcanic ash and clay.

mudstone Very fine-grained sedimentary rock that will not easily split into sheets.

mylonite Rock formed during dynamic metamorphism and characterized by foliation that lies roughly parallel to the fault (shear zone) involved in the shearing process; mylonites have very fine grains formed by the nonbrittle subdivision of larger grains.

native metal A naturally occurring pure mass of a single metal in an ore deposit.

natural arch An arch that forms when erosion along joints leaves narrow walls of rock; when the lower part of the wall erodes while the upper part remains, an arch results.

natural levees A pair of low ridges that appear on either side of a stream and develop as a result of the accumulation of sediment deposited naturally during flooding.

natural selection The process by which the fittest organisms survive to pass on their characteristics to the next generation.

neap tide An especially low tide that occurs when the angle between the direction of the Moon and the direction of the Sun is 90°.

nebula A cloud of gas or dust in space.

nebula theory of planet formation The concept that planets grow out of rings of gas, dust, and ice surrounding a new-born star.

negative anomaly An area where the magnetic field strength is less than expected.

negative feedback Feedback that slows a process down or reverses it.

Neocrystallization The growth of new crystals, not in the protolith, during metamorphism.

Nevadan orogeny A convergent-margin mountain-building event that took place in western North America during the Late Jurassic Period.

nonconformity A type of unconformity at which sedimentary rocks overlie basement (older intrusive igneous rocks and/or metamorphic rocks).

nonflowing artesian well An artesian well in which water rises on its own up to a level that lies below the ground surface.

nonfoliated metamorphic rock Rock containing minerals that recrystallized during metamorphism, but which has no foliation.

nonmetallic mineral resources Mineral resources that do not contain metals; examples include building stone, gravel, sand, gypsum, phosphate, and salt.

nonplunging fold A fold with a horizontal hinge.

nonrenewable resource A resource that nature will take a long time (hundreds to millions of years) to replenish or may never replenish.

nonsystematic joints Short cracks in rocks that occur in a range of orientations and are randomly placed and oriented.

nor'easter A large, mid-latitude North American cyclone; when it reaches the East Coast, it produces strong winds that come out of the northeast.

normal fault A fault in which the hanging-wall block moves down the slope of the fault.

normal force The component of the gravitational force acting perpendicular to a slope.

normal polarity Polarity in which the paleomagnetic dipole has the same orientation as it does today.

normal stress The push or pull that is perpendicular to a surface.

North Atlantic deep-water mass The mass of cold, dense water that sinks in the north polar regions.

northeast tradewinds Surface winds that come out of the northeast and occur in the region between the equator and 30°N.

nuclear fuel Pellets of concentrated uranium oxide or a comparable radioactive material that can provide energy in a nuclear reactor.

nuclear fusion The process by which the nuclei of atoms fuse together, thereby creating new, larger atoms.

nuclear reactor The part of a nuclear power plant where the fission reactions occur.

nuée ardente *Pyroclastic flow.*

oasis A verdant region surrounded by desert, occurring at a place where natural springs provide water at the surface.

oblique-slip fault A fault in which sliding occurs diagonally along the fault plane.

obsidian An igneous rock consisting of a solid mass of volcanic glass.

occluded front A front that no longer intersects the ground surface.

oceanic crust The crust beneath the oceans; composed of gabbro and basalt, overlain by sediment.

oceanic lithosphere Lithosphere topped by oceanic crust; it reaches a thickness of 100 km.

Oil Age The period of human history, including our own, so named because the economy depends on oil.

oil field A region containing a significant amount of accessible oil underground.

oil reserve The known supply of oil held underground.

oil shale Shale containing kerogen.

oil trap A geologic configuration that keeps oil underground in the reservoir rock and prevents it from rising to the surface.

oil window The narrow range of temperatures under which oil can form in a source rock.

olistotrome A large, submarine slump block, buried and preserved.

ophiolite A slice of oceanic crust that has been thrust onto continental crust.

ordinary well A well whose base penetrates below the water table, and can thus provide water.

ore Rock containing native metals or a concentrated accumulation of ore minerals.

ore deposit An economically significant accumulation of ore.

ore minerals Minerals that have metal in high concentrations and in a form that can be easily extracted.

organic carbon Carbon that has been incorporated in an organism.

organic chemical A carbon-containing compound that occurs in living organisms, or that resembles such compounds; it consists of carbon atoms bonded to hydrogen atoms along with varying amounts of oxygen, nitrogen, and other chemicals.

organic coast A coast along which living organisms control landforms along the shore.

organic sedimentary rock Sedimentary rock (such as coal) formed from carbon-rich relicts of organisms.

organic shale Lithified, muddy, organic-rich ooze that contains the raw materials from which hydrocarbons eventually form.

orogen (or orogenic belt) A linear range of mountains.

orogenic collapse The process in which mountains begin to collapse under their own weight and spread out laterally.

orogeny A mountain-building event.

orographic barrier A landform that diverts air flow upward or laterally.

outcrop An exposure of bedrock.

outer core The section of the core, between 2,900 and 5,150 km deep, that consists of liquid iron alloy.

outwash plain A broad area of gravel and sandbars deposited by a braided stream network, fed by the meltwater of a glacier.

overburden The weight of overlying rock on rock buried deeper in the Earth's crust.

overriding plate (or slab) The plate at a subduction zone that overrides the downgoing plate.

oversaturated solution A solution that contains so much solute (dissolved ions) that precipitation begins.

oversized stream valley A large valley with a small stream running through it; the valley formed earlier when the flow was greater.

oxbow lake A meander that has been cut off yet remains filled with water.

oxidation reaction A reaction in which an element loses electrons; an example is the reaction of iron with air to form rust.

ozone O_3, an atmospheric gas that absorbs harmful ultraviolet radiation from the Sun.

ozone hole An area of the atmosphere, over polar regions, from which ozone has been depleted.

pahoehoe A lava flow with a surface texture of smooth, glassy, rope-like ridges.

paleoclimate The past climate of the Earth.

paleomagnetism The record of ancient magnetism preserved in rock.

paleopole The supposed position of the Earth's magnetic pole in the past, with respect to a particular continent.

paleosol Ancient soil preserved in the stratigraphic record.

Paleozoic The oldest era of the Phanerozoic Eon.

Pangaea A supercontinent that assembled at the end of the Paleozoic Era.

Pannotia A supercontinent that may have existed sometime between 800 Ma and 600 Ma.

parabolic dunes Dunes formed when strong winds break through transverse dunes to make new dunes whose ends point upwind.

parallax The apparent movement of an object seen from two different points not on a straight line from the object (e.g., from your two different eyes).

parallax method A trigonometric method used to determine the distance from the Earth to a nearby star.

parent isotope A radioactive isotope that undergoes decay.

partial melting The melting in a rock of the minerals with the lowest melting temperatures, while other minerals remain solid.

passive margin A continental margin that is not a plate boundary.

passive-margin basin A thick accumulation of sediment along a tectonically inactive coast, formed over crust that stretched and thinned when the margin first began.

patterned ground A polar landscape in which the ground splits into pentagonal or hexagonal shapes.

pause An elevation in the atmosphere where temperature stops decreasing and starts increasing, or vice versa.

peat Compacted and partially decayed vegetation accumulating beneath a swamp.

pedalfer soil A temperate-climate soil characterized by well-defined soil horizons and an organic A-horizon.

pediment The broad, nearly horizontal bedrock surface at the base of a retreating desert cliff.

pedocal soil Thin soil, formed in arid climates. It contains very little organic matter, but significant precipitated calcite.

pegmatite A coarse-grained igneous rock containing crystals of up to tens of centimeters across and occurring in dike-shaped intrusions.

pelagic sediment Microscopic plankton shells and fine flakes of clay that settle out and accumulate on the deep-ocean floor.

Pelé's hair Droplets of basaltic lava that mold into long glassy strands as they fall.

Pelé's tears Droplets of basaltic lava that mold into tear-shaped glassy beads as they fall.

peneplain A nearly flat surface that lies at an elevation close to sea level; thought to be the product of long-term erosion.

perched water table A quantity of groundwater that lies above the regional water table because an underlying lens of impermeable rock or sediment prevents the water from sinking down to the regional water table.

percolation The process by which groundwater meanders through tiny, crooked channels in the surrounding material.

peridotite A coarse-grained ultramafic rock.

periglacial environment A region with widespread permafrost but without a blanket of snow or ice.

period An interval of geologic time representing a subdivision of a geologic era.

permafrost Permanently frozen ground.

permanent magnet A special material that behaves magnetically for a long time all by itself.

permanent stream A stream that flows year-round because its bed lies below the water table, or because more water is supplied from upstream than can infiltrate the ground.

permeability The degree to which a material allows fluids to pass through it via an interconnected network of pores and cracks.

permineralization The fossilization process in which plant material becomes transformed into rock by the precipitation of silica from groundwater.

petrified A term used by geologists to describe plant material that has transformed into rock by permineralization.

petroglyph Drawings formed by chipping into the desert varnish of rocks to reveal the lighter rock beneath.

petroleum *Oil.*

phaneritic A textural term used to describe coarse-grained igneous rock.

Phanerozoic Eon The most recent eon, an interval of time from 542 Ma to the present.

phenocryst A large crystal surrounded by a finer-grained matrix in an igneous rock.

photochemical smog Brown haze that blankets a city when exhaust from cars and trucks reacts in the presence of sunlight.

photosynthesis The process during which chlorophyll-containing plants remove carbon dioxide from the atmosphere, form tissues, and expel oxygen back to the atmosphere.

phreatomagmatic eruption An explosive eruption that occurs when water enters the magma chamber and turns into steam.

phyllite A fine-grained metamorphic rock with a foliation caused by the preferred orientation of very fine-grained mica.

phyllitic luster A silk-like sheen characteristic of phyllite, a result of the rock's fine-grained mica.

phylogenetic tree A chart representing the ideas of paleontologists showing which groups of organisms radiated from which ancestors.

physical weathering The process in which intact rock breaks into smaller grains or chunks.

piedmont glacier A fan or lobe of ice that forms where a valley glacier emerges from a valley and spreads out into the adjacent plain.

pillow basalt Glass-encrusted basalt blobs that form when magma extrudes on the sea floor and cools very quickly.

placer deposit Concentrations of metal grains in stream sediment that develop when rocks containing native metals erode and create a mixture of sand grains and metal fragments; the moving water of the stream carries away lighter mineral grains.

planetesimal Tiny, solid pieces of rock and metal that collect in a planetary nebula and eventually accumulate to form a planet.

plankton Tiny plants and animals that float in sea or lake water.

plastic deformation The deformational process in which mineral grains behave like plastic and, when compressed or sheared, become flattened or elongate without cracking or breaking.

plate One of about twenty distinct pieces of the relatively rigid lithosphere.

plate boundary The border between two adjacent lithosphere plates.

plate-boundary earthquakes The earthquakes that occur along and define plate boundaries.

plate-boundary volcano A volcanic arc or mid-ocean ridge volcano, formed as a consequence of movement along a plate boundary.

plate interior A region away from the plate boundaries that consequently experiences few earthquakes.

plate tectonics *Theory of plate tectonics.*

playa The flat, typically salty lake bed that remains when all the water evaporates in drier times; forms in desert regions.

Pleistocene ice age(s) The period of time from about 2 Ma to 14,000 years ago, during which the Earth experienced an ice age.

plunge pool A depression at the base of a waterfall scoured by the energy of the falling water.

plunging fold A fold with a tilted hinge.

pluton An irregular or blob-shaped intrusion; can range in size from tens of m across to tens of km across.

pluvial lake A lake formed to the south of a continental glacier as a result of enhanced rainfall during an ice age.

point bar A wedge-shaped deposit of sediment on the inside bank of a meander.

polar cell A high-latitude convection cell in the atmosphere.

polar easterlies Prevailing winds that come from the east and flow from the polar high to the subpolar low.

polar front The convergence zone in the atmosphere at latitude 60°.

polar glacier *Dry-bottom glacier.*

polar high The zone of high pressure in polar regions created by the sinking of air in the polar cells.

polarity The orientation of a magnetic dipole.

polarity chron The time interval between polarity reversals of Earth's magnetic field.

polarity subchron The time interval between magnetic reversals if the interval is of short duration (less than 200,000 years long).

polarized light A beam of filtered light waves that all vibrate in the same plane.

polar wander The phenomenon of the progressive changing through time of the position of the Earth's magnetic poles relative to a location on a continent; significant polar wander probably doesn't occur—in fact, poles seem to remain fairly fixed, while continents move.

polar-wander path The curving line representing the apparent progressive change in the position of the Earth's magnetic pole, relative to a locality X, assuming that the position of X on Earth has been fixed through time (in fact, poles stay fixed while continents move).

pollen Tiny grains involved in plant reproduction.

polymorphs Two minerals that have the same chemical composition but a different crystal lattice structure.

pore A small open space within sediment or rock.

pore collapse The closer packing of grains that occurs when groundwater is extracted from pores, thus eliminating the support holding the grains apart.

porosity The total volume of empty space (pore space) in a material, usually expressed as a percentage.

porphyritic A textural term for igneous rock that has phenocrysts distributed throughout a finer matrix.

positive anomaly An area where the magnetic field strength is stronger than expected.

positive-feedback mechanism A mechanism that enhances the process that causes the mechanism in the first place.

potentiometric surface The elevation to which water in an artesian system would rise if unimpeded; where there are flowing artesian wells, the potentiometric surface lies above ground.

pothole A bowl-shaped depression carved into the floor of a stream by a long-lived whirlpool carrying sand or gravel.

Precambrian The interval of geologic time between Earth's formation about 4.57 Ga and the beginning of the Phanerozoic Eon 542 Ma.

precession The gradual conical path traced out by Earth's spinning axis; simply put, it is the "wobble" of the axis.

precious metals Metals (like gold, silver, and platinum) that have high value.

precipitation (1) The process by which atoms dissolved in a solution come together and form a solid; (2) rainfall or snow.

preferred mineral orientation The metamorphic texture that exists where platy grains lie parallel to one another and/or elongate grains align in the same direction.

pressure Force per unit area, or the "push" acting on a material in cases where the push is the same in all directions.

pressure gradient The rate of pressure change over a given horizontal distance.

pressure solution The process of dissolution at points of contact, between grains, where compression is greatest, producing ions that then precipitate elsewhere, where compression is less.

prevailing winds Surface winds that generally flow in the same direction for long time periods.

primary porosity The space that remains between solid grains or crystals immediately after sediment accumulates or rock forms.

principal aquifer The geologic unit that serves as the primary source of groundwater in a region.

principle of baked contacts When an igneous intrusion "bakes" (metamorphoses) surrounding rock, the rock that has been baked must be older than the intrusion.

principle of cross-cutting relations If one geologic feature cuts across another, the feature that has been cut is older.

principle of fossil succession In a stratigraphic sequence, different species of fossil organisms appear in a definite order; once a fossil species disappears in a sequence of strata, it never reappears higher in the sequence.

principle of inclusions If a rock contains fragments of another rock, the fragments must be older than the rock containing them.

principle of original continuity Sedimentary layers, before erosion, formed fairly continuous sheets over a region.

principle of original horizontality Layers of sediment, when originally deposited, are fairly horizontal.

principle of superposition In a sequence of sedimentary rock layers, each layer must be younger than the one below, for a layer of sediment cannot accumulate unless there is already a substrate on which it can collect.

principle of uniformitariansim The physical processes we observe today also operated in the past in the same way, and at comparable rates.

prograde metamorphism Metamorphism that occurs as temperatures and pressures are increasing.

Proterozoic The most recent of the Precambrian eons.

protocontinent A block of crust composed of volcanic arcs and hot-spot volcanoes sutured together.

protolith The original rock from which a metamorphic rock formed.

protoplanet A body that grows by the accumulation of planetesimals but has not yet become big enough to be called a planet.

protoplanetary nebula A ring of gas and dust that surrounded the newborn Sun, from which the planets were formed.

protostar A dense body of gas that is collapsing inward because of gravitational forces and that may eventually become a star.

pumice A glassy igneous rock that forms from felsic frothy lava and contains abundant (over 50%) pore space.

punctuated equilibrium The hypothesis that evolution takes place in fits and starts; evolution occurs very slowly for quite a while and then, during a relatively short period, takes place very rapidly.

P-waves Compressional seismic waves that move through the body of the Earth.

P-wave shadow zone A band between 103° and 143° from an earthquake epicenter, as measured along the circumference of the Earth, inside which P-waves do not arrive at seismograph stations.

pycnocline The boundary between layers of water of different densities.

pyroclastic debris Fragmented material that sprayed out of a volcano and landed on the ground or sea floor in solid form.

pyroclastic flow A fast-moving avalanche that occurs when hot volcanic ash and debris mix with air and flow down the side of a volcano.

pyroclastic rock Rock made from fragments blown out of a volcano during an explosion that were then packed or welded together.

quarry A site at which stone is extracted from the ground.

quartzite A metamorphic rock composed of quartz and transformed from a protolith of quartz sandstone.

quenching A sudden cooling of molten material to form a solid.

quick clay Clay that behaves like a solid when still (because of surface tension holding the water-coated clay flakes together), but that flows like a liquid when shaken.

radial network A drainage network in which the streams flow outward from a cone-shaped mountain, and define a pattern resembling spokes on a wheel.

radioactive decay The process by which a radioactive atom undergoes fission or releases particles thereby transforming into a new element.

radioactive isotope An unstable isotope of a given element.

radiometric dating The science of dating geologic events in years by measuring the ratio of parent radioactive atoms to daughter product atoms.

rain band A spiraling arm of a hurricane radiating outward from the eye.

rain shadow The inland side of a mountain range, which is arid because the mountains block rain clouds from reaching the area.

range (for fossils) The interval of a sequence of strata in which a specific fossil species appears.

rapids A reach of a stream in which water becomes particularly turbulent; as a consequence, waves develop on the surface of the stream.

reach A specified segment of a stream's path.

recessional moraine The end moraine that forms when a glacier stalls for a while as it recedes.

recharge area A location where water enters the ground and infiltrates down to the water table.

recrystallization The process in which ions or atoms in minerals rearrange to form new minerals.

rectangular network A drainage network in which the streams join each other at right angles because of a rectangular grid of fractures that breaks up the ground and localizes channels.

recurrence interval The average time between successive geologic events.

red giant A huge red star that forms when Sun-sized stars start to die and expand.

red shift The phenomenon in which a source of light moving away from you very rapidly shifts to a lower frequency; that is, toward the red end of the spectrum.

reef bleaching The death and loss of color of a coral reef.

reflected ray A ray that bounces off a boundary between two different materials.

refracted ray A ray that bends as it passes through a boundary between two different materials.

refraction The bending of a ray as it passes through a boundary between two different materials.

reg A vast stony plain in a desert.

regional metamorphism *Dynamothermal metamorphism;* metamorphism of a broad region, usually the result of deep burial during an orogeny.

regolith Any kind of unconsolidated debris that covers bedrock.

regression The seaward migration of a shoreline caused by a lowering of sea level.

relative age The age of one geologic feature with respect to another.

relative humidity The ratio between the measured water content of air and the maximum possible amount of water the air can hold at a given condition.

relative plate velocity The movement of one lithosphere plate with respect to another.

relief The difference in elevation between adjacent high and low regions on the land surface.

renewable resource A resource that can be replaced by nature within a short time span relative to a human life span.

reservoir rock Rock with high porosity and permeability, so it can contain an abundant amount of easily accessible oil.

residence time The average length of time that a substance stays in a particular reservoir.

residual mineral deposit Soils in which the residuum left behind after leaching by rainwater is so concentrated in metals that the soil itself becomes an ore deposit.

resurgent dome The new mound, or cone, of igneous rock that grows within a caldera as an eruption begins anew.

retrograde metamorphism Metamorphism that occurs as pressures and temperatures are decreasing; for retrograde metamorphism to occur, water must be added.

return stroke An upward-flowing electric current from the ground that carries positive charges up to a cloud during a lightning flash.

reversed polarity Polarity in which the paleomagnetic dipole points north.

reverse fault A steeply dipping fault on which the hanging-wall block slides up.

Richter magnitude scale A scale that defines earthquakes on the basis of the amplitude of the largest ground motion recorded on a seismogram.

ridge axis The crest of a mid-ocean ridge; the ridge axis defines the position of a divergent plate boundary.

right-lateral strike-slip fault A strike-slip fault in which the block on the opposite fault plane from a fixed spot moves to the right of that spot.

rip current A strong, localized seaward flow of water perpendicular to a beach.

riprap Loose boulders or concrete piled together along a beach to absorb wave energy before it strikes a cliff face.

roche moutonnée A glacially eroded hill that becomes elongate in the direction of flow and asymmetric; glacial rasping smoothes the upstream part of the hill into a gentle slope, while glacial plucking erodes the downstream edge into a steep slope.

rock A coherent, naturally occurring solid, consisting of an aggregate of minerals or a mass of glass.

rock burst A sudden explosion of rock off the ceiling or wall of an underground mine.

rock cycle The succession of events that results in the transformation of Earth materials from one rock type to another, then another, and so on.

rock flour Fine-grained sediment produced by glacial abrasion of the substrate over which a glacier flows.

rock glacier A slow-moving mixture of rock fragments and ice.

rock slide A sudden downslope movement of rock.

rocky coast An area of coast where bedrock rises directly from the sea, so beaches are absent.

Rodinia A proposed Precambrian supercontinent that existed around 1 billion years ago.

rotational axis The imaginary line through the center of the Earth around which the Earth spins.

R-waves (*Rayleigh waves*) Surface seismic waves that cause the ground to ripple up and down, like water waves in a pond.

sabkah A region of formerly flooded coastal desert in which stranded seawater has left a salt crust over a mire of mud that is rich in organic material.

salinity The degree of concentration of salt in water.

saltation The movement of a sediment in which grains bounce along their substrate, knocking other grains into the water column (or air) in the process.

salt dome A rising bulbous dome of salt that bends up the adjacent layers of sedimentary rock.

salt wedging The process in arid climates by which dissolved salt in groundwater crystallizes and grows in open pore spaces in rocks and pushes apart the surrounding grains.

sand spit An area where the beach stretches out into open water across the mouth of a bay or estuary.

sandstone Coarse-grained sedimentary rock consisting almost entirely of quartz.

sand volcano (or **sand blow**) A small mound of sand produced when sand layers below the ground surface liquify as a result of seismic shaking, causing the sand to erupt onto the Earth's surface through cracks or holes in overlying clay layers.

saprolite A layer of rotten rock created by chemical weathering in warm, wet climates.

Sargasso Sea The center of North Atlantic Gyre, named for the tropical seaweed sargassum, which accumulates in its relatively noncirculating waters.

saturated solution Water that carries as many dissolved ions as possible under given environmental conditions.

saturated zone The region below the water table where pore space is filled with water.

scattering The dispersal of energy that occurs when light interacts with particles in the atmosphere.

schist A medium-to-coarse-grained metamorphic rock that possesses schistosity.

schistosity Foliation caused by the preferred orientation of large mica flakes.

scientific method A sequence of steps for systematically analyzing scientific problems in a way that leads to verifiable results.

scoria A glassy mafic igneous rock containing abundant air-filled holes.

scouring A process by which running water removes loose fragments of sediment from a stream bed.

sea arch An arch of land protruding into the sea and connected to the mainland by a narrow bridge.

sea-floor spreading The gradual widening of an ocean basin as new oceanic crust forms at a mid-ocean ridge axis and then moves away from the axis.

sea ice Ice formed by the freezing of the surface of the sea.

seal A relatively impermeable rock, such as shale, salt, or unfractured limestone, that lies above a reservoir rock and stops the oil from rising further.

seam A sedimentary bed of coal interlayered with other sedimentary rocks.

seamount An isolated submarine mountain.

seasonal floods Floods that appear almost every year during seasons when rainfall is heavy or when winter snows start to melt.

seasonal well A well that provides water only during the rainy season when the water table rises below the base of the well.

sea stack An isolated tower of land just offshore, disconnected from the mainland by the collapse of a sea arch.

seawall A wall of riprap built on the landward side of a backshore zone in order to protect shore cliffs from erosion.

second The basic unit of time measurement, now defined as the time it takes for the magnetic field of a cesium atom to flip polarity 9,192,631,770 times, as measured by an atomic clock.

secondary enrichment The process by which a new ore deposit forms from metals that were dissolved and carried away from preexisting ore minerals.

secondary porosity New pore space in rocks, created some time after a rock first forms.

secondary recovery technique A process used to extract the quantities of oil that will not come out of a reservoir rock with just simple pumping.

sediment An accumulation of loose mineral grains, such as boulders, pebbles, sand, silt, or mud, that are not cemented together.

sedimentary basin A depression, created as a consequence of subsidence, that fills with sediment.

sedimentary rock Rock that forms either by the cementing together of fragments broken off preexisting rock or by the precipitation of mineral crystals out of water solutions at or near the Earth's surface.

sedimentary sequence A grouping of sedimentary units bounded on top and bottom by regional unconformities.

sediment budget The proportion of sand supplied to sand removed from a depositional setting.

sediment load The total volume of sediment carried by a stream.

sediment maturity The degree to which a sediment has evolved from a crushed-up version of the original rock into a sediment that has lost its easily weathered minerals and become well sorted and rounded.

sediment sorting The segregation of sediment by size.

seep A place where oil-filled reservoir rock intersects the ground surface, or where fractures connect a reservoir to the ground surface, so that oil flows out onto the ground on its own.

seiche Rhythmic movement in a body of water caused by ground motion.

seismic belts (or **zones**) The relatively narrow strips of crust on Earth under which most earthquakes occur.

seismicity Earthquake activity.

seismic-moment magnitude scale A scale that defines earthquake size on the basis of calculations involving the amount of slip, length of rupture, depth of rupture, and rock strength.

seismic ray The changing position of an imaginary point on a wave front as the front moves through rock.

seismic-reflection profile A cross-sectional view of the crust made by measuring the reflection of artificial seismic waves off boundaries between different layers of rock in the crust.

seismic tomography Analysis by sophisticated computers of global seismic data in order to create a three-dimensional image of variations in seismic-wave velocities within the Earth.

seismic velocity The speed at which seismic waves travel.

seismic-velocity discontinuity A boundary in the Earth at which seismic velocity changes abruptly.

seismic (earthquake) waves Waves of energy emitted at the focus of an earthquake.

seismogram The record of an earthquake produced by a seismograph.

seismograph (seismometer) An instrument that can record the ground motion from an earthquake.

semipermanent pressure cell A somewhat elliptical zone of high or low atmospheric pressure that lasts much of the year; it forms because high-pressure zones tend to be narrower over land than over sea.

Sevier orogeny A mountain-building event that affected western North America between about 150 Ma and 80 Ma, a result of convergent margin tectonism; a fold-thrust belt formed during this event.

shale Very fine-grained sedimentary rock that breaks into thin sheets.

shatter cones Small, cone-shaped fractures formed by the shock of a meteorite impact.

shear strain A change in shape of an object that involves the movement of one part of a rock body sideways past another part so that angular relationships within the body change.

shear stress A stress that moves one part of a material sideways past another part.

shear waves Seismic waves in which particles of material move back and forth perpendicular to the direction in which the wave itself moves.

shear zone A fault in which movement has occurred ductilely.

sheetwash A film of water less than a few mm thick that covers the ground surface during heavy rains.

shield An older, interior region of a continent.

shield volcano A subaerial volcano with a broad, gentle dome, formed either from low-viscosity basaltic lava or from large pyroclastic sheets.

shocked quartz Grains of quartz that have been subjected to intense pressure such as occurs during a meteorite impact.

shoreline The boundary between the water and land.

shortening The process during which a body of rock or a region of crust becomes shorter.

short-term climate change Climate change that takes place over hundreds to thousands of years.

Sierran arc A large continental volcanic arc along western North America that initiated at the end of the Jurassic Period and lasted until about 80 million years ago.

silica SiO_2.

silicate minerals Minerals composed of silicon-oxygen tetrahedra linked in various arrangements; most contain other elements too.

silicate rock Rock composed of silicate minerals.

siliceous sedimentary rock Sedimentary rock that contains abundant quartz.

silicic Rich in silica with relatively little iron and magnesium.

sill A nearly horizontal table-top-shaped tabular intrusion that occurs between the layers of country rock.

siltstone Fine-grained sedimentary rock generally composed of very small quartz grains.

sinkhole A circular depression in the land that forms when an underground cavern collapses.

slab-pull force The force that downgoing plates (or slabs) apply to oceanic lithosphere at a convergent margin.

slate Fine-grained, low-grade metamorphic rock, formed by the metamorphism of shale.

slaty cleavage The foliation typical of slate, and reflective of the preferred orientation of slate's clay minerals, that allows slate to be split into thin sheets.

slickensides The polished surface of a fault caused by slip on the fault; lineated slickensides also have groves that indicate the direction of fault movement.

slip face The lee side of a dune, which sand slides down.

slip lineations Linear marks on a fault surface created during movement on the fault; some slip lineations are defined by grooves, some by aligned mineral fibers.

slope failure The downslope movement of material on an unstable slope.

slumping Downslope movement in which a mass of regolith detaches from its substrate along a spoon-shaped sliding surface and slips downward semicoherently.

smelting The heating of a metal-containing rock to high temperatures in a fire so that the rock will decompose to yield metal plus a nonmetallic residue (slag).

snotite A long gob of bacteria that slowly drips from the ceiling of a cave.

snow line The boundary above which snow remains all year.

soda straw A hollow stalactite in which calcite precipitates around the outside of a drip.

soil Sediment that has undergone changes at the surface of the Earth, including reaction with rainwater and the addition of organic material.

soil erosion The removal of soil by wind and runoff.

soil horizon Distinct zones within a soil, distinguished from each other by factors such as chemical composition and organic content.

soil moisture Underground water that wets the surface of the mineral grains and organic material making up soil, but lies above the water table.

soil profile A vertical sequence of distinct zones of soil.

solar wind A stream of particles with enough energy to escape from the Sun's gravity and flow outward into space.

solid-state diffusion The slow movement of atoms or ions through a solid.

solifluction The type of creep characteristic of tundra regions; during the summer, the uppermost layer of permafrost melts, and the soggy, weak layer of ground then flows slowly downslope in overlapping sheets.

solstice A day on which the polar ends of the terminator (the boundary between the day hemisphere and the night hemisphere) lie 23.5° away from the associated geographic poles.

Sonoma orogeny A convergent-margin mountain-building event that took place on the western coast of North America in the Late Permian and Early Triassic periods.

sorting (1) The range of clast sizes in a collection of sediment; (2) the degree to which sediment has been separated by flowing currents into different-size fractions.

source rock A rock (organic-rich shale) containing the raw materials from which hydrocarbons eventually form.

southeast tradewinds Tradewinds in the Southern Hemisphere, which start flowing northward, deflect to the west, and end up flowing from southeast to northwest.

southern oscillation The oscillating of atmospheric pressure cells back and forth across the Pacific Ocean, in association with El Niño.

specific gravity A number representing the density of a mineral, as specified by the ratio between the weight of a volume of the mineral and the weight of an equal volume of water.

speleothem A formation that grows in a limestone cave by the accumulation of travertine precipitated from water solutions dripping in a cave or flowing down the wall of a cave.

sphericity The measure of the degree to which a clast approaches the shape of a sphere.

spreading boundary *Divergent plate boundary.*

spreading rate The rate at which sea floor moves away from a mid-ocean ridge axis, as measured with respect to the sea floor on the opposite side of the axis.

spring A natural outlet from which groundwater flows up onto the ground surface.

spring tide An especially high tide that occurs when the Sun is on the same side of the Earth as the Moon.

stable air Air that does not have a tendency to rise rapidly.

stable slope A slope on which downward sliding is unlikely.

stalactite An icicle-like cone that grows from the ceiling of a cave as dripping water precipitates limestone.

stalagmite An upward-pointing cone of limestone that grows when drips of water hit the floor of a cave.

standing wave A wave whose crest and trough remain in place as water moves through the wave.

star dune A constantly changing dune formed by frequent shifts in wind direction; it consists of overlapping crescent dunes pointing in many different directions.

stick-slip behavior Stop-start movement along a fault plane caused by friction, which prevents movement until stress builds up sufficiently.

stone rings Ridges of cobbles between adjacent bulges of permafrost ground.

stoping A process by which magma intrudes; blocks of wall rock break off and then sink into the magma.

storm An episode of severe weather in which winds, precipitation, and in some cases lightning become strong enough to be bothersome and even dangerous.

storm-center velocity A storm's (hurricane's) velocity along its track.

storm surge Excess seawater driven landward by wind during a storm; the low atmospheric pressure beneath the storm allows sea level to rise locally, increasing the surge.

strain The change in shape of an object in response to deformation (i.e., as a result of the application of a stress).

stratified drift Glacial sediment that has been redistributed and stratified by flowing water.

stratigraphic column A cross-section diagram of a sequence of strata summarizing information about the sequence.

stratigraphic formation A recognizable layer of a specific sedimentary rock type or set of rock types, deposited during a certain time interval, that can be traced over a broad region.

stratigraphic sequence An interval of strata deposited during periods of relatively high sea level, and bounded above and below by regional unconformities.

stratopause The temperature pause that marks the top of the stratosphere.

stratosphere The stable, stratified layer of atmosphere directly above the troposphere.

stratovolcano A large, cone-shaped subaerial volcano consisting of alternating layers of lava and tephra.

stratus cloud A thin, sheet-like, stable cloud.

streak The color of the powder produced by pulverizing a mineral on an unglazed ceramic plate.

stream A ribbon of water that flows in a channel.

stream bed The floor of a stream.

stream capacity The total quantity of sediment a stream carries.

stream capture (or piracy) The situation in which headward erosion causes one stream to intersect the course of another, previously independent stream, so that the intersected stream starts to flow down the channel of the first stream.

stream competence The maximum particle size that a stream can carry.

stream gradient The slope of a stream's channel in the downstream direction.

stream rejuvination The renewed downcutting of a stream into a floodplain or peneplain, caused by a relative drop of the base level.

stress The push, pull, or shear that a material feels when subjected to a force; formally, the force applied per unit area over which the force acts.

stretching The process during which a layer of rock or a region of crust becomes longer.

striations Linear scratches in rock.

strike-slip fault A fault in which one block slides horizontally past another (and therefore parallel to the strike line), so there is no relative vertical motion.

strip mining The scraping off of all soil and sedimentary rock above a coal seam in order to gain access to the seam.

stromatolite Layered mounds of sediment formed by cyanobacteria; cyanobacteria secrete a mucuous-like substance to which sediment sticks, and as each layer of cyanobacteria gets buried by sediment, it colonizes the surface of the new sediment, building a mound upward.

structural control The condition in which geologic structures, such as faults, affect the distribution and drainage of water or the shape of the land surface.

subaerial Pertaining to land regions above sea level (i.e., under air).

subduction The process by which one oceanic plate bends and sinks down into the asthenosphere beneath another plate.

subduction zone The region along a convergent boundary where one plate sinks beneath another.

sublimation The evaporation of ice directly into vapor without first forming a liquid.

submarine canyon A narrow, steep canyon that dissects a continental shelf and slope.

submarine fan A wedge-shaped accumulation of sediment at the base of a submarine slope; fans usually accumulate at the mouth of a submarine canyon.

submarine slump The underwater downslope movement of a semicoherent block of sediment along a weak mud detachment.

submergent coast A coast at which the land is sinking relative to sea level.

subpolar low The rise of air where the surface flow of a polar cell converges with the surface flow of a Ferrel cell, creating a low-pressure zone in the atmosphere.

subsidence The vertical sinking of the Earth's surface in a region, relative to a reference plane.

substrate A general term for material just below the ground surface.

subtropical high (subtropical divergence zone) A belt of high pressure in the atmosphere at 30° latitude formed where the Hadley cell converges with the Ferrel cell, causing cool, dense air to sink.

subtropics Desert climate regions that lie on either side of the equatorial tropics between the lines of 20° and 30° north or south of the equator.

summit eruption An eruption that occurs in the summit crater of a volcano.

sunspot cycle The cyclic appearance of large numbers of sunspots (black spots thought to be magnetic storms on the Sun's surface) every 9 to 11.5 years.

supercontinent cycle The process of change during which supercontinents develop and later break apart, forming pieces that may merge once again in geologic time to make yet another supercontinent.

supernova A short-lived, very bright object in space that results from the cataclysmic explosion marking the death of a very large star; the explosion ejects large quantities of matter into space to form new nebulae.

superplume A huge mantle plume.

superposed stream A stream whose geometry has been laid down on a rock structure and is not controlled by the structure.

surface current An ocean current in the top 100 m of water.

surface waves Seismic waves that travel along the Earth's surface.

surface westerlies The prevailing surface winds in North America and Europe, which come out of the west or southwest.

surf zone A region of the shore in which breakers crash onto the shore.

surge (glacial) A pulse of rapid flow in a glacier.

suspended load Tiny solid grains carried along by a stream without settling to the floor of the channel.

swamp A wetland dominated by trees.

swash The upward surge of water that flows up a beach slope when breakers crash onto the shore.

S-waves Seismic shear waves that pass through the body of the Earth.

S-wave shadow zone A band between 103° and 180° from the epicenter of an earthquake inside of which S-waves do not arrive at seismograph stations.

swelling clay Clay possessing a mineral structure that allows it to absorb water between its layers and thus swell to several times its original size.

symmetry The condition in which the shape of one part of an object is a mirror image of the other part.

syncline A trough-shaped fold whose limbs dip toward the hinge.

systematic joints Long planar cracks that occur fairly regularly throughout a rock body.

tabular intrusions Sheet intrusions that are planar and of roughly uniform thickness.

Taconic orogeny A convergent mountain-building event that took place around 400 million years ago, in which a volcanic island arc collided with eastern North America.

tailings pile A pile of waste rock from a mine.

talus A sloping apron of fallen rock along the base of a cliff.

tar Hydrocarbons that exist in solid form at room temperature.

tarn A lake that forms at the base of a cirque on a glacially eroded mountain.

tar sand Sandstone reservoir rock in which less viscous oil and gas molecules have either escaped or been eaten by microbes, so that only tar remains.

taxonomy The study and classification of the relationships among different forms of life.

tension A stress that pulls on a material and could lead to stretching.

tephra Unconsolidated accumulations of pyroclastic grains.

terminal moraine The end moraine at the farthest limit of glaciation.

terminator The boundary between the half of the Earth that has daylight and the half experiencing night.

terrace The elevated surface of an older floodplain into which a younger floodplain had cut down.

terrestrial A term used to describe the inner, Earth-like planets.

thalweg The deepest part of a stream's channel.

theory A scientific idea supported by an abundance of evidence that has passed many tests and failed none.

theory of plate tectonics The theory that the outer layer of the Earth (the lithosphere) consists of separate plates that move with respect to one another.

thermal metamorphism Metamorphism caused by heat conducted into country rock from an igneous intrusion.

thermocline A boundary between layers of water with differing temperatures.

thermohaline circulation The rising and sinking of water driven by contrasts in water density, which is due in turn to differences in temperature and salinity; this circulation involves both surface and deep-water currents in the ocean.

thermosphere The outermost layer of the atmosphere containing very little gas.

thin section A 3/100-mm-thick slice of rock that can be examined with a petrographic microscope.

thin-skinned deformation A distinctive style of deformation characterized by displacement on faults that terminate at depth along a subhorizontal detachment fault.

thrust fault A gently dipping reverse fault; the hanging-wall block moves up the slope of the fault.

tidal bore A visible wall of water that moves toward shore with the rising tide in quiet waters.

tidal flat A broad, nearly horizontal plain of mud and silt, exposed or nearly exposed at low tide but totally submerged at high tide.

tidal reach The difference in sea level between high tide and low tide at a given point.

tide The daily rising or falling of sea level at a given point on the Earth.

tide-generating force The force, caused in part by the gravitational attraction of the Sun and Moon, and in part by the centrifugal force created by the Earth's spin, that generates tides.

till A mixture of unsorted mud, sand, pebbles, and larger rocks deposited by glaciers.

tillite A rock formed from hardened ancient glacial deposits and consisting of larger clasts distributed through a matrix of sandstone and mudstone.

toe (terminus) The leading edge or margin of a glacier.

tombolo A narrow ridge of sand that links a sea stack to the mainland.

topographical map A map that uses contour lines to represent variations in elevation.

topography Variations in elevation.

topsoil The top soil horizons, which are typically dark and nutrient-rich.

tornado A near-vertical, funnel-shaped cloud in which air rotates extremely rapidly around the axis of the funnel.

tornado swarm Dozens of tornadoes produced by the same storm.

tower karst A karst landscape in which steep-sided residual bedrock towers remain between sinkholes.

transform fault A fault marking a transform plate boundary; along mid-ocean ridges, transform faults are the actively slipping segment of a fracture zone between two ridge segments.

transform plate boundary A boundary at which one lithosphere plate slips laterally past another.

transgression The inland migration of shoreline resulting from a rise in sea level.

transition zone The middle portion of the mantle, from 400 to 670 km deep, in which there are several jumps in seismic velocity.

transpiration The release of moisture as a metabolic byproduct.

transverse dune A simple, wave-like dune that appears when enough sand accumulates for the ground surface to be completely buried, but only moderate winds blow.

travel-time curve A graph that plots the time since an earthquake began on the vertical axis, and the distance to the epicenter on the horizontal axis.

trellis network A drainage system that develops across a landscape of parallel valleys and ridges so that major tributaries flow down the valleys and join a trunk stream that cuts through the ridge; the resulting map pattern resembles a garden trellis.

trench A deep elongate trough bordering a volcanic arc; a trench defines the trace of a convergent plate boundary.

triangulation The method for determining the map location of a point from knowing the distance between that point and three other points; this method is used to locate earthquake epicenters.

tributary A smaller stream that flows into a larger stream.

triple junction A point where three lithosphere plate boundaries intersect.

tropical depression A tropical storm with winds reaching up to 61 km per hour; such storms develop from tropical disturbances, and may grow to become hurricanes.

tropical disturbance Cyclonic winds that develop in the tropics.

tropopause The temperature pause marking the top of the troposphere.

troposphere The lowest layer of the atmosphere, where air undergoes convection and where most wind and clouds develop.

truncated spur A spur (elongate ridge between two valleys) whose end was eroded off by a glacier.

trunk stream The single larger stream into which an array of tributaries flow.

tsunami A large wave along the sea surface triggered by an earthquake or large submarine slump.

tuff A pyroclastic igneous rock composed of volcanic ash and fragmented pumice, formed when accumulations of the debris cement together.

tundra A cold, treeless region of land at high latitudes, supporting only species of shrubs, moss, and lichen capable of living on permafrost.

turbidite A graded bed of sediment built up at the base of a submarine slope and deposited by turbidity currents.

turbidity current A submarine avalanche of sediment and water that speeds down a submarine slope.

turbulence The chaotic twisting, swirling motion in flowing fluid.

typhoon The equivalent of a hurricane in the western Pacific Ocean.

ultimate base level Sea level; the level below which a trunk stream cannot cut.

ultramafic A term used to describe igneous rocks or magmas that are rich in iron and magnesium and very poor in silica.

unconfined aquifer An aquifer that intersects the surface of the Earth.

unconformity A boundary between two different rock sequences representing an interval of time during which new strata were not deposited and/or were eroded.

unconsolidated Consisting of unattached grains.

undercutting Excavation at the base of a slope that results in the formation of an overhang.

undersaturated A term used to describe a solution capable of holding more dissolved ions.

unsaturated zone The region of the subsurface above the water table.

unstable air Air that is significantly warmer than air above and has a tendency to rise quickly.

unstable ground Land capable of slumping or slipping downslope in the near future.

unstable slope A slope on which sliding will likely happen.

updraft Upward-moving air.

upper mantle The uppermost section of the mantle, reaching down to a depth of 400 km.

upwelling zone A place where deep water rises in the ocean, or hot magma rises in the asthenosphere.

U-shaped valley A steep-walled valley shaped by glacial erosion into the form of a U.

vacuum Space that contains very little matter in a given volume (e.g., a region in which air has been removed).

valley A trough with sloping walls, cut into the land by a stream.

valley glacier A river of ice that flows down a mountain valley.

Van Allen radiation belts Belts of solar wind particles and cosmic rays that surround the Earth, trapped by Earth's magnetic field.

varve A pair of thin layers of glacial lake-bed sediment, one consisting of silt brought in during the spring floods, and the other of clay deposited during the winter when the lake's surface freezes over and the water is still.

vascular plant A plant with woody tissue and seeds and veins for transporting water and food.

vein A seam of minerals that forms when dissolved ions carried by water solutions precipitate in cracks.

vein deposit A hydrothermal deposit in which the ore minerals occur in veins that fill cracks in preexisting rocks.

velocity-versus-depth curve A graph that shows the variation in the velocity of seismic waves with increasing depth in the Earth.

ventifact (faceted rock) A desert rock whose surface has been faceted by the wind.

vesicles Open holes in igneous rock formed by the preservation of bubbles in magma as the magma cools into solid rock.

viscosity The resistance of material to flow.

volatiles Elements or compounds such as H_2O and CO_2 that evaporate easily and can exist in gaseous forms at the Earth's surface.

volatility A specification of the ease with which a material evaporates.

volcanic arc A curving chain of active volcanoes formed adjacent to a convergent plate boundary.

volcanic ash Tiny glass shards formed when a fine spray of exploded lava freezes instantly upon contact with the atmosphere.

volcanic bomb A large piece of pyroclastic debris thrown into the atmosphere during a volcanic eruption.

volcanic-danger-assessment map A map delineating areas that lie in the path of potential lava flows, lahars, debris flows, or pyroclastic flows of an active volcano.

volcanic gas Elements or compounds that bubble out of magma or lava in gaseous form.

volcanic island arc The volcanic island chain that forms on the edge of the overriding plate where one oceanic plate subducts beneath another oceanic plate.

volcano (1) A vent from which melt from inside the Earth spews out onto the planet's surface; (2) a mountain formed by the accumulation of extrusive volcanic rock.

V-shaped valley A valley whose cross-sectional shape resembles the shape of a V; the valley probably has a river running down the point of the V.

Wadati-Benioff zone A sloping band of seismicity defined by intermediate- and deep-focus earthquakes that occur in the downgoing slab of a convergent plate boundary.

wadi The name used in the Middle East and North Africa for a dry wash.

warm front A front in which warm air rises slowly over cooler air in the atmosphere.

waste rock Rock dislodged by mining activity yet containing no ore minerals.

waterfall A place where water drops over an escarpment.

water gap An opening in a resistant ridge where a trunk river has cut through the ridge.

watershed The region that collects water that feeds into a given drainage network.

water table The boundary, approximately parallel to the Earth's surface, that separates substrate in which groundwater fills the pores from substrate in which air fills the pores.

wave base The depth, approximately equal in distance to half a wavelength in a body of water, beneath which there is no wave movement.

wave-cut bench A platform of rock, cut by wave erosion, at the low-tide line that was left behind a retreating cliff.

wave-cut notch A notch in a coastal cliff cut out by wave erosion.

wave erosion The combined effects of the shattering, wedging, and abrading of a cliff face by waves and the sediment they carry.

wave front The boundary between the region through which a wave has passed and the region through which it has not yet passed.

wavelength The horizontal difference between two adjacent wave troughs or two adjacent crests.

wave refraction (ocean) The bending of waves as they approach a shore so that their crests make no more than a 5° angle with the shoreline.

weather   Local-scale conditions as defined by temperature, air pressure, relative humidity, and wind speed.

weathered rock   Rock that has reacted with air and/or water at or near the Earth's surface.

weathering   The processes that break up and corrode solid rock, eventually transforming it into sediment.

weather system   A specific set of weather conditions, reflecting the configuration of air movement in the atmosphere, that affects a region for a period of time.

welded tuff   Tuff formed by the welding together of hot volcanic glass shards at the base of pyroclastic flows.

well   A hole in the ground dug or drilled in order to obtain water.

Western Interior Seaway   A north-south-trending seaway that ran down the middle of North America during the Late Cretaceous Period.

wet-bottom (temperate) glacier   A glacier with a thin layer of water at its base, over which the glacier slides.

wetted perimeter   The area in which water touches a stream channel's walls.

wind abrasion   The grinding away at surfaces in a desert by windblown sand and dust.

wind gap   An opening through a high ridge that developed earlier in geologic history by stream erosion, but that is now dry.

xenolith   A relict of wall rock surrounded by intrusive rock when the intrusive rock freezes.

yardang   A mushroom-like column with a resistant rock perched on an eroding column of softer rock; created by wind abrasion in deserts where a resistant rock overlies softer layers of rock.

yazoo stream   A small tributary that runs parallel to the main river in a floodplain because the tributary is blocked from entering the main river by levees.

Younger Dryas   An interval of cooler temperatures that took place 4,500 years ago during a general warming/glacier-retreat period.

zeolite facies   The metamorphic facies just above diagenetic conditions, under which zeolite minerals form.

zone of ablation   The area of a glacier in which ablation (melting, sublimation, calving) subtracts from the glacier.

zone of accumulation   (1) The layer of regolith in which new minerals precipitate out of water passing through, thus leaving behind a load of fine clay; (2) the area of a glacier in which snowfall adds to the glacier.

zone of aeration   *Unsaturated zone.*

zone of leaching   The layer of regolith in which water dissolves ions and picks up very fine clay; these materials are then carried downward by infiltrating water.

Credits

UNNUMBERED PHOTOS AND ART

ii, 1: Stephen Marshak; **8:** R. Williams (ST Scl)/NASA; **16–17:** original artwork by Gary Hincks; **30–31:** original artwork by Gary Hincks; **33:** European Space Agency/NASA; **35:** Johnson Space Center/NASA; **72–73:** original artwork by Gary Hincks; **76:** U.S. Geological Survey; **78:** Richard D. Fisher; **95:** Hal Lott/Corbis; **102:** Visuals Unlimited; **114–15:** original artwork by Gary Hincks; **121:** Stephen Marshak; **148–49:** original artwork by Gary Hincks; **151:** U.S. Geological Survey; **153:** Stephen Marshak; **168–69:** original artwork by Gary Hincks; **173:** Jacques Descloitres, MODIS Land Rapid Response Team, NASA/GSFC; **174:** Stephen Marshak; **180:** U.S. Geological Survey; **188–89:** original artwork by Gary Hincks; **204:** U.S. Geological Survey; **206:** U.S. Geological Survey; **216–17:** original artwork by Gary Hincks; **248:** Stephen Marshak; **264–65:** original artwork by Gary Hincks; **271:** U.S. Geological Survey; **283:** Stephen Marshak; **298–99:** original artwork by Gary Hincks; **306:** Stephen Marshak; **326–27:** original artwork by Gary Hincks; **329:** Stephen Marshak; **348–49:** original artwork by Gary Hincks; **360–61:** original artwork by Gary Hincks; **363:** Stephen Marshak; **364:** JPL/NASA; **368–69:** original artwork by Gary Hincks; **372:** U.S. Geological Survey; **373:** Robert L. Schuster/U.S. Geological Survey; **386–87:** original artwork by Gary Hincks; **391:** Stephen Marshak; **410–11:** original artwork by Gary Hincks; **416:** U.S. Geological Survey; **418:** Stephen Marshak; **440–41:** original artwork by Gary Hincks; **450:** Yann Arthus Bertrand/Corbis; **466–67:** original artwork by Gary Hincks; **473:** Stephen Marshak; **486–87:** original artwork by Gary Hincks; **492:** U.S. Geological Survey; **493:** Stephen Marshak; **510–11:** original artwork by Gary Hincks; **524:** U.S. Geological Survey; **525:** Stephen Marshak; **538–39:** original artwork by Gary Hincks; **544:** U.S. Geological Survey.

NUMBERED PHOTOS AND ART

Prelude: P.2, P.3: Stephen Marshak; **P.4:** AP Images; **P.8:** Courtesy Peter Fiske.

Chapter 1: 1.1A–B: Rare Books Division, The New York Public Library, Astor, Lenox and Tilden Foundations; **1.2A:** NASA; **1.6:** Moonrunner Design; **1.7:** J. Hester and P. Scowen/NASA; **1.8:** NASA; **1.9:** J. Hester and P. Scowen/NASA; **1.11:** Photograph by Pelisson, SaharaMet; **1.12:** JPL/NASA; **1.13B:** Modified from Laing; **1.14A:** Louis A. Frank, The University of Iowa; **1.14B:** Raymond Gehman/Corbis; **1.15A:** Johnson Space Center/NASA; **1.19:** Tate Gallery, London/Art Resource, NY.

Chapter 2: 2.1: Wegener A., *The Origin of the Continents and Oceans* (New York: Dover, 1966; trans. From 1929 German ed.); **2.3:** Modified from American Association of Petroleum Geologists; **2.4:** Modified from Motz; **2.5A:** Modified from Hurley; **2.13:** Modified from McElhinney; **2.20:** Rothé, J.R. 1954. La zone séis-mique mé-dian Indo-Atlantique: Proceedings of the Royal Society of London, Series A, v. 222, p. 388; **2.22B:** Modified from Mason, 1955; **2.25:** Modified from Cox, Dalrymple, and Doell, in Hamblin and Christiansen, 1998; **2.27C:** Modified from Cox, Dalrymple, and Doell; **2.34A:** University of Texas Institute of Geophysics; **2.35:** Photo by Dudley Foster, Woods Hole Oceanographic Institute; **2.36:** Modified from Sloss, NOAA; **2.43B:** R.E. Wallace (228), U.S. Geological Survey; **2.54:** Chris Scotese.

Chapter 3: 3.1: © Crown copyright: Historic Royal Palaces; **3.2a:** © Ken Lucas/Visuals Unlimited; **3.2B:** © G. R. Roberts; **3.3A:** © Jay Schomer; **3.3B:** Wally McNamee/Corbis; **3.4:** Modified from Wicander and Monroe; **3.7D:** Charles O'Rear/Corbis; **3.7F:** Courtesy of John A. Jaszczak, Michigan Technological University; **3.10A–B:** John A. Jaszczak; **3.12A:** © 1996 Jeff Scovil; **3.14:** © Richard P. Jacobs/JLM Visuals; **3.15:** © Richard P. Jacobs/JLM Visuals; **3.16A:** © Breck P. Kent/JLM Visuals; **3.16B:** © Richard P. Jacobs/JLM Visuals; **3.17A:** © 1995-1998 by Amethyst Galleries, Inc., http://mineral.galleries.com; **3.17B:** © 1993 Jeff Scovil; **3.18A:** © Richard P. Jacobs/JLM Visuals; **3.18B:** © 1992 Jeff Scovil; **3.18C:** © Martin Miller; **3.18D–E:** © Richard P. Jacobs/JLM Visuals; **3.19:** © Richard P. Jacobs/JLM; **3.20:** © Richard P. Jacobs/JLM Visuals; **3.23:** © 1996 Smithsonian Institution; **3.24A:** © 1985 Darrel Plowes; **3.24B:** © Ken Lucas/Visuals Unlimited; **3.25:** © A.J. Copley/Visuals Unlimited.

Interlude A: A.1A: © Richard P. Jacobs/JLM Visuals; **A.1B:** Courtesy David W. Houseknecht, U.S. Geological Survey; **A.2A:** © Mark A. Schneider/Visuals Unlimited; **A.2B:** Courtesy of Dr. Kent Ratajeski, Department of Geology and Geophysics, University of Wisconsin, Madison; **A.3A:** © G.R. Roberts; **A.3B–D:** Stephen Marshak; **A.4A:** Corbis; **A.4B:** © John Elk III/Stock, Boston/PictureQuest; **A.4C:** © David Muench/Corbis; **A.5C:** Modified from AGI data sheet; **A.6A:** © Tom Bean; **A.6B:** H. Johnson; **A.8A:** Stephen Marshak; **A.8B:** Courtesy of Bradley Hacker, Department of Geology, University of California, Santa Barbara.

Chapter 4: 4.1A: Jim Sugar Photography/Corbis; **4.1B:** Photo by J.D. Griggs/U.S. Geological Survey; **4.1C:** Roger Ressmeyer/Corbis; **4.8A:** Corbis; **4.11A:** Stephen Marshak; **4.11B:** © 1998 Tom Bean; **4.11D:** Stephen Marshak; **4.13B:** Stephen Marshak; **4.14E:** Stephen Marshak; **4.16A–B:** Courtesy of Dr. Kent Ratajeski, Department of Geology and Geophysics, University of Wisconsin, Madison; **4.16C–D:** © Richard P. Jacobs/JLM Visuals; **4.16E:** © John S. Shelton; **4.16F:** © Richard P. Jacobs/JLM Visuals; **4.16G:** © Dane S. Johnson/Visuals Unlimited; **4.16H:** © Doug Sokell/Visuals Unlimited; **4.17A:** Mineral proportions after Hamblin and Howard.

Chapter 5: 5.2: Stephen Marshak; **5.3:** Stephen Marshak; **5.4A:** © Martin Miller; **5.4B–C:** Stephen Marshak; **5.5B:** Stephen Marshak; **5.5C:** Stephen Marshak; **5.6C:** Stephen Marshak; **5.8B–C:** Stephen Marshak; **5.9A–B:** Stephen Marshak; **5.11A:** © John D. Cunningham/Visuals Unlimited; **5.13:** Jim Richardson/Corbis; **5.16A:** Lloyd Cluff/Corbis; **5.16B:** Stephen Marshak; **5.16C:** Stephen Marshak; **5.16D:** Scottsdale

Community College; **5.16E:** Duncan Heron; **5.16F–G:** Stephen Marshak; **5.16H:** E.R. Degginger/Color-Pic, Inc. **5.16I–J:** Stephen Marshak; **5.16K:** R.L. Kugler and J.C. Pashin, Geological Survey of Alabama; **5.16L:** Earth Sciences Department, Oxford University; **5.17A:** U.S. Geological Survey; **5.17B–C:** Stephen Marshak; **5.18A:** Stephen Marshak; **5.18B:** © Steve McCutcheon/Visuals Unlimited; **5.19B:** Stephen Marshak; **5.19D:** John S. Shelton, courtesy of Morton Salt Company; **5.20A:** © Martin Miller; **5.20B:** © 1998 M.W. Schmidt; **5.21:** Stephen Marshak; **5.23:** Stephen Marshak; **5.24A–B:** Stephen Marshak; **5.25C:** Stephen Marshak; **5.27:** Stephen Marshak; **5.28A:** © G.R. Roberts; **5.28B:** Stephen Marshak; **5.28C:** Martin Miller; **5.29A:** © John S. Shelton; **5.29C:** Stephen Marshak; **5.31A:** Yann Arthus-Bertrand/Corbis; **5.32A:** Scripps Institution of Oceanography, University of California, San Diego; **5.32B:** © G.R. Roberts.

Chapter 6: 6.1A: Courtesy Carlo Giovanella; **6.1B:** L.S. Stepanowicz/Visuals Unlimited; **6.1C:** Stephen Marshak; **6.1D:** Courtesy Kurt Friehauf, Kutztown University of Pennsylvania; **6.6B:** Ludovic Maisant/Corbis; **6.8A–C:** Stephen Marshak; **6.9A:** Kevin Schafer/Corbis; **6.10:** Stephen Marshak; **6.11A–B:** Stephen Marshak; **6.12A:** Stephen Marshak; **6.12B:** Diego Lezama Orezzoli/Corbis; **6.22B:** Stephen Marshak; **6.22C:** David Muench/Corbis.

Chapter 7: 7.1A: Stephen Marshak; **7.1C:** Stephen Marshak; **7.3A:** U.S. Geological Survey; **7.3B–C:** Stephen Marshak; **7.3D:** Stephen Marshak; **7.4:** NOAA; **7.5A:** AP/ World Wide Photos; **7.5C:** Stephen Weaver; **7.6A:** Roger Ressmeyer/Corbis; **7.6B:** Associated Press *Antigua Sun*; **7.7:** U.S. Geological Survey; **7.8:** Stephen Marshak; **7.9C:** N.Banks/U.S. Geological Survey; **7.10D:** Yann Arthus-Bertrand/Corbis; **7.11B:** © 1985 Tom Bean; **7.11D:** Roger Ressmeyer/Corbis; **7.13D:** U.S. Geological Survey; **7.13E:** Gary Braasch/Corbis; **7.16B–C:** Stephen Marshak; **7.18:** Stephen Marshak; **7.21:** Stephen Marshak; **7.22A:** Stephen Marshak; **7.22B:** J.D. Griggs/U.S. Geological Survey; **7.22C:** AP Images; **7.22D:** © Philippe Bourseiller; **7.22E:** AP Images; **7.22F:** Photo by J. Marso, U.S. Geological Survey; **7.23:** Peter Turnley/Corbis; **7.24:** Fact Sheet by Scott, K.M., Wolfe, E.W., and Driedger, C.L., Hawaiian Volcano Observatory, U.S. Geological Survey; **7.25:** photo by Sigurgeir Jonasson/Frank Lane Picture Agency/Corbis; **7.26:** Julian Baum/Photo Researchers; **7.27A–B:** JPL/NASA.

Chapter 8: 8.2: J. Dewey, U.S. Geological Survey; **8.5A:** National Geophysical Data Center/NOAA; **8.9E:** Center for Earthquake Research and Information; **8.12:** Adapted from Bolt, 1978; **8.16:** Bettmann/Corbis; **8.17A:** Courtesy of ABSG Consulting, Inc., Houston, Texas, modified by Stephen Marshak; **8.17B:** Patrick Robert/Corbis Sygma; **8.19B:** Corbis; **8.20A:** Adapted from U.S. Geological Survey map of New Madrid Seismicity; **8.20B:** Courtesy State Historical Society of Missouri, Columbia; **8.23A:** AP Images; **8.23B:** M. Celebi, U.S. Geological Survey; **8.23C:** AP Images; **8.24:** AP Images; **8.25:** NGDC; **8.26C:** Karl V. Steinbrugge Collection, Earthquake Engineering Research Center; **8.27B:** © James Mori, Research Center for Earthquake Prediction, Disaster Prevention Institute, Kyoto University; **8.28:** U.S. Geological Survey; **8.30A:** AFP/Getty Images; **8.30C–D:** Space Imaging; **8.31B–C:** Earth and Space Sciences/University of Washington; **8.32A:** NOAA; **8.33:** Vasily V. Titov, Associate Director, Tsunami Inundation Mapping Efforts (TIME), NOAA/PMEL-UW/JISAO, USA; **8.35A:** Produced by the Global Seismic Hazard Assessment Program; **8.35B:** Earthquake Hazards Program/U.S. Geological Survey; **8.35C:** Adapted from Nishenko, 1989 (U.S. Geological Survey); **8.38:** AP Images.

Interlude C: C.4A: From *Patterns of Nature*, Department of Physics and Astronomy, Arizona State University, Arizona Board of Regents; **C.10B:** Steve Grand, University of Texas; **C.11:** Shijie Zhong, University of Colorado; **C.12A:** © John Q. Thompson, courtesy of Dawson Geophysical Company; **C.12A–B:** Courtesy of Dawson Geophysical Company-Observatory.

Chapter 9: 9.1: Corbis; **9.2:** National Geophysical Data Center/NOAA; **9.5A–B:** Stephen Marshak; **9.8C–D:** Stephen Marshak; **9.12A:** Galen Rowell/Corbis; **9.12B:** Stephen Marshak; **9.13A:** U.S. Geological Survey; **9.13C:** © John S. Shelton; **9.15A:** Stephen Marshak; **9.15C–E:** Stephen Marshak; **9.17A–C:** Stephen Marshak; **9.18D–E:** © John S. Shelton; **9.20C:** Stephen Marshak; **9.21:** Stephen Marshak; **9.23A–B:** Stephen Marshak; **9.24B:** Adapted from Coney et al., 1980; **9.29:** U.S. Geological Survey; **9.30A:** Adapted from Geologic Map of the United States, U.S. Geological Survey.

Interlude D: D.1: Stephen Marshak; **D.2:** Photo by William L. Jones from the Stones & Bones Collection, http://www.stonesbones.com; **D.4A:** © Saint-Petersburg Zoological Museum; **D.4B:** Doug Lundberg (www.americawest.com); **D.4C–D:** Stephen Marshak; **D.4E:** Department of Earth and Space Sciences, UCLA; **D.4F:** Kevin Schafer/Corbis; **D.4G–H:** Stephen Marshak; **D.5A:** Courtesy of Senckenberg, Messel Research Department; **D.5B:** Humboldt-Universität zu Berlin, Museum für Naturkunde. Photo by W. Harre; **D.8:** J.J. Sepkoski, Jr., 1994. *Geotimes*, March 1994, p. 16.

Chapter 10: 10.2A–B: Stephen Marshak; **10.2D:** Stephen Marshak; **10.5A:** Stephen Marshak; **10.7A:** Stephen Marshak; **10.13: (top)** Adapted from Coney, Zion National History Association, 1975; **10.13: (bottom)** Adapted from U.S. Geological Survey; **10.13B–C:** Stephen Marshak; **10.13E:** © Charles Preitner/Visuals Unlimited; **10.13G:** Stephen Marshak; **10.15A–B:** courtesy Jim Connelly, University of Texas at Austin; **10.16A:** Michael Pole/Corbis; **10.16B:** Lonnie G. Thompson, Byrd Polar Research Center, The Ohio State University, Columbus; **10.19:** Courtesy of the Royal Ontario Museum © ROM.

Chapter 11: 11.1: © Bonestell Space Art; **11.4A:** Courtesy of Dr. J. William Schopf/UCLA; **11.4B:** Geological Survey of Newfoundland & Labrador; **11.6:** Modified from Hoffman, 1988; **11.9:** Lisa-Ann Gershwin/U.C. Museum of Palaeontology; **11.11:** Modified from Dott and Prothero; **11.12:** Courtesy Jonathan J. Havens, Irving Materials Inc.; **11.13B:** Modified from Sloss, 1962; **11.13C:** Adapted from Haq et al., 1987; **11.15:** Modified from Dott and Prothero; **11.18:** University of Michigan Exhibit Museum; **11.19:** © The Natural History Museum, London; **11.20:** Modified from Coney et al., 1980; **11.23D:** Earth Observatory/NASA; **11.24:** Image by Don Davis/NASA; **11.26:** Adapted from Atwater, 1989.

Chapter 12: 12.1: Adapted from Skinner and Porter, 1985; **12.3:** Adapted from Skinner and Porter, 1985; **12.6A:** Science Photo Library/Photo Researchers; **12.7:** © G.R. Roberts; **12.8A:** William J. Winters, U.S. Geological Society; **12.8B:** © Craig D. Wood; **12.9:** © Ron Testa/The Field Museum, Chicago, IL; **12.12:** Courtesy University of Melbourne; **12.13:** Adapted from Chernikoff, 1999; **12.14B:** © John D. Cunningham/Visuals Unlimited; **12.15:** Roger Ressmeyer/Corbis; **12.16:** © James Blank/Photophile; **12.18B:** © G.R. Roberts; **12.19:** Stephen Marshak; **12.20A:** Stephen Marshak; **12.20B:** Bettmann/Corbis; **12.23:** © G.R. Roberts; **12.24A:** Layne Kennedy/Corbis; **12.24B:** © John D. Cunningham/ Visuals Unlimited; **12.25:** © Richard P. Jacobs/JLM Visuals; **12.26:** © Richard P. Jacobs/JLM Visuals; **12.31A:** Stephen Marshak; **12.31B:** © Science VU/Visuals Unlimited; **12.34:** © 1997 R.W.

nonmetallic mineral resources, 356–57, *360–61*
common, *358*
nonplunging fold, 259, *259*
nonrenewable resources, 346, 358
Norgay, Tenzing, 248
normal faults, 208, *208, 216,* 228, *257,* **257**
normal polarity, 49, *50*
normal stress, 157, *157*
North America, 311, 313, 316, 322
during ice ages, 494, *516,* 517
Pleistocene climatic belts in, *517*
paleogeographic map of, *317*
North American continental divide, 393, *395*
North American Cordillera, 267, *267, 269, 320, 323*
North American craton, 311, *311*
North American Plate, 64, *65,* 206
North Anatolian Fault, 236, *237*
North Atlantic Current, *424*
North Atlantic Deep Water, 425, *426*
Northridge, California earthquake of 1994, 206, *207,* 210, *228*
fires resulting from, *231*
landslides triggered by, 228, *229*
North Sea, 322
tsunamis in, Storegga Slide and, 377, *377*
Norway, fjords in, 436, *436,* 505
nuclear-bomb testing, earthquakes and, 207
nuclear bonds, **A–4**
nuclear energy, *349*
nuclear fission, 330, 342, *A–9,* **A–9**
nuclear fusion, 14, 347, **A–9,** A–9 to A–10
nuclear power, 330, 341–43, 346
problems with, 343, 540
nuclear power plants
breeder reactors, 346
control rods, 343
safety, 343
nuclear reactor, *342,* **342**
nuclear waste, **343,** 540
nucleus (of an atom), **A–2**
numerical age (absolute age), *283,* **284,** 295–301
geologic column and, 301–2, *303*
other methods of determining, 301, *301*
radiometric dating, 295–301
Nyiragongo, Mt., 198

oasis, ***459,*** *459*
oblique-slip faults, 257, *257*
obsidian, *117,* **118**
oceanic crust, 27, *27–28, 30,* 240, *240*
discoveries about, 46–47
formation at mid-ocean ridges, 57–59
heat flow in, 47, *47*
igneous rock, composed of, 103
oceanic hot-spot volcanoes, 194, *195*
oceanic lithosphere, 32, *32,* 53, *72,* 419, *420*
oceanography, 419

oceans, 22, 418–31, *440–41*
bathymetric provinces of the sea floor, 419, *420*
boundaries, cartographers', 419, *421*
coasts, *see* coasts
composition of seawater, 423, *423*
creation of, *17,* 527
currents, *see* currents
hydrologic cycle and, 367, *368–69*
landscapes beneath the, 419–23
percentage of Earth's surface covered by, 419
temperature of ocean water, 423
tides, 426–27, *427*
wave action, 427–31
ocean tides, 330
offshore bar, **432,** 432–33
oil (petroleum), 330–36
estimated world reserves, *346*
exploration and production, *334–36, 347, 348–49*
formation of, 532
as hydrocarbon, 330–31
as nonrenewable resource, 346
predicted end of Oil Age, 346, *347*
primary sources of, 331, *332,* 348–49
reliance on, 345–46
trans-Alaska pipeline, 516
Oil Age, **346,** *347*
oil refineries, 336
oil reserves, 331–34
distribution of, global, 331, *332*
oil shale, 136, **331,** *332,* 337
oil spills, 347, *347,* 443
oil trap, 334, *335*
oil window, 331, *333*
Old Faithful geyser, *460*
olivine, 79
Olympic Peninsula, beaches by, *431*
Olympus Mons, 203, *204*
On a Piece of Chalk (Huxley), 306
Onondaga, 137
open-pit mines, *329,* 355, 359
ophiolites, 27
orbitals (energy shells), **A–3,** *A–4*
ordered atoms, *81*
ordinary well, **456**
Ordovician Period, 295
ore, **351**
grades of, 351
ore deposits, **351**
formation of, methods of, 351–54, *360–61*
locations of, 354
ore minerals (economic minerals), 79, *79,* 350, 350–51
exploration and production, *355,* 355–56
with metallic luster, 351, *351*
organic chemicals, 24, 81
organic coasts, **443**
organic sedimentary rocks, **131,** 135, 137
Organization of Petroleum-Exporting Countries (OPEC), 345
orientation of structures, *254*

original continuity, principle of, 285, *286*
original horizontality, principle of, **285,** *286*
Origin of Continents and Oceans, The (Wegener), 35
orogenic collapse, 266
orogens (mountain belts), **249,** *250*
orogeny, **249**
see also mountain building
orthoclase, 79, *87,* 91
outcrops, **97,** *98*
studying of, 100
outer core, **244**
overgrazing, 537
overhang, 382, *384*
Owen, Rosa May, 450
oxbow lake, 403, *404, 411*
oxidation
chemical weathering by, 125
oxides, 90
oxygen in the atmosphere, 526, 527
Proterozoic Eon, 312, *312*
oxygen-isotope ratios, study of global climate change and, 530, *532*
Ozark Dome, 270, *270*
ozone, 312
ozone depletion, **540,** *541*
ozone hole, 540, *541*

Pacific Ocean, 324
sea floor age, 59
Pacific Palisades, California, slumping near, 375
Pacific Plate, 64, *65,* 206, 223, 322
pahoehoe lava flow, **182,** *183*
paleoclimate, **530**
paleogeography
of Cenozoic Era, *324,* 324–25, *325*
of Mesozoic Era, *319,* 319–21, *320, 321*
in Paleozoic Era, 313–14, *314, 316,* 316–18, *317*
paleomagnetism, 39–45
apparent polar-wander paths, 43–44, *44–45*
formation of, *42,* 42–43
paleontology, **273**
global climate change, evidence of, 530, *531*
history of, 273
paleopole, **43**
paleosol, 518
Paleozoic Era, **5,** 313–19, *326–27*
Early Paleozoic Era (Cambrian-Ordovician Periods, 542-444 Ma), 313–16
life evolution, 314–16
paleogeography, 313–14, *314*
Middle Paleozoic Era (Silurian-Devonian Periods, 444-359 Ma), 316
Late Paleozoic Era (Carboniferous-Permian Periods, 359-251 Ma), 316–19
life evolution, *318,* 318–19
paleogeography, 316–18, *317, 318*

Paleozoic era, 313–19
Pangaea, **35**, 36, *36*, *75*, **316**, 319, 320, 322, *520*, 528
Pannotia, 311, *312*, 313
parabolic dunes, *488*, 489
"Paradise Lost," 25
Paraná Basin of Brazil, 196
parent isotope, 295
Paricutín, 197
partial melting, **105**, 105–7, *107*, *114*
passive continental margins, **419**, *420*, *422*
passive-margin basins, 147
passive margins, **53**, *55*, 68, 438, *440*
Paterno, 202
patterned ground, *515*, **515**
pearls, 93
peat, **338**, 339
  evolution of coal from, 339, *339*
pebble beaches, 431, *431*
pedalfer soils, 129, *130*
pedestals, *487*
pediments, *487*, **488**
pedocal soils, 129–30, *130*
pegmatites, 93, **116**, 116–17, *172*
Pelée, Mt., *184*, 198, 202
Pelé's hair, 184
"Pelé's tears," 184
peneplains, 405
perched water tables, *454*, **454**, 457, *458*
peridot, 78, 79, *79*
peridotite, 25, 29
periglacial environments, *515*, 515–16
periodic table of the elements, **A–2**, *A–3*
periods, **294**
permafrost, **515**, 515–16
permanent magnet, 40, *40*, 42, 43
permanent streams, **393**, 393–95, *395*
permeability, **333**, *452*, **452**, *453*
permineralization, **275**
Persephone River, *407*
Persian Gulf
  oil reserves, 331, *332*
  subduction of, *271*
Peru-Chile Trench, 62, *62*
petrified wood, 138, **275**, *276*
petroglyphs, 479, *479*
petrographic microscope, 100, *101*
petroleum, *see* oil (petroleum)
phaneritic (coarse-grained) rocks, 116
Phanerozoic Eon, **5**, 294, 313, *327*, *527*
  see also Cenozoic Era; Mesozoic Era; Paleozoic Era
phase change, 154, 156, *156*, 243
phenocrysts, 116
Phoenix, Arizona, canals carrying water to, 415, *416*
photochemical smog, 540
photomicrograph, *101*, **101**
photosynthesis, 310, **330**, 528
  earliest evidence of photosynthetic organisms, 310
  global change and, 526
  K-T boundary event and, 322

photovoltaic cells (solar cells), 344, *345*
phyllite, 159, *160*
  phyllitic luster, 159, *160*
physical weathering, 122, **123**, 123–24, *125*, 383
  combined with chemical weathering, 126–27, *127*
  in deserts, 478
  soil formation and, 128
piedmont glaciers, *494*, 495
pillows, lava, 57, 183, *184*, 194
Pinatubo, Mt., 198, *199*, 200, 201–2, 203
pitch, 10, *11*
placer deposits, *354*, 354
plagioclase, special properties of, 89, 91
planar structures, *254*
planetesimals, *16*, 17, *19*, **19**, 19–20, 307
planets, **15**, *16*
  forming of the, *16*
  gas-giant, 18
  inner or terrestrial, 18
  orbits of, *18*
  relative sizes of, *18*
  volcanoes on other, *203*, 203–4, *204*
plantae, 278, *279*
plasma, **A–7**, *A–8*
plastic deformation, 154, *156*
plate boundaries, **53**, 55–64, *72–73*
  birth and death of, 68–69
  convergent, *see* convergent plate boundaries
  divergent, *see* divergent plate boundaries
  earthquakes and, 55, 56
  hot spots, **65**, 65–68, *66*, *67*
  identifying, *55*, 55–56
  transform, *see* transform plate boundaries
  triple junctions, 64–65, *66*
plate boundaries and, 55, 56
plate-boundary earthquakes, 220, *221*, 221–24
  convergent plate-boundary seismicity, 221–22, *222*
  divergent plate-boundary seismicity, 221, *222*
  transform plate-boundary seismicity, 223–24, *224*
plate interiors, 56
plates, *4*, 36, 53
  map of Earth's principal, *4*, *54*
plate tectonics, 4–5, **36**
  coastal variability and, 438, *440*
  desert distribution and, 478
  mass movements and, 383–84
  metamorphism in context of, 164–70
  plate boundaries, *see* plate boundaries
  ridge-push force, 70, *71*
  sedimentary basins and, 147–50
  shear force, *71*
  slab-pull force, 70, *71*
  space, view from, *35*
  theory of, 4, 36, 53–55, *72–73*

timing of ice ages and, 520
velocity of plate motions, 71–74
volcanic eruptions and, 191, *191*, 194–97
platforms, 268, *269*, 270
playas, *483*, **483**, *487*
Playfair, John, 304
Pleistocene ice ages, 324, *324*, *516*, 516–19
  chronology of, 518, *519*
  life and climate in, *517*, 517–18
  model for, 521–22, *522*
plunge, 254, *254*, 258
plunge poll, 401
plunging fold, 259, *259*
Pluto, 15, 18, *18*
plutons, **110**, *112*, *113*, *114*, 116, 165, *189*, 308
pluvial lakes, *515*, **515**
point bars, *398*, *404*
polar cells, *476*, 476, 477
polar glaciers, 495–96
polarity, **A–7**
  of atoms or molecules, bonds resulting from, A–6 to A–7, *A–7*
  of a magnet, **40**
polarity chrons, 50
Polar Plateau, 1, 35
polar regions, deserts of the, 478
polar wander, 43
polar-wander path, 43, *44*
pollution, **537**
  air, 347
  beach and coastal, 443–44
  global change and, 537–40
  groundwater, 461, 463–64, *464*
  of streams, 414
polymorphs, **83**
Pompeii, 181, *181*
population growth, doubling time and, 535, *536*
pores, **333**, **451**
  collapse of, 461–63, *462*
porosity, **333**, **451**, 451–52
  primary, 451, *452*
  secondary, 451–52, *452*
porphyritic rocks, 116
Portland cement, 357, *357*
potassium feldspar, *see* orthoclase
potential energy, 455, **A–8**, *A–9*
potentiometric surface, 457, *457*
Powell, John Wesley, 283, *284*
Precambrian time, **5**, **294**, 326–27
precipitate, **80**
precipitation, crystals formed by, 84, *85*
predators, extinctions and, 282
preferred mineral orientation, *157*, **157**
preservation potential, of fossils, **277**
pressure, 254, *255*
pressure solution, 154, *156*
primates, 325
Principles of Geology (Lyell), 280
prograde metamorphism, 161–64, *164*
prokaryotes, 278, *279*, 312
prospectors, 355

Proterozoic Eon, **5**, 294, **310**, 310–13, 527
protista, 278, *279*
protocontinents, formation of, 308, *308*
protolith, **153**
    changes in response to the environment, 153–54
    see also metamorphic rocks
protons, **A–2**
protoplanetary disk, 19
protoplanets, **19**, 19–20, 307
protostar, **14**, *16*
proto-Sun, *16*, 19
Pterosaurs, 321
Ptolemy, 9
Puebla, Mexico, deforestation and mass movements at, *384*
Puerto Rico, sand beach on coast of, *431*
pumice, *117*, **118**, 184
punctuated equilibrium, **281**
P-waves, *211*, 212, 213–14, *215*, 227
    velocity of, 241, *241*
P-wave shadow zone, *244*, **244**
pyramids of Egypt, *367*
pyroclastic debris, **104**, *114*, *115*, **184**, 184–85, *185*, *186*, 190
    in major historic eruptions, *192*
pyroclastic flows (*nuée ardente*), **184**, 184–85, *185*, *186*, *189*, 194
    threat of, 198–200, *199*
pyroclastic rocks, **118**
pyroxene, *88*

quark, **A–3**
quarries, 356, *356*
quartz, 80, 91
    colors of, 86, *86*
    conchoidal fractures on, *89*
quartz crystal, *81*
quartzite, **160**, *162*, *169*
Queen Charlotte fault system, 322
quenching, 108

radial drainage networks, 393, *394*
radiation, **A–10**, *A–11*
radioactive elements, heat release and decay of, 104
radioactive isotopes, 295–97
    half-life of, 295–97, *297*
    used in radioactive dating of rocks, 297
radioactivity, discovery of, 302
radiometric dating, *50*, **295**, 295–301, *299*, 302
    meaning of a radiometric date, 300–301
    radioactive decay and half-life, 295–97, *297*
    techniques, 297–300, *300*
rain-forest destruction, consequences for soil of, 131
Rainier, Mt., 201, *202*
rain shadows, **477**
    deserts formed in, 475–77, *477*
range of specific fossils in the sequence, 287

Ranrahica, Peru, 373
rapids, 401, *401*, *410*
Rayleigh waves, *see* R-waves
recessional moraines, 507, *511*, 512
recharge area, **455**
recrystallization, 154, *156*
rectangular drainage networks, 393, *394*
recurrence interval, 235, **235**, **413**
    of a flood, 413–14, *415*
redbeds, *299*
Redoubt Volcano, eruption of, *180*
red shifts, 10–11, *11*
reefs, 145, *146*
    see also coral reefs
reflection, **241**
    of seismic waves, 241, *242*
refraction, **241**
    of seismic waves, 241, *242*
refractory materials, *17*, 18
regional basins, 270
regional domes, 270
regional metamorphism, *166*, 166
regolith, 374, 380, 383, *387*
    saturation with water, effect of, 382, 383, *387*
    as unconsolidated, 380
regression, *150*, **150**
relative age, *283*, **284**
    fossil succession and, 285–87, *289*
    geologic column, *see* geologic column
    physical principles for defining, 284–85, *286*, *287*
    stratigraphic formations and, *see* stratigraphic formations
relative plate velocity, **71**, *74*
relief, **365**, *366*
Renaissance, 9
renewable resources, 346
reptiles, 318, 319, 321
    flying, 319
reservoir rocks, *333*, **333**
residence time (in hydrologic cycle), 370
residual mineral deposits, 353–54, 354, *354*
resistance force, 380–81, *381*
resources, **329**
    population growth and use of, 535
retrograde metamorphism, 164, *164*
reversed polarity, 49, *50*
reverse faults, 208, *208*, *216*, *217*, 228, *257*, **257**
rhyolitic magma, *114*
rhyolite, *117*, 118
rhyolitic lava flows, *182*, 183–84, 194
Richter, Charles, 219
Richter scale, *219*, **219**
ridge, *see* divergent plate boundaries; mid-ocean ridges
ridge-push force, 70, *71*
rift basins, 147
rifting, *68*, **68**, 92
    Basin and Range Province, *see* Basin and Range Province
    East African Rift, *see* East African Rift

Ring of Fire, 194
Rio de Janeiro, Brazil, mudflows in, 376, *378*
Rio Grande Rift, 224, *324*
Rio Lagunillas, 200
rip current, 431
ripple marks, **140**, *141*, 143
ripples, 139–40, *141*, 143
rising tide (flood tide), 426
river environments, 144, *145*
rivers, *391*, 392
    environmental issues, 414–16
    river systems, *410–11*
    see also streams
river valleys, drowned, *430*
road cuts, 97, *98*
roche moutonnées, 505, **505**, *511*
Rochester, New York, drumlins near, *509*
rock cycle, **174**, 174–77, *528*, *539*
    case study of, 175, *176*
    factors driving the, in the Earth system, 177
    rates of movement through, 175–77
    rock-forming environments and, *178–79*
    stages of, *175*
rock falls, *379*, **379**, 379–80, *387*
rock flour, 502, *503*
rocks, **25**, 95–101, *96*
    bonding by natural cement, 96, *96*
    classification of, 25, 97–100
    composition of, 99
    defining characteristics of, 96
    deformation, *see* deformation (rock)
    grain size, 98–99, *100*
    historical record of geological events provided by, 95
    igneous, *see* igneous rocks
    layering in, 99, *100*
    metamorphic, *see* metamorphic rocks
    naming of types of, 99–100
    occurrences, 97
    paleomagnetism, *see* paleomagnetism
    sedimentary, *see* sedimentary rocks
    studying
        with high-tech analytical equipment, 101
        outcrop observations, 100
        thin-section study, 100–101
    texture of, 99
rock slide, **378**, *387*
rocky coasts, *431*, *433*, 434, *434*
Rocky Mountains, 318, 319, 320, *321*, 517
    fault zone, *256*
Rodinia, **311**
Rome, ancient, 357
root wedging, 124, *125*
Rosendale Formation, *357*
Ross Sea, ice tongue protruding into the, *501*
Rotorua, New Zealand, hot springs of, 461

rubies, 93
running water, *see* floods; streams
runoff, 392
R-waves, *211, 212, 213, 227*

Saffir-Simpson scale, 444
sag pond, *216, 217*
Sahara Desert, 473, *474*, 475
    dust cloud originating in, 491, *491*
    oasis in, *459*
Sahel, 489–91, *490*
St. Pierre, Martinique, 184, 198
saline water, 461, *462*, 463
saltation, 397, *397*, **479**, 479–80, *480*
salt crystals, 84
salt dome, **335**
salt-dome trap, *335*
salt lakes, 483
salt marshes, 435, 439
Salton Sea, 461
salt pans, *487*
saltwater evaporation, 136–37, *138*
salt wedging, 124, *125*
San Andreas Fault, 64, *65*, *209*, 210,
    223–24, *224*, 256, 256–57
    formation of, 322
sand beaches, 431, *431*
sand-dune environments, *141*, 144
sand dunes, *see* dunes
sand seas, 488
sand split, **432**
sandstone, *96, 99*, **132**, *134*, 135, 144,
    149
sand volcanoes (sand boils), 229, *230*,
    414
San Francisco earthquakes of 1906, *209,
    210*, 223, *224*, 256
    fire destruction from, 229
San Francisco region, aerial view of previ-
    ous earthquakes and faults in,
    206
San Joaquin Valley, subsidence in, 463
San Juan River
    incised meanders, 406, *406*
Santa River, 373–74
Santorini volcano, 200–201
Saturn, 15, 18
Saudi Arabian oil reserves, 331
sauropod dinosaurs, 319
scarp retreat, *see* cliff retreat
schist, **159**, *160*, *169*, 171
schistosity, 159, *382*
science, **6**
scientific laws, **7**
scientific method, **6**, 6–7
scientists, **6**
scoria, **118**, 184
scorpions, 316
Scott, Walter, 2
scouring, 396
scour marks, 142
sea anemones, 436
sea arches, *433, 434, 435, 440*
sea caves, *433*
sea cliffs, *440*

sea-floor spreading, **36**, **48**, 320, 422
    age of the sea floor and, 59, *60*
    along a divergent plate boundary,
        56–59, *57, 58*
    evidence of
        deep-sea drilling, 53
        marine magnetic anomalies, 48–52
    global change and, 525
    Hess's concept of, 48, *48*
    setting the stage for discovery of, 45–38
        oceanic crust, new observations of,
            46–48
        sea-floor bathymetry, new images
            of, 45–46
    spreading rate, 52
    timing of ice ages and, 520
sea ice, **500**, *502, 517, 523*
sea-level changes, *538*
    coastal variability and, 438, *438, 439*
    contemporary, 439, *439*
    eustatic, 528, *528*
    global warming and, 542, *542*
    ice ages and, *511*, 513, *513*
    stratigraphic sequences and, *315*
sea-level cycle, 528
seamount chains, 46, *46*, 441
seamounts, **423**
seasonal floods, **408**, 408–9, *409*
    case study, 409–12
sea stacks, *433, 434, 435, 440*
seawalls, 442, *443*
SeaWiFS satellite images, *421*
secondary bulge, 427
secondary-enrichment deposits, 352, *353*
sediment, **25**, 37, **122**
    weathering and formation of, *see*
        weathering
    see also soils
sedimentary basins, *147*, 147–50
sedimentary environments, 143–47,
    148–49
    coastal variability and, 439, *440*
    marine, 144–47
    terrestrial (nonmarine), 143–44
sedimentary rocks, 25, **98**, *99*, **122**,
    131–51
    biochemical, 131, 135–36, *136, 137*
    chemical, 131, 136–38, *138, 139*
    classes of, 131–38, *148*
    clastic, *see* clastic sedimentary rocks
    dating of, 302, *302*
        fossils used to date layers of,
            273–74, *299*
    Earth's history contained in, 122
    energy resources contained in, 122
    environments, *see* sedimentary envi-
        ronments
    formation of, *148–49, 178–79*
    fossils found in, *see* fossils
    organic, 131, 135, 136
    rock cycle, 174–77, *178–79*
    structures, *see* sedimentary structures
sedimentary structures, **138**, 138–43
    bedding, **99**, *100*, 139, *140, 149, 158*
    bed-surface markings, 142–43

cross beds, *see* cross beds
dunes, *see* dunes
graded beds, 141, 143
ripples, 139–40, *141*, 143
turbidity currents, *see* turbidity currents
value of studying, 143, *149*
sediment budget, *433*, 433–34
sediment load, 396–97
sediment maturity, 135
seismic belts, 47, *47*, 55, 220, **220**, *221*
seismic gaps, 235, *236*
seismic-hazard assessment, 234
seismic hazard maps, 235, *236*
seismicity, **207**
seismic ray, *240, 241*
    curving of, 243, *243*
seismic-reflection profiling, **245**, 245–46,
    247, 334, *336*
seismic tomography, **245**, *246*
seismic-velocity discontinuities, **243**
seismic waves, 26, *211*, **212**, 216
    discoveries about Earth's layers from,
        240–47
    movement of, through the Earth,
        240–41, *241*
    reflection and refraction of, 241, *242*
    see also L-waves; P-waves; R-waves;
        S-waves
seismic zones, *see* seismic belts
seismograms, **213**, *214*
    digital, 213, *214*
seismographs, **212**
    electronic, 213, *214*
    horizontal-motion, 212–13, *213*
    vertical-motion, 212, *213, 214*
seismologists, 207
seismometer, *see* seismographs
septic tanks, pollution of groundwater
    by, 461, *462*, 463
Sevier orogeny, 320, *320*
shale, *134*, **135**, 143, *149*, 158, 159
shallow-marine clastic deposits, 144–45
shallow-water carbonate environments,
    145, *146*
shatter cones, *7*, **7**
shear force, 71
shear stress, 157, *157*, **253**, 254, *255*
shear waves, *211*, **212**
Sheep Mountain, 382
sheetwash, **392**, *394*
shields, **171**, *171–72*, 268, *269*, 270, **311**
shield volcanoes, **187**, 187–90, *190*, 191,
    195
Shiprock (volcano), *111*, 197
shock metamorphism, 167–70, *170*
shoreline, 426
Siberia, 313, 316, 318
Siccar Point, Scotland, 287, *289*
Sierran arc, 319, *320*
Sierra Nevada Mountains, 320, 502, 517
    rocks of the, 95, *95*, 110, *112*
        exfoliation joints, *124*
silica, 135, 138
    lava composition, 182
silicate rocks, 25

silicates, 25, **89**, 89–91, *90*, *91*
  major groups of, 90–91
silicic magma, *see* felsic magma
silicon-oxygen tetrahedron, **89**, *90*
sill, **110**, *111*, *113*
Silliman, Benjamin
sillimanite, 79
sills, *110*, *112*, *114*, *115*, 116, *189*
siltstone, **135**, *149*
siltstone bed, 139, *140*
sinkholes, **450**, 450–51, *451*, *466*, *467*,
    468–69, *469*, *470*
  collapse, 468
  lake formation in, 469
slab-pull force, 70, *71*
slag, 350
slash-and-burn agriculture, 537, *537*
slate, *158*, 159, *169*
  as roofing material, *158*, 159
  slaty cleavage, 159, *159*
slicken-sides, 258
slip face, 489
slip lineations, 258, *258*
slope stability, mass movements and,
    380–81
  slope failure, factors causing, 381–84
slump, 375, *376*
  identifying conditions leading to, 385,
    385
slumping, 375–76, *376*, *387*
  submarine, 375–76
    tsunamis and, 376, *377*
smelting, 350
Smith, William, 273, 285–87, 290–93
Smithsonian Institution, 91, 93
smog, 540
Snake River Plain, *195*
snotites, 470, *471*
snowball Earth, **313**, **520**
Snowdon, Mt., *124*
snowdrifts, *495*
snowflakes, *495*
soda straw, *466*, 468, *468*
soil erosion, *131*, **131**
soil horizons, **128**
soil moisture, **451**
soil profiles, **128**, 128–29, *129*
soils, 122, **128**, 128–31
  classification schemes, 129–31, *130*,
    131
  destruction of, factors causing, 130–31
  factors affecting makeup of, 129
  formation of, processes involved in,
    128, *128*
solar collector, 344
solar energy, 330, 344, *345*, 346–47, *349*,
    538
  hydrologic cycle and, 370
solar system, our, 9
  age of, 18
  formation of, *16–17*, 18–19
  nature of, 15–18
solar wind, 19, 21, *21*, 23
Solenhofen Limestone, Germany, extra-
    ordinary fossils in, 277–78, *278*

solids, 80, **A–7**
solid-state diffusion, crystals formed by,
    84
solifluction, **375**, *376*, *387*
solution, **80**
solution cavities, 452
sonar, 45, *45*
Sonoran Desert, *473*, *486*
sorting of clasts, *133*, **133**
sound waves
  Doppler effect, 10, *11*
source rock, **331**, *332*, 333, *333*
South America, 311, *312*, 313, 322,
    419–20
South Fork Hunting and Fishing Club,
    391, 392
South Pacific Ocean, lava pillows on floor
    of, *184*
specific gravity of a mineral, **87**
speleothems, 137, *467*, *468*, **468**, *470*
spelunkers, 468
spiders, 316
sponges, 314, 436
spreading boundaries, *see* divergent plate
    boundaries
springs, **456**, **457**, *458*
  artesian, 457, *458*
  formation of, 457, *458*
  hot springs, 457–58, *460*, 461
Sri Lanka, tsunami of 2004 striking, *232*,
    233
stability field, 156
stable slopes, 380
stair-step canyons, 400, *400*
stalactites, *466*, *467*, *468*, 468
stalagmites, *466*, *467*, 468, *468*, 468
star dunes, *486*, *488*, 489
stars
  as element factories, 14–15, *15*
  first, forming of, 13–14, *14*
  length of survival, 14
  low-mass and high-mass, 14
  second generation of, 15
  space, view from, *32*
steady-state condition, 529
steep-walled canyon ("slot canyon), 400,
    400
stegosaurus, 319
stellar nucleosynthesis, 14
stellar wind, 15, *15*, 19
Steno, Nicholaus, 273
stick-slip behavior, **210**
Stone Age, mass extinction by hunting
    in, 536
stony plains, 488
stoping, **110**, *113*
Storegga Slide, *377*, **377**
storm activity, global warming and, 542
storm surge, **445**, 446, *446*
strain, **251**
  deformation and, 249–51, *252*
  shear strain, 251, *252*
  shortening, 251, *252*
  stress distinguished from, 254
  stretching, 251, *252*

strata, 139
stratification, *see* bedding
stratigraphic columns, **289**, *291*
  geologic column constructed from, *see*
    geologic column
stratigraphic formations, **139**, *140*, **289**,
    289–93, *291*
  correlation of, 290–93
  naming of, 290
stratigraphic sequence, **315**
  sea-level change and, *315*
stratigraphic trap, *335*
stratopause, 22
stratosphere, 22, *23*
stratovolcanoes (composite volcanoes),
    190, **190**, 191, 194
streak of a mineral, *86*, **86**
streak plate, 86, *86*
stream cut, 97, *98*
stream piracy, **406**, 406–7, *407*
stream rejuvenation, 405–6, **406**
streams, **392**, 392–408, 410–11
  base level of, 398–400, *399*
  capacity of, 397
  competence of, 397
  discharge, 395, *396*
  drainage, evolution of, 405–8
  environmental issues, 414–16
  ephemeral, *395*, **395**
  erosion processes, 396
    beveling topography, 404–6
  flooding by, *see* floods
  formation of, 392–93, *393*, *410*
  landscape effects of, and their de-
    posits, 400–404
  longitudinal profiles, 398, *399*
  permanent, **393**, 393–95, *395*
  sediment loads, 396–98
  source of, *410*
  turbulence, 396, *396*
stream terraces, *401*, **401**
stress, as cause of deformation, *253*,
    253–54, *255*
strike, *254*, **254**
strike direction, *254*
strike line, 208, *254*
strike-slip faults, *208*, 208–9, *209*, 216,
    217, 221, 223, 228, **257**, *257*, 420
  left-lateral, 257, *257*
  right-lateral, 257, *257*
strip mines, 340, *341*
stromatolites, **309**, *309*, 309–10
Styx Sea, *407*
subduction, **36**, **59**, 59–61, *61*, 92
  global change and, 525
  subduction and, 61–62
subduction zones, *see* convergent plate
    boundaries
sublunar bulge, 427
submarine canyons, 420, *422*
submergent coasts, **438**, *439*
subsidence, **147**, **364**, *462*, 463
  glacial, *512*, *512*
  pore collapse and, 463
  rates of, 366

substrate, **366**
subtropical deserts, 475
subtropics, 475
Sudbury, Ontario, acid smelter smoke in, 359, *359*
sulfates, 91
sulfides, 90
sulfur water, 463
summit eruptions, 186
Sun, 9
    formation of, *17*, 18
    future of the, 543, *543*
    global change and, 525
    mass of the, 15
Sunda plate, 229, *232*
Supai Group, *140*
supercontinent cycle, **528**
supercritical fluid, 158
superimposed streams, *407*, **407**
supernova, 14
supernova explosions, 15, *15*, *16*
superplumes, **320**
superposition, principle of, **284**, 284–85, *286*
surface load, **480**
surface-wave magnitude (M_s), 219
surface waves, **212**
suspended load, **479**, 483
    of a stream, 397, *397*
    of wind, 479, *480*
sustainable growth, **543**
Sutter, Captain John, 350
swamps, 434
    mangrove, *430*, 434–35, *435*
swampy deltas, *430*
swash, *428*, *429*, *441*
S-waves, *211*, 212, 213–14, *215*, 227
    velocity of, 241, *241*
S-wave shadow zone, **244**, *245*
symbiotic relationships, 436
symbol (element), **A–2**
symmetry, **83**
    of crystals, 83–84, *84*
synclines, *259*, **259**, *260*
    nonplunging, 259, *261*

tabular intrusions, *189*
Taconic orogeny, 314, 316, *316*
tailing piles, 356, 359, *359*
Taklimakan Desert, alluvial fan in, *151*
talus, 124, *124*, *379*, **380**
talus aprons, **482**, *483*
Tambora, *192*, 203
T'ang-shan, China, earthquake of 1976 in, 237
tar, fossils preserved in, 274
tarn, 502
tar sands (oil sands), 336–37, **337**, 346
Tavernier, Jean Baptiste, 91
taxonomy, 278
tectonic foliation, 260, 260–61, *262*
temperate glaciers, 495
temperature, **A–10**
    blocking, **301**
    metamorphism and, 155, 156, *156*, *189*

of ocean water, 423
    see also climate; global climate change; global warming
tension, **253**, 254, *255*
Tensleep Formation, *383*
tephra, **184**, 186, *188*, *189*
terminal moraine, **507**, *509*, 512
terraces, *366*, *438*
    uplifted, *430*
terrestrial planets, **18**
Tethys Ocean, 322
theories, **7**, 36
theory of evolution, **281**
    see also evolution
theory of plate tectonics, **4**, **36**, 53–55, 72–73
    see also plate tectonics
thermal energy, 14, **A–10**
thermal expansion, physical weathering of rock by, 124
thermal lance, 356
thermal metamorphism, **165**, *167*
    pottery making analogy, *167*
thermohaline circulation, **425**, *426*, 521, 542
thermonuclear devices, A–9, *A–10*
thermosphere, 22, *23*
thin-section study, **100**, 100–101, *101*
thrust faults, 208, *208*, *257*, **257**
Tibet Plateau, 265, 268, 322
tidal bore, 427
tidal bulge, 427, *427*
tidal flats, *432*, 433
tidal reach, 426, *427*
"tidal waves," 233
    see also tsunamis
tide-generating force, 427, **427**
tides, **426**, 426–27, *427*
tidewater glaciers, 499–500, *501*
Tien Shan Mountains, *265*
till, *see* glacial till; lodgment till
tillites, 519–20
Titan (moon of Saturn), volcanism on, 203
Titanic, 499
Titusville, Pennsylvania, 334
toe, or terminus of a glacier, 498–99, *501*
Tokyo, Japan, earthquake of 1923, 229
tombolos, *433*, 434
topography, **22**, 22–23, *24*
tower karst, 470, *471*
trace fossils, 277
trans-Alaska pipeline, 516
Transantarctic Mountains, 1–2, *2*, *494*
transform faults, 63, 420
    see also transform plate boundaries
transform plate boundaries, *56*, **56**, 63–64, 73
    earthquakes and, 223–24, *224*
    fracture zones and, 63, *64*
transgression, *150*, **150**, 289, 339
transition zone (of the mantle), 29, *30*, **244**
transportation of sediment, 132, *132*
transverse dunes, *488*, 489

traps, **334**, 337
    oil and gas, 334, *335*
travel-time curve, 213–14, *215*
travel time of a seismic ray, **240**, 240–41, *241*
travertine, **137**, *139*
tree rings
    determining numerical age with, 301, *301*
    global climate change and study of, 530
treillis network, 393, *394*
Tremblor Range, *256*
trench axis, *62*
trenches, **45**, *46*, **59**, 420, *441*
    see also convergent plate boundaries
Triassic Period, 295
tributaries, *392*, *410*, 503
trilobites, 279, *279*, 314, *314*, 316
triple junctions, 64–65, *73*
    ridge-ridge-ridge, 65, *66*
    trench-transform-transform, 65, *66*
tropical depression, 444
tropical disturbance, 444
tropical rain forests, 536–37, *537*
tropical storm, 444
tropopause, 22
troposphere, 22, *23*
troughs, *see* channels
trunk glaciers, 503
tsunamis, **231**
    of December 26, 2004, 229–33
    earthquakes and, 229–34, *231–34*
    far-field, 233
    formation of, 231, *231*
    K-T boundary event and, 322
    landslides and, 233
    near-field, 231
    North Sea, Storegga Slide and, 377, *377*
    predicting, 233–34, *234*
    volcanic eruptions and, 200
    width of, 233, *233*
tuff, *117*, **118**, **184**, *189*
turbidites, 420, *422*, 440
turbidity currents, 141, *142*, 143, *149*, 378, 420
turbulence (of a stream), 396, *396*
Turkey, 1999 earthquake in, *228*
turtles, 319, 321
Twain, Mark, 283, 412
Twelve Apostles (sea stacks), *435*
typhoons, *416*, 444
    see also hurricanes
Tyrannosaurus rex, 321

ultramafic magma, **105**
ultramafic rocks, 25
ultraviolet radiation, 540
Uluru (Ayers Rock), Central Australia, 485
unconfined aquifers, 452, *453*
unconformities, **287**, 287–88, *289*, *290*, *291*, *299*
    angular, 287, *290*
    discomformity, 288, *290*
    nonconformity, 288, *290*

undercutting, **382**, *384*
Undersea National Park, Virgin Islands, 436, *437*
uniformitarianism, **284**, *286*
United States
 oil reserves of, 331
 Pleistocene deposits in, 518, *519*
U.S. Army Corps of Engineers, 413
Universe, **8**
 forming of the, 9–12
 geocentric concept of, 9, *9*
 heliocentric concept of, 9, *9*
 modern image of the, 9
unstable slopes, 380
Unzen, Mt., *199*
uplift, 263, **364**, *366*, *538*
 global climate change and, 531–32
 mountain building and, 261–67, *263*
 rates of, 366
uplifted terraces, *430*
upper mantle, 29, *30*, **244**
upwelling, *424*, 424–25, *426*
Ural Mountains, 318
uranium, 342–43, 346, *346*
 deposits, 342–43, *349*
 origins of, 342
 ^{235}U (isotope), 342
Uranus, 15, 18
urbanization, flooding and, 416
U.S. Comprehensive Soil Classification System, 130, *131*
U-shaped valley, **503**, *504*, *511*

vacuum, **20**
Vaiont Dam, 376–78, *378*
Valdez, Alaska, 1964 tsunami hitting, 233
Valley and Ridge Province, *268*
valley glaciers, *494*, 495, *511*, 517
Valley of the Mummies, *459*
valleys, 400, *400*, *410*
 alluvium-filled, 401, *401*
 glacial erosion and, 503
Van Allen belts, 21, *21*
van der Waals, Johannes, A–7
varves, **506**, *507*
vascular plants, 316
vegetation, mass movements and removal of, *383*
vein deposit, 352
veins, *255*, **256**
velocity-versus-depth curve, **246**, *247*
Venice, Italy, *462*, 463
ventifacts, **481**, *482*
Venus, 15, 18
 landscape of, 371, *372*
 volcanoes on, 203
Verne, Jules, 26
vertebrates, 295, 314–16
vesicles, **118**, *186*, **186**
Vesuvius, Mt., 180–81
 destruction of Pompeii, 181, *181*
Vine, Fred, 50
viscosity, movement of magma and lava and, 107–9, *109*, *182*, 182–84

volatile materials, *17*, 18, 23, 25, 104–105
volcanic arcs, **45**, 45–46, *46*, *62*, *62*, *114*, 267, *267*, *441*
volcanic ash, **104**, **184**, *185*, *189*
 threat of, 198, *199*
volcanic breccia, **118**
volcanic danger-assessment maps, 201–2, *202*
volcanic debris flow, 186
volcanic eruptions, 102, 180–205, *188–89*, 520
 airplane travel, affecting, *180*, 198
 conduits for, 186–87, *187*
 controlling hazards of, 201–2, *202*
 destruction and deaths resulting from, 180–81, *181*, 184–85, *186*, *192*, 198–201, *199*, *201*
  protecting against, 201–2, *202*
 diverting flows of, 202, *202*
 earthquakes and, 200, 201, 207
 flank eruptions, 187, *187*
 global climate change and, 527, 531, 533–34
 landslides and, *387*
 major historic, 192–93, *192–93*
 plate tectonics and, 191, *191*, 194–97
 predicting, 201
 products of, 182–86
  lava flows, *see* lava flows
  pyroclastic debris, *see* pyroclastic debris
  volcanic gas, 185–86
 protection from, 201–2
 styles of, *188*, 190–91
 summit eruptions, 186
volcanic eruptions and, 200, 201, 207
volcanic gas, 185–86, *186*, 527, 531
 threat of, 200
volcanic island arc, 62–63, *63*, 194, 267
volcanic neck, *111*
volcano(es), **102**, **180**
 architecture and shape of, 186–90, 197–98, *197–98*
 climate and, 202–3
 fissure, 186, *187*
 geologic settings of, map of, *191*
 hot spot, *see* hot spots
 on other planets, *203*, 203–4, *204*
Voltaire, 225
Voyageurs National Park, *518*
V-shaped valley, 400, *400*, *410*, 503, *504*
Vulcan, god of fire, 180
Vulcano (island), 180

wacke, *134*, 135
Wadati-Benioff zone, *61*, 61, *61*, 61–62, *222*, **222**
wadis, 479, *486*, 488
wall rock, *see* country rock (wall rock)
Wasatch Mountains, Precambrian metamorphic rock in, *153*
water
 atmosphere of the Earth, evolution of, 527–28
 groundwater, *see* groundwater

 hydrologic cycle, *see* hydrologic cycle
 oceans, *see* coasts; oceans
 overuse of, 415, *416*
 streams, *see* streams
waterfalls, *391*, 401, 401–2, *402*, 410, 503, *504*
water gap, *407*
water jet, abrasive, 356
watersheds, **393**, *410–11*
water table, **367**, **452**, 452–54, *453*
 capillary fringe, 453, *453*
 depletion of groundwater supplies and, 461, *461*
 depth of, *453*, 453–54
 perched, 454, *454*
  artesian springs and, 457, *458*
 rising, unwanted effects of, 464, *465*
 saturated zone, 453, *453*, 454
 surface tension and, 453
 topography of, 454, *454*
 unsaturated zone, 453, *453*, 454
wave base, **428**
wave-cut benches, *433*, **434**, *438*
wave-cut notches, *433*, **434**
wave front, **240**, *241*
wavelength, 10, *11*
wave refraction, *429*, **429**, 429–31, 434
waves, *440*
 circular path of water molecules, 428, *428*
 Doppler effect, 10
 frequency of, 10, *11*
 hurricane damage and, 445
waves, ocean, 427–31
weathering, 92, **122**, 122–28, 132, 170
 around joints in rocks, 255, *255*
 chemical, *see* chemical weathering
 combination of physical and chemical, 126–27, *127*
 in deserts, 478–79
 differential, 126, *127–28*
 at edges and corners, 126, *127*
 mass movements and, 380
 physical, *see* physical weathering
 surface area and, 126, *127*
Wegener, Alfred, 35–36, *36*, 520
 continental drift and, 35–39, 53, 75
weight, **A–1**
wells, **456**, 456–57
 artesian, 456–57, *457*
 contamination of, 461, *462*
 ordinary, 456, *456*
West Antarctica, 1, *2*
wetlands
 coastal, 434–35, *435*
  pollution of, 443–44
 restoration of, 413
whirlpools, *see* eddies (whirlpools)
Whitby, England, *125*
White Cliffs of Dover, 306
whitewater, 401, *401*
wildcatters, 334
Wilson, J. Tuzo, 63, *64*, 65–66
wind power, **343**, 343–44, *345*, 347, *348–49*

winds, 22
  convection in the atmosphere and
    generation of prevailing, *476,
    477*
  desert erosion, 479–82
  hurricane, 444, 445
Winston, Harry, 93
Winter Park, Florida, sinkhole, 450–51,
  451
wireline saw, 356
woolly mammoths, fossils of, 274,
  276
work, **A–8**

xenolith, **110**, *113*

Yangtze River, destruction caused by
  flooding of, 408–9
Yellowstone National Park, 65, 67, *183,
  194–96, 195, 460,* 461
  geysers of, 67, *460*
  knob-and-kettle topograhy on a
    moraine in, *509*
Yellowstone River, *195*
Yosemite National Park, *518*
  Half Dome in, 502, *503*
  rock fall of 1996 in, 380
Younger Dryas, 532
Yucca Mountain, storing of nuclear waste
  at, 343
Yungay, Peru, 373–74, *374*

Zabargad Island, *3, 78*
Zabros Mountains, 322
Zion Canyon, 295, *296*
Zion National Park, *125, 141,* 319
  cross beds in sandstone of, *489*
zone of ablation, **498**, *500*
zone of accumulation, 128, **128**, **498**,
  499, 500
zone of leaching, *128,* **128**